$$-\frac{5wl^3}{192EI}$$

$$\frac{5wl^3}{192EI}$$

$$\Delta(x_1) = -\frac{w}{960EI}\left(\frac{16}{l}x_1^5 - 40lx_1^3 + 25l^3x_1\right)$$

$$\Delta(x_2) = -\frac{w}{960EI}\left(-\frac{16}{l}x_2^5 + 40x_2^4 - 40l^2x_2^2 + 8l^4\right)$$

$$x_1 = \frac{l}{2}, \quad x_2 = 0$$

$$-\frac{wl^4}{120EI}$$

$$-\frac{8wl^3}{360EI}$$

$$\frac{7wl^3}{360EI}$$

$$\Delta(x) = -\frac{w}{360EI}\left(-\frac{3x^5}{l} + 15x^4 - 20lx^3 + 8l^3x\right)$$

$$x = 0.481l$$

$$-\frac{0.00652wl^4}{EI}$$

$$\frac{M_o l}{6EI}$$

$$-\frac{M_o l}{3EI}$$

$$\Delta(x) = \frac{M_o}{6EIl}(-x^3 + l^2 x)$$

$$x = \frac{l}{\sqrt{3}}$$

$$\frac{M_o l^2}{9\sqrt{3}EI}$$

Structural Analysis: Skills for Practice

James H. Hanson, PhD, PE
Rose-Hulman Institute of Technology

221 River Street, Hoboken NJ 07030

Senior Vice President Courseware Portfolio Management: Engineering, Computer Science, Mathematics, Statistics, and Global Editions: *Marcia J. Horton*
Director, Portfolio Manager: Engineering, Computer Science, and Global Editions: *Julian Partridge*
Executive Portfolio Manager: *Holly Stark*
Portfolio Management Assistant: *Amanda Perfit*
Managing Producer, ECS and Mathematics: *Scott Disanno*
Senior Content Producer: *Erin Ault*
Project Manager: *Rose Kernan*
Manager, Rights and Permissions: *Ben Ferrini*
Operations Specialist: *Maura Zaldivar-Garcia*
Inventory Manager: *Bruce Boundy*
Product Marketing Manager: *Yvonne Vannatta*
Field Marketing Manager: *Demetrius Hall*
Marketing Assistant: *Jon Bryant*
Cover Design: *Black Horse Designs*
Cover Image: *Thomas Kelley/Alamy*
Composition: *Integra Publishing Services*
Cover Printer: *LSC Communications*
Printer/Binder: *Lake Side Communications, Inc. (LSC)*
Typeface: *TimesTenLTStd 10/12*

Copyright © 2020 Pearson Education, Inc., Hoboken, NJ 07030. **All rights reserved.** Manufactured in the United States of America. This publication is protected by copyright, and permission should be obtained from the publisher prior to any prohibited reproduction, storage in a retrieval system, or transmission in any form or by any means, electronic, mechanical, photocopying, recording, or otherwise. For information regarding permissions, request forms and the appropriate contacts within the Pearson Education Global Rights & Permissions department, please visit www.pearsoned.com/permissions/.

Many of the designations by manufacturers and seller to distinguish their products are claimed as trademarks. Where those designations appear in this book, and the publisher was aware of a trademark claim, the designations have been printed in initial caps or all caps.

The author and publisher of this book have used their best efforts in preparing this book. These efforts include the development, research, and testing of the theories and programs to determine their effectiveness. The author and publisher make no warranty of any kind, expressed or implied, with regard to these programs or the documentation contained in this book. The author and publisher shall not be liable in any event for incidental or consequential damages in connection with, or arising out of, the furnishing, performance, or use of these programs.

Library of Congress Cataloging-in-Publication Data

Names: Hanson, James H., author.
Title: Structural analysis : skills for practice / James H. Hanson, PhD, PE,
 Rose-Hulman Institute of Technology.
Description: First edition. | Pearson, [2019] | Includes
 bibliographical references and index.
Identifiers: LCCN 2018028853 | ISBN 9780133128789 | ISBN 0133128784
Subjects: LCSH: Structural analysis (Engineering)
Classification: LCC TA645 .H374 2019 | DDC 624.1/7—dc23
LC record available at https://lccn.loc.gov/2018028853

ISBN 10: 0-13-312878-4
ISBN 13: 978-0-13-312878-9

Contents

Preface   ix
   Motivation for a New Text   ix
   Homework Problems and Example Structure   xi

0   *Evaluating Results*   1

1   *Loads and Structure Idealization*   2
   1.1   Loads   4
   1.2   Load Combinations   11
   1.3   Structure Idealization   28
   1.4   Application of Gravity Loads   35
   1.5   Application of Lateral Loads   49
   1.6   Distribution of Lateral Loads by Flexible Diaphragm   59
   Homework Problems   67

2   *Predicting Results*   82
   2.1   Qualitative Truss Analysis   84
   2.2   Principle of Superposition   94
   2.3   Bounding the Solution   98
   2.4   Approximating Loading Conditions   102
   Homework Problems   109

3   *Cables and Arches*   118
   3.1   Cables with Point Loads   120
   3.2   Cables with Uniform Loads   140
   3.3   Arches   152
   Homework Problems   163

4   *Internal Force Diagrams*   170
   4.1   Internal Forces by Integration   172
   4.2   Constructing Diagrams by Deduction   192
   4.3   Diagrams for Frames   205
   Homework Problems   216

5   *Deformations*   234
   5.1   Double Integration Method   236
   5.2   Conjugate Beam Method   247
   5.3   Virtual Work Method   257
   Homework Problems   277

6   *Influence Lines*   296
   6.1   The Table-of-Points Method   298
   6.2   The Müller-Breslau Method   312
   6.3   Using Influence Lines   322
   Homework Problems   329

7 Introduction to Computer Aided Analysis   336
    7.1   Computer Results Are Always Wrong   338
    7.2   Identifying Mistakes   340
    7.3   Checking Fundamental Principles   342
    7.4   Checking Features of the Solution   350
    Homework Problems   359

8 Approximate Analysis of Indeterminate Trusses and Braced Frames   372
    8.1   Indeterminate Trusses   374
    8.2   Braced Frames with Lateral Loads   384
    8.3   Braced Frames with Gravity Loads   401
    Homework Problems   417

9 Approximate Analysis of Rigid Frames   436
    9.1   Gravity Load Method   438
    9.2   Portal Method for Lateral Loads   458
    9.3   Cantilever Method for Lateral Loads   473
    9.4   Combined Gravity and Lateral Loads   490
    Homework Problems   497

10 Approximate Lateral Displacements   514
    10.1   Braced Frames—Story Drift Method   516
    10.2   Braced Frames—Virtual Work Method   526
    10.3   Rigid Frames—Stiff Beam Method   542
    10.4   Rigid Frames—Virtual Work Method   550
    10.5   Solid Walls—Single Story   565
    10.6   Solid Walls—Multistory   575
    Homework Problems   583

11 Diaphragms   616
    11.1   Distribution of Lateral Loads by Rigid Diaphragm   618
    11.2   In Plane Shear: Collector Beams   633
    11.3   In Plane Moment: Diaphragm Chords   645
    Homework Problems   661

12 Force Method   674
    12.1   One Degree Indeterminate Beams   676
    12.2   Multi-Degree Indeterminate Beams   691
    12.3   Indeterminate Trusses   699
    Homework Problems   711

13 Moment Distribution Method   726
    13.1   Overview of Method   728
    13.2   Fixed-End Moments and Distribution Factors   730
    13.3   Beams and Sidesway Inhibited Frames   734
    13.4   Sidesway Frames   754
    Homework Problems   777

14 Direct Stiffness Method for Trusses   792
- 14.1 Overview of Method   794
- 14.2 Transformation and Element Stiffness Matrices   795
- 14.3 Compiling the System of Equations   807
- 14.4 Finding Deformations, Reactions, and Internal Forces   815
- 14.5 Additional Loadings   827
- Homework Problems   841

15 Direct Stiffness Method for Frames   854
- 15.1 Element Stiffness Matrix in Local Coordinates   856
- 15.2 Element Stiffness Matrix in Global Coordinates   862
- 15.3 Loads Between Nodes   868
- 15.4 Finding Deformations, Reactions, and Internal Forces   877
- Homework Problems   886

Index   900

Combine this...

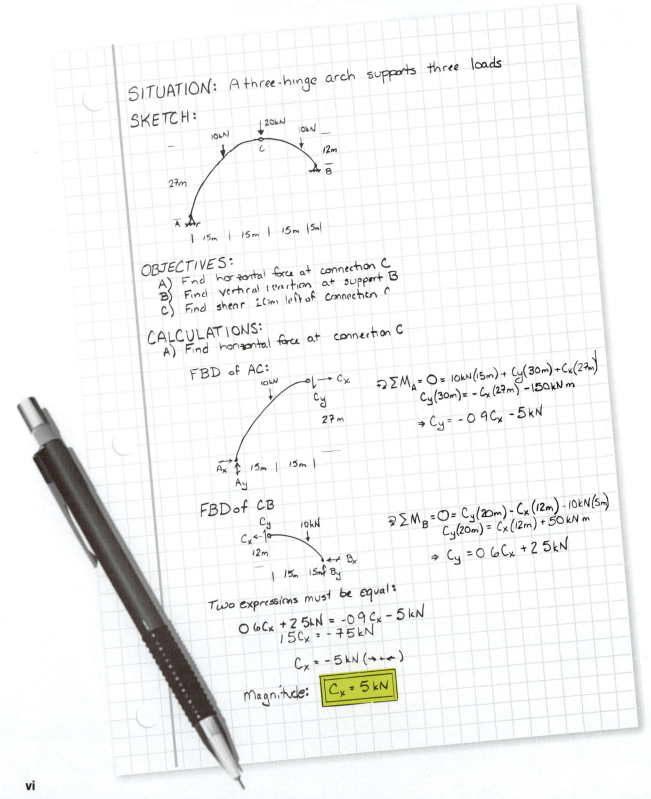

With the Power of Mastering Engineering for Structural Analysis: Skills for Practice

Mastering™ is the teaching and learning platform that empowers every student. By combining trusted authors' content with digital tools developed to engage students and emulate the office hours experience, Mastering personalizes learning and improves results for each student.

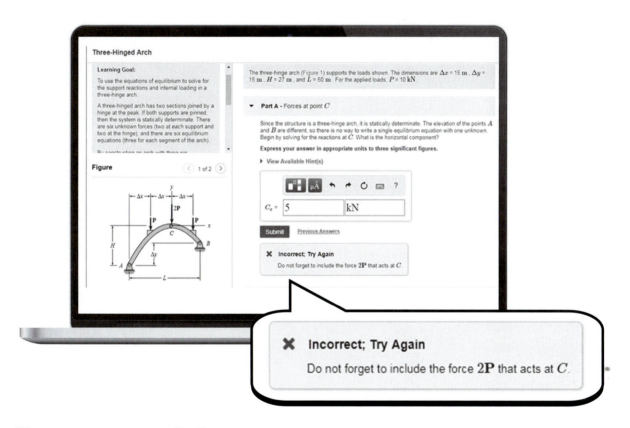

Empower each learner

Each student learns at a different pace. Personalized learning, including optional hints and wrong-answer feedback, pinpoints the precise areas where each student needs practice, giving all students the support they need — when and where they need it — to be successful.

Learn more at www.pearson.com/mastering/engineering

PREFACE

Motivation for a New Text

Let's be realistic, most structural analysis performed in practice is done on a computer. So why do we need another text on how to perform analysis by hand? Because most structural analysis is performed using a computer. That might sound like circular logic, but think about it for a moment. A text on hand methods for structural analysis should be focused on skills needed to complement computer aided analysis, and I couldn't find one of those.

If you ask experienced engineers, there are three practical reasons for performing hand calculations: 1) some problems are faster to solve by hand than by computer, 2) hand methods can be more efficient in the preliminary design phase where we don't yet know the member properties, and 3) hand methods make up many of the tools practitioners use for evaluating detailed analysis results. The topics in this text were carefully chosen to support these three purposes. That meant developing several chapters dedicated to skills used by experienced engineers but not found in other textbooks (e.g., approximate analysis of braced frames, approximating drift, analysis of rigid diaphragms).

Yes, computers have made it possible for us to design structures that could not have been designed before. Nevertheless, even today most structures could still be designed by hand. It is the increase in efficiency that makes computers indispensable in the modern design process. With that increased efficiency, however, also comes the ability to make errors faster than ever before. Therefore, it is especially important that new engineers learn the skills for and develop the habit of evaluating the reasonableness of structural analysis results.

The evaluation skills presented in this text are the result of a ten year project to gather experience from practicing structural engineers and incorporate it in the classroom based on principles from cognitive science. Students following a traditional curriculum and practitioners both took an exam to measure their ability to identify the most reasonable answer and explain why. As expected, practitioners outperformed the students. With the curriculum presented in this text, however, students performed much better on the exam than students following the traditional curriculum. In fact, they cut the gap with practitioners in half.

The curriculum in this text emphasizes developing intuition for reasonable answers and cultivating the habit of predicting results. Intuition allows experienced practitioners to know if a result is not reasonable without giving it conscious thought. The fastest way to develop intuition is to practice evaluation skills routinely and to reflect on the thought process we used. That reflection is called metacognition and is fostered in most of the homework problems in this text. Expert evaluators of results will tell you that they start by predicting results. There are important reasons from cognitive psychology for why it should be in that order, so predicting results before performing detailed analysis is a routine part of the homework and example problems in this text. These skills and habits are valuable not only for students going on to practice structural engineering, but for our students going into any field of engineering.

So while developing a new text, why not address other issues students bring up about their structural analysis textbooks and courses. One such issue is not seeing how the theory connects with the real world. It is difficult for someone new to structural engineering to make the connection between stick figures on a page and real structures they see in the world. To help students make this connection, every example and homework problem is based on a real-world structure with a scenario motivating the requested analysis.

Another issue is the amount of detail in the examples. Students learn a lot by reviewing worked examples and reflecting on why each step is taken. To help in that learning, the examples in this text carry units throughout all calculations and the examples don't skip steps. In addition, the calculations are augmented with comments explaining why different steps were performed and what the results mean.

Organization

Each chapter begins with Motivation: a brief description of why the topics in that chapter are important to practice structural engineering. Most of the sections within the chapters are organized with the following format: Introduction, How-To, Section Highlights (boxed and shaded for easy identification), and Example Problems (boxed for easy identification). The homework problems are grouped at the end of each chapter and are easily identified by a ribbon down the side of the page.

Homework Problems and Example Structure

The homework format is another product of the ten year study. The homework problems are structured to achieve three goals: 1) develop intuition, 2) practice the concept, and 3) accurately evaluate results.

Most homework problems begin with students making a guess about some part of the solution in order to promote development of intuition. It is important to the development process that students make a **guess** without fear of being wrong. Therefore, this part should be graded based on whether it was done or not. If students believe that the quality of the guess will impact their score, they will wait until after they have generated a solution before writing down their guess.

The middle part of each homework problem emphasizes application of the concepts covered in that section of the text. This is the traditional hand calculation portion.

Since many of the hand methods in this text are useful for evaluating computer analysis results, homework problems for about half of the chapters also require that the student calculate the result using structural analysis software. The student is asked to verify fundamental principles for their result (i.e., all equations of equilibrium are satisfied) and features of the graphical solution (e.g., segment of constant shear diagram value where there is no applied load).

The student is then asked to make a comprehensive argument that the computer analysis results are reasonable. For full credit, the student should use all of the previous parts of the homework problem (except the guess) to demonstrate that the computer results are reasonable: hand solution(s), verification of fundamental principles, and verification of features of the graphical solution. In this argument the hand solution part of the homework might have used an approximate analysis method. In those cases, the student should recognize that the hand and computer solutions should not match perfectly. The student will need to decide whether the difference is acceptable or not.

Grading Advice

Each problem that starts with an initial guess ends with reflection on that guess. The student is asked to compare the initial guess with the computer results and reflect on why the two are similar or different. Again, if the instructor wants to successfully promote development of intuition, the students must feel that there is no disadvantage to having an initial guess that does not match the computer result. An example rubric that can be used to score this reflection is shown in the following table:

	Full Credit (10)	**Adequate (7)**	**Marginal (5)**	**Unacceptable (0)**
If the guess and solution generally match:	Explains how previous experience and/or fundamental principles led to a guess that matched.	Identifies previous experienced and/or fundamental principles that guided the guess.	Attempts to explain why the guess matched the solution, but shows little understanding of pertinent fundamental principles and/or features of the solution.	No demonstration of understanding of why the guess matched the solution.
If the guess and solution generally *do not* match:	Explains why guess does not match based on previous experience and/or fundamental principles.	Identifies fundamental principles and/or features of the solution that could be used to explain the difference.	Attempts to explain the difference, but shows little understanding of pertinent fundamental principles and/or features of the solution.	No demonstration of understanding of why the guess did not match the solution.

Examples of how to apply the rubric to score student reflections are also available.

Using Structural Analysis Software

This text is not based on the use of a specific structural analysis software program. Any structural analysis program that can model 2D trusses and frames will be sufficient. Note that in order to model braced frames, the program must allow specification of pinned connections in an otherwise rigid frame.

If students do not already have access to structural analysis software, they can obtain free software via the internet. For example, basic use of the program MASTAN2 can be taught in a single lecture. The program is available for free download from the following website:

www.mastan2.com

Instructor Resources

The single objective of this text is to prepare your students with skills and habits for the practice of engineering, regardless of the specialty. Trust the process. Do all the steps. The organization of the example and homework problems is based on how experienced engineers approach analysis and is supported by cognitive science.

All instructor resources are available for download at www.pearsonhighered.com. If you are in need of a login and password for this site, please contact your local Pearson representative.

Mastering Engineering

This online tutorial and assessment program allows you to integrate dynamic homework with automated grading of the calculation parts of problems and personalized feedback. Mastering^{TM} Engineering allows you to easily track the performance of your entire class on an assignment-by-assignment basis, or the detailed work of an individual student. For more information visit www.masteringengineering.com.

Instructor Solutions Manual

Fully worked-out solutions to the homework problems.

PowerPoint Lecture Images

All figures from the text are available in PowerPoint for your lecture needs. These are used to give students real visual examples of the phenomena.

Learning Catalytics

This "bring your own device" student engagement, assessment and classroom intelligence system enables you to measure student learning during class, and adjust your lectures accordingly. A wide variety of question and answer types allows you to author your own questions, or you can use questions already authored into the system. For more information visit www.learningcatalytics.com or click on the Learning Catalytics link inside Mastering Engineering.

Prerequisite Courses

This text is constructed assuming that students have already completed statics and mechanics of materials courses; therefore, topics such as determinate truss analysis have been omitted.

Acknowledgments

I am blessed to be able to share this approach to teaching structural analysis with you. The old phrase "It takes a village" is so true. All of my students over the years have inspired me and helped me in creating this text. I greatly appreciate their hard work and feedback. My colleagues in Civil and Environmental Engineering at Rose-Hulman have been extremely supportive of me as I focused on making this available to you. There is nothing we won't do for each other to provide a better student experience.

The approach to structural analysis and verification of results unique to this text is a direct result of what I learned from interviews with dozens of experienced structural engineers. Their passion and input really made this text about skills for practice.

The team that Pearson assembled to help me in this process has been stellar. Their unified focus has been to bring my vision to life in order to help you. Part of that team is faculty reviewers, and their feedback made this text so much better. Some of them want to remain anonymous, but others agreed to allow me to thank them publicly:

- Bechara Abboud, Temple University
- Tomasz Arciszewski, George Mason University
- Mikhail Gershfeld, California State Polytechnic University
- Thomas H. Miller, Oregon State University
- Gokhan Pekcan, University of Nevada – Reno
- Hayder A. Rasheed, Kansas State University
- Hung-Liang (Roger) Chen, West Virginia University
- Husam Najm, Rutgers University
- Steven Vukazich, San Jose State University

This text doesn't happen without the support of my friends and family, especially the love of my life Diane. Because they believe in sharing my passion with you, they sacrificed and encouraged me. For example, all of the photos in this text were taken by me. That means my family endured many stops and detours during our travels in order to hunt for those images.

The words "thank you" don't seem adequate, but it is so important for me to thank all these people for joining into the vision for this text. The impact of their contributions permeates every page. Each and every person has my sincere thanks and gratitude!

How You Can Help

This text is meant to help you, both instructor and student. If you see an opportunity to do that better please let me know. As I said, it takes a village.

Thanks!
Prof. Jim Hanson
james.hanson@rose-hulman.edu

Visual Walkthrough

Motivation

Structural analysis, at its most basic level, is predicting the effects of loadings on a structure. A huge variety of methods are used to make those predictions, and those methods are the focus of most of the chapters in this text. Before we can begin implementing any structural analysis methods, however, we need to have a model of the actual structure or the structure we envision.

The model is a representation, an idealization, of reality. It captures the most important attributes of the real structure, but without the full complexity of all the attributes. For example, our structural models typically include information about the cross-sectional properties of the beams and columns. But they typically do not include information about the quantity and placement of reinforcing steel in the concrete or the number and configuration of bolts in a connection. So we need to know how to create a model with sufficient information to perform structural analysis. We call the process of creating the model *idealizing* the structure.

An important part of the model is the loading on the idealized structure. Knowing which loads are significant in the behavior and design of the structure is just as important as knowing which attributes of the real structure are important to capture in the model. For example, a window washer's ladder leaning against the outside wall of a building is generally not significant, but wind during a strong storm is important.

In the modeling process, we also need to convert the loads on the real structure into the resulting loads on the idealized structure. Most loads are actually pressures on surfaces, but those surfaces are often not included in our structural model. Therefore, we need to understand the path that applied loads follow through the real structure in order to predict the loading on the idealized structure.

All of these incredibly important preparatory skills are the focus of this chapter.

Section 1.4 Highlights

Application of Gravity Loads

Assumption: A floor or roof diaphragm is much more flexible out of plane than the members that support it.

Approximations: Each diaphragm panel behaves independently of other panels; we are ignoring continuity between panels.

For panels that are supported on more than two sides, if $S_{long}/S_{short} \geq 2$ we can consider the diaphragm to act one-way.

Distributed Load: $w = \text{pressure} \times \text{tributary width}$

Section Highlights
Section Highlights are boxed and highlighted for easy identification.

Motivations
Motivations start each chapter to provide justification and real world context for what students will be learning about in the chapter and why it's important.

Example 5.8

The manufacturer of freestanding jib cranes has been receiving complaints about how far the tip deflects downward when lifting heavy loads. The manufacturer suspects the problem is poor foundations, but to verify that the problem is not in the design of the cranes, we have been hired to predict the maximum vertical displacement under the rated load, 2 tons.

Figure 5.45

Each jib crane is made of steel, $E = 29,000$ ksi. The arm is a W18 × 40 ($A = 11.8$ in^2, $I = 612$ in^4), and the mast is an HSS 18 × 0.375 ($A = 19.4$ in^2, $I = 754$ in^4). The peak displacement will occur when the hoist is at the far end of the arm.

Evaluation of Results:
Observed Expected Features?
The vertical displacement at A is down, as expected. ✓

Satisfied Fundamental Principles?
The arm bending contribution to the vertical displacement is identical to the approximate solution. We should expect this because the approximate solution considers only bending in the arm. ✓

Approximation Predicted Outcomes?
The predicted displacement is greater than the approximation, as expected. ✓
The detailed prediction is three times larger than the estimate, which might merit closer review. In this case, however, it turns out that the mast does actually contribute a significant amount to the displacement.

Conclusion:
The predicted peak displacement is 2.6 inches. That equates to a relative displacement of $l/83$. The standard of practice for a floor beam is to limit the live load deflection to $l/360$, or $l/180$ for a cantilever. Dead load limits are typically $l/240$ for a beam supported on both ends or $l/120$ for a cantilever. Therefore, it appears reasonable that customers might find the displacement excessive.

Recommendation:
If the crane manufacturer wants to reduce the vertical displacement, our results indicate that increasing the moment of inertia of the mast would be most effective. If that does not provide a sufficient reduction, the manufacturer should increase the moment of inertia of the arm as well. Focusing on the cross-sectional areas of the members will not have a noticeable impact.

Examples
The text emphasizes developing intuition for reasonable answers and cultivating the habit of predicting results.
Evaluation of Results within the Example Problems include icons and headings reinforcing the importance of evaluating results via Observations of Expected Features, Satisfaction of Fundamental Principles, and confirmation that the Approximations Predicted the Outcomes.

Homework Problems

4.9 For ease of construction in the field, a two-span bridge was constructed with a splice and a hinge. The Department of Transportation has received a request to increase the load limit on the bridge. Before acting on that request, they want to know the internal forces due to the dead load. A team member recommends that it will be easier to use deduction rather than integration.

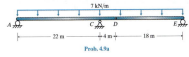

Prob. 4.9a

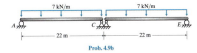

Prob. 4.9b

a. Guess the location of the peak positive moment.
b. Estimate the peak positive moment by considering the compound beam as two simply supported beams. Will this provide an upper bound, a lower bound, or just an approximation for the actual peak positive moment? Justify your answer.
c. Construct the shear and moment diagrams for the original beam. Label values and locations.
d. Identify at least three features of the shear diagram that suggest you have a reasonable answer.
e. Identify at least three features of the moment diagram that suggest you have a reasonable answer.
f. Determine the peak moment due to the dead load and its location.
g. Make a comprehensive argument that the peak moment found in part (f) is reasonable.
h. Comment on why your guess in part (a) was or was not close to the solution in part (f).

Homework Problems
The homework problems are structured to achieve three goals: 1) develop intuition, 2) practice the concept, and 3) accurately evaluate results. Most homework problems and examples begin with students making a guess about some part of the solution in order to promote development of intuition. Each problem that starts with an initial guess ends with a reflection on that guess.

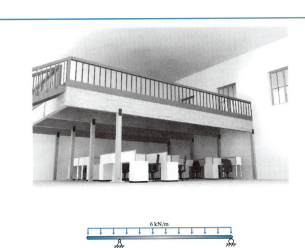

Prob. 2.7

Real-World Connection
Each example and homework problem starts with a real-world scenario to show how the analysis skills apply to practice. With that scenario is a photo or highly detailed photorealistic rendering of the structure to connect the idealization to reality.

CHAPTER 0

EVALUATING RESULTS

The reality is that in the practice of structural engineering we typically rely on complex structural analyses to finalize our designs. Our obligation to hold paramount the safety, health and welfare of the public[1] means that we must *always* evaluate the reasonableness of our results. Practitioners use a wide variety of tools to evaluate their analysis and design results. If we focus specifically on how experienced practitioners evaluate structural analysis results, their tools can be organized into three categories: features of the solution, fundamental principles, and approximations.

Observed Expected Features?

Based on our understanding of mechanics principles and the situation, we expect to see certain features in the results of our analysis. For example, we expect a beam to have a smooth deflected shape as long as there is no internal hinge. Similarly, when we look in the mirror we expect to see certain attributes. Therefore, whenever we have compared features of the solution in the example problems, we have put a small mirror.

Satisfied Fundamental Principles?

Fundamental principles such as equilibrium and compatibility must be satisfied at all times. Just as a compass always points north, these fundamental principles always apply. We typically learn these principles in our statics and mechanics of materials courses. Now we will rely on them to help verify that our results are reasonable. To help you recognize where we used fundamental principles to verify our results in an example problem, we have put a small compass with the check.

Approximations Predicted Outcomes?

We make approximations in order to obtain a solution quickly and with reduced likelihood of error. Approximations are like making a curved road straight; it is not the same journey, but you finish in a similar place. Practitioners use approximations extensively, so we will cover many different approximation tools. We have put a small road sign with each approximation used to verify results.

---

[1] Part of First Fundamental Cannon from the ASCE Code of Ethics, and similar to part of the Code of Professional Conduct of the European Council of Civil Engineers

CHAPTER 1

LOADS AND STRUCTURE IDEALIZATION

In order to convert reality into something we can analyze, we create idealized versions of the structure and loading.

MOTIVATION

Structural analysis, at its most basic level, is predicting the effects of loadings on a structure. A huge variety of methods are used to make those predictions, and those methods are the focus of most of the chapters in this text. Before we can begin implementing any structural analysis methods, however, we need to have a model of the actual structure or the structure we envision.

The model is a representation, an idealization, of reality. It captures the most important attributes of the real structure, but without the full complexity of all the attributes. For example, our structural models typically include information about the cross-sectional properties of the beams and columns. But they typically do not include information about the quantity and placement of reinforcing steel in the concrete or the number and configuration of bolts in a connection. So we need to know how to create a model with sufficient information to perform structural analysis. We call the process of creating the model *idealizing* the structure.

An important part of the model is the loading on the idealized structure. Knowing which loads are significant in the behavior and design of the structure is just as important as knowing which attributes of the real structure are important to capture in the model. For example, a window washer's ladder leaning against the outside wall of a building is generally not significant, but wind during a strong storm is important.

In the modeling process, we also need to convert the loads on the real structure into the resulting loads on the idealized structure. Most loads are actually pressures on surfaces, but those surfaces are often not included in our structural model. Therefore, we need to understand the path that applied loads follow through the real structure in order to predict the loading on the idealized structure.

All of these incredibly important preparatory skills are the focus of this chapter.

1.1 Loads

Introduction

When analyzing and designing a structure, we consider the types of loads that could reasonably act on the structure during its lifetime. For example, it is very unlikely that a structure built on the earth's surface will experience the load of an asteroid impact. Therefore, asteroids are typically not on the list of loads we consider in design. A very strong wind, however, is a reasonable possibility, so we design for that.

The reasonably likely values of load for which we should design are often identified in standards like *Minimum Design Loads for Buildings and Other Structures* (ASCE 2017) or codes like the *AASHTO LRFD Bridge Design Specifications* (AASHTO 2012) and the Eurocode *EN 1991: Actions on Structures* (CEN 2002–2006). In some situations, we anticipate loads that require us to search other resources (e.g., floating ice pushing on a marine structure). In all cases, we can increase the magnitude of loads and add types of loads if, in our judgment, such a change more accurately reflects the risk.

How the load acts on a structure depends on the type of load.

How-To

Dead (D)

Dead load is the self-weight of a structure. The dead load typically includes everything fixed in place, even nonstructural items such as flooring, plumbing, and hand rails. Dead load is considered a gravity load because it acts vertically (due to the effect of gravity).

The self-weight of nonstructural items is often called *superimposed dead load* because the effect is added to the self-weight of the structural members. Examples of superimposed dead load include the self-weights of electrical conduit, light fixtures, air ducts, carpeting, and ceiling tiles.

The total dead load is the product of the density of the material and the volume of the material. Table 1.1 shows typical densities of some common construction materials. For materials with specified or common thicknesses, the dead load is often presented as weight per unit of surface area (Table 1.2). Multiplying the density by the cross-sectional area gives a distributed load that acts along the length of the structure (Figure 1.1).

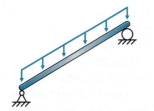

Figure 1.1 Load acting on an inclined structure.

Live (L)

We consider loads associated with the use or occupancy of a structure as live loads. Examples include people, vehicles, furniture, books, merchandise, and partition walls (interior, nonstructural walls that are reasonably likely to be moved over the life of the structure). The two common exceptions are stored liquids and bulk materials (e.g., corn, sand) because those are given their own categories. Live load is also a gravity load, so it acts vertically.

Unlike dead load, live load can act everywhere, somewhere, or nowhere. By this we mean that the full floor space might be used (everywhere), only some of the floor space might be used (somewhere), or at times none of the floor space will have live load (nowhere). We must consider all three possibilities to find the most extreme design case for a structural member.

Table 1.1 Densities of Common Construction Materials (Based on ASCE 2017)

Material	pcf	kN/m^3
Concrete, Reinforced:		
Cinder	111	17.4
Stone	150	23.6
Glass	160	25.1
Masonry, Brick (solid parts of hollow masonry):		
Hard	130	20.4
Soft	100	15.7
Masonry, Concrete (solid parts of hollow masonry):		
Lightweight units	105	16.5
Normal weight units	135	21.2
Plywood	36	5.7
Steel, Cold-Drawn	492	77.3
Wood, Seasoned:		
Ash, commercial white	41	6.4
Fir, Douglas, coast region	34	5.3
Oak, commercial reds and whites	47	7.4
Pine, southern yellow	37	5.8
Redwood	28	4.4

Table 1.2 Dead Load Pressures of Common Construction Materials (Based on ASCE 2017)

Component	psf	kN/m^2
Ceilings:		
Acoustical fiberboard	1	0.05
Mechanical duct allowance	4	0.19
Suspended steel channel system	2	0.10
Coverings, Roof, and Wall Composition:		
Three-ply ready roofing	1	0.05
Four-ply felt and gravel	5.5	0.26
Deck, metal, 18 gauge	3	0.14

(*continued*)

Table 1.2 (Continued)

Component	psf	kN/m²
Insulation, roof boards (per inch or mm thickness):		
Fiberboard	1.5	0.0028
Perlite	0.8	0.0015
Polystyrene foam	0.2	0.004
Urethane foam with skin	0.5	0.009
Waterproofing membranes:		
Bituminous, gravel-covered	5.5	0.26
Liquid applied	1	0.05
Single-ply, sheet	0.7	0.03
Floors and Floor Finishes:		
Ceramic or quarry tile (3/4-in.) on 1/2-in. mortar bed	16	0.77
Hardwood flooring, 7/8-in.	4	0.19
Linoleum or asphalt tile, 1/4-in.	1	0.05
Frame Partitions:		
Movable steel partitions	4	0.19
Wood or steel studs, 1/2-in. gypsum board each side	8	0.38

The magnitude of the live load we expect, and therefore use in design, depends on the use of the area. Table 1.3 lists live load pressures for a variety of building uses. Live load for bridges is typically a uniform lane load typically a uniform lane load and a single point load that represents the design truck. Note that for most building occupancy categories it is extremely unlikely that large areas are experiencing the full live load simultaneously. Therefore, the codes allow us to reduce the average magnitude of the live load based on the total area being supported by a member. This process is called *live load reduction* and is outside the scope of this text.

Most areas that experience live load are flat. But sometimes we deal with inclined areas such as ramps or theater seating. In those cases, we consider the live load to act on the *horizontal projection* (Figure 1.2). The horizontal projection is the area seen from above, the plan view. We do this because a steeper ramp has more surface area but does not hold more people. The number of people who can push together is limited by the length and width as seen from above.

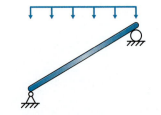

Figure 1.2 Load acting on the horizontal projection of an inclined structure.

Snow (S)

Snow is another gravity load, thus acting down. The amount of snow that falls from the sky onto a given area is not dependent on the surface area but on the length and width as seen from above; therefore, snow acts on the horizontal projection just like live load. Because snow is an environmental load, the weight of snow that accumulates on the ground depends on the geographic region. We modify the ground snow load to account for terrain effects around the structure, thermal behavior of the structure, and potential impact of failure.

Table 1.3 Live Load Pressures (Based on ASCE 2017)

Occupancy or Use	psf	kN/m²
Assembly Areas and Theaters: [a]		
Fixed seats (fastened to floor)	60	2.87
Lobbies	100	4.79
Movable seats	100	4.79
Platforms (assembly)	100	4.79
Stage floors	150	7.18
Balconies and decks [b]	≤ 100	≤ 4.79
Corridors:		
First floor	100	4.79
Other floors [c]		
Dining Rooms and Restaurants [a]	100	4.79
Libraries:		
Reading rooms	60	2.87
Stack rooms [a]	150	7.18
Corridors above first floor	80	3.83
Manufacturing and Storage Warehouses: [a, d]		
Light	125	6.00
Heavy	250	11.97
Office Buildings: [e]		
Lobbies and first-floor corridors	100	4.79
Offices	50	2.40
Corridors above first floor	80	3.83
Roofs:		
Ordinary flat, pitched, and curved roofs	20	0.96
Roofs used for roof gardens	100	4.79

[a] Live load reduction for these uses is not permitted.
[b] Requirement is 1.5 times the live load for the occupancy served. Need not exceed the values shown.
[c] Same as occupancy served except as indicated.
[d] Shall be designed for heavier loads if required for anticipated storage. Live load reduction for these uses is not permitted.
[e] Shall also be designed for a single point load of 2000 lb or 8.90 kN without the pressure.

    Because snow can be moved by wind and might slide on sloped roofs, we typically consider partial loading, unbalanced loading, drifting, and rain-on-snow surcharge. We demonstrate some of these in examples throughout this text.

Wind (W)

Wind is an environmental load that depends on many factors. The wind pressures we use in design are based on geographic location, terrain effects around the structure, shape of the structure, potential impact of failure, and height above the ground surface. The peak wind velocity increases with distance from the ground; therefore, wind pressure on a structure is higher at the top than at the bottom.

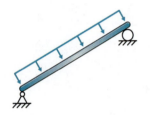

Figure 1.3 Load acting normal to the surface of an inclined structure.

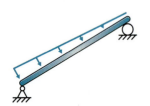

Figure 1.4 Load acting normal to the surface and increasing with depth on an inclined structure.

The wind pressures all act normal to the structure surfaces (Figure 1.3). Therefore, wind load tends to have vertical and lateral components. On the windward side (upwind side), wind typically creates pressure on the surface; on the leeward side (downwind side), it typically creates suction.

Fluid (F) and Soil (H)

In some cases, structures hold fluids in quantities large enough that we need to consider the load (e.g., swimming pool, cistern, chemical vat). Fluids stored in the structure tend to be static. Therefore, the pressure on the structure is hydrostatic and acts normal to the surface. Because the pressure is generated by gravity, the magnitude of the pressure increases with depth (Figure 1.4).

Soil has a similar effect on a structure. If the structure does not move relative to the soil, the at-rest condition, then the soil creates pressure that increases with depth and acts normal to the surface. If the structure moves toward the soil (passive condition) or away from the soil (active condition), there might also be friction stress parallel to the surface.

Because both fluid and soil pressures act normal to surfaces, they might have horizontal and vertical components.

Earthquake (E)

Earthquake loads are fundamentally different from the other loads. All of the other loads are forces that act on the structure. In an earthquake, however, the ground moves underneath the structure. The resulting acceleration of the structural mass has an effect similar to an applied force.

Except for extreme cases, we approximate the dynamic effect on the structure with an equivalent static force. That force is dispersed throughout the structure wherever the largest concentrations of mass are. Therefore, we typically idealize earthquake load as forces applied laterally at the floor and roof levels (Figure 1.5a). Under some circumstances, we also consider the vertical effect. In those cases, we idealize the vertical component of the earthquake load as distributed along the floor and roof levels just like dead load (Figure 1.5b).

Units

Loads act on structures as pressures (e.g., $kPa = kN/m^2$ or $psf = lb/ft^2$). The conversion of those pressures to distributed loads on members is covered in Sections 1.4 and 1.5. The units of distributed loads are force/length (e.g., kN/m or $plf = lb/ft$). Because of the large magnitude of forces used in structural engineering, we often use a unit of force uncommon in other disciplines of engineering: the kip, k. A kip is 1000 lb. In SI units, we typically measure force in kilonewtons, kN.

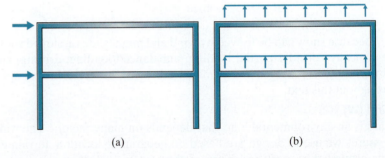

Figure 1.5 Earthquake idealized as equivalent static forces at the levels with the most concentrated mass: the floor and roof; (a) Lateral forces for horizontal seismic effects; (b) Distributed loads for vertical seismic effects.

Section 1.1 Highlights

Loads

Application of loads by type:

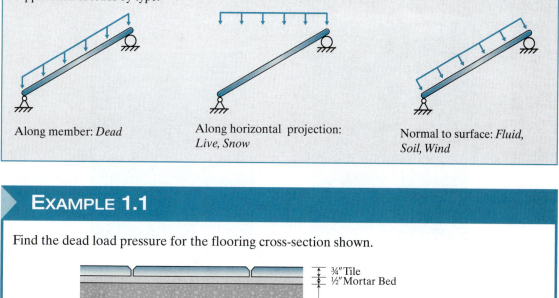

Along member: *Dead*

Along horizontal projection: *Live, Snow*

Normal to surface: *Fluid, Soil, Wind*

Example 1.1

Find the dead load pressure for the flooring cross-section shown.

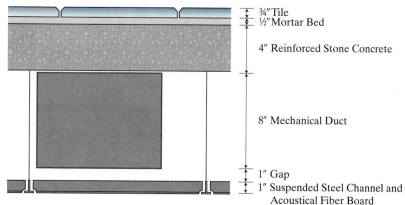

Figure 1.6

Detailed Solution:

From Table 1.2:	
Tile on mortar bed	16 psf
Mechanical duct allowance	4 psf
Suspended steel channel system	2 psf
Acoustical fiberboard	1 psf
From Table 1.1:	
Reinforced stone concrete	
150 pcf (4 in.)/(12 in./ft) =	50 psf
Total:	73 psf

EXAMPLE 1.2

Find the soil pressure along a 1-meter-wide strip of the inclined back of the retaining wall. The soil is a sand-gravel mix. For this type of soil, the typical pressure of the soil on the wall is 5.50 kN/m² per meter of depth.

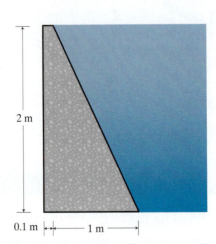

Figure 1.7

Detailed Analysis:

Pressure for a 1-m-strip of wall:	5.50 kN/m per m depth	= 5.50 kN/m²
Pressure at top:	(5.50 kN/m²)(0 m)	= 0 kN/m
Pressure at bottom:	(5.50 kN/m²)(2 m)	= 11 kN/m
Angle:	$\tan^{-1}(1\text{ m}/2\text{ m})$	= 26.6°

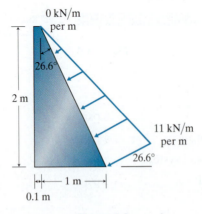

Figure 1.8

1.2 Load Combinations

Introduction

One of the primary goals of structural design is public safety. To have a safe design, we need structural analysis results sufficiently large that there is only a very small probability that the actual loading will exceed those results. To rationally develop those analysis results, we typically amplify the effects of the maximum likely load. The amount of amplification depends on the type of load because the peak magnitude of some types of load is more variable than for others.

Another consideration is that different types of loads often act simultaneously. It would be dangerous to consider the effects of only one load at a time. Therefore, we consider the effects of reasonably likely combinations of loads with load factors. To ensure public safety, we design for the structural analysis results based on the factored and combined effects.

How-To

In Section 1.1, we introduced the concept of maximum likely loads. But the peak load that a structure experiences actually has a probability distribution like the one shown in Figure 1.9. The magnitude that has an acceptably low probability of being exceeded is what we call the *maximum likely load*. The shape of the probability distribution is roughly the same for all types of loads, but the spread of the distribution changes for different types of loads (Figure 1.10).

In Figure 1.10, both distributions lead to the same maximum likely load magnitude, but the range of peak load magnitudes beyond the maximum likely value is much bigger for the dark blue graph. This means that

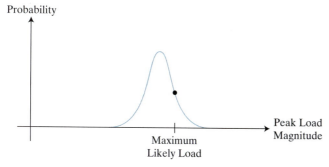

Figure 1.9 Probability distribution of the peak load that a structure experiences in its lifetime.

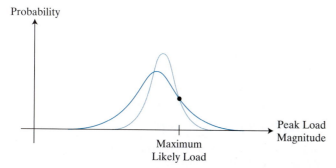

Figure 1.10 Probability distribution of the peak load for two different types of loads with exactly the same maximum likely peak load.

if the maximum likely value is exceeded with the dark blue load, it might reach a much larger value than would the light blue load.

In order to protect the safety, health, and welfare of the public, we use load factors to increase the maximum likely load effects in case the actual peak load exceeds the maximum likely value. We use different load factor values based on the spread of the probability distributions like the ones shown in Figure 1.10. For example, dead load has a narrower distribution, like the light blue curve, so we tend to know the maximum likely self-weight of a structure pretty well. Therefore, we give dead load a load factor of 1.2 in many of the load combinations from ASCE 7. Live load has a broader distribution, like the dark blue curve, so we give live load a load factor of 1.6 in many of the combinations.

The most extreme effect on a structure typically occurs when multiple loads act simultaneously. Therefore, we consider the combined effects of loads, as shown in Table 1.4. The factored and combined effect is called the *design value* (e.g., design moment, design reaction) because that is the value for which we must design the structure. We also call this the *ultimate value*.

The different combinations represent the likelihood of different types of loads reaching their peak simultaneously. For example, it is extremely unlikely that wind and earthquake will reach their maximum likely values at the same moment; therefore, none of the combinations have both wind and earthquake terms.

The combinations use some load factors less than 1 for the same reason. For example, consider combination 3. It is highly unlikely that a structure will experience an extreme live load at the same time as the worst snowstorm in years. People stay home during bad storms like that, but it is reasonable to expect some live load during the snowstorm. Therefore, combination 3 uses a factor of 0.5 on live load effects and a factor of 1.6 on snow effects.

It is important to use judgment with these load factors and combinations. If the effect of a type of load diminishes the effect of another type of load, we must consider the possibility that one or more of the loads does not act at all, a load factor of 0. The only exception is loads that will always act, such as dead. We see from the combinations that it is reasonable to reduce the load factor to 0.9 for a permanent load *if* that would increase the overall effect.

Table 1.4 Load Combinations with Load Factors (Based on ASCE 2017)

ID	Combination
1.	$1.4(D + F)$
2.	$1.2(D + F) + 1.6(L + H) + 0.5S$
3.	$1.2(D + F) + 1.6(S + H) + (0.5L \text{ or } 0.5W)$ [a,b]
4.	$1.2(D + F) + 1.0W + 1.6H + 0.5L + 0.5S$ [a]
5.	$0.9D + 1.0W + 1.6H$
6.	$1.2(D + F) + 1.0E + 1.6H + 0.5L + 0.2S$ [a,c]
7.	$0.9(D + F) + 1.0E + 1.6H$ [c]

[a] For all occupancies with live load pressure greater than 100 psf or classified as "assembly" (Table 1.3), the factor on L is 1.0 rather than 0.5.
[b] The "or" means that only the larger of the two effects need be considered.
[c] When considering vertical seismic load effect, E_v, consider upward or downward to maximize the effect.

Section 1.2 Highlights

Load Combinations
- Consider all relevant combinations.
- Permanent loads can have a load factor ranging from 0.9 to the maximum shown in the combinations.
- All other loads can have a load factor ranging from 0 to the maximum shown.

Example 1.3

Our firm is designing timber roof trusses over the lobby at a new resort. To properly design the members and connections, our team needs to know the design tension and design compression for each member. We have been assigned two members: AB and AF.

Figure 1.11

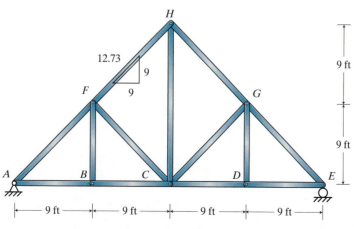

Figure 1.12

For preliminary design, we will consider the left support to be a pin and the right to be a roller. We must consider four types of loads for these trusses: dead, live, snow, and wind. In addition, wind can act from either direction, so we have two mutually exclusive wind cases.

Dead:

One engineer has already estimated the self-weight of the truss members and roof.

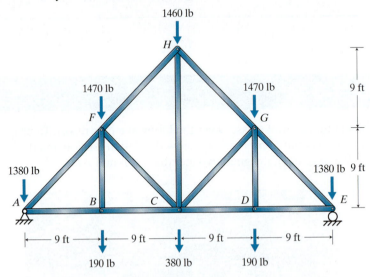

Figure 1.13

Live:

The client wants to be able to hang items to display from the trusses. The senior engineer on the project wants us to consider the possibility of a single point load of 1000 lb that could act at any of the nodes along the bottom chord. For the two members we are assigned, the axial forces will be largest if the live load acts at B.

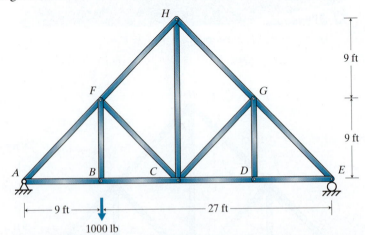

Figure 1.14

Snow:

Another engineer has calculated the maximum likely snow load and the resulting point forces on the truss. These are unfactored.

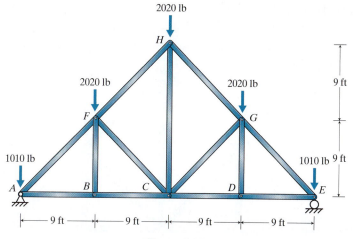

Figure 1.15

Wind Case 1:
The first wind case is wind from the left side.

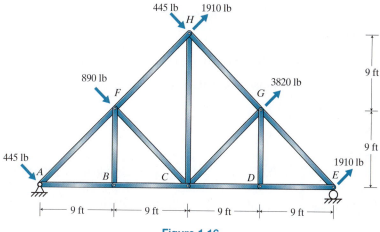

Figure 1.16

Wind Case 2:
The second wind case is the same wind but from the right side.

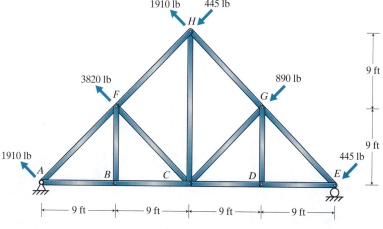

Figure 1.17

16   **Chapter 1**   Loads and Structure Idealization

Find Unfactored Internal Forces:

Dead:

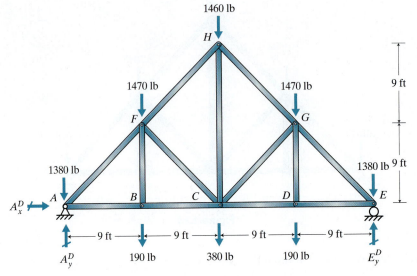

Figure 1.18a

Find the reactions at A:

$$\circlearrowleft \Sigma M_E = 0 = A_y^D(36 \text{ ft}) - 1380 \text{ lb}(36 \text{ ft}) - (1470 \text{ lb} + 190 \text{ lb})(27 \text{ ft})$$
$$- (1460 \text{ lb} + 380 \text{ lb})(18 \text{ ft}) - (1470 \text{ lb} + 190 \text{ lb})(9 \text{ ft})$$

$$A_y^D = 3960 \text{ lb } (+\uparrow)$$

$$\xrightarrow{+} \Sigma F_x = 0 = A_x^D$$

FBD of joint A:

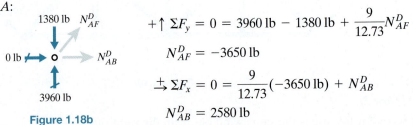

$$+\uparrow \Sigma F_y = 0 = 3960 \text{ lb} - 1380 \text{ lb} + \frac{9}{12.73} N_{AF}^D$$

$$N_{AF}^D = -3650 \text{ lb}$$

$$\xrightarrow{+} \Sigma F_x = 0 = \frac{9}{12.73}(-3650 \text{ lb}) + N_{AB}^D$$

$$N_{AB}^D = 2580 \text{ lb}$$

Figure 1.18b

Live:

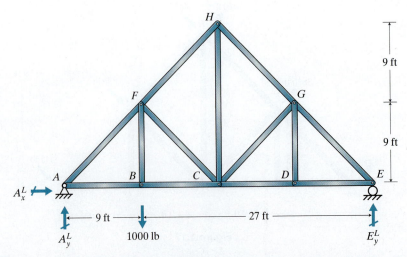

Figure 1.19a

Find the reactions at A:

$$\curvearrowright \Sigma M_E = 0 = A_y^L(36 \text{ ft}) - 1000 \text{ lb}(27 \text{ ft})$$

$$A_y^L = 750 \text{ lb} (+\uparrow)$$

$$\xrightarrow{\pm} \Sigma F_x = 0 = A_x^L$$

FBD of joint A:

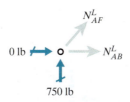

Figure 1.19b

$$+\uparrow \Sigma F_y = 0 = 750 \text{ lb} + \frac{9}{12.73} N_{AF}^L$$

$$N_{AF}^L = -1060 \text{ lb}$$

$$\xrightarrow{\pm} \Sigma F_x = 0 = \frac{9}{12.73}(-1060 \text{ lb}) + N_{AB}^L$$

$$N_{AB}^L = 750 \text{ lb}$$

Snow:

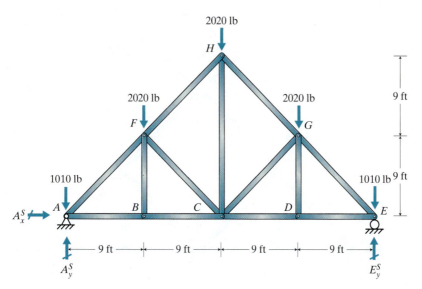

Figure 1.20a

Find the reactions at A:

$$\circlearrowleft \Sigma M_E = 0 = A_y^S(36\text{ ft}) - 1010\text{ lb}(36\text{ ft}) - 2020\text{ lb}(27\text{ ft})$$
$$- 2020\text{ lb}(18\text{ ft}) - 2020\text{ lb}(9\text{ ft})$$
$$A_y^S = 4040\text{ lb}\ (+\uparrow)$$
$$\xrightarrow{+} \Sigma F_x = 0 = A_x^S$$

FBD of joint A:

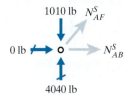

Figure 1.20b

$$+\uparrow \Sigma F_y = 0 = 4040\text{ lb} - 1010\text{ lb} + \frac{9}{12.73}N_{AF}^S$$

$$N_{AF}^S = -4290\text{ lb}$$

$$\xrightarrow{+} \Sigma F_x = 0 = \frac{9}{12.73}(-4290\text{ lb}) + N_{AB}^S$$

$$N_{AB}^S = 3030\text{ lb}$$

Wind Case 1:

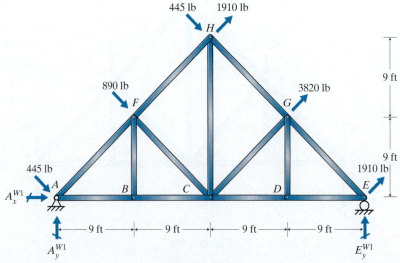

Figure 1.21a

Find the reactions at A:

$$\circlearrowleft \Sigma M_E = 0 = A_y^{W1}(36 \text{ ft}) - 445 \text{ lb}(25.46 \text{ ft}) - 890 \text{ lb}(12.73 \text{ ft})$$
$$+ 1910 \text{ lb}(25.46 \text{ ft}) + 3820 \text{ lb}(12.73 \text{ ft})$$
$$A_y^{W1} = -2070 \text{ lb} \; (+\uparrow)$$

$$\xrightarrow{\pm} \Sigma F_x = 0 = A_x^{W1} + \frac{9}{12.73}(2 \cdot 445 \text{ lb} + 890 \text{ lb}) + \frac{9}{12.73}(2 \cdot 1910 \text{ lb} + 3820 \text{ lb})$$
$$A_x^{W1} = -6660 \text{ lb} \; (\xrightarrow{\pm})$$

FBD of joint A:

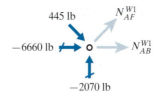

Figure 1.21b

$$+\uparrow \Sigma F_y = 0 = -2070 \text{ lb} - \frac{9}{12.73}(445 \text{ lb}) + \frac{9}{12.73} N_{AF}^{W1}$$
$$N_{AF}^{W1} = 3370 \text{ lb}$$

$$\xrightarrow{\pm} \Sigma F_x = 0 = -6660 \text{ lb} + \frac{9}{12.73}(445 \text{ lb}) + \frac{9}{12.73}(3370 \text{ lb}) + N_{AB}^{W1}$$
$$N_{AB}^{W1} = 3960 \text{ lb}$$

Wind Case 2:

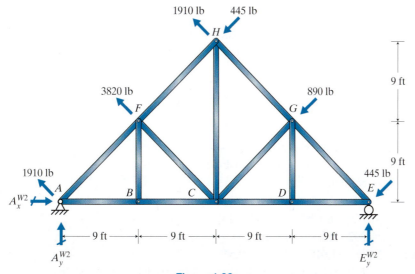

Figure 1.22a

Find the reactions at A:

$$\circlearrowleft \Sigma M_E = 0 = A_y^{W2}(36\text{ ft}) + 1910\text{ lb}(25.46\text{ ft}) + 3820\text{ lb}(12.73\text{ ft})$$
$$- 445\text{ lb}(25.46\text{ ft}) - 890\text{ lb}(12.73\text{ ft})$$
$$A_y^{W2} = -2070\text{ lb}\ (+\uparrow)$$

$$\xrightarrow{+} \Sigma F_x = 0 = A_x^{W2} - \frac{9}{12.73}(2\cdot 1910\text{ lb} + 3820\text{ lb}) - \frac{9}{12.73}(2\cdot 445\text{ lb} + 890\text{ lb})$$
$$A_x^{W2} = 6660\text{ lb}\ (\xrightarrow{+})$$

FBD of joint *A*:

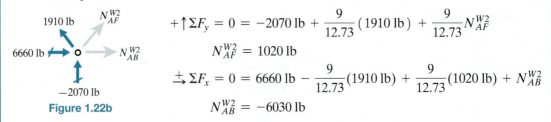

$$+\uparrow \Sigma F_y = 0 = -2070\text{ lb} + \frac{9}{12.73}(1910\text{ lb}) + \frac{9}{12.73}N_{AF}^{W2}$$
$$N_{AF}^{W2} = 1020\text{ lb}$$

$$\xrightarrow{+} \Sigma F_x = 0 = 6660\text{ lb} - \frac{9}{12.73}(1910\text{ lb}) + \frac{9}{12.73}(1020\text{ lb}) + N_{AB}^{W2}$$
$$N_{AB}^{W2} = -6030\text{ lb}$$

Figure 1.22b

Summary of unfactored axial forces:

Load Type	Member *AB*	Member *AF*
D	2580 lb	−3650 lb
L	750 lb	−1060 lb
S	3030 lb	−4290 lb
W1	3960 lb	3370 lb
W2	−6030 lb	1020 lb

Design Axial Forces:

Since the truss is over an assembly area, the lobby, we use a load factor of 1.0 instead of 0.5 on *L* in combinations 3, 4, and 6 (see footnote to Table 1.4).

Member AB: Tension

Of the two wind cases, we use Case 1 because it is the only one that produces tension. The effect of snow is larger than that of live load, so there is no need to try combination 2, since combination 3 emphasizes the larger effect.

Comb 1:   1.4*D*
          = 1.4(2580 lb)
          = 3610 lb

Comb 3a: 1.2*D* + 1.6*S* + 1.0*L*
          = 1.2(2580 lb) + 1.6(3030 lb) + 1.0(750 lb)
          = 8690 lb

Comb 3b: 1.2*D* + 1.6*S* + 0.5*W*
          = 1.2(2580 lb) + 1.6(3030 lb) + 0.5(3960 lb)
          = 9920 lb

Comb 4:   1.2*D* + 1.0*W* + 1.0*L* + 0.5*S*
          = 1.2(2580 lb) + 1.0(3960 lb) + 1.0(750 lb) + 0.5(3030 lb)
          = 9320 lb

Design value:
$$N_{AB}^U = 9920 \text{ lb } (\boldsymbol{T})$$

Member AB: Compression

Of the two wind cases, we use Case 2 because it is the only one that produces compression. Since all other loads produce tension, we use load factors of 0 for the nonpermanent loads L and S. Only one combination maximizes the effect of W and minimizes the effect of the permanent load, D:

Comb 5: $0.9D + 1.0W$
$= 0.9(2580 \text{ lb}) + 1.0(-6030 \text{ lb})$
$= -3710 \text{ lb}$

Design value:
$$N_{AB}^U = 3710 \text{ lb } (\boldsymbol{C})$$

Member AF: Tension

Both wind cases cause tension, so we use the case that causes the largest value, Case 1. Since all other loads produce compression, we use load factors of 0 for the nonpermanent loads L and S. Only one combination maximizes the effect of W and minimizes the effect D:

Comb 5: $0.9D + 1.0W$
$= 0.9(-3650 \text{ lb}) + 1.0(3370 \text{ lb})$
$= 85 \text{ lb}$

Design value:
$$N_{AF}^U = 85 \text{ lb } (\boldsymbol{T})$$

Member AF: Compression

Since neither wind case causes compression, we use a load factor of 0 for W. The effect of snow is larger than that of live load, so there is no need to try combination 2, since combination 3 emphasizes the larger effect.

Comb 1: $1.4D$
$= 1.4(-3650 \text{ lb}) = -5110 \text{ lb}$

Comb 3a: $1.2D + 1.6S + 1.0L$
$= 1.2(-3650 \text{ lb}) + 1.6(-4290 \text{ lb}) + 1.0(-1060 \text{ lb}) = -12{,}300 \text{ lb}$

Comb 3b: $1.2D + 1.6S + 0.5W$
$= 1.2(-3650 \text{ lb}) + 1.6(-4290 \text{ lb}) + 0 = -11{,}240 \text{ lb}$

Design value:
$$N_{AF}^U = 12{,}300 \text{ lb } (\boldsymbol{C})$$

Summary:
$$N_{AB}^U = 9920 \text{ lb } (\boldsymbol{T})$$
$$N_{AB}^U = 3710 \text{ lb } (\boldsymbol{C})$$
$$N_{AF}^U = 85 \text{ lb } (\boldsymbol{T})$$
$$N_{AF}^U = 12{,}300 \text{ lb } (\boldsymbol{C})$$

EXAMPLE 1.4

A continuous beam supports the floor of a house and the carport covering. To properly design the supports at A and B, we need to know the downward design force on the support and the upward design force, if any.

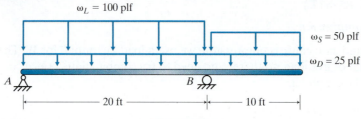

Figure 1.23

Find Unfactored Reactions:

Use equilibrium to find the vertical reactions at A and B due to individual, unfactored loads.

Dead:

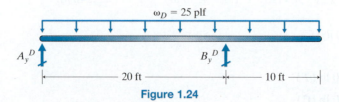

Figure 1.24

$$\curvearrowright \Sigma M_A = 0 = 25 \text{ plf } (30 \text{ ft})(15 \text{ ft}) - B_y^D(20 \text{ ft})$$
$$B_y^D = +562 \text{ lb } (+\uparrow)$$
$$+\uparrow \Sigma F_y = 0 = A_y^D - 25 \text{ plf } (30 \text{ ft}) + 562 \text{ lb}$$
$$A_y^D = +188 \text{ lb } (+\uparrow)$$

Live:

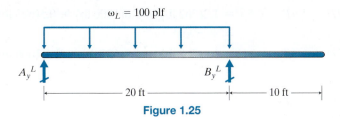

Figure 1.25

$$\circlearrowright \Sigma M_A = 0 = 100 \text{ plf } (20 \text{ ft})(10 \text{ ft}) - B_y^L (20 \text{ ft})$$
$$B_y^L = +1000 \text{ lb } (+\uparrow)$$
$$+\uparrow \Sigma F_y = 0 = A_y^L - 100 \text{ plf } (20 \text{ ft}) + 1000 \text{ lb}$$
$$A_y^L = +1000 \text{ lb } (+\uparrow)$$

Snow:

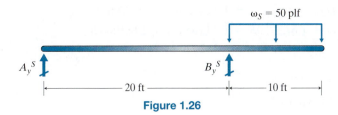

Figure 1.26

$$\circlearrowright \Sigma M_A = 0 = 50 \text{ plf } (10 \text{ ft})(25 \text{ ft}) - B_y^S (20 \text{ ft})$$
$$B_y^S = +625 \text{ lb } (+\uparrow)$$
$$+\uparrow \Sigma F_y = 0 = A_y^S + 625 \text{ lb } - 50 \text{ plf } (10 \text{ ft})$$
$$A_y^S = -125 \text{ lb } (+\uparrow)$$

Summary of unfactored reactions:

Load Type	A_y	B_y
D	188 lb	562 lb
L	1000 lb	1000 lb
S	−125 lb	625 lb

Design Reaction By:

All reactions are up; therefore, the design value will occur when as many of the forces act as can reasonably be expected.

Note: From Table 1.3, the closest occupancy to a home is probably an office (live load of 50 psf). From Table 1.4, the live load is less than 100 psf and is not considered an "assembly" area; therefore, the factor on the live load need *not* be increased to 1.0.

Comb 1: 1.4D = 1.4(562 lb) = +787 lb (+↑)
Comb 2: 1.2D + 1.6L + 0.5S = 1.2(562 lb) + 1.6(1000 lb) + 0.5(625 lb)
           = +2590 lb (+↑)
Comb 3: 1.2D + 1.6S + 0.5L = 1.2(562 lb) + 1.6(625 lb) + 0.5(1000 lb)
           = +2170 lb (+↑)

Design value is the largest:

$$B_y^U = +2590 \text{ lb} (+\uparrow)$$

There is no design downward reaction at B.

Design Reaction A_y:

There are upward and downward reactions that counteract each other; therefore, the design values will occur when only some of the forces act.

Combinations for upward reaction:

We do not include snow because it causes a downward reaction.

Comb 1: 1.4D = 1.4 (188 lb) = +263 lb (+↑)
Comb 2: 1.2D + 1.6L + 0S = 1.2 (188 lb) + 1.6(1000 lb)
           = +1826 lb (+↑)

Design value is the largest:

$$A_y^{U+} = +1826 \text{ lb} (+\uparrow)$$

Combinations for downward reaction:

We do not include live because it causes an upward reaction.

Comb 3: 1.2D + 1.6S + 0L = 1.2(188 lb) + 1.6(−125 lb) = +26 lb (+↑)

This combination does not result in a downward reaction.

We must include the dead load because it will always act, but it is reasonable to consider that the dead load is actually less than the maximum likely value.

Comb 3a: 0.9D + 1.6S + 0L = 0.9(188 lb) + 1.6(−125 lb) = −31 lb (+↑)

Design value is the largest negative:

$$A_y^{U-} = -31 \text{ lb} (+\uparrow)$$

Summary:

$$A_y^{U+} = 1826 \text{ lb} (+\uparrow)$$
$$A_y^{U-} = -31 \text{ lb} (+\uparrow)$$
$$B_y^{U+} = 2590 \text{ lb} (+\uparrow)$$
$$B_y^{U-} = \text{none}$$

Example 1.5

Our firm is designing a new monorail to be built in Canada. The system will carry two trains on adjacent rails, and the rails will be carried on single-mast piers. We are in the preliminary stages of design, but the geotechnical engineers need a sense of the vertical force and any overturning moment that the pier must carry. The engineers want us to determine factored and combined values for the force and moment.

Although we are still very early in the design, a team member has been able to estimate that the maximum likely, unfactored dead load from each rail will be 250 kN. The maximum likely, unfactored live load from each rail will probably be 500 kN.

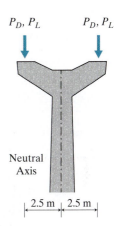

Figure 1.27

Because the pier is symmetric, we need to find only the largest-magnitude design moment rather than both positive and negative values. Remember that the moment is calculated at the neutral axis.

Note that although *Canadian Highway Bridge Design Code* (CSA Group, 2014) load factors and combinations would govern this design, we will use the ASCE 7 load factors and combinations for this preliminary analysis while we wait for a copy of the Canadian code to arrive.

Find Unfactored Reactions:

We use equilibrium to find the vertical and moment reactions at the base due to each load individually.

Dead on left side:

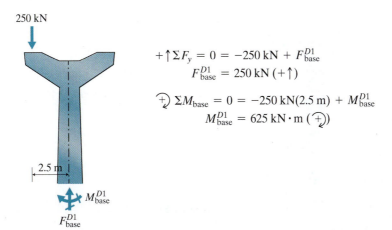

$+\uparrow \Sigma F_y = 0 = -250 \text{ kN} + F_{\text{base}}^{D1}$
$F_{\text{base}}^{D1} = 250 \text{ kN} (+\uparrow)$

$\circlearrowright \Sigma M_{\text{base}} = 0 = -250 \text{ kN}(2.5 \text{ m}) + M_{\text{base}}^{D1}$
$M_{\text{base}}^{D1} = 625 \text{ kN} \cdot \text{m} (\circlearrowright)$

Figure 1.28

Dead on right side:

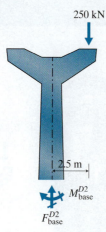

$$+\uparrow \Sigma F_y = 0 = -250 \text{ kN} + F_{base}^{D2}$$
$$F_{base}^{D2} = 250 \text{ kN} (+\uparrow)$$

$$\overset{+}{\curvearrowleft} \Sigma M_{base} = 0 = 250 \text{ kN} (2.5 \text{ m}) + M_{base}^{D2}$$
$$M_{base}^{D2} = -625 \text{ kN} \cdot \text{m} (\overset{+}{\curvearrowright})$$

Figure 1.29

Live on left side:

$$+\uparrow \Sigma F_y = 0 = -500 \text{ kN} + F_{base}^{L1}$$
$$F_{base}^{L1} = 500 \text{ kN} (+\uparrow)$$

$$\overset{+}{\curvearrowleft} \Sigma M_{base} = 0 = -500 \text{ kN} (2.5 \text{ m}) + M_{base}^{L1}$$
$$M_{base}^{L1} = 1250 \text{ kN} \cdot \text{m} (\overset{+}{\curvearrowleft})$$

Figure 1.30

Live on right side:

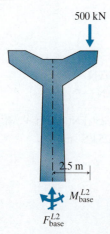

$$+\uparrow \Sigma F_y = 0 = -500 \text{ kN} + F_{base}^{L2}$$
$$F_{base}^{L2} = 500 \text{ kN} (+\uparrow)$$

$$\overset{+}{\curvearrowleft} \Sigma M_{base} = 0 = 500 \text{ kN} (2.5 \text{ m}) + M_{base}^{L2}$$
$$M_{base}^{L2} = -1250 \text{ kN} \cdot \text{m} (\overset{+}{\curvearrowright})$$

Figure 1.31

Summary of unfactored reactions:

Load Type	F_{base}	M_{base}
D1	250 kN	625 kN·m
D2	250 kN	−625 kN·m
L1	500 kN	1250 kN·m
L2	500 kN	−1250 kN·m

Design Reactions:

Note that the dead loads are not mutually exclusive; in fact, they will both act on the pier at all times. The live loads are not mutually exclusive either; trains can be on both sides simultaneously, but they do not have to be.

Vertical:

The largest reaction will occur if both dead cases occur and if both live cases occur.

Comb 1: $1.4(\Sigma D)$
$= 1.4(250 \text{ kN} + 250 \text{ kN})$
$= 700 \text{ kN } (+\uparrow)$

Comb 2: $1.2(\Sigma D) + 1.6(\Sigma L)$
$= 1.2(250 \text{ kN} + 250 \text{ kN}) + 1.6(500 \text{ kN} + 500 \text{ kN})$
$= 2200 \text{ kN } (+\uparrow)$

Design value is largest

$F_{base}^U = 2200 \text{ kN } (+\uparrow)$

Moment:

Left-side loads create clockwise moment; right-side loads create counterclockwise moment. Therefore, simultaneous loads tend to cancel. Since a train might be on only one side, consider just L1.

Comb 2: $1.2(\Sigma D) + 1.6 L1$
$= 1.2(625 \text{ kN·m} - 625 \text{ kN·m}) + 1.6(1250 \text{ kN·m})$
$= 2000 \text{ kN·m } (\curvearrowleft)$

But the dead load might not be the same for each girder. Worst case would be a heavy girder on one side and a light girder on the other.

Comb 2a: $1.2 D1 + 0.9 D2 + 1.6 L1$
$= 1.2(625 \text{ kN·m}) + 0.9(-625 \text{ kN·m}) + 1.6(1250 \text{ kN·m})$
$= 21{,}900 \text{ kN·m } (\curvearrowleft)$

Design value is largest

$M_{base}^U = 2190 \text{ kN·m } (\curvearrowleft)$

Summary:

$F_{base}^U = 2200 \text{ kN } (+\uparrow)$
$M_{base}^U = 2190 \text{ kN·m } (\curvearrowleft) \text{ or } (\curvearrowright)$

1.3 Structure Idealization

Introduction
Structural analysis is the process of creating a mathematical model of a structure and using that model to predict the structure's response. Reality is very complex; therefore, we typically simplify reality when we create the model of the structure and its loading. The process of converting our vision for the real structure into the model we use for analysis is called *structure idealization*.

How-To
There are three key components of the structure we need to idealize: the members themselves, the supports, and the connections. The actual members are three-dimensional, but we idealize them as thin lines or flat areas. We typically idealize structural members that have cross-sections with small aspect ratios as lines (Figure 1.32a) and those with large aspect ratios as flat areas (Figure 1.32b).

The key to idealization is where to put the line or area to most reasonably represent the three-dimensional member. The answer comes from topics learned in mechanics of materials. Consider the concrete T-beam in Figure 1.33a. If the structural member experiences uniform compression, the force resultant would act at the centroid (Figure 1.33b). If the same structural member experiences only bending moment, the strain (and therefore the axial deformation) would be zero at the neutral axis (Figure 1.33c). In most structural members, the centroid and neutral axis occur at the same point in the cross-section. That is the place most appropriate for us to put the idealization of the structural member (Figure 1.33d).

The various parts of the structural system are shown in Figure 1.34. Those parts are often given the following unofficial names and descriptions:

① *Beam:* Typically supports floors or roofs; oriented primarily horizontally; carries primarily shear and moment, and in some cases axial force.
② *Girder:* Typically supports beams; oriented primarily horizontally; carries primarily shear and moment, and in some cases axial force.

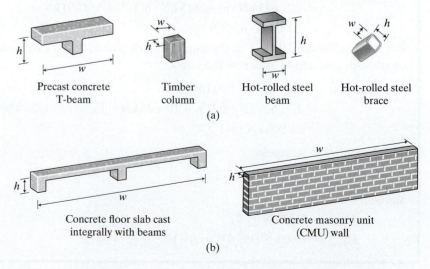

Figure 1.32 (a) Structural members with small aspect ratios; (b) Structural members with large aspect ratios.

③ *Column:* Typically supports beams and girders; oriented primarily vertically; carries primarily axial force, and in some cases shear and moment.
④ *Brace:* Typically resists lateral movement; oriented primarily on a diagonal; carries primarily axial force.
⑤ *Structural wall:* Typically resists lateral movement using in plane stiffness; large aspect ratio; oriented primarily vertically; carries primarily axial force but also shear and moment.
⑥ *Floor diaphragm:* Typically resists gravity loads using out of plane stiffness; large aspect ratio; oriented primarily horizontally; carries primarily shear and moment, and in some cases axial force; also used as roofs.

The key to idealizing the supports and connections is to identify which motions are restrained. Restrained translations result in transferred forces, and restrained rotations result in transferred moments. If a member is held by something that prevents it from moving in at least one direction, we call that a *support*. If two or more members are attached such that they can only move together, we call that a *connection*.

Table 1.5 shows a few idealized supports with descriptions of which motions are restrained. The idealizations we use for supports in our models rarely match reality exactly, so we choose the idealization that most closely matches reality.

Unlike supports, members at a connection can move in all directions (translations and rotations). But with a connection, the structural members are linked such that at least one motion (translation or rotation) is the same for all the members. Consider the two photos in Figure 1.35. The photo on the left shows a pin between a large member and the concrete foundation. The foundation prevents the member from moving left/right or up/down, but it allows the member to rotate. The photo on the right shows a pin that prevents the two members from moving away from each other, but they are not prevented from both moving left/right or up/down. They just have to move together. Therefore, the photo on the right is of a connection.

Table 1.6 shows a few idealized connections with descriptions of which motions are linked. The most difficult distinction to make is often between pinned and rigid connections. If we construct different steel connections and test them in the lab, we would obtain results similar to the

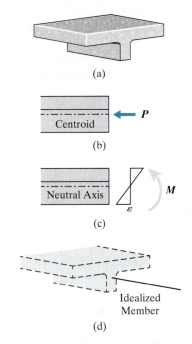

Figure 1.33 (a) Cross-section of a concrete T-beam; (b) Axial force resultant acts at the centroid of the member; (c) Strain and axial deformation are zero at the neutral axis when subjected to only bending; (d) Idealization of member at the location of the centroid/neutral axis.

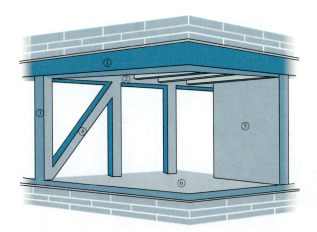

Figure 1.34 Schematic of part of a structural system with different types of structural members labeled.

30 Chapter 1 Loads and Structure Idealization

Table 1.5 Typical Idealizations of Structural Supports

Name	Motions Assumed Completely Restrained	Resulting Reactions	Symbol of Idealized Support	Example
Pin	All displacements			
Roller	Normal to surface[a]			
Fixed	All displacements and rotations			

[a] We assume that a roller restrains motion into and away from the surface. If results of analysis show that it must restrain away from the surface, we can design the support to do so.

Figure 1.35 Example of (a) pin support and (b) pin connection.

Table 1.6 Typical Idealizations of Structural Connections

Name	Motions Assumed Completely Restrained	Symbol of Idealized Connection (Elevation)	Symbol of Idealized Connection (Plan)	Example
Pin	All displacements			
Roller	Normal to longitudinal axis			
Rigid	All displacements and rotations			

behaviors shown in Figure 1.36. The graph is of the moment applied, M, and the relative rotation of one member to the other, θ. Our idealized connections would act along the axes.

If we have laboratory data for the specific connection we plan to use, we can idealize the connection as pinned with a rotational spring (Figure 1.37). Rarely is knowing the spring constant worth the expense of laboratory testing, however. Instead, we idealize the connection as one of the extremes: pinned or rigid.

32 Chapter 1 Loads and Structure Idealization

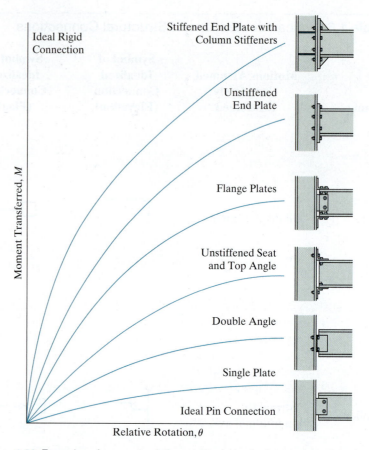

Figure 1.36 Examples of moment–relative rotation behavior for a variety of steel connections.

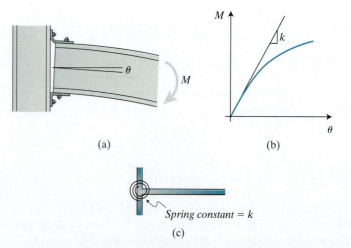

Figure 1.37 Idealizing a connection as semirigid: (a) Actual behavior has some relative rotation; (b) Laboratory data showing moment–relative rotation relationship; (c) Idealized connection with rotational spring.

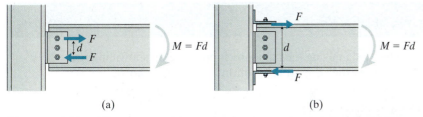

Figure 1.38 The ability of a connection to transfer moment is governed primarily by where the cross-section is connected: (a) Connection at the web only can transfer little moment; (b) Connecting the outer edges creates the largest possible moment.

In design, we choose the actual connection to have the behavior we used in our idealization. When analyzing an existing structure, we must choose the idealization that most closely matches the actual connection. To help us make that choice, consider how moment is transferred at the connection in Figure 1.38. The attempted rotation of the connection creates two forces that are equal in magnitude and opposite in direction. Those two forces create a couple moment. The forces are located where the two members are connected. The greater the distance from the neutral axis, the larger the couple moment; therefore, the better moment is transferred. As a standard practice, we will consider a connection rigid if the members are connected at the outer edges. Otherwise, we will consider the connection pinned.

SECTION 1.3 HIGHLIGHTS

Structure Idealization

Connections:

- When relative motion between members is restrained, force or moment is transferred.
- Connections that grab the member far from its neutral axis (e.g., flanges) are idealized as rigid.
- Connections that grab the member close to its neutral axis (e.g., web) are idealized as pinned.

Supports:

- When motion is restrained, there is a force or moment reaction.

Members:

- Idealize members along their centroid/neutral axis.

Example 1.6

We want to design a new highway sign using the standard support detail shown in Figure 1.39. What is a reasonable idealization of the support for motion in the plane shown?

Figure 1.39

Idealized Support:

The bolts between the sign post and the concrete foundation prevent vertical and lateral motion. The narrowly spaced bolts will not provide much resistance to rotation in this plane though. Therefore, the support is most like a pin in this plane.

Pin: ⚬⟋⟍

Example 1.7

The same standard support detail has a very different look from the side. What is a reasonable idealization of the support for motion in the side plane shown in Figure 1.40?

Figure 1.40

Idealized Support:

The bolts between the sign post and the concrete foundation still prevent vertical and lateral motion. The bolts are now widely spaced and will provide resistance to rotation in this plane. Therefore, the support is most like a fixed support in this plane.

Fixed: ////

EXAMPLE 1.8

In order to design the connection shown in Figure 1.41, we need to know the forces and/or moments transferred. Which internal forces are transferred for this connection, and how would you idealize this connection?

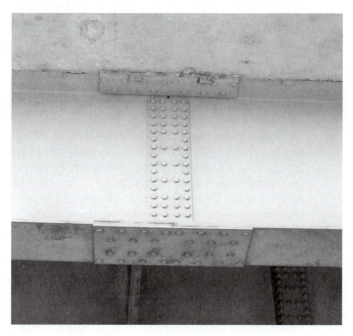

Figure 1.41

Idealized Connection:

The entire cross-section is connected. The bolted plate in the middle ensures that there is no relative translation left-right or up-down. The bolted plates on the top and bottom flanges ensure that there is no relative rotation between the two beams. This connection is best idealized as *rigid*.

1.4 Application of Gravity Loads

Introduction

Before we can begin to analyze an idealized structure, we need to know how the loads act on the structure. Pressures on surfaces (e.g., snow on a roof or live load on a floor) are carried to supporting members. Predicting how those pressures are distributed to the supporting members is called idealizing the loading.

Because of the way structures are typically made, we use different assumptions and approximations to idealize pressures due to gravity loads than we use to idealize lateral pressures.

How-To

Our goal is to determine the idealized gravity loading on the structural members that support floors and roofs. As we saw in Section 1.1, gravity loads act vertically on the structure. Downward loads include dead, live, snow, and others. Structures also experience some upward gravity loads. Suction on a roof due to wind acts upward, and water pressure due to a high water table acts upward under a basement floor. With the exception of self-weight, gravity loads on a structure rarely act directly on the beams or columns. Instead, the loads act as pressures on floors and roofs. Therefore, we need to convert those pressures into loads on the supporting structural members before we can analyze the supporting structure. To perform that conversion, we need to make some assumptions or approximations; therefore, we call this process idealizing the loading.

The key to idealizing any loading is to understand the load path. The load path is the sequence of structural members through which load travels from where it acts on the structure to the supports (the locations prevented from moving—Section 1.3). To understand what approximations we make and why, consider the floor system shown in Figure 1.42. Uniform pressure from a gravity load acts on the floor diaphragm. Note that pressure on a surface is also called an *area load*. The diaphragm is in direct contact with only the three beams, and all three beams are in direct contact with columns that we presume are supported by foundations. The load must go from the diaphragm to the foundations, so we find the load path by looking for a continuous sequence of connected pieces. In this case, the gravity load acts on the floor diaphragm, then travels to the beams, which carry the force to the columns, which carry the force to the foundations.

To idealize the loading on the beams, we need to make some approximations. Look at the floor system from the side (Figure 1.43a). As the load from the floor diaphragm pushes down on the beams, the beams deflect downward except at the columns. Therefore, the best idealization of the beams as supports would be springs (Figure 1.43b). If we know the spring constants, we can find the force in each spring (i.e., the force on each beam), but that requires structural analysis skills outside the scope of this text.

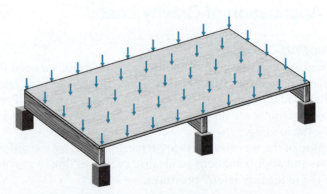

Figure 1.42 Gravity load acting on a floor diaphragm supported by three beams and six columns.

However, *if** the floor diaphragm is very flexible out of plane compared to the supporting beams (Figure 1.43c), we can reasonably assume that the beams do not move. Therefore, we can idealize the beams as a pin and two rollers under the floor diaphragm (Figure 1.43d). This idealization leads to an indeterminate structure that we can analyze with skills from Chapters 12 to 15. However, a common approximation we make is that each span of the floor diaphragm between supporting beams acts independently, without continuity (Figure 1.43e). Since each span is simply supported, we can determine how much load goes to each support using statics. The result is that uniform load is carried by the nearest supporting member (Figure 1.43f). Note that this last statement is true only when the pressure is uniform; for cases of nonuniform pressure (e.g., snow drifting on a roof), we still often approximate the diaphragm as simply supported, thus enabling us to use statics to determine how much load is carried by each supporting member.

If a diaphragm experiences uniform pressure, our assumption and approximation result in the statement that load is carried by the nearest supporting member. We call the width of the diaphragm that contributes load to a supporting member the *tributary width*. For our example floor diaphragm, the tributary widths for each supporting beam are shown in Figure 1.44. The resulting loads on the supporting beams are distributed loads, also called *line loads* (Figure 1.45). The magnitude of the distributed load, w, on any supporting member is the uniform gravity pressure, p, multiplied by the tributary width:

$$w = p \times \text{tributary width}$$

When the panels of a diaphragm are supported on only two sides, the pressure on the diaphragm travels in one direction to the supported sides (e.g., see the arrows in Figure 1.44). We call this behavior *one-way action* and the diaphragm a *one-way diaphragm*. But when a panel of a diaphragm is supported on three or four sides, the pressure travels in two directions to the supported sides. We call this behavior *two-way action* and the diaphragm a *two-way diaphragm*. Again, *if* the gravity pressure is uniform, we assume the load is carried by the nearest supporting member. The distributed load is still calculated as the product of the uniform

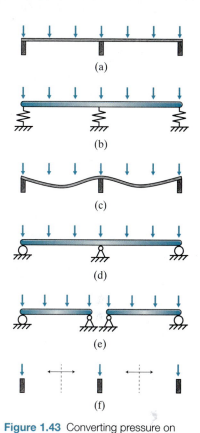

Figure 1.43 Converting pressure on a floor or roof into idealized loading on the supporting members: (a) Uniform pressure on diaphragm; (b) Best idealization of beams supporting the diaphragm; (c) Behavior of diaphragm if much less stiff out of plane than the supporting beams; (d) Reasonable idealization if diaphragm is much less stiff than supporting beams; (e) Common approximation that each span of diaphragm is simply supported; (f) Resulting idealized loads on each supporting beam.

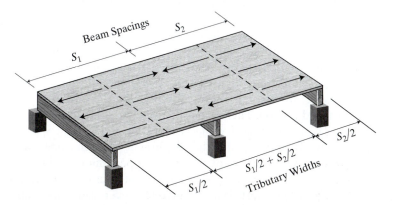

Figure 1.44 Floor diaphragm supported by three beams showing the beam spacing and the resulting tributary widths each beam supports.

---

* The *if* in italics indicates that we are making an assumption or approximation.

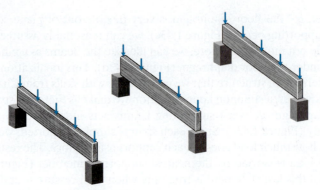

Figure 1.45 Idealized distributed loads on supporting beams due to uniform pressure on the floor diaphragm (not shown).

pressure and the tributary width, but the tributary width is not constant along the supporting member.

Consider the square diaphragm in Figure 1.46a supported on all four sides. In Figure 1.46b, we start to draw the dividing line between the area that sends load to beam AB and the area that sends load to beam AC. Because we assume that the uniform pressure will go to the nearest supporting member, the dividing line bisects the corner angle. Since the corner angle of our square diaphragm is 90°, the dividing line is at 45°. We draw similar dividing lines for each corner and extend them until they all meet, as in Figure 1.46c. For beam AC, the tributary width at ends A and C is zero. The tributary width is greatest at midspan and changes linearly in between. The resulting distributed load, therefore, is also zero at the ends and greatest at midspan; it changes linearly in between (Figure 1.46d).

Determining how gravity pressure on a nonsquare diaphragm panel distributes to the supporting beams is done the same way; see Figure 1.47a. We begin by drawing a boundary line from a corner and bisecting the angle of that corner (Figure 1.47b). We repeat the process for all corners, and extend the boundary lines until they meet (Figure 1.47c). In this case, the four lines do not meet at one point. We draw one more boundary line that bisects the large area in the middle, since we assume the uniform gravity load goes to the nearest supporting member (Figure 1.47d). Again, the distributed load on beam AC is the product of the gravity pressure, p, and the tributary width, but the tributary width is trapezoidal (Figure 1.47e). The resulting distributed load on beam AB is triangular (Figure 1.47f).

If the aspect ratio, S_{long}/S_{short}, of a two-way diaphragm panel becomes sufficiently large, we can reasonably approximate the behavior as one-way in the short direction. There is no standard threshold for "sufficiently large," but a commonly used value is 2. Therefore, if $S_{long}/S_{short} \geq 2$, we can reasonably approximate the diaphragm as acting one-way. By making this approximation, we overpredict the peak moment in the long beam by only 1%. The approximation does result in us completely ignoring the moment in the short beam though. However, remember our underlying assumption: The diaphragm is much more flexible out of plane than the supporting beams. If we design the short beams to be flexible because we ignored any load coming to them, they will actually attract less load. Therefore, approximating diaphragms with large aspect ratios as one-way diaphragms is reasonable.

Section 1.4 Application of Gravity Loads **39**

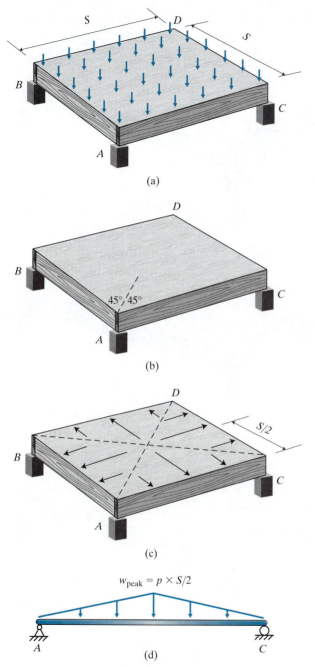

Figure 1.46 Converting gravity pressure on a square diaphragm supported on all four sides into idealized loading on the supporting members: (a) Uniform pressure on a square diaphragm; (b) Dividing the surface into regions supported by beams *AB* and *AC*; (c) Tributary areas for each supporting beam identified on the diaphragm along with arrows showing the two-way action; (d) Idealized loading on beam *AC*.

40 **Chapter 1** Loads and Structure Idealization

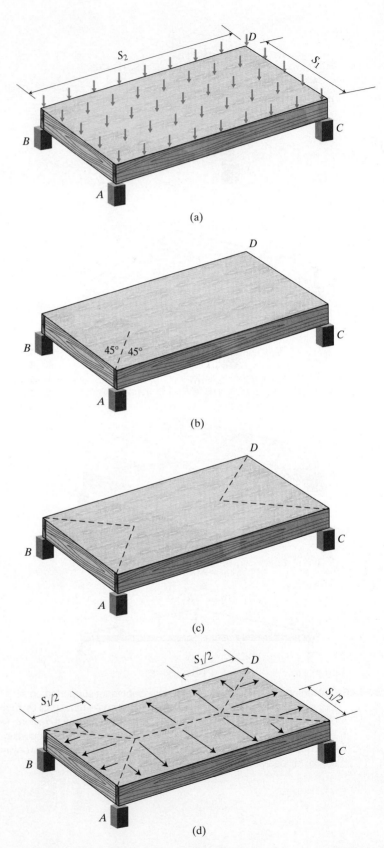

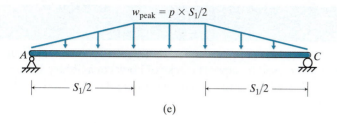

(e)

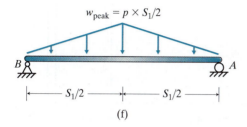

(f)

Figure 1.47 Converting gravity pressure on a rectangular diaphragm supported on all four sides into idealized loading on the supporting members: (a) Uniform pressure on a rectangular diaphragm; (b) Dividing the surface into regions supported by beams *AB* and *AC*; (c) Joining the boundary lines from the four corners leaves a large, undivided area in the middle; (d) Tributary areas for each supporting beam identified on the diaphragm along with arrows showing the two-way action; (e) Idealized loading on beam *AC* is trapezoidal, since the tributary width is trapezoidal; (f) Idealized loading on beam *AB* is triangular, since the tributary width is triangular.

SECTION 1.4 HIGHLIGHTS

Application of Gravity Loads

Assumption: A floor or roof diaphragm is much more flexible out of plane than the members that support it.

Approximations: Each diaphragm panel behaves independently of other panels; we are ignoring continuity between panels.

For panels that are supported on more than two sides, if $S_{long}/S_{short} \geq 2$ we can consider the diaphragm to act one-way.

Distributed Load: $w = $ pressure $\times$ tributary width

Example 1.9

The floor shown in Figure 1.48 is made of cast-in-place concrete (i.e., continuous) and is supported by three beams. The floor will be used to support stacks of books in a library. Before we can determine the design moment in each beam, we need to determine the idealized live load on each beam.

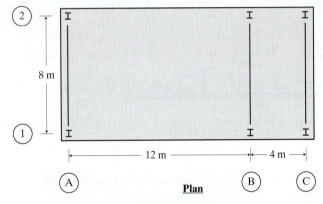

Figure 1.48

Assumptions and Approximations:
- The floor diaphragm is much more flexible out of plane than the beams that support it.
- Each diaphragm panel behaves independently of the other one (i.e., discontinuous).

Idealized Live Load:

Live load:
From Table 1.3, for library stacks, $p_L = 7.18$ kN/m².

Tributary widths:
Because beams are on only two sides of each floor panel, the floor must act one-way.

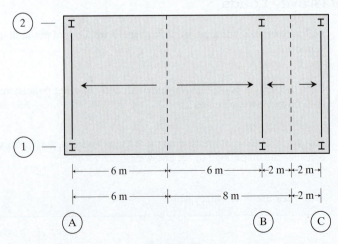

Figure 1.49

Calculate the resulting distributed loads:

$w = p \times$ tributary width

$w_{LA} = 7.18 \text{ kN/m}^2(6 \text{ m}) = 43.1 \text{ kN/m}$
$w_{LB} = 7.18 \text{ kN/m}^2(8 \text{ m}) = 57.4 \text{ kN/m}$
$w_{LC} = 7.18 \text{ kN/m}^2(2 \text{ m}) = 14.4 \text{ kN/m}$

Idealized loading:

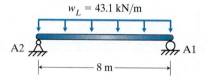

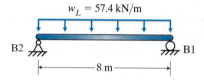

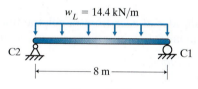

Figure 1.50

Evaluation of Results:
Satisfied Fundamental Principles?

Total load applied $= p_L \times$ area $= 7.18 \text{ kN/m}^2(8 \text{ m})(16 \text{ m}) = 919 \text{ kN}$

Total load on beams $= \Sigma \int w_L \cdot d\text{length} = 43.1 \text{ kN/m}(8 \text{ m}) + 57.4 \text{ kN/m}(8 \text{ m})$
$+ 14.4 \text{ kN/m}(8 \text{ m}) = 919 \text{ kN}$ ✓

EXAMPLE 1.10

For architectural reasons, the columns at line B from Example 1.9 must be removed. The redesigned structural system for the library floor is shown in Figure 1.51. What is the idealized live load on beam B1-B2 and on girder A1-C1?

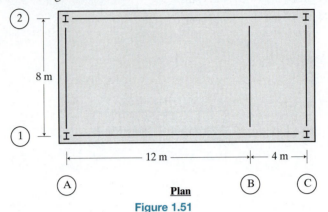

Plan
Figure 1.51

Assumptions and Approximations:
- The floor diaphragm is much more flexible out of plane than the beams that support it.
- Each diaphragm panel behaves independently of the other one (i.e., discontinuous).
- For the panel with an aspect ratio equal to 2, consider the floor to act one-way.

Idealized Live Load:

Live load:
From Table 1.3, for library stacks, $p_L = 7.18$ kN/m².

Tributary widths:
Because there are beams on all sides of each floor panel, the floor will act two-way, but we can approximate the panel with the large aspect ratio as acting one-way.

Calculate the resulting distributed loads:
$w = p \times$ tributary width
$w_{LA\,max} = 7.18 \text{ kN/m}^2(4 \text{ m}) = 28.7 \text{ kN/m}$
$w_{LB\,ends} = 7.18 \text{ kN/m}^2(2 \text{ m}) = 14.4 \text{ kN/m}$
$w_{LB\,max} = 7.18 \text{ kN/m}^2(4 \text{ m} + 2 \text{ m}) = 43.1 \text{ kN/m}$
$w_{LC} = 7.18 \text{ kN/m}^2(2 \text{ m}) = 14.4 \text{ kN/m}$
$w_{LAB\,max} = 7.18 \text{ kN/m}^2(4 \text{ m}) = 28.7 \text{ kN/m}$

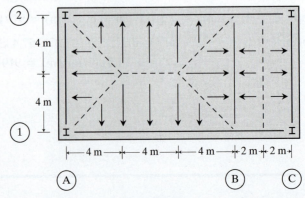

Figure 1.52

Idealized loading on beam B1-B2:

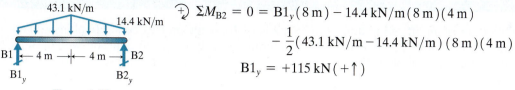

Figure 1.53

$$\circlearrowleft \Sigma M_{B2} = 0 = B1_y(8\text{ m}) - 14.4\text{ kN/m}(8\text{ m})(4\text{ m})$$
$$- \frac{1}{2}(43.1\text{ kN/m} - 14.4\text{ kN/m})(8\text{ m})(4\text{ m})$$
$$B1_y = +115\text{ kN}(+\uparrow)$$

Idealized loading on girder A1-C1:

Note that beams A1-A2 and C1-C2 attach directly to the columns; therefore, they do not put load onto the girders.

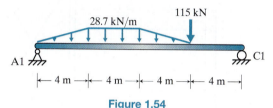

Figure 1.54

Evaluation of Results:

Approximation Predicted Outcomes?

We approximate the structural system as if there were no beam along line B. This results in one floor panel with an aspect ratio of 2. If we approximate the floor behavior as one-way, we get a uniform distributed load on girder A1-C1.

The total approximated load on the girder should overpredict the total load on the idealized girder because the approximation does not account for any load going to the end beams.

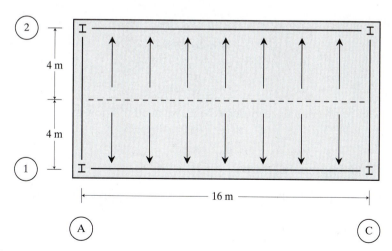

Figure 1.55

$w_{L\text{ approx}} = 7.18\text{ kN/m}^2(4\text{ m}) = 28.7\text{ kN/m}$

Total approx. load on girder AC = $28.7\text{ kN/m}(16\text{ m}) = 459\text{ kN}$

Total actual load on girder AC = $\frac{1}{2}(12\text{ m} + 4\text{ m})(28.7\text{ kN/m}) + 115\text{ kN}$
= $345\text{ kN} < 459\text{ kN}$ ✓

Example 1.11

The floor of an office building is to be made of 4-ft-wide precast concrete planks on a steel frame. The planks weigh 60 psf. For a preliminary analysis, assume the self-weight of each beam and girder is 50 plf. The box with an × in it in Figure 1.56 represents an opening in the floor.

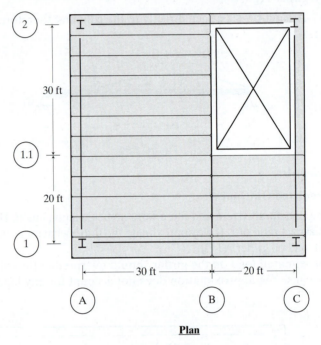

Plan

Figure 1.56

Before we can begin to analyze girder A2-C2, we need to know the idealized dead load on the girder.

Assumptions and Approximations:

- For any plank supported on more than two sides, consider it to act one-way.

Idealized Dead Load:

Floor behavior:

Because the planks act as individual members, they behave one-way in the long direction.

Tributary widths:

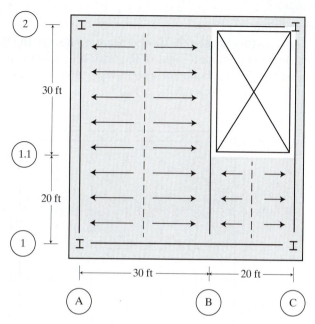

Figure 1.57

Resulting distributed loads on beam B1-B2:

$w_{DB1-B1.1} = 60 \text{ psf}(15 \text{ ft} + 10 \text{ ft}) + 50 \text{ plf} = 1550 \text{ plf}$
$w_{DB1-B1} = 60 \text{ psf}(15 \text{ ft} + 10 \text{ ft}) + 50 \text{ plf} = 950 \text{ plf}$

Idealized loading on beam B1-B2:

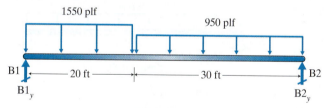

Figure 1.58

$\circlearrowleft^+ \Sigma M_{B1} = 0 = 1550 \text{ plf}(20 \text{ ft})(10 \text{ ft}) + 950 \text{ plf}(30 \text{ ft})(35 \text{ ft}) - B2_y(50 \text{ ft})$
$B2_y = 26{,}150 \text{ lb } (+\uparrow)$

Idealized loading on girder A2-C2:

The only distributed load on the girder is from its self-weight.

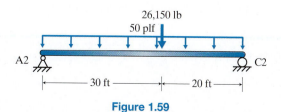

Figure 1.59

Evaluation of Results:

Approximation Predicted Outcomes?
We can approximate the structural system as if there were no hole in the floor and no beam along line B. Although the planks run left-right in the floor plan and even though the floor is square, we can approximate the floor behavior as one-way running up-down. This allows us to approximate the actual effect on girder A2-C2. The result is a uniform distributed load on girder A2-C2.

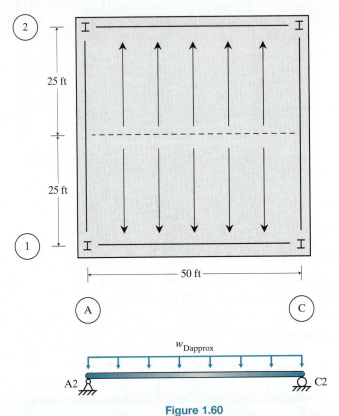

Figure 1.60

The total approximated load on the girder should overpredict the total load on the idealized girder because the approximation does not account for the hole in the floor or the load carried by beams A1-A2 and C1-C2.

$w_{Dapprox}$ = 60 psf × 25 ft + 50 plf = 1550 plf

Total approx. load on girder A2-C2 = 1550 plf × 50 ft = 77,500 lb = 77.5 k

Total actual load on girder A2-C2 = 50 plf × 50 ft + 26,150 lb = 28,650 lb
= 28.6 k < 77.5 k ✓

Note that the hole in the floor results in 36 k less dead load than the approximation we used. Therefore, the difference between our approximation and the actual total load we obtained is reasonable.

1.5 Application of Lateral Loads

Introduction

Lateral area loading, pressure, on the sides of a structure typically comes from wind, soil, or fluids. In order to perform structural analysis, we need to convert those pressures into idealized loadings on the structure. The process is similar to idealizing gravity loads, but there are some important differences.

Notice that seismic is not on this list. Seismic loading is not a lateral pressure on the outside of the structure. Instead, it is a load distributed across the interior of the structure. We discuss seismic loading more in Chapter 11: Diaphragms.

How-To

The key to idealizing any load is to understand the load path. The key to understanding the load path is to know where and how pieces of the structure are connected.

Pressure on a wall diaphragm typically transfers one-way because of how the diaphragm is attached to the rest of the structure. For example, the tilt-up concrete panels in Figure 1.61a do not experience continuity along their vertical edges. Therefore, the load can travel in only one way, vertically. These panels are supported out of plane by the slab-on-grade at the bottom and the roof near the top.

As another example, consider the precast concrete panels in Figure 1.61b. The panels do not experience continuity along their horizontal edges, so the load travels one way. But in this case, the load travels horizontally to the supporting columns.

The resulting loading in both cases is distributed load on the supporting members. To determine the magnitude, we make the following assumption: *If* the wall diaphragm is much more flexible out of plane than the supporting members, we can reasonably assume that the supporting members do not move.

For the tilt-up example, the wall diaphragm is idealized as simply supported with a cantilevered end (Figure 1.62a). Because this idealization is determinate, we can calculate the force on the roof and the slab-on-grade using only the equations of equilibrium, without needing to make another assumption.

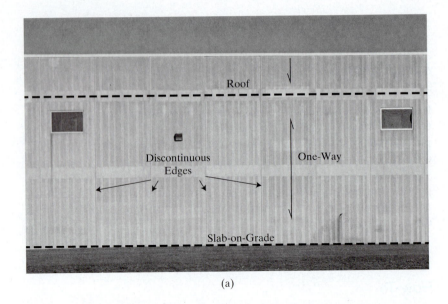

(a)

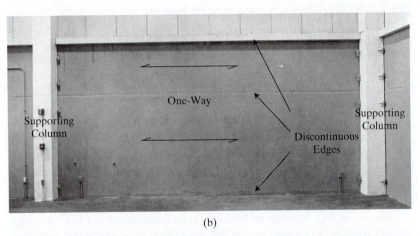

(b)

Figure 1.61 Examples of wall diaphragms that experience wind load out of plane (in or out of the page): (a) Wall made of tilt-up panels that are discontinuous along their vertical edges; (b) Wall made of precast panels that are discontinuous along their horizontal edges.

For the precast panel example, each panel is simply supported by the columns (Figure 1.62b). Again, this is a determinate problem, so we do not need any additional assumptions to calculate the force on each column.

In some cases, we design structures with continuous walls, such as concrete masonry unit (CMU) walls (Figure 1.63). Then we typically design the supporting structure to be the floor and roof diaphragms; the walls are often attached directly to the floor and roof diaphragms rather than columns. As a result, the wall diaphragm acts one-way. But note that the idealized wall might not be determinate.

Consider the three-story CMU wall in Figure 1.63. The idealized wall diaphragm is shown in Figure 1.64a. The idealized wall is indeterminate.

Section 1.5 Application of Lateral Loads

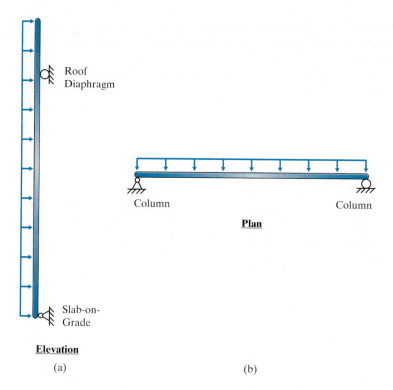

Figure 1.62 Idealized wall diaphragms with idealized load due to uniform lateral pressure: (a) Elevation view of idealization of tilt-up panel; (b) Plan view of idealization of precast panel.

With skills we cover in Chapters 12, 13, and 15, we could determine the load carried by each supporting diaphragm. However, it is common practice to approximate the wall as simply supported between each supporting diaphragm (Figure 1.64b). As a result of this approximation, uniform pressure on a wall will gather at the nearest supporting member.

Figure 1.63 Example of a continuous wall diaphragm that transfers lateral wind pressure to floor and roof diaphragms.

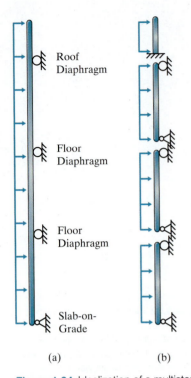

Therefore, we can convert uniform pressure on a wall into an idealized distributed load on a floor or roof diaphragm by multiplying the lateral pressure by the tributary height.

In cases where the lateral pressure is not uniform, we calculate the idealized distributed load on the floor or roof diaphragm by considering a vertical strip of wall of unit width and calculating the approximate reactions provided by the floor and roof diaphragms (Figure 1.65).

Figure 1.64 Idealization of a multistory continuous wall subjected to uniform lateral pressure: (a) Elevation of idealization; (b) Approximating each segment as simply supported.

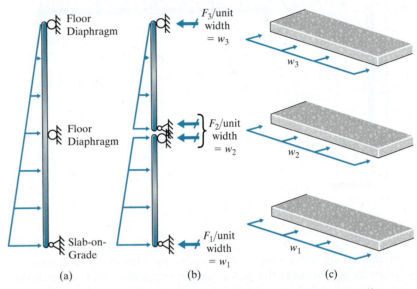

Figure 1.65 Idealization of a two-story continuous wall subjected to nonuniform lateral pressure: (a) Elevation of idealization of unit width strip; (b) Approximating each segment as simply supported with resulting reactions; (c) Reactions are the distributed loads on the edges of the supporting floor diaphragms.

SECTION 1.5 HIGHLIGHTS

Application of Lateral Loads

Assumption: A wall diaphragm is much more flexible out of plane than the members that support it.

Approximation: If a wall diaphragm is supported in more than two locations because of continuity, consider the wall to behave as simply supported between each supporting member.

Distributed Load: If the pressure on the wall is uniform,

$$w = \text{pressure} \times \text{tributary height}$$

If the pressure on the wall is nonuniform, calculate the distributed load considering a unit width vertical strip of wall.

EXAMPLE 1.12

A one-story warehouse is built with tilt-up concrete panels. Before we can design the lateral load-resisting system, we need to know how much wind load gathers at the roof diaphragm and how much gathers at the slab-on-grade.

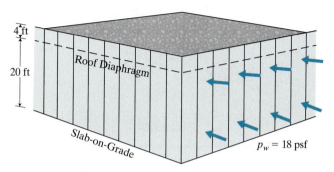

Figure 1.66

Assumptions and Approximations:

- Joints between tilt-up panels are discontinuous; therefore, load travels one-way to the diaphragms.

Idealized Lateral Load:

Idealized 1-foot strip of wall:

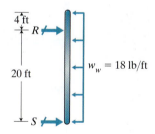

Figure 1.67

Distributed loads:

Apply equilibrium to find R and S:

$$\circlearrowleft^+ \Sigma M_S = 0 = R(20 \text{ ft}) - (18 \text{ plf})(24 \text{ ft})(12 \text{ ft})$$
$$R = 259 \text{ lb} (\rightarrow)$$
$$\xrightarrow{+} \Sigma F_x = 0 = 259 \text{ lb} - (18 \text{ plf})(24 \text{ ft}) + S$$
$$S = 173 \text{ lb} (\rightarrow)$$

Roof = 259 lb/ft
Slab = 173 lb/ft

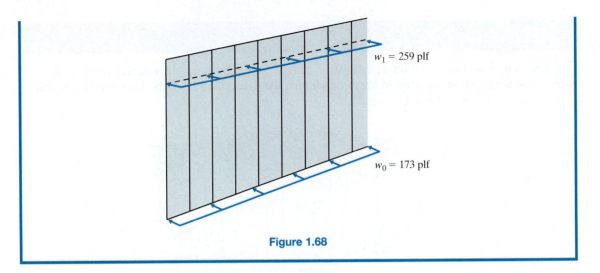

Figure 1.68

Example 1.13

The tall wall of a sports facility is made of precast concrete panels laid horizontally and vertical metal panels. The horizontal members are called girts. All wind load on that wall is ultimately transferred to the columns. To perform a structural analysis of column A4, we need to know the distributed and point loads on the column. The unfactored wind pressure is 1.0 kPa (1.0 kN/m²) over the entire wall surface.

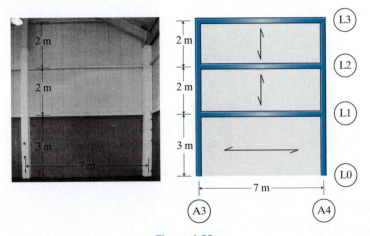

Figure 1.69

Assumptions and Approximations:

- Joints between the concrete panels are discontinuous; therefore, load travels one-way to the columns.
- If the top concrete panel presses against the girt at level 1, the long aspect ratio means we can approximate its behavior as one-way in the long direction.
- Consider each level of metal panels to be simply supported between girts.

Idealized Lateral Load:

Tributary widths:

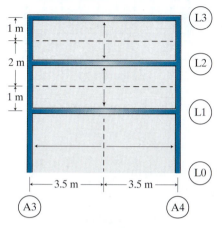

Figure 1.70

The concrete panels carry their load laterally to the columns; therefore, the tributary width is 3.5 m to each column. The metal panels carry their load vertically to the girts; therefore, the tributary width is 1 m to L1, 2 m to L2, and 1 m to L3.

Distributed loads on girts:

Multiply pressure by tributary width:
$w_{L3} = 1.0 \text{ kN/m}^2 (1 \text{ m}) = 1.0 \text{ kN/m}$
$w_{L2} = 1.0 \text{ kN/m}^2 (2 \text{ m}) = 2.0 \text{ kN/m}$
$w_{L1} = 1.0 \text{ kN/m}^2 (1 \text{ m}) = 1.0 \text{ kN/m}$

Point loads on column:

Girt level 3

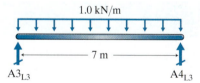

Figure 1.71

$\circlearrowleft^{+} \Sigma M_{A3} = 0 = (1.0 \text{ kN/m})(7 \text{ m})(3.5 \text{ m}) - A4_{L3}(7 \text{ m})$
$A4_{L3} = 3.5 \text{ kN } (+\uparrow)$

Girt level 2

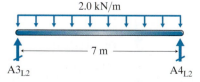

Figure 1.72

$\circlearrowleft^{+} \Sigma M_{A3} = 0 = (2.0 \text{ kN/m})(7 \text{ m})(3.5 \text{ m}) - A4_{L2}(7 \text{ m})$
$A4_{L2} = 7.0 \text{ kN } (+\uparrow)$

Girt level 1

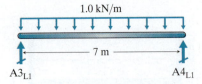

Figure 1.73

$$+\circlearrowright \Sigma M_{A3} = 0 = (1.0 \text{ kN/m})(7 \text{ m})(3.5 \text{ m}) - A4_{L1}(7 \text{ m})$$
$$A4_{L1} = 3.5 \text{ kN } (+\uparrow)$$

Distributed load on column:
$$w_{\text{col}} = 1.0 \text{ kN/m}^2 (3.5 \text{ m}) = 3.5 \text{ kN/m}$$

Summary of loads on column A4:

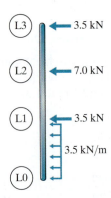

Figure 1.74

Example 1.14

A single-story building has CMU walls above ground and cast-in-place concrete walls for the basement level. In order to analyze the lateral load-resisting system, we need to know the distributed load generated at each diaphragm level by the soil pressure. The effective lateral soil pressure for this site is 60 psf per foot of depth.

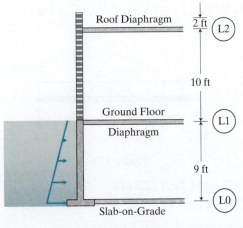

Figure 1.75

Assumptions and Approximations:

- Consider each story of wall to be simply supported between diaphragm levels.

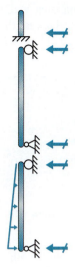

Figure 1.76

Idealized Lateral Load:

Earth pressure:
We multiply density by depth to obtain pressure:

$$p_{L1} = 60 \text{ psf/ft}(0 \text{ ft}) = 0 \text{ psf}$$
$$p_{L0} = 60 \text{ psf/ft}(9 \text{ ft}) = 540 \text{ psf}$$

Idealized 1-foot strip of wall:

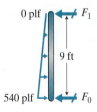

Figure 1.77

$$\curvearrowright \Sigma M_{L0} = 0 = \frac{1}{2}(540 \text{ plf})(9 \text{ ft})(3 \text{ ft}) - F_1(9 \text{ ft})$$

$$F_1 = 810 \text{ lb } (\leftarrow)$$

$$\xrightarrow{+} \Sigma F_x = 0 = -810 \text{ lb} + \frac{1}{2}(540 \text{ plf})(9 \text{ ft}) - F_0$$

$$F_0 = 1620 \text{ lb } (\leftarrow)$$

Distributed loads:

$$w_{L2} = 0 \text{ plf}$$
$$w_{L1} = 810 \text{ plf}$$
$$w_{L0} = 1620 \text{ plf}$$

Example 1.15

A two-story building with a continuous masonry wall will experience wind load. The building is shown with only one wall in Figure 1.78 to reveal the floor and roof diaphragms. In order to choose the lateral load-resisting system, it will help to know the magnitude of the distributed load generated at each diaphragm level.

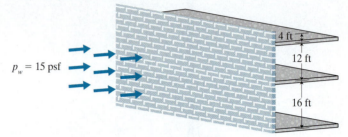

Figure 1.78

Assumptions and Approximations:

- Consider each story of the wall to be simply supported between diaphragm levels.

Idealized Lateral Load:

Tributary heights:

Although the wall rests on a foundation, that foundation is typically designed to carry the vertical load from the wall. The lateral load at the base of the wall will typically be carried by the slab-on-grade.

At the top of the wall is a parapet. Based on our approximation, we consider all of the parapet to transfer its lateral load to the roof diaphragm.

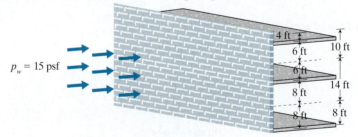

Figure 1.79

Distributed loads:

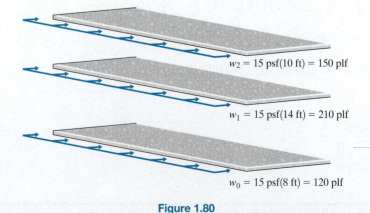

$w_2 = 15 \text{ psf}(10 \text{ ft}) = 150 \text{ plf}$

$w_1 = 15 \text{ psf}(14 \text{ ft}) = 210 \text{ plf}$

$w_0 = 15 \text{ psf}(8 \text{ ft}) = 120 \text{ plf}$

Figure 1.80

1.6 Distribution of Lateral Loads by Flexible Diaphragm

Introduction

For most structures, lateral pressure on a wall is carried first to the floor or roof diaphragms. From there the load is transferred to the structural frame. Determining how much of the load along the edge of the floor or roof diaphragm is transferred to the various lateral load-resisting elements of the structural frame depends on the in plane stiffness of the floor or roof diaphragm.

If the floor or roof is much more flexible in plane than the structural frame, we call that floor or roof a *flexible diaphragm*. The lateral load carried in plane by a flexible diaphragm distributes to the structural frame very differently than if it was a rigid diaphragm. This section of the chapter covers only flexible diaphragms. We deal with rigid diaphragms in Section 11.1: Distribution of Lateral Loads by Rigid Diaphragm.

How-To

Only certain parts of the structural frame are lateral load resisting, and the floor or roof diaphragm transfers load only to lateral load-resisting elements. The three types of lateral load-resisting elements are rigid frames, braced frames, and shear walls (Figure 1.81). These elements are much stiffer in plane than out of plane; therefore, we consider them to resist only lateral load applied in the direction of the plane of the lateral load-resisting elements (Figure 1.82). A frame that consists of only pin-connected members and no diagonal members (Figure 1.83) does not have lateral stiffness and, therefore, does not receive lateral load from the diaphragms.

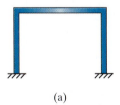
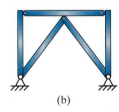
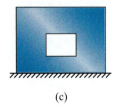

(a)   (b)   (c)

Figure 1.81 Idealizations of lateral load-resisting elements: (a) Rigid frame has rigid connections and/or fixed bases; (b) Braced frame has diagonal members to stiffen the frame; (c) Shear wall is solid but might have openings for windows or doors.

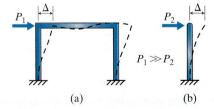

Figure 1.82 Lateral stiffness is defined as the force required to generate a unit lateral displacement: (a) Lateral stiffness of a rigid frame (or braced frame or shear wall) in plane is significant; (b) Lateral stiffness of a rigid frame (or braced frame or shear wall) out of plane is negligible in comparison.

60   Chapter 1   Loads and Structure Idealization

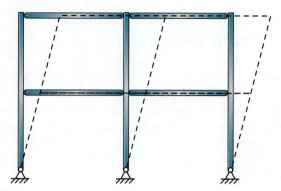

Figure 1.83 Example of a frame unable to support lateral load: pinned at base, all connections pinned, and no diagonal members.

When a floor or roof diaphragm is very flexible in plane compared to the lateral load-resisting elements, we call it a *flexible* diaphragm. *If* the floor or roof diaphragm is much more flexible in plane than the supporting elements, we can reasonably assume that the lateral load-resisting elements do not move. Figure 1.84a illustrates this idealization. Because the distributed load on the edge of the flexible diaphragm acts in the north-south direction, none of the rigid frames in the east-west direction carry any of the load. With our assumption that the lateral load-resisting elements do not move compared to the diaphragm, we can idealize the diaphragm as a beam supported by a pin and rollers (Figure 1.84b). Note that column line 3 does

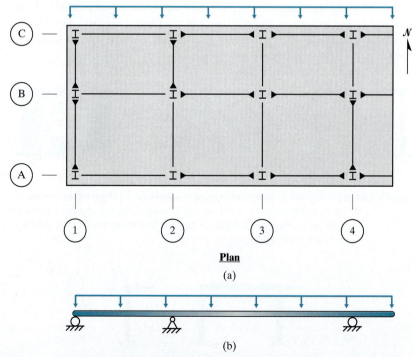

Figure 1.84 Converting lateral load along a floor or roof diaphragm into point loads on the lateral load-resisting elements: (a) Flexible diaphragm with uniform lateral load; (b) Reasonable idealization of the diaphragm if it is much less stiff than the lateral load-resisting elements.

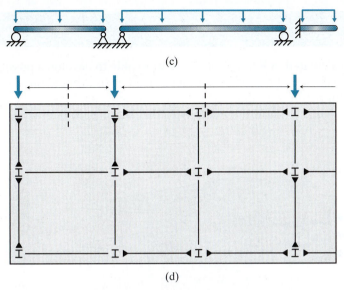

Figure 1.84 (*Continued*) (c) Common approximation that each span of diaphragm is simply supported; (d) Resulting idealized point loads on each lateral load-resisting element.

not act as a support because it is not a lateral load-resisting element. We call this idealization of a flexible diaphragm the *Equivalent Beam Model*.

If the equivalent beam is determinate, we use equilibrium to find the reactions. The reactions are the forces that act on the lateral load-resisting elements. If the equivalent beam is indeterminate, like the one in Figure 1.84b, we can use methods from Chapter 12: Force Method or Chapter 13: Moment Distribution Method to perform analysis, or we can approximate the beam as simply supported between supports (Figure 1.84c). If the idealized lateral load on the diaphragm is uniformly distributed, our approximation results in the load being carried by the nearest lateral load-resisting element (Figure 1.84d).

Section 1.6 Highlights

Distribution of Lateral Loads by Flexible Diaphragm

Assumption: A floor or roof diaphragm is much more flexible in plane than the lateral load-resisting elements that support it.

Approximation: If a flexible floor or roof diaphragm is supported in more than two locations, consider the diaphragm to behave as simply supported between each supporting element.

Distributed Load: If the load along the edge of the diaphragm is uniformly distributed,

$$P = \text{distributed load} \times \text{tributary width}$$

If the load along the edge of the diaphragm is nonuniformly distributed, calculate the forces on the lateral load-resisting elements using approximation and equilibrium.

EXAMPLE 1.16

With the distributed loads we calculated in Example 1.15, we were able to develop a possible structural system. We want to make a preliminary selection of member sizes based on approximate analysis results. But before we can perform an approximate analysis on the lateral load-resisting system, we need to know the point loads that act on each lateral load-resisting element.

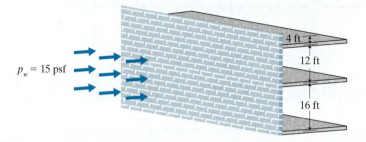

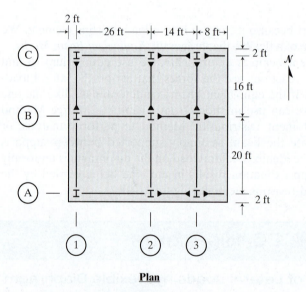

Plan

Figure 1.85

Assumptions and Approximations:
- For preliminary analysis, consider the floor and roof diaphragms to be relatively flexible. We can update our analysis based on rigid diaphragms after a preliminary selection of member sizes.
- Consider the floor and roof diaphragms to be simply supported between lateral load-resisting elements.

Idealized Lateral Load:

Distributed loads on diaphragms:

We calculated the distributed loads on each diaphragm in Example 1.15.

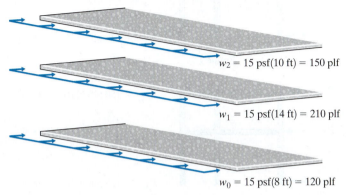

Figure 1.86

Note that knowing the total lateral force applied to the wall will help us check our final results.

$$\text{Total force} = 15\,\text{psf}\,(32\,\text{ft})\,(50\,\text{ft}) = 24{,}000\,\text{lb}$$

Tributary widths:

In the north-south direction, there are three lateral load-resisting elements: rigid frames at column lines 1, 2, and 3. Therefore, each of those column lines receives some load based on its tributary width.

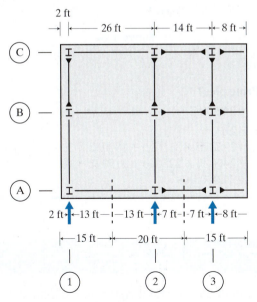

Figure 1.87

Point loads on lateral load-resisting elements:

Note that the slab-on-grade probably transfers most of its load directly to the ground through friction. However, some will likely transfer to the foundation at the base of the lateral load-resisting frames. Therefore, we will calculate the point forces at the bases of the frames as an upper bound on what the foundation might need to carry.

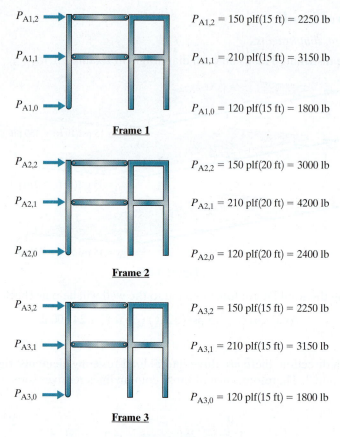

$P_{A1,2} = 150 \text{ plf}(15 \text{ ft}) = 2250 \text{ lb}$

$P_{A1,1} = 210 \text{ plf}(15 \text{ ft}) = 3150 \text{ lb}$

$P_{A1,0} = 120 \text{ plf}(15 \text{ ft}) = 1800 \text{ lb}$

Frame 1

$P_{A2,2} = 150 \text{ plf}(20 \text{ ft}) = 3000 \text{ lb}$

$P_{A2,1} = 210 \text{ plf}(20 \text{ ft}) = 4200 \text{ lb}$

$P_{A2,0} = 120 \text{ plf}(20 \text{ ft}) = 2400 \text{ lb}$

Frame 2

$P_{A3,2} = 150 \text{ plf}(15 \text{ ft}) = 2250 \text{ lb}$

$P_{A3,1} = 210 \text{ plf}(15 \text{ ft}) = 3150 \text{ lb}$

$P_{A3,0} = 120 \text{ plf}(15 \text{ ft}) = 1800 \text{ lb}$

Frame 3

Figure 1.88

Evaluation of Results:

Satisfied Fundamental Principles?

Total applied force = 24,000 lb

Total load on frames = ΣP
= 2250 lb + 3150 lb + 1800 lb + 3000 lb + 4200 lb
+ 2400 lb + 2250 lb + 3150 lb + 1800 lb
= 24,000 lb ✓

EXAMPLE 1.17

We want to compare the cost of the structural system in Example 1.16 with an alternative system. The alternative system uses chevron braces rather than rigid connections to provide lateral load resistance.

Assumptions and Approximations:

- For preliminary analysis, consider the floor and roof diaphragms to be relatively flexible. We can update our analysis based on rigid diaphragms after we have a preliminary selection of member sizes.

Section 1.6 Distribution of Lateral Loads by Flexible Diaphragm

Idealized Lateral Load:

Distributed loads on diaphragms:

The distributed loads on the diaphragms are not affected by the position of the lateral load-resisting elements.

The total force is also unchanged:

$$\text{Total force} = 15 \text{ psf}\,(32 \text{ ft})\,(50 \text{ ft}) = 24{,}000 \text{ lb}$$

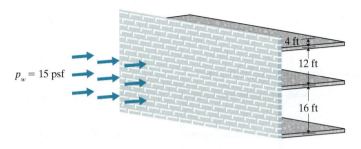

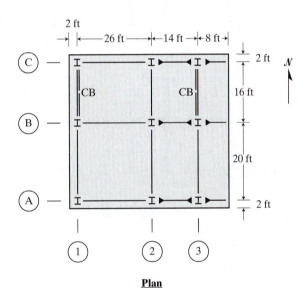

Plan

Figure 1.89

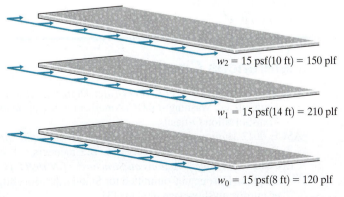

$w_2 = 15 \text{ psf}(10 \text{ ft}) = 150 \text{ plf}$

$w_1 = 15 \text{ psf}(14 \text{ ft}) = 210 \text{ plf}$

$w_0 = 15 \text{ psf}(8 \text{ ft}) = 120 \text{ plf}$

Figure 1.90

Chapter 1 Loads and Structure Idealization

Equivalent Beam Model:

In the north-south direction, there are now only two lateral load-resisting elements: braced frames at column lines 1 and 3. Therefore, we apply equilibrium to the Equivalent Beam Model.

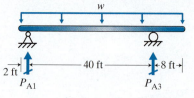

$$\circlearrowleft \Sigma M_{A1} = 0 = w(50 \text{ ft})(25 \text{ ft} - 2 \text{ ft}) - P_{A3}(40 \text{ ft})$$
$$P_{A3} = w(28.8 \text{ ft}) \; (+\uparrow)$$
$$+\uparrow \Sigma F_y = 0 = P_{A1} - w(50 \text{ ft}) + w(28.8 \text{ ft})$$
$$P_{A1} = w(21.2 \text{ ft}) \; (+\uparrow)$$

Figure 1.91

Point loads on lateral load-resisting elements:

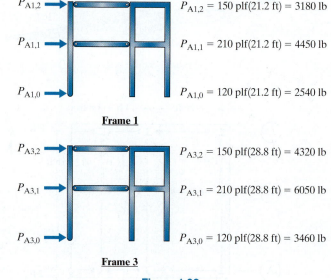

$P_{A1,2} = 150 \text{ plf}(21.2 \text{ ft}) = 3180 \text{ lb}$
$P_{A1,1} = 210 \text{ plf}(21.2 \text{ ft}) = 4450 \text{ lb}$
$P_{A1,0} = 120 \text{ plf}(21.2 \text{ ft}) = 2540 \text{ lb}$

Frame 1

$P_{A3,2} = 150 \text{ plf}(28.8 \text{ ft}) = 4320 \text{ lb}$
$P_{A3,1} = 210 \text{ plf}(28.8 \text{ ft}) = 6050 \text{ lb}$
$P_{A3,0} = 120 \text{ plf}(28.8 \text{ ft}) = 3460 \text{ lb}$

Frame 3

Figure 1.92

Evaluation of Results:

Satisfied Fundamental Principles?

Total applied force = 24,000 lb

Total load on frames = ΣP
  = 3180 lb + 4450 lb + 2540 lb + 4320 lb
  + 6050 lb + 3460 lb
  = 24,000 lb ✓

References Cited

AASHTO. 2012. *AASHTO LRFD Bridge Design Specifications, Customary U.S. Units*, 6th ed. Washington, DC: American Association of State Highway and Transportation Officials.

ASCE. 2017. *Minimum Design Loads for Buildings and Other Structures (7–16)*. Reston, VA: American Society of Civil Engineers.

CEN. 2002. *Eurocode 1: Actions on Structures (EN 1991)*, TC250. Brussels, Belgium: European Committee for Standardization. http://eurocodes.jrc.ec.europa.eu/showpage.php?id=131

CSA Group. 2014. *Canadian Highway Bridge Design Code (S6-14)*. Mississauga, Ontario: Canadian Standards Association.

Homework Problems

1.1 A mechanical engineer and an architect have designed the roof cross-section that results in the materials and depths shown.

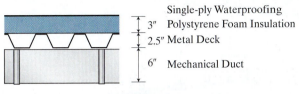

Prob. 1.1

What is the unfactored dead load pressure for the anticipated roof cross-section?

1.2 Based on preliminary estimates and the architect's design, a team member has developed the anticipated cross-section for an office floor. The plan is to use cinder concrete, which is also called lightweight concrete.

What is the unfactored dead load pressure for the anticipated floor cross-section?

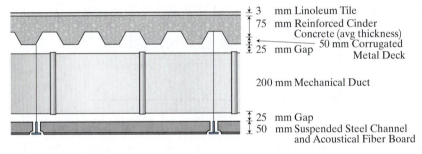

Prob. 1.2

1.3 To design the floor of a building, we need to know the load created by the walls. For this building, the walls are considered fixed partitions (they will not be moved around by renovations).

What is the unfactored line load generated by the self-weight of the wall?

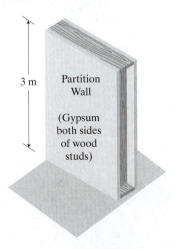

Prob. 1.3

1.4 To design the floor of a building, we need to know the load created by the walls. For this building, the walls are considered fixed partitions (they will not be moved around by renovations).

What is the unfactored line load generated by the self-weight of the wall?

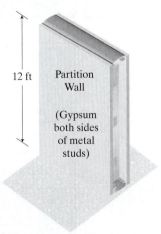

Prob. 1.4

68 **Chapter 1** Loads and Structure Idealization

1.5 The architect on our project has designed the second-floor layout for an office building to have two corridors as shown, but the interior walls are not structural and are likely to be moved over time, resulting in new corridor locations.

a. What are the minimum unfactored live loads for the offices and corridors?
b. How can we account for the corridors moving over time? Why is that a reasonable choice?

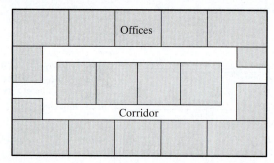

Prob. 1.5

1.6 The architect has laid out the anticipated functions for a new library: offices, reading areas, stack areas, and corridors. The open layout of the upper floors, however, means that the entire layout could be reconfigured at a later time.

a. What are the minimum unfactored live loads for the offices, reading areas, stack areas, and upper floor corridors?
b. How can we account for the layout changing over time? Why is that a reasonable choice?

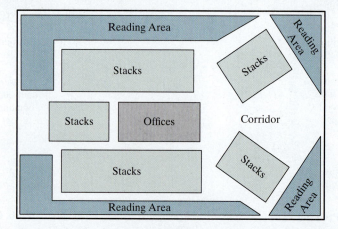

Prob. 1.6

1.7 A construction company wants to use a truss to hold equipment over the edge of a building during construction. The maximum likely, unfactored dead and live loads have been calculated. To properly design the locations where the truss will attach to the building, we need to know the design reactions.

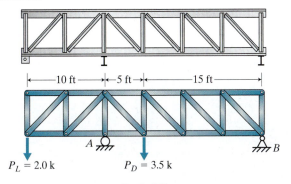

Prob. 1.7

a. What is the design vertical reaction at A? Provide values for both upward and downward reactions.
b. What is the design vertical reaction at B? Provide values for both upward and downward reactions.

1.8 We are designing a concrete retaining wall for a dolphin tank at the new zoo. We need to know the peak design moments at the base of the wall in order to begin design of the wall. The geotechnical engineer recommends that we consider this soil pressure to be based on a density of 45 pcf. Caretakers at the zoo have given us the ideal salinity of the water for the dolphins; with a bit of research, we estimate the resulting density of the water to be 64.1 pcf.

a. What is the design counterclockwise moment? What load factors and combination give that design value?
b. What is the design clockwise moment? What load factors and combination give that design value?

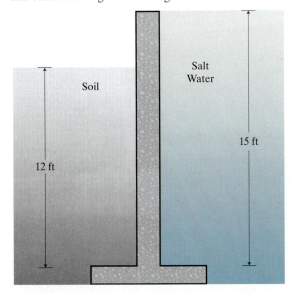

Prob. 1.8

1.9 We want to design a reinforced concrete beam that overhangs the supports on both ends. To design that beam, we need to know the design moments, both positive and negative, at midspan (point B). Remember that live load can act everywhere, somewhere, or nowhere.

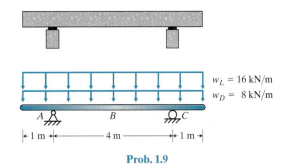

Prob. 1.9

1.10 Our team is designing a new highway overpass. Although we have not completed the design of this overpass, the geotechnical engineer needs a preliminary estimate of the design vertical force and the design moment that the foundation must carry. At this time, we estimate that the maximum likely, unfactored dead load from each girder is 40 kips and the live load is 20 kips.

For this situation, we need only the largest-magnitude design moment rather than both positive and negative moments. Remember that the moment is calculated at the neutral axis. Note that although AASHTO load factors and combinations would govern this design, we will use the ASCE 7 load factors and combinations for this preliminary analysis, since they are handy.

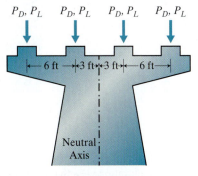

Prob. 1.10

1.11 Two steel beams are connected at the corner of a cantilevered balcony. The web of one member is bolted to a web stiffener of the other member.

1.13 The connection shown in the photo is used in a bridge girder. One reason the connection is used is to allow for thermal expansion and contraction of the girder.

What is a reasonable idealization of the connection? Draw the idealization in an elevation view that matches the photo.

Prob. 1.11

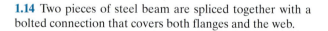

Prob. 1.13

What is a reasonable idealization of the connection? Draw the idealization in an elevation view that matches the photo. Also draw the idealization in a plan view.

1.12 A concrete beam-column connection was cast with notches on two sides. The notches are black in the photo. The concrete beam is continuous over the column.

1.14 Two pieces of steel beam are spliced together with a bolted connection that covers both flanges and the web.

Prob. 1.12

Prob. 1.14

What is a reasonable idealization of the connection? Draw the idealization in an elevation view that matches the photo.

What is a reasonable idealization of the connection? Draw the idealization in a plan view that matches the photo.

1.15 A reinforced concrete column in a parking garage supports two inverted-T girders. The extensions from the column are called corbels. The girders sit on elastomeric pads on the corbels. The pads allow the girders to expand and contract as well as rotate.

What is a reasonable idealization of the connection between the column and one of the girders? Draw the idealization of the column and both girders in an elevation view that matches the photo. Also draw the idealization in a plan view.

Prob. 1.15

1.16 A rocking element is a common support for bridge girders. The top and bottom are slightly rounded to allow the support to rotate.

Prob. 1.16

What is a reasonable idealization of the support? Draw the idealization in an elevation view that matches the photo.

1.17 In a parking lot, an inattentive driver can seriously damage a steel column; therefore, we typically use reinforced concrete to a height above bumper level. The steel column then attaches to the concrete column just as it would have attached to the foundation.

Prob. 1.17

What is a reasonable idealization of the support shown? Draw the idealization in an elevation view that matches the photo.

1.18 A railroad bridge rests on a support anchored to the concrete abutment and hinged at the top.

Prob. 1.18

What is a reasonable idealization of the support? Draw the idealization in an elevation view that matches the photo.

1.19 A bridge girder is supported by a short steel block anchored to the concrete pier. The top of the block is slightly rounded to allow rotation.

Prob. 1.19

What is a reasonable idealization of the support? Draw the idealization in an elevation view that matches the photo.

1.20 Our firm is designing a platform for part of a theatrical stage. The platform can be built in either of two ways: beams on girders, or beams and girders at the same level. We want to perform analyses of both options to help decide which to choose. But in order to perform those analyses, we need to know the idealized loading on beam BE and girder ABC.

We anticipate that the floor diaphragm will be made of 18-mm-thick plywood on 38 × 190-mm yellow pine lumber. The floor on the second option has been partially cut away to show how the beams and girders connect; the plywood actually covers the entire frame.

a. What are the idealized live loads on beam BE and girder ABC for each option?
b. What are the idealized dead loads on beam BE and girder ABC for each option?

1.21 and 1.22 Our firm is evaluating an existing facility that is to be used as an overflow area for a library's book collection. The facility has a cast-in-place concrete floor supported by steel beams. The box with an × in the figure represents an opening in the floor where the elevators are.

1.21 To assess whether the existing members are adequate for the new use, the team needs to know the idealized loads on the members. We have been tasked with finding the idealized live load on certain members. Assume that the entire floor, except the elevator space, will be used for stacks of books.

a. Find the idealized live load on girder A1-C1.
b. Find the idealized live load on girder C1-C3.

1.22 To assess whether the existing members are adequate for the new use, the team needs to know the idealized loads on the members. We have been tasked with finding the idealized dead load on certain members. The concrete floor weighs 150 psf and carries a superimposed dead load of 10 psf for utilities and flooring. The beams all weigh about 100 plf.

a. Find the idealized dead load on beam A2-C2.
b. Find the idealized dead load on girder A1-A3.

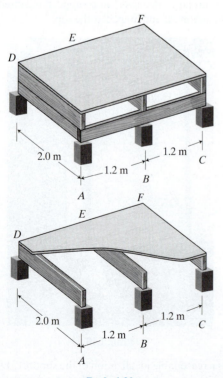

Prob. 1.20

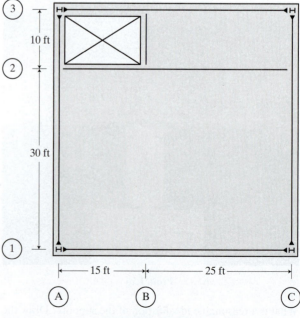

Probs. 1.21 and 1.22

1.23 through 1.25 Our firm is designing an office building for a narrow site. The plan is to use a cast-in-place slab for the floor. Architectural requirements have driven the floor plan. To begin preliminary design of the beams and girders, the team needs to know the loading on each member due to each unfactored load.

The floor will be a combination of offices and corridors. Since the locations of the corridors can change with each new tenant, the project leader decided we should use a uniform live load of 2.9 kN/m² over the entire floor. That is larger than the live load for offices, but smaller than the live load for corridors.

A more experienced engineer on the team has provided approximate self-weights for the various members. The beams are running up-down in the plan view, and the girders are running left-right.

$$p_o^{slab} = 5 \text{ kN/m}^2$$
$$w_o^{beam} = 1 \text{ kN/m}$$
$$w_o^{girder} = 2 \text{ kN/m}$$

1.23 We have been tasked with finding some of the idealized live loads.

a. Find the idealized live load on beam D2-D3.
b. Find the idealized live load on girder C2-E2.

1.24 We have been tasked with finding some of the idealized live loads.

a. Find the idealized live load on beam B2-B3.
b. Find the idealized live load on girder A2-C2.

1.25 We have been tasked with finding some of the idealized dead loads.

a. Find the idealized dead load on beam D2-D3.
b. Find the idealized dead load on girder C2-E2.

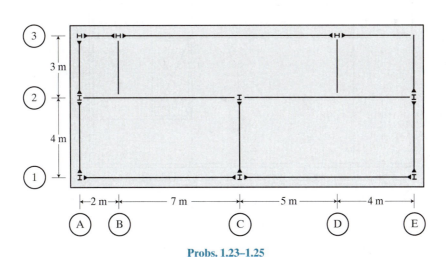

Probs. 1.23–1.25

74   Chapter 1   Loads and Structure Idealization

1.26 through 1.28 The senior engineer working on a two-story office building project has laid out a framing plan to be used for the first floor and roof level. The box with an × in the plan view represents an opening in the floor and roof for the elevator. The senior engineer anticipates using a concrete slab supported by steel beams and girders. Before we can begin to design the members, we need to know the idealized load on each member due to each type of load.

1.26 We are tasked with finding the idealized live load. For preliminary design of the members, we can ignore the corridors and consider the entire floor to be offices. Let's start with the idealized live load on girder B3-B4 for level 1 (first floor).

1.27 The maximum likely snow load on the roof is 20 psf. Our task is to find the idealized snow load on beam A2-B2 for level 2 (the roof).

1.28 A more experienced engineer on the team anticipates the floor and roof to be 12 inches thick. The superimposed dead load (weight of flooring, ceilings, lights, ducts, etc.) will be about 10 psf on the floor and roof. Initially let's assume each beam and girder has a self-weight of 75 plf. Our task is to find the idealized dead load on girder A3-B3 for level 2, including its own self-weight.

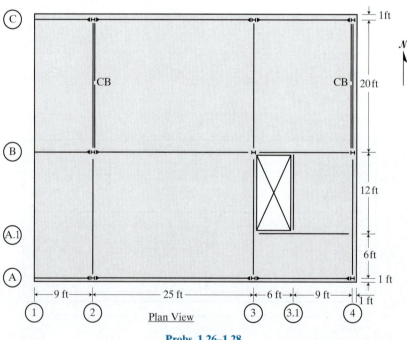

Probs. 1.26–1.28

1.29 We are to design the lateral load-resisting system for a building with a high roof. The senior engineer has already decided that the outer walls perpendicular to the wind will be the lateral load-resisting elements (e.g., the beige walls will resist wind from the left in the photo). The senior engineer has also concluded that the best way to transfer the wind pressure to the outer walls is by purlins. Therefore, the outer skin of the wall touches only the slab-on-grade and the purlins.

Before we can analyze and design the purlins, we need to determine the idealized line load on each purlin due to the wind. The wind generates a pressure of 950 Pa.

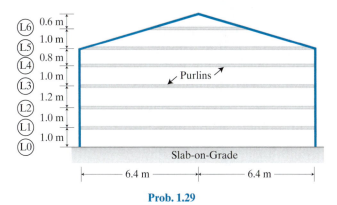

Prob. 1.29

1.30 The contractor for a new high-rise building must temporarily brace the large excavation for the basement. Our firm has been selected to design the bracing. The geotechnical report recommends that we design for an effective lateral soil pressure of 2.5 kN/m² per meter of depth.

The senior engineer on the team recommends that we use braces spaced every 5 m along the length of the excavation. Our job is to find the unfactored force generated on one brace at each level. Another team member will convert those applied forces into axial forces in the braces; the two will not be the same, since each brace is on at least a slight angle.

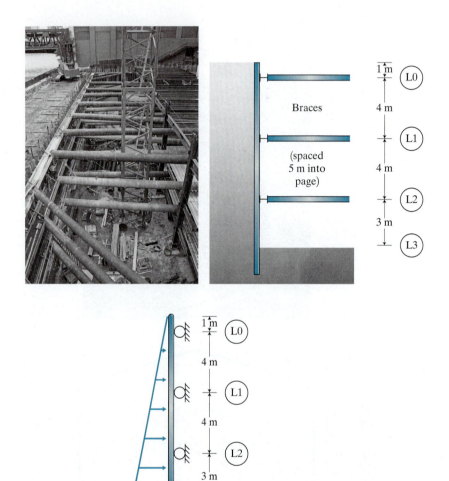

Prob. 1.30

1.31 A two-story building will be made with thin floors that will behave as flexible diaphragms. The lateral load-resisting system will be rigid frames. For preliminary analysis, consider the wind from the south to cause uniform pressure on the entire wall. Note that the south and east walls have been cut away so that we can see the system supporting them.

a. Our job is to convert the applied wind pressure into line loads that act at each diaphragm level on the windward side.
b. Then we are to convert the line loads into point loads that act at the locations where the diaphragms meet the lateral load-resisting elements.

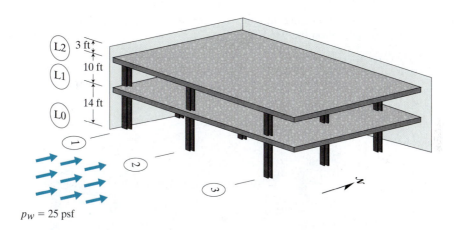

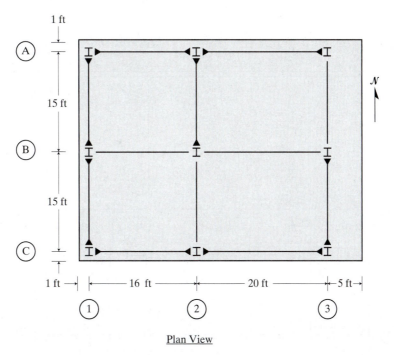

Plan View

Prob. 1.31

Chapter 1 Loads and Structure Idealization

1.32 and 1.33 Our team is designing the lateral load-resisting system for a two-story office building. To do so, we need to determine the wind loads on the lateral load-resisting elements. At this preliminary stage, let's consider the floor and roof diaphragms to be flexible, and consider the wind pressure to be uniform. Note that the south and east walls have been cut away so that we can see the system supporting them.

1.32 When the wind blows from the south, it creates a uniform pressure of 25 psf on the south elevation.

a. Our job is to convert the applied wind pressure into line loads that act at each diaphragm level on the windward side.

b. Then we are to convert the line loads into point loads that act at the locations where the diaphragms meet the lateral load-resisting elements.

1.33 When the wind blows from the east, it creates a uniform pressure of 30 psf on the east elevation.

a. Our job is to convert the applied wind pressure into line loads that act at each diaphragm level on the windward side.

b. Then we are to convert the line loads into point loads that act at the locations where the diaphragms meet the lateral load-resisting elements.

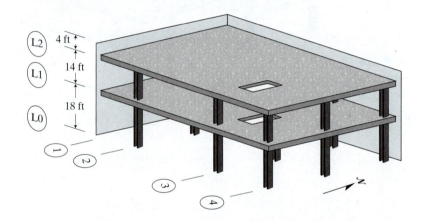

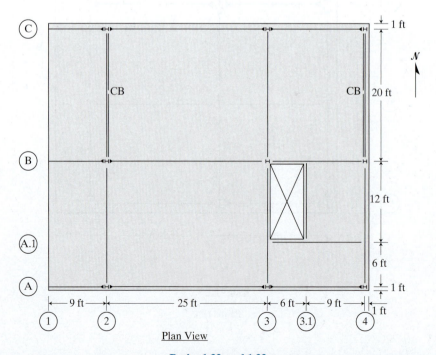

Plan View

Probs. 1.32 and 1.33

1.34 and 1.35 Our firm is designing an office building for a narrow site. Due to architectural requirements, the columns along line C have to be moved for the second level. The plan is to use a cast-in-place slab for the floor and roof. Although the result will likely be rigid diaphragms, for preliminary design let's consider the floor and roof diaphragms to be flexible.

Although we will ultimately design the lateral system for nonuniform wind pressure, we can use uniform pressure for this preliminary analysis. Note that the south and east walls have been cut away so that we can see the system supporting them.

1.34 When the wind blows from the west, it creates a uniform pressure of 0.9 kN/m² on the west elevation.

a. Our job is to convert the applied wind pressure into line loads that act at each diaphragm level on the windward side.
b. Then we are to convert the line loads into point loads that act at the locations where the diaphragms meet the lateral load-resisting elements.

1.35 When the wind blows from the south, it creates a uniform pressure of 1.0 kN/m² on the south elevation.

a. Our job is to convert the applied wind pressure into line loads that act at each diaphragm level on the windward side.
b. Then we are to convert the line loads into point loads that act at the locations where the diaphragms meet the lateral load-resisting elements.

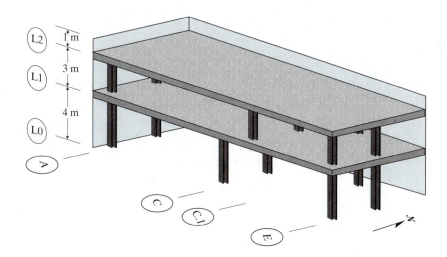

80   **Chapter 1**   Loads and Structure Idealization

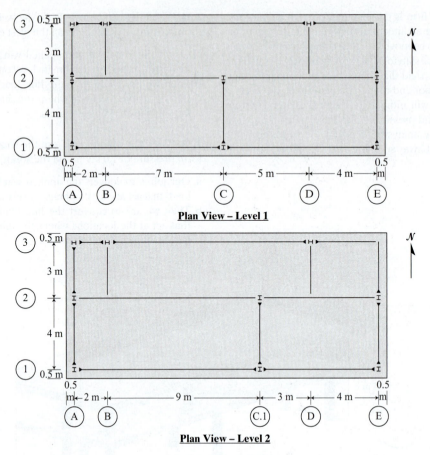

Probs. 1.34 and 1.35

CHAPTER 2

PREDICTING RESULTS

The more complex a structure, the more important it is to predict analysis results.

MOTIVATION

Experienced practitioners have a saying: "Don't ask the computer to solve a problem that you don't already know the solution to." Actually, practitioners live by a broader mantra: "Don't perform a detailed analysis on a structure until you know roughly what the results will be." It does not matter whether you are performing the detailed analysis by hand or with the assistance of a computer.

There are two primary reasons seasoned practitioners start with a prediction of the results rather than start with the detailed analysis and then confirm the results. The first is very practical. Detailed analysis is time consuming, and the results depend on the structural system and the member properties used in the idealization. But we choose the member properties based on the results of analysis; therefore, structural design is an iterative process. Rather than spend significant time performing a detailed analysis with guessed member properties, we can use other analysis methods to quickly estimate the required properties. We then use those properties as the starting point in the detailed analysis. The result is that we typically arrive at a reasonable design faster.

The second reason is rooted in cognitive psychology. In the phenomenon called *inconsistency discounting*, we place more trust in the first information we have and we tend to discount inconsistencies with the later information we obtain. Since complex analysis has many opportunities for errors, it is dangerous to give more credence to the results of complex analysis than to the results of a simplified analysis. Another contributing cognitive phenomenon is *information overload*. Computer aided analysis produces volumes of information about the behavior of a structure. It is difficult and impractical to evaluate all this information, so we pick and choose. When overwhelmed with information, we often pick information that looks familiar (i.e., reasonable). Therefore, we tend to be biased toward checking the reasonable parts of the results. If we start with a simplified analysis, we start with much less information and therefore focus on evaluating that information in the detailed analysis.

Many parts of this text focus on tools for performing simplified analysis. This chapter presents several quick methods as an introduction.

2.1 Qualitative Truss Analysis

Introduction

Knowing the sense of the resulting axial force (tension or compression) in each member of a truss helps us determine how to design the members even when we don't know their magnitudes. Members in compression tend to fail by buckling; therefore, they need larger cross-sectional properties than tension members that carry the same force.

When we apply the equations of equilibrium over and over again in the Method of Joints or the Method of Sections, we increase the possibility of making a math error. When we perform computer aided analysis on complex trusses, we are not likely to make a math error, but there is a good possibility of making an error in the idealization (e.g., support conditions, direction of loading). Knowing the sense of the axial forces before performing those detailed calculations gives us a quick check on our detailed results.

How-To

For our detailed truss analysis, we will use the Method of Joints or the Method of Sections, which we learned in an earlier course. For qualitative analysis, we will use the same methods but without the numbers. In most cases, we can identify the required sense of an internal force (tension or compression) in order to keep that joint or section in equilibrium.

Although we can use qualitative analysis to figure out the sense of most members in a truss, for some members we can only identify their sense with quantitative analysis.

Zero-Force Members

A skill that you might have covered previously but that is worth repeating here is identifying zero-force members. There are two situations in which we can identify members with no axial force. The first is shown in Figure 2.1: two members meet at a joint without an applied force. In that case, both members are zero force. Note that no applied force includes no force due to support reactions.

The second situation is shown in Figure 2.2: three members meet at a joint without an applied force and two are colinear. In this case, the member that is not colinear is zero force.

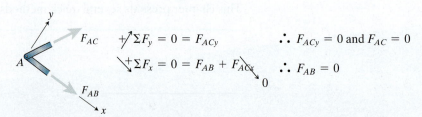

Figure 2.1 Identifying zero-force members when only two members meet at a joint.

Section 2.1 Qualitative Truss Analysis

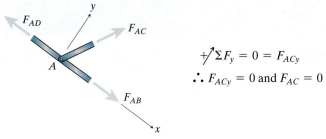

$$+\nearrow \Sigma F_y = 0 = F_{ACy}$$
$$\therefore F_{ACy} = 0 \text{ and } F_{AC} = 0$$

Figure 2.2 Identifying zero-force members when three members meet at a joint and two are colinear.

SECTION 2.1 HIGHLIGHTS

Qualitative Truss Analysis

Approach:

Apply the Method of Joints or the Method of Sections without numbers to determine the sense of the internal force in each truss member:

- Tension
- Compression
- Zero
- Can't be determined

Zero-Force Members:

- Two members
- No applied force

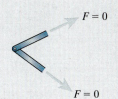

Both are zero force.

- Three members
- Two colinear
- No applied force

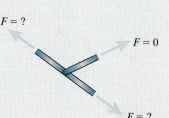

Non-colinear member is zero force.

EXAMPLE 2.1

Our firm has designed a large pipeline along the edge of a canal. We are tasked with designing a structure to support the pipe. Before we make a preliminary selection of members for the truss, it will help to know which members are in tension and which are in compression.

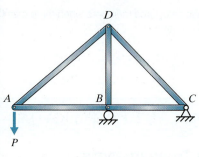

Figure 2.3

Qualitative Analysis:

Find the reaction directions:

Regardless of what the actual dimensions are, let's assume that the spacings between joints A, B, and C are equal. This allows us to determine the direction of the reactions but not the exact magnitudes.

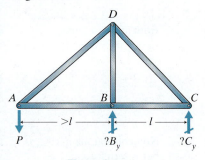

Figure 2.4a

$$\circlearrowleft^+ \Sigma M_C = 0 = P(2^+l) - B_y(l)$$
$$B_y = 2^+P > 0$$

$$+\uparrow \Sigma F_y = 0 = -P + 2^+P + C_y$$
$$C_y = -P^+ < 0$$

Method of Joints:

FBD of joint A:

Figure 2.4b

$$+\uparrow \Sigma F_y = 0 = -P + F_{ADy}$$
$$F_{ADy} = P > 0$$
$$\therefore F_{AD} = \textbf{T}$$

$$\xrightarrow{+} \Sigma F_x = 0 = +F_{ADx} + F_{AB}$$
$$F_{AB} = -F_{ADx} < 0$$
$$\therefore F_{AB} = \textbf{C}$$

Section 2.1 Qualitative Truss Analysis **87**

FBD of joint C:

Figure 2.4c

$+\uparrow \Sigma F_y = 0 = -C_y + F_{CDy}$
$F_{CDy} = C_y > 0$
$\therefore F_{CD} = \mathbf{T}$

$\xleftarrow{+} \Sigma F_x = 0 = +F_{CDx} + F_{CB}$
$F_{CB} = -F_{CDx} < 0$
$\therefore F_{CB} = \mathbf{C}$

FBD of joint B:

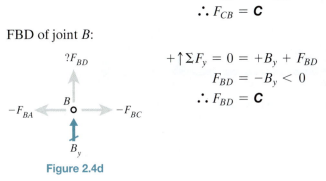

Figure 2.4d

$+\uparrow \Sigma F_y = 0 = +B_y + F_{BD}$
$F_{BD} = -B_y < 0$
$\therefore F_{BD} = \mathbf{C}$

Evaluation of Results:
Satisfied Fundamental Principles?

FBD of joint B:

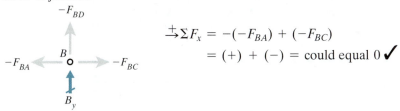

Figure 2.4e

$\xrightarrow{+} \Sigma F_x = -(-F_{BA}) + (-F_{BC})$
$\quad = (+) + (-) =$ could equal 0 ✓

FBD of joint D:

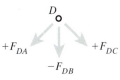

Figure 2.4f

$\xrightarrow{+} \Sigma F_x = -F_{DAx} + F_{DCx}$
$\quad = (-) + (+) =$ could equal 0 ✓

$+\uparrow \Sigma F_y = -(F_{DAy}) - (-F_{DB}) + (-F_{DCy})$
$\quad = (-) + (+) + (-) =$ could equal 0 ✓

Summary:

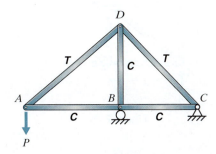

Figure 2.5

Example 2.2

We are going to analyze a truss that has three different point loads. In order to check the results of our detailed analysis, we want to know the sense of the axial force in some of the members. Let's choose BC, BH, GH, and CH.

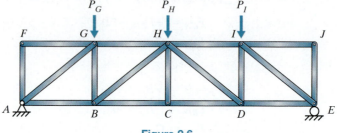

Figure 2.6

Qualitative Analysis:

Find the reaction directions:

Even though the applied loads are different, let's assume that the loads P_G, P_H, and P_I are all equal. This will allow us to determine the direction of the reaction we need.

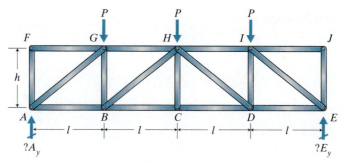

Figure 2.7a

$$\circlearrowleft \Sigma M_E = 0 = A_y(4l) - P(3l) - P(2l) - P(l)$$

$$A_y = 1.5P > 0$$

Method of Sections:
FBD of left section:

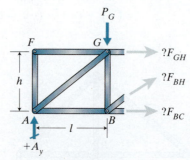

Figure 2.7b

$+\uparrow \Sigma F_y = 0 = +A_y - P_G + F_{BHy}$
$\quad F_{BHy} = -(A_y - P_G)$
$\quad$ Since $A_y > P_G$, $F_{BHy} < 0$
$\quad \therefore F_{BH} =$ **C**

$\curvearrowright \Sigma M_B = 0 = A_y(l) + F_{GH}(h)$
$\quad F_{GH} = -A_y(l/h) < 0$
$\quad \therefore F_{GH} =$ **C**

$\curvearrowright \Sigma M_H = 0 = A_y(2l) - P_G(l) - F_{BC}(h)$
$\quad F_{BC} = [A_y(2l) - P_G(l)]/h > 0$
$\quad \therefore F_{BC} =$ **T**

Method of Joints:
$\quad$ FBD of joint C:

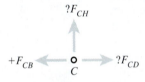

Figure 2.7c

$+\uparrow \Sigma F_y = 0 = F_{CH}$
$\quad F_{CH} = 0$
$\quad \therefore F_{CH} =$ **0**

Summary:

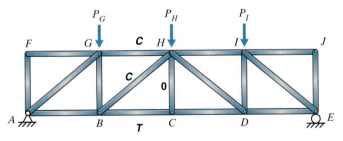

Figure 2.8

EXAMPLE 2.3

The most cost-effective lateral load-resisting system for a particular storage building is diagonal rods. Two load cases we must consider are wind from the left and wind from the right. Before we begin preliminary selection of member sizes, we want to know which members experience tension and which experience compression due to each of the two wind load cases.

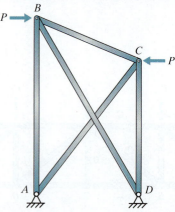

Figure 2.9

Because each rod is very thin compared to its length, the rod will buckle elastically if we try to load it in compression. Elastic buckling means that the rod will return to its original shape when we stop pushing on it, but it also means that the rods carry only tension.

Qualitative Analysis for Wind from the Left:

Find the sense of the axial force in the rods:

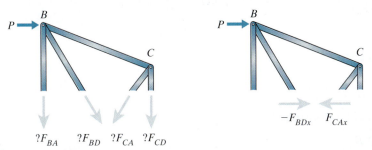

Figure 2.10

If both diagonal members helped to resist the lateral load, the horizontal components of the axial forces would be to the left to oppose the load.

Since rod BD cannot be in compression, the structure behaves as if member BD does not exist (under this loading).

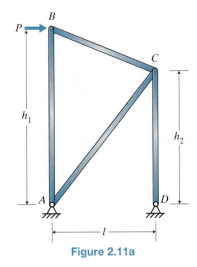

Figure 2.11a

FBD of top section:

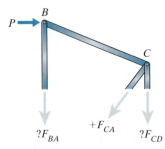

Figure 2.11b

$+\circlearrowleft \Sigma M_A = 0 = P(h_1) + F_{CD}(l)$
$ F_{CD} = -P(h_1/l) < 0$
$ \therefore F_{CD} = \mathbf{C}$

$+\circlearrowleft \Sigma M_C = 0 = P(h_1 - h_2) - F_{BA}(l)$
$ F_{BA} = P(h_1 - h_2)/l > 0$
$ \therefore F_{BA} = \mathbf{T}$

FBD of joint B:

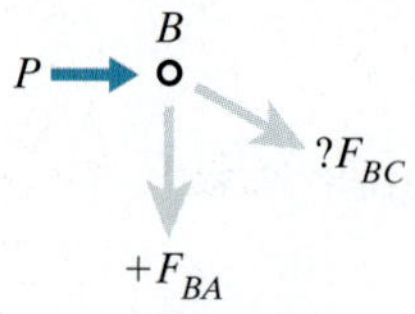

$\pm \rightarrow \Sigma F_x = 0 = P + F_{BCx}$
$F_{BCx} = -P < 0$
$\therefore F_{BC} = \mathbf{C}$

Figure 2.11c

Evaluation of Results:
Satisfied Fundamental Principles?

FBD of joint B:

$+\uparrow \Sigma F_y = -F_{BA} - (-F_{BCy})$
$= (-) + (+) = $ could equal 0 ✓

Figure 2.11d

Qualitative Analysis for Wind from the Right:

Find the sense of the axial force in the rods:

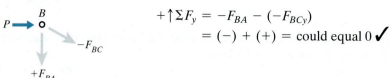

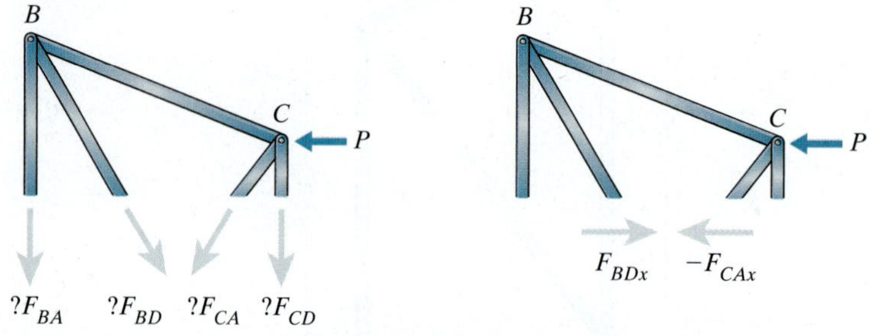

Figure 2.12

Under this loading, rod AC does not have axial force.

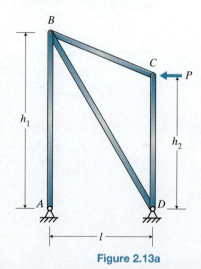

Figure 2.13a

Section 2.1 Qualitative Truss Analysis

FBD of top section:

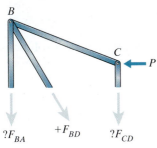

Figure 2.13b

$$\circlearrowleft \Sigma M_D = 0 = P(h_2) + F_{BA}(l)$$
$$F_{BA} = -P(h_2/l) < 0$$
$$\therefore F_{BA} = \mathbf{C}$$

$$\circlearrowleft \Sigma M_B = 0 = P(h_1 - h_2) + F_{CD}(l)$$
$$F_{CD} = -P(h_1 - h_2)/l < 0$$
$$\therefore F_{CD} = \mathbf{C}$$

FBD of joint C:

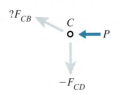

Figure 2.13c

$$\xrightarrow{+} \Sigma F_x = 0 = F_{CBx} + P$$
$$F_{CBx} = -P < 0$$
$$\therefore F_{BC} = \mathbf{C}$$

Evaluation of Results:
Satisfied Fundamental Principles?

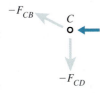

Figure 2.13d

$$+\uparrow \Sigma F_y = (-F_{CBy}) - (-F_{CD})$$
$$= (-) + (+) = \text{could equal } 0 \checkmark$$

Summary:

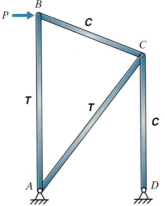 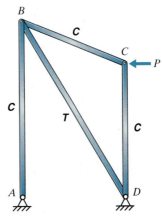

Figure 2.14

Observation: It is not a suprise that the tension diagonal switches when the load direction switches. Nor is it a suprise that the sense in member AB switches with the load. We might not anticipate that member CD would be in compression for both load cases, but it must be true for equilibrium to be satisfied.

2.2 Principle of Superposition

Introduction
Some structures are relatively easy to analyze or their solutions are readily available. For other structures, sometimes we can reproduce a complex situation as the addition of two or more simple situations. In those cases, we can find the solution to the complex situation more quickly by adding the solutions from the simple situations.

How-To
Under certain conditions, the deformation of a structure due to the total load is the same as the sum of the deformations due to each individual load (Figure 2.15). Two conditions are required for that to be true: the material must remain linear elastic, and the second-order effects should be minimal. If either condition is not met, the load-versus-displacement plot is not linear (Figure 2.15d).

Although there are situations where we want to capture the nonlinear behavior of a material during structural analysis, in many cases assuming linear elastic behavior provides reasonable results. Second-order effects, also called P-Δ effects, are the result of applied loads on the deformed shape of the structure. For most structures, the second-order effects are relatively small, so we ignore them. Analyzing a structure considering the effects of loads on the undeformed shape is called first-order analysis and is the basis of all structural analysis methods in this text. First-order analysis will always produce a linear load-versus-displacement plot as long as the material remains linear elastic. Therefore, it is reasonable to assume linear behavior for many structures.

It turns out that reactions and internal forces also add *if* the structure remains linear. This means we can find the reactions or internal forces for a structure with complex loading by adding the reactions or internal forces due to each loading individually. We call this the *principle of superposition*.

This ability can be very helpful when the reaction, deformation, or internal force has been precalculated for one or more of the individual loads. For example, the inside front cover of this text has a list of precalculated expressions for displacement and slope for certain combinations of beam supports and loadings. Initially, we will use the principle of superposition to add only the effects of loads, but in Chapters 11 and 12 we will add different support conditions to replicate indeterminate structures.

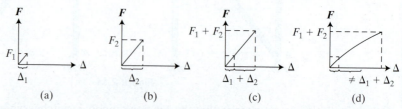

Figure 2.15 Load versus displacement behavior of a structure under different situations: (a) Loaded with only force F_1; (b) Loaded with only force F_2; (c) Loaded with both forces F_1 and F_2; (d) Loaded with forces F_1 and F_2 but material is nonlinear.

Section 2.2 Highlights

Principle of Superposition

Assumption:
Linear behavior

Concept:
The response of a structure to multiple loads is equal to the summation of the responses of that structure to individual loads.

Applies to:
- Reactions
- Internal forces
- Deformations

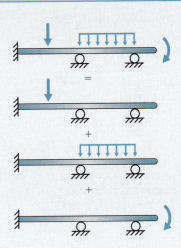

Example 2.4

In Example 1.10 we determined the idealized live load on beam B1-B2. Now we are ready to begin the design process. One of the design criteria for a beam is the live load deflection. We typically want to limit the unfactored live load displacement to no more than the span length divided by 360. Therefore, we want to know the minimum moment of inertia, I, for a steel beam that will not exceed this live load displacement criterion. Note that the modulus of elasticity, E, for steel is 200 GPa (200×10^9 N/m^2).

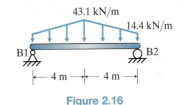

Figure 2.16

Detailed Analysis:

In Chapter 5 we learn several different methods for predicting displacement. For now, we need to use the displacement formulas given inside the front cover of this text. This distribution of loading is not one of the options given, but we can create this loading by superimposing two load cases that are provided.

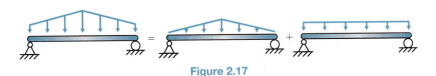

Figure 2.17

Calculate the individual displacements:

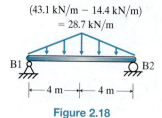

Figure 2.18

$$\Delta_{midspan1} = \frac{wl^4}{120EI} = \frac{(28.7\,\text{kN/m})(8\,\text{m})^4}{120(200 \times 10^6\,\text{kN/m}^2)I}$$

$$= \frac{4.90 \times 10^{-6}\,\text{m}^5}{I}$$

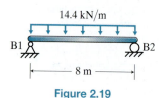

Figure 2.19

$$\Delta_{midspan2} = \frac{5wl^4}{384EI} = \frac{5(14.4\,\text{kN/m})(8\,\text{m})^4}{384(200 \times 10^6\,\text{kN/m}^2)I}$$

$$= \frac{3.84 \times 10^{-6}\,\text{m}^5}{I}$$

Superimpose the results:

$$\Delta_{midspan} = \frac{4.90 \times 10^{-6}\,\text{m}^5}{I} + \frac{3.84 \times 10^{-6}\,\text{m}^5}{I} = \frac{8.74 \times 10^{-6}\,\text{m}^5}{I}$$

Determine the minimum moment of inertia:

Set the deflection equal to the limiting deflection, $l/360$.

$$\frac{8.74 \times 10^{-6}\,\text{m}^5}{I} \leq \frac{8\,\text{m}}{360}$$

$$I \geq 3.93 \times 10^{-4}\,\text{m}^4 = 3.93 \times 10^8\,\text{mm}^4$$

Example 2.5

We will be designing a three-span continuous beam to meet live load displacement criteria. Before we perform detailed calculations to predict the displacement, we should determine the placement of the live load such that displacement is maximized. Remember from Section 1.1 that the live load can act everywhere, somewhere, or nowhere.

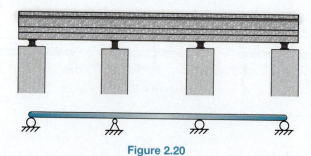

Figure 2.20

Qualitative Analysis:

We should consider the possibility that the largest displacement is caused by the live load on only one or two spans. If we can visualize the displaced shape due to the distributed load on each span individually, we can superimpose the results to maximize the displacement.

To visualize the displaced shape, we can build a physical model of the beam rather quickly. Putting a thin peice of cardboard on four pens works well. Using our fingers, we apply a distributed load to one span of the beam. Since each roller support restrains both downward and upward motion, we might need to help the model at supports where the cardboard beam tries to lift off.

Figure 2.21

Displaced shapes:

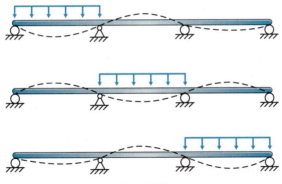

Figure 2.22

Loading to maximize displacements:

Outer spans:
Loading either of the outer spans causes both outer spans to displace down, but loading the middle span causes both outer spans to displace up. Therefore, the largest displacement in the outer spans will be due to live load on just the outer spans.

Figure 2.23

Middle span:
To maximize the displacement of the middle span, the beam needs to experience live load on *only* the middle span.

Figure 2.24

2.3 Bounding the Solution

Introduction

In many cases, superposition does not help us analyze a complex situation more quickly. The structure's shape or support conditions might not have precalculated solutions, or the loading might be so complex that superposition does not save time. However, small changes in the loading, the shape of the structure, or the boundary conditions might result in a structure that does have a precalculated solution or is easy to analyze. Making those changes means that we do not obtain the detailed solution we ultimately want, but we obtain an estimate of the solution. In addition, we can often identify whether that estimate is a guaranteed upper or lower bound to the detailed solution. Knowing the bounds to a detailed solution helps us more accurately evaluate the reasonableness of that solution.

How-To

The key to bounding the solution is making a change that simplifies the structural situation to something we can easily analyze. By considering the impact of the change, we can often determine whether we have obtained an over- or underestimate of part of the solution. If we know that we have overpredicted the solution, we have identified an upper bound to the detailed solution. An underprediction of the solution means we have a lower bound.

It is important to note that because we have changed the original situation, we are only able to compare parts of the solution. The simplified situation is not an exact scale replica of the original; some quantities will be larger and some will be smaller.

It is also important to understand that the approximate answers might be an order of magnitude larger or smaller than the detailed result. Knowing the bounds is still quite helpful because it helps us identify order-of-magnitude errors in the results of the detailed analysis. Order-of-magnitude errors happen all the time in the design process; however, if left uncorrected, they can be costly.

To simplify the situation, we might change the loading, structure shape, or boundary conditions. Figure 2.25 shows an example of each type of change.

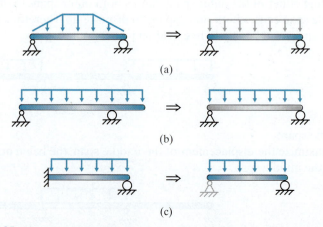

Figure 2.25 Examples of simplifying a structural situation in order to bound part of the solution: (a) Changing loading; (b) Changing structure shape; (c) Changing boundary conditions.

Section 2.3 Highlights

Bounding the Solution

Objective:

Change the structural situation to something that we can easily analyze and that we know will overpredict or underpredict some aspect of the original situation.

Keys:

- May change loading, structure shape, or boundary conditions.
- Make only one change if possible (or it may become difficult to know whether we will overpredict or underpredict).
- Will only be able to compare one or a few aspects of behavior due to the change.

Bounds:

- Lower bound = Underpredicts
- Upper bound = Overpredicts

Example 2.6

A simply supported beam with a cantilevered end is to support a point load on the cantilevered end. We want to be able to predict the vertical displacement of the cantilevered end in order to ensure that it does not exceed tolerances. Before we perform detailed calculations to predict the displacement, we want an estimate. Since we have not yet chosen the material or section properties for the beam, our answer will be in terms of the modulus of elasticity, E, and the moment of inertia, I.

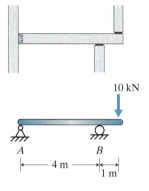

Figure 2.26

Estimated Solution:

The displaced shape should rise up between the supports and slope down over the roller support.

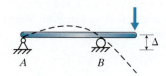

Figure 2.27

Approximate the beam:
Consider the beam to be fixed at B, leaving only the cantilevered end.

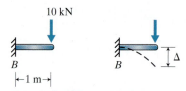

Figure 2.28

Since the actual beam has a downward slope at B, the actual beam will displace *more* than the approximation. Therefore, this should be a lower bound.

We can find the displacement of the approximate beam on the inside front cover of this text:

$$\Delta_{end} = \frac{Pl^3}{3EI} = \frac{10 \text{ kN}(1 \text{ m})^3}{3EI} = \frac{3.33 \text{ kN} \cdot \text{m}^3}{EI} \text{ (lower bound)}$$

Evaluation of Results:

Approximation Predicted Outcomes?

Compare with the results of detailed analysis:
Using methods we learn later in this text, we obtain the following displacement:

$$\Delta_{end} = \frac{16.67 \text{ kN} \cdot \text{m}^3}{EI}$$

The detailed result is greater than the approximation, as expected. ✓

EXAMPLE 2.7

During a job interview with a design firm, the interviewer draws a simply supported beam with a cantilevered end (Figure 2.29).

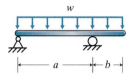

Figure 2.29

The interviewer then asks you for a quick estimate of the peak displacement of the cantilevered end and whether that estimate will be an upper or lower bound for the actual displacement. The interviewer asks that you "think aloud" in order to understand your thought process.

Estimated Solution:

Approximate the beam:
We approximate the beam as being length b and fixed at the left end.

Figure 2.30

From the formula inside the front cover of this text (and readily available from many references), we find the displacement of the end:

$$\Delta_{end} = \frac{wb^4}{8EI}$$

Determine the kind of bound on the solution:
We have essentially made two changes. Let's consider each individually and superimpose the effects.

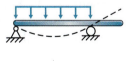

The load between the supports causes the cantilevered end to rise; therefore, ignoring this effect (which we did) will *overpredict* Δ.

+

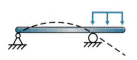

The beam will rotate over the roller support. Ignoring this effect will *underpredict* Δ.

Figure 2.31

One change causes us to overpredict Δ and one causes us to underpredict Δ. We don't know which effect is larger; therefore, we **can't predict** whether the approximation is an upper bound or a lower bound. We only know it is an estimate.

This is a trick question by the interviewer; good thing we expected that.

2.4 Approximating Loading Conditions

Introduction

If we take a complex loading situation and replace it with simpler loading, we call the results of our analysis an approximate solution or an estimate of the solution. In many cases, we can analyze the structure with simple loading much faster than with complex loading. Therefore, we can make a quick prediction by using an approximation of the actual loading. Usually, we can also identify whether this approximation will give an upper or lower bound on key parts of the detailed solution.

How-To

The goal of approximating the loading conditions is to create a situation that is easier to analyze. In preliminary design, we might want an estimate of the peak moment or displacement so that we can make initial design decisions before performing a more detailed analysis. Most often, we want a solution in which we have high confidence, so that we can verify our detailed results.

The simplest loading conditions are a uniform distributed load or a point load. How we choose the magnitude of the approximate load determines whether we will obtain an upper bound, a lower bound, or just an estimate of the solution. The best estimates typically come from using a statically equivalent load. That means the total force of the approximate loading is the same as the original loading, and the location of the equivalent resultant force is unchanged. For illustration, consider the simply supported beam in Figure 2.32. The original loading is trapezoidal with a nonzero magnitude at the ends. If we use a uniform distributed load with the smaller magnitude, we will obtain a lower bound solution. If we use a uniform distributed load with the larger magnitude, we will obtain an upper bound. But if we take the total applied load, P_{equiv}, and spread it uniformly over the beam, we have $P_{equiv}/l = w_{equiv}$. Our prediction using w_{equiv} will be most accurate because it is closest to the original loading.

Sometimes, but not often, we are able to predict whether the equivalent loading gives an upper or lower bound for the peak moment and displacement. To be able to make that determination, the approximate situation must clearly move the load either toward the supports, which results in a lower bound solution, or away from the supports, which results in an upper bound (Figure 2.33). By the way, the effect on peak shear and reactions

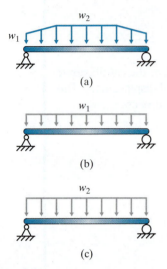

Figure 2.32 Simply supported beam: (a) Original loading is trapezoidal with nonzero values at the ends; (b) Uniform loading using the smaller end value produces a lower bound solution; (c) Uniform loading using the peak distributed load value produces an upper bound solution.

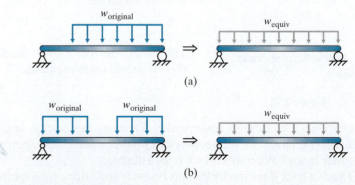

Figure 2.33 Predicting the effect of using an equivalent loading: (a) Moving the load toward the supports, away from midspan, results in a lower bound solution on peak moment and displacement; (b) Moving the load away from the supports results in an upper bound solution on peak moment and displacement.

Situation:

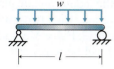

FBD:

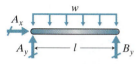

Apply equilibrium:

$$\overset{+}{\curvearrowleft}\Sigma M_B = 0 = A_y(l) - wl(l/2)$$

$$A_y = +\frac{wl}{2} \,(+\uparrow)$$

$$V_{max} = +\frac{wl}{2} \,(\downarrow+\uparrow)$$

Cut at midspan:

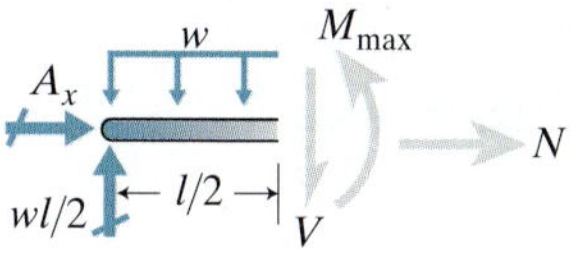

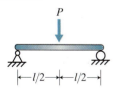

Apply equilibrium:

$$\overset{+}{\curvearrowleft}\Sigma M_{cut} = 0 = \frac{wl}{2}\left(\frac{l}{2}\right) - w\left(\frac{l}{2}\right)\left(\frac{l}{4}\right) - M_{max}$$

$$M_{max} = +\frac{wl^2}{8} \,(\curvearrowright+\curvearrowleft)$$

Figure 2.34 Derivation of peak shear, V_{max}, and peak moment, M_{max}, for a simply supported, uniformly loaded beam.

is exactly the opposite. Moving the load toward the supports results in an upper bound on peak shear and reactions, while moving load away from the supports results in a lower bound on those two quantities.

Simply supported beams are extremely common in structural engineering. Therefore, it will be helpful to know the peak shear and moment in a simply supported beam due to a uniform distributed load (Figure 2.34) and due to a single point load at midspan (Figure 2.35). In Chapter 4 we cover how to know *where* to find V_{max} and M_{max}.

Situation:

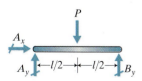

FBD:

Apply equilibrium:

$$\overset{+}{\curvearrowleft}\Sigma M_B = 0 = A_y(l) - P(l/2)$$

$$A_y = +\frac{P}{2} \,(+\uparrow)$$

$$V_{max} = +\frac{P}{2} \,(\downarrow+\uparrow)$$

Cut at point of applied load:

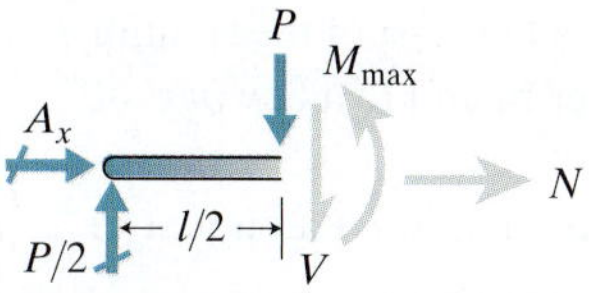

Apply equilibrium:

$$\overset{+}{\curvearrowleft}\Sigma M_{cut} = 0 = \frac{P}{2}\left(\frac{l}{2}\right) - M_{max}$$

$$M_{max} = +\frac{Pl}{4} \,(\curvearrowright+\curvearrowleft)$$

Figure 2.35 Derivation of peak shear, V_{max}, and peak moment, M_{max}, for a simply supported beam with a single point load at midspan.

Note that the benefits of using approximate loading are not limited to simply supported beams. For example, consider the indeterminate girder in Figure 2.36. The loading comes from uniform live load on a two-way floor with intermediate beams. As we will see in Chapter 12: Force Method, analyzing the beam with a uniform distributed load will be much easier than with the original loading.

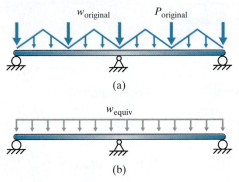

Figure 2.36 Live loading on an indeterminate girder due to intermediate beams and two-way action from the floor: (a) Original loading; (b) Equivalent uniform loading used as an approximation.

SECTION 2.4 HIGHLIGHTS

Approximating Loading Conditions

Equivalent Loads:

$$w_{equiv} = \frac{\Sigma \text{Force applied}}{\text{Length}} \qquad P_{equiv} = \Sigma \text{Force applied}$$

Peak Internal Forces for Special Cases:

$$V_{max} = \frac{wl}{2} \qquad M_{max} = \frac{wl^2}{8} \qquad V_{max} = \frac{P}{2} \qquad M_{max} = \frac{Pl}{4}$$

Key Questions:
- Can we determine whether equivalent loading will produce an upper or lower bound to a key part of the solution?
- Is there a way to simplify the loading (no longer equivalent) to obtain an upper or lower bound to a key part of the solution?
- Will superposition help us analyze the structure or help us determine whether the approximation is an upper or lower bound?

Example 2.8

In order to design a beam, we need to know the peak moment generated by the loading on the beam. Before we perform a detailed analysis, it will be helpful to know the upper and lower bounds on the solution. These bounds will allow us to evaluate the results of our detailed analysis.

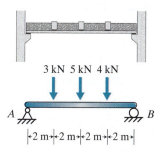

Figure 2.37

Approximate the Loading as a Single Point Load at an Equivalent Location:

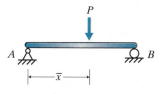

Figure 2.38

Equivalent system:

$P = \Sigma F = 3 \text{ kN} + 5 \text{ kN} + 4 \text{ kN} = 12 \text{ kN}$

$P \cdot \bar{x} = \circlearrowleft \Sigma M_A = 3 \text{ kN}(2 \text{ m}) + 5 \text{ kN}(4 \text{ m}) + 4 \text{ kN}(6 \text{ m}) = 50 \text{ kN} \cdot \text{m}$

$\bar{x} = (50 \text{ kN} \cdot \text{m})/12 \text{ kN} = 4.17 \text{ m}$

Note that this should be an upper bound because we moved most of the load farther from the supports.

Find the reaction:

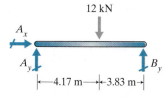

Figure 2.39

$\circlearrowleft \Sigma M_B = 0 = A_y(8 \text{ m}) - 12 \text{ kN}(3.83 \text{ m})$
$A_y = +5.74 \text{ kN } (+\uparrow)$

Find the peak moment:

We cover how to know where the peak moment is in Chapter 4: Internal Force Diagrams. For now, we will trust a more experienced designer who says that the peak moment for this beam is at the location of the single point load.

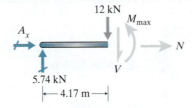

Figure 2.40

$+\circlearrowleft \Sigma M_{cut} = 0 = 5.74 \text{ kN}(4.17 \text{ m}) - M_{max}$

$M_{max} = +23.9 \text{ kN} \cdot \text{m} \ (\circlearrowleft + \circlearrowright) \text{ (upper bound)}$

Approximate the Loading as Equal, Average Point Loads:

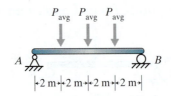

Figure 2.41

Average load:

$$P_{avg} = \frac{\Sigma F}{\# \text{ Forces}} = \frac{3 \text{ kN} + 5 \text{ kN} + 4 \text{ kN}}{3} = 4 \text{ kN}$$

Note that this should be a lower bound because we moved the load closer to the supports. And, it should be more accurate because we moved less load and we moved it a shorter distance.

Find the reaction:

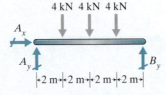

Figure 2.42

$+\circlearrowleft \Sigma M_B = 0 = A_y(8 \text{ m}) - 4 \text{ kN}(6 \text{ m}) - 4 \text{ kN}(4 \text{ m}) - 4 \text{ kN}(2 \text{ m})$

$A_y = +6 \text{ kN} \ (+\uparrow)$

Find the peak moment:
The peak moment occurs at midspan for this beam under this loading.

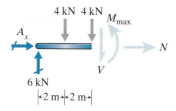

Figure 2.43

$\overset{+}{\curvearrowleft}\Sigma M_{cut} = 0 = 6\text{ kN}(4\text{ m}) - 4\text{ kN}(2\text{ m}) - M_{max}$
$M_{max} = +16.0\text{ kN}\cdot\text{m}\ (\curvearrowleft+\curvearrowright)$ (lower bound)

Evaluation of Results:
Approximation Predicted Outcomes?
Using methods we learn later in this text, we obtain the following peak moment:

$$M_{max} = +17.0\text{ kN}\cdot\text{m}\ (\curvearrowleft+\curvearrowright)$$

The detailed result is between the upper and lower bounds from our approximations, $+23.9\text{ kN}\cdot\text{m} \geq +17.0\text{ kN}\cdot\text{m} \geq +16.0\text{ kN}\cdot\text{m}$, and the detailed result is closer to the lower bound, as expected. ✓

EXAMPLE 2.9

A 20-foot-long girder will support part of a two-way slab and a supporting beam. One of the design criteria for selecting the girder cross-section is a sufficient moment of inertia to keep the peak displacement due to live loads below a certain limit. Therefore, we need to predict the displacement as a function of the moment of inertia.

Because of the complex loading, a detailed analysis of the displacement will be time consuming. Therefore, let's approximate the loading in order to use the displacement formulas inside the front cover of this text.

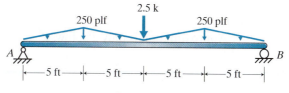

Figure 2.44

2.3 Open-web joists are a common choice for supporting floors and roofs because they are lightweight and relatively easy to install.

Dead load will tend to push down on the joists, and wind will typically cause uplift. To learn more about the behavior, we want to find the sense of the axial force in members *FG*, *PF*, *PG*, and *PQ* due to each type of load separately.

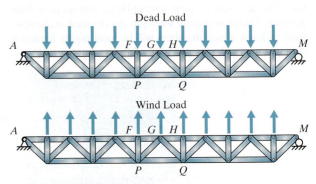

Prob. 2.3

2.4 If the dead load deflection of a beam is expected to be significant, we can prebend the beam upward to counteract the expected downward deflection. This prebending is called camber. The industry standard is that for beams less than 50 feet long, it is not cost effective to camber less than ¾ inch.

To have the desired flow of customers in a retail space, the architect wants to use a transfer girder to carry load from a second-story column out to other columns on the first floor. The girder will also carry floor load.

A member of the engineering team has chosen a steel W14 × 233 ($E = 29{,}000$ ksi, $I = 3010$ in^4) to carry the dead loads shown. Our job is to determine whether the beam should be cambered. To make that determination, we need to predict the displacement of the beam chosen. Although the peak displacement might not be exactly at midspan, the midspan displacement will be close enough. The fastest way to predict the displacement will be to use the formulas from the inside front cover of this book.

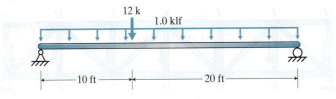

Prob. 2.4

Find the peak moment:
The peak moment occurs at midspan for this beam under this loading.

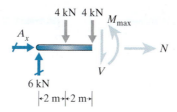

Figure 2.43

$$+\circlearrowleft \Sigma M_{cut} = 0 = 6 \text{ kN}(4 \text{ m}) - 4 \text{ kN}(2 \text{ m}) - M_{max}$$
$$M_{max} = +16.0 \text{ kN} \cdot \text{m } (\circlearrowleft+\circlearrowright) \text{ (lower bound)}$$

Evaluation of Results:

Approximation Predicted Outcomes?

Using methods we learn later in this text, we obtain the following peak moment:

$$M_{max} = +17.0 \text{ kN} \cdot \text{m } (\circlearrowleft+\circlearrowright)$$

The detailed result is between the upper and lower bounds from our approximations, $+23.9 \text{ kN} \cdot \text{m} \geq +17.0 \text{ kN} \cdot \text{m} \geq +16.0 \text{ kN} \cdot \text{m}$, and the detailed result is closer to the lower bound, as expected. ✓

EXAMPLE 2.9

A 20-foot-long girder will support part of a two-way slab and a supporting beam. One of the design criteria for selecting the girder cross-section is a sufficient moment of inertia to keep the peak displacement due to live loads below a certain limit. Therefore, we need to predict the displacement as a function of the moment of inertia.

Because of the complex loading, a detailed analysis of the displacement will be time consuming. Therefore, let's approximate the loading in order to use the displacement formulas inside the front cover of this text.

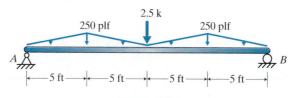

Figure 2.44

Use superposition:

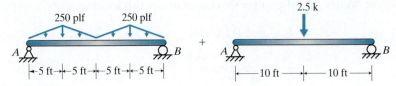

Figure 2.45

Approximate the loading:

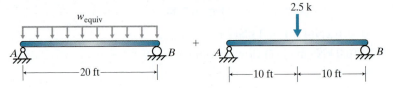

Figure 2.46

Note that we will not be able to predict whether the approximation is an upper or lower bound because we moved an equal amount of load toward the supports as well as away from the supports.

Find the equivalent loading:

$w_{equiv} = \Sigma F/l$ 　　　　　　　No change for point load

$= [2(1/2)(250 \text{ plf})(10 \text{ ft})]/20 \text{ ft}$

$= 125 \text{ plf}$

Find the peak displacement:
We use formulas from inside the front cover of this text.

$$\Delta_{max}^{w} = \frac{-5wl^4}{384EI} \qquad\qquad \Delta_{max}^{P} = \frac{-Pl^3}{48EI}$$

$$= \frac{-5(0.125 \text{ klf})(20 \text{ ft})^4}{384EI} \qquad\qquad = \frac{-2.5 \text{ k}(20 \text{ ft})^3}{48EI}$$

$$= \frac{-260 \text{ k} \cdot \text{ft}^3}{EI} \qquad\qquad\qquad = \frac{-417 \text{ k} \cdot \text{ft}^3}{EI}$$

$$\Delta_{max}^{Total} = \frac{-260 \text{ k} \cdot \text{ft}^3}{EI} + \frac{-417 \text{ k} \cdot \text{ft}^3}{EI} = \frac{-677 \text{ k} \cdot \text{ft}^3}{EI}$$

Evaluation of Results:

Approximation Predicted Outcomes?

Compare with the results of detailed analysis:
Using methods we learn later in this text, we obtain the following peak deflection:

$$\Delta_{max} = \frac{-693 \text{ k} \cdot \text{ft}^3}{EI} \qquad \therefore \text{ Approximation is only 2\% low.} \checkmark$$

Homework Problems

2.1 A conservation group purchased an old power transmission tower with the goal of turning the lower part into a platform for watching wildlife in a local park. We will need to consider the effect of gravity loads and lateral loads on the truss. The members are most susceptible to damage from compression. Therefore, before we begin our detailed analysis, we want to know which members will be most highly loaded in compression by the combination of gravity and lateral loads.

a. Considering only the lateral load, determine the sense of the axial force in each member.
b. Considering only the gravity loads, determine the sense of the axial force in each member.
c. Use superposition to determine which members are most highly loaded in compression. Also determine which members are in tension under the combined loading.

Prob. 2.1

2.2 We are designing a truss to support a pedestrian walkway over a downtown street. The live load displacement will be an important quantity to control so that the glass remains weatherproof (large displacements lead to gaps between the glass panels and weather strips). In preparation for calculating the live load displacement, we need to determine which members are in tension, which are in compression, and which are zero force.

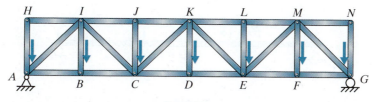

Prob. 2.2

2.3 Open-web joists are a common choice for supporting floors and roofs because they are lightweight and relatively easy to install.

Dead load will tend to push down on the joists, and wind will typically cause uplift. To learn more about the behavior, we want to find the sense of the axial force in members *FG*, *PF*, *PG*, and *PQ* due to each type of load separately.

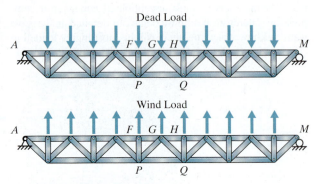

Prob. 2.3

2.4 If the dead load deflection of a beam is expected to be significant, we can prebend the beam upward to counteract the expected downward deflection. This prebending is called camber. The industry standard is that for beams less than 50 feet long, it is not cost effective to camber less than ¾ inch.

To have the desired flow of customers in a retail space, the architect wants to use a transfer girder to carry load from a second-story column out to other columns on the first floor. The girder will also carry floor load.

A member of the engineering team has chosen a steel W14 × 233 ($E = 29{,}000$ ksi, $I = 3010$ in^4) to carry the dead loads shown. Our job is to determine whether the beam should be cambered. To make that determination, we need to predict the displacement of the beam chosen. Although the peak displacement might not be exactly at midspan, the midspan displacement will be close enough. The fastest way to predict the displacement will be to use the formulas from the inside front cover of this book.

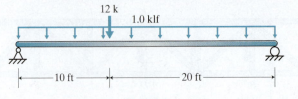

Prob. 2.4

2.5 Our design firm is working on a large round floor supported by beams that extend radially out from a central ring. The resulting idealized live load on each beam is a linear distributed load.

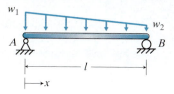

Prob. 2.5

We do not yet know how many beams will be in the final design; therefore, we do not yet know the distributed load values w_1 or w_2.

a. To help in the design process, the team wants us to develop an expression for the vertical displacement along the entire length of the beam. Before doing so, we want to identify some bounds to help verify our expression. The location of the peak displacement will change with the ratio w_1/w_2. We can consider two extremes for that ratio: $w_1/w_2 = 1$ (uniform distributed load) and $w_1/w_2 = \infty$ (triangular distributed load). Using the formulas inside the front cover of this text, find the location of the peak displacement for each of these extremes measured from end A.

b. Determine the expression for the vertical displacement in terms of x measured from end A. Although the displacement can be derived from the loading (we get to that in Chapter 5: Deformations), it is probably easier in this case to use the formulas from the inside front cover of this text.

c. Using a spreadsheet or another computer tool, plot the displaced shapes for the following ratios of w_1/w_2: 1.0, 2.5, 5.0. Show all three plots on one graph. To make the plot, a team member recommends setting $E, I, l,$ and w_2 all equal to 1. Use your results from part (a) to verify these results.

d. From your results in parts (a) and (c), what do you conclude about the sensitivity of the peak displacement location for this project?

112   **Chapter 2**   Predicting Results

2.6 An industrial client contacted our firm with a problem for us to fix: a beam is displacing upward and damaging fragile equipment above it. As we review photos from the field and the original structural drawings, we discover that the beam can be idealized as simply supported with no substantial load except a couple moment at one end due to how the façade is attached.

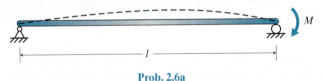

Prob. 2.6a

A variety of solutions are available (e.g., add another beam to share the load, stiffen the beam), but the simplest might be to hang a weight at midspan of the beam.

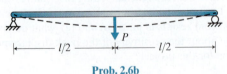

Prob. 2.6b

a. Using the formulas for peak vertical displacement on the inside front cover of this text, determine the load P required to cancel the upward deflection due to the moment M. Since we do not yet know the magnitude of the moment, an expression in terms of M and l is most helpful.

b. While reviewing our calculations, a coworker points out that we have made an approximation/assumption that causes P to be an estimate (beyond the idealization process). What was that approximation or assumption?

2.7 An engineer on the team was assigned the task of designing a simply supported beam with a cantilevered end. That engineer calculated the peak positive moment to be 57 kN·m. Another team member points out that wherever the beam bends concave up (a smile) the moment is positive, and wherever the beam bends concave down (a frown) the moment is negative.

Before moving forward with the design, our coworker has asked us to verify the result.

a. Calculate a bound on the peak moment by considering the beam to be simply supported with supports at only A and C (i.e., move the support from B to A). Should this be an upper or lower bound? Justify your answer about the bound.

b. Calculate a bound on the peak moment by ignoring the cantilevered end (i.e., there is only a simply supported beam from B to C). Should this be an upper or lower bound? Justify your answer about the bound.

c. Using your results from parts (a) and (b), tell your coworker whether or not you believe the original result is reasonable.

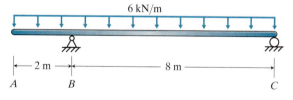

Prob. 2.7

2.8 We will be designing the frame to support a pedestrian walkway over a parking area. We want to estimate the peak moment before beginning a detailed analysis of the frame. The walkway applies two point loads to the beam of the frame. One of the point loads is very close to the support at the wall (point A). A more experienced engineer on the team teaches us that the point load near the support is important to consider when calculating the peak shear or designing the support, but that the load plays a very small role in the peak moment or displacement. That engineer also says that the peak positive moment in this beam will occur under the middle point load.

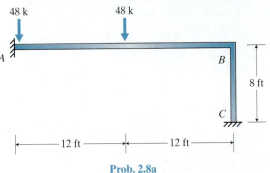

Prob. 2.8a

a. The column provides some restraint to the beam, but point B is not free to rotate, nor is it completely restrained from rotating. If the column was extremely flexible, it would not restrain rotation at B. The beam would still be indeterminate, but by checking several reference manuals, we find the reactions for these support conditions.

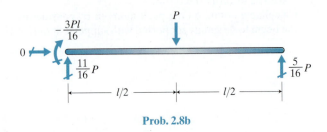

Prob. 2.8b

Use these reactions to find an estimate of the peak positive moment in the beam. Will this estimate be an upper or lower bound on the peak moment? Justify your answer about the bound.

b. If the column was extremely stiff, it would completely stop rotation at B. We can find the reactions to those support conditions inside the back cover of this book.

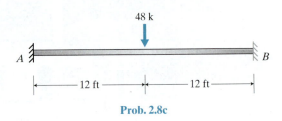

Prob. 2.8c

Estimate the peak positive moment in the beam using the reactions. Will this estimate be an upper or lower bound on the peak moment? Justify your answer about the bound.

2.9 Our team is designing the structure to support a corridor. Several of the beams are simply supported, but they support live load over only part of their length.

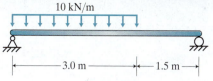

Prob. 2.9a

A member of the team has designed each of the beams to be steel ($E = 200$ GPa) HSS sections and to have a moment of inertia, I, of 6.55×10^6 mm^4. That person predicts the peak vertical displacement will be 1.6 mm. We have been tasked with evaluating the reasonableness of the team member's results.

a. Approximate the beam as being only as long as the distributed load.

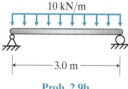

Prob. 2.9b

Estimate the peak vertical displacement based on this configuration. Will this estimate be an upper or lower bound on the peak displacement, or can't it be determined? Justify your answer about the bound.

b. Approximate the beam as if the distributed load extended the full length of the beam.

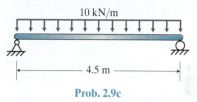

Prob. 2.9c

Estimate the peak vertical displacement based on this configuration. Will this estimate be an upper or lower bound on the peak displacement, or can't it be determined? Justify your answer about the bound.

c. Using your results from parts (a) and (b), inform the team member whether or not you believe the original result is reasonable.

2.10 We are tasked with predicting the dead load displacement of a beam that supports four equally spaced open-web joists. The other two joists are supported directly by the columns. At this stage of the design process, we will ignore the contribution from the self-weight of the steel beam ($E = 29{,}000$ ksi). The preliminary choice for the beam is a W12 × 65 that has a moment of inertia, I, of 533 in^4.

a. Approximate the four point loads with an equivalent uniform distributed load. Should this be an upper or lower bound on the solution, or can't it be determined? What is the predicted approximate vertical displacement?

b. Because of the symmetry of the problem, we know that the peak displacement will be at midspan. Using superposition and the formulas from the inside front cover of this text, make a more accurate prediction of the peak vertical displacement.

c. Using your results from parts (a) and (b), assess how reasonable it is to use an equivalent uniform load to approximate the four point loads.

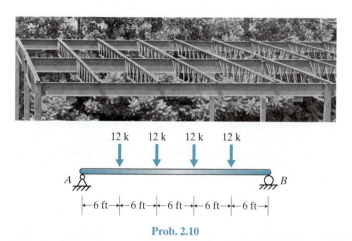

Prob. 2.10

2.11 The senior engineer on our team decided that a precast, prestressed concrete beam is the best choice to support a uniform distributed load and two point loads.

To begin preliminary design of the beam, we need an estimate of the peak moment.
a. Determine the peak moment considering only the uniform distributed load. Should this be an upper or lower bound on the solution, or can't it be determined?
b. Determine the peak moment after converting all of the applied load into an equivalent point load at midspan. Should this be an upper or lower bound on the solution, or can't it be determined?
c. Determine the peak moment after converting all of the applied load into an equivalent uniform distributed load. Should this be an upper or lower bound on the solution, or can't it be determined?
d. Which of the peak moments determined in parts (a), (b), and (c) should be the most accurate? Justify your answer.

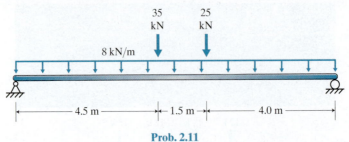

Prob. 2.11

2.12 Our team is designing a high-rise building where snow is a concern. The roof has a penthouse for mechanical equipment. A roof beam to the left of the penthouse must be designed for snow drifting off the penthouse roof.

In order to design the roof beam, we will need to know the peak moment due to snow.

a. Estimate the peak moment by considering the beam to support a uniform distributed load of 250 plf. Will this estimate be an upper or lower bound on the peak moment, or can't it be determined? Justify your answer about the bound.

b. Estimate the peak moment by considering the beam to have an equivalent uniform distributed load (i.e., same total load). Will this estimate be an upper or lower bound on the peak moment, or can't it be determined? Justify your answer about the bound.

c. Estimate the peak moment by considering the beam to support a uniform distributed load of 700 plf. Will this estimate be an upper or lower bound on the peak moment, or can't it be determined? Justify your answer about the bound.

d. Which of the peak moments determined in parts (a), (b), and (c) should be the most accurate? Justify your answer.

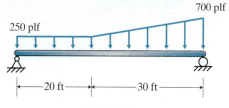

Prob. 2.12

CHAPTER 3

CABLES AND ARCHES

Cables and arches can be used in a wide variety of situations, but they are most commonly used to support gravity loads.

MOTIVATION

Cables can be a great way to support loads because they are very lightweight. They are lightweight because cables can experience only tension, not shear or moment, and most materials are strongest in tension. The light weight required to support loads typically results in cables that are much less stiff than beams or frames. That is not necessarily a problem for the behavior of the structure, but it makes analysis more of a challenge.

Because cables have low stiffness, we need to apply equilibrium to the displaced shape in order to find the internal forces. That means we must analyze the cable, predict the deformations, update the geometry, and iterate in order to find the equilibrium geometry. We call this *second-order analysis*, and it is typically done using structural analysis software. But in some situations we start the process knowing what we want the final cable geometry to be. In these cases, we can calculate the tension all along the cable.

Arches work much the same way as cables. If we design the arch geometry based on the load it will carry, the arch experiences only compression. However, if the loading is different from the pattern used to design the arch shape, the arch also experiences shear and moment. Because arches tend to be much stiffer than cables, first-order analysis (applying equilibrium to the undisplaced shape) is typically adequate.

Most arches are indeterminate structures, and we leave detailed analysis of indeterminate arches to other references. One kind of arch is determinate though: the three-hinge arch. Treating an indeterminate arch like a three-hinge arch allows us to generate approximate predictions of reactions and internal forces.

3.1 Cables with Point Loads

Introduction
Although we will use the term *cable* throughout this chapter, there are several types of structural members that behave the same way: ropes, chains, and thin rods. The key is that the member can experience only axial tension.

Analyzing cables is challenging, but if we assume to know the final geometry of the loaded cable, we can analyze it by hand. One of the typical loadings of cables is point loads.

How-To
Cables can carry only tension, so in some ways they are very similar to trusses. In a previous course, we learned the Method of Joints and Method of Sections for truss analysis. Before we can start applying these methods to cables, we need to understand a few important differences from trusses.

Difference 1: Connection Types
The unknowns for a cable system depend on the way the load is connected to the cable. There are two types of connections: pulley and pin.

(a)

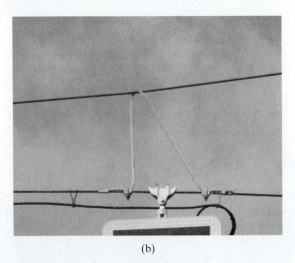

(b)

Figure 3.1 Example of a connection that acts like a pulley: (a) Cable used to support traffic signals; (b) Close-up of loop attached to the cable such that it cannot transfer force along the axis of the cable.

Figure 3.2 Cable with a single applied load with a pulley connection.

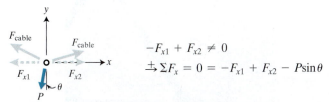

$$-F_{x1} + F_{x2} \neq 0$$
$$\xrightarrow{+} \Sigma F_x = 0 = -F_{x1} + F_{x2} - P\sin\theta$$

Figure 3.3 When there is a pulley connection, the direction of the applied load is determined by the requirements of equilibrium.

Pulley Connection
In some cases, the connection just sits on the cable (Figure 3.1). If we neglect any friction, this connection acts like a pulley. The pulley does not change the tension in the cable; there would have to be transfer along the axis of the cable to do that. The pulley only transfers force normal to the cable, thus resulting in a change in direction for the cable.

To identify the resulting unknowns, let's consider a cable that has one point load attached with a pulley connection (Figure 3.2).

Because a pulley changes only the direction of the cable, the cable force is unchanged at the pulley. If we draw a free body diagram of the point where the load is applied, we can write the equations of equilibrium (Figure 3.3). Since the tension in the cable segments is the same on both sides, we see that horizontal equilibrium can only be satisfied if the applied load is not vertical. Therefore, the two unknowns are the cable force and the angle of the applied load.

Pin Connection
If the connection grabs the cable (Figure 3.4), force can be transferred normal to and along the cable. Therefore, the magnitude of the tension can change. Based on our definitions in Section 1.3: Structure Idealization, this is a pin connection.

To identify the resulting unknowns, consider the cable in Figure 3.5. We can draw a free body diagram of the point where the load is applied (Figure 3.6). Since this is a pin connection, we know the direction of the applied load. Typically the load just hangs on the cable, so the direction is straight down. In order to have equilibrium, the horizontal components of the cable forces must be equal. The cable segments are at different angles, so the resulting tension in each cable segment will be different. Therefore, with pin connections the force in each cable segment is unknown.

Difference 2: Geometric Nonlinearity
Another important difference between cables and trusses is that cables are geometrically nonlinear. We will define that term in a moment, but

122 **Chapter 3** Cables and Arches

(a)

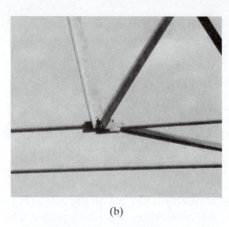

(b)

Figure 3.4 Example of a connection that can change the tension in the cable: (a) Cables used to support a watering pipe system; (b) Close-up of the connection showing that the members grab the cable.

Figure 3.5 Cable with a single applied load with a pin connection.

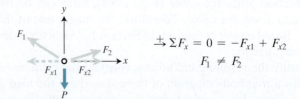

Figure 3.6 When there is a pin connection, the tension in each cable segment is different to satisfy equilibrium.

first we'll illustrate the issue. Consider a weightless cable extending between two supports and carrying a point load at midspan (Figure 3.7a). Because the cable can carry only axial force, no moment, the point under the load acts just like a hinge. The way the load is attached is called a pulley or pin connection, but the cable itself acts as if it has a hinge or pin connection between its segments (Figure 3.7b).

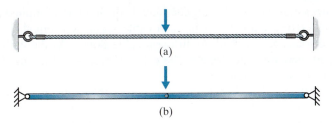

Figure 3.7 Idealizing cables: (a) Straight cable with a point load; (b) Idealization of the cable has a hinge or pin connection between segments under the load.

If we draw a free body diagram of the point under the load, we discover that it is not in vertical equilibrium (Figure 3.8).

The only way the point can be in equilibrium is if the cable force has a vertical component. The only way the cable force has a vertical component is if the point moves down relative to the supports (Figure 3.9).

For the cable to move down it must get slightly longer, and for the cable to get longer there must be axial force. So if we know how far the cable displaces, we can calculate the internal force. If we must know the displaced shape of a structure before we can accurately calculate the internal forces, we call that structure *geometrically nonlinear*.

For most structures (e.g., trusses, beams, frames), we find the internal forces by applying equilibrium to the undisplaced shape. This is called *first-order analysis*, and it is typically accurate enough for design purposes for trusses, beams, and frames. As we just saw from considering a horizontal cable with a point load, first-order analysis is typically not adequate for analysis of cables.

Applying equilibrium to the displaced shape in order to find internal forces is called *second-order analysis* or *geometric nonlinear analysis*. This level of analysis produces more accurate results than first-order analysis for all structures, but the difference in results between first-order and second-order analysis is sufficiently small for most trusses, beams, and frames. For cables, however, we need to perform second-order analysis.

So how do we perform second-order analysis? Before we can answer that question, consider the free body diagram of the point in Figure 3.9b. The point could be in equilibrium for any vertical displacement other than zero. Vertical equilibrium means that the vertical components of the two

Figure 3.8 If the cable remains horizontal, the point below the load is not in equilibrium.

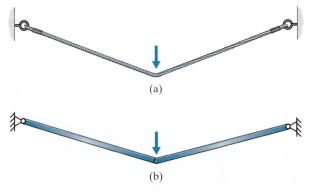

Figure 3.9 The point below the point load can be in equilibrium only if the cable displaces down: (a) Deformed cable under a point load; (b) The point below the load is in equilibrium.

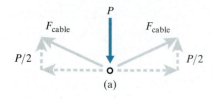

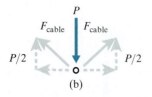

Figure 3.10 Net force in the cable depends on the slope of the cables because the magnitude of the vertical component is set: (a) Smaller slope results in a larger net force; (b) Larger slope results in a smaller net force.

cable forces must add to the applied load. If the load, P, is applied at midspan, the vertical component in each cable segment is equal. Different vertical displacements result in different cable slopes. As the vertical displacement increases, the slope of the cable increases and the net force in the cable decreases (Figure 3.10).

But as the vertical displacement increases, the cable length increases, so the required axial force in the cable increases. This means that only one configuration satisfies equilibrium and compatibility of deformations. To find that configuration, we can write equations of equilibrium and compatibility. However, for most cables this becomes a very complex system of nonlinear equations to solve. Rather than predict the displaced shape by hand, we typically model the cable using computer software and analyze it iteratively until the conditions of equilibrium and compatibility are satisfied. How to perform second-order analysis is outside the scope of this text.

Performing second-order analysis requires numerical methods that sometimes do not converge or sometimes converge to invalid solutions. Therefore, having a sense of the answer beforehand is incredibly important. Fortunately, there are a couple options for analysis without using structural analysis software.

Difference 3: Additional Unknowns

For trusses, the unknowns are the internal forces in each of the members and the reactions. If the truss is determinate, the number of equations of equilibrium found from the Method of Joints will equal the total number of unknowns.

For cables, we can still determine the total number of equations of equilibrium from the Method of Joints. There will be two equations for each place the cable experiences force, either support or applied. For example, consider the cable in Figure 3.11. It is supported at the two ends and has a single point load; therefore, we have six equations of equilibrium. If the load is applied with a pin connection, we have six unknowns: four reactions and two cable tensions (Figure 3.11a). If the load is applied with a pulley connection, we still have six unknowns: four reactions,

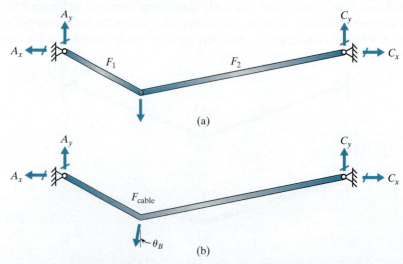

Figure 3.11 All six of the unknowns for a cable with one applied load: (a) Pin connection; (b) Pulley connection.

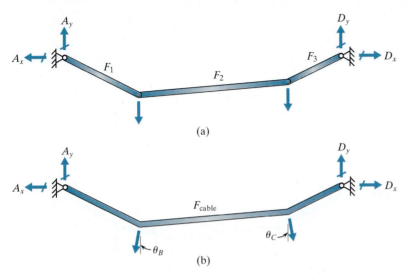

Figure 3.12 Seven of the unknowns for a cable with two applied loads: (a) Pin connections; (b) Pulley connections.

the cable tension, and the applied load direction (Figure 3.11b). In either case, the number of equations matches the number of unknowns.

The difference between cables and trusses shows up when we have more than one applied force. Consider the cable in Figure 3.12. The cable is supported at two ends and has two applied forces; therefore, we have eight equations of equilibrium. Figure 3.12a shows the cable if we use pin connections. In this case, there are seven unknowns: four reactions and three cable tensions. Figure 3.12b shows the cable if we use pulley connections. In this case, there are still seven unknowns: four reactions, one cable tension, and two applied load directions. These observations show that we have eight equations and only seven unknowns regardless of the type of connections.

Mathematically, when we have more equations than unknowns, the system of equations is overdetermined. That means there is no solution. For cables, this means equilibrium will probably not be enforced somewhere along the cable. Since a lack of equilibrium would mean the cable is moving, we need an additional unknown to perform static analysis. The additional unknown is the vertical displacement at one of the points of load application (Figure 3.13). For cables, we call that vertical displacement *sag*.

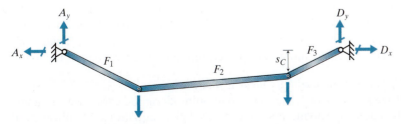

Figure 3.13 All eight unknowns labeled for a cable with two applied loads with pin connections.

126 Chapter 3 Cables and Arches

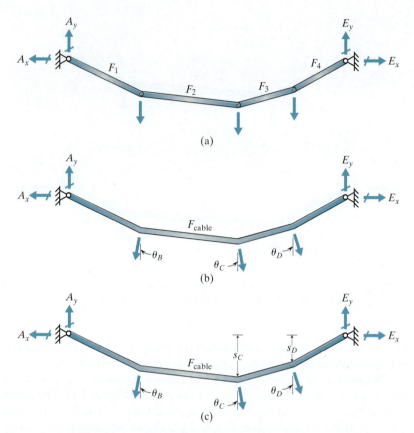

Figure 3.14 Cable with three applied loads: (a) Eight of the unknowns for pin connections; (b) Eight of the unknowns for pulley connections; (c) All ten unknowns labeled for a cable with pulley connections.

Before we can create a rule, we need to consider another situation. Figure 3.14 shows three applied loads. There are ten equations of equilibrium. If the loads are applied with pin connections, we have eight unknowns: four reactions and four cable tensions (Figure 3.14a). If the loads are applied with pulley connections, we still have eight unknowns: four reactions, one cable tension, and three applied load directions (Figure 3.14b). In both cases, we need two additional unknowns. The additional unknowns are the sags at two of the points of load application (Figure 3.14c).

Therefore, our rule is that we can specify the sag at only one point of load application. For all other points of load application, the sag must be unknown.

Analysis Methods
Although cables are geometrically nonlinear, there are still a couple ways to analyze them without using second-order analysis. These methods are typically approximations, however, because we are ignoring axial elongation of the cable and we are prescribing the sag under one of the loads.

Option 1: System of Nonlinear Equations
If we know or pick the sag at one point along the cable, we can analyze the cable by solving a system of nonlinear equations. To illustrate this approach, consider the cable in Figure 3.15. There are two applied loads

with pulley connections, so we don't know their angles θ_B or θ_C. Let's say we chose the left sag, so the other sag, s_C, is unknown.

We have the same number of unknowns as equations of equilibrium; therefore, we can use the Method of Joints to develop the equations. The free body diagrams for the four joints are presented in Figure 3.16. The unknowns are shown in bold. The resulting system of eight equations has unknowns that are trig functions and unknowns that are part of square roots in denominators.

Even if the cable uses pin connections, we end up with a system of nonlinear equations. Consider the same cable but with pin connections (Figure 3.17). Now we know the directions of the applied forces, down in this case, but each cable segment has a different tension.

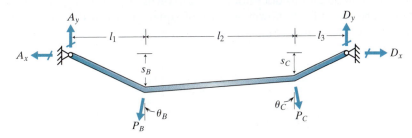

Figure 3.15 Cable with two point loads applied with pulley connections.

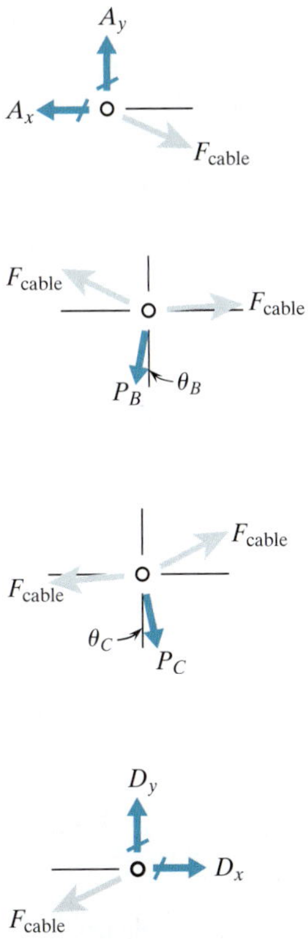

$$\xrightarrow{\pm} \Sigma F_x = 0 = -\mathbf{A_x} + \frac{l_1}{\sqrt{l_1^2 + s_B^2}} F_{cable}$$

$$+\uparrow \Sigma F_y = 0 = \mathbf{A_y} - \frac{s_B}{\sqrt{l_1^2 + s_B^2}} F_{cable}$$

$$\xrightarrow{\pm} \Sigma F_x = 0 = -\frac{l_1}{\sqrt{l_1^2 + s_B^2}} F_{cable} + \frac{l_2}{\sqrt{l_2^2 + (s_B - s_C)^2}} F_{cable} - P_B \sin \boldsymbol{\theta_B}$$

$$+\uparrow \Sigma F_y = 0 = \frac{s_B}{\sqrt{l_1^2 + s_B^2}} F_{cable} + \frac{s_B - s_C}{\sqrt{l_2^2 + (s_B - s_C)^2}} F_{cable} - P_B \cos \boldsymbol{\theta_B}$$

$$\xrightarrow{\pm} \Sigma F_x = 0 = -\frac{l_2}{\sqrt{l_2^2 + (s_B - s_C)^2}} F_{cable} + \frac{l_3}{\sqrt{l_3^2 + s_C^2}} F_{cable} + P_C \sin \boldsymbol{\theta_C}$$

$$+\uparrow \Sigma F_y = 0 = -\frac{s_B - s_C}{\sqrt{l_2^2 + (s_B - s_C)^2}} F_{cable} + \frac{s_C}{\sqrt{l_3^2 + s_C^2}} F_{cable} - P_C \cos \boldsymbol{\theta_C}$$

$$\xrightarrow{\pm} \Sigma F_x = 0 = -\frac{l_3}{\sqrt{l_3^2 + s_C^2}} F_{cable} + \mathbf{D_x}$$

$$+\uparrow \Sigma F_y = 0 = -\frac{s_C}{\sqrt{l_3^2 + s_C^2}} F_{cable} + \mathbf{D_y}$$

Figure 3.16 Free body diagrams and equations of equilibrium for the four joints with pulley connections.

128 **Chapter 3** Cables and Arches

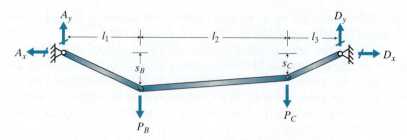

Figure 3.17 Cable with two point loads applied with pin connections.

Again we apply the Method of Joints to develop the system of equations (Figure 3.18). The unknowns are in bold. The resulting system of eight equations does not have unknowns as part of trig functions, but there are still unknowns as part of square roots in denominators. Therefore, the system of equations is still nonlinear.

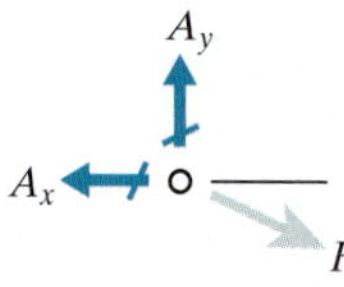

$$\xrightarrow{\pm} \Sigma F_x = 0 = -\mathbf{A_x} + \frac{l_1}{\sqrt{l_1^2 + s_B^2}} \mathbf{F_1}$$

$$+\uparrow \Sigma F_y = 0 = \mathbf{A_y} - \frac{s_B}{\sqrt{l_1^2 + s_B^2}} \mathbf{F_1}$$

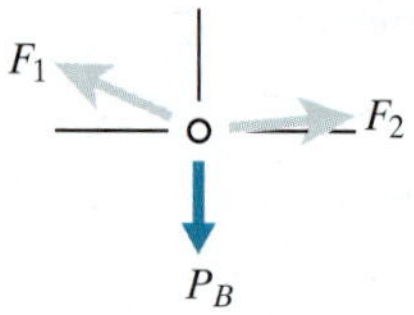

$$\xrightarrow{\pm} \Sigma F_x = 0 = -\frac{l_1}{\sqrt{l_1^2 + s_B^2}} \mathbf{F_1} + \frac{l_2}{\sqrt{l_2^2 + (s_B - s_C)^2}} \mathbf{F_2}$$

$$+\uparrow \Sigma F_y = 0 = \frac{s_B}{\sqrt{l_1^2 + s_B^2}} \mathbf{F_1} + \frac{s_B - s_C}{\sqrt{l_2^2 + (s_B + s_C)^2}} \mathbf{F_2} - P_B$$

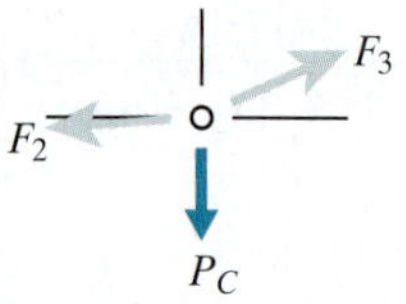

$$\xrightarrow{\pm} \Sigma F_x = 0 = -\frac{l_2}{\sqrt{l_2^2 + (s_B - s_C)^2}} \mathbf{F_2} + \frac{l_3}{\sqrt{l_3^2 + s_C^2}} \mathbf{F_3}$$

$$+\uparrow \Sigma F_y = 0 = -\frac{s_B - s_C}{\sqrt{l_2^2 + (s_B - s_C)^2}} \mathbf{F_2} + \frac{s_C}{\sqrt{l_3^2 + s_C^2}} \mathbf{F_3} - P_C$$

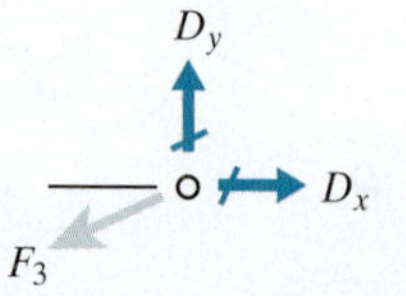

$$\xrightarrow{\pm} \Sigma F_x = 0 = -\frac{l_3}{\sqrt{l_3^2 + s_C^2}} \mathbf{F_3} + \mathbf{D_x}$$

$$+\uparrow \Sigma F_y = 0 = -\frac{s_C}{\sqrt{l_3^2 + s_C^2}} \mathbf{F_3} + \mathbf{D_y}$$

Figure 3.18 Free body diagrams and equations of equilibrium for the four joints with pin connections.

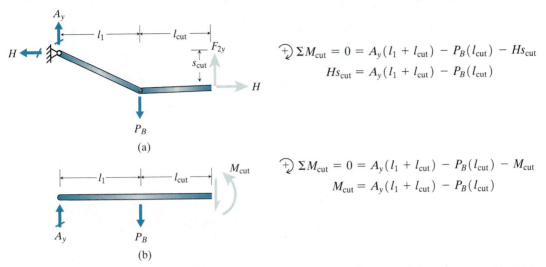

Figure 3.19 Demonstration of the General Cable Theorem: (a) FBD of cable segment showing moment equilibrium; (b) FBD of simply supported beam segment showing moment equilibrium.

To solve either system of nonlinear equations, we need to use iterative numerical methods outside the scope of this text. We typically use mathematics computer software to find the solution to a system of nonlinear equations. Take note that this approach is not the same as performing second-order analysis using structural analysis software. In second-order analysis, we don't need to know or choose any of the sags; we calculate all of the displaced shape with second-order analysis.

Option 2: General Cable Theorem
If all applied forces are vertical, we can benefit from the *General Cable Theorem* to simplify the analysis. The key is that all of the applied forces must be vertical; therefore, we must treat pulley connections as pin connections in order to use this analysis method.

The General Cable Theorem follows:

> *If a cable carries only vertical loads, then at any point along the cable, the product of the horizontal component of the cable force and the sag at that point is equal to the moment in a simply supported beam that spans the same horizontal distance and is subjected to the same vertical forces.*

This theorem is possible because the horizontal component of the cable force is constant along the cable. We saw that in Figure 3.6. To illustrate the theorem, let's use the Method of Sections on a cable. Figure 3.19a is a free body diagram of the left piece of a cable cut at a random location. If we apply moment equilibrium, we see how to calculate the product of the horizontal force and the sag at the cut. Figure 3.19b is a free body diagram of a simply supported beam with the same applied forces at the same horizontal spacings. We can apply moment equilibrium to find the moment at the cut.

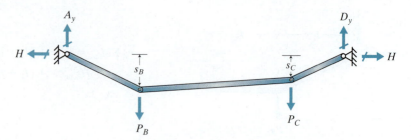

Figure 3.20 Cable with ends at the same height.

Because the expressions on the right sides of the equations are the same, we know that the product of horizontal force and sag must equal the moment in the simply supported beam.

So why is this helpful? If the cable ends are at the same height (Figure 3.20), we can use moment equilibrium to calculate one of the vertical reactions. We can then use that quantity with our free body diagram of the cut cable (Figure 3.19a) to solve for the horizontal force or the sag. Now we see that the General Cable Theorem might be quicker, but it is not required.

If the ends of the cable are offset (Figure 3.21), we cannot calculate the vertical reaction with moment equilibrium. The unknown horizontal component contributes moment.

The General Cable Theorem still applies in this situation, however. In the case of a cable with offset ends, the sag is measured as the vertical distance to the straight line that connects the two ends. If you would like the full proof for a cable with offset ends, proofs are available in other references. Here we will just accept that it is true.

The other reason the General Cable Theorem is helpful is that it allows us to anticipate the displaced shape of a cable. In Chapter 4: Internal Force Diagrams, we get a lot of practice developing the moment diagram for beams. The General Cable Theorem results in the following observation: the displaced shape of a cable matches the shape of the moment diagram of a simply supported beam with the same loads at the same spacings. Of course, the caveat to this observation is that if the cable ends are offset, we are measuring the sag from the line that connects the ends.

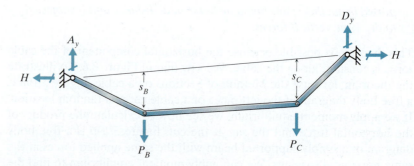

Figure 3.21 Cable with ends offset vertically.

SECTION 3.1 HIGHLIGHTS

Cables with Point Loads

Definitions:

Geometrically nonlinear—A structure for which we must know the displaced shape before we can accurately calculate the internal forces.

First-order analysis—Calculating internal forces due to loads on the undisplaced shape.

Second-order analysis, or geometric nonlinear analysis—Calculating internal forces due to loads on the displaced shape.

Sag—Vertical displacement of a cable from the straight line that connects the two ends.

Connection Types:

Pulley
- Does not transfer force along the cable, only perpendicular.
- Unknowns: cable force, angle of load application.

Pin
- Transfers force along and perpendicular to the cable.
- Unknowns: two cable forces (one on each side).

Analysis Methods:

We have two by hand methods for approximate analysis of cables.

Assumptions:
- The effect of axial deformation of the cable segments is negligible.
- We know the final sag at one applied load.

1. System of Nonlinear Equations
   - Can be used with either type of connection or a combination.
   - Use Method of Joints to formulate the equations of equilibrium.
   - System of equations will be nonlinear, so use a numerical method to solve for unknowns.

2. General Cable Theorem
   - Theorem:
     If a cable carries only vertical loads, at any point along the cable the product of the horizontal component of the cable force and the sag at that point is equal to the moment in a simply supported beam that spans the same horizontal distance and is subjected to the same vertical forces.
   - Can be used only if all pin connections and vertical loads.
   - Use theorem to find the horizontal component of cable force and sags at applied loads.
   - Use Method of Joints or Method of Sections to find the remaining unknowns.

Example 3.1

The owners of a zip line attraction want to know if they can increase the weight limit for riders from 100 kg to 120 kg. They recently purchased new equipment that is designed for the increase in rider weight. The question is whether the zip line and anchorages are strong enough.

Figure 3.22

Our team is analyzing the cable considering the live load at various locations along the zip line. The team has collectively decided that we should use a load factor of 2.0 for the live load rather than the code factor because of the lack of redundancy in the system. The resulting factored live load is 2.4 kN.

To perform a preliminary analysis, the team has decided to make a couple of simplifying assumptions:

- Although the problem is geometrically nonlinear, we are going to superimpose the results of analysis of the cable subjected to self-weight with the results of the cable with a live load.
- Although the rider is not in equilibrium, we are going to conservatively treat the problem as static. In reality, the rider's weight vector remains vertical and the unresolved horizontal component at the pulley connection is what accelerates and decelerates the rider.

We have been assigned to analyze the cable with a rider halfway across the zip line. An experienced member of the team tells us that for this particular situation, the cable should drop by 3 m from the start point to the rider. Our task is to find the resulting force in the cable.

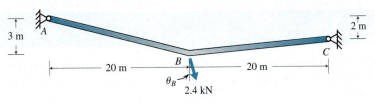

Figure 3.23

Analysis by System of Nonlinear Equations:

Since there is only one applied load, we can solve for both the load angle and the cable force by hand without needing numerical methods.

Calculate the cable segment lengths:

$$l_{AB} = \sqrt{(20\,\text{m})^2 + (3\,\text{m})^2} = 20.22\,\text{m}$$
$$l_{BC} = \sqrt{(20\,\text{m})^2 + (3\,\text{m} - 2\,\text{m})^2} = 20.02\,\text{m}$$

FBD of B:

$$\xrightarrow{+}\Sigma F_x = 0 = -\frac{20}{20.22}F_{\text{cable}} + \frac{20}{20.02}F_{\text{cable}} + 2.4\,\text{kN}\sin\theta_B$$
$$-9.88\times 10^{-3} F_{\text{cable}} = 2.4\,\text{kN}\sin\theta_B$$
$$-4.12\times 10^{-3}/\text{kN} \cdot F_{\text{cable}} = \sin\theta_B$$

$$+\uparrow\Sigma F_y = 0 = \frac{3}{20.22}F_{\text{cable}} + \frac{1}{20.02}F_{\text{cable}} - 2.4\,\text{kN}\cos\theta_B$$
$$-0.1983 F_{\text{cable}} = -2.4\,\text{kN}\cos\theta_B$$
$$8.26\times 10^{-2}/\text{kN} \cdot F_{\text{cable}} = \cos\theta_B$$

Divide the two equations:

$$\frac{-4.12\times 10^{-3}/\text{kN} \cdot F_{\text{cable}}}{8.26\times 10^{-2}/\text{kN} \cdot F_{\text{cable}}} = \frac{\sin\theta_B}{\cos\theta_B}$$
$$-4.99\times 10^{-2} = \tan\theta_B$$
$$\theta_B = -2.86°$$

Substitute to find the cable force:

$$8.26\times 10^{-2}/\text{kN} \cdot F_{\text{cable}} = \cos(-2.86°)$$
$$F_{\text{cable}} = 12.09\,\text{kN}$$

EXAMPLE 3.2

Our firm has been hired to design the rigging for a concert tour. The set designer has presented us with some unique challenges. We need to support three large props in front of a draped, glowing canvas. The senior engineer has worked out with the set designer that we will support the large props from a single cable that does not interfere with the draped canvas.

Figure 3.24

In order to ensure enough room for the canvas and enough height for the large props, the senior engineer has set the maximum drop of the suspended props to be 15 ft. We initially assume it will be the middle prop, but we must check that.

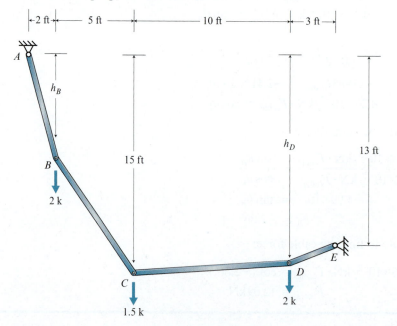

Figure 3.25

Our task is to perform preliminary analysis of the cable to determine how far down the props are from the top frame, and to determine the tension in each segment of the cable.

Section 3.1 Cables with Point Loads

Analysis by General Cable Theorem:

Because there is more than one applied load, this is the most efficient approximate analysis method.

Equivalent beam:

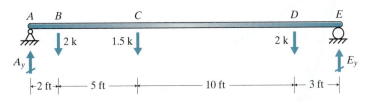

Figure 3.26

$+\circlearrowleft \Sigma M_E = 0 = A_y(20\,\text{ft}) - 2\text{k}(18\,\text{ft}) - 1.5\text{k}(13\,\text{ft}) - 2\text{k}(3\,\text{ft})$
$A_y = 3.08\,\text{k}$

Find the internal moments:

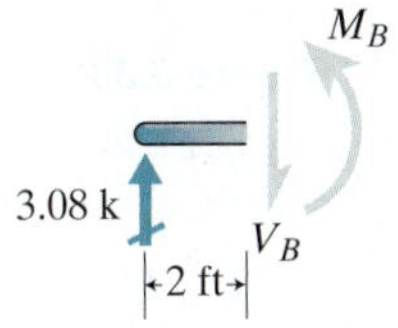

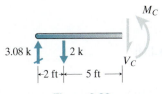

Figure 3.27

$+\circlearrowleft \Sigma M_{\text{cut}} = 3.08\text{k}(2\,\text{ft}) - M_B$
$M_B = 6.16\,\text{k}\cdot\text{ft}$

Figure 3.28

$+\circlearrowleft \Sigma M_{\text{cut}} = 3.08\text{k}(7\,\text{ft}) - 2\text{k}(5\,\text{ft}) - M_C$
$M_C = 11.56\,\text{k}\cdot\text{ft}$

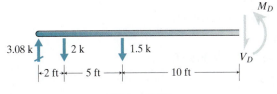

Figure 3.29

$+\circlearrowleft \Sigma M_{\text{cut}} = 3.08\text{k}(17\,\text{ft}) - 2\text{k}(15\,\text{ft}) - 1.5\text{k}(10\,\text{ft}) - M_D$
$M_D = 7.36\,\text{k}\cdot\text{ft}$

Cable geometry:

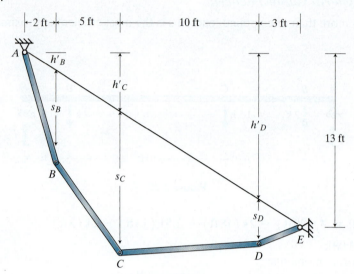

Figure 3.30

From similar triangles:

$$\frac{h'_B}{13\,\text{ft}} = \frac{2\,\text{ft}}{20\,\text{ft}} \qquad h'_B = 1.30\,\text{ft}$$

$$\frac{h'_C}{13\,\text{ft}} = \frac{7\,\text{ft}}{20\,\text{ft}} \qquad h'_C = 4.55\,\text{ft}$$

$$\frac{h'_D}{13\,\text{ft}} = \frac{17\,\text{ft}}{20\,\text{ft}} \qquad h'_D = 11.05\,\text{ft}$$

Calculate the sag at C:

$$s_C = 15\,\text{ft} - 4.55\,\text{ft} = 10.45\,\text{ft}$$

General Cable Theorem:

Horizontal component of cable times sag at C is moment from equivalent beam.

$$10.45\,\text{ft} \cdot H = 11.56\,\text{k} \cdot \text{ft}$$

$$H = 1.106\,\text{k}$$

Find the vertical positions of the other loads:

General Cable Theorem:

$$s_B \cdot H = s_B(1.106\,\text{k}) = 6.16\,\text{k} \cdot \text{ft}$$

$$S_B = 5.57\,\text{ft}$$

$$s_D \cdot H = s_D(1.106\,\text{k}) = 7.36\,\text{k} \cdot \text{ft}$$

$$s_D = 6.65\,\text{ft}$$

Vertical positions:

$$h_B = s_B + h'_B = 5.57\,\text{ft} + 1.30\,\text{ft}$$

$$h_B = 6.87\,\text{ft}$$

$$h_D = s_D + h'_D = 6.65\,\text{ft} + 11.05\,\text{ft}$$

$$h_D = 17.7\,\text{ft}$$

This configuration results in point D being 2.7 ft too low, since the maximum drop of any of the suspended props is 15 ft.

Calculate the resulting cable forces anyway for comparison:

Lengths of cable segments:

$$l_{AB} = \sqrt{(2\,\text{ft})^2 + (6.87\,\text{ft})^2} = 7.16\,\text{ft}$$
$$l_{BC} = \sqrt{(5\,\text{ft})^2 + (15\,\text{ft} - 6.87\,\text{ft})^2} = 9.54\,\text{ft}$$
$$l_{CD} = \sqrt{(10\,\text{ft})^2 + (17.7\,\text{ft} - 15.0\,\text{ft})^2} = 10.36\,\text{ft}$$
$$l_{DE} = \sqrt{(3\,\text{ft})^2 + (17.7\,\text{ft} - 13.0\,\text{ft})^2} = 5.58\,\text{ft}$$

Cable forces:

Since the horizontal component is known for all segments,

$$F_{\text{cable}} = H\left(\frac{l_{\text{segment}}}{w_{\text{segment}}}\right)$$

$$F_{AB} = 1.106\,\text{k}\left(\frac{7.16\,\text{ft}}{2\,\text{ft}}\right) = 3.96\,\text{k}$$

$$F_{BC} = 1.106\,\text{k}\left(\frac{9.54\,\text{ft}}{5\,\text{ft}}\right) = 2.11\,\text{k}$$

$$F_{CD} = 1.106\,\text{k}\left(\frac{10.36\,\text{ft}}{10\,\text{ft}}\right) = 1.15\,\text{k}$$

$$F_{DE} = 1.106\,\text{k}\left(\frac{5.58\,\text{ft}}{3\,\text{ft}}\right) = 2.06\,\text{k}$$

Summary:

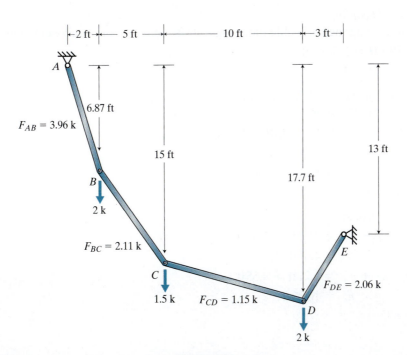

Figure 3.31

138 Chapter 3 Cables and Arches

Analysis with Maximum Sag at D:

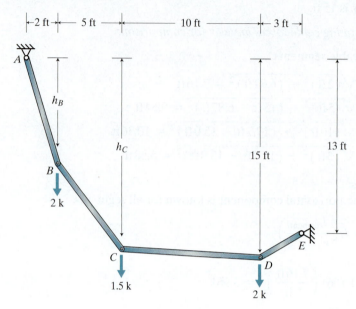

Figure 3.32

Cable geometry:
Since the start and end points are not changed,
$$h'_B = 1.30 \text{ ft}$$
$$h'_C = 4.55 \text{ ft}$$
$$h'_D = 11.05 \text{ ft}$$

Calculate the sag at D:
$$s_D = 15 \text{ ft} - 11.05 \text{ ft} = 3.95 \text{ ft}$$

General Cable Theorem:
Horizontal component of cable times sag at D is moment from equivalent beam.
$$3.95 \text{ ft} \cdot H \equiv 7.36 \text{ k} \cdot \text{ft}$$
$$H = 1.863 \text{ k}$$

Find the vertical positions of the other loads:
General Cable Theorem:
$$s_B \cdot H = s_B(1.863 \text{ k}) \equiv 6.16 \text{ k} \cdot \text{ft}$$
$$s_B = 3.31 \text{ ft}$$
$$s_C \cdot H = s_C(1.863 \text{ k}) \equiv 7.36 \text{ k} \cdot \text{ft}$$
$$s_C = 6.21 \text{ ft}$$

Vertical positions:
$$h_B = s_B + h'_B = 3.31 \text{ ft} + 1.30 \text{ ft}$$
$$h_B = 4.61 \text{ ft}$$
$$h_C = s_C + h'_C = 6.21 \text{ ft} + 4.55 \text{ ft}$$
$$h_C = 10.76 \text{ ft}$$

Lengths of cable segments:
$$l_{AB} = \sqrt{(2\,\text{ft})^2 + (4.61\,\text{ft})^2} = 5.03\,\text{ft}$$
$$l_{BC} = \sqrt{(5\,\text{ft})^2 + (10.76\,\text{ft} - 4.61\,\text{ft})^2} = 7.93\,\text{ft}$$
$$l_{CD} = \sqrt{(10\,\text{ft})^2 + (15\,\text{ft} - 10.76\,\text{ft})^2} = 10.86\,\text{ft}$$
$$l_{DE} = \sqrt{(3\,\text{ft})^2 + (13\,\text{ft} - 15\,\text{ft})^2} = 3.61\,\text{ft}$$

Cable forces:
$$F_{AB} = 1.863\,\text{k}\left(\frac{5.03\,\text{ft}}{2\,\text{ft}}\right) = 4.69\,\text{k}$$

$$F_{BC} = 1.863\,\text{k}\left(\frac{7.93\,\text{ft}}{5\,\text{ft}}\right) = 2.95\,\text{k}$$

$$F_{CD} = 1.863\,\text{k}\left(\frac{10.86\,\text{ft}}{10\,\text{ft}}\right) = 2.02\,\text{k}$$

$$F_{DE} = 1.863\,\text{k}\left(\frac{3.61\,\text{ft}}{3\,\text{ft}}\right) = 2.24\,\text{k}$$

Summary:

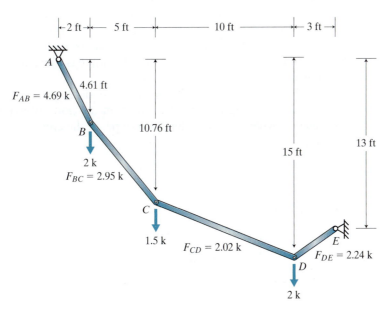

Figure 3.33

Conclusion:

The configuration that meets the concert designer's requirements has point D hanging at 15 ft. The peak cable force in this configuration is 4.69 k.

Observations:

The final configuration results in a 79% increase in tension in one of the cable segments and an increase of 0.73 k in another segment compared with the initially assumed configuration.

∴ The choice of the sag can have a significant effect on the resulting axial forces in the cable segments.

3.2 Cables with Uniform Loads

Introduction
We often use cables to carry uniform loads. Sometimes the load hangs from the cable, and sometimes the load is the cable's self-weight. The analysis is similar to that for a cable with point loads, but an infinite number of point loads act along the cable. Therefore, we need to modify our approach to analysis.

How-To
Just like cables with point loads, the most accurate results for cables with distributed loads will come from second-order analysis that incorporates the stretch of the cable. However, if the distributed load is uniform, we can perform a reasonably accurate approximate analysis by hand. The key is that we need to specify the change in elevation at some point along the cable. If we specify that change where the cable is horizontal, the calculations become much simpler.

Consider the cable in Figure 3.34. The location where the cable is horizontal, zero slope, will always be where the maximum change in elevation occurs. Notice that the ends of the cable don't need to be at the same elevation.

Our goal is to predict the displaced shape of the cable and find the axial force anywhere along the cable. To do that, we start with an observation we made in Section 3.1: *if* the load is applied only vertically, then the horizontal component of the cable force is constant everywhere along the cable.

We can find the reactions at the ends of the cable by constructing a free body diagram of the cable from the point of maximum elevation change, where the cable has zero slope, to the end (Figure 3.35) and then applying equilibrium.

Figure 3.34 Cable carrying uniform distributed load with ends at different elevations.

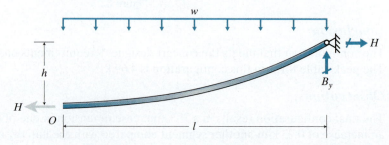

Figure 3.35 Free body diagram of a uniformly loaded cable from the point of zero slope to the end.

Apply vertical equilibrium:

$$+\uparrow \Sigma F_y = 0 = -wl + B_y$$

$$B_y = wl$$

Apply moment equilibrium:

$$\overset{+}{\curvearrowleft} \Sigma M_O = 0 = wl\left(\frac{l}{2}\right) - wl(l) + Hh$$

$$Hh = -\frac{wl^2}{2} + wl^2 = \frac{wl^2}{2}$$

$$H = \frac{wl^2}{2h}$$

So we can find the horizontal component of the cable force if we know where the cable is horizontal and what the maximum change in elevation is.

We can find where the cable is horizontal by using a bit of algebra. Consider the cable in Figure 3.36. The point where the cable is horizontal is labeled O. We saw in Section 3.1 that we must know an elevation change, so let's assume we know the peak elevation change. Relative to the left support, that peak elevation change is h_1; relative to the right support, that peak elevation change is h_2.

We know that the horizontal component of the cable force is the same on both sides of the cable, so the following must be true:

$$\frac{wl_1^2}{2h_1} = \frac{wl_2^2}{2h_2}$$

$$\frac{l_1^2}{l_2^2} = \frac{h_1}{h_2}$$

$$\frac{l_1}{l_2} = \sqrt{\frac{h_1}{h_2}}$$

Since we know the total span of the cable, we can find the location where the cable is horizontal.

To find the vertical position of the cable anywhere along its length, let's draw a free body diagram of a piece of the cable. Specifically, we want a piece cut at the point where the cable is horizontal, since there will

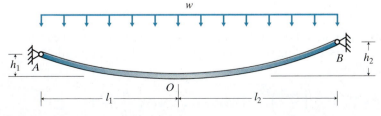

Figure 3.36 The point where the cable is horizontal will not be at midspan of a cable that has supports at different elevations.

142 Chapter 3 Cables and Arches

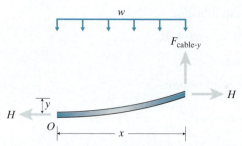

Figure 3.37 Free body diagram of a piece of uniformly loaded cable.

be no vertical component of force at that end. The other end of the piece will be an arbitrary distance x from the horizontal point (Figure 3.37).

First we apply vertical equilibrium; then we apply moment equilibrium.

$$+\uparrow \Sigma F_y = 0 = -wx + F_{\text{cable-}y}$$
$$F_{\text{cable-}y} = wx$$

$$\circlearrowleft \Sigma M_O = 0 = wx\left(\frac{x}{2}\right) - F_{\text{cable-}y}x + Hy$$

$$Hy = -\frac{wx^2}{2} + F_{\text{cable-}y}x$$

We can substitute in for $F_{\text{cable-}y}$.

$$Hy = -\frac{wx^2}{2} + wx^2 = \frac{wx^2}{2}$$

$$y = \frac{w}{2H}x^2$$

But we know how to relate the horizontal component, H, to the geometry of the cable, so we can substitute that into the expression.

$$y = \frac{w}{2} \cdot \frac{2h}{wl^2}x^2$$

$$y = \frac{h}{l^2}x^2$$

Now to find the cable force along the cable. At any point along the cable, the cable force is the vector resultant of the horizontal and vertical components.

$$F_{\text{cable}} = \sqrt{H^2 + F_{\text{cable-}y}^2}$$

$$F_{\text{cable}} = \sqrt{\left(\frac{wl^2}{2h}\right)^2 + (wx)^2}$$

This expression shows us that the maximum cable tension occurs at a support, where x is largest. Calculating the cable force at position l gives us the following:

$$F_{\text{cable}}^{\max} = \sqrt{\left(\frac{wl^2}{2h}\right)^2 + (wl)^2} = \sqrt{(wl)^2\left[\frac{l^2}{4h^2} + 1\right]}$$

$$F_{\text{cable}}^{\max} = wl\sqrt{\frac{l^2}{4h^2} + 1}$$

Point Loads with Uniform Loads

If a cable experiences a uniform load and a point load, we can superimpose the effects *if* the uniform load is the dominant load. If the uniform load dominates, then we know the displaced shape is quadratic and we know the location where the cable is horizontal. That means we can find the effect of just the point load. Consider the cable in Figure 3.38a. The free body diagram in Figure 3.38b is from the point of zero slope to the end and includes only the point load.

If we apply moment equilibrium to the piece of cable in Figure 3.38b, we can find the horizontal component of the cable due to just the point force.

$$+\circlearrowleft \Sigma M_B = 0 = Hh_2 - P(l_2 - x_{load})$$

$$H = \frac{P(l_2 - x_{load})}{h_2}$$

To find the vertical reactions at the ends of the cable, we need to consider a free body diagram of the whole cable with just the point load (Figure 3.38c). Moment equilibrium gives us the vertical reaction at one end of the cable.

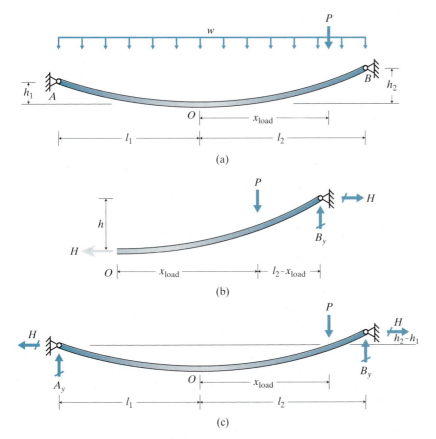

Figure 3.38 Cable experiencing a uniform load and a point load: (a) If the uniform load dominates, the displaced shape is quadratic; (b) Free body diagram of part of the cable considering only the point load; (c) Free body diagram of the cable with quadratic shape but considering only the point load.

$$\stackrel{+}{\curvearrowleft} \Sigma M_B = 0 = A_y(l_1 + l_2) + H(h_2 - h_1) - P(l_2 - x_{\text{load}})$$

$$A_y = P\frac{(l_2 - x_{\text{load}})}{(l_1 + l_2)} - H\frac{(h_2 - h_1)}{(l_1 + l_2)}$$

Vertical equilibrium gives us the vertical reaction at the other end of the cable.

$$+\uparrow \Sigma F_y = 0 = P\frac{(l_2 - x_{\text{load}})}{(l_1 + l_2)} - H\frac{(h_2 - h_1)}{(l_1 + l_2)} - P + B_y$$

$$B_y = P\frac{(l_1 + l_2)}{(l_1 + l_2)} - P\frac{(l_2 - x_{\text{load}})}{(l_1 + l_2)} + H\frac{(h_2 - h_1)}{(l_1 + l_2)}$$

$$B_y = P\frac{(l_1 + x_{\text{load}})}{(l_1 + l_2)} + H\frac{(h_2 - h_1)}{(l_1 + l_2)}$$

To find the net effect on the cable due to the uniform and point loads, we superimpose the horizontal components and vertical components in the cable.

Self-Weight of Cables

So far we have considered the distributed load to act uniformly over the horizontal projection. For situations like suspension bridges (Figure 3.39a), the weight of the bridge is much greater than the weight of the cable. In those cases, uniform load over the horizontal projection is a reasonable idealization. But in some cases, like power lines (Figure 3.39b), the uniform load acts over the length of the cable.

Section 3.2 Cables with Uniform Loads   **145**

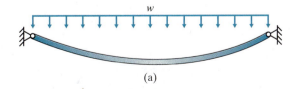

(a)

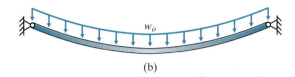

(b)

Figure 3.39 Examples of cables carrying uniform loads: (a) The dominant load for the main cable of a suspension bridge acts over the horizontal projection; (b) The dominant load for a power line cable is its self-weight, which acts along its length.

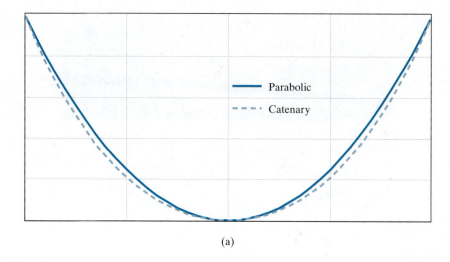

(a)

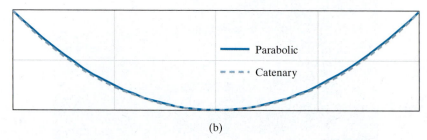

(b)

Figure 3.40 Comparison of parabolic and catenary curves for cables: (a) Sag-to-span ratio of 0.5; (b) Sag-to-span ratio of 0.25.

We determined a few pages ago that when the uniform load acts over the horizontal projection the cable shape is parabolic, but a cable that experiences self-weight has a slightly different shape. That shape is called *catenary*. You can find information on the catenary shape in other resources. For most sag-to-span ratios that we use for structural purposes, however, we can reasonably approximate the effect of self-weight as acting over the horizontal projection. Figure 3.40a shows a comparison of the parabolic and catenary curves when the sag-to-span ratio is 0.5 (peak sag is one-half the overall span length). That is a much larger ratio than we typically use for structural purposes. Figure 3.40b makes the same comparison, but for a sag-to-span ratio of 0.25. This ratio is still typically larger than we use for structural purposes, but the curves are almost identical.

SECTION 3.2 HIGHLIGHTS

Cables with Uniform Loads

Displaced Shape:

- Cables subjected to a uniform load over the horizontal projection have a parabolic shape.
- Cables subjected to a uniform load over their length have a catenary shape, but we can typically approximate the situation as a uniform load over the horizontal projection.

Uniform Load:

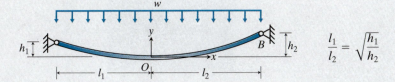

$$\frac{l_1}{l_2} = \sqrt{\frac{h_1}{h_2}}$$

All along cable:

$$y = \frac{h_i}{l_i^2} x^2$$

$$H = \frac{w l_i^2}{2 h_i}$$

$$F_{\text{cable-}y} = wx$$

$$F_{\text{cable}} = \sqrt{\left(\frac{w l_i^2}{2 h_i}\right)^2 + (wx)^2}$$

At end of cable:

$$F_{\text{cable}}^{\max} = w l_i \sqrt{\frac{l_i^2}{4 h_i^2} + 1}$$

Uniform Load Plus Point Load:

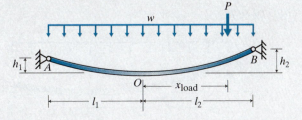

If the uniform load dominates, we can superimpose effects.
Reactions due to P:

$$H = \frac{P(l_i - x_{\text{load}})}{h_i}$$

$$A_y = P \frac{(l_2 - x_{\text{load}})}{(l_1 + l_2)} - H \frac{(h_2 - h_1)}{(l_1 + l_2)}$$

$$B_y = P \frac{(l_1 + x_{\text{load}})}{(l_1 + l_2)} + H \frac{(h_2 - h_1)}{(l_1 + l_2)}$$

Assumptions and Limitations:

- All loads are applied vertically.
- The effect of cable elongation is negligible.

Example 3.3

The owners of a zip line attraction want to know if they can increase the weight limit for riders from 100 kg to 120 kg. They recently purchased new equipment that is designed for the increase in rider weight. The question is whether the zip line and anchorages are strong enough.

Figure 3.41

Part of the assessment is to determine the effect of the self-weight of the zip line. Our team leader wants us to find the peak force generated in the cable by its self-weight. The maximum elevation change was set by the owners so that riders slow down adequately as they approach the landing.

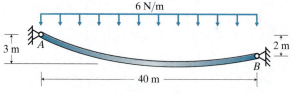

Figure 3.42

Analysis:

Find the location of zero slope:

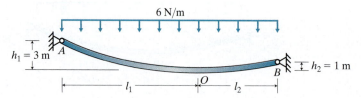

Figure 3.43

$$\frac{l_1}{l_2} = \sqrt{\frac{h_1}{h_2}} = \sqrt{\frac{3\,\text{m}}{1\,\text{m}}} = 1.732$$

$$l_1 = 1.732\, l_2$$

$$40\,\text{m} = l_1 + l_2 = 1.732\, l_2 + l_2 = 2.732\, l_2$$

$$l_2 = 14.64\,\text{m};\quad l_1 = 40\,\text{m} - 14.64\,\text{m} = 25.36\,\text{m}$$

Find the peak cable force: It will occur at the left end, since $l_1 > l_2$.

$$F_{\text{cable}}^{\max} = w l_1 \sqrt{\frac{l_1^2}{4 h_1^2} + 1} = 6\,\text{N/m}\,(25.36\,\text{m}) \sqrt{\frac{(25.36\,\text{m})^2}{4\,(3\,\text{m})^2} + 1}$$

$$F_{\text{cable}}^{\max} = 625\,\text{N} = 0.625\,\text{kN}$$

Section 3.2 Cables with Uniform Loads

> ### EXAMPLE 3.4

Our firm has been hired to evaluate the anchorages for the main cables of a historic suspension bridge.

Figure 3.44

Part of the assessment is to determine the force generated in one main cable due to the dead load of the bridge. A team member has already calculated the uniform distributed load, and we have the bridge dimensions from the Port Authority.

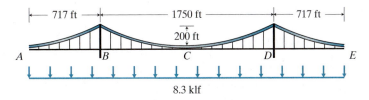

Figure 3.45

Our task is to find the cable force due to the dead load as it arrives at the anchorage. A team member is developing a computer model to perform a nonlinear analysis of the cable. We are providing an approximate solution to use for comparison.

Analysis:

If we assume that the main cable passes smoothly over the pier, H will be constant along the cable. Since w is also constant along the cable, the parabolic shape of the end spans will be the same as the shape of the main span.

Find the horizontal component from the main span:

$$H = \frac{wl^2}{2h} = \frac{8.3\,\text{klf}\,(1750\,\text{ft}/2)^2}{2(200\,\text{ft})}$$

$$H = 15{,}890\,\text{k}$$

Find the vertical component at the anchorage:

The end span has the same parabolic shape as the main span.

$$y = \frac{h}{l^2}x^2 = \frac{200\,\text{ft}}{(1750\,\text{ft}/2)^2}x^2$$

$$y = 2.612 \times 10^{-4} x^2$$

But the end span does not start at zero slope.

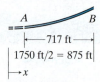

Figure 3.46

At the anchorage,

$$x = l_{\text{main span}} - l_{\text{end span}}$$
$$= 875\,\text{ft} - 717\,\text{ft}$$
$$x = 158\,\text{ft}$$

The vertical component is based on x.

$$F_{\text{cable-}y} = wx = 8.3\,\text{klf}\,(158\,\text{ft})$$
$$F_{\text{cable-}y} = 1311\,\text{k}$$

Combine the components to find the cable force at the anchorage:

$$F_{\text{cable}} = \sqrt{H^2 + F_{\text{cable-}y}^2}$$
$$= \sqrt{(15{,}890\,\text{k})^2 + (1311\,\text{k})^2}$$
$$F_{\text{cable}} = 15{,}940\,\text{k}$$

EXAMPLE 3.5

The Port Authority for the bridge in Example 3.4 has received a request for a special permit for a heavy load to traverse the bridge. The gross vehicle weight is 300 kips, well above the 80-kip maximum for trucks without permits.

Before the Port Authority will authorize the permit, they want to make sure the load will not damage the bridge.

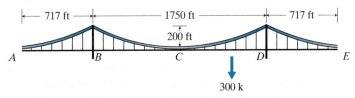

Figure 3.47

Our team is checking the strength of all components due to the large load. Our task is to determine the largest effect of the vehicle on the forces in the main cable. Before performing a nonlinear analysis of the cable, we want an approximate solution.

Analysis:

We assume that the cable shape remains parabolic because the dead load dominates. The location of the load that maximizes the effect on the cable force is

$$H = \frac{P(l - x_{\text{load}})}{h}$$

The horizontal component is largest when $x_{\text{load}} = 0$, at the middle of the main span.

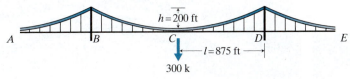

Figure 3.48

Find the force components at the tower, where the impact is greatest.

$$H = \frac{300\,\text{k}\,(875\,\text{ft} - 0)}{200\,\text{ft}}$$

$$H = 1312\,\text{k}$$

$$A_y = P\frac{(l_2 - x_{\text{load}})}{(l_1 + l_2)} - H\frac{(h_2 - h_1)}{(l_1 + l_2)}$$

$$= 300\,\text{k}\frac{(875\,\text{ft} - 0)}{(1750\,\text{ft})} - 1312\,\text{k}\frac{(200\,\text{ft} - 200\,\text{ft})}{(1750\,\text{ft})}$$

$$A_y = 150\,\text{k}\;(+\uparrow)$$

$$B_y = P - A_y$$
$$= 300\,\text{k} - 150\,\text{k}$$

$$B_y = 150\,\text{k}\;(+\uparrow)$$

3.3 Arches

Introduction

Arches were developed long ago to carry gravity loads. If the shape of the arch is made just right for the loading, the arch will experience only axial compression, no shear or moment. Therefore, we can think of an arch as the mirror image of a cable with the same loading.

How-To

The concept of the arch is essentially a cable upside down. If the shape of the arch matches the shape that a cable would take under the same loads, the arch experiences only axial compression. However, if the shape of the arch and the loading pattern do not match, the arch does not change shape like a cable does; instead, the arch develops shear and moment in addition to compression.

Unlike with cables, there are a few options for supports at the ends of an arch. The ends can be fix supported (Figure 3.49a), pin supported (Figure 3.49b), or simply supported (Figure 3.49c). The fix- and pin-supported arches are called *thrust* arches because the horizontal components of the end forces push outward. The reactions of a simply supported

(b)

(a)

(c)

Figure 3.49 Types of continuous arches: (a) Thrust arch with fixed supports; (b) Thrust arch with pinned supports; (c) Tied arch with simple supports.

arch cannot resist that thrust. Therefore, we connect the ends with a tension member. We call these *tied* arches.

Fix-supported arches are not common because the foundations must be substantially stiffer than for pin-supported arches. If the arch is the correct shape for the dominant load, there is little need for fixed supports. Tied arches are typically preferred when it is okay to have the tie pass across the space below the arch. When that is not practical, pin-supported arches are most common.

Arches tend to be much stiffer than cables, so first-order analysis is typically adequate. The challenge for the analysis is that all three of these types of arches are indeterminate. The only determinate arch is a *three-hinge* arch (Figure 3.50). The internal hinge makes the arch determinate as long as it is a thrust arch with pinned supports or a tied arch with simple supports. A three-hinge arch lacks the redundancy that an

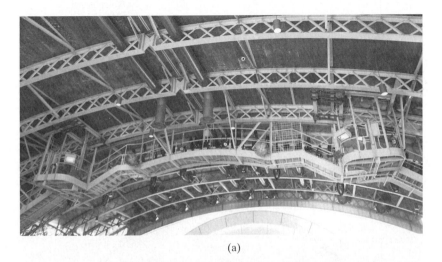

(a)

(b)

Figure 3.50 Three-hinge arch: (a) Example of a three-hinge arch supporting a theater roof; (b) A three-hinge arch is not continuous but has an internal hinge.

indeterminate arch has, but the resulting benefit is that the three-hinge arch does not develop internal forces due to temperature change, creep, or shrinkage.

Performing detailed analyses of indeterminate arches is outside the scope of this text, but many references show how to derive expressions for internal forces and deformations for indeterminate arches (e.g., Karnovsky 2012). There are also references that provide formulas for reactions for indeterminate arches subjected to certain load patterns (e.g., Mikhelson and Hicks 2013).

We can also perform approximate analysis by treating an indeterminate arch as a three-hinge arch. That allows us to approximate the reactions and internal forces.

Analysis of Three-Hinge Arches

For a three-hinge arch, the internal hinge is typically at the apex of the arch. The other two hinges are at the supports (i.e., pin supports). If we cut the arch at all three hinges, we get two free body diagrams with a total of six unknowns (Figure 3.51). Using the six equations of equilibrium, we can find the internal forces at the hinges. Once we have those, we can calculate internal forces anywhere along the arch.

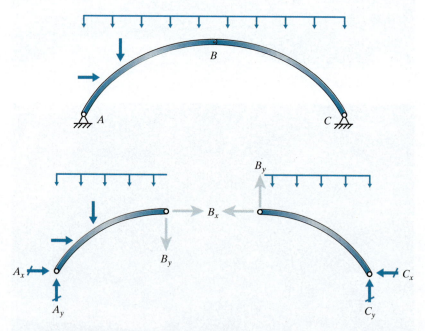

Figure 3.51 Analysis of a three-hinge arch: (a) The arch can experience any loading; (b) Two free body diagrams with six unknowns at the three hinges.

Section 3.3 Highlights

Arches

Characteristics:

- *If* the arch is a mirror image of the shape a cable takes under the same loading, the arch experiences only axial compression.
- Arches can also carry shear and moment.

Indeterminate arches:

Can approximate as a three-hinge arch (determinate):

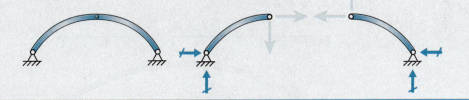

Example 3.6

Our firm is designing a bridge to cross a large creek. The bridge will be a tied arch.

Figure 3.52

In order to make a preliminary selection of member properties, the team needs to determine approximate internal forces. We have been tasked with finding the tie force and the vertical reactions at the abutments. A more experienced member of the team has estimated the combined dead and live load on the arch.

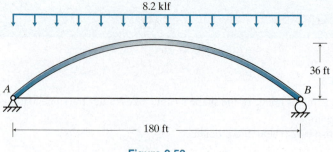

Figure 3.53

Analysis:

Find the reactions:

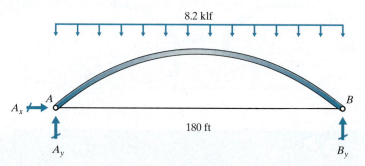

Figure 3.54

$$\circlearrowright \Sigma M_A = 0 = 8.2\,\text{klf}\,(180\,\text{ft})(180\,\text{ft}/2) - B_y(180\,\text{ft})$$

$$B_y = 738\,\text{k}\ (+\uparrow)$$

$$+\uparrow \Sigma F_y = 0 = A_y - 8.2\,\text{klf}\,(180\,\text{ft}) + 738\,\text{k}$$

$$A_y = 738\,\text{k}\ (+\uparrow)$$

$$\xrightarrow{+} \Sigma F_x = 0 = A_x$$

Note: The result for A_x confirms that the tie carries all the horizontal force.

Although the deck hangs from the arch by diagonal cables, we can use the equivalent downward load since we are trying to find only the tie force.

Section 3.3 Arches   157

Approximate the tied arch as a three-hinge arch:

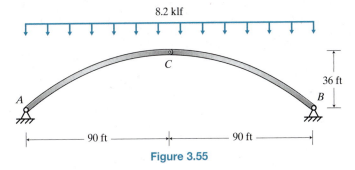

Figure 3.55

FBD of *AC*:

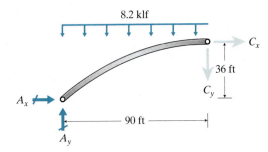

Figure 3.56

$\xrightarrow{\pm} \Sigma F_x = 0 = A_x + C_x$

$\qquad A_x = -C_x$

$+\uparrow \Sigma F_y = 0 = A_y - 8.2\,\text{klf}\,(90\,\text{ft}) - C_y$

$\qquad A_y = 738\,\text{k} + C_y$

$\curvearrowright \Sigma M_A = 0 = 8.2\,\text{klf}\,(90\,\text{ft})\,(90\,\text{ft}/2) + C_x(36\,\text{ft}) + C_y(90\,\text{ft})$

$\qquad C_x(36\,\text{ft}) = -33{,}210\,\text{k}\cdot\text{ft} - C_y(90\,\text{ft})$

$\qquad C_x = -922\,\text{k} - 2.5C_y$

FBD of *CB*:

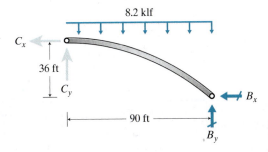

Figure 3.57

$\pm \Sigma F_x = 0 = C_x + B_x$

$B_x = -C_x$

$+\uparrow \Sigma F_y = 0 = C_y - 30\,\text{kN/m}\,(34\,\text{m}) + B_y$

$B_y = 1020\,\text{kN} - C_y$

$\circlearrowright \Sigma M_B = 0 = -C_x(36\,\text{ft}) + C_y(90\,\text{ft}) - 8.2\,\text{klf}\,(90\,\text{ft})(90\,\text{ft}/2)$

$C_x(36\,\text{ft}) = -33{,}210\,\text{k}\cdot\text{ft} + C_y(90\,\text{ft})$

$C_x = -922\,\text{k} + 2.5 C_y$

Combine the C_x expression with the result from AC:

$-922\,\text{k} - 2.5 C_y = -922\,\text{k} + 2.5 C_y$

$5.0 C_y = 0$

$C_y = 0$

$\therefore A_y = 738\,\text{k}\ (+\uparrow)$

$B_y = 738\,\text{k}\ (+\uparrow)$

Evaluation of Results:

Satisfied Fundamental Principles?

These reactions match what we calculated at the beginning of the problem. ✓

Find tie force:

Substitute C_y into the expression for C_x:

$C_x = -922\,\text{k}$

$\therefore A_x = 922\,\text{k}\ (\xrightarrow{+})$

$B_x = 922\,\text{k}\ (\xleftarrow{+})$

Note: This is the tie force.

Summary:

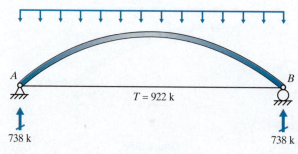

Figure 3.58

Example 3.7

We are designing a new pedestrian bridge to cross a major highway. The architect and senior engineer have decided on a thrust arch with fixed supports. The deck widens from one end to the other, so the dead load is not uniform.

Figure 3.59

Our team is still in the preliminary design phase, but the geotechnical engineers need order-of-magnitude estimates of the forces on the foundations. Each team member is performing an approximate analysis of the arch for different unfactored loads. A team member has calculated the seismic loads based on the assumed self-weight of the bridge. Although the static equivalent seismic load will be distributed along the arch, we can simplify that as five point loads for this preliminary analysis. Our task is to use approximate analysis to estimate the vertical and horizontal forces on the foundations due to seismic load in the plane of the arch.

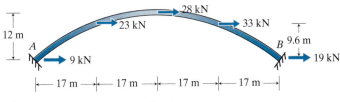

Figure 3.60

Analysis:

Approximate the fix-supported arch as a three-hinge arch. This will not predict moment reactions, but it will predict all other reactions.

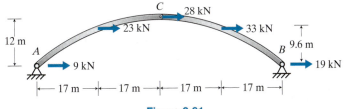

Figure 3.61

FBD of AC:

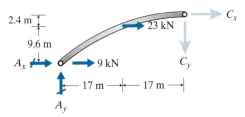

Figure 3.62

160 Chapter 3 Cables and Arches

$$\xrightarrow{+} \Sigma F_x = 0 = A_x + 9\,\text{kN} + 23\,\text{kN} + C_x$$
$$A_x = -C_x - 32\,\text{kN}$$
$$+\uparrow \Sigma F_y = 0 = A_y - C_y$$
$$A_y = C_y$$
$$\circlearrowright \Sigma M_A = 0 = 23\,\text{kN}(9.6\,\text{m}) + C_x(12\,\text{m}) + C_y(34\,\text{m})$$
$$C_x(12\,\text{m}) = -C_y(34\,\text{m}) - 221\,\text{kN}\cdot\text{m}$$
$$C_x = -2.83 C_y - 18.41\,\text{kN}$$

FBD of CB:

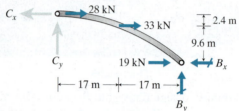

Figure 3.63

$$\xleftarrow{+} \Sigma F_x = 0 = -C_x + 28\,\text{kN} + 33\,\text{kN} + 19\,\text{kN} - B_x$$
$$B_x = -C_x + 80\,\text{kN}$$
$$+\uparrow \Sigma F_y = 0 = C_y + B_y$$
$$B_y = -C_y$$
$$\circlearrowright \Sigma M_B = 0 = -C_x(12\,\text{m}) + C_y(34\,\text{m}) + 28\,\text{kN}(12\,\text{m}) + 33\,\text{kN}(9.6\,\text{m})$$
$$C_x(12\,\text{m}) = C_y(34\,\text{m}) + 653\,\text{kN}\cdot\text{m}$$
$$C_x = 2.83 C_y + 54.4\,\text{kN}$$

Combine the C_x expression with the result from AC:

$$2.83 C_y + 54.4\,\text{kN} = -2.83 C_y - 18.41\,\text{kN}$$
$$5.66 C_y = -72.8\,\text{kN}$$
$$C_y = -12.86\,\text{kN}$$

$$\therefore A_y = -12.86\,\text{kN}\ (+\uparrow)$$
$$B_y = 12.86\,\text{kN}\ (+\uparrow)$$

Substitute C_y into the expression for C_x:

$$C_x = 2.83(-12.86\,\text{kN}) + 54.4\,\text{kN} = 18.01\,\text{kN}$$

Use C_x to find the horizontal reactions:

$$A_x = -18.01\,\text{kN} - 32\,\text{kN} \qquad B_x = -18.01\,\text{kN} + 80\,\text{kN}$$
$$A_x = -50.0\,\text{kN}\ (\xrightarrow{+}) \qquad B_x = 62.0\,\text{kN}\ (\xleftarrow{+})$$

Summary:

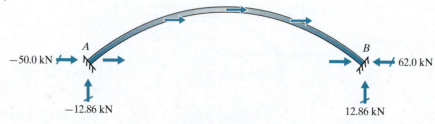

Figure 3.64

Section 3.3 Arches   161

Example 3.8

Our company designs and manufactures pre-engineered steel buildings. The structural system uses tapered steel members to create a rigid frame with a pitched roof. The members become thinner near the top, so the frame behaves like a three-hinge arch.

Figure 3.65

To design the connection between the vertical and diagonal members, we need to know the moment transferred at that point. One of the load cases we need to check is heavy snow on only one side of the roof.

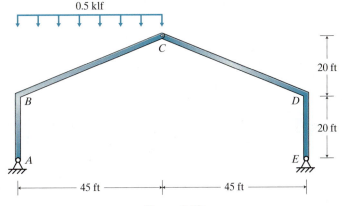

Figure 3.66

Analysis:

FBD of *AC*:

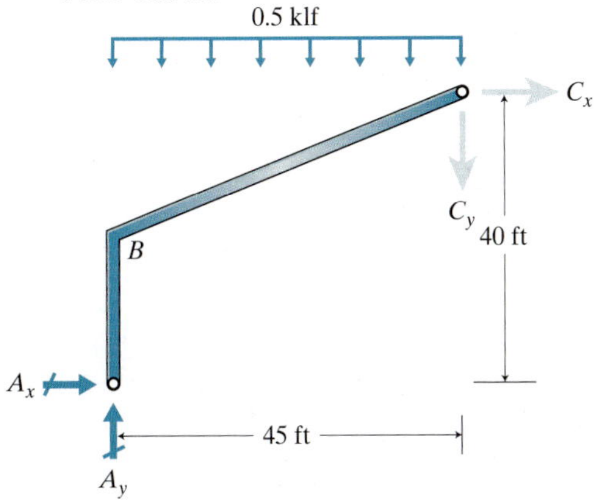

$$\circlearrowleft \Sigma M_A = 0 = 0.5\,\text{klf}\,(45\,\text{ft})\,(45\,\text{ft}/2)$$
$$+ C_x(40\,\text{ft}) + C_y(45\,\text{ft})$$
$$C_x(40\,\text{ft}) = -C_y(45\,\text{ft}) - 506\,\text{k}\cdot\text{ft}$$
$$C_x = -1.125 C_y - 12.65\,\text{k}$$

Figure 3.67

162   Chapter 3   Cables and Arches

FBD of *CE*:

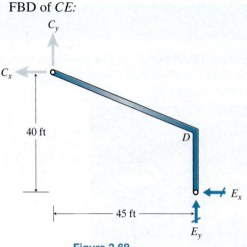

$$+\circlearrowleft \Sigma M_E = 0 = -C_x(40\,\text{ft}) + C_y(45\,\text{ft})$$
$$C_x(40\,\text{ft}) = C_y(45\,\text{ft})$$
$$C_x = 1.125 C_y$$

Figure 3.68

Combine the C_x expression with the result from *AC*:

$$1.125 C_y = -1.125 C_y - 12.65\,\text{k}$$
$$2.25 C_y = -12.65\,\text{k}$$
$$C_y = -5.62\,\text{k}$$

Substitute C_y into the expression for C_x:

$$C_x = 1.125(-5.62\,\text{k}); \quad C_x = -6.32\,\text{k}$$

FBD of *BC*:

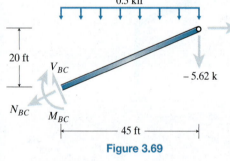

$$+\circlearrowleft \Sigma M_B = 0 = M_{BC} + 0.5\,\text{klf}(45\,\text{ft})(45\,\text{ft}/2)$$
$$+ (-6.32\,\text{k})(20\,\text{ft}) + (-5.62)(45\,\text{ft})$$
$$M_{BC} = -127.0\,\text{k}\cdot\text{ft}\;(\curvearrowleft)$$

Figure 3.69

FBD of *CD*:

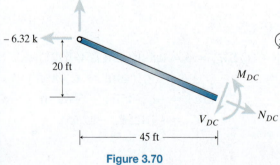

$$+\circlearrowright \Sigma M_D = 0 = (-6.32\,\text{k})(20\,\text{ft})$$
$$- (-5.62)(45\,\text{ft}) + M_{DC}$$
$$M_{DC} = -126.5\,\text{k}\cdot\text{ft}\;(\curvearrowright)$$

Figure 3.70

Summary: $M_{BC} = -127.0\,\text{k}\cdot\text{ft}\;(\curvearrowleft)$
$\phantom{\text{Summary: }}M_{DC} = -126.5\,\text{k}\cdot\text{ft}\;(\curvearrowright)$

References Cited

Karnovsky, I. A. 2012. *Theory of Arched Structures: Strength, Stability, Vibration.*
    New York: Springer-Verlag.
Mikhelson, I., and T. G. Hicks. 2013. *Structural Engineering Formulas*, 2nd ed.
    New York: McGraw-Hill Education.

Homework Problems

3.1 Our team is analyzing a zip line cable to explore the feasibility of increasing the maximum rider weight. The new factored live load is 2.4 kN.

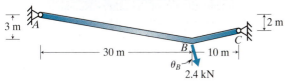

Prob. 3.1

Our team is analyzing the cable by considering the live load at various locations along the zip line. We have been assigned to analyze the rider three-quarters of the way across the zip line. A more experienced member of the team tells us that for this particular situation, the cable should drop by 3 m from the start point to the rider.

To perform the preliminary analysis, the team has decided to make a couple simplifying assumptions:

- Although the problem is geometrically nonlinear, we are going to superimpose the results of the analysis of the cable subjected to self-weight on the results of the analysis of the cable with a live load.

- Although the rider is not in equilibrium, we are going to conservatively treat the problem as static. In reality, the rider's weight vector remains vertical and the unresolved horizontal component at the pulley connection is what accelerates and decelerates the rider.

  a. Guess whether the load will tilt toward end A or end C or be exactly vertical for this analysis.
  b. Find the cable force and the load angle for this situation.
  c. Compare your cable force from part (b) with the result from Example 3.1. Which rider location causes a larger cable force?
  d. Comment on why your guess in part (a) was or was not the same as what you discovered in part (b).

3.2 The Transit Authority in our city uses electric buses that get their power from an overhead line. The Transit Authority is performing its annual inspection of the lines and has concerns about one of the support cables. They have contracted our firm to verify the adequacy of the cable. The cable appears to be more than adequate to support the dead load of the power line, but there is concern about what will happen if the power line becomes coated with ice. A team member has calculated the factored combination of dead load plus ice.

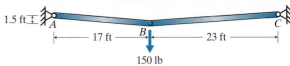

Prob. 3.2

a. Guess which side will have the larger cable force.
b. Find the peak cable force by calculating the cable force in each segment.
c. Comment on why your guess in part (a) was or was not the same as what you discovered in part (b).

3.3 The Transit Authority is considering adding a second powered bus line along an existing route. They want to hang the new power line on the existing cable.

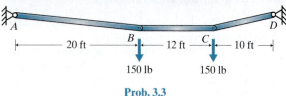

Prob. 3.3

Because of height restrictions in the area, the Transit Authority wants us to limit the maximum sag to 2 feet. A team member has calculated the design load from each power line.

a. Guess which point will have the largest sag.
b. Choose one of the load points to cause the maximum sag. Calculate the resulting sag of the other load point. If the second load point sags more than the maximum allowable, reduce that sag to the maximum and recalculate the sag of the first.
c. Using the configuration that keeps the largest sag at the maximum allowable, find the peak cable force by calculating the cable force in each segment.
d. Comment on why your guess in part (a) was or was not the same as what you discovered in part (b).

3.4 We work for the City Engineer's office, and the city is getting ready to install a new set of traffic signals across a wide road. Since this span will be the longest the office has placed, the City Engineer wants us to verify that the standard cable will be strong enough.

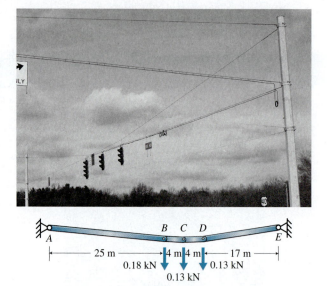

Prob. 3.4

The maximum allowable sag is 2 m due to vehicle clearance and standard pole height. The weights of the different traffic signals come from the manufacturer's catalog.

a. Guess which cable segment will have the largest force.
b. Choose one of the load points to cause the maximum allowable sag. Calculate the resulting sag of the other two load points. If a different load point sags more than the maximum allowable, reduce that sag to the maximum and recalculate the sag of the others. Repeat if necessary in order to make all three sags at or below the maximum allowable.
c. Using the configuration that keeps the largest sag at the maximum allowable, find the peak cable force by calculating the cable force in each segment.
d. Comment on why your guess in part (a) was or was not the same as what you discovered in part (c).

3.5 Our team is analyzing zip line cables to explore the feasibility of increasing the maximum rider weight for a regional attraction. Different members of the team have been assigned different lines and different load conditions.

3.6 Our firm has been contacted by a local manufacturer that has suspended pipes between two buildings from cables. The cables are anchored into the block walls of the two buildings. The block near one of the anchorages is beginning to spall, so the manufacturer is concerned that the suspension cable will pull out.

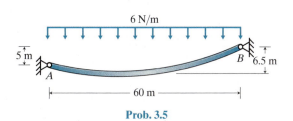

Prob. 3.5

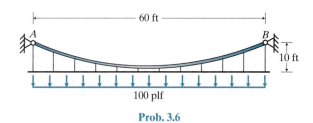

Prob. 3.6

One member of our firm is in the field checking on the capacity of the anchorage. Our task is to determine the horizontal and vertical components of the pull being generated by the cable. We consider the components separately because they are associated with different failure mechanisms.

The weight of the pipes effectively acts as a uniform load on the suspension cables. Our coworker has sent back the approximate weight of the pipes, which we have already factored.

We have been assigned the longest line. Our task is to find the peak cable force due to the self-weight of the cable. The maximum drop of the cable was set by the client to give the riders a safe but exciting experience.

a. Guess which end will experience the largest cable force.
b. Find the largest cable force.
c. Comment on why your guess in part (a) was or was not the same as what you discovered in part (b).

a. Guess which component will be larger: horizontal or vertical.
b. Find the components of the cable force at the anchorage.
c. Comment on why your guess in part (a) was or was not the same as what you discovered in part (b).

3.7 Our firm is designing a shade structure for outdoor concerts. The roof will be fabric held up by a cable down the middle. In order to keep the fabric spread out over the full length of the structure, our senior engineer wants to put a pair of stiffeners at midspan.

3.8 Maintenance is scheduled for a long suspension bridge. The contractor has requested that materials be stockpiled near one of the piers. Before the Department of Transportation will okay that request, they want our firm to perform an assessment of the impact.

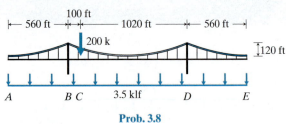

Prob. 3.8

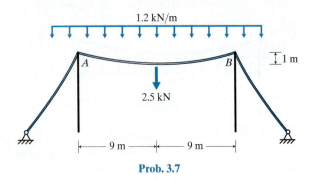

Prob. 3.7

Before performing a computer aided analysis of the structure, we want to have a sense of the solution. Therefore, let's find the tension generated in the cable right before it passes over the mast.

The Department of Transportation sent us information about the self-weight of the bridge and the proposed stockpile of materials. As another team member begins to create a computer model of the bridge, we will predict the peak tension in the main cable.

a. Guess which load will cause greater tension: distributed load due to the self-weight of the fabric, or point load due to the self-weight of the stiffeners.
b. Find the horizontal and vertical components of the cable force just right of point A due to just the distributed load. Combine the components to find the cable force.
c. Assuming that the shape of the cable is set by the distributed load, find the horizontal and vertical components of the cable force just right of point A due to just the point load. Combine the components to make an estimate of the cable force.
d. Combine the components from the two loadings and find the resultant. The resultant is our approximation of the cable force generated by both loads.
e. Comment on why your guess in part (a) was or was not the same as what you discovered in parts (b) and (c).

a. Guess which load will cause greater tension: distributed load due to the self-weight of the bridge, or point load due to the stockpiled materials.
b. Find the horizontal and vertical components of the cable force just right of point B due to just the distributed load. Combine the components to find the cable force.
c. Assuming that the shape of the cable is set by the distributed load, find the horizontal and vertical components of the cable force just right of point B due to just the point load. Combine the components to make an estimate of the cable force.
d. Combine the components from the two loadings and find the resultant. The resultant is our approximation of the cable force generated by both loads.
e. Comment on why your guess in part (a) was or was not the same as what you discovered in parts (b) and (c).

3.9 Our client is a chemical company that wants to expand its operations across a street and rail line. The expansion requires connecting the two facilities by several pipes. The site is just below an elevated highway, so the company cannot run the pipes underground. Working with the client, our principal-in-charge has decided to use an arch to carry the pipes over the road and rail line. Because of the tight geometry of the site and the close proximity to the road and rail line, the senior engineer has chosen to use a semicircular arch even though that is not the ideal shape for the loads.

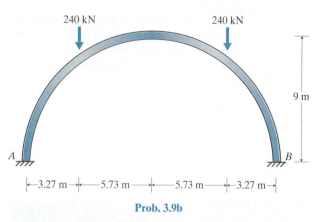

Prob. 3.9b

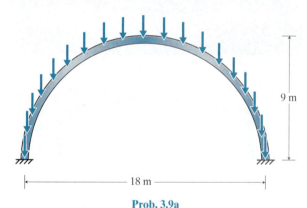

Prob. 3.9a

Another team member will be performing computer-aided analysis of the arch. Our task is to generate approximate results that we can use to validate the computer results. The primary load will be the self-weight of the truss and pipes, so we will analyze the arch subjected to dead load. Because the load is distributed along the length, it is not uniform along the horizontal axis.

For this comparison we want approximate reactions, therefore we can consider the resultant of the distributed load on each half of the arch. A team member has calculated the forces and their locations.

a. Guess whether the reactions will have a larger vertical or horizontal magnitude.
b. Approximate the arch as a three-hinge arch. Find the vertical and horizontal reactions.
c. Comment on why your guess in part (a) was or was not the same as what you discovered in part (b).

3.10 A small river runs adjacent to a city. To expand its business opportunities, the city wants to extend a street across that river. The senior engineer on the project thinks a tied arch will be the best structural system. To help confirm that choice, our team is going to perform an approximate analysis of the arch subjected to the largest load types.

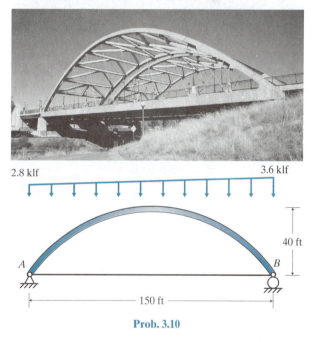

Prob. 3.10

To accommodate a turn lane at one end of the bridge, the bridge will taper in width. Because of the taper, the dead load on the arch is nonuniform. A team member has estimated the dead load on one side of the bridge.

a. Guess which of the following will have the largest magnitude and which will have the smallest: left vertical reaction, right vertical reaction, tie force.
b. Calculate the vertical reactions due to the dead load.
c. Approximate the arch as a three-hinge arch. Calculate the resulting tie force.
d. Comment on why your guess in part (a) was or was not the same as what you discovered in parts (b) and (c).

168 Chapter 3 Cables and Arches

3.11 A small river runs adjacent to a city. To expand its business opportunities, the city wants to extend a street across that river. The senior engineer on the project thinks a tied arch will be the best structural system. To help confirm that choice, our team is going to perform an approximate analysis of the arch subjected to the largest load types.

One of the load cases for the bridge is the axle loads from the design truck. The team must consider the effect of the truck as it moves from one side to the other. An experienced engineer in the firm suggests we start by calculating the effect of the truck as the semi passes midspan of the bridge.

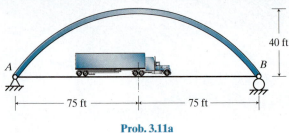

Prob. 3.11a

The design truck has the following axle loads and dimensions:

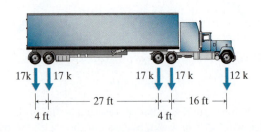

Prob. 3.11b

a. Guess which of the following will have the largest magnitude and which will have the smallest: left vertical reaction, right vertical reaction, tie force.
b. Calculate the vertical reactions due to the live load.

c. Approximate the arch as a three-hinge arch. Calculate the resulting tie force.
d. Comment on why your guess in part (a) was or was not the same as what you discovered in parts (b) and (c).

3.12 Our firm is designing a small warehouse with a pitched roof to be constructed in British Columbia. The senior engineer has chosen to use tapered members that will act as a three-hinge arch.

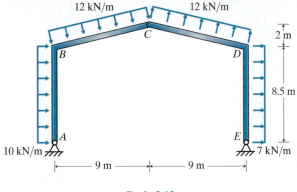

Prob. 3.12

Our team needs to analyze the structure for a variety of loads. Our task is to find the reactions and internal forces due to wind. One of our team members has calculated the wind loads.

a. Guess which of the four reactions will have the largest magnitude.
b. Calculate the reactions due to the wind load.
c. Calculate the moment at connections B and D.
d. Comment on why your guess in part (a) was or was not the same as what you discovered in part (b).

CHAPTER 4

INTERNAL FORCE DIAGRAMS

Internal forces along members affect our design choices.

Motivation

Internal force diagrams are graphical representations of the internal forces along a member. These diagrams present significant information (peak positive value and location, peak negative value and location, and location of minimum value) in a concise way. Therefore, practitioners frequently use these diagrams for evaluating the results of their analyses (see Section 7.3: Checking Features of the Solution) and for making design decisions.

Typically, internal force diagrams are created using structural analysis software. However, the by hand skills covered in this chapter serve several important purposes:

- Creating internal force diagrams by hand builds an understanding of the features of an accurate internal force diagram so that we can evaluate diagrams created by hand or computer.
- When we design a member, the goal is often to find the peak internal force and its location. For determinate members, this is often done faster by hand than by creating and running a computer model.
- When designing steel flexure members, we often want to calculate the moment gradient modification factor. To do so, we need to know the moment value at specific points along the member. In those cases, the equations for the moments are often more helpful than the diagram.
- Predicting deformations (slope and displacement) using the Double Integration Method (see Section 5.1) or the Virtual Work Method (see Section 5.3) requires equations for the moments along the member.

4.1 Internal Forces by Integration

Introduction
An insightful way to construct internal force diagrams is to use integration to obtain equations for the internal forces along the member. Those equations can be plotted to give the diagrams, and they can be used to find deformations (see Section 5.1: Double Integration Method and Section 5.3: Virtual Work Method).

How-To
We can demonstrate how to integrate applied loads to obtain equations for internal forces by cutting out a small section and using equations of equilibrium.

To begin, we need to establish a sign convention. Let's cut a member and examine the effects of the exposed internal forces. An axial force that tends to stretch the end by pulling on it is a positive axial force, N (Figure 4.1a). We define the shear, V, that tends to rotate the piece clockwise as positive (Figure 4.1b). We define the moment, M, that tends to bend the piece into a smile as positive (Figure 4.1c). Note that it is impossible to define positive as an absolute direction, as we do for a support reaction, because the direction is opposite on the other side of the cut (compliments of Newton's 3rd law).

Distributed Loads Normal to a Member
Consider a member with a distributed load that is a function of position, $w(x)$. We define a positive distributed load as acting up, away from the

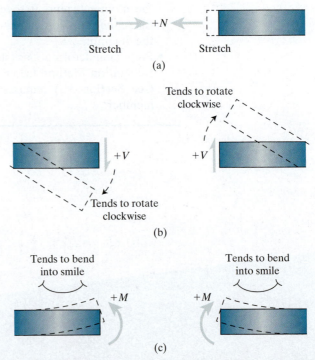

Figure 4.1 Sign conventions for positive internal forces: (a) Axial force that tends to stretch the ends of the cut; (b) Shear that tends to rotate the piece clockwise; (c) Moment that tends to bend the piece into a smile.

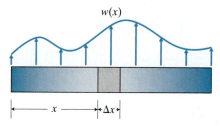

Figure 4.2 Member with a distributed load that is a function of position.

member. Therefore, $w(x)$ due to gravity is typically negative. We also define x to increase from left to right (Figure 4.2). Consistency is essential in these conventions.

We can cut out an extremely small piece of width Δx. Over such a short distance, $w(x)$ is essentially constant, so the resulting applied load is $w(x)\Delta x$. If we draw only the shears at the cut, we see that the shear at one side is not the same magnitude as the shear at the other side of the piece (Figure 4.3).

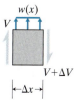

Figure 4.3 Extremely short piece of the member showing only shears at the cut ends.

When we apply vertical equilibrium to the piece, we obtain

$+\uparrow \Sigma F = 0 = V + w(x)\Delta x - V - \Delta V$

$\Delta V = w(x)\Delta x$

In the limit, as Δx gets infinitely small, we obtain an expression with derivatives:

$dV = w(x)dx$

This expression leads to two important pieces of information:

$\dfrac{dV}{dx} = w(x)$ The slope of the shear diagram is the value of the distributed load at that point.

$V(x) = \displaystyle\int w(x)dx$ The equation for the shear is the integral of the distributed load.

Since the last expression is an indefinite integral (no limits of integration), it produces an equation $V(x)$ rather than a value. Note, though, that the indefinite integration process produces a constant of integration. To find the value of the constant, we must use known boundary conditions.

Consider the same piece of member from Figure 4.2 but now showing both shears and moments at the cut ends (Figure 4.4).

Applying moment equilibrium about the left edge produces

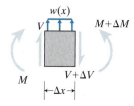

Figure 4.4 Extremely short piece of the member showing only shears and moments at the cut ends.

$\curvearrowleft \Sigma M_{\text{left edge}} = 0 = M - w(x)\Delta x (\Delta x/2) + (V + \Delta V)\Delta x - (M + \Delta M)$

$\Delta M = -\dfrac{w(x)(\Delta x)^2}{2} + V\Delta x + \Delta V \Delta x$

But $\Delta V = w(x)\Delta x$, so

$\Delta M = -\dfrac{w(x)(\Delta x)^2}{2} + V\Delta x + w(x)(\Delta x)^2$

In the limit as Δx gets infinitely small, $(\Delta x)^2$ is essentially zero and we obtain the following expression with derivatives:

$dM = V(x)dx$

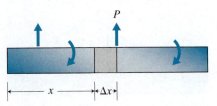

Figure 4.5 Member experiencing only point forces and applied moments.

This expression leads to two important pieces of information:

$$\frac{dM}{dx} = V(x)$$
The slope of the moment diagram is the value of the shear diagram at that point.

$$M(x) = \int V(x)\,dx$$
The equation for the moment is the integral of the shear equation.

This expression for the moment is also an indefinite integral that produces a constant that must be found from known boundary conditions.

Point Loads Normal to a Member and Applied Moments

We have derived a process for finding $V(x)$ and $M(x)$ if there is a distributed load, but what about when there are point loads or applied moments?

Consider a member with applied moments and point forces (Figure 4.5). Our sign convention is that upward forces are positive and downward forces are negative. Clockwise applied moments are positive, so counterclockwise moments are negative. We can cut out a very narrow piece that contains one point force at its right edge (Figure 4.6).

Applying vertical equilibrium to the piece, we obtain the expression

$$+\uparrow \Sigma F = 0 = V + P - (V + \Delta V)$$
$$\Delta V = P$$

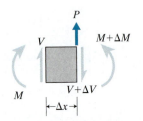

Figure 4.6 Extremely short piece of the member that contains one applied point force.

We discover that there is a jump in the shear value equal to the value of the applied force. Therefore, a force that pulls down on the member causes a negative jump, and a force that pushes up on the member causes a positive jump in the shear diagram.

Now we apply moment equilibrium:

$$\curvearrowright \Sigma M_{\text{left edge}} = 0 = M - P\Delta x + (V + \Delta V)\Delta x - (M + \Delta M)$$
$$\Delta M = -P\Delta x + V\Delta x + \Delta V \Delta x$$

But $\Delta V = P$, so

$$\Delta M = -P\Delta x + V\Delta x - P\Delta x$$
$$\Delta M = V\Delta x$$

which is no different from what we found before. There is no jump in the moment value due to an applied force.

Now we can cut out a very narrow piece of member that contains only one applied moment (Figs. 4.7 and 4.8).

Applying vertical equilibrium to the piece, we obtain the expression

$$+\uparrow \Sigma F = 0 = V - (V + \Delta V)$$
$$\Delta V = 0$$

From this, we discover that there is no direct effect (i.e., no jump) in the shear value where there is an applied moment.

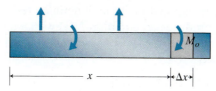
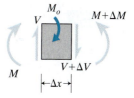

Figure 4.7 Member experiencing only point forces and applied moments.

Figure 4.8 Extremely short piece of the member that contains one applied moment.

Applying moment equilibrium to the piece, we obtain the expression

$$\curvearrowright \Sigma M_{\text{left edge}} = 0 = M + M_o + (V + \Delta V)\Delta x - (M + \Delta M)$$
$$\Delta M = M_o + V\Delta x + \Delta V \Delta x$$

But $\Delta V = 0$, so

$$\Delta M = M_o + V\Delta x$$

This is the same expression we found earlier but with one additional term, M_o. As Δx gets very small,

$$\Delta M = M_o$$

Therefore, there is a jump in the moment value where a moment is applied. Note that a moment applied clockwise causes a positive jump in the moment diagram, and a counterclockwise moment causes a negative jump. This is a result of the sign convention we adopted.

Distributed Loads Parallel to a Member

Loads parallel to a member cause an axial force. The equation for axial force in a member is found in a similar manner as for shear. Consider a member with a distributed axial load that is a function of position (Figure 4.9). Let's define x to increase from bottom to top (again, consistency in this is essential).

We can cut out an extremely short piece of length Δx. Over such a length, $w(x)$ is essentially constant. Drawing only axial forces at the cut ends, we see that the axial force at one end is not the same as at the other end (Figure 4.10).

When we apply vertical equilibrium to the piece, we obtain

$$+\uparrow \Sigma F = 0 = -N - w(x)\Delta x + (N + \Delta N)$$
$$\Delta N = w(x)\Delta x$$

In the limit, as Δx gets infinitely small, we obtain an expression with derivatives:

$$dN = w(x)\,dx$$

This expression leads to two important pieces of information:

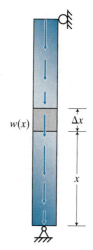

Figure 4.9 Member with a distributed load that is parallel to the axis of the member and is a function of position.

$\dfrac{dN}{dx} = w(x)$ The slope of the axial force diagram is the value of the applied load at that point.

$N(x) = \int w(x)\,dx$ The equation for the axial force is the integral of the distributed load.

Note that the sign convention assumes the distributed load acts down. If the distributed load acts up (e.g., skin friction on a pile), the sign on $w(x)$ is reversed.

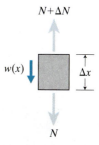

Figure 4.10 Extremely short piece of the member showing only axial forces at the cut ends.

176   Chapter 4   Internal Force Diagrams

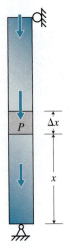

Figure 4.11 Member that experiences only point forces parallel to the axis of the member.

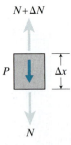

Figure 4.12 Extremely short piece of the member that contains only one point force and shows only axial forces at the cut ends.

Again, the last expression is an indefinite integral (no limits of integration), so it produces an equation $N(x)$ with a constant of integration. To find the value of the constant, we must use known axial force boundary conditions.

Point Loads Parallel to a Member
To derive the effect of a point load parallel to the axis of the member, we consider a member with only applied point loads (Figure 4.11).

Let's cut out a very short piece that contains only one of the point loads (Figure 4.12).

Applying vertical equilibrium to the piece, we obtain the expression

$$+\uparrow \Sigma F = 0 = -N - P + (N + \Delta N)$$
$$\Delta N = P$$

There is a jump in the axial force value that is opposite the value of the applied force. Therefore, a force that pulls down on the member causes a negative jump, and a force that pushes up on the member causes a positive jump in the axial force diagram.

Equation Intervals
It is important to note that the equations describing the shear, moment, and axial force might be discontinuous (different equations are required over different regions of the member). Wherever the equation describing the applied load changes, wherever there is an applied force, or wherever there is an applied moment, there will be a new equation for the shear, moment, and/or axial force. We deal with this by using a new x variable at each of these locations (Figure 4.13).

Remember that each time we integrate an equation, we obtain a constant of integration that must be found with a known boundary condition. If we were to find the five equations for the moments that define the moment diagram for the beam in Figure 4.13, we would have 10 constants, requiring 10 boundary conditions. Some of these conditions are known at the ends of the member once the reactions are known. However, this does not provide enough boundary conditions. To obtain the rest of the needed boundary conditions, we can cut the member at the boundary of an equation ($x_i = 0$), calculate the internal forces using equilibrium or the previous equation (x_{i-1} = end of range) adding any jump, and solve for the constant.

Finding Peak Values
From calculus we know that a relative maximum or minimum occurs where the slope is zero, but the peak might occur at the boundary of an interval. Therefore, we need to check both. If we plot the expressions, we can typically reduce the number of potential locations we need to check.

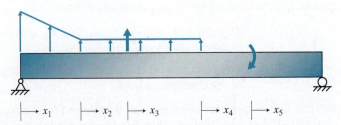

Figure 4.13 Intervals of new x variables for a member that has discontinuous shear and moment equations.

When checking for relative maxima or minima, we can use information from our derivations earlier in this section. The slope of the shear expression is the magnitude of the distributed load; therefore, a relative maximum or minimum for the shear occurs where the distributed load, w, is zero. The slope of the moment expression is the magnitude of the shear; therefore, the moment reaches a relative maximum or minimum where $V = 0$. And the slope of the axial force expression is the magnitude of the distributed load parallel to the member; therefore, the axial force has a relative maximum or minimum where $w = 0$. Remember that if the location of zero slope based on the expressions lies outside the interval, there is no relative maximum or minimum in that interval.

In most cases, the peak shear and axial force occur at a support or where two members connect. The peak moment, however, often occurs inside an interval.

Section 4.1 Highlights

Internal Forces by Integration

Sign Convention:

Defining Intervals:

- Change in $w(x)$ equation (including changing to $w(x) = 0$)
- Applied point load, P (including support reactions)
- Applied moment, M (including support reactions)

Shear:

$$\text{Slope} = w(x) \qquad V(x) = \int w(x)\,dx \qquad \Delta V = P$$

Moment:

$$\text{Slope} = V(x) \qquad M(x) = \int V(x)\,dx \qquad \Delta M = M_o$$

Axial Force:

$$\text{Slope} = w(x) \qquad N(x) = \int w(x)\,dx \qquad \Delta N = P$$

Boundary Conditions (for Constants of Integration):
- Find at the ends using equations of equilibrium.
- Find at the start of an interval using the previous equation and adding any jump.

Finding Peak Values:

- Check ends of intervals.
- Check where the slope is zero:
  – If outside the interval, there is none in that interval.
  – For $V_{\max}$, where $w = 0$
  – For $M_{\max}$, where $V = 0$
  – For $N_{\max}$, where $w = 0$

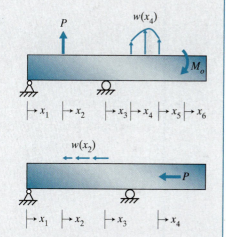

EXAMPLE 4.1

A beam will be supporting equipment that has an eccentric center of gravity. The result is unequal point forces on the supporting beam. The process engineers specify the placement of the equipment; they want it placed off center in this case.

We will be designing the beam to be made of steel. We have not yet decided where along the length we will brace the beam, so we do not yet know where we will need to calculate the moment magnitude in order to obtain the moment gradient modification factor. Therefore, we want to find the equations for the moment along the entire length of the beam.

To build our experience base for evaluating internal force diagrams, let's also construct the diagrams for this beam.

Figure 4.14

Estimated Solution:

We can approximate the loading conditions as a symmetric loading on the simply supported beam so that we can quickly find an approximation of the maximum moment, which will be at midspan due to the symmetry.

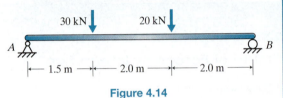

Figure 4.15

Find the left reaction:

$$\curvearrowleft^+ \Sigma M_B = 0 = A_y(5.5 \text{ m}) - 25 \text{ kN}(2.0 \text{ m}) - 25 \text{ kN}(3.5 \text{ m})$$
$$A_y = +25 \text{ kN} (+\uparrow)$$

Find the approximate M_{max}:

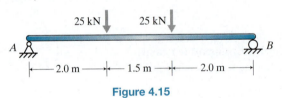

Figure 4.16

$$\curvearrowleft^+ \Sigma M_{cut} = 0 = 25 \text{ kN}(2.75 \text{ m}) - 25 \text{ kN}(0.75 \text{ m}) - M_{max}$$
$$M_{max} \approx +50 \text{ kN} \cdot \text{m} \, (\smile + \curvearrowleft)$$

Implication for evaluating the results:

Since we moved the total load closer to midspan (farther from the supports) for this approximation, we have overestimated the maximum moment. Therefore, the peak moment we obtain in our detailed analysis should be somewhat smaller.

Detailed Analysis:

Find the reactions:

$\circlearrowright^+ \Sigma M_A = 0 = 30 \text{ kN}(1.5 \text{ m}) + 20 \text{ kN}(3.5 \text{ m}) - B_y(5.5 \text{ m})$
$\quad B_y = +20.9 \text{ kN}(+\uparrow)$

$+\uparrow \Sigma F_y = 0 = A_y - 30 \text{ kN} - 20 \text{ kN} + 20.9 \text{ kN}$
$\quad A_y = +29.1 \text{ kN}(+\uparrow)$

Define the intervals:

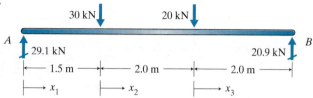

Figure 4.17

Integrate:

$w(x_1) = 0, \quad w(x_2) = 0, \quad w(x_3) = 0$

$V(x_1) = \int w(x_1)dx_1 = \int 0 dx_1 = C_1$
$M(x_1) = \int V(x_1)dx_1 = \int C_1 dx_1 = C_1 x_1 + C_2$

$V(x_2) = \int w(x_2)dx_2 = \int 0 dx_2 = C_3$
$M(x_2) = \int V(x_2)dx_2 = \int C_3 dx_2 = C_3 x_2 + C_4$

$V(x_3) = \int w(x_3)dx_3 = \int 0 dx_3 = C_5$
$M(x_3) = \int V(x_3)dx_3 = \int C_5 dx_3 = C_5 x_3 + C_6$

Boundary conditions:

At $x_1 = 0$, $V(x_1) = 29.1 \text{ kN}$
Substitute into the expression for $V(x_1)$:
$29.1 \text{ kN} = C_1$ $\Rightarrow C_1 = 29.1 \text{ kN}$

At $x_1 = 0$, $M(x_1) = 0 \text{ kN} \cdot \text{m}$
Substitute into the expression for $M(x_1)$:
$0 \text{ kN} \cdot \text{m} = 29.1 \text{ kN}(0 \text{ m}) + C_2$ $\Rightarrow C_2 = 0 \text{ kN} \cdot \text{m}$

At $x_2 = 0$, $x_1 = 1.5 \text{ m}$, $V(x_2) = V(x_1) - 30 \text{ kN}$
Substitute into the expressions for $V(x_1)$ and $V(x_2)$:
$C_3 = 29.1 \text{ kN} - 30 \text{ kN}$ $\Rightarrow C_3 = -0.9 \text{ kN}$

At $x_2 = 0$, $x_1 = 1.5 \text{ m}$, $M(x_2) = M(x_1)$
Substitute into the expressions for $M(x_1)$ and $M(x_2)$:
$-0.9 \text{ kN}(0 \text{ m}) + C_4 = 29.1 \text{ kN}(1.5 \text{ m})$ $\Rightarrow C_4 = 43.6 \text{ kN} \cdot \text{m}$

At $x_3 = 0$, $x_2 = 2.0 \text{ m}$, $V(x_3) = V(x_2) - 20 \text{ kN}$
Substitute into the expressions for $V(x_2)$ and $V(x_3)$:
$C_5 = -0.9 \text{ kN} - 20 \text{ kN}$ $\Rightarrow C_5 = -20.9 \text{ kN}$

At $x_3 = 0$, $x_2 = 2.0 \text{ m}$, $M(x_3) = M(x_2)$
Substitute into the expressions for $M(x_2)$ and $M(x_3)$:
$-20.9 \text{ kN} \cdot \text{m}(0 \text{ m}) + C_6 = -0.9 \text{ kN}(2 \text{ m}) + 43.6 \text{ kN} \cdot \text{m}$ $\Rightarrow C_6 = 41.8 \text{ kN} \cdot \text{m}$

Summary of expressions:

$V(x_1) = 29.1 \text{ kN}$
$M(x_1) = 29.1 \text{ kN} \cdot x_1$

$V(x_2) = -0.9 \text{ kN}$
$M(x_2) = -0.9 \text{ kN} \cdot x_2 + 43.6 \text{ kN} \cdot \text{m}$
$V(x_3) = -20.9 \text{ kN}$
$M(x_3) = -20.9 \text{ kN} \cdot x_3 + 41.8 \text{ kN} \cdot \text{m}$

Diagrams:

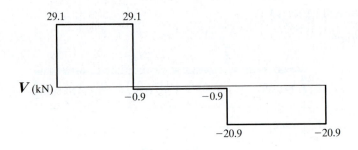

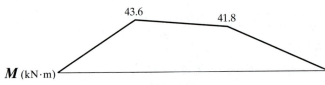

Figure 4.18

Evaluation of Results:

Observed Expected Features?
Since there is no distributed load in any of the intervals, the shear expressions should be constant. They are. ✓

Observed Expected Features?
Since the shear expressions are constant, the moment expressions in all of the intervals should be linear. They are. ✓

Satisfied Fundamental Principles?
At $x_3 = 2$ m, $V(x_3) = -20.9$ kN because of the reaction.
  Substitute into the expression for $V(x_3)$ and check the result:
$V(x_3) = -20.9$ kN ✓

Satisfied Fundamental Principles?
At $x_3 = 2$ m, $M(x_3) = 0$ kN $\cdot$ m.
  Substitute into the expression for $M(x_3)$ and check the result:
$M(x_3) = -20.9 \text{ kN}(2 \text{ m}) + 41.8 \text{ kN} \cdot \text{m} = 0$ ✓

Approximation Predicted Outcomes?
The maximum moment occurs at $x_2 = 0$ and is 43.6 kN $\cdot$ m.
  Our approximate maximum moment was about 15% different; that is a reasonable difference for an approximation.
  As expected, the value of the detailed analysis result is smaller than the approximation. ✓

Conclusion:
Together these observations strongly suggest that we have reasonable detailed analysis results.

EXAMPLE 4.2

A short cantilever beam in a manufacturing plant is to be extended as part of the preparation for a new product line. Because of access limitations, the best solution will be to attach the extension with a hinged connection.

There will be tight tolerances on the displacement at several potential interference points. Therefore, we want to find the equations for the shear and moment along the beam so that later we can find the equations for the displacement along the entire length of the beam.

To build our experience base for evaluating internal force diagrams, let's also construct the diagrams for this beam.

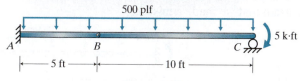

Figure 4.19

Estimated Solution:

Upper bound for positive moment:

We can approximate the geometry as a solid member and change the reaction at A to a pin. This is equivalent to moving the point of zero moment to A instead of B.

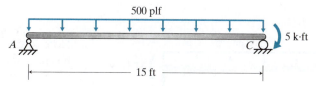

Figure 4.20

If we use superposition, we see that the distributed load causes positive moment (i.e., displaced shape will smile) and the couple moment at C causes negative moment (i.e., displaced shape will frown).

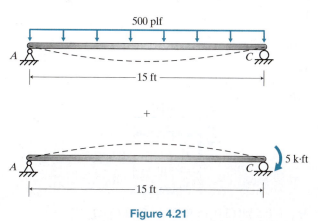

Figure 4.21

To estimate the peak moment, we can consider the contribution of the distributed load only. In Section 2.4: Approximating Loading Conditions, we derived the peak moment in a simply supported beam subjected to a uniform distributed load:

$$M_{max}^+ \approx 0.5 \text{ klf}(15 \text{ ft})^2/8$$
$$M_{max}^+ = +14.06 \text{ k} \cdot \text{ft} \ (\circlearrowright+\circlearrowleft)$$

This is an upper bound on the positive moment. Moving the point of zero moment from B to A increases the peak positive moment. Ignoring the couple moment in this case also increases the peak positive moment.

Upper bound for negative moment:

To find an upper bound limit on the negative moment, we can move the point of zero moment to end C. That makes the approximated beam a cantilever. Such a dramatic change means that we might significantly overpredict the moment, but if our detailed solution is bigger, we can be certain that we made a mistake.

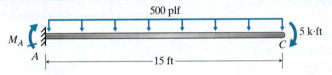

Figure 4.22a

$$\circlearrowleft+\Sigma M_A = 0 = M_A + 0.5 \text{ klf}(15 \text{ ft})(15 \text{ ft}/2) + 5 \text{ k} \cdot \text{ft}$$
$$M_A \approx M_{max}^- = -61.25 \text{ k} \cdot \text{ft} \ (\circlearrowright+\circlearrowleft)$$

Detailed Analysis:

Find the reactions:

FBD of member BC:

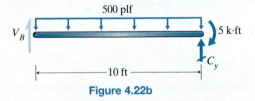

Figure 4.22b

Note: since the axial force has no effect on the shear or moment diagram, we have not included it in these FBDs.

$$\circlearrowleft+\Sigma M_C = 0 = V_B(10 \text{ ft}) - 0.5 \text{ klf}(10 \text{ ft})(10 \text{ ft}/2) + 5 \text{ k} \cdot \text{ft}$$
$$V_B = +2.0 \text{ k}(+\uparrow)$$

$$+\uparrow \Sigma F_y = 0 = 2.0 \text{ k} - 0.5 \text{ klf}(10 \text{ ft}) + C_y$$
$$C_y = +3.0 \text{ k}(+\uparrow)$$

FBD of member AB:

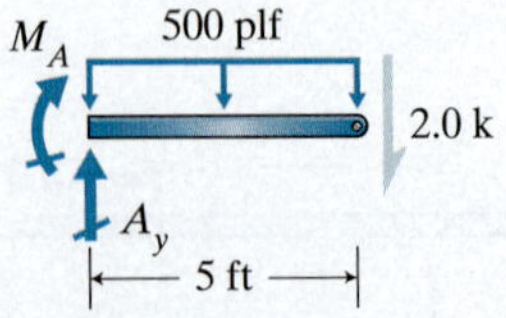

Figure 4.22c

$$\circlearrowleft+\Sigma M_A = 0 = M_A + 0.5 \text{ klf}(5 \text{ ft})(5 \text{ ft}/2) + 2.0 \text{ k}(5 \text{ ft})$$

$$M_A = -16.25 \text{ k} \cdot \text{ft} \ (\circlearrowleft+)$$

$+\uparrow \Sigma F_y = 0 = A_y - 0.5\text{ klf}(5\text{ ft}) - 2.0\text{ k}$
$A_y = +4.5\text{ k}(+\uparrow)$

Define the intervals:

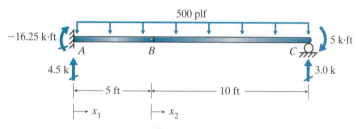

Figure 4.23

Note that according to our rules for starting a new interval, we do not need a second interval. If we intended to stop with only the internal force equations, one interval would be sufficient in this case. However, we will see in Section 5.1: Double Integration Method that we need a new interval at the hinge because the slope at the end of x_1 is not the same as the slope at the start of x_2.

Integrate:
$$w(x_1) = -0.5\text{ klf}, \qquad w(x_2) = -0.5\text{ klf}$$
$$V(x_1) = \int w(x_1)\,dx_1 = \int -0.5\text{ klf}\cdot dx_1 = -0.5\text{ klf}\cdot x_1 + C_1$$
$$M(x_1) = \int V(x_1)\,dx_1 = \int(-0.5\text{ klf}\cdot x_1 + C_1)\,dx_1 = -0.25\text{ klf}\cdot x_1^2 + C_1 x_1 + C_2$$
$$V(x_2) = \int w(x_2)\,dx_2 = \int -0.5\text{ klf}\cdot dx_2 = -0.5\text{ klf}\cdot x_2 + C_3$$
$$M(x_2) = \int V(x_2)\,dx_2 = \int(-0.5\text{ klf}\cdot x_2 + C_3)\,dx_2 = -0.25\text{ klf}\cdot x_2^2 + C_3 x_2 + C_4$$

Boundary conditions:
 At $x_1 = 0$, $V(x_1) = 4.5\text{ k}$
 Substitute into the expression for $V(x_1)$:
 $4.5\text{ k} = -0.5\text{ klf}(0\text{ ft}) + C_1$ $\Rightarrow C_1 = 4.5\text{ k}$

 At $x_1 = 0$, $M(x_1) = -16.25\text{ k}\cdot\text{ft}$
 Substitute into the expression for $M(x_1)$:
 $-16.25\text{ k}\cdot\text{ft} = -0.25\text{ klf}(0\text{ ft})^2 + 4.5\text{ k}(0\text{ ft}) + C_2$ $\Rightarrow C_2 = -16.25\text{ k}\cdot\text{ft}$

 At $x_2 = 0$, $x_1 = 5\text{ ft}$, $V(x_2) = V(x_1)$
 Substitute into the expressions for $V(x_1)$ and $V(x_2)$:
 $-0.5\text{ klf}(0\text{ ft}) + C_3 = -0.5\text{ klf}(5\text{ ft}) + 4.5\text{ k}$ $\Rightarrow C_3 = 2.0\text{ k}$

 At $x_2 = 0$, $x_1 = 5\text{ ft}$, $M(x_2) = M(x_1)$
 Substitute into the expressions for $M(x_1)$ and $M(x_2)$:
 $-0.25\text{ klf}(0\text{ ft})^2 + 2.0\text{ k}(0\text{ ft}) + C_4 = -0.25\text{ klf}(5\text{ ft})^2 + 4.5\text{ k}(5\text{ ft}) - 16.25\text{ k}\cdot\text{ft} \Rightarrow C_4 = 0\text{ k}\cdot\text{ft}$

Summary of expressions:
$$V(x_1) = -0.5\text{ klf}\cdot x_1 + 4.5\text{ k}$$
$$M(x_1) = -0.25\text{ klf}\cdot x_1^2 + 4.5\text{ k}\cdot x_1 - 16.25\text{ k}\cdot\text{ft}$$
$$V(x_2) = -0.5\text{ klf}\cdot x_2 + 2.0\text{ k}$$
$$M(x_2) = -0.25\text{ klf}\cdot x_2^2 + 2.0\text{ k}\cdot x_2$$

Peak values of the moment:
   The peak moment occurs where $V = 0$ or at one of the ends of the interval.

In the first interval,
$$-0.5 \text{ klf} \cdot x_1 + 4.5 \text{ k} = 0$$
$$x_1 = 9 \text{ ft}$$

Since this is outside the first interval, the peak must be at one of the ends.
   At $x_1 = 0$,
$$M(x_1) = -0.25 \text{ klf}(0)^2 + 4.5 \text{ k}(0) - 16.25 \text{ k} \cdot \text{ft}$$
$$M(x_1) = -16.25 \text{ k} \cdot \text{ft}$$

At $x_1 = 5$ ft,
$$M(x_1) = -0.25 \text{ klf}(5 \text{ ft})^2 + 4.5 \text{ k}(5 \text{ ft}) - 16.25 \text{ k} \cdot \text{ft}$$
$$M(x_1) = 0$$

In the second interval,
$$-0.5 \text{ klf} \cdot x_2 + 2.0 \text{ k} = 0$$
$$x_2 = 4 \text{ ft}$$
$$M(x_2) = -0.25 \text{ klf}(4 \text{ ft})^2 + 2.0 \text{ k}(4 \text{ ft})$$
$$M(x_2) = +4.0 \text{ k} \cdot \text{ft}$$

We also check the ends of the interval.
   At $x_2 = 0$,
$$M(x_2) = -0.25 \text{ klf}(0)^2 + 2.0 \text{ k}(0)$$
$$M(x_2) = 0$$

At $x_2 = 10$ ft,
$$M(x_2) = -0.25 \text{ klf}(10 \text{ ft})^2 + 2.0 \text{ k}(10 \text{ ft})$$
$$M(x_2) = -5.0 \text{ k} \cdot \text{ft}$$

Peak moments:
$$M_{\text{max}}^- = -16.25 \text{ k} \cdot \text{ft}$$
$$M_{\text{max}}^+ = +4.00 \text{ k} \cdot \text{ft}$$

Diagrams:

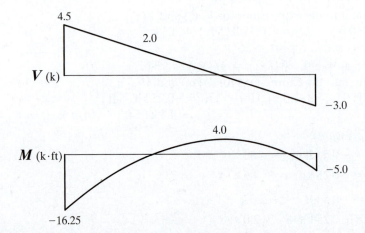

Figure 4.24

Evaluation of Results:

Observed Expected Features?
There is a moment reaction at A; therefore, the moment expression in the first interval should have a constant term. It does. ✓

Observed Expected Features?
There is uniform distributed load along both intervals; therefore, the shear expressions should be linear. They are. ✓

Observed Expected Features?
Since the shear expressions are linear, the moment expressions in both intervals should be quadratic. They are. ✓

Satisfied Fundamental Principles?
At $x_1 = 5$ ft, $M(x_1) = 0$ k·ft because of the hinge.
   We saw above that substituting into the expression for $M(x_1)$ results in zero moment. ✓

Satisfied Fundamental Principles?
At $x_2 = 10$ ft, $V(x_2) = -3$ k because of the reaction.
   We can substitute into the expression for $V(x_2)$ and check the result:
   $V(x_2) = -0.5 \text{ klf}(10 \text{ ft}) + 2.0 \text{ k} = -3 \text{ k}$ ✓

Satisfied Fundamental Principles?
At $x_2 = 10$ ft, $M(x_2) = -5$ k·ft because of the applied moment.
   We can substitute into the expression for $M(x_2)$ and check the result:
   $M(x_2) = -0.25 \text{ klf}(10 \text{ ft})^2 + 2.0 \text{ k}(10 \text{ ft}) = -5 \text{ k·ft}$ ✓

Approximation Predicted Outcomes?
The peak negative moment is -16.25 k·ft.
   As expected, this detailed analysis result is smaller than the upper bound limit of -61.25 k·ft we calculated. ✓

Approximation Predicted Outcomes?
The peak positive moment is $+4.0$ k·ft.
   As expected, this detailed analysis result is smaller than the upper bound limit of $+14.06$ k·ft we calculated. ✓

Conclusion:
Together these observations strongly suggest that we have reasonable detailed analysis results.

Example 4.3

In order to have the desired flow of customers in a retail space, the architect wants us to use a transfer girder to carry load from a second-story column out to other columns on the first floor. The girder will also carry the floor load.

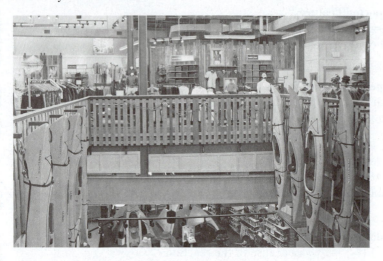

Figure 4.25

We want to know the peak moment for the transfer girder, but its location is not obvious. Therefore, let's find the equations for the shear and moment along the entire length of the beam.

To build our experience base for evaluating internal force diagrams, let's also construct the diagrams for this beam.

Estimated Solution:

We can approximate the loading conditions as one equivalent uniform distributed load along the entire beam.

Equivalent uniform load:

$$P_{total} = 2.0 \text{ klf}(30 \text{ ft}) + 12 \text{ k} = 72 \text{ k}$$
$$w_{equiv} = 72 \text{ k}/30 \text{ ft} = 2.4 \text{ klf}$$

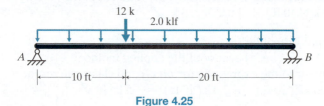

Figure 4.26

Section 4.1 Internal Forces by Integration

Peak moment:
For a uniformly loaded, simply supported beam, the peak moment occurs at midspan and can be calculated by the following formula:

$$M_{max} = \frac{wl^2}{8} = \frac{2.4\text{ klf}(30\text{ ft})^2}{8} = 270\text{ k}\cdot\text{ft}$$

We cannot tell whether this is an upper or lower bound for the actual peak moment because distributing the point load moves some load toward the middle of the beam (increases peak moment) and moves some load farther away from the middle (decreases peak moment). But this is still a reasonable estimate.

Detailed Analysis:

Find the reactions:

$$\curvearrowright \Sigma M_A = 0 = 12\text{ k}(10\text{ ft}) + 2.0\text{ klf}(30\text{ ft})(30\text{ ft}/2) - B_y(30\text{ ft})$$
$$B_y = +34\text{ k}(+\uparrow)$$
$$+\uparrow \Sigma F_y = 0 = A_y - 12\text{ k} - 2.0\text{ klf}(30\text{ ft}) + 34\text{ k}$$
$$A_y = +38\text{ k}(+\uparrow)$$

Define the intervals:

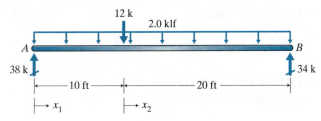

Figure 4.27

Integrate:

$$w(x_1) = -2.0\text{ klf}, \qquad w(x_2) = -2.0\text{ klf}$$

$$V(x_1) = \int w(x_1)\,dx_1 = \int -2.0\text{ klf}\cdot dx_1 = -2.0\text{ klf}\cdot x_1 + C_1$$
$$M(x_1) = \int V(x_1)\,dx_1 = \int (-2.0\text{ klf}\cdot x_1 + C_1)\,dx_1 = -1.0\text{ klf}\cdot x_1^2 + C_1 x_1 + C_2$$

$$V(x_2) = \int w(x_2)\,dx_2 = \int -2.0\text{ klf}\cdot dx_2 = -2.0\text{ klf}\cdot x_2 + C_3$$
$$M(x_2) = \int V(x_2)\,dx_2 = \int (-2.0\text{ klf}\cdot x_1 + C_3)\,dx_2 = -1.0\text{ klf}\cdot x_2^2 + C_3 x_2 + C_4$$

Boundary conditions:
At $x_1 = 0, V(x_1) = 38\text{ k}$
  Substitute into the expression for $V(x_1)$:
  $38\text{ k} = -2.0\text{ klf}(0\text{ ft}) + C_1$ $\Rightarrow C_1 = 38\text{ k}$

At $x_1 = 0, M(x_1) = 0\text{ k}\cdot\text{ft}$
  Substitute into the expression for $M(x_1)$:
  $0\text{ k}\cdot\text{ft} = -1.0\text{ klf}(0\text{ ft})^2 + 38\text{ k}(0\text{ ft}) + C_2$ $\Rightarrow C_2 = 0\text{ k}\cdot\text{ft}$

At $x_2 = 0, x_1 = 10\text{ ft}, V(x_2) = V(x_1) - 12\text{ k}$
  Substitute into the expressions for $V(x_1)$ and $V(x_2)$:
  $-2.0\text{ klf}(0\text{ ft}) + C_3 = -2.0\text{ klf}(10\text{ ft}) + 38\text{ k} - 12\text{ k}$ $\Rightarrow C_3 = 6\text{ k}$

At $x_2 = 0$, $x_1 = 10$ ft, $M(x_2) = M(x_1)$
Substitute into the expressions for $M(x_1)$ and $M(x_2)$:
$-1.0 \text{ klf}(0 \text{ ft})^2 + 6 \text{ k}(0 \text{ ft}) + C_4 = -1.0 \text{ klf}(10 \text{ ft})^2 + 38 \text{ k}(10 \text{ ft}) \Rightarrow C_4 = 280 \text{ k} \cdot \text{ft}$

Summary of expressions:

$$V(x_1) = -2.0 \text{ klf} \cdot x_1 + 38 \text{ k}$$
$$M(x_1) = -1.0 \text{ klf} \cdot x_1^2 + 38 \text{ k} \cdot x_1$$

$$V(x_2) = -2.0 \text{ klf} \cdot x_2 + 6 \text{ k}$$
$$M(x_2) = -1.0 \text{ klf} \cdot x_2^2 + 6 \text{ k} \cdot x_2 + 280 \text{ k} \cdot \text{ft}$$

Find the peak moment:
The peak moment occurs where $V = 0$ or at one of the ends of the interval.

In the first interval,

$$-2.0 \text{ klf} \cdot x_1 + 38 \text{ k} \equiv 0$$
$$x_1 = 19 \text{ ft}$$

Since this is outside the first interval, the peak must be at one of the ends.
At $x_1 = 0$,

$$M(x_1) = -1.0 \text{ klf}(0)^2 + 38 \text{ k}(0)$$
$$M(x_1) = 0$$

At $x_1 = 10$ ft,

$$M(x_1) = -1.0 \text{ klf}(10 \text{ ft})^2 + 38 \text{ k}(10 \text{ ft})$$
$$M(x_1) = +280 \text{ k} \cdot \text{ft}$$

In the second interval,

$$-2.0 \text{ klf} \cdot x_2 + 6 \text{ k} \equiv 0$$
$$x_2 = 3 \text{ ft}$$
$$M(x_2) = -1.0 \text{ klf}(3 \text{ ft})^2 + 6 \text{ k}(3 \text{ ft}) + 280 \text{ k} \cdot \text{ft}$$
$$M(x_2) = +289 \text{ k} \cdot \text{ft}$$

Also we check the ends of the interval.
At $x_2 = 0$,

$$M(x_2) = -1.0 \text{ klf}(0)^2 + 6 \text{ k}(0) + 280 \text{ k} \cdot \text{ft}$$
$$M(x_2) = 280 \text{ k} \cdot \text{ft}$$

At $x_2 = 20$ ft,

$$M(x_2) = -1.0 \text{ klf}(20 \text{ ft})^2 + 6 \text{ k}(20 \text{ ft}) + 280 \text{ k} \cdot \text{ft}$$
$$M(x_2) = 0$$

Peak moment:

$$M_{\max}^+ = +289 \text{ k} \cdot \text{ft}$$

There is no negative moment.

Diagrams:

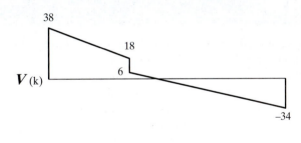
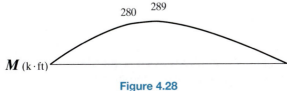

Figure 4.28

Evaluation of Results:

Observed Expected Features?
There is a uniform distributed load along both intervals; therefore, the shear expressions should be linear. They are. ✓

Observed Expected Features?
Since the shear expressions are linear, the moment expressions in both intervals should be quadratic. They are. ✓

Satisfied Fundamental Principles?
At $x_2 = 20$ ft, $V(x_2) = -34$ k because of the reaction.
  Substitute into the expression for $V(x_2)$ and check the result:
  $V(x_2) = -2.0$ klf $(20$ ft$) + 6$ k $= -34$ k ✓

At $x_2 = 20$ ft, $M(x_2) = 0$ k·ft because there is no applied moment.
  We saw above that substituting into the expression for $M(x_2)$ results in zero moment. ✓

Approximation Predicted Outcomes?
The maximum moment occurs at $x_2 = 3.0$ ft and is 289 k·ft.
  Our approximate maximum moment was about 7% lower; that is a reasonable difference for an approximation. ✓

Conclusion:
Together these observations strongly suggest that we have reasonable detailed analysis results.

Example 4.4

An artist has developed a sculpture that resembles a tree for the front lawn of an art museum. The artist would like our help designing the main support member in the middle of the sculpture. She would like the member to be tapered like a tree trunk.

The idealized self-weight is a linearly decreasing distributed axial load as shown. The buckling capacity depends on the distribution of the axial force in the column. Therefore, we need to find the equation for the axial force along the length of the column.

To build our experience base for evaluating internal force diagrams, let's also construct the diagram.

Estimated Solution:

We can approximate the loading conditions as one equivalent uniform distributed load along the entire column. Then we use the value from mid-height.
Approximate the reaction:

$$w_{\text{midht}} = w(2.5 \text{ m}) = -0.5 \text{ kN/m}^2(2.5 \text{ m}) + 2.5 \text{ kN/m} = 1.25 \text{ kN/m}$$
$$+\uparrow \Sigma F_x = 0 = A_x - 1.25 \text{ kN/m}(5 \text{ m})$$
$$A_x = +6.25 \text{ kN}(+\uparrow)$$

Figure 4.29

This approximation of the loading should produce the same vertical reaction as the detailed analysis because we used the average magnitude of the applied distributed load (the middle of a linearly varying distribution).

Detailed Analysis:

Find the reactions:

$$+\uparrow \Sigma F_x = 0 = A_x + \int_{0\text{m}}^{5\text{m}} w(x) \cdot dx$$

$$A_x = -\int_{0\text{m}}^{5\text{m}} (-0.5 \text{ kN/m}^2 \cdot x + 2.5 \text{ kN}) \, dx$$

$$= -(-0.25 \text{ kN/m}^2 \cdot x^2 + 2.5 \text{ kN} \cdot x)\big|_{0\text{ m}}^{5\text{ m}}$$

Section 4.1 Internal Forces by Integration **191**

$$= -[-0.25 \text{ kN/m}^2(5\text{ m})^2 + 2.5 \text{ kN}(5\text{ m})] + [-0.25 \text{ kN/m}^2(0\text{ m})^2 + 2.5 \text{ kN}(0\text{ m})]$$
$$A_x = +6.25 \text{ kN}(+\uparrow)$$

Define the intervals:
Only one interval is needed for this member, x.

Integrate:
$$w(x) = -0.5 \text{ kN/m}^2 \cdot x + 2.5 \text{ kN/m}$$
$$N(x) = \int w(x)\,dx = \int(-0.5 \text{ kN/m}^2 \cdot x + 2.5 \text{ kN/m})\,dx$$
$$= -0.25 \text{ kN/m}^2 \cdot x^2 + 2.5 \text{ kN/m} \cdot x + C$$

Boundary condition:
At $x = 0, N(x) = -6.25$ kN
Note that the sign is reversed from the reaction because the reaction direction is the opposite of the positive internal axial force.

Substitute into the expression for $N(x)$:
$$-6.25 \text{ kN} = -0.25 \text{ kN/m}^2(0\text{ m})^2 + 2.5 \text{ kN/m}(0\text{ m}) + C \implies C = -6.25 \text{ kN}$$

Expression:
$$N(x) = -0.25 \text{ kN/m}^2 \cdot x^2 + 2.5 \text{ kN/m} \cdot x - 6.25 \text{ kN}$$

Diagram:

N (kN) -6.25
Figure 4.30

Evaluation of Results:

Satisfied Fundamental Principles?
At $x = 5$ m, $N(x) = 0$ kN
Substitute into the expression for $N(x)$ and check the result:
$N(x) = -0.25 \text{ kN/m}^2(5\text{ m})^2 + 2.5 \text{ kN/m}(5\text{ m}) - 6.25 \text{ kN} = 0$ ✓

Approximation Predicted Outcomes?
The vertical reaction at A is 6.25 kN as predicted. ✓

Conclusion:
These observations suggest that we have reasonable detailed analysis results.

4.2 Constructing Diagrams by Deduction

Introduction

Knowing that internal force diagrams are related to applied loads by integration allows us to identify rules that the diagrams will follow. These rules can guide us in constructing most internal force diagrams without the need for equations.

Approaching internal force diagrams this way reinforces the fundamental principles needed to evaluate them and the features the diagrams should have. Therefore, this section emphasizes the skills practitioners use to assess the reasonableness of internal force diagrams.

How-To

Shear and Moment Diagrams

In the previous section, we saw that using an indefinite integral results in the equation for the internal force. If we apply limits to the integral, we have a definite integral that produces a value. That value is the change in the internal force value.

For example, consider a member with a distributed load perpendicular to its axis, $w(x)$. The change in the shear value from point A to point B (Figure 4.31) is given by the following definite integral:

$$\int_{x_A}^{x_B} w(x)\,dx = V_B - V_A = \Delta V_{A-B}$$

where
$$\begin{aligned} x_A &= \text{Location of point } A \\ V_A &= \text{Shear value at point } A \\ x_B &= \text{Location of point } B \\ V_B &= \text{Shear value at point } B \\ \Delta V_{A-B} &= \text{Change in shear value from point } A \text{ to point } B \end{aligned}$$

Similarly, the change in the moment value from point A to point B (Figure 4.32) is the area under the shear diagram:

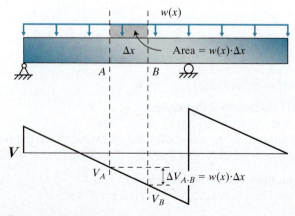

Figure 4.31 Beam experiencing distributed load and the resulting shear diagram. The change in the shear value from point A to point B is the area under the distributed load.

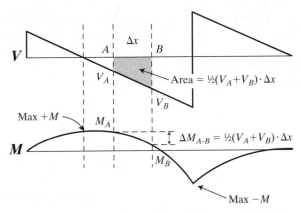

Figure 4.32 Shear diagram for a beam and the resulting moment diagram. The change in the moment value from point A to point B is the area under the shear diagram. Note that the locations where the shear is zero correspond to the locations of local maximums (+/−) in the moment diagram.

$$\int_{x_A}^{x_B} V(x)\,dx = M_B - M_A = \Delta M_{A-B}$$

where

M_A = Moment value at point A
M_B = Moment value at point B
ΔM_{A-B} = Change in moment value from point A to point B

The indefinite integrals discussed in the previous section lead to several observations that will help us construct shear and moment diagrams without developing the equations.

The equation for the shear is given by the expression

$$V(x) = \int w(x)\,dx$$

If there is no distributed load, $w(x) = 0$, the indefinite integral results in just the constant of integration. Therefore, if there is no distributed load, that portion of the V diagram is constant; it is a horizontal line.

If there is a uniform distributed load, $w(x)$ = constant, the indefinite integral results in a linear equation for that portion of the V diagram:

$$V(x) = w \cdot x + \text{constant}$$

From this equation we see that the slope of that line is w. Based on our sign convention, this means that if the distributed load acts down, the shear is decreasing (becoming less positive); if the distributed load acts up, the shear is increasing (becoming more positive).

The equation for the moment is given by

$$M(x) = \int V(x)\,dx$$

If there is no distributed load, that portion of the shear diagram is a horizontal line, V = constant, and the moment is linear in that portion.

If there is a uniform distributed load, that portion of the shear diagram is linear and the moment diagram is quadratic in that portion:

$$M(x) = \frac{w}{2} \cdot x^2 + \text{constant}_1 \cdot x + \text{constant}_2$$

The sign of the quadratic term, x^2, tells us the concavity of the moment diagram. Based on our sign convention, if the distributed load acts down, the moment diagram is concave down; if the distributed load acts up, the moment diagram is concave up.

In the previous section, we proved that an applied point load results in a jump in the shear diagram, and an applied moment results in a jump in the moment diagram. With these last two rules, we have all the information we need to construct shear and moment diagrams without deriving their equations.

If we include any jumps due to applied loads, applied moments, reaction forces, or reaction moments, the shear and moment diagrams must close to zero. Therefore, we have a way to check the diagrams we construct.

Axial Force Diagrams

Axial force diagrams possess properties similar to those of shear diagrams. If a member experiences a distributed axial load $w(x)$, the change in the axial force value from point A to point B (Figure 4.33) is given by the following definite integral:

$$\int_{x_A}^{x_B} w(x)\,dx = N_B - N_A = \Delta N_{A-B}$$

where

$\quad x_A =$ Location of point A
$\quad N_A =$ Axial force value at point A
$\quad x_B =$ Location of point B
$\quad N_B =$ Axial force value at point B
$\quad \Delta N_{A-B} =$ Change in axial force value from point A to point B

Therefore, we can quantify the change in the axial force by calculating the area under the distributed axial load.

If there is no distributed axial load, $w(x) = 0$, the indefinite integral results in just the constant of integration. Therefore, if there is

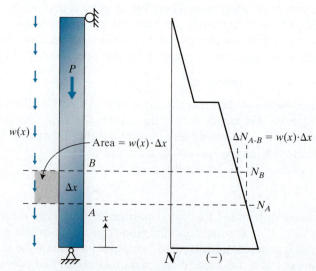

Figure 4.33 Axial force diagram for a column that experiences a distributed load and a point load. The change in the axial force value from point A to point B is the area under the distributed load.

no distributed load, that portion of the N diagram is constant; it is a vertical line.

If there is a uniform distributed load, $w(x) = $ constant, the indefinite integral results in a linear equation for that portion of the N diagram:

$$N(x) = w \cdot x + \text{constant}$$

From this equation we see that the slope of that line is $+w$. Based on our sign convention, this means that if the distributed axial load acts down, the axial force is increasing; if the distributed axial load acts up, the axial force is decreasing.

If there is an applied point load parallel to the axis of the member, the axial force diagram jumps by the value of the point load. Based on our sign convention, if the point load acts down, the axial force increases; if the point load acts up, the axial force is decreased.

If we include any jumps due to applied point loads or reactions, the axial force diagram must also close to zero. Therefore, we have a way to check the axial force diagrams we construct as well.

SECTION 4.2 HIGHLIGHTS

Constructing Diagrams by Deduction

Sign Convention (unchanged):

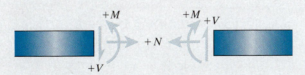

Shear Diagram (building left to right):

- No distributed load → Constant V
- Constant distributed load → Linear V
  - Upward acting load → Increasing V at rate w
  - Downward acting load → Decreasing V at rate w
- Area under $w(x)$ diagram → Change in V
- Point load → Jump in V
  - Upward acting load → Increase in V
  - Downward acting load → Decrease in V

Moment Diagram (building left to right):

- No distributed load → Linear M increase/decrease at rate V
- Constant distributed load → Quadratic M
  - Upward acting load → Concave up M diagram
  - Downward acting load → Concave down M diagram
- Area under V diagram → Change in M
  - Positive V area → Increase in M
  - Negative V area → Decrease in M
- Applied moment → Jump in M
  - Clockwise M_o → Increase in M
  - Counterclockwise M_o → Decrease in M
- Location where $V = 0$ → Relative max/min for M

Axial Force Diagram (building bottom to top):

- No distributed load → Constant N
- Constant distributed load → Linear N
  - Upward acting load → Decreasing N at rate w
  - Downward acting load → Increasing N at rate w
- Area under $w(x)$ diagram → Change in N
- Point load → Jump in N
  - Upward acting load → Decrease in N
  - Downward acting load → Increase in N

Checking Diagrams:

- All diagrams (V, M, and N) must close to zero, including jumps at the ends due to applied loads, applied moments, and/or reactions.

Example 4.5

A beam will be supporting equipment that has an eccentric center of gravity. The result is unequal point forces on the supporting beam. Process engineers specify the placement of the equipment; they want it placed off center in this case.

In order to make design decisions later, we would like to know the shear and moment along the length of the beam. Therefore, we will construct the internal force diagrams for the entire length of the beam.

Note that since we do not need the equations for the shear and moment, we can create the diagrams from equations (see Example 4.1) *or* by deduction, as we will do in this example.

Figure 4.34

Estimated Solution:

The approach is the same as it was in Example 4.1.

Detailed Analysis:

Find the reactions:

$$\circlearrowleft^+ \Sigma M_A = 0 = 30 \text{ kN}(1.5 \text{ m}) + 20 \text{ kN}(3.5 \text{ m}) - B_y(5.5 \text{ m})$$
$$B_y = +20.9 \text{ kN}(+\uparrow)$$

$$+\uparrow \Sigma F_y = 0 = A_y - 20 \text{ kN} - 30 \text{ kN} + 20.9 \text{ kN}$$
$$A_y = +29.1 \text{ kN}(+\uparrow)$$

Shear diagram:
- Starts at zero.
- Positive jump of 29.1 kN at A.
- Constant until the location of the first point load, since there is no distributed load.
- Negative jump of 30 kN at 1.5 m from A (location of the point load). New shear value is 29.1 kN − 30 kN = −0.9 kN.
- Constant until the location of the second point load, since there is no distributed load.
- Negative jump of 20 kN at 3.5 m from A (location of the second point load). New shear value is −0.9 kN − 20 kN = −20.9 kN.
- Constant until B, since there is no distributed load.

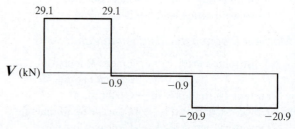

Figure 4.35

Moment diagram:
- Starts at zero.
- Increases, since V is positive. Linear, since V is constant.
- At 1.5 m from A, the new moment value is $(29.1 \text{ kN})(1.5 \text{ m}) = 43.6 \text{ kN} \cdot \text{m}$.
- Then decreases, since V is negative. Still linear, since V is constant.
- At 3.5 m from A, the new moment value is $43.6 \text{ kN} \cdot \text{m} + (-0.9 \text{ kN})(2.0 \text{ m}) = 41.8 \text{ kN} \cdot \text{m}$.
- Decreases at faster rate, since V is more negative. Still linear.

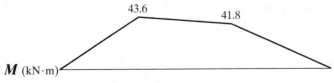

Figure 4.36

Evaluation of Results:
Shear diagram:
Satisfied Fundamental Principles?
At B, there is a positive jump of 20.9 kN. The new shear value is $-20.9 \text{ kN} + 20.9 \text{ kN} = 0$. ✓

Observed Expected Features?
Since there is no distributed load in any of the intervals, the shear segments should be constant. They are. ✓

Observed Expected Features?
The diagram jumps up where there are upward reactions and down where there are applied forces. ✓

Conclusion:
These observations suggest that we have reasonable detailed analysis results. As added confirmation, this diagram matches the shear diagram in Example 4.1.

Moment diagram:
Satisfied Fundamental Principles?
At B, the new moment value is $41.8 \text{ kN} \cdot \text{m} + (-20.9 \text{ kN})(2.0 \text{ m}) = 0$. ✓

Observed Expected Features?
Since there is no distributed load in any of the intervals, all of the moment segments should be linear. They are. ✓

Observed Expected Features?
Since there are no applied moments, there should not be any jumps in the moment diagram. There are none. ✓

Approximation Predicted Outcomes?
The maximum moment is $43.6 \text{ kN} \cdot \text{m}$.
   Our approximate maximum moment was about 15% different; that is a reasonable difference for an approximation.
   As expected, the detailed analysis result is smaller than the approximation. ✓

Conclusion:
Together these observations strongly suggest that we have reasonable detailed analysis results. As added confirmation, this diagram matches the moment diagram in Example 4.1.

EXAMPLE 4.6

Sometimes we can make a problem easier to solve by transforming it into a superposition of simple problems. As a demonstration of this, let's redo Example 4.5 by considering the effects of each point load separately and then adding the results.

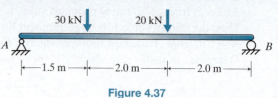

Figure 4.37

Estimated Solution:
The approach is the same as it was in Example 4.1.

Detailed Analysis:
Just the left applied load:

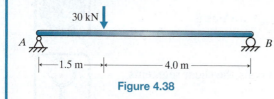

Figure 4.38

Find the reactions:
$$\circlearrowleft \Sigma M_A = 0 = 30 \text{ kN}(1.5 \text{ m}) - B_y(5.5 \text{ m})$$
$$B_y = +8.2 \text{ kN}(+\uparrow)$$

$$+\uparrow \Sigma F_y = 0 = A_y - 30 \text{ kN} + 8.2 \text{ kN}$$
$$A_y = +21.8 \text{ kN}(+\uparrow)$$

Shear diagram:
- Starts at zero.
- Positive jump of 21.8 kN at A.
- Constant until the point load, since there is no distributed load.
- Negative jump of 30 kN at 1.5 m from A (location of the point load). New shear value is $21.8 \text{ kN} - 30 \text{ kN} = -8.2 \text{ kN}$.
- Constant until B, since there is no distributed load.

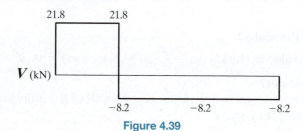

Figure 4.39

Evaluation of Results:

Satisfied Fundamental Principles?
At B, there is a positive jump of 8.2 kN. The new shear value is $-8.2 \text{ kN} + 8.2 \text{ kN} = 0$. ✓

Moment diagram:
- Starts at zero.
- Increases, since V is positive. Linear, since V is constant.
- At 1.5 m from A, the new moment value is $(21.8 \text{ kN})(1.5 \text{ m}) = 32.7 \text{ kN} \cdot \text{m}$
- Then decreases, since V is negative. Still linear, since V is constant.

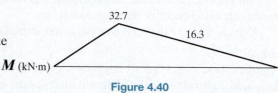

Figure 4.40

Evaluation of Results:
Satisfied Fundamental Principles?
At B, the new moment value is $32.7 \text{ kN} \cdot \text{m} + (-8.2 \text{ kN})(4.0 \text{ m}) = -0.1 \text{ kN} \cdot \text{m} \approx 0.$ ✓

Note that to determine whether the residual is approximately zero, we compare it with the other terms in the equation: 32.7 and 32.8. The residual is about 0.3% of these terms. By redoing the calculations with more digits, we could demonstrate that this difference is due to roundoff.

Just the right applied load:

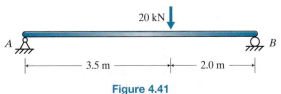

Figure 4.41

Find the reactions:
$$\curvearrowright \Sigma M_A = 0 = 20 \text{ kN}(3.5 \text{ m}) - B_y(5.5 \text{ m})$$
$$B_y = +12.7 \text{ kN}(+\uparrow)$$

$$+\uparrow \Sigma F_y = 0 = A_y - 20 \text{ kN} + 12.7 \text{ kN}$$
$$A_y = +7.3 \text{ kN}(+\uparrow)$$

Shear diagram:
– Starts at zero.
– Positive jump of 7.3 kN at A.
– Constant until the point load, since there is no distributed load.
– Negative jump of 20 kN at 3.5 m from A (location of the point load). New shear value is $7.3 \text{ kN} - 20 \text{ kN} = -12.7 \text{ kN}$.
– Constant until B, since there is no distributed load.

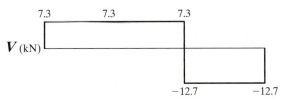

Figure 4.42

Evaluation of Results:
Satisfied Fundamental Principles?
At B, there is a positive jump of 12.7 kN. The new shear value is $-12.7 \text{ kN} + 12.7 \text{ kN} = 0.$ ✓

Moment diagram:
– Starts at zero.
– Increases, since V is positive. Linear, since V is constant.
– At 3.5 m from A, the new moment value is $(7.3 \text{ kN})(3.5 \text{ m}) = 25.6 \text{ kN} \cdot \text{m}$.
– Then decreases, since V is negative. Still linear, since V is constant.

Figure 4.43

Evaluation of Results:
Satisfied Fundamental Principles?
At B, the new moment value is $25.6 \text{ kN} \cdot \text{m} + (-12.7 \text{ kN})(2.0 \text{ m}) = 0.2 \text{ kN} \cdot \text{m} \approx 0$. ✓
Again, we consider this residual close to zero because it is approximately 0.8% of the terms. This difference is also due to roundoff.

Superimpose diagrams:
Shear diagram:

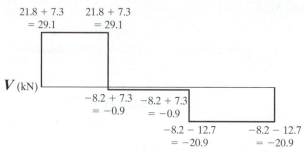

Figure 4.44

Moment diagram:

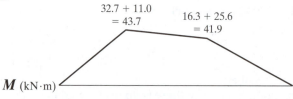

Figure 4.45

Evaluation of Results:

Observed Expected Features?
Since there is no distributed load in any of the intervals, the shear segments should be constant, even though they were created by superposition. They are. ✓

Observed Expected Features?
The shear diagram jumps up where there are upward reactions and down where there are applied forces. ✓

Observed Expected Features?
Since there is no distributed load in any of the intervals, all the moment segments should be linear. They are. ✓

Observed Expected Features?
Since there are no applied moments, there should be no jumps in the moment diagram. There are none. ✓

Conclusion:
These observations suggest that we have reasonable detailed analysis results. As added confirmation, these diagrams match the ones in Examples 4.1 and 4.5 to within roundoff error.

We decide on a case-by-case basis whether it is advantageous to use superposition and divide the problem into a series of simpler problems, or whether it is better to analyze the structure with all the loads applied at one time.

EXAMPLE 4.7

A short cantilever beam in a manufacturing plant is to be extended as part of the preparation for a new product line. Because of access limitations, the best solution will be to attach the extension with a hinged connection.

To confirm the strength capacity of the original beam and the extension, shear and moment diagrams will be very helpful.

Note that since we do not need the equations for the shear and moment, we can create the diagrams from equations (see Example 4.2) or by deduction, as we will do in this example.

Figure 4.46a

Estimated Solution:

The approach is the same as it was in Example 4.2.

Detailed Analysis:

Find the reactions:
    FBD of member BC:

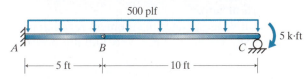

Figure 4.46b

Note: since the axial force has no effect on the shear or moment diagram, we have not included it in these FBDs.

$$+\circlearrowleft \Sigma M_C = 0 = V_B(10\text{ ft}) - 0.5\text{ klf}(10\text{ ft})(10\text{ ft}/2) + 5\text{ k} \cdot \text{ft}$$
$$V_B = +2.0\text{ k}(+\uparrow)$$

$$+\uparrow \Sigma F_y = 0 = 2.0\text{ k} - 0.5\text{ klf}(10\text{ ft}) + C_y$$
$$C_y = +3.0\text{ k}(+\uparrow)$$

FBD of member AB:

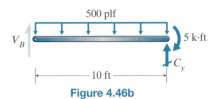

Figure 4.46c

$$+\circlearrowleft \Sigma M_A = 0 = M_A + 0.5\text{ klf}(5\text{ ft})(5\text{ ft}/2) + 2.0\text{ k}(5\text{ ft})$$
$$M_A = -16.25\text{ k} \cdot \text{ft}\;(\circlearrowleft)$$

$$+\uparrow \Sigma F_y = 0 = A_y - 0.5\text{ klf}(5\text{ ft}) - 2.0\text{ k}$$
$$A_y = +4.5\text{ k}(+\uparrow)$$

Shear diagram:
- Starts at zero.
- Positive jump of 4.5 k at A.
- Decreases linearly at a rate of 0.5 klf, since it is a uniform distributed load.
- At 5 ft from A, the location of the hinge, the new value of the shear is $4.5\text{ k} - (0.5\text{ klf})(5\text{ ft}) = 2.0\text{ k}$.
- Continues to decrease linearly at a rate of 0.5 klf until C.
- At C, just before the reaction, the shear has dropped to $2.0\text{ k} - (0.5\text{ klf})(10\text{ ft}) = -3.0\text{ k}$.

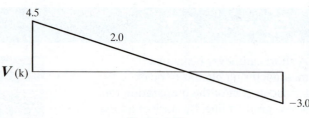

Figure 4.47

Moment diagram:
- Starts at zero.
- Negative jump, since the moment reaction is counterclockwise, of 16.25 k·ft at A.
- Initially increases, since V is initially positive. Quadratic, since V is linear. Concave down, since V is decreasing.
- Peak where $V = 0$. From similar triangles, $x/4.5\text{ k} = 15\text{ ft}/(4.5\text{ k} + 3.0\text{ k})$. So, the peak moment is at $x = 9$ ft from A (4 ft from B).
- Moment increases from A by $\frac{1}{2}(4.5\text{ k})(9.0\text{ ft}) = 20.25\text{ k}\cdot\text{ft}$. Therefore, the peak moment is $-16.25\text{ k}\cdot\text{ft} + 20.25\text{ k}\cdot\text{ft} = 4.0\text{ k}\cdot\text{ft}$.
- At C, just before the applied moment, the new moment value is $4.0\text{ k}\cdot\text{ft} + \frac{1}{2}(-3.0\text{ k})(6\text{ ft}) = -5.0\text{ k}\cdot\text{ft}$.

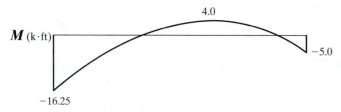

Figure 4.48

Evaluation of Results:

Shear diagram:
Satisfied Fundamental Principles?
At C, there is a positive jump of 3.0 k. The new shear value is $-3.0\text{ k} + 3.0\text{ k} = 0$. ✓

Observed Expected Features?
There is a uniform distributed load along the entire member; therefore, the shear diagram should be linear. It is. ✓

Observed Expected Features?
The shear diagram jumps up at the ends where there are upward reactions but not in between where there are no applied forces. ✓

Moment diagram:
Satisfied Fundamental Principles?
At 5 ft from A (point B, the location of the hinge), the new moment value is $-16.25\text{ k}\cdot\text{ft} + \frac{1}{2}(4.5\text{ k} + 2.0\text{ k})(5\text{ ft}) = 0$. ✓

Satisfied Fundamental Principles?
At C, there is a positive jump of 5.0 k·ft, since the applied moment is clockwise. The new moment value is $-5.0\,\text{k}\cdot\text{ft} + 5.0\,\text{k}\cdot\text{ft} = 0$. ✓

Observed Expected Features?
There is a uniform distributed load along the entire member; therefore, the moment diagram should be quadratic. It is. ✓

Observed Expected Features?
The uniform load is downward; therefore, the moment diagram should be concave down. It is. ✓

Observed Expected Features?
Since there are applied moments at both ends, there should be jumps in the moment diagram at these locations. There are. ✓

Approximation Predicted Outcomes?
The peak negative moment is $-16.25\,\text{k}\cdot\text{ft}$. As expected, this detailed analysis result is smaller than the upper bound limit of $-61.25\,\text{k}\cdot\text{ft}$ we calculated. ✓

Approximation Predicted Outcomes?
The peak positive moment is $+4.0\,\text{k}\cdot\text{ft}$. As expected, this detailed analysis result is smaller than the upper bound limit of $+14.06\,\text{k}\cdot\text{ft}$ we calculated. ✓

Conclusion:
Together these observations strongly suggest that we have reasonable detailed analysis results. As added confirmation, these diagrams match the ones in Example 4.2.

EXAMPLE 4.8

In order to have the desired flow of customers in a retail space, the architect wants us to use a transfer girder to carry load from a second-story column out to other columns on the first floor. The girder will also carry the floor load.

We want to know the peak moment for the beam shown, but its location is not obvious. We can find the location and magnitude by creating the internal force diagrams by deduction.

Note that this is a different approach to the same problem in Example 4.3. Which approach is better is a matter of personal preference.

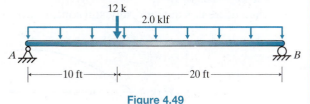

Figure 4.49

204 Chapter 4 Internal Force Diagrams

Estimated Solution:

The approach for estimating the solution is the same as it was in Example 4.3.

Detailed Analysis:

Find the reactions:

$$+\circlearrowleft \Sigma M_A = 0 = 12\,k\,(10\,ft) + 2.0\,klf\,(30\,ft)\,(30\,ft/2) - B_y(30\,ft)$$
$$B_y = +34\,k\,(+\uparrow)$$

$$+\uparrow \Sigma F_y = 0 = A_y - 12\,k - (2.0\,klf)\,(30\,ft) + 34\,k$$
$$A_y = +38\,k\,(+\uparrow)$$

Shear diagram:
- Starts at zero.
- Positive jump of 38 k at A.
- Decreases until the location of the point load, since there is a downward distributed load. Linearly decreasing, at a rate of 2.0 klf, since there is a uniform distributed load.
- At 10 ft from A, right before the point load, the new value of the shear is $38\,k - 2.0\,klf\,(10\,ft) = 18\,k$.
- Negative jump of 12 k at 10 ft from A (the location of the point load). The new shear value is $18\,k - 12\,k = 6\,k$.
- Continues to decrease linearly, at a rate of 2.0 klf, until B.
- At B, just before the reaction, the new shear value is $6\,k - 2.0\,klf\,(20\,ft) = -34\,k$.

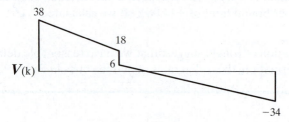

Figure 4.50

Moment diagram:
- Starts at zero.
- Initially increases, since V is initially positive. Quadratic, since V is linear. Concave down, since V is decreasing.
- At 10 ft from A, the location of the jump in V, the new moment value is $\frac{1}{2}(38\,k + 18\,k)(10\,ft) = 280\,k\cdot ft$.
- After the jump in V, continues to decrease initially but at a slower rate. Still quadratic and concave down.
- Peak where $V = 0$. From similar triangles, $x/6\,k = 20\,ft/(6\,k + 34\,k)$. So, there is a peak moment at $x = 3.0\,ft$ from the location of the applied point load.
- Moment increases by $\frac{1}{2}(6\,k)(3.0\,ft) = 9\,k\cdot ft$. Therefore, peak M is $280\,k\cdot ft + 9\,k\cdot ft = 289\,k\cdot ft$.
- Then decreases quadratically, still concave down.

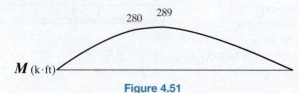

Figure 4.51

Evaluation of Results:

Shear diagram:
Satisfied Fundamental Principles?
At B, there is a positive jump of 34 k. The new shear value is $-34\,\text{k} + 34\,\text{k} = 0$. ✓

Observed Expected Features?
There is a uniform distributed load along the entire member; therefore, the shear diagram should be linear. It is. ✓

Observed Expected Features?
The shear diagram jumps up at the ends where there are upward reactions and down where the column transfers its point load. ✓

Moment diagram:
Satisfied Fundamental Principles?
At B, the moment has decreased by $\frac{1}{2}(34\,\text{k})(20\,\text{ft} - 3\,\text{ft}) = 289\,\text{k}\cdot\text{ft}$. The new moment value is $289\,\text{k}\cdot\text{ft} - 289\,\text{k}\cdot\text{ft} = 0$. ✓

Observed Expected Features?
There is a uniform distributed load along the entire member; therefore, the moment diagram should be quadratic. It is. ✓

Observed Expected Features?
The uniform load is downward; therefore, the moment diagram should be concave down. It is. ✓

Observed Expected Features?
Since there are no applied moments, there should be no jumps in the moment diagram. There are none. ✓

Approximation Predicted Outcomes?
The maximum moment is $289\,\text{k}\cdot\text{ft}$.
  Our approximate maximum moment was about 7% lower; that is a reasonable difference for an approximation. ✓

Conclusion:
Together these observations strongly suggest that we have reasonable detailed analysis results. As added confirmation, these diagrams match the ones in Example 4.3.

4.3 Diagrams for Frames

Introduction

The first two sections of this chapter deal with developing internal force diagrams for individual structural members. However, most structures are assemblies of many members.
  Constructing internal force diagrams for rigid frames presents some unique challenges because the shear or axial force at the end of one member is not necessarily the shear or axial force for the attached member.

How-To

The principles and procedures presented earlier in this chapter are still relevant, even essential, for constructing internal force diagrams for rigid frames. The challenge lies in converting internal forces at the intersection of members into components normal and parallel to the axis of each of the members in turn.

Finding the internal forces at the intersection of members might be possible using equations of equilibrium (e.g., if the frame is determinate). For indeterminate frames, the internal forces can be found approximately using the methods described in Chapter 9: Approximate Analysis of Rigid Frames. Otherwise, the internal forces must be found using the methods in Chapter 12: Force Method, Chapter 13: Moment Distribution Method, or Chapter 15: Direct Stiffness Method for Frames.

Once the internal forces have been determined at the intersection of two members, we can focus on converting them into the proper format to help create the internal force diagrams for each of the members. Throughout this process it is essential that we maintain the sign convention presented in Section 4.1, including the direction for increasing x (left to right and bottom to top).

We use vector mechanics to convert internal forces at the intersection into components normal to the axis of the member (these will affect the shear) and parallel to the axis of the member (these will affect the axial force) (Figure 4.52). The moment need not be converted when we are dealing with a planar (two-dimensional) frame.

When constructing the diagrams, we treat the converted internal forces at the ends of the member as if they were applied loads or reactions. Therefore, they should cause jumps in the internal force diagrams. If the internal forces are at the trailing end of the member, the end of the x-direction, they should cause the diagrams to close to zero.

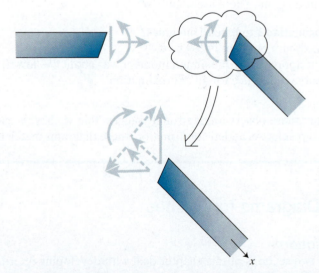

Figure 4.52 Cut at the intersection of two members of a rigid frame showing the equal and opposite internal forces at the ends of the members. Focus on the right member shows how the internal forces are divided into components normal and parallel to the axis of the member.

Section 4.3 Highlights

Diagrams for Frames

Step 1: Find the reactions.
Step 2: Cut the frame into members (straight segments) and calculate the internal forces at the cuts.
Step 3: Convert the internal forces at the cuts and the support forces everywhere into components normal and parallel to the axis of each member.
Step 4: Use the methods from Section 4.1 or 4.2 to construct the internal force diagrams. Note that internal forces at the cut are treated as applied point forces or moments.
Step 5: Check that the diagrams close to zero.

Example 4.9

The architect's vision of the entrance to a building is an asymmetric triangle. In order to design the members and connection for the rigid frame, we need to know the internal forces that are generated throughout the frame. Therefore, we want to construct the axial force, shear, and moment diagrams for the frame. For this analysis, the frame experiences only the dead load shown. Our preliminary plan is to make the left end pinned and allow the right end to behave as a roller.

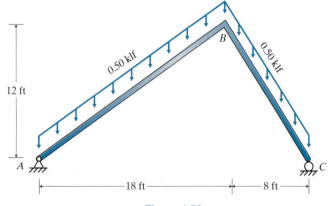

Figure 4.53

Lengths of members:

$$L_{AB} = \sqrt{(18\,\text{ft})^2 + (12\,\text{ft})^2} = 21.63\,\text{ft}$$

$$L_{BC} = \sqrt{(12\,\text{ft})^2 + (8\,\text{ft})^2} = 14.42\,\text{ft}$$

Estimated Solution:

Since all of the load is down (no lateral loads), we can approximate the frame as a simply supported beam from A to C. Let's also approximate the loading conditions as one equivalent uniform distributed load along the entire beam.

Equivalent uniform load:

$$P_{\text{total}} = 0.5\,\text{klf}\,(21.63\,\text{ft}) + 0.5\,\text{klf}\,(14.42\,\text{ft}) = 18.02\,\text{k}$$

$$w_{\text{equiv}} = 18.02\,\text{k}/26\,\text{ft} = 0.693\,\text{klf}$$

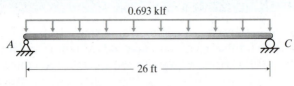

Figure 4.54

Peak moment:

For a uniformly loaded, simply supported beam, the peak moment occurs at midspan and can be calculated from the following formula which we derived in Section 2.4: Approximating Loading Conditions.

$$M_{\max} = \frac{wl^2}{8} = \frac{0.693\,\text{klf}\,(26\,\text{ft})^2}{8} = 58.6\,\text{k}\cdot\text{ft}$$

The equivalent load actually moves more load to the middle part of the beam, so we would expect this approximation to overpredict the peak moment.

In general, using an approximate geometry can cause major changes to what we would predict for maximum internal forces. However, since we only changed the geometry in the direction of the applied load (we did not make it longer or shorter from A to C), our approximation of the geometry should not affect our prediction of peak moment in this case.

Detailed Analysis:

Find the reactions:

$$\circlearrowleft \Sigma M_A = 0 = 0.50\,\text{klf}\,(21.63\,\text{ft})\,(18\,\text{ft}/2)$$
$$+ 0.50\,\text{klf}\,(14.42\,\text{ft})\,(18\,\text{ft} + 8\,\text{ft}/2) - C_y(26\,\text{ft})$$

$$C_y = +9.84\,\text{k}\,(+\uparrow)$$

$$+\uparrow \Sigma F_y = 0 = A_y - 0.50\,\text{klf}\,(21.63\,\text{ft}) - 0.50\,\text{klf}\,(14.42\,\text{ft}) + 9.84\,\text{k}$$

$$A_y = +8.18\,\text{k}\,(+\uparrow)$$

$$\xrightarrow{+} \Sigma F_x = 0 = A_x$$

$$A_x = 0$$

Section 4.3 Diagrams for Frames

Cut the frame into its individual members.
Member AB:

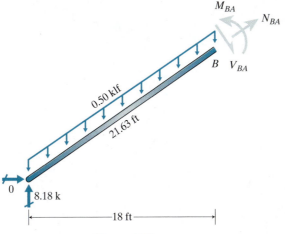

Figure 4.55

We could find the internal forces at the cut by dividing the internal forces into horizontal and vertical components. But to construct the internal force diagrams, we want the reaction at A and the applied load to be in components along and perpendicular to the member.

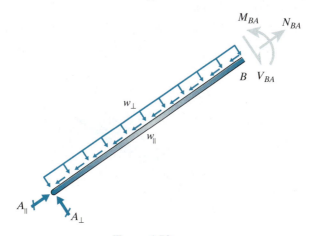

Figure 4.56

Convert the applied load:

Figure 4.57

$$w_\parallel = \frac{12}{21.63}(0.50 \text{ klf}) = 0.277 \text{ klf} \ (\nearrow)$$

$$w_\perp = \frac{18}{21.63}(0.50 \text{ klf}) = 0.416 \text{ klf} \ (\searrow)$$

Convert the reactions at A:

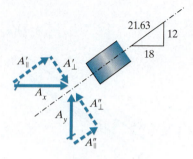

Figure 4.58

$A_x = 0 \; (\xrightarrow{+})$
$\quad A'_{\parallel} = 0 \; (+\nearrow)$
$\quad A'_{\perp} = -0 \; (\nwarrow)$
$A_y = 8.18 \text{ k} \; (+\uparrow)$
$\quad A''_{\parallel} = \dfrac{12}{21.63}(8.18 \text{ k}) = 4.54 \text{ k} \; (+\nearrow)$
$\quad A''_{\perp} = \dfrac{18}{21.63}(8.18 \text{ k}) = 6.81 \text{ k} \; (\nwarrow)$
$A_{\parallel} = A'_{\parallel} + A''_{\parallel} = 0 \text{ k} + 4.54 \text{ k} = 4.54 \text{ k} \; (+\nearrow)$
$A_{\perp} = A'_{\perp} + A''_{\perp} = 0 \text{ k} + 6.81 \text{ k} = 6.81 \text{ k} \; (\nwarrow)$

FBD of member AB:

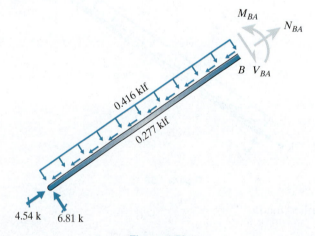

Figure 4.59

Apply equilibrium:

$\circlearrowleft \Sigma M_{\text{cut}} = 0 = 6.81 \text{ k} (21.63 \text{ ft}) - 0.416 \text{ klf} (21.63 \text{ ft})(21.63 \text{ ft}/2) - M_{BA}$
$\quad M_{BA} = +50.0 \text{ k} \cdot \text{ft} \; (\curvearrowright)$

$\nwarrow \Sigma F_y = 0 = 6.81 \text{ k} - 0.416 \text{ klf} (21.63 \text{ ft}) - V_{BA}$
$\quad V_{BA} = -2.19 \text{ k} \; (\searrow)$

$+\nearrow \Sigma F_y = 0 = 4.54 \text{ k} - 0.277 \text{ klf} (21.63 \text{ ft}) + N_{BA}$
$\quad N_{BA} = +1.45 \text{ k} \; (\nearrow)$

Member BC:

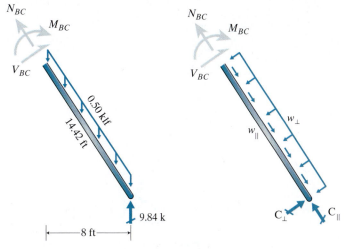

Figure 4.60

Convert the applied load:

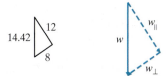

Figure 4.61

$$w_\| = \frac{12}{14.42}(0.50 \text{ klf}) = 0.416 \text{ klf} (\nwarrow)$$

$$w_\perp = \frac{8}{14.42}(0.50 \text{ klf}) = 0.277 \text{ klf} (\swarrow)$$

Convert the reaction at C:

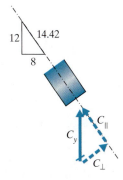

Figure 4.62

$$C_y = 9.84 \text{ k} (+\uparrow)$$

$$C_\| = \frac{12}{14.42}(9.84 \text{ k}) = 8.19 \text{ k} (\nwarrow)$$

$$C_\perp = \frac{8}{14.42}(9.84 \text{ k}) = 5.46 \text{ k} (\nearrow)$$

FBD of member *BC*:

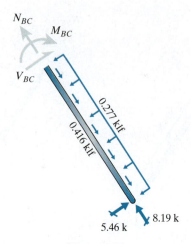

Figure 4.63

Apply equilibrium:

$$\curvearrowleft \Sigma M_{cut} = 0 = M_{BC} + 0.277 \text{ klf}(14.42 \text{ ft})(14.42 \text{ ft}/2) - 5.46 \text{ k}(14.42 \text{ ft})$$
$$M_{BC} = +49.9 \text{ k} \cdot \text{ft} \;(\curvearrowleft +)$$

$$\nwarrow \Sigma F_y = 0 = 8.19 \text{ k} - 0.416 \text{ klf}(14.42 \text{ ft}) + N_{BC}$$
$$N_{BC} = -2.19 \text{ k} \;(\nearrow +)$$

$$\nearrow \Sigma F_x = 0 = 5.46 \text{ k} - 0.277 \text{ klf}(14.42 \text{ ft}) + V_{BC}$$
$$V_{BC} = -1.46 \text{ k} \;(\nwarrow +)$$

Evaluation of Results:
Before spending effort to create the diagrams, we can check our results by verifying the equilibrium of point *B*.

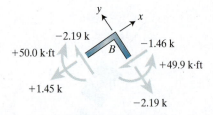

Figure 4.64

Satisfied Fundamental Principles?

$$\curvearrowleft \Sigma M_{cut} = 50.0 \text{ k} \cdot \text{ft} - 49.9 \text{ k} \cdot \text{ft} = -0.1 \text{ k} \cdot \text{ft} \approx 0 \quad \checkmark$$
$$\nwarrow \Sigma F_y = -2.19 \text{ k} - (-2.19 \text{ k}) = 0 \quad \checkmark$$
$$\nearrow \Sigma F_x = -1.45 \text{ k} - (-1.46 \text{ k}) = 0.01 \text{ k} \approx 0 \quad \checkmark$$

Equilibrium is satisfied to within roundoff error.

Internal force diagrams for AB:
FBD of member AB:

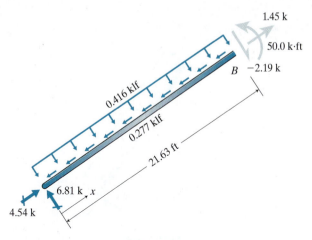

Figure 4.65

Axial force diagram:
– Starts at zero.
– Negative jump of 4.54 k at $x = 0$.
– Linearly increases at a rate of 0.277 klf. At the other end, has increased to $-4.54 \text{ k} + 0.277 \text{ klf}(21.63 \text{ ft}) = 1.45 \text{ k}$.
– At end, jump of -1.45 k.

Evaluation of Results:
Satisfied Fundamental Principles?
– *Check*: 1.45 k − 1.45 k = 0 ✓

Shear diagram:
– Starts at zero.
– Positive jump of 6.81 k at $x = 0$.
– Linearly decreases at a rate of 0.416 klf. At the other end, has decreased to $6.81 \text{ k} - 0.416 \text{ klf}(21.63 \text{ ft}) = -2.19 \text{ k}$.
– At end, jump of $-(-2.19 \text{ k})$.

Evaluation of Results:
Satisfied Fundamental Principles?
– *Check*: −2.19 k − (−2.19 k) = 0 ✓

Moment diagram:
– Starts at zero.
– Initially increases, since V is initially positive. Quadratic, since V is linear. Concave down, since V is decreasing.
– Peak where $6.81 \text{ k} - 0.416 \text{ klf} \cdot x = 0$. Therefore, peak at $x = 16.37$ ft.
– Moment increases by $\frac{1}{2}(6.81 \text{ k})(16.37 \text{ ft}) = 55.7 \text{ k} \cdot \text{ft}$. Therefore, the peak M is 55.7 k·ft.
– Then M decreases quadratically but is still concave down.
– Moment decreases by $\frac{1}{2}(-2.19 \text{ k})(21.63 \text{ ft} - 16.37 \text{ ft}) = -5.8 \text{ k} \cdot \text{ft}$. Therefore, the moment decreases to 55.7 k·ft − 5.8 k·ft = 49.9 k·ft.
– At end, a jump of −50.0 k·ft.

Evaluation of Results:
Satisfied Fundamental Principles?
– *Check*: 49.9 k·ft − 50.0 k·ft = −0.1 k·ft ≈ 0 ✓

Internal force diagrams for BC:
FBD of member *BC*:

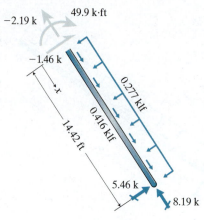

Figure 4.66

Axial force diagram:
Since we are constructing the axial force diagram from the top, our rules from Section 4.2 are reversed.
– Starts at zero.
– Positive jump of -2.19 k at $x = 0$. Therefore decreases, since negative.
– Linearly decreases at a rate of 0.416 klf. At the other end, has decreased to -2.19 k $-$ 0.416 klf (14.42 ft) $= -8.19$ k.
– At end, a positive jump of 8.19 k.

Evaluation of Results:
Satisfied Fundamental Principles?
– *Check*: -8.19 k $+ 8.19$ k $= 0$ ✓

Shear diagram:
– Starts at zero.
– Positive jump of -1.46 k at $x = 0$. Therefore decreases, since negative.
– Linearly decreases at a rate of 0.277 klf. At the other end, has decreased to -1.46 k $-$ 0.277 klf (14.42 ft) $= -5.45$ k.
– At end, a positive jump of 5.46 k.

Evaluation of Results:
Satisfied Fundamental Principles?
– *Check*: -5.45 k $+ 5.46$ k $= +0.01$ k ≈ 0 ✓

Moment diagram:
– Starts at zero.
– Positive jump of 49.9 k·ft.
– Decreases, since *V* is negative. Quadratic, since *V* is linear. Concave down, since *V* is decreasing.
– Peak at $x = 0$, since $V \neq 0$ everywhere along the member.

Evaluation of Results:
Satisfied Fundamental Principles?
– *Check*: At the other end, has decreased to
49.9 k · ft $+ \frac{1}{2}(-1.46$ k $- 5.45$ k$)(14.42$ ft$) = +0.08$ k · ft ≈ 0 ✓

Internal force diagrams for entire frame:

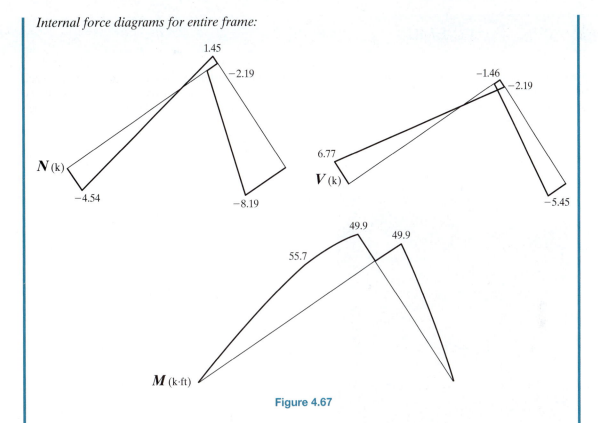

Figure 4.67

Evaluation of Results:

Satisfied Fundamental Principles?
When taking apart the frame and finding the internal forces at the cuts, we confirmed equilibrium to within roundoff error at point B. ✓

Satisfied Fundamental Principles?
In the construction of the diagrams, we checked that each diagram closed to zero to within roundoff error. ✓

Approximation Predicted Outcomes?
The maximum moment is 55.7 k · ft.
  Our approximate maximum moment was about 5% different; that is a reasonable difference for an approximation.
  As expected, the detailed analysis result is smaller than the approximation.

Conclusion:
Together these observations suggest that we have reasonable detailed analysis results.

Homework Problems

4.1 The preliminary design of a roof has the deck running one way, radially, to beams that distribute the load to the girders that run radially. The senior engineer leading our team would also like to explore the effect of running the deck one way toward the girders, thus eliminating the need for the beams. The result will be a triangular distributed load on the girder. Our assignment is to determine the peak moment caused by full snow load.

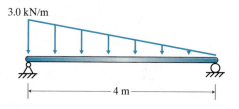

Prob. 4.1a

a. Guess the location of the peak moment.
b. Estimate the peak moment by replacing the linear distributed load with an equivalent uniform distributed load. Will this provide an upper bound, a lower bound, or just an approximation for the actual peak moment? Justify your answer.

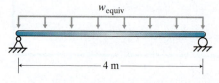

Prob. 4.1b

c. Derive the equations for the shear and moment for the beam with the triangular distributed load.
d. Construct the shear and moment diagrams for the beam. Label values and locations.
e. Identify at least two features of the shear diagram that suggest you have a reasonable answer.
f. Identify at least two features of the moment diagram that suggest you have a reasonable answer.
g. Determine the peak moment due to snow and its location.
h. Make a comprehensive argument that the peak moment found in part (g) is reasonable.
i. Comment on why your guess in part (a) was or was not close to the solution in part (g).

4.2 An elevated walkway will connect a pair of facilities on one side of a street to a single building on the other side. As a result, it was decided that the walkway will be wider at one end and narrower at the other. The resulting idealized dead load is trapezoidal. Our job is to determine the peak moment in the girder due to the dead load.

Prob. 4.2a

a. Guess the location of the peak moment.
b. Estimate the peak moment by replacing the linear distributed load by an equivalent uniform distributed load. Will this provide an upper bound, a lower bound, or just an approximation for the actual peak moment? Justify your answer.
c. Derive the equations for the shear and moment for the beam with the trapezoidal load.
d. Construct the shear and moment diagrams for the beam. Label values and locations.
e. Identify at least two features of the shear diagram that suggest you have a reasonable answer.
f. Identify at least two features of the moment diagram that suggest you have a reasonable answer.
g. Determine the peak moment due to the dead load and its location.
h. Make a comprehensive argument that the peak moment found in part (g) is reasonable.
i. Comment on why your guess in part (a) was or was not close to the solution in part (g).

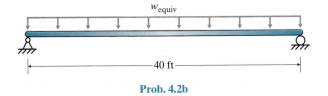

Prob. 4.2b

4.3 A simply supported beam supports an elevated walkway. The walkway covers only part of the beam, as shown in the idealization. In order to design the steel beam, a member of the team will need expressions for the moment in order to calculate the moment gradient modification factor. That person will also need to know the shear along the beam.

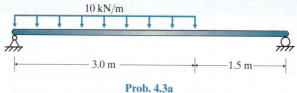

Prob. 4.3a

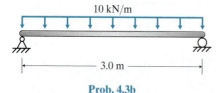

Prob. 4.3b

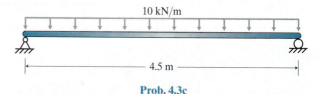

Prob. 4.3c

a. Guess the location of the peak moment.
b. Estimate the peak moment by considering the beam to be uniformly loaded over its entire length with 10 kN/m, but only 3 m long. Will this provide an upper bound, a lower bound, or just an approximation for the actual peak moment? Justify your answer.
c. Estimate the peak moment by considering the beam to be uniformly loaded over its entire length of 4.5 m with 10 kN/m. Will this provide an upper bound, a lower bound, or just an approximation for the actual peak moment? Justify your answer.
d. Derive the equations for the shear and moment for the original beam.
e. Construct the shear and moment diagrams for the beam. Label values and locations.
f. Identify at least three features of the shear diagram that suggest you have a reasonable answer.
g. Identify at least three features of the moment diagram that suggest you have a reasonable answer.
h. Determine the peak moment due to the live load and its location.
i. Make a comprehensive argument that the peak moment found in part (h) is reasonable.
j. Comment on why your guess in part (a) was or was not close to the solution in part (h).

4.4 Our firm has been hired to design a sunscreen to protect cars on the top level of a parking garage. The design team has determined the ideal structural system: two beams will extend from a central column, pass over exterior columns, and cantilever over the parking spots. To design the beam, we will need to know the internal forces along the beam. Also, we will need expressions for the moment in order to predict deflections later.

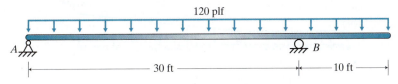

Prob. 4.4a

a. Guess the location of the peak moment.
b. Estimate the peak moment by ignoring the distributed load on the cantilevered end. With no load on the cantilevered end, we can ignore the presence of that end. Will this provide an upper bound, a lower bound, or just an approximation for the actual peak moment? Justify your answer.
c. Derive the equations for the shear and moment for the original beam.
d. Construct the shear and moment diagrams for the beam. Label values and locations.
e. Identify at least three features of the shear diagram that suggest you have a reasonable answer.
f. Identify at least three features of the moment diagram that suggest you have a reasonable answer.
g. Determine the peak moment and its location.
h. Make a comprehensive argument that the peak moment found in part (g) is reasonable.
i. Comment on why your guess in part (a) was or was not close to the solution in part (g).

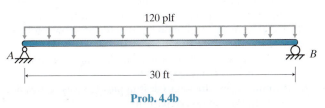

Prob. 4.4b

4.5 The roof of a tall building will be supported by beams that extend from the core to the perimeter. Because of the mechanical penthouse, snow will drift, causing a non-uniform load on the beams. Our team needs to know the internal forces along the beam in order to design it. A more experienced designer on the team points out that since the distributed load is not uniform, it will be easier to use integration rather than deduction.

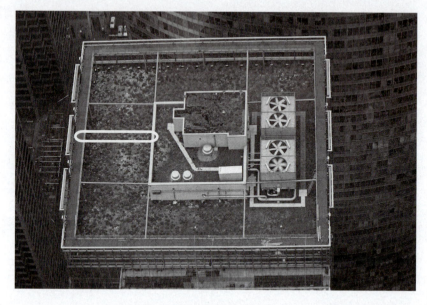

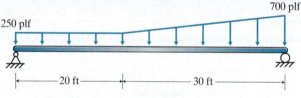

Prob. 4.5a

a. Guess the location of the peak moment.
b. Estimate the peak moment by approximating the loading as uniform along the entire length without the drifting. Will this provide an upper bound, a lower bound, or just an approximation for the actual peak moment? Justify your answer.
c. Estimate the peak moment by replacing the loading by an equivalent uniform distributed load. Will this provide an upper bound, a lower bound, or just an approximation for the actual peak moment? Justify your answer.
d. Derive the equations for the shear and moment for the beam with the non-uniform load.
e. Construct the shear and moment diagrams for the beam. Label values and locations.
f. Identify at least three features of the shear diagram that suggest you have a reasonable answer.
g. Identify at least three features of the moment diagram that suggest you have a reasonable answer.
h. Determine the peak moment due to the snow load and its location.
i. Make a comprehensive argument that the peak moment found in part (h) is reasonable.
j. Comment on why your guess in part (a) was or was not close to the solution in part (h).

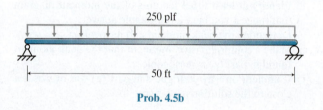

Prob. 4.5b

4.6 Because of the placement of ductwork below an office floor, we needed to move beam A2-B2 away from the middle of the floor. The resulting live load on girder A1-A3 has been calculated. In order to design the girder, we need to know the peak moment generated by the live load. A more experienced designer on the team points out that since the distributed load is not uniform, it will be easier to use integration rather than deduction.

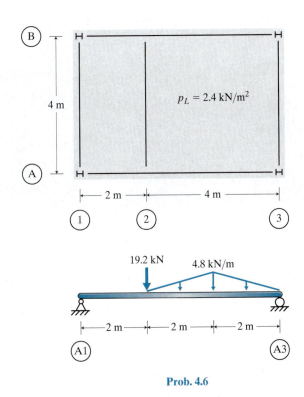

Prob. 4.6

a. Guess the location of the peak moment.
b. Estimate the peak moment by replacing the point and distributed loads by an equivalent uniform distributed load. Will this provide an upper bound, a lower bound, or just an approximation for the actual peak moment? Justify your answer.
c. Estimate the peak moment by replacing the point and distributed loads by an equivalent single point load at midspan. Will this provide an upper bound, a lower bound, or just an approximation for the actual peak moment? Justify your answer.
d. Derive the equations for the shear and moment for the beam with the original loading.
e. Construct the shear and moment diagrams for the beam. Label values and locations.
f. Identify at least three features of the shear diagram that suggest you have a reasonable answer.
g. Identify at least three features of the moment diagram that suggest you have a reasonable answer.
h. Determine the peak moment due to the live load and its location.
i. Make a comprehensive argument that the peak moment found in part (h) is reasonable.
j. Comment on why your guess in part (a) was or was not close to the solution in part (h).

4.7 The air conditioning unit shown in the photo is supported by a beam that can be idealized as simply supported. The unit applies a uniform distributed load on the beam. That load can be considered "dead" because the weight of the unit is known reasonably well and it will not be moved. The self-weight of the beam is so small in comparison to the unit that we can ignore the effect of the self-weight for now. To design the beam, we need to know the internal forces. Since there are three intervals and a uniform distributed load, it will probably be easier to use deduction rather than integration.

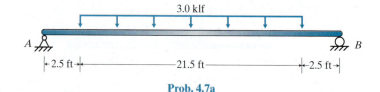

Prob. 4.7a

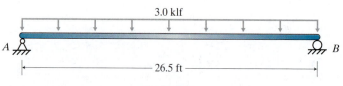

Prob. 4.7b

a. Guess the shape of the moment diagram.
b. Estimate the peak moment by extending the distributed load over the entire length of the beam. Will this provide an upper bound, a lower bound, or just an approximation for the actual peak moment? Justify your answer.
c. Construct the shear and moment diagrams for the beam with the actual load. Label values and locations.
d. Identify at least three features of the shear diagram that suggest you have a reasonable answer.
e. Identify at least three features of the moment diagram that suggest you have a reasonable answer.
f. Determine the peak moment due to the air conditioning unit and its location.
g. Make a comprehensive argument that the moment diagram found in part (c) is reasonable.
h. Comment on why your guess in part (a) was or was not close to the solution in part (c).

4.8 Due to the potential for snow drifting next to a roof parapet, the open-web joists cause uneven point loads on the center steel beam. The open-web joists at the ends of the span are supported directly by the columns; therefore, they do not carry load to the beam. In order to design the steel beam, we will need to calculate the peak moment considering all loads. That peak might be in a different location than the peak moment for each of the types of load individually. Therefore, we need to know the moment caused by snow along the entire beam. An experienced team member recommends using deduction in this case.

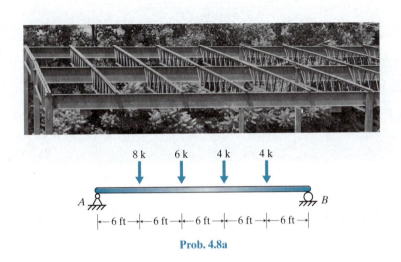

Prob. 4.8a

a. Guess the shape of the moment diagram.
b. Estimate the peak moment by replacing the four point loads by an equivalent uniform distributed load. Will this provide an upper bound, a lower bound, or just an approximation for the actual peak moment? Justify your answer.

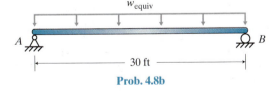

Prob. 4.8b

c. Construct the shear and moment diagrams for the beam with the point loads. Label values and locations.
d. Identify at least three features of the shear diagram that suggest you have a reasonable answer.
e. Identify at least three features of the moment diagram that suggest you have a reasonable answer.
f. Determine the peak moment due to snow and its location.
g. Make a comprehensive argument that the moment diagram found in part (c) is reasonable.
h. Comment on why your guess in part (a) was or was not close to the solution in part (c).

4.9 For ease of construction in the field, a two-span bridge was constructed with a splice and a hinge. The Department of Transportation has received a request to increase the load limit on the bridge. Before acting on that request, they want to know the internal forces due to the dead load. A team member recommends that it will be easier to use deduction rather than integration.

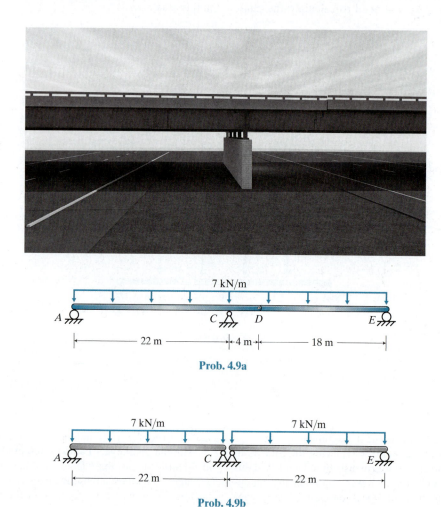

Prob. 4.9a

Prob. 4.9b

a. Guess the location of the peak positive moment.
b. Estimate the peak positive moment by considering the compound beam as two simply supported beams. Will this provide an upper bound, a lower bound, or just an approximation for the actual peak positive moment? Justify your answer.
c. Construct the shear and moment diagrams for the original beam. Label values and locations.
d. Identify at least three features of the shear diagram that suggest you have a reasonable answer.
e. Identify at least three features of the moment diagram that suggest you have a reasonable answer.
f. Determine the peak moment due to the dead load and its location.
g. Make a comprehensive argument that the peak moment found in part (f) is reasonable.
h. Comment on why your guess in part (a) was or was not close to the solution in part (f).

4.10 Wood beams cantilevered out from a house to form an elevated deck. The owners wanted a bigger deck, so they extended the beams and added columns at the far ends. The addition was not engineered, so the owners have no idea whether the deck is safe. Now that they want to sell the house, they have hired our firm to evaluate the capacity of the deck in its current condition. As part of that evaluation, we need to know the internal forces generated by the full live load. Since the load is uniform, it is probably more efficient to use deduction rather than integration for this problem.

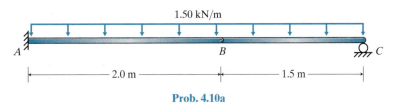

Prob. 4.10a

a. Guess the location of the peak positive moment.
b. Estimate the moment reaction at A by considering the original configuration of the deck: uniformly loaded, cantilevered segment AB. Will this provide an upper bound, a lower bound, or just an approximation for the actual moment reaction at A? Justify your answer.

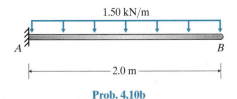

Prob. 4.10b

c. Construct the shear and moment diagrams for the extended beam. Label values and locations.
d. Identify at least three features of the shear diagram that suggest you have a reasonable answer.
e. Identify at least three features of the moment diagram that suggest you have a reasonable answer.
f. Determine the peak positive moment due to the live load and its location.
g. Make a comprehensive argument that the peak positive moment found in part (f) is reasonable.
h. Comment on why your guess in part (a) was or was not close to the solution in part (f).

4.11 The roof covering a walkway in a northern city is slanted to allow melting snow to drain off. Since the slope of the beam is very small, −0.05, we can reasonably approximate the beam as horizontal for our calculations. Our team is to determine the internal forces so that they can select the beam section properties. A more experienced designer on the team points out that since the distributed load is uniform, it will be easier to use deduction rather than integration.

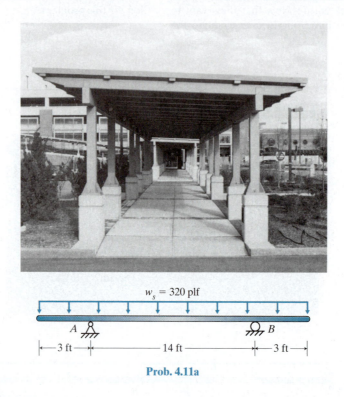

Prob. 4.11a

a. Guess the shape of the moment diagram.
b. Estimate the peak midspan moment by considering only the segment between the supports, AB. Will this provide an upper bound, a lower bound, or just an approximation for the actual peak midspan moment? Justify your answer.

c. Construct the shear and moment diagrams for the full beam. Label values and locations.
d. Identify at least three features of the shear diagram that suggest you have a reasonable answer.
e. Identify at least three features of the moment diagram that suggest you have a reasonable answer.
f. Determine the peak midspan moment due to snow.
g. Make a comprehensive argument that the moment diagram found in part (c) is reasonable.
h. Comment on why your guess in part (a) was or was not close to the diagram in part (c).

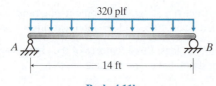

Prob. 4.11b

4.12 Our firm has been hired to design the support beams for an HVAC unit. The facility might expand to need an additional unit, so the structure should be able to handle both conditions: one unit and two. Our team is exploring the effect of just one unit. The beam is indeterminate, but one member of the team has calculated the reactions due to the single HVAC unit. In order to design the beam, we need to know the internal forces along the entire beam. A team member points out that since there is only a uniform distributed load, it is probably faster to use deduction rather than integration.

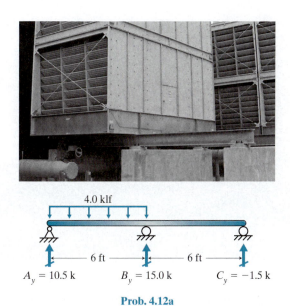

Prob. 4.12a

a. Guess the shape of the moment diagram.
b. Estimate the peak positive moment by considering only the loaded segment, AB. Will this provide an upper bound, a lower bound, or just an approximation for the actual peak positive moment? Justify your answer.
c. Construct the shear and moment diagrams for the full beam. Label values and locations.
d. Identify at least three features of the shear diagram that suggest you have a reasonable answer.
e. Identify at least three features of the moment diagram that suggest you have a reasonable answer.
f. Determine the peak positive moment due to the HVAC unit.
g. Make a comprehensive argument that the moment diagram found in part (c) is reasonable.
h. Comment on why your guess in part (a) was or was not close to the diagram in part (c).

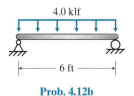

Prob. 4.12b

4.13 Our team is designing the beam to support the bleacher seating at a sports stadium. The beam is simply supported and cantilevers past the upper support. A team member reminds us that a live load acts over the horizontal projection.

To design the beam, we need to know the axial force, shear, and moment along its entire length. Note that we need to convert the distributed load over the horizontal projection into an equivalent distributed load along the length of the beam; we will then need to break the equivalent distributed load along the beam into vector components parallel to the beam and normal to the beam.

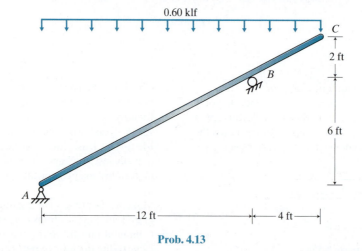

Prob. 4.13

a. Guess the shape of the moment diagram.
b. Imagine that the beam is simply supported at A and C (roller at C, no support at B). Calculate the peak moment. Will this provide an upper bound, a lower bound, or just an approximation for the actual peak moment? Justify your answer.
c. Construct the axial force, shear, and moment diagrams for the beam with the original supports. Label values and locations.
d. Identify at least two features of the axial force diagram that suggest you have a reasonable answer.
e. Identify at least three features of the shear diagram that suggest you have a reasonable answer.
f. Identify at least three features of the moment diagram that suggest you have a reasonable answer.
g. Determine the peak moment and its location.
h. Make a comprehensive argument that the peak moment found in part (g) is reasonable.
i. Comment on why your guess in part (a) was or was not close to the diagram in part (c).

4.14 Our team is designing the structural members for a portable gantry crane. The manufacturer wants the crane to have a rated capacity of 25 kN. Ultimately we will need to design the members to carry the maximum load at any point along the top member, but for preliminary design, our team leader wants us to consider only the maximum load at midspan. We need to know the internal forces throughout the frame in order to perform preliminary design.

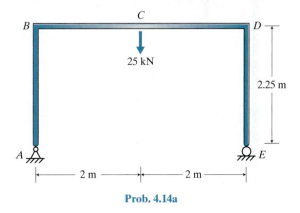

Prob. 4.14a

a. Guess the shape of the moment diagram.
b. Imagine that the top member, BD, is simply supported. Calculate the peak moment. Will this provide an upper bound, a lower bound, or just an approximation for the actual peak moment? Justify your answer.

c. Construct the axial force, shear, and moment diagrams for the full frame. Label values and locations.
d. Identify at least two features of the axial force diagram that suggest you have a reasonable answer.
e. Identify at least three features of the shear diagram that suggest you have a reasonable answer.
f. Identify at least three features of the moment diagram that suggest you have a reasonable answer.
g. Determine the peak moment and its location.
h. Make a comprehensive argument that the peak moment found in part (g) is reasonable.
i. Comment on why your guess in part (a) was or was not close to the diagram in part (c).

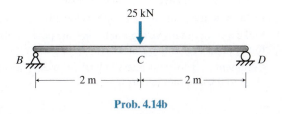

Prob. 4.14b

4.15 Our team is designing this simple frame to support a roof for parked cars. The geotechnical engineer has provided us with the point of fixity (i.e., the depth where the foundation behaves as though it is perfectly fixed), A. In order to design the members, we need to know the internal forces due to each type of load. Our task is to determine the internal forces due to snow on the roof. Although the roof load transfers to the beam at four points, it will be a reasonable approximation to consider the snow as a uniform distributed load.

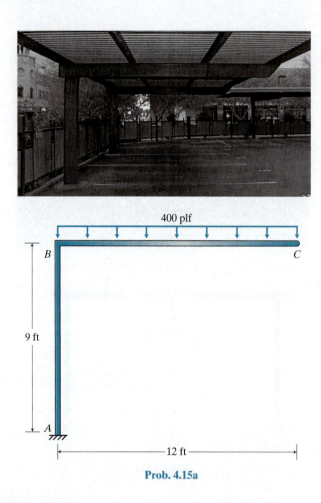

Prob. 4.15a

a. Guess the shape of the moment diagram.
b. Estimate the peak moment by considering the beam to be cantilevered out from a fixed support (i.e., no rotation at B). Will this provide an upper bound, a lower bound, or just an approximation for the actual peak moment? Justify your answer.
c. Construct the axial force, shear, and moment diagrams for the full frame. Label values and locations.
d. Identify at least two features of the axial force diagram that suggest you have a reasonable answer.
e. Identify at least three features of the shear diagram that suggest you have a reasonable answer.
f. Identify at least three features of the moment diagram that suggest you have a reasonable answer.
g. Determine the peak moment and its location.
h. Make a comprehensive argument that the peak moment found in part (g) is reasonable.
i. Comment on why your guess in part (a) was or was not close to the diagram in part (c).

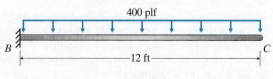

Prob. 4.15b

4.16 Our company is getting into the business of designing traffic signal structures. In order to design the cross-member and the post, we will need to know the internal forces due to the self-weight.

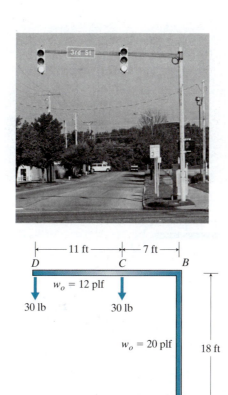

Prob. 4.16

a. Guess the shape of the moment diagram.
b. Calculate the moment reaction at A, ignoring the weights of the lights (the 30-lb forces). Will this provide an upper bound, a lower bound, or just an approximation for the actual moment reaction at A? Justify your answer.
c. Construct the axial force, shear, and moment diagrams for the frame with full self-weight. Label values and locations.
d. Identify at least two features of the axial force diagram that suggest you have a reasonable answer.
e. Identify at least three features of the shear diagram that suggest you have a reasonable answer.
f. Identify at least three features of the moment diagram that suggest you have a reasonable answer.
g. Determine the peak moment due to the full self-weight and its location.
h. Make a comprehensive argument that the peak moment found in part (g) is reasonable.
i. Comment on why your guess in part (a) was or was not close to the diagram in part (c).

4.17 The owners of a building with a below-grade window well would like to put on a removable roof in order to turn the space into a greenhouse. The roof will attach to the outer wall and lean on the inner wall. In order to design the frame, we need to know the internal forces along the length of the frame due to self-weight.

Prob. 4.17a

a. Guess the shape of the axial force diagram.
b. Approximate the frame as a simply supported beam with an equivalent uniform distributed load. Calculate the peak moment. Will this provide an upper bound, a lower bound, or just an approximation for the actual peak moment? Justify your answer.
c. Construct the axial force, shear, and moment diagrams for the leaning frame. Label values and locations.
d. Identify at least three features of the axial force diagram that suggest you have a reasonable answer.
e. Identify at least three features of the shear diagram that suggest you have a reasonable answer.
f. Identify at least three features of the moment diagram that suggest you have a reasonable answer.
g. Determine the peak moment due to the self-weight and its location.
h. Make a comprehensive argument that the peak moment found in part (g) is reasonable.
i. Comment on why your guess in part (a) was or was not close to the diagram in part (c).

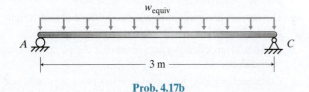

Prob. 4.17b

4.18 The owners of an amusement park want to add a shelter from the sun over their outdoor eating areas. The architect would like the shelter to rise at an angle and then cantilever out like a leaning tree. For preliminary design, we can focus on the self-weight of the sun screen, ignoring the self-weight of the frame. Our team needs to know the internal forces throughout the frame in order to select the members.

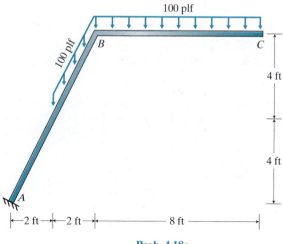

Prob. 4.18a

a. Guess the shape of the axial force diagram.
b. Approximate the frame as a cantilevered beam with an equivalent uniform distributed load. Calculate the peak moment. Will this provide an upper bound, a lower bound, or just an approximation for the actual peak moment? Justify your answer.

c. Construct the axial force, shear, and moment diagrams for the original frame. Label values and locations.
d. Identify at least three features of the axial force diagram that suggest you have a reasonable answer.
e. Identify at least three features of the shear diagram that suggest you have a reasonable answer.
f. Identify at least three features of the moment diagram that suggest you have a reasonable answer.
g. Determine the peak moment due to the self-weight and its location.
h. Make a comprehensive argument that the peak moment found in part (g) is reasonable.
i. Comment on why your guess in part (a) was or was not close to the diagram in part (c).

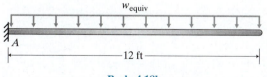

Prob. 4.18b

CHAPTER 5

DEFORMATIONS

Loading causes deformations in structures that can interfere with use of the structure or can cause damage to nonstructural elements; therefore, we need to predict those deformations in advance.

Motivation

The three products of structural analysis are reactions, internal forces, and deformations. Deformations are the displacements and slopes of the members of a structure. Displacement is the translation (i.e., left or right, up or down) of a point in the structure. Because we typically deal with relatively small deformations, the change in the slope of a point is essentially also the rotation of that point.

We want to know deformations for a variety of reasons. We predict dead load displacement to determine how much to camber a beam (i.e., prebend upward). We limit peak live load displacement to prevent non-structural damage such as cracked drywall and jammed windows. We predict lateral displacement due to wind or earthquake to know whether a structure is likely to bump into an adjacent structure.

In addition, some structural analysis methods are based on our ability to predict slopes or displacements. We cover those methods (e.g., Force Method, Moment Distribution Method) in later chapters.

Notice that we use the word *predict* when we discuss deformations. Because the deformations for most structures are very small relative to the dimensions of the structures, a small variance in deformation might be large relative to the prediction. Any difference between reality and the idealized structure can cause the measured deformation in the field to differ perceptibly from our prediction. Connections typically have some play or slack in them, and as we saw in Section 1.3: Structure Idealization, connections rarely behave as frictionless pins or perfectly rigid.

Fortunately, the limits we typically impose on deformations are based on the successful performance of many other structures. That means these limits already account for most variations between reality and our predicted deformations.

5.1 Double Integration Method

Introduction

Of all the tools we use for predicting deformations, only one gives us expressions for deformations throughout the structure: the Double Integration Method. Having expressions for deformations allows us to create visualizations of displaced shapes. Those expressions also let us find the peak slope or displacement when we do not know in advance where the peak is located.

This method considers deformations due to bending only (i.e., no axial deformation). Although the Double Integration Method can be used for frames, that process is cumbersome and prone to error. Therefore, the Double Integration Method is a tool most practical for beams.

How-To

Deriving the Governing Equation

The Double Integration Method gets its name because we can integrate the moment to obtain the slope, and we can integrate the slope to obtain the displacement. Proving that requires us to draw upon our knowledge from mechanics of materials and calculus and to make a few simplifying assumptions.

There are four key steps in the derivation: finding a link between the horizontal strain in the beam and the radius of curvature, finding a link between the horizontal strain and the moment at the cross-section, linking the radius of curvature and the moment, and expressing a change in the vertical position in terms of the moment.

1. Finding a Link Between the Horizontal Strain and the Radius of Curvature

Consider a beam subjected to some downward loading (Figure 5.1a). Vertical displacement is associated with bending, so there is curvature in the deformed beam (Figure 5.1b). One way to quantify that curvature is

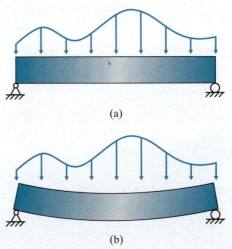

Figure 5.1 Beam subjected to some downward load: (a) Undisplaced shape; (b) Displaced shape.

the radius of curvature, ρ. If we take a small slice of the beam before loading, the faces are parallel (Figure 5.2a), but after bending, the extensions of the faces meet at a point (Figure 5.2b). The distance from the neutral axis to that point is called the *radius of curvature*. A very slightly bent beam has little curvature, and as a result ρ is very large. For an undeformed beam, ρ is infinity. The tighter the curve, the smaller the ρ.

To find the relationship between the horizontal strain, ϵ, and the radius of curvature, ρ, we need to consider a narrow slice from the beam (Figure 5.3a). Let's call the width of the slice dx. Once bent, the widths become arcs because of the curvature (Figure 5.3b). The arc length is different from dx everywhere but at the neutral axis. Remember that the neutral axis is defined as the location in the cross-section that does not change length when bent. Let's call these arc lengths ds. If we are looking at a very narrow slice, the extensions of the faces make a small angle $d\theta$. Therefore, we can reasonably approximate the arc length with the horizontal width (Figure 5.3c).

If plane sections remain plane, we can use trigonometry to obtain the following relationships:

$$\tan(d\theta) = \frac{dx'}{\rho} \approx \frac{dx}{\rho} \quad \text{and} \quad \tan(d\theta) = \frac{ds'}{(\rho - y)} \approx \frac{ds}{(\rho - y)}$$

Since the angle is very small, we can use the small-angle approximation:

$$\tan(d\theta) \approx d\theta$$

We can combine and rearrange these expressions to get expressions for the widths:

$$dx = \rho\, d\theta$$
$$ds = (\rho - y)\, d\theta$$

Now we use principles from mechanics of materials. Engineering strain is the change in length divided by the original length. When we calculate the strain at a distance y from the neutral axis, we get a relationship between the horizontal strain and the radius of curvature:

$$\epsilon = \frac{ds - dx}{dx} = \frac{(\rho - y)\, d\theta - \rho\, d\theta}{\rho\, d\theta} = -\frac{y}{\rho}$$

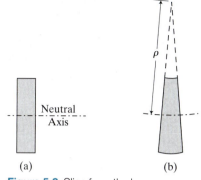

Figure 5.2 Slice from the beam: (a) Before loading, faces are parallel; (b) Under loading, faces turn in.

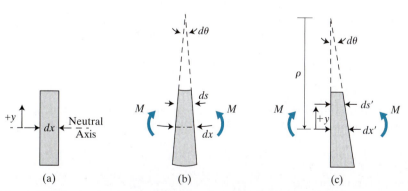

Figure 5.3 Narrow slice from a point along the beam: (a) Shape before load is applied; (b) After load is applied, slice experiences moment and segments arc; (c) For a very small angle, the arc lengths are approximated as the horizontal lengths.

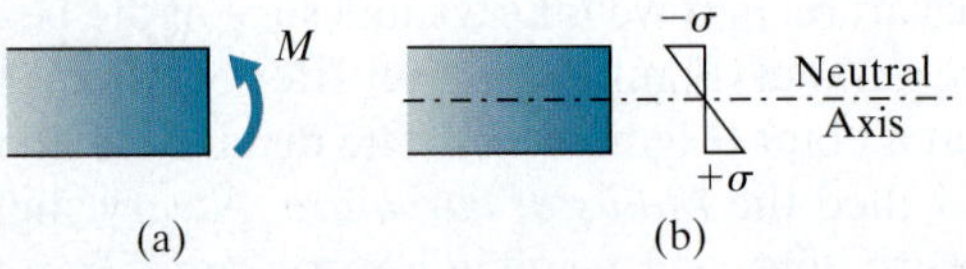

Figure 5.4 Beam under load cut at an arbitrary point: (a) Resulting internal moment; (b) Resulting horizontal stress distribution.

2. Finding a Link Between the Horizontal Strain and the Moment

From mechanics of materials we learned the relationship between the moment at a cross-section and the horizontal stress. *If* plane sections remain plane, we have the following relationship, where the moment and stress are defined in Figure 5.4:

$$\sigma = \frac{-My}{I}$$

If the material remains linear elastic, Hooke's law applies:

$$\sigma = E\epsilon$$

Combining these results, we have a relationship between the horizontal strain and the moment at that cross-section:

$$\epsilon = \frac{-My}{EI}$$

3. Linking the Radius of Curvature and the Moment

When we combine the results of the first two steps, we observe the following:

$$-\frac{y}{\rho} = -\frac{My}{EI}$$

We can simplify this expression. Note that the moment typically changes along the length of the beam (we saw that in Chapter 4: Internal Force Diagrams); therefore, it is a function of the horizontal position, x. That means the radius of curvature is also a function of x:

$$\frac{1}{\rho(x)} = \frac{M(x)}{EI}$$

4. Expressing the Change in the Vertical Position in Terms of the Moment

From calculus we can find an expression for $1/\rho$ in terms of x- and y-coordinates. The y-value in this case is vertical displacement.

$$\frac{1}{\rho} = \frac{d^2y/dx^2}{[1 + (dy/dx)^2]^{\frac{3}{2}}}$$

Combining this with the result of the previous step gives us a second-order, nonlinear, ordinary differential equation (ODE):

$$\frac{d^2y/dx^2}{[1 + (dy/dx)^2]^{\frac{3}{2}}} = \frac{M(x)}{EI}$$

There is not a simple analytical solution to this ODE. Fortunately, we can simplify the ODE for most structural situations we will encounter. Note that dy/dx is the slope at one point of the deformed beam. *If* the slope is relatively small, $(dy/dx)^2$ will be extremely small—essentially zero. In these cases, the ODE simplifies to an expression that is much easier to solve:

$$\frac{d^2y}{dx^2} = \frac{M(x)}{EI}$$

Using the Governing Equation
There are primarily three steps to the Double Integration Method: finding expressions for slope, finding expressions for displacement, and solving for the constants of integration.

1. Finding Expressions for Slope
Integrating the governing equation once results in dy/dx on the left side. As we pointed out earlier, that is the slope.

$$\theta(x_i) = \int \frac{M(x_i)}{EI} dx_i$$

This is an indefinite integral, so we obtain an expression with a constant of integration rather than a value. We will determine the values of the constants in the third step. Note that the expression is valid only in the range of $M(x_i)$. Therefore, if a beam needs three intervals to describe M [e.g., $M(x_1)$, $M(x_2)$, and $M(x_3)$], we will obtain three expressions for the slope [e.g., $\theta(x_1)$, $\theta(x_2)$, and $\theta(x_3)$].

2. Finding Expressions for Displacement
Integrating dy/dx results in only y on the left side: the vertical displacement.

$$\Delta(x_i) = \int \theta(x_i) dx_i$$

This is also an indefinite integral that will generate a constant of integration.

3. Solving for the Constants of Integration
To find the constants, we need boundary conditions. In fact, we need the same number of boundary conditions as the number of constants of integration we have in a problem. Boundary conditions are found at places where we know the magnitude of θ or Δ.

We will use two types of boundary conditions: fixed and compatibility. Every support provides at least one fixed boundary condition (Figure 5.5). Since we might not have enough fixed boundary conditions to solve for all of the constants, we may also need compatibility boundary conditions. Compatibility means that the deformation calculated immediately to the left of a point is the same immediately right of the point. This is helpful at the boundaries between intervals of x_i (Figure 5.6).

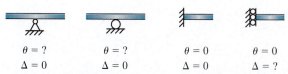

Figure 5.5 Fixed boundary conditions due to supports.

240　Chapter 5　Deformations

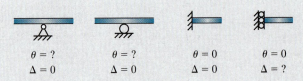

Figure 5.6 Compatibility boundary conditions: (a) Two boundary conditions at a continuous point in the beam; (b) One boundary condition at an internal hinge.

SECTION 5.1 HIGHLIGHTS

Double Integration Method

Typical Uses:
- Beams only.
- Any loading, especially distributed loads.
- For finding expressions for slope and displacement (e.g., to show displaced shape, to determine peak θ or Δ when location is not known in advance).

Procedure:
1. Integrate moment expressions to obtain expressions for slope.
$$\theta(x_i) = \int \frac{M(x_i)}{EI} dx_i$$
2. Integrate slope expressions to obtain expressions for displacement.
$$\Delta(x_i) = \int \theta(x_i) dx_i$$
3. Find constants of integration using boundary conditions.

Boundary Conditions:
- Need same number as constants of integration.
- Fixed boundary conditions:

$\theta = ?$	$\theta = ?$	$\theta = 0$	$\theta = 0$
$\Delta = 0$	$\Delta = 0$	$\Delta = 0$	$\Delta = ?$

- Compatibility boundary conditions:

  (a) $\theta(x_2 = l) = \theta(x_3 = 0)$, $\Delta(x_2 = l) = \Delta(x_3 = 0)$

  (b) $\theta(x_2 = l) \neq \theta(x_3 = 0)$, $\Delta(x_2 = l) = \Delta(x_3 = 0)$

Assumptions:
- Plane sections remain plane.
- Material remains linear elastic.
- Slopes are relatively small.

Peak:
- Peak θ will be at a free end, internal hinge, or where $M(x) = 0$.
- Peak Δ will be at a free end, internal hinge, or where $\theta(x) = 0$.

Section 5.1 Double Integration Method 241

EXAMPLE 5.1

To verify our understanding of a new structural analysis tool, we should use it to find a known solution. This helps ensure that we are applying the tool properly and, in the case of computer software, that the tool is working properly. Therefore, let's find the expression for the vertical displacement of a simply supported, uniformly loaded beam and then compare that expression with the one inside the front cover of this text.

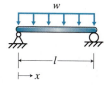

Figure 5.7a

Detailed Analysis:

Find the reactions:
FBD:

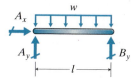

Figure 5.7b

Apply equilibrium:

$$\circlearrowleft \Sigma M_B = 0 = A_y(l) - wl(l/2)$$

$$A_y = +\frac{wl}{2} \ (+\uparrow)$$

Find the moment expression:
FBD of beam cut at x:

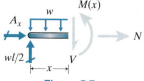

Figure 5.7c

Apply equilibrium:

$$\circlearrowleft \Sigma M_{cut} = 0 = \frac{wl}{2}(x) - w(x)\left(\frac{x}{2}\right) - M(x)$$

$$M(x) = -\frac{wx^2}{2} + \frac{wlx}{2}$$

Find expressions for θ and Δ:
Integrate $M(x)$:

$$\theta(x) = \int \frac{M(x)}{EI} dx = \frac{1}{EI} \int \left(-\frac{wx^2}{2} + \frac{wlx}{2}\right) dx$$

$$\theta(x) = \frac{1}{EI}\left[-\frac{wx^3}{6} + \frac{wlx^2}{4} + C_1\right]$$

242   Chapter 5   Deformations

Integrate $\theta(x)$:

$$\Delta(x) = \int \theta(x)\,dx = \frac{1}{EI}\int \left(-\frac{wx^3}{6} + \frac{wlx^2}{4} + C_1\right)dx$$

$$\Delta(x) = \frac{1}{EI}\left[-\frac{wx^4}{24} + \frac{wlx^3}{12} + C_1 x + C_2\right]$$

Boundary conditions:
Since there are two constants, we need two boundary conditions.

At $x = 0$, $\Delta(x) = 0$

$$\Delta(0) = \frac{1}{EI}[-0 + 0 + 0 + C_2] \equiv 0 \qquad \Rightarrow C_2 = 0$$

At $x = l$, $\Delta(x) = 0$

$$\Delta(l) = \frac{1}{EI}\left[-\frac{wl^4}{24} + \frac{wl^4}{12} + C_1 l\right] \equiv 0$$

$$C_1 l = \frac{wl^4}{24} - \frac{wl^4}{12} = -\frac{wl^4}{24} \qquad \Rightarrow C_1 = -\frac{wl^3}{24}$$

Summary of expressions:

$$\theta(x) = \frac{1}{EI}\left[-\frac{wx^3}{6} + \frac{wlx^2}{4} - \frac{wl^3}{24}\right]$$

$$\Delta(x) = \frac{1}{EI}\left[-\frac{wx^4}{24} + \frac{wlx^3}{12} - \frac{wl^3 x}{24}\right]$$

Comparison with Precalculated Expressions:

At $x = 0$, $\theta(x) = -\dfrac{wl^3}{24EI}$

At $x = l$, $\theta(x) = \dfrac{1}{EI}\left[-\dfrac{wl^3}{6} + \dfrac{wl^3}{4} - \dfrac{wl^3}{24}\right] = \dfrac{wl^3}{24EI}$

$$\Delta(x) = -\frac{w}{24EI}\left[x^4 - 2lx^3 + l^3 x\right]$$

The expressions all match the precalculated expressions found inside the front cover of this text. Therefore, it appears that we have correctly applied the Double Integration Method.

Example 5.2

We are tasked with designing a beam to support a moving hoist. The track for the hoist attaches to the beam at midspan. Since the hoist will only sometimes load the beam, it is a live load. In order to allow the hoist to move smoothly along the track, we have tight limits on the live load deflection. One option is to make the beam simply supported and use a sufficiently large moment of inertia, I. Another option is to make the end against the wall a fixed support. The fixed support will probably cost more than a pinned support, but we might be able to save even more money on a smaller beam. To help choose between the options, we want to know how much of an impact adding a fixed support has compared to simple supports.

Figure 5.8

Using methods we learn later in this text, one member of our team has calculated the reactions if we use a fixed support on the wall end.

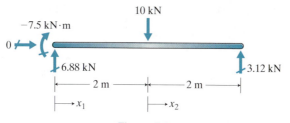

Figure 5.9

Simply Supported Beam Solution:

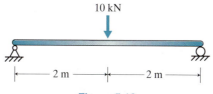

Figure 5.10

From inside the front cover of this text, we find

$$\Delta_{max} = -\frac{Pl^3}{48EI} = -\frac{10\text{ kN}(4\text{ m})^3}{48EI} = -\frac{13.33\text{ kN}\cdot\text{m}^3}{EI}$$

244 Chapter 5 Deformations

Fixed-Roller Beam Solution:
Find expressions for shear and moment:
  Integrate $w(x)$:
  $$V(x_1) = \int w \, dx_1 = \int 0 \, dx_1 = C_1$$
  At $x_1 = 0$, $V(x_1) = 6.88 \text{ kN}$ $\Rightarrow C_1 = 6.88 \text{ kN}$

  $$V(x_2) = \int w \, dx_2 = \int 0 \, dx_2 = C_2$$
  At $x_2 = 0$, $V(x_2) = V(x_1 = 2 \text{ m}) - 10 \text{ kN}$
  $C_2 = 6.88 \text{ kN} - 10 \text{ kN}$ $\Rightarrow C_2 = -3.12 \text{ kN}$

  Integrate $V(x)$:
  $$M(x_1) = \int V(x_1) \, dx_1 = \int 6.88 \text{ kN} \, dx_1 = 6.88 \text{ kN} \cdot x_1 + C_3$$
  At $x_1 = 0$, $M(x_1) = -7.50 \text{ kN} \cdot \text{m}$ $\Rightarrow C_3 = -7.50 \text{ kN} \cdot \text{m}$
  $$M(x_2) = \int V(x_2) \, dx_2 = \int -3.12 \text{ kN} \, dx_2 = -3.12 \text{ kN} \cdot x_2 + C_4$$

  At $x_2 = 0$, $M(x_2) = M(x_1 = 2 \text{ m})$
  $C_4 = 6.88 \text{ kN} \cdot 2 \text{ m} - 7.50 \text{ kN} \cdot \text{m}$ $\Rightarrow C_4 = 6.26 \text{ kN} \cdot \text{m}$

Evaluation of Results:
Satisfied Fundamental Principles?
  At $x_2 = 2 \text{ m}$, $M(x_2) = -3.12 \text{ kN} \cdot 2 \text{ m} + 6.26 \text{ kN} \cdot \text{m}$
    $= 0.02 \text{ kN} \cdot \text{m} \approx 0$ ✓

The difference is due to roundoff error.

Summary:
  $M(x_1) = 6.88 \text{ kN} \cdot x_1 - 7.50 \text{ kN} \cdot \text{m}$ $M(x_2) = -3.12 \text{ kN} \cdot x_2 + 6.26 \text{ kN} \cdot \text{m}$

Find expressions for θ and Δ:
  Integrate $M(x)$:
  $$\theta(x_1) = \int \frac{M(x_1)}{EI} \, dx_1$$
  $$= \frac{1}{EI} \int (6.88 \text{ kN} \cdot x_1 - 7.50 \text{ kN} \cdot \text{m}) \, dx_1$$
  $$= \frac{1}{EI} (3.44 \text{ kN} \cdot x_1^2 - 7.50 \text{ kN} \cdot \text{m} \cdot x_1 + C_5)$$

  $$\theta(x_2) = \int \frac{M(x_2)}{EI} \, dx_2$$
  $$= \frac{1}{EI} \int (-3.12 \text{ kN} \cdot x_2 + 6.26 \text{ kN} \cdot \text{m}) \, dx_2$$
  $$= \frac{1}{EI} (-1.56 \text{ kN} \cdot x_2^2 + 6.26 \text{ kN} \cdot \text{m} \cdot x_2 + C_6)$$

  Integrate $\theta(x)$:
  $$\Delta(x_1) = \int \theta(x_1) \, dx_1$$

Section 5.1 Double Integration Method

$$= \frac{1}{EI}\int (3.44 \text{ kN} \cdot x_1^2 - 7.50 \text{ kN} \cdot \text{m} \cdot x_1 + C_5)\, dx_1$$

$$= \frac{1}{EI}(1.147 \text{ kN} \cdot x_1^3 - 3.75 \text{ kN} \cdot \text{m} \cdot x_1^2 + C_5 \cdot x_1 + C_7)$$

$$\Delta(x_2) = \int \theta(x_2)\, dx_2$$

$$= \frac{1}{EI}\int (-1.56 \text{ kN} \cdot x_2^2 + 6.26 \text{ kN} \cdot \text{m} \cdot x_2 + C_6)\, dx_1$$

$$= \frac{1}{EI}(-0.52 \text{ kN} \cdot x_2^3 + 3.13 \text{ kN} \cdot \text{m} \cdot x_2^2 + C_6 \cdot x_2 + C_8)$$

Boundary conditions:
Since there are four constants, we need four boundary conditions.
At $x_1 = 0$, $\theta(x_1) = 0$

$$\theta(0) = \frac{1}{EI}(0 - 0 + C_5) = 0 \qquad \Rightarrow C_5 = 0$$

At $x_1 = 0$, $\Delta(x_1) = 0$

$$\Delta(0) = \frac{1}{EI}(0 - 0 + 0 + C_7) = 0 \qquad \Rightarrow C_7 = 0$$

At $x_1 = 2$ m, $x_2 = 0$, $\theta(x_1) = \theta(x_2)$

$$\frac{1}{EI}\left[3.44 \text{ kN}(2 \text{ m})^2 - 7.50 \text{ kN} \cdot \text{m}(2 \text{ m})\right] = \frac{1}{EI}(-0 + 0 + C_6)$$

$$\Rightarrow C_6 = -1.24 \text{ kN} \cdot \text{m}^2$$

At $x_2 = 2$ m, $\Delta(x_2) = 0$

$$\Delta(2 \text{ m}) = \frac{1}{EI}\left[-0.52 \text{ kN}(2 \text{ m})^3 + 3.13 \text{ kN} \cdot \text{m}(2 \text{ m})^2\right.$$
$$\left. - 1.24 \text{ kN} \cdot \text{m}^2(2 \text{ m}) + C_8\right] = 0$$
$$\Rightarrow C_8 = -5.88 \text{ kN} \cdot \text{m}^3$$

Evaluation of Results:
Satisfied Fundamental Principles?
At $x_1 = 2$ m, $x_2 = 0$, $\Delta(x_1) = \Delta(x_2)$

$$\Delta(x_1 = 2 \text{ m}) = \frac{1}{EI}\left[1.147 \text{ kN}(2 \text{ m})^3 - 3.75 \text{ kN} \cdot \text{m}(2 \text{ m})^2\right]$$

$$= -\frac{5.82 \text{ kN} \cdot \text{m}^3}{EI}$$

$$\Delta(x_2 = 0) = \frac{1}{EI}\left[-0 + 0 - 0 - 5.88 \text{ kN} \cdot \text{m}^3\right]$$

$$= -\frac{5.88 \text{ kN} \cdot \text{m}^3}{EI}$$

$$-\frac{5.82 \text{ kN} \cdot \text{m}^3}{EI} \approx -\frac{5.88 \text{ kN} \cdot \text{m}^3}{EI} \checkmark$$

The difference is due to roundoff error.

Summary:

$$\theta(x_1) = \frac{1}{EI}(3.44 \text{ kN} \cdot x_1^2 - 7.50 \text{ kN} \cdot \text{m} \cdot x_1)$$

$$\theta(x_2) = \frac{1}{EI}(-1.56 \text{ kN} \cdot x_2^2 + 6.26 \text{ kN} \cdot \text{m} \cdot x_2 - 1.24 \text{ kN} \cdot \text{m}^2)$$

$$\Delta(x_1) = \frac{1}{EI}(1.147 \text{ kN} \cdot x_1^3 - 3.75 \text{ kN} \cdot \text{m} \cdot x_1^2)$$

$$\Delta(x_2) = \frac{1}{EI}(-0.52 \text{ kN} \cdot x_2^3 + 3.13 \text{ kN} \cdot \text{m} \cdot x_2^2 - 1.24 \text{ kN} \cdot \text{m}^2 \cdot x_2 - 5.88 \text{ kN} \cdot \text{m}^3)$$

Find $\Delta_{\max}$:

Check the range of x_1:

$$\frac{1}{EI}(3.44 \text{ kN} \cdot x_1^2 - 7.50 \text{ kN} \cdot \text{m} \cdot x_1) \equiv 0$$

$$3.44 \text{ kN} \cdot x_1 - 7.50 \text{ kN} \cdot \text{m} = 0$$

$$\Rightarrow x_1 = 2.18 \text{ m} > 2.0 \text{ m} \qquad \therefore \text{ Not in range of } x_1$$

Check the range of x_2:

$$\frac{1}{EI}(-1.56 \text{ kN} \cdot x_2^2 + 6.26 \text{ kN} \cdot \text{m} \cdot x_2 - 1.24 \text{ kN} \cdot \text{m}^2) \equiv 0$$

$$-1.56 \text{ kN} \cdot x_2^2 + 6.26 \text{ kN} \cdot \text{m} \cdot x_2 - 1.24 \text{ kN} \cdot \text{m}^2 = 0$$

$$x_2 = \frac{-6.26 \text{ kN} \cdot \text{m} \pm \sqrt{(6.26 \text{ kN} \cdot \text{m})^2 - 4(-1.56 \text{ kN})(-1.24 \text{ kN} \cdot \text{m}^2)}}{2(-1.56 \text{ kN})}$$

$$= \frac{-6.26 \text{ kN} \cdot \text{m} \pm 5.61 \text{ kN} \cdot \text{m}}{-3.12 \text{ kN}}$$

$$\Rightarrow x_2 = 3.80 \text{ m} \quad \text{or} \quad 0.21 \text{ m} \qquad \therefore x_2 = 0.21 \text{ m}$$

Calculate $\Delta_{\max}$:

$$\Delta_{\max} = \frac{1}{EI}\big[-0.52 \text{ kN}(0.21 \text{ m})^3 + 3.13 \text{ kN} \cdot \text{m}(0.21 \text{ m})^2$$
$$- 1.24 \text{ kN} \cdot \text{m}^2(0.21 \text{ m}) - 5.88 \text{ kN} \cdot \text{m}^3\big]$$

$$\Delta_{\max} = -\frac{6.01 \text{ kN} \cdot \text{m}^3}{EI}$$

Evaluation of Results:

Observed Expected Features?
The slope should start down, negative, once we move away from the support.
Using a position of $x_1 = 0.1$ m, we do get a negative slope. ✓

Observed Expected Features?
The slope should be positive, upward, at the right end. At $x_2 = 2$ m,
$\theta(x_2) = +5.04 \text{ kN} \cdot \text{m}^2/EI$. ✓

Satisfied Fundamental Principles?
At the right end, the displacement should be zero because of the support.
Substituting $x_2 = 2$ m into the expression for $\Delta(x_2)$ gives $\Delta(x_2) = 0$. ✓

Approximation Predicted Outcomes?
This Δ_{max} is smaller than the Δ_{max} for a simply supported beam, as expected. ✓

Conclusion:
Together with the two checks made earlier, these observations strongly suggest that we have a reasonable result.

Compare displacements:

Simply supported $\quad \Delta_{max} = -\dfrac{13.33 \text{ kN} \cdot \text{m}^3}{EI}$

Fixed-roller supported $\quad \Delta_{max} = -\dfrac{6.01 \text{ kN} \cdot \text{m}^3}{EI}$

The impact of changing the support at the wall end from pinned to fixed is a 55% reduction in live load deflection. Therefore, we could use a beam with a 55% lower moment of inertia. The cost savings for beam material will likely be greater than the cost increase for the fixed support, so making the supports fixed-roller appears to be a good choice.

5.2 Conjugate Beam Method

Introduction
For engineers who are comfortable constructing shear and moment diagrams by deduction, the Conjugate Beam Method can be a quicker alternative to the Double Integration Method for predicting deformations. Because the Conjugate Beam Method is fundamentally the same as the Double Integration Method, the Conjugate Beam Method is practical only for beams. Furthermore, it is most efficient when there are no distributed loads on the beams, just point loads and/or couple moments.

How-To
The basic concept for the Conjugate Beam Method comes from what we learned in the Double Integration Method (Section 5.1): slope is the integral of M/EI, and displacement is the integral of slope. This is analogous to how we found internal force diagrams in Chapter 4: shear is the integral of the distributed load, and moment is the integral of shear. So, the Conjugate Beam Method is really just another way to integrate M/EI twice.

The basic idea is that if we apply the M/EI diagram to the beam as if it was the distributed load, then the shear calculated anywhere along this conjugate beam will be the slope anywhere along the actual beam. Likewise, the moment calculated anywhere along the conjugate beam will be the displacement along the actual beam. To preserve our standard sign conventions for positive slope and displacement, we must consider positive M/EI acting up and negative M/EI acting down.

The analogy is straightforward, so why do we use a "conjugate" beam? The key is boundary conditions. If you remember from Chapter 4: Internal Force Diagrams, we used indefinite integrals to obtain expressions for shear and moment. Indefinite integrals produce expressions with constants of integration. We need boundary conditions to find the unique value of the constant for a certain beam. Those boundary conditions come from reactions at the supports (Figure 5.11). Notice that the expressions are identical for both beams in Figure 5.11 because the

248   **Chapter 5**   Deformations

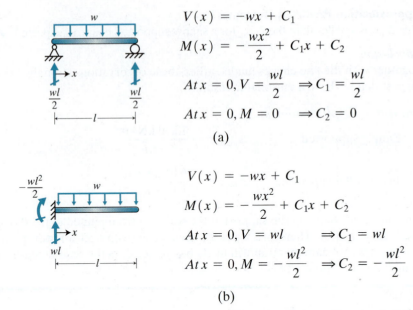

Figure 5.11 Uniformly loaded beam with different support conditions results in the same expressions for V and M but different constants of integration: (a) Simply supported conditions; (b) Cantilever conditions.

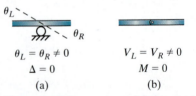

Figure 5.12 Comparison of real support condition and the required condition for a conjugate beam: (a) Real beam supported on a roller; (b) Conjugate beam has an internal hinge.

loading and span length are identical. Only the support conditions affect the values of the constants.

To obtain the proper values of the constants in the Conjugate Beam Method, we need support conditions for the conjugate beam that match the slope and displacement behavior of the original beam. For example, consider part of a beam sitting on an interior roller (Figure 5.12a). The real beam can have a slope (first integral) over the roller, but its displacement (second integral) will be zero. In the analogous conjugate beam, we need a condition that can have shear (first integral) but no moment (second integral). That requires a hinge (Figure 5.12b).

Figure 5.13 shows the support conditions we need to use on the conjugate beam in order to replicate the behavior of the real beam. Since *conjugate* means "inversely or oppositely related," the method is appropriately named.

To illustrate the transformation, Figure 5.14 shows a variety of real beams and their conjugates. Note that some of the conjugate beams are unstable. No worries; the M/EI diagrams that we apply as distributed loads will keep the conjugate beams in equilibrium.

Once we have the conjugate beam and we apply the M/EI diagram from the real beam, we can use either of the methods from Chapter 4: Internal Force Diagrams to obtain the shear and moment diagrams for the conjugate beam. The shear diagram shows the values of slope everywhere along the real beam, and the moment diagram shows the values of displacement.

Section 5.2 Conjugate Beam Method

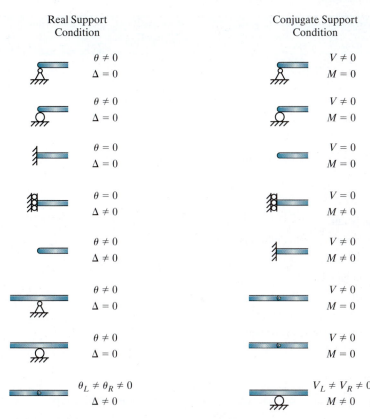

Figure 5.13 Converting support conditions for a real beam into support conditions for the conjugate beam.

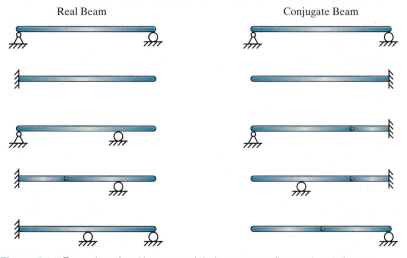

Figure 5.14 Examples of real beams and their corresponding conjugate beams.

Section 5.2 Highlights

Conjugate Beam Method

Typical Uses:
- Beams only.
- Point loads and/or couple moments.
- For finding the displaced shape or determining the peak θ or Δ when the location is not known in advance.

Procedure:
1. Construct the moment diagram for the real beam.
2. Create the conjugate beam with the appropriate support conditions.
3. Divide the real moment by EI to create an M/EI diagram.
4. Apply M/EI as a distributed load to the conjugate beam:
   $+M/EI$ acts upward, $-M/EI$ acts downward.
5. Construct the shear diagram for the conjugate beam (θ for real beam).
6. Construct the moment diagram for the conjugate beam (Δ for real beam).

Conjugate Conditions:

Real Support Condition

Conjugate Support Condition

Assumptions (same as for Double Integration Method):
- Plane sections remain plane.
- Material remains linear elastic.
- Slopes are relatively small.

Peak:
- Peak θ will be where M/EI applied to the conjugate beam is zero.
- Peak Δ will be where V for the conjugate beam is zero.

Example 5.3

To verify our understanding of a new structural analysis tool, we should use it to find a known solution. In order to confirm that we are properly applying the Conjugate Beam Method, let's find the peak vertical displacement of a simply supported beam with a single point load at midspan. We can compare our result with the solution inside the front cover of this book.

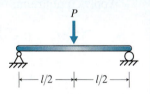

Figure 5.15a

Detailed Analysis:

Find the reactions:
    FBD:

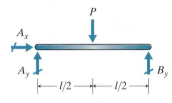

Figure 5.15b

Apply equilibrium:
$$\circlearrowleft \Sigma M_B = 0 = A_y(l) - P(l/2)$$
$$A_y = +\frac{P}{2}(+\uparrow)$$

Construct diagrams for the real beam by deduction:

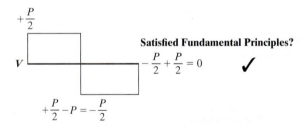

Satisfied Fundamental Principles?
$$-\frac{P}{2} + \frac{P}{2} = 0 \quad \checkmark$$

Satisfied Fundamental Principles?
$$\frac{Pl}{4} + \left(-\frac{P}{2}\right)\cdot\frac{l}{2} = 0 \quad \checkmark$$

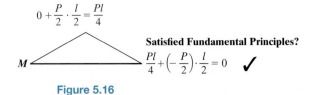

Figure 5.16

Conjugate beam:
For these supports, the conjugate beam is exactly the same as the real beam.

Figure 5.17a

Apply the M/EI diagram:
The modulus of elasticity, E, and moment of inertia, I, are uniform along the beam.

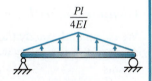

Figure 5.17b

252 Chapter 5 Deformations

FBD:

Apply equilibrium:

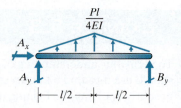

Figure 5.17c

$$\curvearrowright \Sigma M_B = 0 = A_y \cdot l + \frac{1}{2}\left(\frac{Pl}{4EI} \cdot l\right)\left(\frac{l}{2}\right)$$

$$A_y = -\frac{Pl^2}{16EI} \quad (+\uparrow)$$

$$+\uparrow \Sigma F_y = 0 = -\frac{Pl^2}{16EI} + \frac{1}{2}\left(\frac{Pl}{4EI} \cdot l\right) + B_y$$

$$B_y = -\frac{Pl^2}{16EI} \quad (+\uparrow)$$

Construct the shear diagram for the conjugate beam by deduction:

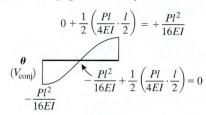

Figure 5.18

Calculate the peak displacement:
FBD of the conjugate beam cut at $V = 0$:

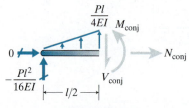

Figure 5.19

Apply equilibrium:

$$\curvearrowright \Sigma M_{cut} = 0 = -\frac{Pl^2}{16EI} \cdot \frac{l}{2} + \frac{1}{2}\left(\frac{Pl}{4EI} \cdot \frac{l}{2}\right)\left(\frac{1}{3} \cdot \frac{l}{2}\right) - M_{conj}$$

$$M_{conj} = -\frac{Pl^3}{32EI} + \frac{Pl^3}{96EI} = -\frac{Pl^3}{48EI}$$

$$\Delta_{max} = -\frac{Pl^3}{48EI} \quad (+\uparrow)$$

Comparison with Precalculated Displacement:

We calculated exactly the same peak displacement as the value we find inside the front cover of this text. Now we are ready to use the Conjugate Beam Method to analyze other beams.

Example 5.4

A steel beam is fixed at one end and cantilevers out past a roller support. For aesthetic reasons, the architect requested that the beam change depths for the cantilevered end. Now the owner wants to hang a sign from the tip of the cantilevered end. A team member has confirmed that the beam will be strong enough. Our assignment is to confirm that the deflection in each span due to the added load does not exceed $l/240$. Remember that for steel, $E = 29{,}000$ ksi. Our teammate has provided us with the reactions.

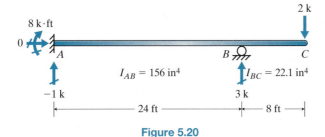

Figure 5.20

Estimated Solution:

In order to estimate the vertical displacement at end C, we can consider segment BC as a cantilever. This should provide a lower bound on the vertical displacement at C because the actual beam will rotate at B.

Figure 5.21

From the inside front cover of this text, we have the following formula for the displacement of a cantilevered beam with a point load at the end:

$$\Delta_{end} = -\frac{P}{6EI}(2l^3) = -\frac{Pl^3}{3EI}$$

Estimated displacement:

$$\Delta_{Cy} = -\frac{2\,\text{k}\,(8\,\text{ft} \cdot 12\,\text{in./ft})^3}{3(29{,}000\,\text{ksi})(22.1\,\text{in}^4)} = -0.92\,\text{in.}\ (+\uparrow)$$

Detailed Analysis:

Construct diagrams for the real beam by deduction:

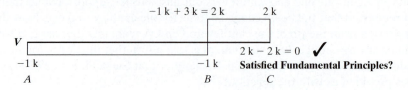

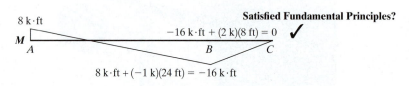

Figure 5.22

Conjugate beam:
For these supports, the conjugate beam is the following:

Figure 5.23a

Apply the M/EI diagram:
Calculate the ordinates of the M/EI diagram:

$$\frac{8 \text{ k} \cdot \text{ft}\,(12 \text{ in./ft})}{156 \text{ in}^4\,(29{,}000 \text{ ksi})} = 2.12 \times 10^{-5} \text{ in}^{-1} \quad \text{Acts up since } +$$

$$\frac{-16 \text{ k} \cdot \text{ft}\,(12 \text{ in./ft})}{156 \text{ in}^4\,(29{,}000 \text{ ksi})} = -4.24 \times 10^{-5} \text{ in}^{-1} \quad \text{Acts down since } -$$

$$\frac{-16 \text{ k} \cdot \text{ft}\,(12 \text{ in./ft})}{22.1 \text{ in}^4\,(29{,}000 \text{ ksi})} = -3.00 \times 10^{-4} \text{ in}^{-1} \quad \text{Acts down since } -$$

Apply M/EI:

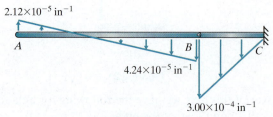

Figure 5.23b

Find the location of zero M/EI using similar triangles:

$$\frac{x}{24 \text{ ft}} = \frac{2.12}{(4.24 + 2.12)} \quad \Rightarrow x = 8 \text{ ft}$$

FBD of AB:

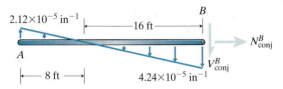

Figure 5.23c

Apply equilibrium:

$$+\uparrow \Sigma F_y = 0 = \frac{1}{2}(2.12\times 10^{-5}\text{ in}^{-1})(8\text{ ft}\cdot 12\text{ in./ft})$$
$$-\frac{1}{2}(4.24\times 10^{-5}\text{ in}^{-1})(16\text{ ft}\cdot 12\text{ in./ft}) - V_{\text{conj}}^B$$

$$V_{\text{conj}}^B = -3.05\times 10^{-3}$$

Note: We can verify that this insufficiently constrained conjugate beam is still in equilibrium.

$$\circlearrowleft \Sigma M_B = \frac{1}{2}(2.12\times 10^{-5}\text{ in}^{-1})(8\text{ ft})\left(\frac{2}{3}\cdot 8\text{ ft} + 16\text{ ft}\right)$$
$$-\frac{1}{2}(4.24\times 10^{-5}\text{ in}^{-1})(16\text{ ft})\left(\frac{1}{3}\cdot 16\text{ ft}\right) = 0 \checkmark$$

FBD of BC:

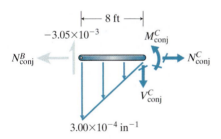

Figure 5.23d

Apply equilibrium:

$$+\uparrow \Sigma F_y = 0 = -3.05\times 10^{-3} - \frac{1}{2}(3.00\times 10^{-4}\text{ in}^{-1})(8\text{ ft}\cdot 12\text{ in./ft}) - V_{\text{conj}}^C$$

$$V_{\text{conj}}^C = -0.01745$$

$$\circlearrowleft \Sigma M_C = 0 = 3.05\times 10^{-3}(8\text{ ft}\cdot 12\text{ in./ft})$$
$$+ \frac{1}{2}(3.00\times 10^{-4}\text{ in}^{-1})(8\text{ ft}\cdot 12\text{ in./ft})\left(\frac{2}{3}\cdot 8\text{ ft}\cdot 12\text{ in./ft}\right) + M_{\text{conj}}^C$$

$$M_{\text{conj}}^C = -1.214\text{ in.}$$

256 **Chapter 5** Deformations

Construct the shear diagram for the conjugate beam by deduction:

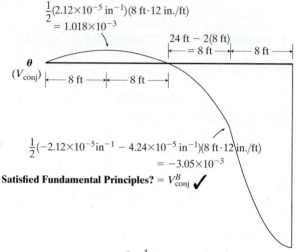

Figure 5.24

Peak displacement in span AB:
FBD of conjugate beam cut at $V_{conj} = 0$:

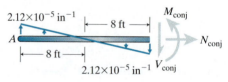

Figure 5.25

Apply equilibrium:

$$+\circlearrowleft \Sigma M_{cut} = 0 = \frac{1}{2}(2.12\times10^{-5}\text{ in}^{-1})(8\text{ ft})\left(\frac{2}{3}\cdot 8\text{ ft} + 8\text{ ft}\right)(12\text{ in./ft})^2$$

$$+ \frac{1}{2}(-2.12\times10^{-5}\text{ in}^{-1})(8\text{ ft})\left(\frac{1}{3}\cdot 8\text{ ft}\right)(12\text{ in./ft})^2 - M_{conj}$$

$M_{conj} = 0.1303$ in.
$\Delta^{AB}_{max} = +0.1303$ in. $(+\uparrow)$

Peak displacement in span BC:
Peak occurs where $V_{conj} = 0$, which is at C.

$M^C_{conj} = -1.214$ in.
$\Delta^{BC}_{max} = -1.214$ in. $(+\uparrow)$

Evaluation of Results:

Observed Expected Features?
The slope at the left end should be zero because of the fixed support. At that end $V_{conj} = 0$; therefore, it is. ✓

Observed Expected Features?
Just past the support at A, the beam should slope up, positive. V_{conj} is positive there. ✓

Observed Expected Features?
At the far right, C, the beam should slope down, negative. V_{conj} is negative there. ✓

Approximation Predicted Outcomes?
Our prediction of ΔC_y is -1.21 in., which is larger than our estimate of -0.92 in., as expected. ✓

Approximation Predicted Outcomes?
Our estimated displacement at C was only 24% low, which is reasonable for this type of estimate. ✓

Conclusion:

Together with the two checks made earlier, these observations strongly suggest that we have a reasonable result.

Conclusion:

Compare the peak displacement in AB with the limit:

$$\Delta^{AB}_{limit} = \frac{24 \text{ ft } (12 \text{ in./ft})}{240} = 1.2 \text{ in.}$$

$$\Delta^{AB}_{max} = +0.13 \text{ in.} (+\uparrow)$$

$$\Delta^{AB}_{max} < \Delta^{AB}_{limit} \therefore \text{ okay}$$

Compare the peak displacement in BC with the limit:

$$\Delta^{BC}_{limit} = \frac{8 \text{ ft}(12 \text{ in./ft})}{240} = 0.4 \text{ in.}$$

$$\Delta^{BC}_{max} = +1.21 \text{ in.} (+\downarrow)$$

$$\Delta^{BC}_{max} > \Delta^{BC}_{limit} \therefore \text{ NOT okay}$$

The cantilevered end will deflect too much. The team will need to explore alternatives.

5.3 Virtual Work Method

Introduction

The Virtual Work Method is our most versatile tool for predicting deformations. We can use it with trusses, beams, and frames. The basic premise of the method is that we can predict deformation based on the energy stored in the structure due to the internal forces. Therefore, the method tells us how much of the deformation is due to each internal force in each member of the structure. For example, we can know what percentage of the lateral displacement (i.e., drift) of a frame is due to the axial load in the columns. Knowing which structural members are the major contributors to a particular deformation allows us to make efficient design choices when we want to limit or reduce that deformation.

The key limitation to this method is that it produces the displacement or slope at only one point in the structure at a time. If we want to know both, we need to apply the method twice. If we want to know the displacement in multiple locations, we need to apply the method multiple times. Because of this limitation, the method is not very useful for predicting the peak displacement if we don't know where that will occur.

How-To

Deriving the Governing Equation

The Virtual Work Method is based on conservation of energy. To develop the governing equation, we go through four steps: equivalency of work terms, strain energy due to the axial load, strain energy due to bending, and generation of the governing equation. Although the method is valid for nonlinear material behavior, we will limit our derivation and application to materials that behave linear elastically (i.e., E is constant at all loads).

1. Equivalency of Work Terms

Our derivation starts by calculating the external work, U_e, performed as we slowly load a rod axially. As the rod in Figure 5.26a is loaded from zero to P_1, the deformation grows from zero to Δ_1. The external work is the area under the force-displacement curve, $\frac{1}{2}P_1\Delta_1$. Notice that the work performed is not $P_1\Delta_1$ because the full force, P_1, was not acting throughout the entire deformation, Δ_1. If we apply a different load, P_2, to the same rod, we get a final deformation of Δ_2 (Figure 5.26b). The external work is $\frac{1}{2}P_2\Delta_2$.

Now consider that we apply both loads to the rod. In Figure 5.27a, we apply P_1 and then P_2. In Figure 5.27b, we apply P_2 and then P_1. The total applied load is the same, $P_1 + P_2$, and the resulting total deformation in both cases is the same, $\Delta_1 + \Delta_2$. Therefore, the total external work should be the same for both cases:

$$\frac{1}{2}P_1\Delta_1 + \frac{1}{2}P_2\Delta_2 + P_1\Delta_2 = \frac{1}{2}P_2\Delta_2 + \frac{1}{2}P_1\Delta_1 + P_2\Delta_1$$

Once we cancel the like terms on each side, we obtain an interesting result that is central to the Virtual Work Method:

$$P_1\Delta_2 = P_2\Delta_1$$

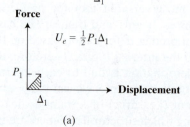
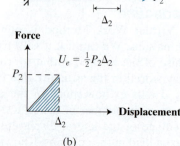

(a)   (b)

Figure 5.26 Rod loaded axially and the resulting force-displacement behavior: (a) Rod subjected to load P_1; (b) Rod subjected to a different load P_2.

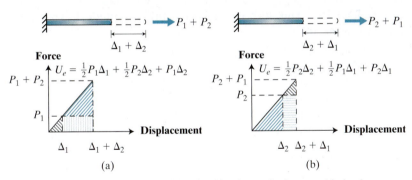

Figure 5.27 Rod loaded axially and the resulting force-displacement behavior: (a) Rod subjected to load P_1 and then P_2; (b) Rod subjected to load P_2 and then P_1.

These two terms are the light blue rectangles in Figure 5.27. From just looking at the rectangles, it is not obvious that they are the same size, but we have shown that they must be equal. What makes this result interesting is that it is true for any load P_2. So that load does not need to really exist; it can be *virtual*.

It is important to remember that work is a dot product; the load and deformation must be at the same location and in the same direction (Figure 5.28).

The energy equality we just derived also applies to systems of loads (Figure 5.29). The only requirement is that the loads in the two systems be applied at the same locations and in the same directions. The same direction is from a vector perspective; reversing signs is okay. Note that the second system does not need to be a scaled version of the first system. That means some of the loads can be zero. Since we can make the second system anything we want, let's make all of the loads zero except one; we call the

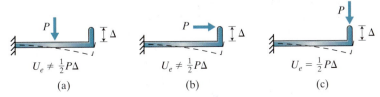

Figure 5.28 Frame loaded with a single point force and the resulting deformed shape:
(a) Displacement measured at different location from applied load;
(b) Displacement measured in different direction than applied load;
(c) Displacement measured at location and in direction of applied load.

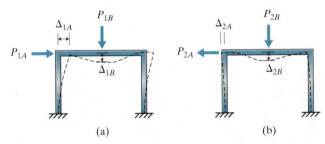

Figure 5.29 Frame loaded with system of loads and the resulting work-related deformations: (a) First system of loads; (b) Second system of loads.

nonzero load P_{2k}. And, let's give load P_{2k} the magnitude 1. The energy equality reduces to

$$\Sigma P_{1j}\Delta_{2j} = 1 \cdot \Delta_{1k}$$

In this expression, the P_{1j}'s are all the real loads and the Δ_{2j}'s are the deformations due to the virtual load, P_{2k}. The right-hand side, Δ_{1k}, is the real deformation where the virtual load was applied.

This means we can calculate one real Δ by applying a virtual unit load at the location where we want to measure Δ. All we need in order to finish the calculation are the deflections due to the virtual load, Δ_{2j}'s. But if we knew how to calculate those deflections, we would know how to calculate Δ_{1k}.

To get past this conundrum, we have to use conservation of energy. The external work performed on the structural system is energy, and this energy must go somewhere. If we limit ourselves to static conditions, none of the energy becomes kinetic, so it must all be stored internally as strain energy:

$$U_e = U_i$$

Therefore, all we need is a way to calculate the total strain energy stored in a structure.

2. Strain Energy Due to the Axial Load

One of the forms of internal work is strain energy stored due to the axial load. We can calculate the strain energy in a piece of the structure by integrating the product of the axial force and the axial deformation along the piece. Note that in our field, we often refer to the pieces as *elements*.

$$U_i = \int_0^l N(x)\,d\Delta$$

Remember that for our energy equality, the loads are from the real system and the deformations are from the virtual system. Therefore, in this integral, the axial force is due to the real loads and the axial deformation is due to the virtual load.

To express $d\Delta$ as a function of position, x, we need to use mechanics of materials. Axial stress, σ, is related to axial force. To distinguish axial force due to the virtual load from axial force due to the real loads, let's use lowercase letters for the internal forces due to the virtual load (i.e., n for virtual axial force and m for virtual moment):

$$\sigma = \frac{n}{A}$$

If the material remains linear elastic, Hooke's law applies:

$$\sigma = E\epsilon \quad \Rightarrow \quad \frac{n}{A} = E\epsilon$$

Engineering strain is defined as the ratio of the change in length, Δ, to the original length, l:

$$\epsilon = \frac{\Delta}{l} \quad \Rightarrow \quad \frac{n}{A} = E\frac{\Delta}{l} \quad \Rightarrow \quad \Delta = \frac{nl}{AE}$$

Next, we need to take the derivative of each side of the equation. *If* the cross-sectional area, A, and the modulus of elasticity, E, are constant along the length of the element, and *if* the virtual axial load is constant along the length of the element, we obtain

$$d\Delta = \frac{n}{AE}dx$$

When we substitute this expression into the integral for the strain energy, we arrive at a way to calculate the axial strain energy in an element of the structure due to real loads acting through virtual deformations. The j's denote that each quantity is unique to element j of the structure.

$$U_i = \int_0^{l_j} \frac{N_j(x_j)n_j(x_j)}{A_j E_j} dx_j$$

If the axial forces due to both the real and virtual loads are constant along the element, the term simplifies to

$$U_i = \frac{N_j n_j}{A_j E_j}\int_0^{l_j} dx_j = \frac{N_j n_j l_j}{A_j E_j}$$

3. Strain Energy Due to Bending

Although strain energy is stored due to shear, we typically ignore its contribution because it is much smaller than the strain energy stored due to bending. The strain energy stored due to bending is the integral of the internal moment acting through rotation along the element.

$$U_i = \int_0^l M(x)\, d\theta$$

From our derivation of the Double Integration Method in Section 5.1, we know how to express $d\theta$ as a function of x. Since the deformation is due to the virtual load, we use lowercase m for the moment. Note that in most cases the moment changes along the element; therefore, both M and m are functions of position:

$$d\theta = \frac{m(x)}{EI}dx$$

When we substitute this expression into the integral, we obtain

$$U_i = \int_0^{l_j} \frac{M(x_j)m(x_j)}{E_j I_j} d\theta_j$$

Again, the j's denote that each quantity is unique to element j of the structure. *If* the material and section properties are uniform along the element, EI will be a constant for the piece.

4. Generation of the Governing Equation

To obtain our energy equality, we applied a unit force, $P_{2k} = 1$, to measure the displacement, Δ_{1k}. But we can also measure the slope. Similar to force acting through displacement, moment acting through rotation is external work. Therefore, to measure the slope θ_{1k}, we apply a unit moment, $M_{2k} = 1$.

To properly account for all of the energy stored in the structural system, we must consider the strain energy in *every* element of the structure. That includes elements far from the locations where the real and virtual loads are applied. We must also consider the strain energy stored due to *all* of the internal forces. For two-dimensional problems, there can be three internal forces: axial force, shear, and moment. As long as we are predicting deformations of relatively long and narrow elements, the contribution from shear is negligible. Therefore, we have ignored the strain energy term due to shear.

The resulting governing equation for the Virtual Work Method is

$$\begin{matrix} 1 \cdot \Delta \\ \text{or} \\ 1 \cdot \theta \end{matrix} = \sum_{j=1}^{\#\text{Elem}} \int_0^{l_j} \frac{N_j(x_j) n_j(x_j)}{A_j E_j} dx_j + \sum_{j=1}^{\#\text{Elem}} \int_0^{l_j} \frac{M(x_j) m(x_j)}{E_j I_j} dx_j$$

If the axial force is constant along the length of each element (i.e., N and n are constants), the governing equation simplifies to

$$\begin{matrix} 1 \cdot \Delta \\ \text{or} \\ 1 \cdot \theta \end{matrix} = \sum_{j=1}^{\#\text{Elem}} \frac{N_j n_j l_j}{A_j E_j} + \sum_{j=1}^{\#\text{Elem}} \int_0^{l_j} \frac{M(x_j) m(x_j)}{E_j I_j} dx_j$$

Using the Governing Equation

To use the governing equation, we need to divide the structure into elements. The number of elements and the start and end of each element are dictated by several factors:

- A, I, and E are constant along each element.
- One expression can describe N for the entire element.
- One expression can describe n for the entire element.
- One expression can describe M for the entire element.
- One expression can describe m for the entire element.

Because the governing equation is based on calculating work, we must apply the virtual load at the location where we want to measure the deformation. If we want to measure the displacement, we must apply the virtual force along the line that represents the direction of the displacement we want to measure. If we want to measure the rotation (i.e., slope), we must apply the virtual moment about the axis whose rotation we want to measure. If the deformation we calculate is negative, it means that the resulting deformation is in the opposite direction as the applied virtual load.

A unique benefit of this method is that we can identify how much of the predicted deformation is due to the energy stored in each element of the structure. Typically, the elements that contribute the most to the deformation will be most cost effective to change if we want to reduce the deformation.

Section 5.3 Highlights

Virtual Work Method

Typical Uses:

- Trusses, beams, and frames.
- Any loading.
- For finding the slope or displacement at one location at a time.

Governing Equation:

$$\begin{matrix} 1 \cdot \Delta \\ \text{or} \\ 1 \cdot \theta \end{matrix} = \sum_{j=1}^{\#\,\text{Elem}} \int_0^{l_j} \frac{N_j(x_j)\,n_j(x_j)}{A_j E_j}\,dx_j + \sum_{j=1}^{\#\,\text{Elem}} \int_0^{l_j} \frac{M(x_j)\,m(x_j)}{E_j I_j}\,dx_j$$

If the axial force is constant along each element,

$$\begin{matrix} 1 \cdot \Delta \\ \text{or} \\ 1 \cdot \theta \end{matrix} = \sum_{j=1}^{\#\,\text{Elem}} \frac{N_j n_j l_j}{A_j E_j} + \sum_{j=1}^{\#\,\text{Elem}} \int_0^{l_j} \frac{M(x_j)\,m(x_j)}{E_j I_j}\,dx_j$$

Procedure:

1. Find $N(x)$ and $M(x)$ throughout the structure due to the real loads.
2. Determine the appropriate virtual load (force for Δ or moment for θ), its location and direction.
3. Find $n(x)$ and $m(x)$ throughout the structure due to the virtual load.
4. Evaluate the governing equation.

   + result = in the direction of the virtual load
   − result = in the opposite direction

Assumptions:

- Material and section properties (i.e., A, I, E) are constant along the length of each element.
- Contribution due to shear is negligible (will result in plane sections remaining plane).
- Material remains linear elastic.
- Slopes are relatively small.

Example 5.5

Our firm has designed a large pipeline along the edge of a canal in Singapore. Our team was tasked with designing a structure to support the pipe. The truss members have been selected for strength, but we need to verify that the truss does not deflect too much when the pipeline is full. To minimize the possibility of leaks in the pipeline, the truss must not deflect more than 15 mm when the pipeline changes from empty to full.

Member properties:

Diagonals: Double angle $120 \times 120 \times 8$ $A = 37.6 \text{ cm}^2$
All others: Universal columns $250 \times 250 \times 66.5$ $A = 84.7 \text{ cm}^2$
Steel: $E = 200 \text{ GPa} = 20{,}000 \text{ kN/cm}^2$

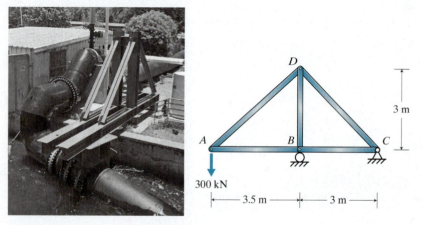

Figure 5.30

Estimated Solution:

We can estimate the vertical displacement at A by considering only the vertical component of the elongation of member AD. This will provide a lower bound for the actual ΔA_y because member AB will get shorter, thus causing rotation of member AD. Deformations of the other members will also contribute to increasing ΔA_y. Because of all the effects we have ignored, this approximation might give us only an order-of-magnitude estimate.

$$l_{AD} = \sqrt{(3.5 \text{ m})^2 + (3.0 \text{ m})^2} = 4.610 \text{ m}$$

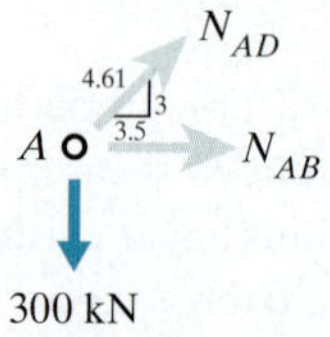

FBD of joint A:

$$+\uparrow \Sigma F_y = 0 = -300 \text{ kN} + \left(\frac{3}{4.61}\right) N_{AD}$$

$$N_{AD} = +461 \text{ kN}$$

Figure 5.31

Elongation of member AD using mechanics of materials:

$$\Delta_{AD} = \frac{N_{AD} l_{AD}}{E A_{AD}} = \frac{461 \text{ kN}(4.61 \text{ m})(1000 \text{ mm/m})}{200 \text{ kN/mm}^2 (3760 \text{ mm}^2)}$$

$$= 2.83 \text{ mm}$$

Vertical component of elongation:

$$\Delta_{ADy} = \frac{3}{4.61}\Delta_{AD} = \frac{3(2.83\text{ mm})}{4.61} = 1.8\text{ mm}$$

Detailed Analysis:

Since the members of a truss do not experience moment, the governing equation reduces to

$$1 \cdot \Delta_{Ay} = \sum_{j=1}^{5} \frac{N_j n_j l_j}{A_j E_j}$$

Length of member:

$$l_{CD} = \sqrt{(3.0\text{ m})^2 + (3.0\text{ m})^2} = 4.243\text{ m}$$

Analysis Using Real Load:

Apply equilibrium to find the reactions:

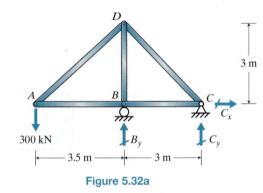

Figure 5.32a

$$\circlearrowleft \Sigma M_C = 0 = 300\text{ kN}(6.5\text{ m}) - B_y(3\text{ m})$$
$$B_y = 650\text{ kN}\ (+\uparrow)$$
$$+\uparrow\Sigma F_y = 0 = -300\text{ kN} + 650\text{ kN} + C_y$$
$$C_y = -350\text{ kN}\ (+\uparrow)$$
$$\pm\!\!\!\rightarrow \Sigma F_x = 0 = C_x$$
$$C_x = 0\ (\pm\!\!\!\rightarrow)$$

FBD of joint A:

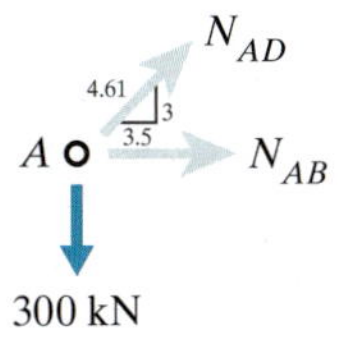

Figure 5.32b

$$+\uparrow\Sigma F_y = 0 = -300\text{ kN} + \left(\frac{3}{4.61}\right)N_{AD}$$

$$N_{AD} = +461\text{ kN}$$

$$\pm\!\!\!\rightarrow \Sigma F_x = 0 = 461\text{ kN}\left(\frac{3.5}{4.61}\right) + N_{AB}$$

$$N_{AB} = -350\text{ kN}$$

266 Chapter 5 Deformations

FBD of joint B:

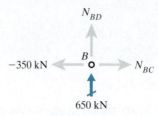

Figure 5.32c

$+\uparrow \Sigma F_y = 0 = 650 \text{ kN} + N_{BD}$
$\qquad N_{BD} = -650 \text{ kN}$
$\xrightarrow{+} \Sigma F_x = 0 = -(-350 \text{ kN}) + N_{BC}$
$\qquad N_{BC} = -350 \text{ kN}$

FBD of joint C:

$+\uparrow \Sigma F_y = 0 = -350 \text{ kN} + \left(\dfrac{3}{4.243}\right) N_{CD}$

$\qquad N_{CD} = +495 \text{ kN}$

Figure 5.32d

Evaluation of Results:

Satisfied Fundamental Principles?

$\xleftarrow{+} \Sigma F_x = 0 = 495 \text{ kN}\left(\dfrac{3}{4.243}\right) - 350 \text{ kN} = 0$ ✓

Analysis Using Virtual Load:

Since we want to find the vertical displacement at A, we apply a virtual force vertically at A. Apply equilibrium to find the reactions:

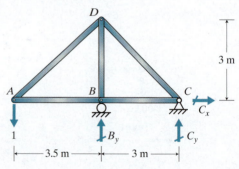

Figure 5.33a

$\curvearrowleft \Sigma M_C = 0 = 1(6.5 \text{ m}) - B_y(3 \text{ m})$
$\qquad B_y = 2.17 (+\uparrow)$
$+\uparrow \Sigma F_y = 0 = -1 + 2.17 + C_y$
$\qquad C_y = -1.17 (+\uparrow)$
$\xrightarrow{+} \Sigma F_x = 0 = C_x$
$\qquad C_x = 0 \;\; (\xrightarrow{+})$

FBD of joint A:

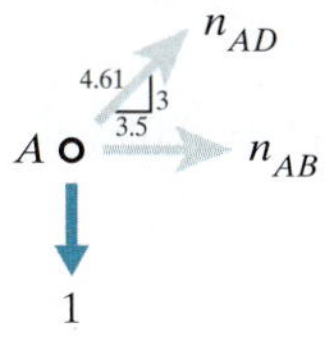

Figure 5.33b

$$+\uparrow \Sigma F_y = 0 = -1 + \left(\frac{3}{4.61}\right)n_{AD}$$

$$n_{AD} = +1.537$$

$$\xrightarrow{+} \Sigma F_x = 0 = 1.537\left(\frac{3.5}{4.61}\right) + n_{AB}$$

$$n_{AB} = -1.167$$

FBD of joint B:

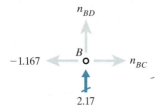

Figure 5.33c

$$+\uparrow \Sigma F_y = 0 = 2.17 + n_{BD}$$
$$n_{BD} = -2.17$$
$$\xrightarrow{+} \Sigma F_x = 0 = -(-1.167) + n_{BC}$$
$$n_{BC} = -1.167$$

FBD of joint C:

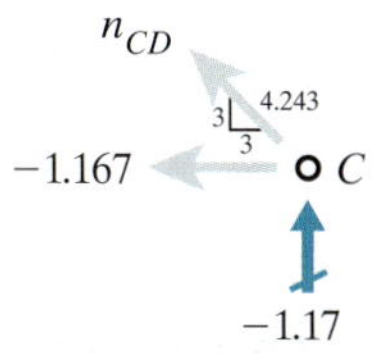

Figure 5.33d

$$+\uparrow \Sigma F_y = 0 = -1.17 + \left(\frac{3}{4.243}\right)n_{CD}$$
$$n_{CD} = +1.655$$

Evaluation of Results:
Satisfied Fundamental Principles?

$$\xrightarrow{+} \Sigma F_x = 0 = 1.655\left(\frac{3}{4.243}\right) - 1.167 = 0.003 \approx 0 \ \checkmark$$

Observation:
The virtual results are directly proportional to the real results. This is not a coincidence. Because the real system was exactly one force applied downward at A and the virtual system was exactly one force applied downward at A, the results must be proportional. Recognizing this can save us effort on future problems.

Calculate displacement:

$$\Delta_{Ay} = \sum_{j=1}^{5} \frac{N_j n_j l_j}{A_j E_j}$$

$$= \frac{461 \text{ kN}(1.537)(4610 \text{ mm})}{(3760 \text{ mm}^2)(200 \text{ kN/mm}^2)} + \frac{495 \text{ kN}(1.655)(4243 \text{ mm})}{(3760 \text{ mm}^2)(200 \text{ kN/mm}^2)}$$

$$+ \frac{-350 \text{ kN}(-1.167)(3500 \text{ mm})}{(8470 \text{ mm}^2)(200 \text{ kN/mm}^2)} + \frac{-350 \text{ kN}(-1.167)(3000 \text{ mm})}{(8470 \text{ mm}^2)(200 \text{ kN/mm}^2)}$$

$$+ \frac{-650 \text{ kN}(-2.17)(3000 \text{ mm})}{(8470 \text{ mm}^2)(200 \text{ kN/mm}^2)}$$

$$= 4.34 \text{ mm} + 4.62 \text{ mm} + 0.84 \text{ mm} + 0.72 \text{ mm} + 2.50 \text{ mm} = 13.0 \text{ mm} \ (+\downarrow)$$

Evaluation of Results:
Observed Expected Features?
The vertical displacement of A is down, as we expect. ✓

Approximation Predicted Outcomes?
Our prediction of Δ_{Ay} is 13.0 mm, which is larger than our lower bound estimate of 1.8 mm, as expected. ✓

Approximation Predicted Outcomes?
Our prediction of Δ_{Ay} is within an order of magnitude of our estimate.

Conclusion:
Together with the two evaluations made earlier, these observations suggest that we have a reasonable result.

Evaluation of Design:

The predicted live load deflection, 13.0 mm, is less than the allowable limit of 15 mm; therefore, the design should be adequate.

Example 5.6

To verify our understanding of a new structural analysis tool, we should use it to find a known solution. This helps ensure that we are applying the tool properly and, in the case of computer software, that the tool is working properly. Therefore, let's find the formula for the midspan vertical displacement of a simply supported, uniformly loaded beam and then compare that formula with the one inside the front cover of this text.

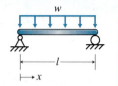

Figure 5.34

Detailed Analysis:

The beam does not experience an axial force; therefore, we only need to consider the bending term. Since we are finding displacement at midspan, we will need to consider two intervals.

The governing equation is

$$1 \cdot \Delta_{midspan} = \sum_{j=1}^{2} \int_{0}^{l_j} \frac{M(x_j)m(x_j)}{E_j I_j} dx_j$$

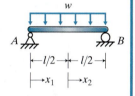

Figure 5.35

Analysis Using Real Load:
Apply equilibrium to find the reactions:

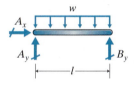

Figure 5.36

$$\circlearrowright \Sigma M_B = 0 = A_y(l) - wl(l/2)$$

$$A_y = +\frac{wl}{2}(+\uparrow)$$

Cut the beam within the range of x_1 and apply equilibrium to find the expression for the moment:

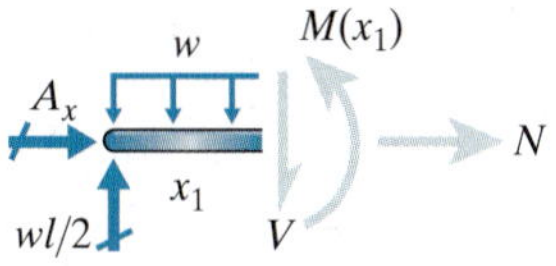

Figure 5.37

$$\circlearrowright \Sigma M_{cut} = 0 = \frac{wl}{2}(x_1) - w \cdot x_1\left(\frac{x_1}{2}\right) - M(x_1)$$

$$M(x_1) = -\frac{wx_1^2}{2} + \frac{wlx_1}{2}$$

Cut the beam within the range of x_2 and apply equilibrium to find the expression for the moment:

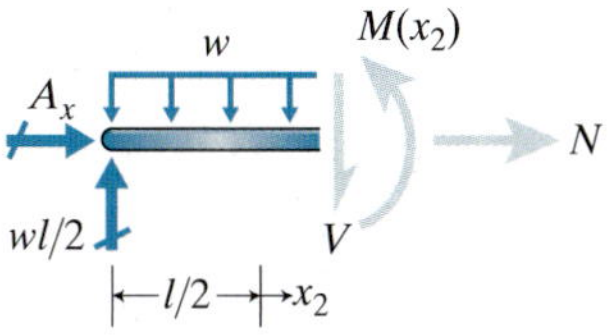

Figure 5.38

$$\circlearrowright \Sigma M_{cut} = 0 = \frac{wl}{2}\left(\frac{l}{2} + x_2\right)$$
$$- w\left(\frac{l}{2} + x_2\right)\left(\frac{1}{2}\right)\left(\frac{l}{2} + x_2\right) - M(x_2)$$

$$M(x_2) = -\frac{wx_2^2}{2} + \frac{wl^2}{8}$$

Analysis Using Virtual Load:
Since we want to find the vertical displacement at midspan, we need to apply a virtual force vertically at midspan. In order to match the formula inside the front cover of this text, we want up to be a positive displacement; therefore, we must apply the virtual force up.

Apply equilibrium to find the reactions:

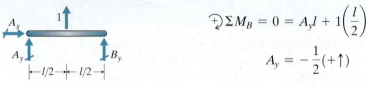

Figure 5.39

$$+\circlearrowleft \Sigma M_B = 0 = A_y l + 1\left(\frac{l}{2}\right)$$

$$A_y = -\frac{1}{2}(+\uparrow)$$

Cut the beam within the range of x_1 and apply equilibrium to find the expression for the moment:

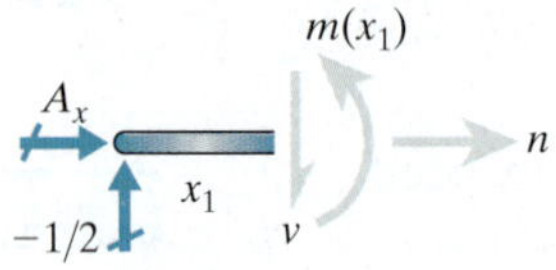

Figure 5.40

$$+\circlearrowleft \Sigma M_{\text{cut}} = 0 = -\frac{1}{2}(x_1) - m(x_1)$$

$$m(x_1) = -\frac{x_1}{2}$$

Cut the beam within the range of x_2 and apply equilibrium to find the expression for the moment:

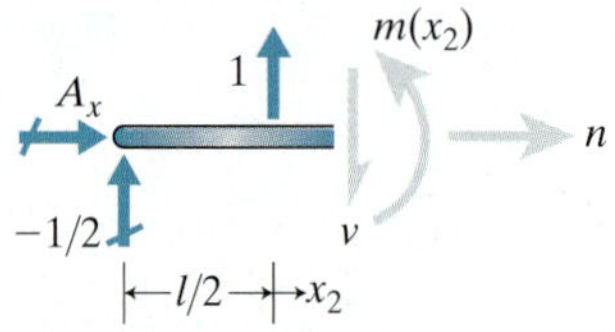

Figure 5.41

$$+\circlearrowleft \Sigma M_{\text{cut}} = 0 = -\frac{1}{2}\left(\frac{l}{2} + x_2\right) + 1 \cdot x_2 - m(x_2)$$

$$m(x_2) = \frac{x_2}{2} - \frac{l}{4}$$

Calculate displacement:

$$1 \cdot \Delta_{\text{midspan}} = \int_0^{l/2} \frac{M(x_1)m(x_1)}{EI} dx_1 + \int_0^{l/2} \frac{M(x_2)m(x_2)}{EI} dx_2$$

$$= \int_0^{l/2} \frac{\left(-\frac{wx_1^2}{2} + \frac{wlx_1}{2}\right)\left(-\frac{x_1}{2}\right)}{EI} dx_1 + \int_0^{l/2} \frac{\left(-\frac{wx_2^2}{2} + \frac{wl^2}{8}\right)\left(\frac{x_2}{2} - \frac{l}{4}\right)}{EI} dx_2$$

$$= \frac{1}{EI} \int_0^{l/2} \left(\frac{wx_1^3}{4} - \frac{wlx_1^2}{4}\right) dx_1 + \frac{1}{EI} \int_0^{l/2} \left(-\frac{wx_2^3}{4} + \frac{wlx_2^2}{8} + \frac{wl^2 x_2}{16} - \frac{wl^3}{32}\right) dx_2$$

$$= \frac{1}{EI}\left[\frac{wx_1^4}{16} - \frac{wlx_1^3}{12}\right]_0^{l/2} + \frac{1}{EI}\left[-\frac{wx_2^4}{16} + \frac{wlx_2^3}{24} + \frac{wl^2 x_2^2}{32} - \frac{wl^3 x_2}{32}\right]_0^{l/2}$$

$$= \frac{1}{EI}\left[\frac{wl^4}{256} - \frac{wl^4}{96}\right] + \frac{1}{EI}\left[-\frac{wl^4}{256} + \frac{wl^4}{192} + \frac{wl^4}{128} - \frac{wl^4}{64}\right]$$

$$= \frac{wl^4}{EI}\left[\frac{1}{256} - \frac{4}{384} - \frac{1}{256} + \frac{2}{384} + \frac{3}{384} - \frac{6}{384}\right]$$

$$= \frac{-5wl^4}{384EI}(+\uparrow)$$

Comparison with Precalculated Displacement:

We calculated exactly the same midspan displacement as the value we find inside the front cover of this text. Now we appear ready to use the Virtual Work Method to predict displacements.

Section 5.3 Virtual Work Method   271

EXAMPLE 5.7

To ensure that we understand how to properly apply the Virtual Work Method to find slope, we should use the method to predict a known slope. Let's find the formula for the slope at the right end of a simply supported, uniformly loaded beam and then compare that formula with the one inside the front cover of this text.

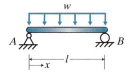

Figure 5.42

Detailed Analysis:

The beam does not experience an axial force; therefore, we only need to consider the bending term. Since we are finding the slope at the end of the member and there is a uniform distributed load, we need only one interval.

The governing equation is

$$1 \cdot \theta_B = \int_0^l \frac{M(x)m(x)}{EI} dx$$

Analysis Using Real Load:

From Example 5.1, we have

$$M(x) = -\frac{wx^2}{2} + \frac{wlx}{2}$$

Analysis Using Virtual Load:

Since we want to find the slope at B, we need to apply a virtual moment at B. In order to match the formula inside the front cover of this text, we want counterclockwise to be a positive slope; therefore, we must apply the virtual moment counterclockwise.

Apply equilibrium to find the reactions:

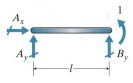

Figure 5.43

$$\circlearrowleft \Sigma M_B = 0 = A_y l + (-1)$$

$$A_y = +\frac{1}{l} \quad (+\uparrow)$$

$$+\uparrow \Sigma F_y = 0 = +\frac{1}{l} + B_y$$

$$B_y = -\frac{1}{l} \quad (+\uparrow)$$

Cut the beam and apply equilibrium to find the expression for the moment:

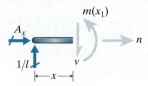

Figure 5.44

$$\overset{+}{\curvearrowleft} \Sigma M_{\text{cut}} = 0 = \frac{1}{l}(x) - m(x)$$

$$m(x) = +\frac{x}{l}$$

Calculate displacement:

$$1 \cdot \theta_B = \int_0^l \frac{M(x)m(x)}{EI} dx$$

$$= \int_0^l \frac{\left(-\frac{wx^2}{2} + \frac{wlx}{2}\right)\left(\frac{x}{l}\right)}{EI} dx$$

$$= \frac{1}{EI}\int_0^l \left(-\frac{wx^3}{2l} + \frac{wx^2}{2}\right) dx$$

$$= \frac{1}{EI}\left[-\frac{wx^4}{8l} + \frac{wx^3}{6}\right]_0^l = \frac{1}{EI}\left[-\frac{3wl^3}{24} + \frac{4wl^3}{24}\right]$$

$$= \frac{wl^3}{24EI} \; (\measuredangle +)$$

Comparison with Precalculated Slope:

We calculated exactly the same slope as the value we find inside the front cover of this text, including the sign. We are ready to use the Virtual Work Method to predict slopes as well.

Example 5.8

The manufacturer of freestanding jib cranes has been receiving complaints about how far the tip deflects downward when lifting heavy loads. The manufacturer suspects the problem is poor foundations, but to verify that the problem is not in the design of the cranes, we have been hired to predict the maximum vertical displacement under the rated load, 2 tons.

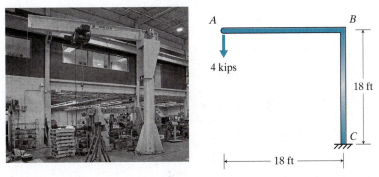

Figure 5.45

Each jib crane is made of steel, $E = 29{,}000$ ksi. The arm is a W18 × 40 ($A = 11.8$ in^2, $I = 612$ in^4), and the mast is an HSS 18 × 0.375 ($A = 19.4$ in^2, $I = 754$ in^4). The peak displacement will occur when the hoist is at the far end of the arm.

Estimated Solution:

To develop a sense of the answer, we will predict the vertical displacement by considering the arm to be a cantilever that is fixed against rotation at B. Since B will rotate, this will be a lower bound for our detailed analysis result.

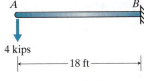

Figure 5.46

From the inside front cover of this text, we have the following formula for the displacement of a cantilevered beam with a point load at the end:

$$\Delta_{end} = -\frac{P}{6EI}(2l^3) = -\frac{Pl^3}{3EI}$$

Estimate the displacement:

$$\Delta_{end} = -\frac{4\,\text{k}\,(18\,\text{ft} \cdot 12\,\text{in./ft})^3}{3\,(29{,}000\,\text{ksi})(612\,\text{in}^4)} = -0.76\,\text{in.}\,(+\uparrow)$$

Detailed Analysis:

Consider the contribution from both the axial force and bending:

$$1 \cdot \Delta_{Ay} = \sum_{j=1}^{2} \frac{N_j n_j l_j}{A_j E_j} + \sum_{j=1}^{2} \int_0^{l_j} \frac{M(x_j)m(x_j)}{E_j I_j} dx_j$$

Analysis Using Real Load:

Apply equilibrium to find the reactions:

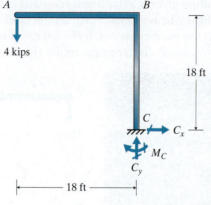

Figure 5.47

$$\circlearrowleft \Sigma M_C = 0 = -4\,\text{k} \cdot 18\,\text{ft} + M_C$$
$$M_C = +72\,\text{k}\cdot\text{ft}\ (\circlearrowleft)$$
$$+\uparrow \Sigma F_y = 0 = -4\,\text{k} + C_y$$
$$C_y = 4\,\text{k}\ (+\uparrow)$$
$$\xrightarrow{\pm} \Sigma F_x = 0 = C_x$$
$$C_x = 0$$

Cut the arm and apply equilibrium to find expressions for the axial force and moment:

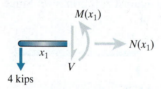

Figure 5.48

$$\circlearrowright \Sigma M_{\text{cut}} = 0 = 4\,\text{k} \cdot x_1 + M(x_1)$$
$$M(x_1) = -4\,\text{k} \cdot x_1$$
$$\xrightarrow{\pm} \Sigma F_x = 0 = N(x_1)$$

Cut the mast and apply equilibrium to find expressions for the axial force and moment:

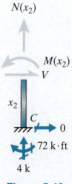

Figure 5.49

$$+\circlearrowright \Sigma M_{cut} = 0 = 72 \text{ k} \cdot \text{ft} - M(x_2)$$
$$M(x_2) = 72 \text{ k} \cdot \text{ft}$$
$$+\uparrow \Sigma F_y = 0 = 4 \text{ k} + N(x_2)$$
$$N(x_2) = -4 \text{ k}$$

Analysis Using Virtual Load:

To find the vertical displacement at A, we need to apply a unit force vertically at A.

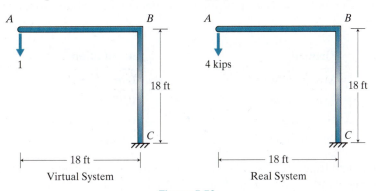

Figure 5.50

Since the virtual system and real system are identical except for the magnitude of the load, the internal forces must be proportional.

Summary of internal forces due to the virtual load:

$$m(x_1) = -x_1$$
$$n(x_1) = 0$$
$$m(x_2) = 18 \text{ ft}$$
$$n(x_2) = -1$$

Calculate displacement:
Caution about units: If we convert the limits of integration to inches, we must be sure to convert all of the lengths to inches, including the moments. Rather than risk error, it is typically easier to wait until after integration to make the unit conversion.

$$1 \cdot \Delta_{Ay} = \frac{0(0)(18 \text{ ft})}{(11.8 \text{ in}^2)(29{,}000 \text{ ksi})} + \frac{-1(-4 \text{ k})(18 \text{ ft})}{(19.4 \text{ in}^2)(29{,}000 \text{ ksi})}$$

$$+ \int_0^{18 \text{ ft}} \frac{(-4 \text{ k} \cdot x_1)(-x_1)}{(29{,}000 \text{ ksi})(612 \text{ in}^4)} dx_1$$

$$+ \int_0^{18 \text{ ft}} \frac{(72 \text{ k} \cdot \text{ft})(18 \text{ ft})}{(29{,}000 \text{ ksi})(754 \text{ in}^4)} dx_2$$

$$= 0 + 1.280 \times 10^{-4} \text{ ft} (12 \text{in.}/\text{ft}) + \left[\frac{4 \text{ k} \cdot x_1^3}{3(29{,}000 \text{ ksi})(612 \text{ in}^4)} \right]_0^{18 \text{ ft}}$$

$$+ \left[\frac{1296 \text{ k} \cdot \text{ft}^2 \cdot x_2}{(29{,}000 \text{ ksi})(754 \text{ in}^4)} \right]_0^{18 \text{ ft}}$$

$$= 0 + 1.536 \times 10^{-3} \text{ in.} + 4.38 \times 10^{-4} \frac{\text{ft}^3}{\text{in}^2}(12 \text{ in.}/\text{ft})^3$$

$$+ 1.067 \times 10^{-3} \frac{\text{ft}^3}{\text{in}^2}(12 \text{ in.}/\text{ft})^3$$

$$= 0 + 0.0015 \text{ in.} + 0.757 \text{ in.} + 1.844 \text{ in.}$$

$$= 2.60 \text{ in.} \ (+\downarrow)$$

Observation:
Each term of the summation tells us how much that element contributes to the overall response of the structure.

Element	Contribution
Arm axial force	0 in.
Arm bending	0.757 in.
Mast axial force	0.0015 in.
Mast bending	1.844 in.

Note that in this rigid frame, the axial deformation contributes very little to the overall response. This is generally true for rigid frames. Therefore, we typically neglect the axial terms for rigid frames.

Evaluation of Results:

Observed Expected Features?
The vertical displacement at A is down, as expected. ✓

Satisfied Fundamental Principles?
The arm bending contribution to the vertical displacement is identical to the approximate solution. We should expect this because the approximate solution considers only bending in the arm. ✓

Approximation Predicted Outcomes?
The predicted displacement is greater than the approximation, as expected. ✓

The detailed prediction is three times larger than the estimate, which might merit closer review. In this case, however, it turns out that the mast does actually contribute a significant amount to the displacement.

Conclusion:

The predicted peak displacement is 2.6 inches. That equates to a relative displacement of $l/83$. The standard of practice for a floor beam is to limit the live load deflection to $l/360$, or $l/180$ for a cantilever. Dead load limits are typically $l/240$ for a beam supported on both ends or $l/120$ for a cantilever. Therefore, it appears reasonable that customers might find the displacement excessive.

Recommendation:
If the crane manufacturer wants to reduce the vertical displacement, our results indicate that increasing the moment of inertia of the mast would be most effective. If that does not provide a sufficient reduction, the manufacturer should increase the moment of inertia of the arm as well. Focusing on the cross-sectional areas of the members will not have a noticeable impact.

Homework Problems

5.1 To verify our understanding of a new structural analysis tool, we should use it to find a known solution. This helps ensure that we are applying the tool properly. To be sure that we understand how to apply the Double Integration Method, let's find the expressions for the slope and vertical displacement of a simply supported beam with a triangular distributed load. We can compare those expressions with information inside the front cover of this text.

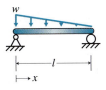

Prob. 5.1

a. Guess the location of the peak vertical displacement.

b. Derive the expressions for the shear and moment for the beam.

c. Derive the expressions for the slope and displacement.

d. Calculate the slope at both ends of the beam. Determine the location and value of the peak vertical displacement. Note: the peak is not at midspan. Compare those values with the values from inside the front cover of this text.

e. Comment on why your guess in part (a) was or was not close to the solution in part (d).

5.2 To verify our understanding of a new structural analysis tool, we should use it to find a known solution. This helps ensure that we are applying the tool properly. To be sure that we understand how to apply the Double Integration Method with multiple regions, let's find the expressions for the slope and vertical displacement of a simply supported beam with a point load not at midspan. We can compare those expressions with information inside the front cover of this text.

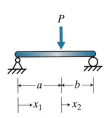

Prob. 5.2

a. Guess the location of the peak vertical displacement.

b. Derive the expressions for the shear and moment for the beam.

c. Derive the expressions for the slope and displacement.

d. Calculate the slope at both ends of the beam. Determine the location and value of the peak vertical displacement. Note: the peak is not at midspan. Compare those values with the values from inside the front cover of this text.

e. Comment on why your guess in part (a) was or was not close to the solution in part (d).

5.3 A simply supported beam carries a floor over only part of its length. Part of the design process is ensuring that the live load deflection is not excessive. The design team has made a preliminary selection of a steel section based on strength requirements: HSS 254 × 50.8 × 3.2 (A = 1740 mm², I = 1.186×10⁷ mm⁴, E = 200 GPa = 200 kN/mm²).

a. Guess the location of the peak vertical displacement (e.g., near the left support, middle of the unloaded region).

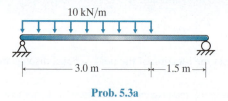

Prob. 5.3a

b. Estimate the peak vertical displacement by approximating the beam as uniformly loaded and only 3 m long. Will this provide an upper bound, a lower bound, or just an approximation for the actual peak vertical displacement? Justify your answer.

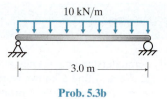

Prob. 5.3b

c. Estimate the peak vertical displacement by approximating the loading as uniform along the entire length. Will this provide an upper bound, a lower bound, or just an approximation for the actual peak vertical displacement? Justify your answer.

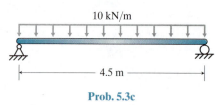

Prob. 5.3c

d. Derive the expressions for the shear and moment for the planned configuration of the beam.

e. Derive the expressions for the slope and displacement.

f. Identify at least four features of the expressions for the slope and displacement that suggest you have a reasonable answer.

g. Determine the location and value of the peak vertical displacement.

h. Evaluate the displacement prediction using your results from parts (b), (c), and (f).

i. Comment on why your guess in part (a) was or was not close to the solution in part (g).

5.4 The roof of a tall building will be supported by beams that extend from the core to the perimeter. Because of the mechanical penthouse, snow will drift, causing a nonuniform load on the beams.

a. Guess the location of the peak vertical displacement (e.g., near the right support, middle of the uniform distributed load region).

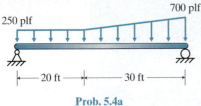

Prob. 5.4a

b. Estimate the peak vertical displacement by approximating the loading as uniform along the entire length without the drifting. Will this provide an upper bound, a lower bound, or just an approximation for the actual peak vertical displacement? Justify your answer.

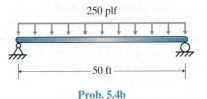

Prob. 5.4b

c. Estimate the peak vertical displacement by approximating the loading as the equivalent uniform load. Will this provide an upper bound, a lower bound, or just an approximation for the actual peak vertical displacement? Justify your answer.

d. Derive the expressions for the shear and moment for the beam with the nonuniform load.

e. Derive the expressions for the slope and displacement.

f. Identify at least four features of the expressions for the slope and displacement that suggest you have a reasonable answer.

g. Determine the location and value of the peak vertical displacement.

h. Evaluate the displacement prediction using your results from parts (b), (c), and (f).

i. Comment on why your guess in part (a) was or was not close to the solution in part (g).

5.5 Our firm has been hired to design a sun screen to protect cars on the top level of a parking garage. The design team has determined the ideal structural system: two beams will extend from a central column, pass over exterior columns, and cantilever over the parking spots. For aesthetic reasons, the architect wants us to limit the dead load deflection of the beam to no more than 1 inch. Determine the peak vertical deflection for one of the beams given our current choice of beam section. The distributed load shown already includes the self-weight of the beam.

a. Guess the anticipated deflected shape.
b. Estimate the peak deflection by ignoring the distributed load on the cantilevered end. Will this provide an upper bound, a lower bound, or just an approximation for the actual peak vertical displacement? Justify your answer.
c. Derive the expressions for the slope and displacement due to dead load along the entire beam. Determine the peak displacement and its location (creating a table of values is acceptable for this).
d. Plot the expressions for the slope and displacement.
e. Identify at least four features of the expressions for the slope and displacement that suggest you have a reasonable answer.
f. Evaluate the displacement prediction using your results from parts (b) and (e).
g. Comment on why your guess in part (a) was or was not close to the solution in part (d).

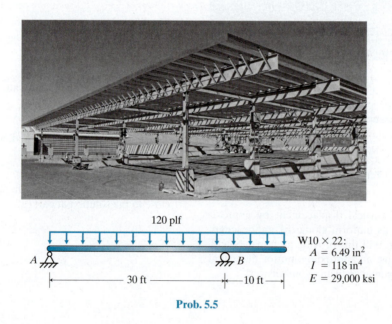

Prob. 5.5

5.6 The roof covering a walkway in a northern city is slanted to allow melting snow to drain off. If the roof deflects too much under a heavy snow load, the water will not be able to run off, thus causing maintenance problems. We are tasked with determining the minimum moment of inertia for the steel support beam ($E = 29{,}000$ ksi) in order to prevent water from ponding on the roof.

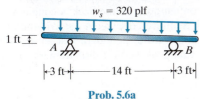

Prob. 5.6a

Because this is a serviceability issue rather than a strength issue, we use the maximum likely unfactored snow load. Since the slope of the beam is very small, -0.05, we can reasonably approximate the beam as horizontal for all of our calculations. Note that our answers will be in terms of the moment of inertia, I, since we have not yet chosen the beam cross-section.

a. Guess the displaced shape under uniform snow load.

b. Estimate the peak slope by considering only the span between the supports, AB. Will this provide an upper bound, a lower bound, or just an approximation for the actual peak vertical displacement? Justify your answer.

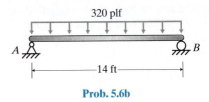

Prob. 5.6b

c. Derive the expressions for the shear and moment for the full beam under a uniform load along its entire length.

d. Derive the expressions for the slope and displacement for the full beam under a uniform load along its entire length.

e. Plot the expressions for the displacement to show the displaced shape.

f. Identify at least three features of the displaced shape that suggest you have a reasonable answer.

g. Determine the peak positive slope and its location.

h. Evaluate the slope prediction using your results from parts (b) and (f).

i. As long as the peak positive slope does not exceed the initial negative slope of the roof, -0.05, the roof will always have a downward slope and the water will drain. Determine the minimum moment of inertia required to ensure this.

j. Comment on why your guess in part (a) was or was not close to the solution in part (e).

5.7 To verify our understanding of a new structural analysis tool, we should use it to find a known solution. This helps ensure that we are applying the tool properly. To be sure that we understand how to apply the Conjugate Beam Method, let's find the slope and peak vertical displacement of a simply supported beam with a couple moment applied at one end. We can compare our results with information inside the front cover of this text.

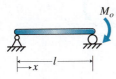

Prob. 5.7

a. Guess the location of the peak vertical displacement.

b. Construct the moment diagram for the beam.

c. Calculate the slope at both ends of the beam. Compare those values with the values inside the front cover of this text.

d. Determine the location and value of the peak vertical displacement. Note: the peak is not at midspan. Compare those values with the values from inside the front cover of this text.

e. Comment on why your guess in part (a) was or was not close to the solution in part (d).

5.8 To verify our understanding of a new structural analysis tool, we should use it to find a known solution. This helps ensure that we are applying the tool properly. To be sure that we understand how to apply the Conjugate Beam Method, let's find the slope and vertical displacement at the free end of a cantilever with a point load. We can compare our results with information inside the front cover of this text.

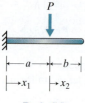

Prob. 5.8

a. Guess the location of the peak slope.

b. Construct the moment diagram for the beam.

c. Calculate the slope along the beam. Identify where the slope is largest. Compare the slope at the free end with the value from inside the front cover of this text.

d. Determine the vertical displacement at the free end of the beam. Compare this value with the value inside the front cover of this text.

e. Comment on why your guess in part (a) was or was not close to the solution in part (c).

5.9 Wood beams cantilever out from a house to form an elevated deck. The owners wanted a bigger deck, so they extended the beams and added columns at the far ends. The addition was not engineered, so the owners have no idea whether the deck is safe. Now that they want to sell the house, they have hired our firm to evaluate the capacity of the deck in its current condition.

Although this is not a strength issue, the senior engineer on the team wants us to determine the peak vertical deflection of the extended beam due to a lateral load on the handrail. The handrails are rigidly connected to the ends of some beams rather than connecting to the columns. Each beam is 50 × 300 lumber ($I = 9.63 \times 10^7$ mm^4, $E = 10$ GPa $= 10$ kN/mm^2). Because there is only a couple moment acting on the extended beam, the Conjugate Beam Method is probably more efficient than the Double Integration Method.

a. Guess the location of the peak vertical displacement (e.g., middle of span AB, near C).

b. Estimate the peak vertical displacement considering the original configuration of the deck: a couple moment at the end of cantilevered segment AB. Will this provide an upper bound, a lower bound, or just an approximation for the actual peak vertical displacement? Justify your answer.

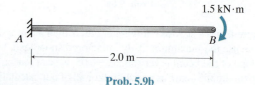

Prob. 5.9b

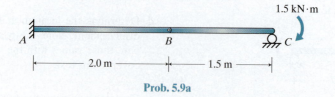

Prob. 5.9a

c. Construct the moment diagram for the deck beam as currently constructed. Identify at least three features of the moment diagram that suggest you have a reasonable answer.
d. Determine the location and value of the peak vertical displacement for the beam.

e. Evaluate the displacement prediction using your results from part (b).
f. Comment on why your guess in part (a) was or was not close to the solution in part (d).

5.10 The owners of a manufacturing plant want to hang a piece of equipment from a retrofit beam. They are very concerned about additional deflection of the beam because it is closely surrounded by sensitive equipment. If the additional deflection anywhere along the beam exceeds $1/4$ inch, the system must be redesigned. The beam is a steel W8 × 15 ($I = 48.0$ in^4, $E = 29,000$ ksi). A team member points out that since the beam experiences only a point load, the Conjugate Beam Method is probably faster than the Double Integration Method in this case.

a. Guess the location of the peak vertical displacement (e.g., near A, just right of B).
b. Estimate the peak vertical displacement considering segment BC as simply supported. Will this provide an upper bound, a lower bound, or just an approximation for the actual peak vertical displacement? Justify your answer.

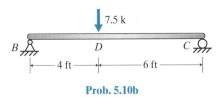

Prob. 5.10b

c. Construct the moment diagram for the full beam. Identify at least three features of the moment diagram that suggest you have a reasonable answer.
d. Determine the location and value of the peak vertical displacement.
e. Evaluate the displacement prediction using your results from part (b).
f. Evaluate whether the beam will meet the deflection limit or must be redesigned.
g. Comment on why your guess in part (a) was or was not close to the solution in part (d).

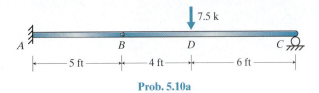

Prob. 5.10a

5.11 A beam will be supporting equipment at two points. The equipment mixes product before discharging it into a vat below, so there are live loads on the beam. The beam will be made of steel, $E = 200$ GPa $= 200$ kN/mm^2, but we have not yet selected the section. The manufacturer of the mixing equipment has recommended limits on the tilt of the machine; they equate to a live load displacement limit of $l/400$ in this case. Although we could use formulas from inside the front cover with superposition, we don't know the location of the peak displacement. Therefore, we would need to use a spreadsheet. It is probably faster to use the Conjugate Beam Method in this case.

a. Guess the location of the peak vertical displacement (e.g., just to the left of the 6-kN load, near B).

b. Estimate the peak vertical displacement considering the loading to be symmetric. With this approximation, the peak vertical displacement will be at midspan, so you can use formulas from inside the front cover of this text along with superposition. Will this provide an upper bound, a lower bound, or just an approximation for the actual peak vertical displacement? Justify your answer.

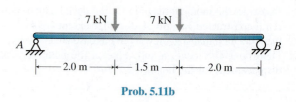

Prob. 5.11b

c. Construct the moment diagram for the beam with the unequal loads. Identify at least three features of the moment diagram that suggest you have a reasonable answer.

d. Determine the location and value of the peak vertical displacement. The answer will be in terms of I.

e. Evaluate the displacement prediction using your results from part (b).

f. Determine the minimum moment of inertia, I, in order to meet the displacement limit for this 5.5-m beam.

g. Comment on why your guess in part (a) was or was not close to the solution in part (d).

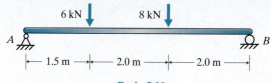

Prob. 5.11a

5.12 Our team is redesigning the beam that supports the track for a moving hoist. The original plan was for the track to run midspan of the beam, but new requirements have moved the track away from the middle. A team member has calculated the new reactions. In order to allow the hoist to move smoothly along the track, we have tight limits on the live load deflection of the support beam. Because the beam is loaded only with a point load, the Conjugate Beam Method is probably fastest in this case.

a. Guess the location of the peak vertical displacement (e.g., near A, midspan of BC).

b. Estimate the peak vertical displacement considering the beam to be simply supported. The answer will be in terms of EI. Will this provide an upper bound, a lower bound, or just an approximation for the actual peak vertical displacement? Justify your answer.

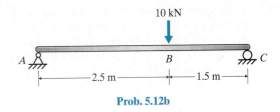

Prob. 5.12b

c. Construct the moment diagram for the beam with the fixed support. Identify at least three features of the moment diagram that suggest you have a reasonable answer.

d. Determine the location and value of the peak vertical displacement. The answer will be in terms of EI.

e. Evaluate the displacement prediction using your results from part (b).

f. Comment on why your guess in part (a) was or was not close to the solution in part (d).

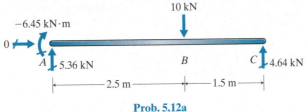

Prob. 5.12a

5.13 To verify our understanding of a new structural analysis tool, we should use it to find a known solution. This helps ensure that we are applying the tool properly. To be sure that we understand how to apply the Virtual Work Method, let's find the slope at one end of a simply supported beam with a triangular distributed load. We can compare our results with information inside the front cover of this text.

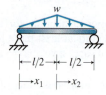

Prob. 5.13a

a. Guess whether the slope at the right end will be greater or less than the slope at the same end of a simply supported beam with an equivalent uniform load.

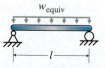

Prob. 5.13b

b. Calculate the slope at the right end of the beam with the equivalent uniform load using formulas from inside the front cover of this text.
c. Derive the expressions for the moment for the beam with a triangular distributed load.
d. Calculate the slope at the right end of the beam using the Virtual Work Method. Compare that value with the value inside the front cover of this text.
e. Comment on why your guess in part (a) was or was not close to the solution in parts (b) and (d).

5.14 To verify our understanding of a new structural analysis tool, we should use it to find a known solution. This helps ensure that we are applying the tool properly. To be sure that we understand how to apply the Virtual Work Method, let's find the vertical displacement at midspan of a simply supported beam with a triangular distributed load. We can compare our results with information inside the front cover of this text.

a. Guess whether the midspan displacement will be greater or less than the midspan displacement of a simply supported beam with an equivalent uniform load.

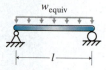

Prob. 5.14b

b. Calculate the midspan displacement of the beam with the equivalent uniform load using formulas from inside the front cover of this text.
c. Derive the expressions for the moment for the beam with a triangular distributed load.
d. Calculate the midspan displacement of the beam using the Virtual Work Method. Compare that value with the value inside the front cover of this text.
e. Comment on why your guess in part (a) was or was not close to the solution in parts (b) and (d).

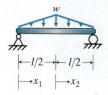

Prob. 5.14a

5.15 Our team has been tasked with designing the truss to support a large billboard. The team has decided to use steel angles for the members of the truss ($E = 29{,}000$ ksi). For preliminary design, a team member has chosen L4 × 4 × 3/8's: $A = 2.86$ in^2. One truss carries 3 kips of billboard weight, resulting in the forces shown on the idealized truss. If the displacement of the truss under the weight of the sign is too large, we can adjust the design to compensate. Therefore, we want to know how much the truss moves.

a. Guess which direction the top of the truss moves under the self-weight of the sign.
b. Estimate the vertical displacement at C by ignoring the lateral loads and assuming that all vertical load is carried only in the vertical members, AB and BC. Will this provide an upper bound, a lower bound, or just an approximation for the actual vertical displacement? Justify your answer.
c. Calculate the horizontal displacement of point C considering all components of the load on the full truss.
d. Calculate the vertical displacement of point C.
e. Evaluate the vertical displacement prediction using your results from part (b).
f. If you wanted to reduce the horizontal displacement, you could increase the cross-sectional area of a member. Which single member should you increase? Why that one?
g. Comment on why your guess in part (a) was or was not close to the solutions in parts (c) and (d).

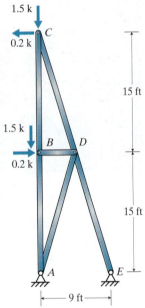

Prob. 5.15

5.16 Our team is designing a pedestrian footbridge. Preliminary design resulted in steel section properties for the chord and diagonal members ($E = 29{,}000$ ksi). The chords will be MC 18 × 42.7 channels, $A = 12.6$ in^2, and the diagonals will be round HSS 4 × 0.220, $A = 2.44$ in^2. Now our job is to confirm that the live load deflection remains within the tolerances. In order to maintain the seals around the glass panels, the manufacturer of this system recommends that the live load deflection remain below $l/500$ (i.e., 1.7 in.).

a. Guess which members will be in tension, which will be in compression, and which are zero force (if any).
b. Estimate the peak vertical displacement by approximating the truss as a beam with an equivalent uniform distributed load. You can approximate the moment of inertia using the parallel axis theorem, considering only the chord areas. Will this provide an upper bound, a lower bound, or just an approximation for the actual vertical displacement? Justify your answer.
c. Calculate the peak vertical displacement of the actual truss. Because of the symmetry of the structure and loading, the peak will be at midspan, point C.
d. Evaluate the peak displacement prediction using your results from part (b).
e. Is the truss adequately designed to limit the peak live load deflection?
f. If you wanted to reduce the peak vertical displacement, regardless of what you determine in part (e), you could increase the cross-sectional area of certain members. Should you increase the chords or the diagonals? Why?
g. Comment on why your guess in part (a) was or was not close to what you discovered while working on part (c).

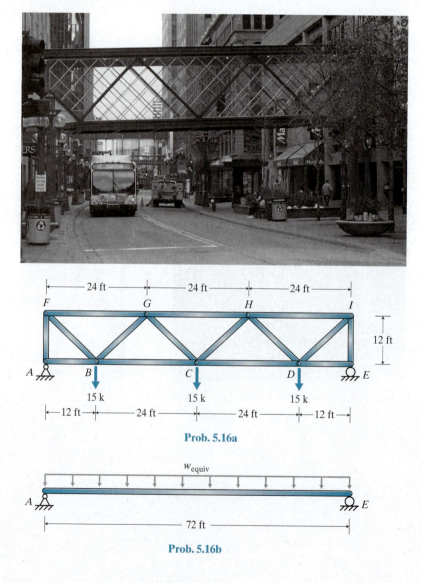

Prob. 5.16a

Prob. 5.16b

5.17 Our team is designing a pedestrian bridge using steel hollow structural sections (HSS). Other team members completed a preliminary design and chose HSS 152.4 × 152.4 × 6.4 to use for all of the elements ($A = 3380$ mm^2, $E = 200$ GPa $= 200$ kN/mm^2).

Full live load will cause seven point forces of 20 kN each along the bottom chord of the bridge. In order to minimize the likelihood of damage to the windows, we want to keep the live load deflection below $l/360$; for this 22-m span, that means keeping the peak deflection below 60 mm. Therefore, we need to know how large the peak deflection will likely be.

a. Guess which members will be in tension, which will be in compression, and which are zero force (if any).
b. Estimate the peak vertical displacement of the actual truss by approximating the truss as a beam with an equivalent uniform distributed load. You can approximate the moment of inertia using the parallel axis theorem, considering only the chord areas. Will this provide an upper bound, a lower bound, or just an approximation for the actual vertical displacement? Justify your answer.
c. Calculate the peak vertical displacement of the actual truss. Because of the symmetry of the structure and loading, the peak will be at midspan, point E.
d. Evaluate the vertical displacement prediction using your results from part (b).
e. Is the truss adequately designed to limit the peak live load deflection?
f. If you wanted to reduce the peak vertical displacement, regardless of what you determined in part (e), you could increase the cross-sectional area of certain members. Should you increase the chords, diagonals, or verticals? Why?
g. Comment on why your guess in part (a) was or was not close to what you discovered while working on part (c).

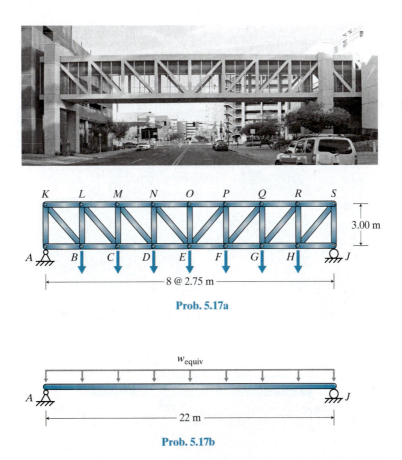

Prob. 5.17a

Prob. 5.17b

5.18 A continuous beam supports the floor of a house and the carport covering. If there is a heavy snowfall while the room is empty, the beam will slope at B such that the snow melt will drain away from the house. But if the room has a live load, the beam might slope such that the snow melt will pond next to the house. Based on our initial design, we plan to use 14-inch-deep glulam wood joists ($I = 90 \text{ in}^4$, $E = 1500 \text{ ksi}$).

a. Guess which way the beam will slope at B under this loading.

b. Estimate the slope at B by considering the live load only. You can approximate the geometry as simply supported between A and B. Will this provide an upper bound, a lower bound, or just an approximation for the actual slope? Justify your answer.

c. Calculate the slope at B due to both unfactored loads acting at the same time.

d. Evaluate the slope prediction using your results from part (b).

e. Under this loading, will the snow melt gather next to the house or will it drain away?

f. Which loading contributed more to the slope? Would increasing the moment of inertia of the beam change the result in part (e)? Why or why not?

g. Comment on why your guess in part (a) was or was not close to what you discovered while working on part (c).

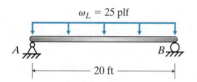

Prob. 5.18b

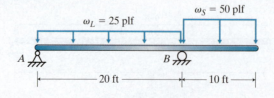

Prob. 5.18a

5.19 We work for a jib crane company. A jib crane has a single arm cantilevered out from a single mast. The arm supports a hoist that can move along the length of the arm. The newest model of crane for our company has a hoist that needs a relatively flat arm in order to move smoothly. Based on tests done in our shop, we know that the arm should have a slope no greater than 0.5% for the hoist to travel smoothly. This model of crane is designed to lift 8 kN. The steel arm and mast have been designed for strength already ($E = 200$ GPa $= 200$ kN/mm^2).

a. Guess which component will contribute most to the slope at A: axial force in the arm, bending in the arm, axial force in the mast, or bending in the mast.

b. Estimate the slope at A by considering the arm to be cantilevered out from a fixed support (i.e., no rotation at B). Will this provide an upper bound, a lower bound, or just an approximation for the actual slope? Justify your answer.

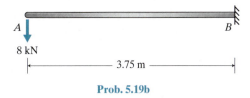

Prob. 5.19b

c. Calculate the slope at A for the full jib crane.

d. Evaluate the slope prediction using your results from part (b).

e. Is the jib crane adequately designed to limit the peak slope?

f. Which contributed most to the slope: axial force in the arm, bending in the arm, axial force in the mast, or bending in the mast? If you wanted to reduce the peak slope, regardless of what you determined in part (e), what cross-sectional property do you recommend be increased? Why that one?

g. Comment on why your guess in part (a) was or was not close to your answer in part (f).

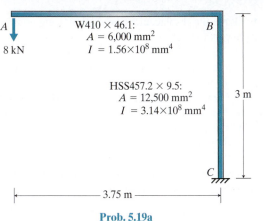

Prob. 5.19a

5.20 Our team is designing this simple frame to support a roof for parked cars. We will have the frame fabricated with a slight upward angle so that even under a full snow load, the roof will drain toward the column side. In order to do that, we want to know how much the tip of the frame deflects downward, ΔC_y. Although the roof load transfers to the beam at four points, it will be a reasonable approximation to consider the snow as a uniform distributed load. The geotechnical engineer has provided us with the point of fixity (i.e., the depth where the foundation behaves as though it is perfectly fixed), A.

a. Guess which component will contribute most to the vertical displacement at C: axial force in the beam, bending in the beam, axial force in the column, or bending in the column.

b. Estimate the vertical displacement at C by considering the beam to be cantilevered out from a fixed support (i.e., no rotation at B). Will this provide an upper bound, a lower bound, or just an approximation for the actual displacement? Justify your answer.

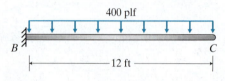

Prob. 5.20b

c. Calculate the vertical displacement at C for the frame due to a full snow load. We will fabricate the frame to have an initial upward lean slightly larger than this displacement.

d. Evaluate the displacement prediction using your results from part (b).

e. Which contributed most to the displacement: axial force in the beam, bending in the beam, axial force in the column, or bending in the column? If you wanted to reduce the vertical displacement at the end, what cross-sectional property do you recommend be increased? Why that one?

f. Comment on why your guess in part (a) was or was not close to your answer in part (e).

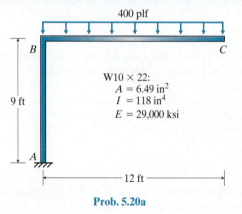

Prob. 5.20a

5.21 Our company is getting into the business of designing traffic signal structures. A coworker has performed a preliminary design of the members for strength. The members are made of galvanized steel ($E = 29{,}000$ ksi). Before finalizing the design, the company owner wants to know how much we can expect the tip of the structure to displace under dead load.

a. Guess which component will contribute most to the vertical displacement at D: axial force in the cross-member, bending in the cross-member, axial force in the post, or bending in the post.

b. Estimate the vertical displacement at D by considering the cross-member to be cantilevered out from a fixed support (i.e., no rotation at B). You can use the formulas inside the front cover of this text along with superposition. Will this provide an upper bound, a lower bound, or just an approximation for the actual displacement? Justify your answer.

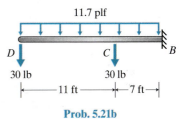

Prob. 5.21b

c. Calculate the vertical displacement at D for the full structure due to the self-weight of the structure and signals.

d. Evaluate the displacement prediction using your results from part (b).

e. Which contributed most to the displacement: axial force in the cross-member, bending in the cross-member, axial force in the post, or bending in the post? If you wanted to reduce the vertical displacement at the end, what cross-sectional property do you recommend be increased? Why that one?

f. Comment on why your guess in part (a) was or was not close to your answer in part (e).

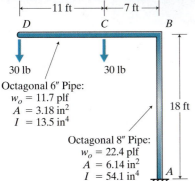

Prob. 5.21a

5.22 Owners of a building that has a below-grade window well would like to put on a removable roof in order to turn the space into a greenhouse. The roof will attach to the outer wall and lean on the inner wall. The primary loading will be the self-weight. A preliminary design resulted in wood 45×145 members ($A = 62.5 \text{ cm}^2$, $I = 1143 \text{ cm}^4$, $E = 10 \text{ GPa} = 1000 \text{ kN/cm}^2$). One of our design concerns is that water will pond where the roof leans against the wall, point A. Therefore, we will build the frame slightly elevated at A to ensure that water drains away from the building. In order to fabricate the frame correctly, we need to know how much the frame is likely to displace at A under the self-weight.

a. Guess which component will contribute most to the vertical displacement at A: axial force in the top member, bending in the top member, axial force in the diagonal member, or bending in the diagonal member.

b. Estimate the vertical displacement at A by considering the frame to be a flat, cantilevered beam with an equivalent uniform distributed load. Will this provide an upper bound, a lower bound, or just an approximation for the actual displacement? Justify your answer.

c. Calculate the vertical displacement at A for the frame due to dead load.

d. Evaluate the displacement prediction using your results from part (b).

e. Which contributed most to the displacement: axial force in the top member, bending in the top member, axial force in the diagonal member, or bending in the diagonal member? If you wanted to reduce the vertical displacement at A, what cross-sectional property do you recommend be increased? Why that one?

f. Comment on why your guess in part (a) was or was not close to your answer in part (e).

Prob. 5.22a

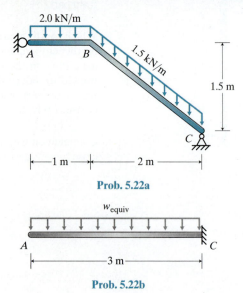

Prob. 5.22b

CHAPTER 6

INFLUENCE LINES

The internal force generated at a specific location in a structure depends on the location of the live load.

MOTIVATION

We saw in Section 1.1 that live load can act everywhere, somewhere, or nowhere. We must consider all three options when determining the design requirements for a member, connection, or support. Checking the effects of live load everywhere and nowhere is straightforward. But knowing where to put the "somewhere" live load to maximize the effect might not be intuitive. To help us with this issue, we use influence lines.

An influence line is a graphical representation of the internal force or reaction at one location in a structure due to a load applied at different locations along the structure. Although an influence line gives us information at only one point in the structure, that point might be very important. We often splice members together in the field because of length limitations, so we need to know the maximum internal forces that the splice must carry.

We can use influence lines to determine where to place load to maximize the effect (i.e., reaction or internal force). An influence line shows us the critical location to place a point force and the regions where distributed load should be placed in order to generate the largest effect. Once we know the maximum effect, we can adequately design the support or member.

Another benefit of influence lines is that we can use them to calculate the resulting reaction or internal force given the load placement. Therefore, we can use influence lines to calculate the effect due to any loading, including dead or snow.

6.1 The Table-of-Points Method

Introduction
An influence line visually illustrates the effect of a single, unit point load placed at different locations along the structure. One way to construct the influence line is to first create a table showing locations of the unit load and the resulting effects. We then connect the points to draw the influence line.

How-To
First a word of caution: When initially introduced to influence lines, engineers often confuse them with internal force diagrams (Chapter 4). The only similarity between the two, however, is that both have graphical representations. The influence line for the moment at a point is **not** the integral of the influence line for the shear at that point.

An influence line is a graphical representation of a single reaction or internal force generated by a single point load with unit magnitude. We call that reaction or internal force the *effect* that the influence line is displaying. The key attribute of an influence line is that it shows the effect generated by the unit load at all locations along the structure. We often speak of the unit load "moving" across the structure, much as a person walks along a hallway or a car moves along a bridge.

For determinate structures, all influence lines are made of straight segments. This is not true of indeterminate structures. In this text, however, we focus on determinate structures.

Since the influence line is made of straight segments, if we know points along the influence line, we can complete the graphical representation by connecting the dots. A subtle but important idea is that we will not create the actual influence line if we don't have points at the right locations.

To make sure we have enough information, we need to calculate the effect due to the unit load at each of the following locations:

For axial force in a truss member
- Each joint where the live load might act

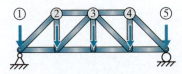

Figure 6.1a

For force or moment reaction
- Each support location
- Each internal hinge
- Free end if the member has any

Figure 6.1b

For internal shear
- Each support location
- Each internal hinge
- Free end if the member has any
- Immediately to the left of the point where we want to know the shear
- Immediately to the right of the point

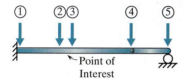

Figure 6.1c

For internal moment
- Each support location
- Each internal hinge
- Free end if the member has any
- Point where we want to know the moment

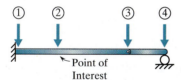

Figure 6.1d

We can generate an influence line using a unit load in any direction we want. Almost always, we apply the unit load in the direction of gravity. Therefore, it is safe to assume that unless otherwise noted, an influence line is based on a downward unit load. However, there are times when we might create an influence line with a lateral unit load. We must design handrails for live load that acts in any direction, including sideways; we design stadiums for live load that acts parallel and perpendicular to the rows of seating.

The influence line is made of the magnitudes of the effect of the unit force. Because the unit force has no units, an influence line for a force reaction, axial force, or shear is unitless. An influence line for an internal moment or moment reaction has units of length.

An influence line provides information for only one internal force or one reaction at only one point in a structure. If we want to display information for a different internal force or reaction, or if we want to display information for a different location, we need to construct another influence line.

SECTION 6.1 HIGHLIGHTS

The Table-of-Points Method

Table headings:
- Location of unit force.
- Effect generated (e.g., reaction, shear, moment).

Locations where we need to apply force:

For axial force in a truss member
- Each joint where a live load might act

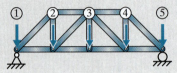

For force or moment reaction
- Each support location
- Each internal hinge
- Free end if the member has any

For internal shear
- Each support location
- Each internal hinge
- Free end if the member has any
- Immediately to the left of the point where we want to know the shear
- Immediately to the right of the point

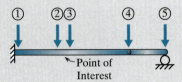

For internal moment
- Each support location
- Each internal hinge
- Free end if the member has any
- Point where we want to know the moment

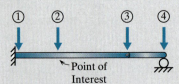

Creating the influence line:
- Plot each point from the table to scale on a graph.
- Connect each point with a straight line (for determinate structures).

EXAMPLE 6.1

A continuous beam supports the floor of a house and the carport covering. In order to properly design the support at *A*, we need to know which loads generate an upward reaction at *A* and which generate a downward reaction. An influence line for the vertical reaction at *A* will provide us that information.

Figure 6.2a

Detailed Analysis:

Identify locations for the unit load:
  Reaction locations: A, B
  Free end: C

Calculate the reaction, A_y, due to each unit load:
  Unit load at A:
    FBD:

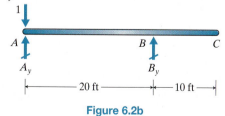

Figure 6.2b

  Apply equilibrium:

$$\circlearrowleft \Sigma M_B = 0 = A_y(20 \text{ ft}) - 1(20 \text{ ft})$$
$$A_y = 1 \; (+\uparrow)$$

302   **Chapter 6**   Influence Lines

Unit load at B:
FBD:

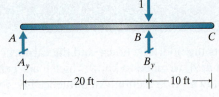

Figure 6.2c

Apply equilibrium:

$$\circlearrowleft^+ \Sigma M_B = 0 = A_y(20\text{ ft})$$

$$A_y = 0$$

Unit load at C:
FBD:

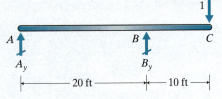

Figure 6.2d

Apply equilibrium:

$$\circlearrowleft^+ \Sigma M_B = 0 = A_y(20\text{ ft}) + 1(10\text{ ft})$$

$$A_y = -0.50\ (+\uparrow)$$

Table of points:

Location of Unit Load	A_y
A	1.0
B	0.0
C	−0.5

Create the influence line:

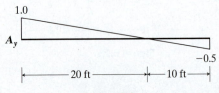

Figure 6.3

Example 6.2

Tractor trailers beyond a certain length require special permits. As a result, long bridge girders are often brought to the site in multiple parts and spliced together. In order to design the splice for a particular single-span bridge, we need to know the peak shear generated at that splice. To know where to put the traffic load to generate the largest shear, we can construct an influence line.

Figure 6.4

Detailed Analysis:

Identify locations for the unit load:
  Reaction locations: A, C
  Points of interest: B^-, B^+

Calculate the shear, V_B, due to each unit load:
  Unit load at A:
    FBD:

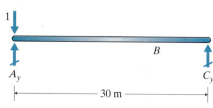

Figure 6.5a

Apply equilibrium:
$$\circlearrowleft \Sigma M_C = 0 = A_y(30\,\text{m}) - 1(30\,\text{m})$$
$$A_y = 1\ (+\uparrow)$$

Cut at B:

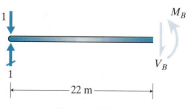

Figure 6.5b

Apply equilibrium:
$$+\uparrow \Sigma F_y = 0 = 1 - 1 - V_B$$
$$V_B = 0\ (\downarrow+\uparrow)$$

Unit load at B^-:
FBD:

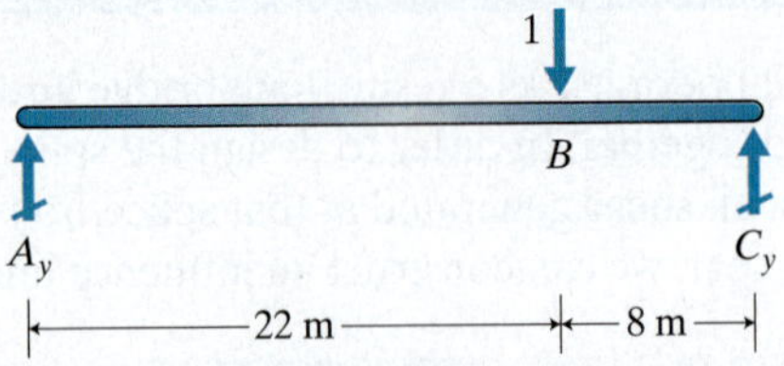

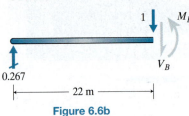
Figure 6.6a

Apply equilibrium:
$$\circlearrowleft^+ \Sigma M_C = 0 = A_y(30\,\text{m}) - 1(8\,\text{m})$$
$$A_y = 0.267 \; (+\uparrow)$$

Cut at B:

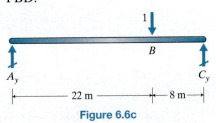
Figure 6.6b

Apply equilibrium:
$$+\uparrow \Sigma F_y = 0 = 0.267 - 1 - V_B$$
$$V_B = -0.733 \; (\downarrow + \uparrow)$$

Unit load at B^+:
FBD:

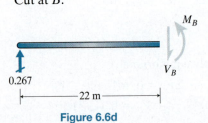
Figure 6.6c

Apply equilibrium:
$$\circlearrowleft^+ \Sigma M_C = 0 = A_y(30\,\text{m}) - 1(8\,\text{m})$$
$$A_y = 0.267 \; (+\uparrow)$$

Cut at B:

Figure 6.6d

Apply equilibrium:
$$+\uparrow \Sigma F_y = 0 = 0.267 - V_B$$
$$V_B = +0.267 \; (\downarrow + \uparrow)$$

Unit load at C:
FBD:

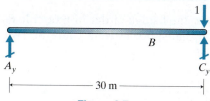

Figure 6.7a

Apply equilibrium:

$\circlearrowleft +\ \Sigma M_C = 0 = A_y(30\text{ m})$
$A_y = 0\ (+\uparrow)$

Cut at B:

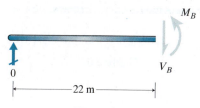

Figure 6.7b

Apply equilibrium:
$+\uparrow \Sigma F_y = 0 = 0 - V_B$
$V_B = 0\ (\downarrow+\uparrow)$

Table of points:

Location of Unit Load	V_B
A	0.000
B^-	−0.733
B^+	0.267
C	0.000

Create the influence line:

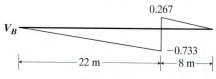

Figure 6.8

EXAMPLE 6.3

In order to design the splice for the bridge girder in Example 6.2, we need to know the peak moment generated at that splice as well. Therefore, we need to construct another influence line.

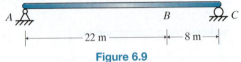

Figure 6.9

Detailed Analysis:

Identify locations for the unit load:
    Reaction locations: A, C
    Point of interest: B

Calculate the moment, M_B, due to each unit load:
    Unit load at A:
        FBD:

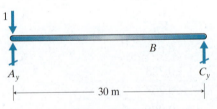

Apply equilibrium:
$$\circlearrowleft^+ \Sigma M_C = 0 = A_y(30 \text{ m}) - 1(30 \text{ m})$$
$$A_y = 1 \ (+\uparrow)$$

Figure 6.10a

Note that the reaction is the same as when we generated the shear influence line. The reaction is not a function of where we calculate the internal force.

Cut at B:

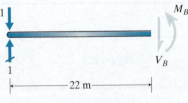

Figure 6.10b

Apply equilibrium:

$$+\circlearrowleft \Sigma M_{cut} = 0 = 1(22\text{ m}) - 1(22\text{ m}) - M_B$$
$$M_B = 0 \;\; (\circlearrowleft+)$$

Unit load at B:
Cut at B:
Note that we actually cut the beam at either B^- or B^+, but it does not matter to the value of the moment at the cut.

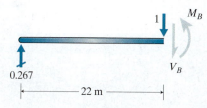

Figure 6.11

Apply equilibrium:

$$+\circlearrowleft \Sigma M_{cut} = 0 = 0.267(22\text{ m}) - M_B$$
$$M_B = 5.87\text{ m} \;\; (\circlearrowleft+)$$

Unit load at C:
Cut at B:

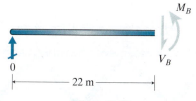

Figure 6.12

Apply equilibrium:

$$+\circlearrowleft \Sigma M_{cut} = 0 = 0(22\text{ m}) - M_B$$
$$M_B = 0 \;\; (\circlearrowleft+)$$

Table of points:

Location of Unit Load	M_B
A	0.00 m
B	5.87 m
C	0.00 m

Create the influence line:

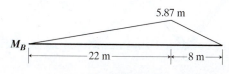

Figure 6.13

EXAMPLE 6.4

Our team is designing a pedestrian footbridge. Before we can select member sizes, we need to know the peak axial force for each member. An influence line will show where the live load must act to generate the peak force in a certain member and will allow us to calculate the resulting axial force. Let's focus on member BH for now.

Figure 6.14

Detailed Analysis:

Identify locations for the unit load:
   Joints where the live load might act: A, B, C, D, E

Calculate the axial force, F_{BH}, due to each unit load:
   Unit load at A:
      FBD:

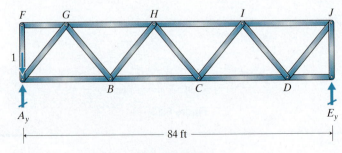

Figure 6.15a

Apply equilibrium:

$$\stackrel{+}{\curvearrowleft}\Sigma M_E = 0 = A_y(84\text{ ft}) - 1(84\text{ ft})$$
$$A_y = 1\ (+\uparrow)$$

Cut section through BH:

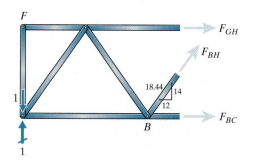

Apply equilibrium:

$$+\uparrow \Sigma F_y = 0 = 1 - 1 + \frac{14}{18.44}F_{BH}$$

$$F_{BH} = 0$$

Figure 6.15b

Unit load at B:
FBD:

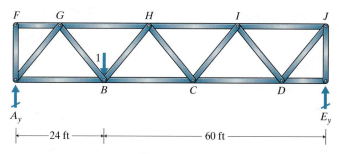

Figure 6.16a

Apply equilibrium:

$$\stackrel{+}{\curvearrowleft}\Sigma M_E = 0 = A_y(84\text{ ft}) - 1(60\text{ ft})$$
$$A_y = 0.714\ (+\uparrow)$$

Cut section through BH:

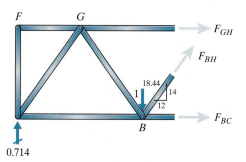

Apply equilibrium:

$$+\uparrow \Sigma F_y = 0 = 0.714 - 1 + \frac{14}{18.44}F_{BH}$$

$$F_{BH} = 0.377$$

Figure 6.16b

310   Chapter 6   Influence Lines

Unit load at C:
FBD:

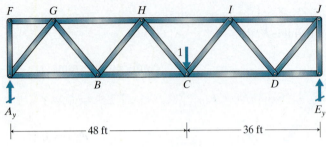

Figure 6.17a

Apply equilibrium:

$$\circlearrowright \Sigma M_E = 0 = A_y(84 \text{ ft}) - 1(36 \text{ ft})$$
$$A_y = 0.429 \; (+\uparrow)$$

Cut section through BH:

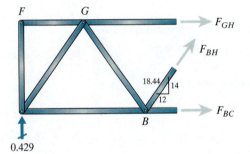

Apply equilibrium:

$$+\uparrow \Sigma F_y = 0 = 0.429 + \frac{14}{18.44} F_{BH}$$
$$F_{BH} = -0.565$$

Figure 6.17b

Unit load at D:
FBD:

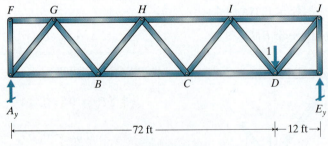

Figure 6.18a

Apply equilibrium:

$$\circlearrowright \Sigma M_E = 0 = A_y(84 \text{ ft}) - 1(12 \text{ ft})$$
$$A_y = 0.143 \; (+\uparrow)$$

Section 6.1 The Table-of-Points Method **311**

Cut section through BH:

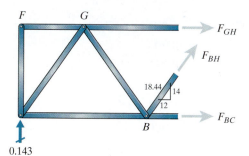

Apply equilibrium:

$$+\uparrow \Sigma F_y = 0 = 0.143 + \frac{14}{18.44}F_{BH}$$

$$F_{BH} = -0.188$$

Figure 6.18b

Unit load at E:
FBD:

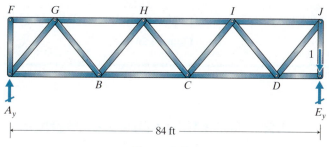

Figure 6.19a

Apply equilibrium:

$$\curvearrowleft \Sigma M_E = 0 = A_y(84 \text{ ft}) - 1(0 \text{ ft})$$

$$A_y = 0 \;(+\uparrow)$$

Cut section through BH:

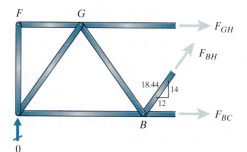

Apply equilibrium:

$$+\uparrow \Sigma F_y = 0 = 0 + \frac{14}{18.44}F_{BH}$$

$$F_{BH} = 0$$

Figure 6.19b

Table of points:

Location of Unit Load	F_{BH}
A	0.000
B	0.377
C	−0.565
D	−0.188
E	0.000

Create the influence line:

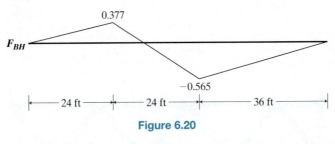

Figure 6.20

6.2 The Müller-Breslau Method

Introduction
Another method for creating influence lines was developed by a Prussian engineer, Heinrich Müller-Breslau, in the late 1800s. He showed that we can visualize the influence line by releasing the structure at the location of interest, applying the desired internal force or reaction, and sketching the displaced shape. This method is limited to beams and rigid frames, however.

How-To
Müller-Breslau developed his method using insights gained from the Virtual Work Method of predicting deformations (Section 5.3). The Müller-Breslau Method is applicable to beams and frames but not to trusses.

To construct the shape of the influence line, we release the structure so that it cannot resist the internal force or reaction of interest. If the structure is determinate, releasing it results in an unstable structure. We then apply the internal force or reaction to the released structure (Figure 6.21). To avoid confusion, we should apply the internal forces in their positive directions using the sign conventions from Section 4.1. Since the released structure is unstable, the pieces move without bending. The resulting shape is the exact shape of the influence line for the original structure.

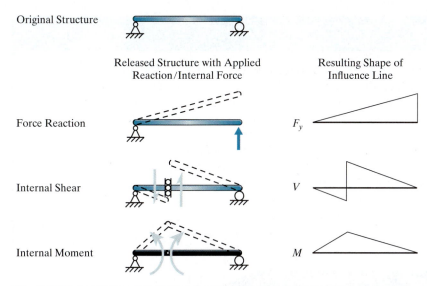

Figure 6.21 Application of the Müller-Breslau Method to a simply supported beam.

Using this method, we don't know the exact magnitudes on the influence line, but we do know the relative magnitudes because of the straight line segments. Therefore, to complete the influence line, we need just one value. We can find all the other values once we know the one. To find that one value, we use the Table-of-Points Method for one location of the unit load. Remember that we use the original structure to find this one value.

Section 6.2 Highlights

The Müller-Breslau Method

For a determinate beam or frame:

1. Release the structure so that it cannot resist the internal force or reaction of interest.

2. Apply the internal force or reaction to the released structure in the positive direction.

3. The displaced shape of the released structure is the shape of the influence line to scale.

4. Find one nonzero value on the influence line using the Table-of-Points Method.

5. Find all other values on the influence line using similar triangles.

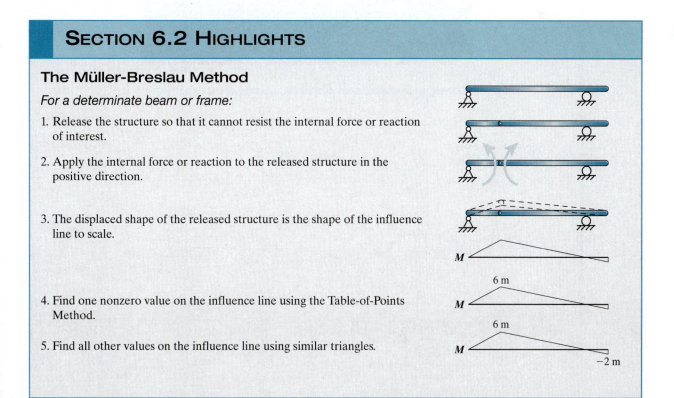

Example 6.5

Wood beams cantilevered out from a house to form an elevated deck. The owners wanted a bigger deck, so they extended the beams and added columns at the far ends. The addition was not engineered, so the owners have no idea whether the deck is safe. Now that they want to sell the house, they have hired our firm to evaluate the capacity of the deck in its current condition.

Our team is to determine the required capacity of the moment reaction at A. To help in that process, our assignment is to develop the influence line for M_A. This is a great opportunity for us to practice the Müller-Breslau Method.

Figure 6.22

Construct Influence Line:

Release the structure and apply the moment at A:

Replace the moment reaction with a pin and apply a positive moment reaction:

Figure 6.23

The unstable beam deforms in straight line segments.

Figure 6.24

The displaced shape is the influence line to the correct scale.

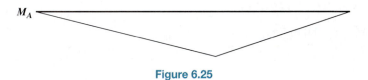

Figure 6.25

Find the value at B:

Unit load just to the left of *B* (although it doesn't matter which side):

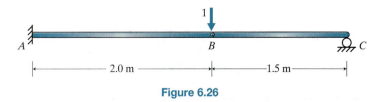

Figure 6.26

Apply equilibrium to BC:

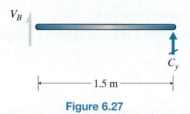

Figure 6.27

$$\circlearrowleft^{+}\Sigma M_C = 0 = V_B(1.5\text{ m})$$
$$V_B = 0$$

Apply equilibrium to AB:

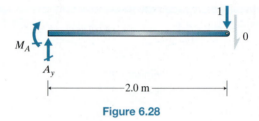

Figure 6.28

$$\circlearrowleft^{+}\Sigma M_A = 0 = M_A + 1(2.0\text{ m})$$
$$M_A = -2.0\text{ m }(\circlearrowleft^{+})$$

Label the influence line for the moment reaction at A:

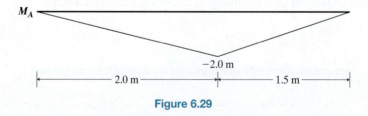

Figure 6.29

Section 6.2 The Müller-Breslau Method

EXAMPLE 6.6

A field inspection of the deck in Example 6.5 reveals a patch of rotting wood in one of the beams. The damage is localized, so before deciding to replace it, our team will estimate the residual capacity of the member and compare that with the design internal forces.

Our job is to construct an influence line for the internal shear at the damaged location, point *D*.

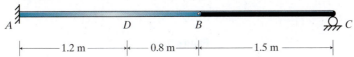

Figure 6.30

Construct Influence Line:

Release the shear carrying capability at D and apply the shear at D:
  Insert the constrained roller at *D* and apply a positive internal shear:

Figure 6.31

The unstable beam deforms in straight line segments. The constrained roller does not allow relative rotation, so segment AD does not deform and segment DB remains horizontal.

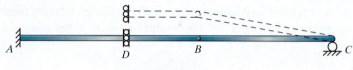

Figure 6.32

The displaced shape is the influence line to the correct scale.

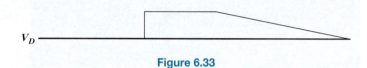

Figure 6.33

Find the value at D:
Unit load just to the right of D:

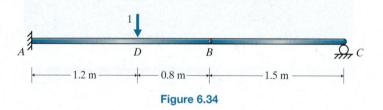

Figure 6.34

Apply equilibrium to BC:

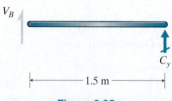

Figure 6.35

$$+\circlearrowleft \Sigma M_C = 0 = V_B(1.5 \text{ m})$$
$$V_B = 0$$

Apply equilibrium to DB:

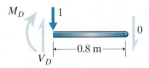

Figure 6.36

$$+\uparrow \Sigma F_y = 0 = V_D - 1$$
$$V_D = 1.0 \; (\downarrow + \uparrow)$$

Label the influence line for the shear at D:

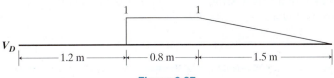

Figure 6.37

Observation: The load between the wall, A, and the location of the damage, D, has no effect on the shear generated at D.

EXAMPLE 6.7

We have also been tasked with developing the influence line for the internal moment at the damaged location, D.

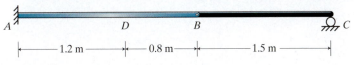

Figure 6.38

Construct Influence Line:

Release the moment carrying capability at D and apply the moment at D:

Insert a pin at D and apply a positive internal moment:

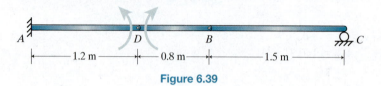

Figure 6.39

The unstable beam deforms in straight line segments. Segment AD cannot deform, but segment DB turns down.

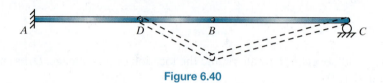

Figure 6.40

The displaced shape is the influence line to the correct scale.

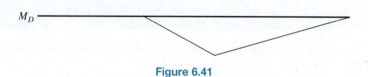

Figure 6.41

Find the value at B:

Unit load just to the right of B (although it doesn't matter which side):

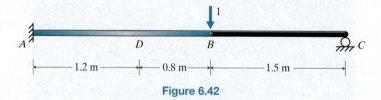

Figure 6.42

Apply equilibrium to BC:

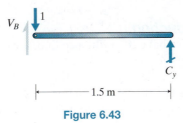

Figure 6.43

$+\circlearrowleft \Sigma M_C = 0 = V_B(1.5\text{ m}) - 1(1.5\text{ m})$

$V_B = 1\ (\downarrow + \uparrow)$

Apply equilibrium to DB:

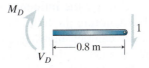

Figure 6.44

$+\circlearrowleft \Sigma M_D = 0 = M_D + 1(0.8\text{ m})$

$M_D = -0.8\text{ m}\ (\circlearrowleft)$

Label the influence line for the moment at D:

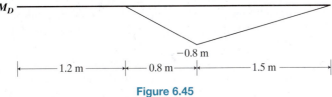

Figure 6.45

Observation: The load between the wall, A, and the location of the damage, D, has no effect on the moment generated at D either.

6.3 Using Influence Lines

Introduction
Once we have the influence line, it can help us in a couple ways. We can quickly see where to place a live point load or a distributed load to maximize the effect. That location is not always apparent from just looking at the structure. We can also use the influence line to quantitatively predict the effect of a set combination of point loads and distributed loads.

How-To
An influence line shows us the effect of a unit point load. As long as the material remains linear elastic (an assumption), we can scale the effect of the unit load by the magnitude of an applied load. For example, if the shear generated by the unit load at a certain location is 0.4, then the shear due to a 12-kN point load at that same location is $0.4 \times 12 \text{ kN} = 4.8 \text{ kN}$.

We can also use the influence line to identify where to place a live point load to generate the largest positive or negative effect. Remember that live loads can act anywhere, so knowing the critical placement is essential for predicting the reactions or internal forces that we need to design for. The location of the largest positive magnitude on the influence line is the location where a point force will generate the largest positive effect (Figure 6.46a). Likewise, we can place a point force at the location of the largest negative value on the influence line in order to generate the largest negative effect (Figure 6.46b). Depending on the situation, we might be concerned about both the largest positive and the largest negative effects.

In most cases, a live load results in a distributed load on a structure. The influence line still shows us where to place a distributed load to maximize the effect. If we want to know the peak positive effect, we place the live distributed load only along the parts of the structure where the influence line is positive (Figure 6.47a). For the peak negative effect, we place the distributed load along the parts where the influence line is negative (Figure 6.47b).

A distributed load is like an infinite set of small point loads next to one another. Therefore, we should be able to use influence lines to determine the effect of a distributed load as well.

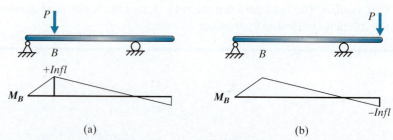

Figure 6.46 Using an influence line to determine where to place a point load to generate the largest effect: (a) Location to apply point force to generate largest positive M_B; (b) Location to apply point force to generate largest negative M_B.

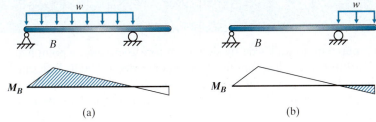

Figure 6.47 Using an influence line to determine where to place a distributed load to generate the largest effect: (a) Location to apply distributed load to generate largest positive M_B; (b) Location to apply distributed load to generate largest negative M_B.

To find the total applied load, we integrate the distributed load. But to find the total effect, we need to integrate the product of the distributed load and the influence line value:

$$\text{Effect} = \int_0^l Infl(x) \cdot w(x)\,dx$$

Fortunately, we often deal with a uniform distributed load, so the integration simplifies:

$$\text{Effect} = w \int_0^l Infl(x)\,dx$$

$$\text{Effect} = w \times \text{Area under influence line}$$

SECTION 6.3 HIGHLIGHTS

Using Influence Lines

Point loads:

- Location of largest magnitude on the influence line is the location to put a point load to maximize the effect.
- Effect is the product of the influence line magnitude where the point load is applied and the point load magnitude:

$$\text{Effect} = \text{Influence} \times P$$

Distributed loads:

- Regions of positive magnitude on the influence line are the locations to put a distributed load to maximize the positive effect.
- Regions of negative magnitude on the influence line are where to put a distributed load to maximize the negative effect.
- Effect is the product of the area under the influence line and the uniform distributed load magnitude:

$$\text{Effect} = \text{Area under influence line} \times \text{Distributed load}$$

EXAMPLE 6.8

People walking together in a group tend to walk side-by-side. Therefore, they cause a point live load on a pedestrian bridge. For this bridge, the point live load on one truss is 600 lb. In Example 6.4, we developed the influence line for the axial force in member BH. In order to calculate the peak tension and peak compression in member BH due to the point live load, let's use the influence line.

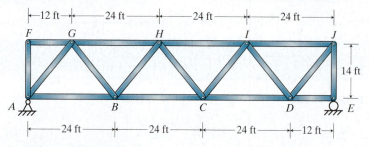

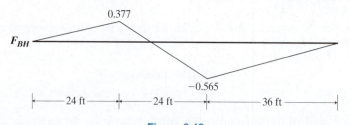

Figure 6.48

Detailed Analysis:

Peak tension:

Location of peak positive influence line value: B

Calculate tension when the live load is applied at B:

$$F_{BH} = 0.377\,(600\text{ lb}) = 226 \text{ lb } (\textbf{\textit{T}})$$

Peak compression:
   Location of peak negative influence line value: C
   Calculate compression when the live load is applied at C:
   $$F_{BH} = -0.565(600 \text{ lb}) = 339 \text{ lb (C)}$$

Summary:

Desired Peak Effect	Location of Point Load	F_{BH}
Tension	B	226 lb
Compression	C	−339 lb

EXAMPLE 6.9

The largest live load effect on a pedestrian bridge is likely due to distributed load over part of the bridge. In Example 6.4, we developed the influence line for the axial force in member BH. The maximum likely live load for this bridge results in a distributed load of 500 plf that can act anywhere along the bottom chord of the truss. Our job is to calculate the peak tension and peak compression that can develop in member BH due to this uniform distributed live load.

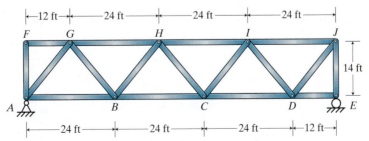

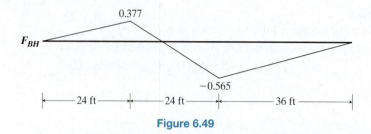

Figure 6.49

Detailed Analysis:

Location where the influence line is zero:
To find where the influence line crosses 0, use similar triangles:

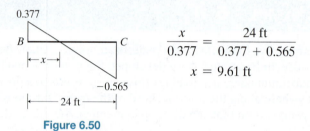

$$\frac{x}{0.377} = \frac{24 \text{ ft}}{0.377 + 0.565}$$

$$x = 9.61 \text{ ft}$$

Figure 6.50

Peak tension:
Apply the distributed load where the influence line is positive: A to 9.61 ft past B:

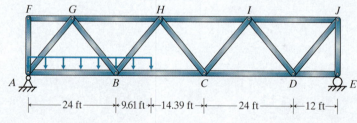

Figure 6.51

Calculate the resulting peak tension:

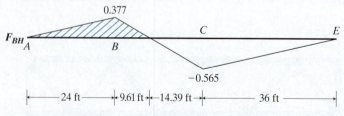

Figure 6.52

$$\text{Area} = \tfrac{1}{2}(0.377)(24 \text{ ft} + 9.61 \text{ ft}) = 6.34 \text{ ft}$$
$$F_{BH} = 6.34 \text{ ft}(500 \text{ plf}) = 3170 \text{ lb } (\boldsymbol{T})$$

Peak compression:
Apply the distributed load where the influence line is negative: 9.61 ft past B to E:

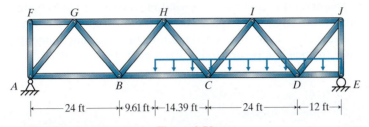

Figure 6.53

Calculate the resulting peak compression:

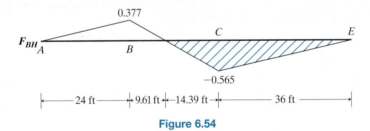

Figure 6.54

Area $= \frac{1}{2}(-0.565)(14.39 \text{ ft} + 36 \text{ ft}) = -14.24 \text{ ft}$
$F_{BH} = -14.24 \text{ ft}(500 \text{ plf}) = 7120 \text{ lb}$ (**C**)

Summary:

Desired Peak Effect	Location of Distributed Load	F_{BH}
Tension	0–33.6 ft	3170 lb
Compression	33.6–84 ft	−7120 lb

EXAMPLE 6.10

Repairs need to be made to the left side of the bridge. Therefore, the contractors want to store materials and equipment on the right half. The resulting distributed load on the girder will be rather large. Before authorizing the contractor to use the bridge for storage, we should check the adequacy of the splice. To do that, we need to know the resulting moment generated at the splice. Let's use the influence line we created in Example 6.3.

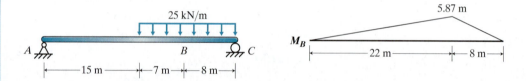

Figure 6.55

Detailed Analysis:

Magnitude of the influence line at the start of the distributed load:
 To find the value of the influence line where the distributed load starts, use similar triangles:

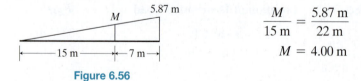

$$\frac{M}{15 \text{ m}} = \frac{5.87 \text{ m}}{22 \text{ m}}$$

$$M = 4.00 \text{ m}$$

Figure 6.56

Calculate the moment due to the distributed load:

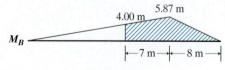

Figure 6.57

$$\text{Area} = \tfrac{1}{2}(7 \text{ m})(4.00 \text{ m} + 5.87 \text{ m}) + \tfrac{1}{2}(8 \text{ m})(5.87 \text{ m})$$

$$= 58.0 \text{ m}^2$$

$$M_B = 58.0 \text{ m}^2 (25 \text{ kN/m}) = 1450 \text{ kN} \cdot \text{m}$$

Homework Problems

6.1 and 6.2 A portable gantry crane has a rated capacity of 25 kN. It was recently dented when hit by a load carried by the rail-mounted gantry. As a result, the owner wants us to evaluate whether the portable gantry can still carry the rated load. A member of the team is evaluating the dent and calculating the shear and moment capacity of the beam at the location of the damage.

6.1 Our role is to calculate the peak shear generated at the damage site, B, if the crane is used to lift the full 25 kN. Creating an influence line will help us determine where to place the load.

a. Guess where to place a single point load to generate the peak shear at B (positive or negative).
b. Develop the influence line for the internal shear at B. Because this is idealized as a simply supported beam, the Table-of-Points Method is probably the fastest.
c. Identify where a single point load must act to generate the largest shear at B (positive or negative).
d. What is the peak shear generated at B (positive or negative) due to a live point load of 25 kN?
e. Comment on why your guess in part (a) was or was not close to your conclusion in part (c).

6.2 Our role is to calculate the peak moment generated at the damage site, B, if the crane is used to lift the full 25 kN. Creating an influence line will help us determine where to place the load.

a. Guess where to place a single point load to generate the peak moment at B (positive or negative).
b. Develop the influence line for the internal moment at B. Because this is idealized as a simply supported beam, the Table-of-Points Method is probably the fastest.
c. Identify where a single point load must act to generate the largest moment at B (positive or negative).
d. What is the peak moment generated at B (positive or negative) due to a live point load of 25 kN?
e. Comment on why your guess in part (a) was or was not close to your conclusion in part (c).

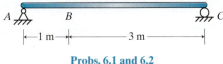

Probs. 6.1 and 6.2

6.3 and 6.4 Crews are preparing to perform maintenance on the roof of a covered walkway. The maintenance contractor wants to stack supplies on the outer edges of the roof, but the owner is concerned about overloading the structure.

Our team leader would like us to explore two options for stacking the supplies. The contractor would like to stack the supplies on the outer edges of the roof because it is easier to put them there; this option would result in a distributed load of 350 plf from the supports to the outer edges. The other option is to distribute the same total weight of supplies between the supports of the beam; this option would result in a distributed load of 150 plf between the supports. Influence lines will help us in the evaluation.

To promote drainage, the roof drops 1 foot over its entire length. Because the live load does not act normal to the support beam, it will be easier to construct the influence lines using the Table-of-Points Method.

6.3 One of our concerns is the possibility that the loading will cause uplift at support D that exceeds the connection design strength.

a. Guess where to put the distributed load to generate the largest uplift at support D.
b. Develop the influence line for the vertical reaction at D.
c. Determine the largest uplift that can be generated at support D by a distributed live load of 350 plf. This live load can be placed on either or both ends of the beam.
d. Determine the largest uplift that can be generated at support D by a distributed live load of 150 plf. This live load can only be placed between the supports, B to D.
e. Determine the vertical reaction generated at support D by the self-weight of 15 plf.
f. Which live load situation generates the largest uplift at support D? When combined with the self-weight effect, is there still uplift? If so, the team will need to verify that the connection is sufficiently strong.
g. Comment on why your guess in part (a) was or was not the same as what you discovered in part (f).

6.4 Another of our concerns is exceeding the bending capacity of the beam. The critical location will be midspan, C. Because of the material choice and the way the roof is constructed, the capacity of the beam is the same for both positive and negative moments.

a. Guess whether the largest midspan moment generated by the roofing materials will be positive or negative.
b. Develop the influence line for the moment at midspan of the support beam, point C.
c. Determine the midspan moment generated by a distributed live load on both outer ends of 350 plf.
d. Determine the midspan moment generated by a distributed live load between the supports of 150 plf.
e. Which situation generates the largest midspan moment? We should encourage the maintenance contractor to place supplies at the location that generates the smallest moment.
f. Comment on why your guess in part (a) was or was not the same as what you discovered in part (e).

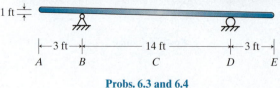

Probs. 6.3 and 6.4

6.5 and 6.6 A pedestrian bridge connects two parts of a city's convention center. The convention committee wants to move a heavy display across the bridge from one building to the other. The effect will be essentially a large point load of 15 kips traveling across the bridge. Our firm has been hired to evaluate the ability of the bridge to safely support the display as it is moved across. In order to do that, we need to know the peak axial force generated in each member due to the live load. Influence lines will help.

6.5 The team divided up the responsibility for different members. We have been tasked with members CK and DK.

a. Guess where to apply the unit load to generate the largest force in member CK.
b. Develop the influence line for the axial force in member CK.
c. Determine the largest tension generated in member CK, if any, due to the live point load. Determine the largest compression generated, if any, by the same point load.
d. Develop the influence line for the axial force in member DK.
e. Determine the largest tension generated in member DK, if any, due to the live point load. Determine the largest compression generated, if any, by the same point load.
f. Comment on why your guess in part (a) was or was not the same as what you discovered in part (c).

6.6 The team divided up the responsibility for different members. We have been tasked with members JK and JC.

a. Guess where to apply the unit load to generate the largest force in member JK.
b. Develop the influence line for the axial force in member JK.
c. Determine the largest tension generated in member JK, if any, due to the live point load. Determine the largest compression generated, if any, by the same point load.
d. Develop the influence line for the axial force in member JC.
e. Determine the largest tension generated in member JC, if any, due to the live point load. Determine the largest compression generated, if any, by the same point load.
f. Comment on why your guess in part (a) was or was not the same as what you discovered in part (c).

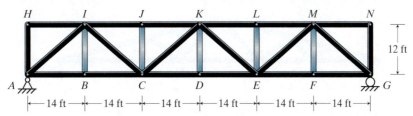

Probs. 6.5 and 6.6

332 **Chapter 6** Influence Lines

6.7 and 6.8 The interior trusses for a shelter will be supported at the ends only. As designers, we must consider the possibility that the owner will hang equipment or other items from the joints of the truss. Our sales team reports that for a truss this size, clients typically want to be able to support up to a 4-kN point load. To help us account for this in the design, influence lines will be extremely helpful.

6.7 The design team divided up the responsibility for different members. We have been tasked with members DJ and JK.

a. Guess where to apply the unit load to generate the largest force in member DJ.
b. Develop the influence line for the axial force in member DJ.
c. Determine the largest tension that can be generated in member DJ, if any, due to a live point load of 4 kN. Determine the largest compression, if any, that can be generated by the same point load.
d. Develop the influence line for the axial force in member JK.
e. Determine the largest tension that can be generated in member JK, if any, due to a live point load of 4 kN. Determine the largest compression, if any, that can be generated by the same point load.
f. Comment on why your guess in part (a) was or was not the same as what you discovered in part (c).

6.8 The design team divided up the responsibility for different members. We have been tasked with members DL and DE.

a. Guess where to apply the unit load to generate the largest force in member DL.
b. Develop the influence line for the axial force in member DL.
c. Determine the largest tension that can be generated in member DL, if any, due to a live point load of 4 kN. Determine the largest compression, if any, that can be generated by the same point load.
d. Develop the influence line for the axial force in member DE.
e. Determine the largest tension that can be generated in member DE, if any, due to a live point load of 4 kN. Determine the largest compression, if any, that can be generated by the same point load.
f. Comment on why your guess in part (a) was or was not the same as what you discovered in part (c).

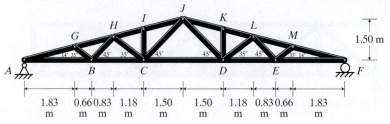

Probs. 6.7 and 6.8

6.9 and 6.10 Our team is designing a pedestrian footbridge. In order to do that, we need to know the peak axial force generated in each member due to a distributed live load of 625 plf. Influence lines will help.

6.9 The design team divided up the responsibility for different members. We have been tasked with member *CD*.

a. Guess whether the largest force generated in member *CD* by a distributed live load will be tension or compression.
b. Develop the influence line for the axial force in member *CD*.
c. Determine the largest tension that can be generated in member *CD*, if any, due to a 625-plf live distributed load. Determine the largest compression, if any, that can be generated by the same live load.
d. Comment on why your guess in part (a) was or was not the same as what you discovered in part (c).

6.10 The design team divided up the responsibility for different members. We have been tasked with member *CH*.

a. Guess whether the largest force generated in member *CH* by a distributed live load will be tension or compression.
b. Develop the influence line for the axial force in member *CH*.
c. Determine the largest tension that can be generated in member *CH*, if any, due to a 625-plf live distributed load. Determine the largest compression, if any, that can be generated by the same live load.
d. Comment on why your guess in part (a) was or was not the same as what you discovered in part (c).

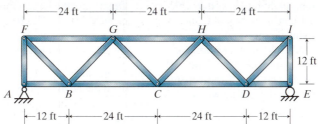

Probs. 6.9 and 6.10

6.11 and 6.12 Our team is designing a pedestrian footbridge. In order to do that, we need to know the peak axial force generated in each member due to a distributed live load of 6.36 kN/m. Influence lines will help.

6.11 The design team divided up the responsibility for different members. We have been tasked with member DL.
a. Guess whether the largest force generated in member DL by a distributed live load will be tension or compression.
b. Develop the influence line for the axial force in member DL.
c. Determine the largest tension that can be generated in member DL, if any, due to a 6.36-kN/m live distributed load. Determine the largest compression, if any, that can be generated by the same live load.
d. Comment on why your guess in part (a) was or was not the same as what you discovered in part (c).

6.12 The design team divided up the responsibility for different members. We have been tasked with member DM.
a. Guess whether the largest force generated in member DM by a distributed live load will be tension or compression.
b. Develop the influence line for the axial force in member DM.
c. Determine the largest tension that can be generated in member DM, if any, due to a 6.36-kN/m live distributed load. Determine the largest compression, if any, that can be generated by the same live load.
d. Comment on why your guess in part (a) was or was not the same as what you discovered in part (c).

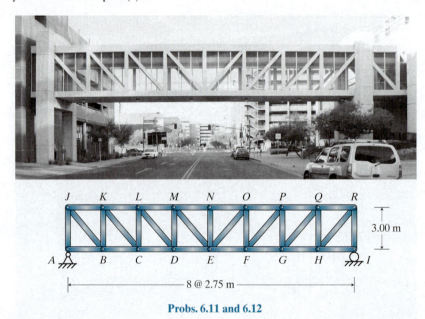

Probs. 6.11 and 6.12

6.13 and 6.14 Retrofits are being planned for a manufacturing facility in preparation for a new product line. A short beam will be extended using a pinned connection at B. The extended beam will need to support a point load of 2 kips, but the team is not yet sure where the load will be attached. In order to move forward with the retrofit design, we can use influence lines to determine worst-case scenarios.

6.13 One of the concerns is the impact of the new load on the moment reaction at A.
a. Guess where to place a single point load to generate the peak moment reaction at A (positive or negative).
b. Develop the influence line for the moment reaction at A. The Müller-Breslau Method is probably the quickest.
c. Identify where a single point load must act to generate the largest moment reaction at A (positive or negative).
d. Determine the largest moment reaction at A (positive or negative) that could be generated by a 2-kip point force.
e. Comment on why your guess in part (a) was or was not close to your conclusion in part (c).

6.14 The mechanical design team wants to run a pipe through the beam at point D. In order to help our structural team evaluate the request, we are tasked with generating an influence line for the shear at D.
a. Guess whether the peak shear at D due to a point live load will be positive or negative.
b. Develop the influence line for the shear at D. The Müller-Breslau Method is probably the quickest.
c. Identify where a single point load must act to generate the largest shear at D (positive or negative).
d. Determine the largest shear at D (positive or negative) that could be generated by a 2-kip point force.
e. Comment on why your guess in part (a) was or was not the same as your conclusion in part (c).

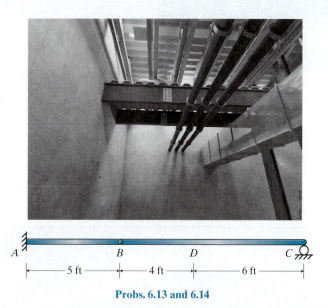

Probs. 6.13 and 6.14

6.15 and 6.16 For ease of construction in the field, a two-span bridge was built with a splice and a hinge. The Department of Transportation has received a request to increase the load limit on the bridge that will result in a live load on each girder of 5 kN/m. Before deciding, they need to know the impact on the internal forces at the splice.

6.15 One of the concerns is the shear generated in the splice, point B.

a. Guess where to place a distributed load to generate the peak shear at the splice B (positive or negative).
b. Develop the influence line for the shear at splice B. The Müller-Breslau Method is probably the quickest.
c. Identify where a distributed load must act to generate the largest shear at splice B (positive or negative).
d. What is the peak shear generated at splice B due to a uniform distributed live load of 5 kN/m?
e. Comment on why your guess in part (a) was or was not close to your conclusion in part (c).

6.16 One of the concerns is the moment generated in the splice, point B.

a. Guess where to place a distributed load to generate the peak moment at the splice B (positive or negative).
b. Develop the influence line for the moment at splice B. The Müller-Breslau Method is probably the quickest.
c. Identify where a distributed load must act to generate the largest moment at splice B (positive or negative).
d. What is the peak moment generated at splice B due to a uniform distributed live load of 5 kN/m?
e. Comment on why your guess in part (a) was or was not close to your conclusion in part (c).

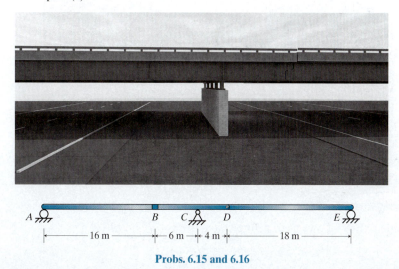

Probs. 6.15 and 6.16

CHAPTER 7

Introduction to Computer Aided Analysis

Computer aided analysis allows us to refine calculations for complex structures.

MOTIVATION

Experienced engineers around the world all have a similar complaint: new engineers run to the computer too quickly and trust the results far too much. The experienced engineers cringe when they hear "That's what the computer told me." This chapter is all about helping to change that.

Computer aided analysis can be a very helpful tool. The structures shown in the chapter-opening photos could not be designed practically without the assistance of computer analysis. Computers have opened up a huge range of complex designs that we can now bring to reality. Even for noncomplex structures, computer aided analysis allows us to determine the impact of changes to the design relatively quickly (i.e., parameter studies).

The common problem new engineers fall into is the false sense of accuracy that computer results give. Modern computer programs allow complex representations of reality, and the computer presents results with up to 10 digits of precision. However, the models are still only approximations of reality, so the computer results never match reality exactly. Our responsibility as engineers is not to produce "correct" analysis results; it is to produce reasonable results.

One of the ways we evaluate the reasonableness of results is by predicting the solution *before* we use computer aided analysis. This gives us a sense of the answer before we become buried in output data. After we have the computer results, our first priority should be to validate the results. We do that with several tools: fundamental principles, features of the solution, and approximations.

7.1 Computer Results Are Always Wrong

Introduction
Okay, so the title of this section is a bit misleading. Actually, all structural analysis results are wrong: both by hand and computer. None of them predicts the real behavior of a structure exactly. Understanding the sources of error, however, helps us estimate how far from the real behavior we might be.

How-To
We can define *error* as anything that causes our structural analysis results to be different from the structure's actual behavior. There are many sources of error, but we lump them into three categories: errors in the inputs to our structural analysis, errors in the analysis process, and human mistakes.

Input Errors
Dr. A. R. Dykes is credited with the following description of Structural Engineering:

> *Structural Engineering is the Art of moulding materials we do not wholly understand into shapes we cannot precisely analyse, so as to withstand forces we cannot really assess, in such a way that the community at large has no reason to suspect the extent of our ignorance. (Dykes 1978)*

The quote keenly hits three important areas of uncertainty in the inputs to our structural analysis: material behavior, structural idealization, and loading.

1. Material Behavior
When performing structural analysis, we almost always use material property values as if we know them precisely. However, if we test several samples of the same material, we will not obtain the same results from all the samples. We know that there is variability in material property results even with standardized test methods. Adding to the complexity of the issue, materials have certain behaviors that change over time. For example:

- Concrete gets stronger and develops a higher modulus of elasticity.
- Steel in tension relaxes (elongates).
- Masonry expands.
- Concrete masonry units (CMU) shrink.
- Timber creeps (sags) under flexural loading.

Under service loads, most materials remain in their linear-elastic range. But when we design structural members for strength (e.g., Ultimate Strength Design, Load and Resistance Factor Design), we are considering extreme loads and typically count on the material's inelastic capacity. Therefore, when we analyze structures in order to design members with adequate strength, the materials are probably beyond the linear range. This leads to the following results:

- Superposition (Section 2.2) is not actually valid.
- Deformations (Chapter 5) are underpredicted.
- Analysis based on compatibility (Chapter 12) or stiffness (Chapters 13–15) is not completely accurate.

Often our analysis methods are sufficiently accurate when we use constant material properties, but sometimes we need more accurate representation of the nonlinear material behavior. We leave that advanced topic for other references.

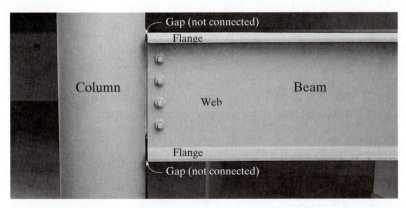

Figure 7.1 Beam-column connection where the beam is connected to the column by its web; we idealize the connection as pinned.

2. Structural Idealization

As a simple example of how structural idealization can affect our analysis results, consider the connection in Figure 7.1. Since the beam is connected to the column only at the web, we idealize this connection as pinned (Section 1.3: Structure Idealization). In reality, the connection is not pinned because the beam cannot rotate freely relative to the column, and it is not rigid because the angle between the beam and the column does change some.

Fortunately, our idealization is typically a good enough representation of reality to provide reasonable results. However, if we are concerned that our idealization does not adequately model reality, we can change the idealization to a rotational spring, but that requires laboratory testing to determine an accurate spring constant. Another option is to bound the solution (Section 2.3: Bounding the Solution) by performing two analyses: one with the connection idealized as pinned, and another with it idealized as rigid.

3. Loading

We don't know the actual loading on a structure. Consider wind loading as an example. The pressures proscribed in the building codes are higher for components and cladding (i.e., individual members) than for the main wind force-resisting system. This is because wind near the surface of the earth has very turbulent flow. The effect on even a small building is that there will be patches of intense pressure scattered around the surface. Based on field data, we know that those pockets move around the structure surface. Therefore, there is no way for us to predict the exact wind pressure distribution on a structure at a given moment in time. Instead, we use averages and envelopes of peak pressure. We consider localized intensity when designing individual pieces, and we use averages or envelopes of peak pressure when designing the whole system.

Analysis Errors

The two key contributors to errors in analysis methods are assumptions made in the method and roundoff error. Like the input errors, these are always present.

1. Assumptions in the Method

As we introduce each structural analysis method in this text, we identify the assumptions made in using the method. Some of those assumptions are overt (e.g., assuming the location of inflection points in Chapter 9: Approximate Analysis of Rigid Frames), and some are embedded (e.g., assuming relatively small deformations in Chapter 5: Deformations). It is our responsibility as engineers to be on the lookout for when we might be violating those assumptions enough to make our analysis results unreasonable. Knowing when we have gone too far typically requires comparing our results with more complex analysis that relaxes those assumptions. How to perform some of those analyses is outside the scope of this text. Instead, we focus on knowing when we need to check.

2. Roundoff Error
When we consider the variability of our structural analysis inputs, roundoff error is rarely significant enough to be of concern. In certain instances, roundoff error can play an important role in structural analysis, but those rare cases are typically associated with nonlinear analysis methods outside the scope of this text. Carrying three or four significant digits will almost always be sufficient for the problems we analyze by hand.

Human Mistakes
In most regions of the world, structural design is regulated by design codes. These codes are written by teams of experienced structural engineers who recognize all the sources of error we have identified. In general, the design methodology in our codes accounts for the input errors and analysis errors described in this section. The one source of error that is *not* accounted for is human mistakes. Therefore, the codes implicitly count on us engineers to identify and remove the human mistakes.

We use the term *mistake* to refer to any error we introduce. For example, mistakes include unintentionally changing an input, using an analysis method for something it was not intended, or misreading a calculator.

SECTION 7.1 HIGHLIGHTS

Computer Results Are Always Wrong

All structural analysis results differ from reality due to the following:

Input errors:
- Nonconstant material behavior.
- Imperfect idealization of the structure.
- Inexact prediction of loading.

Analysis errors:
- Assumptions inherent in the analysis method.
- Roundoff error in the calculations.

Human mistakes

We can and must control certain errors, namely, human mistakes.

7.2 Identifying Mistakes

Introduction
Finding a mistake can be a long, involved process. Although we discuss some common mistakes, the real focus of this chapter is identifying the presence of a mistake. Recognizing that a mistake exists, or likely exists, is more

important for a structural engineer than being able to find the mistake. We can ask for help or start all over if we recognize the presence of a mistake. But if we don't know it is there, our design might get someone hurt.

How-To

Based on conversations with many experienced structural engineers, we recommend the following procedure for performing structural analysis and evaluating the reasonableness of the results:

Step 1. Predict the Results

We discussed in the Motivation in Chapter 2: Predicting Results that there are important reasons to predict the results *before* we perform a detailed analysis. One reason is the way our mind works. Two phenomena from cognitive psychology reinforce the need to predict results first.

First: Inconsistency discounting is the phenomenon of placing more trust in the first information we have and tending to discount inconsistencies with information we receive later. Because our predictions tend to be simplified analysis results, it is easier for us to ensure that those results warrant the trust we inherently put on our first information.

Second: The other phenomenon that supports the need to predict results is *information overload*. For even a medium-size structure, the volume of information created by computer aided analysis can be huge (e.g., over a hundred pages of numbers about element forces, nodal displacements, etc.). It is impractical to evaluate all of the information, so we pick and choose. Unfortunately, when overwhelmed with information, we tend to pick information that looks familiar. What looks familiar will be the parts that are reasonable. Therefore, if we start with the detailed analysis, we tend to be biased toward checking what we know is reasonable. But if we start by using a simplified analysis to predict results, those are the results that look familiar. So we tend to check those parts of the computer aided analysis results, regardless of whether they look reasonable or not.

Step 2. Perform the Analysis

This is the step most new engineers want to jump right into. Here we apply the detailed analysis, either by computer or by hand.

Step 3. Evaluate the Results

As we discussed in the frontmatter, we are using three categories of tools throughout this text to identify whether our results are reasonable or we suspect a mistake. In addition, a fourth category is sometimes used: comparing duplicate results.

- Satisfy fundamental principles (Section 7.3).
- Observe expected features of the solution (Section 7.4).
- Approximate outcomes (Chapters 8–10).
- Compare duplicate analysis results (Chapters 12 and 13).

Throughout this text, we show how to use fundamental principles and features of the solution to evaluate results. By hand methods often provide only some information about what is going on in the structure. Computer aided analysis, however, provides reactions, internal forces, and deformations every time. Therefore, the rest of this chapter is dedicated to showing how we can use those tools to evaluate the results specifically of computer aided analysis. Note that the comparison to our prediction from Step 1 is typically part of verifying the features and comparing with approximate solutions. Our evaluation might go much further than just comparison with the prediction, but at the very least we make that comparison.

In Section 2.3: Bounding the Solution and Section 2.4: Approximating Loading Conditions, we started learning ways to develop approximate solutions in order to predict results. We add to those skills in Chapters 8–10 when we discuss specific approximate analysis methods. Those methods are excellent tools for developing an estimate of the solution that can be used to evaluate computer aided analysis results.

For complex structures, many structural design firms choose to have two different people or teams create the model and perform the analysis. Ideally, the two use different computer programs or structural analysis methods. It might seem, initially, that performing duplicate analyses is an inefficient use of time. However, comparing independent results can be one of the most reliable methods of verification. But even when we use duplicate analyses, each analysis result needs to be evaluated using the first three categories of tools. Although we do not focus on duplicate analyses using multiple software programs, we use this tool in Chapter 12: Force Method and Chapter 13: Moment Distribution Method. Those structural analysis methods are essentially as accurate as the Direct Stiffness Method (Chapters 14 and 15), which is the method used in computer aided analysis. Therefore, they can provide duplicate analysis results to compare with computer aided analysis results.

Section 7.2 Highlights

Identifying Mistakes

Mistakes are errors made by humans and are not accounted for in design codes.

The process for identifying the presence, or likely presence, of mistakes:
1. Predict results.
2. Perform analysis.
3. Evaluate results.
   - Check fundamental principles.
   - Verify features of the solution (including prediction).
   - Compare with approximate solutions (including prediction).
   - Compare duplicate analysis results.

7.3 Checking Fundamental Principles

Introduction

Unless we are performing dynamic analysis, which is outside the scope of this text, the equations of equilibrium always apply. Period. Although that might be obvious, verifying equilibrium for computer aided analysis results helps identify the presence or absence of a multitude of common mistakes.

How-To

Indeterminate structures have redundant load paths. As we see in later chapters, how much load travels through each path is a function of relative stiffness. But regardless of how the load is distributed throughout the structure, equilibrium must always be satisfied. This is true for the entire structure and for any section of the structure we choose to examine.

Each equation of equilibrium helps identify different types of mistakes. Therefore, we should verify all equations of equilibrium for each structure. In two dimensions there are three equations, and in three dimensions there are six. We will focus on 2D problems, but the process is the same in 3D.

2D	3D
$\Sigma F_x = 0$	$\Sigma F_x = 0$
$\Sigma F_y = 0$	$\Sigma F_y = 0$
$\Sigma M_z = 0$	$\Sigma F_z = 0$
	$\Sigma M_x = 0$
	$\Sigma M_y = 0$
	$\Sigma M_z = 0$

We cannot create a comprehensive list of mistakes that can be detected by checking equilibrium. However, experience helps us identify several common mistakes. The lists of possibilities included here are intended to help you start your hunt for the error once you discover that equilibrium is not satisfied. Note that there might be multiple errors and that the error might be completely different from any item on these lists.

As we create the lists, our assumption is that there are no mistakes in the check itself. Note that a common mistake is made when we check moment equilibrium: forgetting the moment reactions at fixed supports. Making this mistake gives a potentially false indication of a mistake in the analysis.

Before going to the lists, we will share some insights into one specific error common in computer aided analysis: *inconsistent units*. Practically all structural analysis software programs are based on the Direct Stiffness Method, which we cover in some detail in Chapters 14 and 15. As shown in those chapters, the Direct Stiffness Method depends on consistent units. For example, if the geometry is specified in feet, then the cross-sectional areas of members must be in ft^2. Likewise, if the modulus of elasticity is in kN/mm^2, then any distributed load must be in kN/mm. Some structural analysis programs are written with the capability of converting units. In those programs, we can create the idealized structure's geometry in meters and input the moment of inertia in in^4. A word of caution though: in those cases, the computer program still requires us to distinguish the units. The program cannot anticipate a switch in units.

Horizontal Equilibrium
If horizontal equilibrium is not satisfied, consider the possibility of one or more of these mistakes:

- We used inconsistent units if there is a horizontal distributed load.
- We applied an unintended magnitude for one or more lateral loads (e.g., software added loads rather than changing the load, forgot to deselect the previous load location when applying the next load magnitude).
- In 3D, the applied force was in the other horizontal direction than we intended.

Vertical Equilibrium

If vertical equilibrium is not satisfied, consider the possibility of one or more of these mistakes:

- We used inconsistent units if there is a vertical distributed load.
- Software included self-weight, but that was not our intent (e.g., we already applied it, we were considering only live load).
- Results were based on an unintended load combination or load factor if the software has load combination capability.
- We used the occupancy live load on the roof as well.
- We applied the load in the wrong direction.
- We used an unintended magnitude for one or more vertical loads.

Moment Equilibrium

If moment equilibrium is not satisfied, consider the possibility of one or more of these mistakes:

- We used inconsistent units if there is a distributed load or moment reactions.
- We applied an unintended magnitude or direction of force.
- Results were based on an unintended load combination or load factor.
- We switched the magnitudes of two forces applied to the structure in the same direction.
- The self-weight we intended to include was missing.

SECTION 7.3 HIGHLIGHTS

Checking Fundamental Principles

Check force equilibrium in each direction:

- Consider all forces applied in that direction, including self-weight if we intended to include it.
- Compare with all support reactions in that direction.
- Temperature effects and other self-straining loads are not part of the summation, but equilibrium still applies.

Check moment equilibrium about each axis:

- Pick a reference point.
- Sum the moments about the reference point due to the applied forces and couple moments.
- Sum the moments about the reference point due to the support reactions; remember to include the moment reactions.
- Compare the two summations.

Section 7.3 Checking Fundamental Principles **345**

EXAMPLE 7.1

As part of new employee training at our firm, the structural engineers are asked to design a rigid frame to withstand wind load.

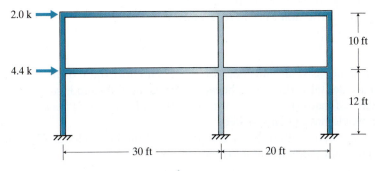

Figure 7.2

One of the new engineers obtained the following results from computer aided analysis. Our job is to verify that the results satisfy the fundamental principles.

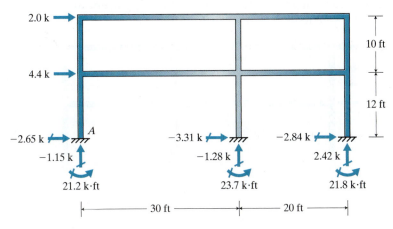

Figure 7.3

Check Equilibrium:
Satisfied Fundamental Principles?

$\xrightarrow{+} \Sigma F_x = 2.0 \text{ k} + 4.4 \text{ k} - 2.65 \text{ k} - 3.31 \text{ k} - 2.84 \text{ k}$

$\qquad = -2.4 \text{ k} \neq 0$ ✗

∴ There is a mistake.

$+\uparrow \Sigma F_y = -1.15 \text{ k} - 1.28 \text{ k} + 2.42 \text{ k}$

$\qquad = -0.01 \text{ k} \approx 0$ ✓

$$\underset{\curvearrowright}{+)} \Sigma M_A^{\text{applied}} = 2.0\,\text{k}\,(22\,\text{ft}) + 4.4\,\text{k}\,(12\,\text{ft}) = 96.8\,\text{k} \cdot \text{ft}$$

$$\underset{\curvearrowright}{+)} \Sigma M_A^{\text{reactions}} = -(-1.28\,\text{k})(30\,\text{ft}) - 2.42\,\text{k}\,(50\,\text{ft}) - 21.2\,\text{k} \cdot \text{ft} - 23.7\,\text{k} \cdot \text{ft}$$

$$-21.8\,\text{k} \cdot \text{ft} = -149.3\,\text{k} \cdot \text{ft}$$

$96.8\,\text{k} \cdot \text{ft} - 149.3\,\text{k} \cdot \text{ft} = -52.5\,\text{k} \cdot \text{ft} \neq 0$ ✗

∴ More confirmation that there is a mistake.

Identify Mistakes:

- Not an issue of inconsistent units because there is no horizontal distributed load.
- Can't be an accidental vertical load because vertical equilibrium is satisfied.
- Can't be switched loads (i.e., 4.4 k at top and 2.0 k at lower level) because horizontal equilibrium would have been satisfied.
- Likely an unintended magnitude for one or more of the lateral loads. This would show in both horizontal and moment equilibrium.

EXAMPLE 7.2

As part of new employee training at our firm, the structural engineers are asked to design a rigid frame to withstand wind load.

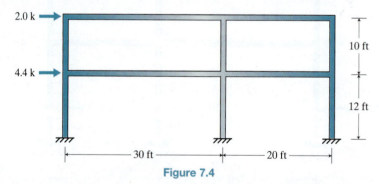

Figure 7.4

These are the results from the second new engineer. Our job is still to verify that the results satisfy the fundamental principles.

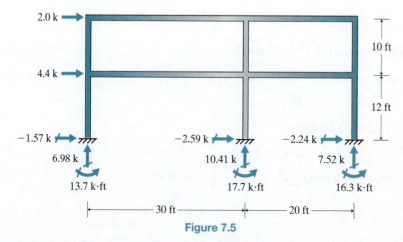

Figure 7.5

Check Equilibrium:

Satisfied Fundamental Principles?

$\xrightarrow{\pm} \Sigma F_x = 2.0\,k + 4.4\,k - 1.57\,k - 2.59\,k - 2.24\,k = 0$ ✓

$+\uparrow \Sigma F_y = 6.98\,k + 10.41\,k + 7.52\,k = 24.9\,k \neq 0$ ✗

∴ There is a mistake.

$\circlearrowright \Sigma M_A^{applied} = 2.0\,k\,(22\,ft) + 4.4\,k\,(12\,ft) = 96.8\,k \cdot ft$

$\circlearrowright \Sigma M_A^{reactions} = -10.41\,k\,(30\,ft) - 7.52\,k\,(50\,ft) - 13.7\,k \cdot ft$

$\qquad -17.7\,k \cdot ft - 16.3\,k \cdot ft = -736\,k \cdot ft$

$96.8\,k \cdot ft - 736\,k \cdot ft = -639\,k \cdot ft \neq 0$ ✗

∴ More confirmation that there is a mistake.

Identify Mistakes:

- Not an issue of inconsistent units because there is no vertical distributed load.
- Can't be switched loads (i.e., 4.4 k at top and 2.0 k at lower level) because vertical equilibrium would not be affected.
- Could be an accidental vertical load, but 24.9 k is an odd number to add.
- Likely that self-weight was included. There is 166 ft of members, which would be 150 plf. That is a reasonable self-weight for structural members. Self-weight would affect both vertical and moment equilibrium.

EXAMPLE 7.3

As part of new employee training at our firm, the structural engineers are asked to design a rigid frame to withstand wind load.

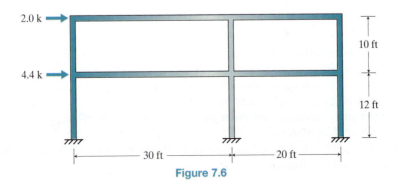

Figure 7.6

Here is the third submission:

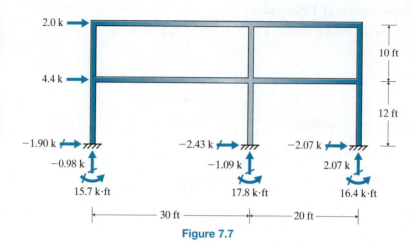

Figure 7.7

Check Equilibrium:
Satisfied Fundamental Principles?

$\xrightarrow{\pm} \Sigma F_x = 2.0\,\text{k} + 4.4\,\text{k} - 0.98\,\text{k} - 2.43\,\text{k} - 2.07\,\text{k}$
$= 0$ ✓

$+\uparrow \Sigma F_y = -0.98\,\text{k} - 1.09\,\text{k} + 2.07\,\text{k}$
$= 0$ ✓

$\circlearrowleft \Sigma M_A^{\text{applied}} = 2.0\,\text{k}\,(22\,\text{ft}) + 4.4\,\text{k}\,(12\,\text{ft}) = 96.8\,\text{k}\cdot\text{ft}$

$\circlearrowleft \Sigma M_A^{\text{reactions}} = -(-1.09\,\text{k})\,(30\,\text{ft}) - 2.07\,\text{k}\,(50\,\text{ft}) - 15.7\,\text{k}\cdot\text{ft} - 17.8\,\text{k}\cdot\text{ft}$
$-16.4\,\text{k}\cdot\text{ft} = -120.7\,\text{k}\cdot\text{ft}$

$96.8\,\text{k}\cdot\text{ft} - 120.7\,\text{k}\cdot\text{ft} = -23.9\,\text{k}\cdot\text{ft} \neq 0$ ✗

∴ There is a mistake.

Identify Mistakes:

- Can't be an unintended magnitude for one or more of the loads because horizontal equilibrium is satisfied.
- Can't be an accidental load because both horizontal and vertical equilibrium are satisfied.
- Could be an accidental couple moment applied.
- Could be an issue of inconsistent units. Since there are no distributed loads, only moment equilibrium would be affected.
- Could be switched loads (i.e., 4.4 k at top and 2.0 k at lower level) because only moment equilibrium would be affected.

EXAMPLE 7.4

As part of new employee training at our firm, the structural engineers are asked to design a rigid frame to withstand wind load.

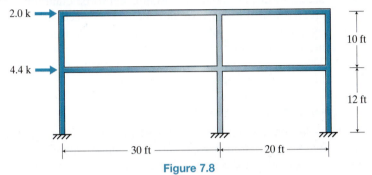

Figure 7.8

The fourth submission:

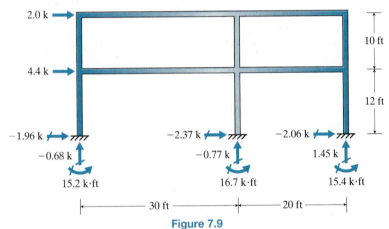

Figure 7.9

Check Equilibrium:

Satisfied Fundamental Principles?

$$\xrightarrow{\pm} \Sigma F_x = 2.0 \text{ k} + 4.4 \text{ k} - 1.96 \text{ k} - 2.37 \text{ k} - 2.06 \text{ k}$$
$$= 0.01 \text{ k} \approx 0 \checkmark$$

$$+\uparrow \Sigma F_y = -0.68 \text{ k} - 0.77 \text{ k} + 1.45 \text{ k}$$
$$= 0 \checkmark$$

$$\circlearrowright \Sigma M_A^{\text{applied}} = 2.0 \text{ k} (22 \text{ ft}) + 4.4 \text{ k} (12 \text{ ft}) = 96.8 \text{ k} \cdot \text{ft}$$

$$\circlearrowright \Sigma M_A^{\text{reactions}} = -(-0.77 \text{ k})(30 \text{ ft}) - 1.45 \text{ k}(50 \text{ ft}) - 15.2 \text{ k} \cdot \text{ft} - 16.7 \text{ k} \cdot \text{ft}$$
$$- 15.4 \text{ k} \cdot \text{ft} = -96.7 \text{ k} \cdot \text{ft}$$

$$96.8 \text{ k} \cdot \text{ft} - 96.7 \text{ k} \cdot \text{ft} = -0.1 \text{ k} \cdot \text{ft} \approx 0 \checkmark$$

Conclusions:

- Satisfies fundamental principles.
- Review internal force diagrams and compare with approximations or a duplicate analysis before declaring the results are reasonable.

7.4 Checking Features of the Solution

Introduction

One of the advantages of computer aided analysis is the relatively quick access to displaced shape and internal force diagrams once the model is created and the analysis completed. Displaced shape and internal force diagrams present lots of information in a concise way: graphically. The human mind is quite adept at identifying features or patterns in images. Therefore, reviewing these graphical results can be an effective way to identify the presence or absence of mistakes.

How-To

From our understanding of the idealized structure and mechanics, we can anticipate certain aspects, features, of the analysis results. We learned an example of this in Section 2.1: Qualitative Truss Analysis. There we practiced identifying whether truss members would be in tension or compression. In Section 4.2: Constructing Diagrams by Deduction, we learned how to predict the shapes of internal force diagrams based on the loading and support conditions.Section 5.1: Double Integration Method showed us how the displaced shape is related to the internal moment. We also practiced identifying deformation boundary conditions based on the types of supports. Together these skills allow us to anticipate features of the graphical solutions we obtain from computer aided analysis.

The first thing practitioners typically review is the displaced shape, which can identify many mistakes. If the displaced shape appears reasonable, then we review the internal force diagrams. These reviews typically involve asking the following questions. Note that the list of potential mistakes that could cause the phenomenon is not exhaustive; it is only a sample of common mistakes to consider.

Review of Displaced Shape

1. Do the axial-force-only members (i.e., pin connected at both ends with load applied only along the axis of the member or no load) remain straight?
    - Self-weight will cause non-vertical members to bend.
    - Rigid connections on either or both ends will cause the member to bend.
    - Analyzing a truss as a frame will cause the members to bend.
2. Does the angle between pin-connected members change if needed?
    - Rigid connections will prevent one member from rotating relative to another.
3. Does the angle between rigidly connected members remain unchanged while the connection rotates?
    - Pinned connections will allow one member to rotate relative to the other.
4. Do members displace in the anticipated direction (e.g., up, down, left, right)?
    - Reversing the sign of the applied load will cause members to displace in the opposite direction.

- Forgetting to apply a load can leave a member to displace in the opposite direction than anticipated.
- Incorrectly designating a pin connection as rigid, or designating a rigid connection as a pin, can cause members to displace in unanticipated directions.
- Adding unwanted loads can cause members to displace in unanticipated directions.
- Adding unwanted supports will stop displacement where there should be some and will cause some parts to move in unanticipated directions.
- Forgetting a support will allow motion where there should be none and will cause some parts to move in unanticipated directions.

5. Do the support reactions behave as intended?
   - Switching the restrained direction of a roller will allow the structure to move in an unintended direction.
   - Forgetting the rotational restraint will allow the joint to rotate when it should remain in its original orientation.

Review of Internal Force Diagrams

1. Is the sense of the axial force (i.e., tension or compression) as expected?
   - Reversing the sign of the applied load will reverse the sense of the axial force.
   - Using two pin supports instead of a pin and roller (i.e., simply supported) will tend to reduce the tension in truss members, causing some to switch to compression.
   - Adding extra supports will tend to reverse the sense of some members, typically those near the extra support.
2. Are the features consistent with the loading (see the Section Highlights in Section 4.2: Constructing Diagrams by Deduction)?
   - Adding unwanted restraints or missing a desired restraint will change the jumps in the diagrams.
   - Reversing the sign of an applied load will reverse the direction of the axial force and/or shear diagrams.
   - Forgetting a distributed load will change the shape of the diagrams.
   - Unwanted self-weight will change the shape of the diagrams where there is no other applied distributed load.

If we have a completely symmetric structure, there is an easily identifiable feature that will help us evaluate the results. If the structure, its supports, member properties, and loading are symmetric, then the displaced shape and internal force diagrams will be symmetric. Note that *all* aspects of the idealized structure must mirror about the center line for this to be true.

In the chapters that follow this one, we develop even more skills for anticipating the features of structural analysis results. When we get there, we add those skills to the skills from this chapter.

Sometimes we see an unanticipated feature or a feature that is the opposite of what we expected. That does not automatically mean there is a mistake. It means that our results merit closer review. For example, consider the idealized beam in Figure 7.10. Predict which way the far

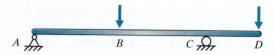

Figure 7.10 Simply supported beam with a cantilevered end and two point forces applied.

right end, point *D*, will displace: up or down. Take a moment and think through your prediction.

No matter which prediction you make, I can make *D* do the opposite. The actual behavior depends on the relative magnitudes of the two forces and the relative lengths of spans *AC* and *CD*. If *CD* is relatively short or the force at *D* is relatively small, *D* will go up. If *CD* is sufficiently long or the force at *D* is sufficiently large, *D* will go down. When in doubt about whether your prediction and the obtained solution are reasonably close, take some time to give the situation a thorough review. Sometimes that requires us to identify why our original prediction was flawed. Sometimes it requires us to seek assistance from a fresh set of eyes. The key is that we are not done until we are convinced that we have reasonable results. Nothing should be dismissed as too difficult to explain.

SECTION 7.4 HIGHLIGHTS

Checking Features of the Solution

Review the displaced shape:

- Do the axial-force-only members remain straight?
- Does the angle between pin-connected members change if needed?
- Does the angle between rigidly connected members remain unchanged while the connection rotates?
- Do members displace in the anticipated direction (e.g., up, down, left, right)?
- Do the support reactions behave as intended?

Review the internal force diagrams:

- Is the sense of the axial force (i.e., tension or compression) as expected?
- Are the features consistent with the loading (see the Section Summary in Section 4.2: Constructing Diagrams by Deduction)?

Section 7.4 Checking Features of the Solution   353

EXAMPLE 7.5

Our firm has been contracted to design a truss to support rigging in a theater. A member of our team has used computer aided analysis to analyze the truss. That member has verified equilibrium but would like our help evaluating the reasonableness of the results.

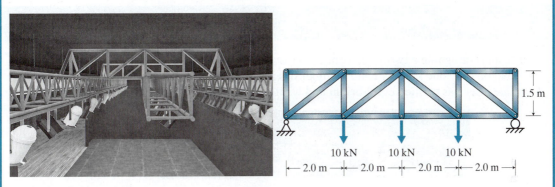

Figure 7.11

Review the Displaced Shape:
Observed Expected Features?

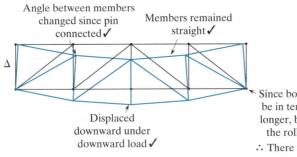

Figure 7.12

Review the Axial Force Diagram:
Observed Expected Features?

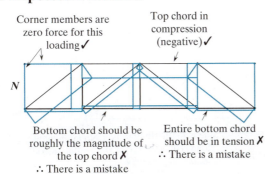

Figure 7.13

Identify Mistakes:
- There can't be an extra horizontal load pushing the right end in because horizontal equilibrium was satisfied.
- There is likely a pin support on both sides. That change would make the structure perfectly symmetric, thus resulting in these perfectly symmetric results.

EXAMPLE 7.6

Our team is designing a two-span, continuous overpass bridge. One of the critical loadings is live load on only one span. A team member has modeled this and has obtained results. The team member has verified equilibrium and wants our help verifying the features of the solution.

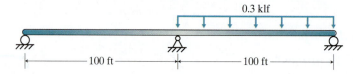

Figure 7.14

Review the Displaced Shape:

Observed Expected Features?

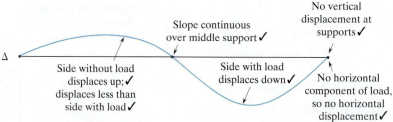

Figure 7.15

Review the Moment Diagram:

Observed Expected Features?

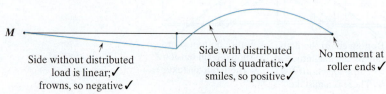

Figure 7.16

Conclusions:

- The displaced shape and moment diagram both have only features we expect.
- Equilibrium was already verified.
- The results are probably reasonable. To make a final judgment, we should check the moment and displacement magnitudes with approximate predictions.

EXAMPLE 7.7

At the interview for a job as a structural engineer, the principal gives us a test. She sketches a two-story, diagonally braced frame and shows us a printout of the displaced shape. The test is to determine whether the displaced shape is reasonable.

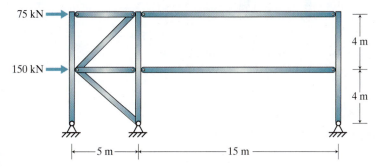

Figure 7.17

Review the Displaced Shape:

Observed Expected Features?

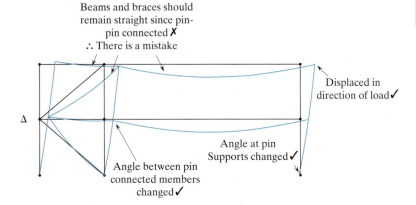

Figure 7.18

Identify Mistakes:

- There can't be rigid connections because the angle between pin-connected members changed.
- There could be distributed load applied downward.
- Self-weight is probably included. Every member including the braces are displaced downward.

Example 7.8

Apparently the test is not over. The principal shows us another printout of the displaced shape for the same structure.

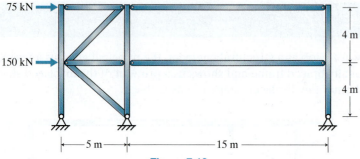

Figure 7.19

Review the Displaced Shape:

Observed Expected Features?

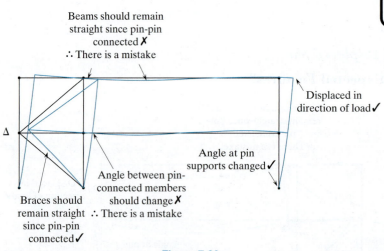

Figure 7.20

Identify Mistakes:
- There can be no self-weight because the braces remained straight.
- There are probably rigid connections between the beams and columns, since all members at the joint are rotating.

EXAMPLE 7.9

We must be doing well because the principal gives us a third printout of the displaced shape for the same structure.

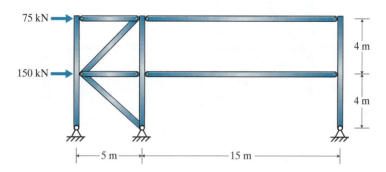

Figure 7.21

Review the Displaced Shape:
Observed Expected Features?

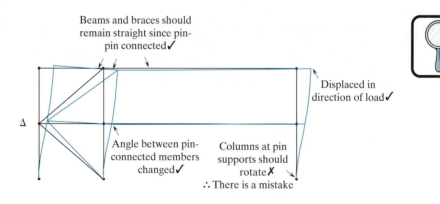

Figure 7.22

Identify Mistakes:
- There can't be rigid connections because the angle between members changed.
- Fixed supports are probably used at the bases of the columns.

Example 7.10

The principal congratulates us on doing so well on the test. One last displaced shape remains to be evaluated.

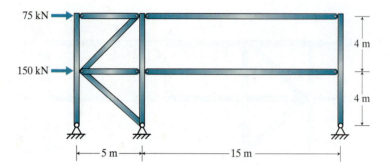

Figure 7.23

Review the Displaced Shape:

Observed Expected Features?

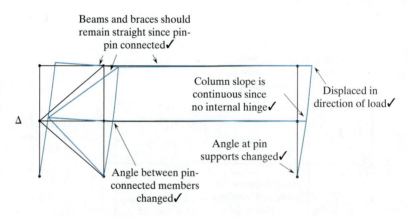

Figure 7.24

Conclusion:

- The displaced shape has only features we expect.
- This displaced shape appears reasonable.

Reference Cited

Dykes, A. R. 1975. "Bees in His Bonnet." *The Structural Engineer: The Monthly Journal of the Institution of Structural Engineers* 56A (May, No. 5): 150–151.

Homework Problems

7.1 Our team is responsible for analyzing a new gantry rail support beam. The gantry travels on a pair of wheels 1 meter apart. The extreme live load case occurs when the capacity load is closest to that rail. One of the conditions we need to consider is the gantry at midspan of one of the spans. A member of the team completed the analysis; our task is to verify the reasonableness of the results.

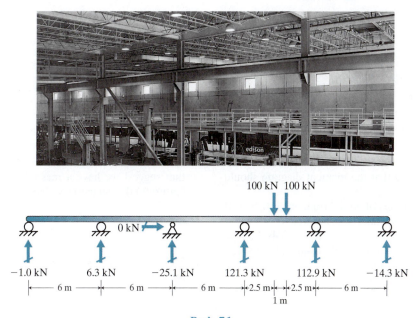

Prob. 7.1

a. Predict three features that the displaced shape should have.
b. Predict three features that the moment diagram should have.
c. Verify that all equations of equilibrium are satisfied. If one or more shows a mistake, indicate which.
d. Identify at least three features of the displaced shape that suggest we have a reasonable result. Identify any feature(s) that suggests we have a mistake.

e. Identify at least three features of the moment diagram that suggest we have a reasonable result. Identify any feature(s) that suggests we have a mistake.

f. Make a comprehensive argument that the computer aided analysis results are reasonable, if you believe they are. If there appears to be a mistake, suggest the most likely cause based on your observations. Justify your hypothesis.

7.2 Our firm has been contracted by a state Department of Transportation to design a three-span overpass. For the live load, the peak negative moment will occur if we load only adjacent spans, not the entire girder. Therefore, a team member has used computer aided analysis to determine the peak negative moment in this girder. Our job is to verify the results.

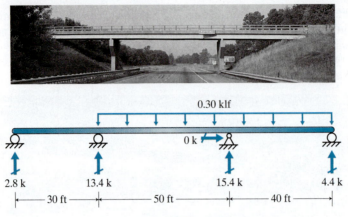

Prob. 7.2

a. Predict three features that the displaced shape should have.
b. Predict three features that the moment diagram should have.
c. Verify that all equations of equilibrium are satisfied. If one or more shows a mistake, indicate which.
d. Identify at least three features of the displaced shape that suggest we have a reasonable result. Identify any feature(s) that suggests we have a mistake.

e. Identify at least three features of the moment diagram that suggest we have a reasonable result. Identify any feature(s) that suggests we have a mistake.

f. Make a comprehensive argument that the computer aided analysis results are reasonable, if you believe they are. If there appears to be a mistake, suggest the most likely cause based on your observations. Justify your hypothesis.

7.3 The cantilevered walkway at a hotel is showing signs of distress. Our firm has been contracted to ensure that the walkway does not collapse as a retrofit is designed and implemented. One possibility is to brace the walkway at the end only; that would reduce the peak moment. If that option does not reduce the moment enough, we could add two braces: one at the end and one midspan. We have explored both options using computer aided analysis. Our job is to evaluate the reasonableness of the results for the second option: two supports.

a. Predict three features that the displaced shape should have.
b. Predict three features that the moment diagram should have.
c. Verify that all equations of equilibrium are satisfied. If one or more shows a mistake, indicate which.

Continued on next page

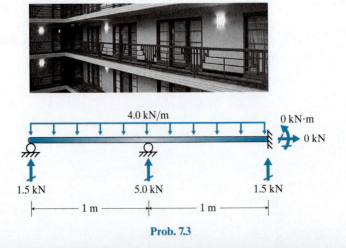

Prob. 7.3

d. Identify at least three features of the displaced shape that suggest we have a reasonable result. Identify any feature(s) that suggests we have a mistake.

e. Identify at least three features of the moment diagram that suggest we have a reasonable result. Identify any feature(s) that suggests we have a mistake.

f. Make a comprehensive argument that the computer aided analysis results are reasonable, if you believe they are. If there appears to be a mistake, suggest the most likely cause based on your observations. Justify your hypothesis.

7.4 A pedestrian walkway must be suspended over a maintenance corridor. To do that, our firm has chosen to support the walkway on a rigid frame fixed to the adjacent building and to the foundation under the column. One of our team members has completed the analysis of the frame under full live load. Our firm's quality assurance procedures require that another person verify the results.

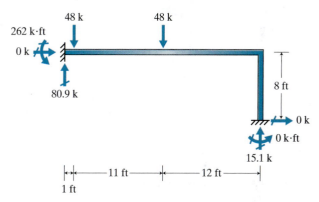

Prob. 7.4

a. Predict three features that the displaced shape should have.
b. Predict three features that the shear diagram should have.
c. Predict three features that the moment diagram should have.
d. Verify that all equations of equilibrium are satisfied. If one or more shows a mistake, indicate which.
e. Identify at least three features of the displaced shape that suggest we have a reasonable result. Identify any feature(s) that suggests we have a mistake.

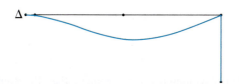

f. Identify at least three features of the shear diagram that suggest we have a reasonable result. Identify any feature(s) that suggests we have a mistake.

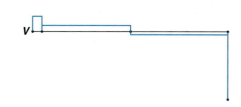

g. Identify at least three features of the moment diagram that suggest we have a reasonable result. Identify any feature(s) that suggests we have a mistake.

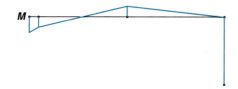

h. Make a comprehensive argument that the computer aided analysis results are reasonable, if you believe they are. If there appears to be a mistake, suggest the most likely cause based on your observations. Justify your hypothesis.

7.5 We were recently hired by a company that designs and fabricates metal frame buildings. To help introduce us to the company's design process, we have been tasked with evaluating the computer aided analysis results for one of the wind load conditions.

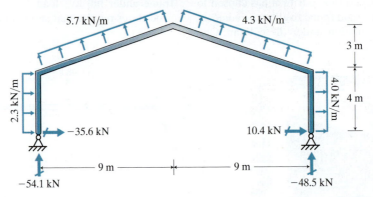

Prob. 7.5

a. Predict three features that the displaced shape should have.
b. Predict three features that the shear diagram should have.
c. Predict three features that the moment diagram should have.
d. Verify that all equations of equilibrium are satisfied. If one or more shows a mistake, indicate which.
e. Identify at least three features of the displaced shape that suggest we have a reasonable result. Identify any feature(s) that suggests we have a mistake.

g. Identify at least three features of the moment diagram that suggest we have a reasonable result. Identify any feature(s) that suggests we have a mistake.

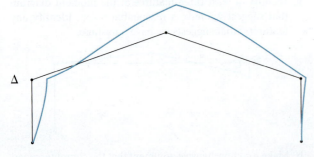

f. Identify at least three features of the shear diagram that suggest we have a reasonable result. Identify any feature(s) that suggests we have a mistake.

h. Make a comprehensive argument that the computer aided analysis results are reasonable, if you believe they are. If there appears to be a mistake, suggest the most likely cause based on your observations. Justify your hypothesis.

7.6 Our team is designing a bridge pier to be constructed in the median of a highway. The design calls for deep foundations that will provide fixed supports. The geotechnical engineer determined that the point of fixity is 10 feet below the ground surface. That means the columns are effectively 30 feet tall. We completed a computer aided analysis and need to verify that the results are reasonable.

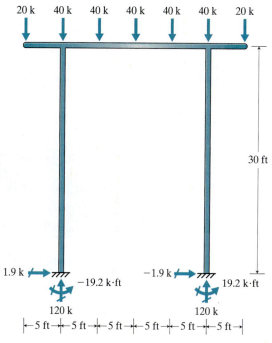

Prob. 7.6

a. Predict three features that the displaced shape should have.
b. Predict three features that the shear diagram should have.
c. Predict three features that the moment diagram should have.
d. Verify that all equations of equilibrium are satisfied. If one or more shows a mistake, indicate which.
e. Identify at least three features of the displaced shape that suggest we have a reasonable result. Identify any feature(s) that suggests we have a mistake.

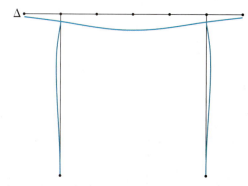

f. Identify at least three features of the shear diagram that suggest we have a reasonable result. Identify any feature(s) that suggests we have a mistake.

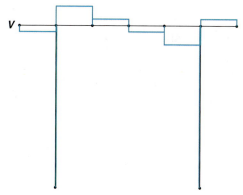

g. Identify at least three features of the moment diagram that suggest we have a reasonable result. Identify any feature(s) that suggests we have a mistake.

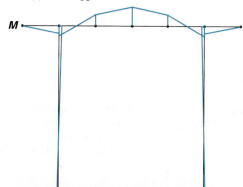

364   **Chapter 7**   Introduction to Computer Aided Analysis

h. Make a comprehensive argument that the computer aided analysis results are reasonable, if you believe they are. If there appears to be a mistake, suggest the most likely cause based on your observations. Justify your hypothesis.

7.7 An inspection of an existing railroad bridge revealed significant corrosion in some members. Our firm has been hired to evaluate whether the members still have sufficient strength. To do that, we need to find the internal forces in all truss members due to the various loads. An analysis of the truss with full live load has been completed. Our task is to evaluate the results before we use them to make decisions about the corroded members.

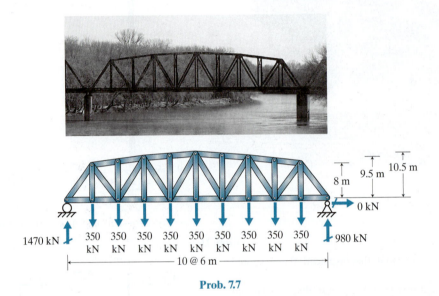

Prob. 7.7

a. Predict three features that the displaced shape should have.
b. Predict three features that the axial diagram should have.
c. Verify that all equations of equilibrium are satisfied. If one or more shows a mistake, indicate which.
d. Identify at least three features of the displaced shape that suggest we have a reasonable result. Identify any feature(s) that suggests we have a mistake.

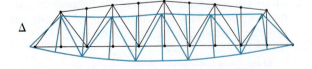

e. Identify at least three features of the axial diagram that suggest we have a reasonable result. Identify any feature(s) that suggests we have a mistake.

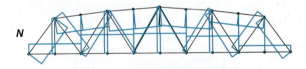

f. Make a comprehensive argument that the computer aided analysis results are reasonable, if you believe they are. If there appears to be a mistake, suggest the most likely cause based on your observations. Justify your hypothesis.

7.8 Large trusses will support the roof over an athletic facility. Our team is to design the trusses. Our current task is to analyze the truss with full snow load. We have completed the computer aided analysis and need to evaluate the results.

a. Predict three features that the displaced shape should have.
b. Predict three features that the axial diagram should have.
c. Verify that all equations of equilibrium are satisfied. If one or more shows a mistake, indicate which.

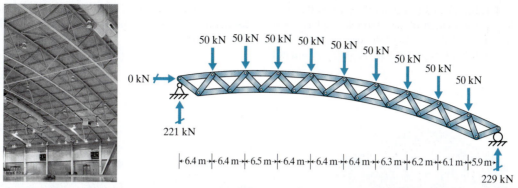

Prob. 7.8

d. Identify at least three features of the displaced shape that suggest we have a reasonable result. Identify any feature(s) that suggests we have a mistake.

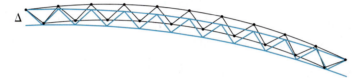

Close-up of the displaced shape of the left end:

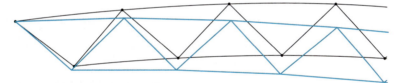

Close-up of the displaced shape of the right end:

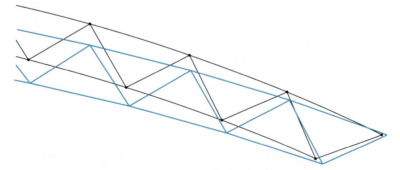

e. Identify at least three features of the axial diagram that suggest we have a reasonable result. Identify any feature(s) that suggests we have a mistake.

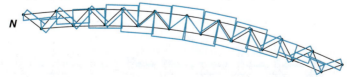

f. Make a comprehensive argument that the computer aided analysis results are reasonable, if you believe they are. If there appears to be a mistake, suggest the most likely cause based on your observations. Justify your hypothesis.

7.9 Our firm has been hired to design the trusses that will support the roof for a chain of gas stations. We have performed a computer aided analysis of the truss with the roof self-weight applied. For this preliminary analysis, we did not include the self-weight of the truss itself. Before passing the results to the next member of the team, we need to verify that they are reasonable.

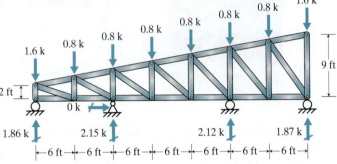

Prob. 7.9

a. Predict three features that the displaced shape should have.
b. Predict three features that the axial diagram should have.
c. Verify that all equations of equilibrium are satisfied. If one or more shows a mistake, indicate which.
d. Identify at least three features of the displaced shape that suggest we have a reasonable result. Identify any feature(s) that suggests we have a mistake.

e. Identify at least three features of the axial diagram that suggest we have a reasonable result. Identify any feature(s) that suggests we have a mistake.

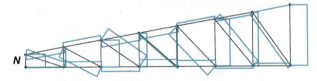

f. Make a comprehensive argument that the computer aided analysis results are reasonable, if you believe they are. If there appears to be a mistake, suggest the most likely cause based on your observations. Justify your hypothesis.

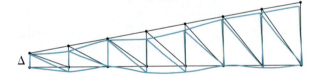

7.10 Our firm has been hired to design a new athletic facility. A member of the team has analyzed one of the frame lines with wind load. Before finalizing the design, we need to evaluate the reasonableness of the results.

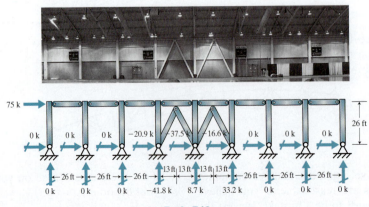

Prob. 7.10

a. Predict three features that the displaced shape should have.
b. Predict three features that the axial diagram should have.
c. Verify that all equations of equilibrium are satisfied. If one or more shows a mistake, indicate which.
d. Identify at least three features of the displaced shape that suggest we have a reasonable result. Identify any feature(s) that suggests we have a mistake.

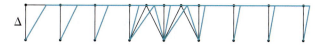

e. Identify at least three features of the axial diagram that suggest we have a reasonable result. Identify any feature(s) that suggests we have a mistake.

f. Make a comprehensive argument that the computer aided analysis results are reasonable, if you believe they are. If there appears to be a mistake, suggest the most likely cause based on your observations. Justify your hypothesis.

7.11 Our team was selected to design a new four-story apartment building. A member of the team has completed a computer aided analysis of one of the frame lines, considering all of the dead load *except* the self-weight of the frame. The plan is to use these results to make our initial selection of member properties. Before we take that step, we need to verify the reasonableness of these results.

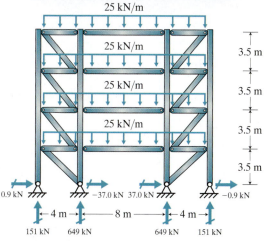

Prob. 7.11

a. Predict three features that the displaced shape should have.
b. Predict three features that the moment diagram should have.

c. Verify that all equations of equilibrium are satisfied. If one or more shows a mistake, indicate which.
d. Identify at least three features of the displaced shape that suggest we have a reasonable result. Identify any feature(s) that suggests we have a mistake.

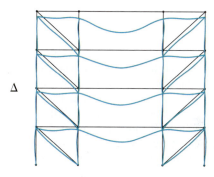

e. Identify at least three features of the axial diagram that suggest we have a reasonable result. Identify any feature(s) that suggests we have a mistake.

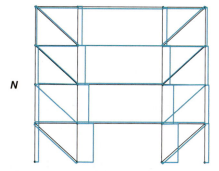

f. Make a comprehensive argument that the computer aided analysis results are reasonable, if you believe they are. If there appears to be a mistake, suggest the most likely cause based on your observations. Justify your hypothesis.

7.12 Our firm was selected to design a large warehouse facility. In one direction, the main wind force-resisting system is diagonal bracing. The bracing is helped by rigid connections between some of the columns and the open-web joists. Note that the horizontal members do not help resist wind load in this plane; they gather wind normal to this plane. A team member completed an analysis of the frame with unfactored wind load. That team member has asked us to evaluate the reasonableness of the results.

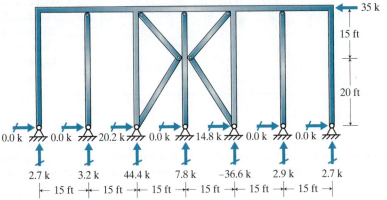

Prob. 7.12

a. Predict three features that the displaced shape should have.
b. Predict three features that the axial diagram should have.
c. Verify that all equations of equilibrium are satisfied. If one or more shows a mistake, indicate which.
d. Identify at least three features of the displaced shape that suggest we have a reasonable result. Identify any feature(s) that suggests we have a mistake.
e. Identify at least three features of the axial diagram that suggest we have a reasonable result. Identify any feature(s) that suggests we have a mistake.

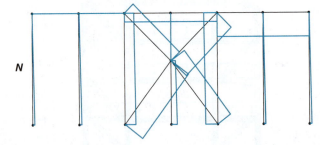

f. Make a comprehensive argument that the computer aided analysis results are reasonable, if you believe they are. If there appears to be a mistake, suggest the most likely cause based on your observations. Justify your hypothesis.

7.13 Our team is designing a new office building. For architectural reasons, our team leader chose rigid frames as the lateral load-resisting system in the long direction. Now we are checking the frame under gravity loads. The unfactored live load on each beam of each floor except the roof is 10 kN/m. We have the results of our computer aided analysis of the rigid frame with live load. Our task is to evaluate the reasonableness of those results.

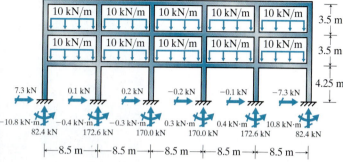

Prob. 7.13

a. Predict three features that the displaced shape should have.
b. Predict three features that the axial diagram should have.
c. Predict three features that the shear diagram should have.
d. Predict three features that the moment diagram should have.
e. Verify that all equations of equilibrium are satisfied. If one or more shows a mistake, indicate which.
f. Identify at least three features of the displaced shape that suggest we have a reasonable result. Identify any feature(s) that suggests we have a mistake.

h. Identify at least three features of the shear diagram that suggest we have a reasonable result. Identify any feature(s) that suggests we have a mistake.

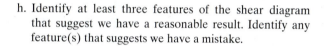
V

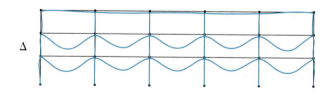
Δ

i. Identify at least three features of the moment diagram that suggest we have a reasonable result. Identify any feature(s) that suggests we have a mistake.

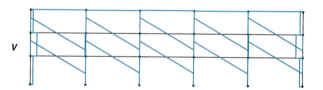
M

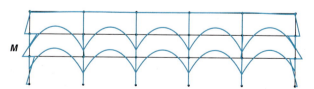

g. Identify at least three features of the axial diagram that suggest we have a reasonable result. Identify any feature(s) that suggests we have a mistake.

j. Make a comprehensive argument that the computer aided analysis results are reasonable, if you believe they are. If there appears to be a mistake, suggest the most likely cause based on your observations. Justify your hypothesis.

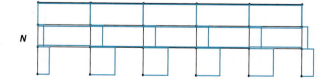
N

7.14 To give a new office building a unique look, the architect chose to expose the structural system. The system is a rigid frame that looks like a grid. A team member has determined the equivalent static seismic load for each level of the building and has performed a computer aided analysis. Consistent with company policy, verification of the results is passed to another person: us.

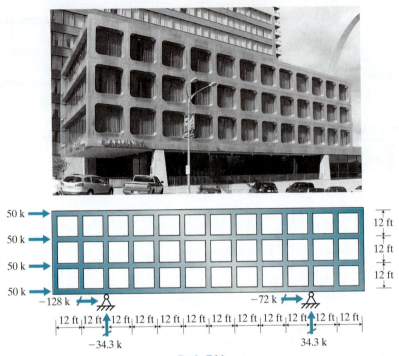

Prob. 7.14

a. Predict three features that the displaced shape should have.
b. Predict three features that the moment diagram should have.
c. Verify that all equations of equilibrium are satisfied. If one or more shows a mistake, indicate which.
d. Identify at least three features of the displaced shape that suggest we have a reasonable result. Identify any feature(s) that suggests we have a mistake.

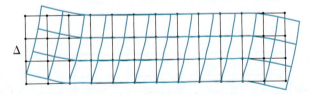

e. Identify at least three features of the axial diagram that suggest we have a reasonable result. Identify any feature(s) that suggests we have a mistake.

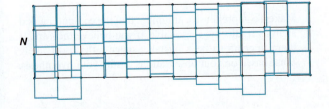

f. Identify at least three features of the shear diagram that suggest we have a reasonable result. Identify any feature(s) that suggests we have a mistake.

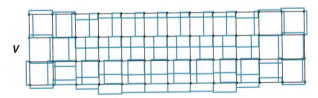

g. Identify at least three features of the moment diagram that suggest we have a reasonable result. Identify any feature(s) that suggests we have a mistake.

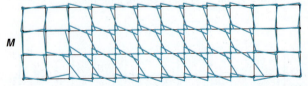

h. Make a comprehensive argument that the computer aided analysis results are reasonable, if you believe they are. If there appears to be a mistake, suggest the most likely cause based on your observations. Justify your hypothesis.

CHAPTER 8

APPROXIMATE ANALYSIS OF INDETERMINATE TRUSSES AND BRACED FRAMES

Braced frames are a popular lateral load resisting system, and they behave similarly to trusses.

MOTIVATION

Indeterminate trusses can be a great choice when only one or two trusses carry the load (e.g., a bridge) or when rigid connections are not practical (e.g., architectural requirements). Their advantage is the redundancy they provide. If one member or connection fails, another load path is typically available, thus preventing catastrophic collapse. Indeterminate trusses also tend to have smaller diagonal members, which might be important for aesthetic, construction, or weight considerations.

We frequently choose braced frames for the lateral load-resisting system for a structure. They are much stiffer than rigid frames for the same amount of material, and they are much lighter than shear walls.

Although we tend to use computer software to help us analyze the final design, approximate analysis helps us in important ways:

1. Approximate results allow us to make informed decisions without the cost of detailed analysis when choosing between different lateral load-resisting systems for a particular structure.
2. We can use approximate results to make our initial selection of member properties. The better our initial choice of member properties, the faster we converge to an effective solution.
3. Approximate results can be quickly obtained and then used to verify the reasonableness of results from computer aided analysis.

An indeterminate structure has more unknowns than equations of equilibrium. When we perform approximate analysis, we close the gap between equations and unknowns by making assumptions. It turns out that the assumptions we make to analyze indeterminate trusses are very similar to those for braced frames.

Chapter 8 Approximate Analysis of Indeterminate Trusses and Braced Frames

8.1 Indeterminate Trusses

Introduction
When we first learned how to analyze a truss, we probably used the Method of Sections and the Method of Joints. Those two approaches still form the basis for how we perform approximate analysis of an indeterminate truss. The difference is that with an indeterminate truss, we make assumptions in order to start the process.

How-To
Approximate analysis is based on making assumptions until the number of unknowns equals the number of equations of equilibrium. This section focuses on the assumptions we make for an indeterminate truss that has extra members (Figure 8.1), not extra supports (Figure 8.2).

The assumption we make for trusses with extra members is that the net shear at a section through the truss is shared equally as components of the axial forces in the diagonal members at that section. We illustrate this in Figure 8.3.

Figure 8.1 Indeterminate truss due to extra members.

Figure 8.2 Indeterminate truss due to extra supports.

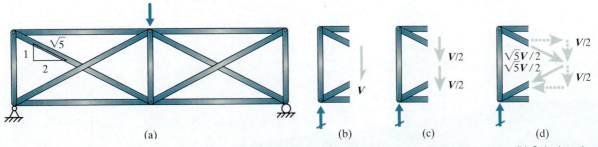

Figure 8.3 Approximate analysis assumption for indeterminate trusses: (a) Cut a section through the truss; (b) Calculate the net shear at the section; (c) The net shear is shared equally as the vertical components of the axial forces in the diagonal members; (d) Calculate the resulting axial forces in the diagonal members.

Note, however, that some trusses appear to be indeterminate but are not. This can be the case when crossing diagonal members are very thin, such as rods or cables (Figure 8.4). One of the diagonal members is in tension while the other tries to go into compression (Figure 8.5a). In Section 2.1: Qualitative Truss Analysis, we learned how to determine which members are in tension and which are in compression. Because the diagonal members have such a small cross-section, they have a very low buckling strength. Therefore, the members' capacity in compression is essentially zero. In these cases, the truss behaves as if the thin compression member does not exist. The effect is a determinate truss (Figure 8.5b).

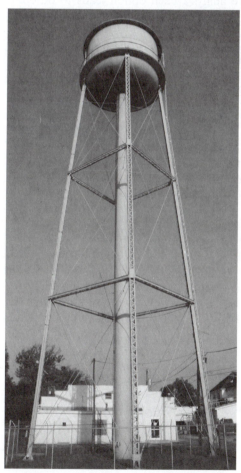

Figure 8.4 Truss appears to be indeterminate due to extra members; however, the diagonals are too thin to carry compression.

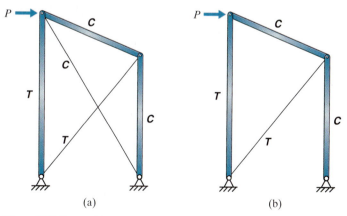

Figure 8.5 Truss with extra members that are too thin to carry compression: (a) Lateral load results in tension and compression in the diagonals; (b) Diagonals cannot carry compression, so the effective truss is determinate.

Dealing with More Than Two Diagonals

In rare cases, we design trusses such that a panel has more than two diagonals. Then we have choices in how to apply the assumptions. For example, consider the truss in Figure 8.6.

We can calculate the degree of indeterminacy by comparing the number of unknowns and the number of equations of equilibrium:

$$\begin{aligned}
\text{Unknowns} &= 3 \text{ reactions} + 16 \text{ members} = 19 \\
\text{Equations} &= 2 \times 8 \text{ joints} = 16 \\
\text{Degrees indeterminate} &= 3
\end{aligned}$$

Since the truss is third-degree indeterminate, we need to make three assumptions in order to perform approximate analysis. Figure 8.7 shows three different sets of assumptions we could make.

Because this is approximate analysis, each option leads to different results. For example, we can apply each of the three sets of assumptions and find the forces in members *EB, EC,* and *ED* (Figures 8.8–8.10). Table 8.1 summarizes the results with a comparison to the results of computer aided analysis (the Direct Stiffness Method).

Approximate Analysis Process

Here are the steps to follow for approximate analysis of an indeterminate truss:

Step 1: Find the reactions. With this method, we are focusing on trusses that have extra members; therefore, we should be able to find all of the reactions using only the equations of equilibrium.

Step 2: Use our assumption. We can apply our assumption as many times as needed up to the degree of indeterminacy. If we are only trying to find the force in certain members, we might need to apply this assumption only once or twice. If each panel has no more than two crossing members, there is one possible solution. If there is a panel that has more than two crossing members, each of the options for applying the assumption leads to different results.

Step 3: Use the Method of Joints or Method of Sections to find any other member forces we want. Once we apply our assumption in Step 2, the number of unknown member forces drops to the number of equations of equilibrium that remain.

Section 8.1 Indeterminate Trusses 377

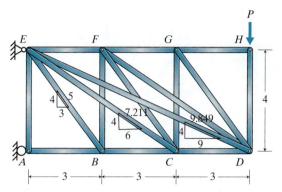

Figure 8.6 An indeterminate truss with more than two diagonals in each panel.

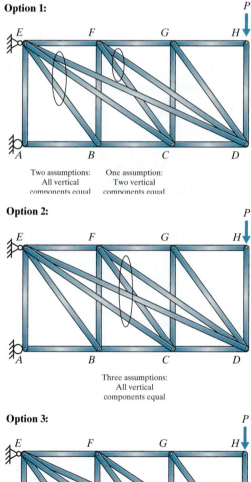

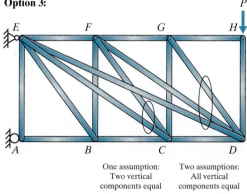

Figure 8.7 Three options for making the three assumptions for approximate analysis.

378   Chapter 8   Approximate Analysis of Indeterminate Trusses and Braced Frames

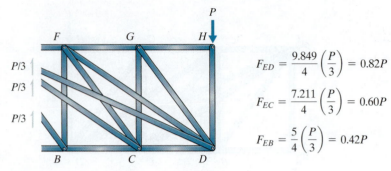

Figure 8.8 Approximate analysis using Option 1: All three vertical components in the left panel are equal.

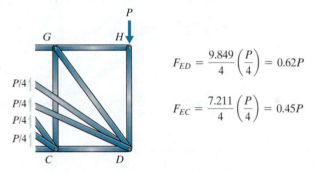

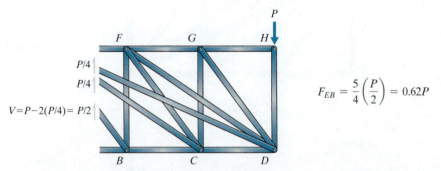

Figure 8.9 Approximate analysis using Option 2: All four vertical components in the middle panel are equal.

Section 8.1 Indeterminate Trusses

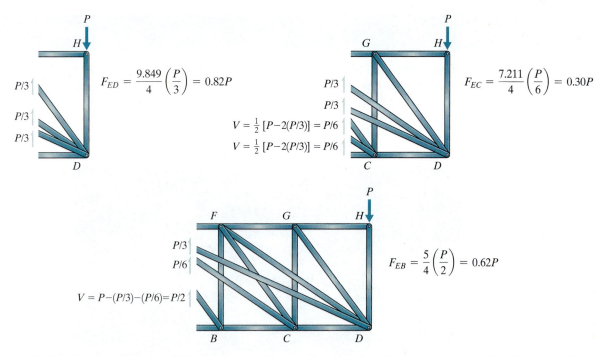

Figure 8.10 Approximate analysis using Option 3: All three vertical components in the right panel are equal, and the other two vertical components in the middle panel are equal.

Table 8.1 Summary of Approximate Analysis Results Using the Three Different Options and Comparing With Results of Computer Aided Analysis

Member	Option 1	Option 2	Option 3	Direct Stiffness
F_{EB}	0.42P	0.62P	0.62P	0.68P
F_{EC}	0.60P	0.45P	0.30P	0.51P
F_{ED}	0.82P	0.62P	0.82P	0.44P

SECTION 8.1 HIGHLIGHTS

Indeterminate Trusses

This approximate analysis method is limited to trusses that have extra members, not extra supports.
Assumption: The net shear at a section is shared equally as components of the axial forces in the diagonal members at that section.

Analysis Steps:

1. Find the reactions.
2. Use our assumption in each panel as needed.
3. Use the Method of Joints or Method of Sections to find any additional member forces desired.

EXAMPLE 8.1

Our firm has been contracted to design a truss to support rigging in a theater. To provide redundancy, the team leader has chosen to use an indeterminate truss. In order to make preliminary selection of member properties, we have been asked to perform approximate analysis of the truss under an unbalanced live load.

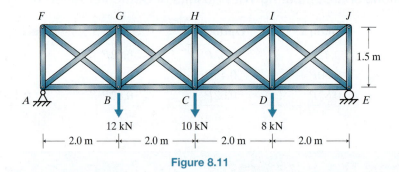

Figure 8.11

Initially, let's find the approximate axial forces in members BC, BG, BH, CG, and GH.

Predict the Sense of the Axial Forces:

Find the reaction directions:

We need only the left support reaction, since we will use the Method of Sections with the left section.

Section 8.1 Indeterminate Trusses **381**

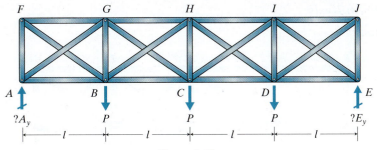

Figure 8.12a

$$+\circlearrowleft \Sigma M_E = 0 = A_y(4l) - P(3l) - P(2l) - P(l)$$
$$A_y = 1.5P > 0$$

Use the Method of Sections:
FBD of left section:

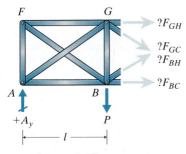

Figure 8.12b

Since $A_y > P$, the diagonals must point down.

$$\therefore F_{GC} = T \quad F_{BH} = C$$

Since $A_y > P$, the net moment at the cross-section is clockwise, so the chords must provide a counterclockwise moment.

$$\therefore F_{GH} = C \quad F_{BC} = T$$

Summary:

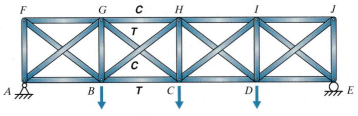

Figure 8.13

Approximate Analysis:
Find the reactions:

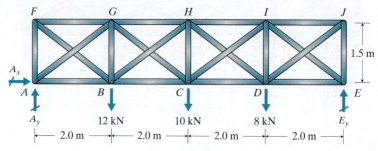

Figure 8.14a

$$\xrightarrow{\pm} \Sigma F_x = 0 = A_x$$
$$\circlearrowright \Sigma M_E = 0 = A_y(8\text{ m}) - 12\text{ kN}(6\text{ m}) - 10\text{ kN}(4\text{ m}) - 8\text{ kN}(2\text{ m})$$
$$A_y = +16\text{ kN} \;(+\uparrow)$$

Cut in the second panel:
Assume the shear in the panel is carried equally by the diagonals.

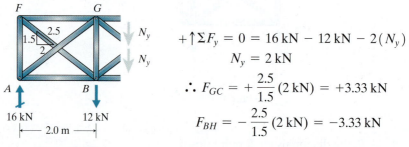

Figure 8.14b

$$+\uparrow \Sigma F_y = 0 = 16\text{ kN} - 12\text{ kN} - 2(N_y)$$
$$N_y = 2\text{ kN}$$
$$\therefore F_{GC} = +\frac{2.5}{1.5}(2\text{ kN}) = +3.33\text{ kN}$$
$$F_{BH} = -\frac{2.5}{1.5}(2\text{ kN}) = -3.33\text{ kN}$$

Find the chord forces:

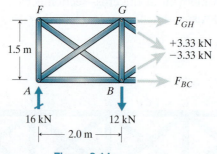

Figure 8.14c

$$\circlearrowright \Sigma M_B = 0 = 16\text{ kN}(2\text{ m}) + F_{GH}(1.5\text{ m}) + \frac{2.0}{2.5}(3.33\text{ kN})(1.5\text{ m})$$
$$F_{GH} = -24\text{ kN}$$

$$\xrightarrow{\pm} \Sigma F_x = 0 = -24\text{ kN} + \frac{2.0}{2.5}(3.33\text{ kN}) - \frac{2.0}{2.5}(3.33\text{ kN}) + F_{BC}$$
$$F_{BC} = +24\text{ kN}$$

Cut in the first panel:
Assume the shear in the panel is carried equally by the diagonals.

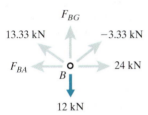

$+\uparrow \Sigma F_y = 0 = 16 \text{ kN} - 2(N_y)$
$N_y = 8 \text{ kN}$

$\therefore F_{FB} = +\dfrac{2.5}{1.5}(8 \text{ kN}) = +13.33 \text{ kN}$

Figure 8.14d

FBD of joint B:

[Figure showing joint B with forces: 13.33 kN, F_{BG}, -3.33 kN, F_{BA}, 24 kN, 12 kN]

Figure 8.14e

$+\uparrow \Sigma F_y = 0 = \dfrac{1.5}{2.5}(13.33 \text{ kN}) + F_{BG} + \dfrac{1.5}{2.5}(-3.33 \text{ kN}) - 12 \text{ kN}$

$F_{BG} = +6 \text{ kN}$

Summary:

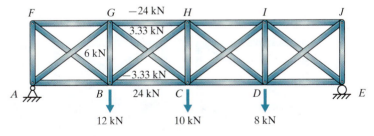

Figure 8.15

Evaluation of Results:

Observed Expected Features?
The results match the predicted senses from our qualitative analysis. ✓

8.2 Braced Frames with Lateral Loads

Introduction

Braced frames have diagonal members to provide lateral stability. Because the diagonals tend to make the braced frame stiff, there is rarely an advantage to using rigid connections. Therefore, braced frames tend to have pinned connections. Pinned connections and diagonal members mean that braced frames behave very similarly to trusses, so we make similar assumptions for the analysis of braced frames with lateral loads as we did with indeterminate trusses.

How-To

There are basically three types of braced frames (Figure 8.16). Diagonally braced frames have only one member bracing each bay. Cross braced frames have two members that cross in each bay. Chevron braced frames have two members that meet at the beam.

Because braced frames are very efficient for carrying lateral load, we typically need only one or two bays of bracing even for very wide frames (Figure 8.17). Notice that we typically use pinned connections between all of the members. Adding rigid connections rarely changes the behavior enough to reduce the sizes of the members; therefore, the additional cost is not warranted.

Assumptions

Even though the members are all connected with pins, braced frames taller than one story are indeterminate. Also, all chevron braced frames are indeterminate regardless of the number of stories. Therefore, we need to make assumptions in order to perform approximate analysis. To see where the first assumption comes from, let's draw a free body diagram of the top half of a diagonally braced frame (Figure 8.18).

The brace is a two-force member[1], so it carries only axial force. The two columns can carry axial force, shear, and moment, though. Horizontal equilibrium gives us an expression with too many unknowns:

$$\xrightarrow{+} \Sigma F_x = 0 = P - V_{Col1} - N_{Brace} - V_{Col2}$$
$$P = V_{Col1} + N_{Brace} + V_{Col2}$$

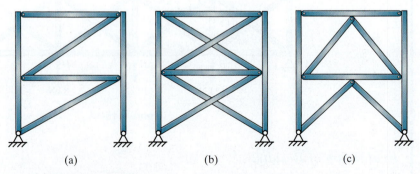

Figure 8.16 Three types of braced frames: (a) Diagonally braced; (b) Cross braced; (c) Chevron braced.

---

[1] Two-force members are pin connected at both ends and have no load applied except through the connections. Equilibrium shows that there is only one internal force: axial.

Figure 8.17 One bay of diagonal bracing is sufficient for this multibay frame.

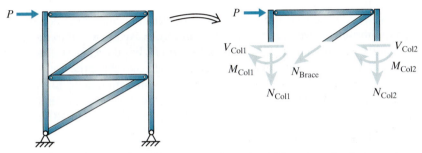

Figure 8.18 Diagonally braced frame with a lateral load. We can cut the frame and create a free body diagram.

In later chapters, we prove that stiffness attracts load in an indeterminate structure. If we compare the stiffness of a column shearing to a brace stretching, the brace is much stiffer. You can do an experiment yourself to demonstrate this; all you need is a strip of cardboard (Figure 8.19).

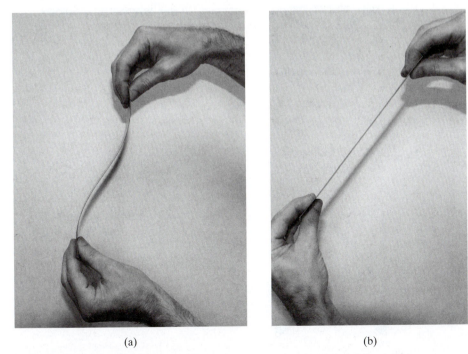

Figure 8.19 Experiment to show relative stiffness: (a) Shearing a member is relatively easy; (b) Stretching the same member is much more difficult.

Chapter 8 Approximate Analysis of Indeterminate Trusses and Braced Frames

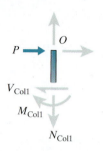

Figure 8.20 Free body diagram of the top piece of column from the braced frame.

Since the brace is significantly stiffer than the columns, our first assumption is that the braces carry all the shear at a level; no shear is generated in the columns. This assumption results in zero moment in the columns as well. To demonstrate that, let's draw a free body diagram of the piece of column from Figure 8.18 (Figure 8.20). If we sum the moments about the top, only the internal moment contributes, since there is no shear. So the moment must be zero.

This one assumption is enough for a single-bay diagonally braced frame, but for all other cases we still have an indeterminate structure. To demonstrate that, let's draw a free body diagram of the top half of a cross braced frame (Figure 8.21).

Now we have four unknown internal forces but only three equations of equilibrium, so we need one more assumption. Again, we go to the idea that stiffness attracts load. Each of the two braces has the same stiffness because they are the same length, at the same angle, and share the same properties; therefore, they should each carry the same amount of horizontal force. Formally, our second assumption is that the shear at a level is shared equally as the horizontal components of the brace forces (Figure 8.22).

When Braces Are Not Identical

If the braces at a level are not the same length, at the same inclination, and made with the same material and cross-sectional area, their stiffnesses are different (Figure 8.23). In these cases, our assumption about the distribution of shear to the braces is not as accurate, but for now we continue to make the assumption. Fortunately, these situations do not happen very often. In Section 10.1: Braced Frames—Story Drift Method, we see how to account for these situations more accurately if desired.

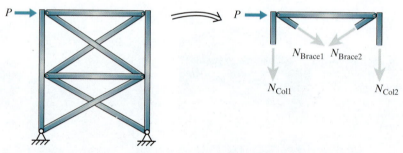

Figure 8.21 Cross braced frame with a lateral load. We can cut the frame to create a free body diagram with four unknowns.

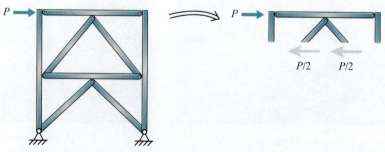

Figure 8.22 Chevron braced frame with a lateral load. We can cut the frame to see that the lateral load is carried by the two braces equally.

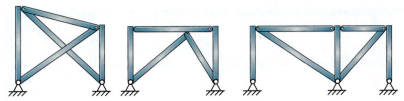

Figure 8.23 Examples of braced frames where the braces are not identical: different lengths and angles of inclination.

Axial Force Members

The beams in diagonally and cross braced frames are two-force members; therefore, they carry only axial force. Beams in symmetric chevron braced frames are not two-force members; however, they also carry only axial force based on our assumptions. We can see this by applying moment equilibrium to the free body diagram of a beam from a chevron braced frame (Figure 8.24).

Because the braces are set at the same angle, if the horizontal components are equal, then the vertical components must be equal in magnitude and opposite in direction. Therefore, the vertical components cancel each other in the moment equilibrium equation:

$$\circlearrowright \Sigma M_2 = 0 = V_1(l) - N_{y1}(l/2) + N_{y2}(l/2)$$
$$V_1(l) = 0$$

That leaves only the shear at the end of the beam. So the shear must be zero. From Chapter 4: Internal Force Diagrams, we learned that if there is no distributed load, the shear must be constant. Therefore, the shear throughout the beam is zero. We also learned in Chapter 4 that if the shear has a constant value of zero, the moment is constant. Since the moment at the pinned end is zero, the moment throughout the beam must be zero. So the beam carries only axial force.

We have shown that with our assumptions for approximate analysis, all members of a laterally loaded braced frame carry only axial force.

Purpose of Each Member

Each member of a braced frame has a specific purpose when the frame carries lateral load. To understand the purpose, consider the free body diagram of the top of a chevron braced frame (Figure 8.25).

We already know that each of the members carries only axial load. Because of our first assumption, we know that the purpose of the braces is to carry the lateral load down to the foundation. From the free body diagram, we can see that the beam delivers the lateral load to the

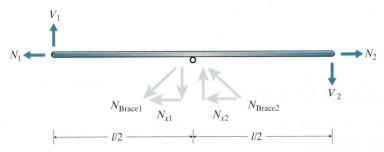

Figure 8.24 Free body diagram of a beam from a chevron braced frame.

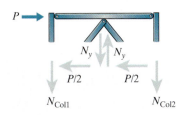

Figure 8.25 Free body diagram of the top part of a chevron braced frame.

388 **Chapter 8** Approximate Analysis of Indeterminate Trusses and Braced Frames

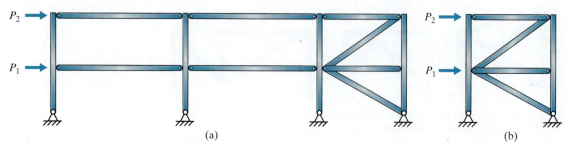

Figure 8.26 Lateral load acting on a column line with lateral load-resisting elements: (a) Load applied to a multibay frame; (b) Lateral load is carried through the bays without lateral bracing to the braced bay.

braces. To understand the purpose of the columns, first we apply vertical equilibrium:

$$+\downarrow \Sigma F_y = 0 = N_{Col1} + N_{Col2} + N_y - N_y$$
$$N_{Col1} = -N_{Col2}$$

So the two columns have parallel forces that are equal and opposite. Therefore, they create a couple moment. The purpose of the columns is to resist the overturning moment of the applied lateral load.

Transfer of Lateral Force

We saw in Section 1.5: Application of Lateral Load that lateral load typically acts on the outer envelope of the structure and gathers at the diaphragm levels. In Section 1.6: Distribution of Lateral Loads by Flexible Diaphragm, we saw that the load at the diaphragm level gathers at lateral load-resisting frames like the one in Figure 8.17. Since there is no shear in any of the columns, the lateral force must pass through the column into the beams or floor diaphragms. The beams or diaphragms just carry the lateral load to the bay with the braced frame. Therefore, we can simplify the problem from the full frame in Figure 8.17 to only the braced bay in Figure 8.26.

Approximate Analysis Process

Because our assumptions result in only axial force members in braced frames, we can approach the analysis as for indeterminate trusses.

Step 1: Deliver lateral load to the braced frame. Using the methods we learned in Chapter 1: Loads and Structure Idealization, determine the magnitude of the force applied to each level of the braced frame.

Step 2: At each level, cut the braced frame and find the brace forces. We apply the assumptions here.

Step 3: Use the Method of Joints or Method of Sections to find any other axial forces desired.

SECTION 8.2 HIGHLIGHTS

Braced Frames with Lateral Loads

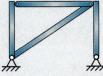
Diagonally Braced

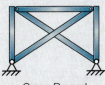
Cross Braced

Chevron Braced

Role of Members:

Beams:	Deliver lateral force to braces
Braces:	Carry lateral force to foundation
Columns:	Balance overturning moment caused by lateral force

In approximate analysis, all members carry only axial force.

Assumptions:

1. All shear at a level is carried in the braces; none is carried in the columns.
2. Shear at a level is shared equally as the horizontal components of the brace forces.

Analysis Steps:

1. Deliver lateral load to the braced frames.
2. At each level, cut the braced frame and find the brace forces.
3. Use the Method of Sections or Method of Joints to find any other axial forces desired.

390 Chapter 8 Approximate Analysis of Indeterminate Trusses and Braced Frames

EXAMPLE 8.2

Our firm is designing a two-story research facility. The team leader has chosen to use diagonal braces as the lateral load-resisting system. In order to make a preliminary selection of member properties, we have been tasked to perform approximate analysis of the frame subjected to wind load. We are to find the axial force requirement in each of the braces and columns.

Figure 8.27

Deliver Lateral Load to the Braced Frame:

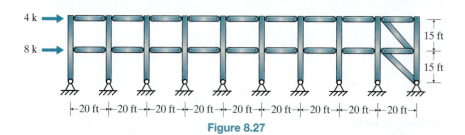

Figure 8.28

Qualitative Analysis:
Use the Method of Joints:
  FBD of joint F:

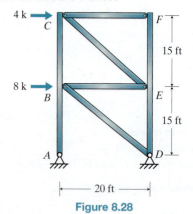

$$\xleftarrow{\pm} \Sigma F_x = 0 = N_{FC}$$
$$+\downarrow \Sigma F_y = 0 = N_{FE}$$

Figure 8.29a

Section 8.2 Braced Frames with Lateral Loads

FBD of joint C:

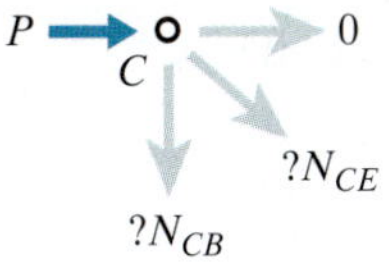

$\pm \Sigma F_x = 0 = P + N_{CEx}$ $\quad N_{CE} < 0$
$+\downarrow \Sigma F_y = 0 = N_{CB} + (-N_{CEy})$ $\quad N_{CB} > 0$

Figure 8.29b

FBD of joint E:

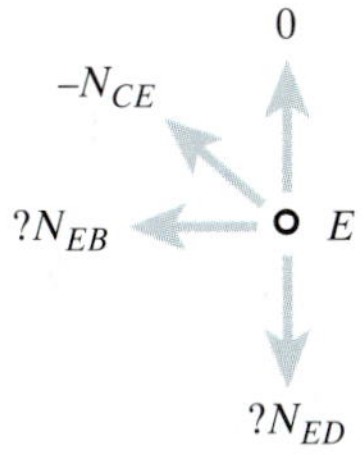

$\pm \Sigma F_x = 0 = -N_{CEx} + N_{EB}$ $\quad N_{EB} > 0$
$+\uparrow \Sigma F_y = 0 = -N_{CEy} + (-N_{ED})$ $\quad N_{ED} < 0$

Figure 8.29c

FBD of joint B:

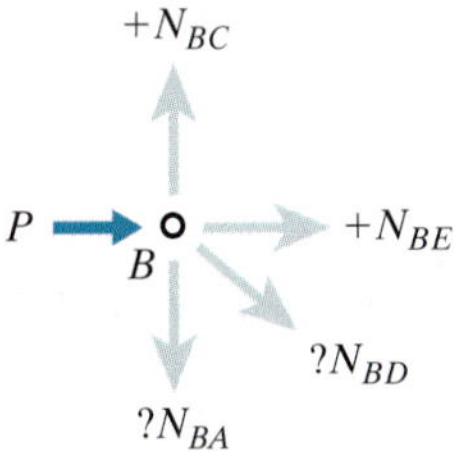

$\pm \Sigma F_x = 0 = P + N_{BE} + N_{BDx}$ $\quad N_{BD} < 0$
$+\uparrow \Sigma F_y = 0 = N_{BC} - (-N_{BDy}) - N_{BA}$ $\quad N_{BA} > 0$

Figure 8.29d

Summary:

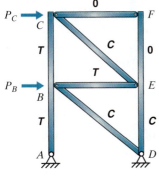

Figure 8.30

Approximate Analysis:

Cut the top story:
  Find the axial force in the brace:
    Assume the shear is carried only by the brace.

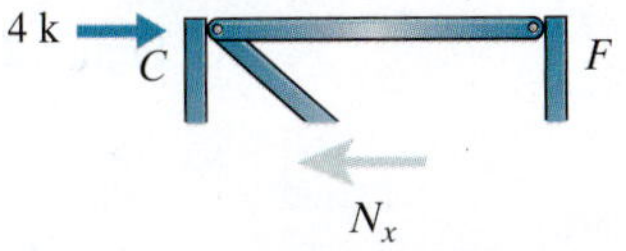

$\pm \Sigma F_x = 0 = 4\,\text{k} - N_x$
$N_x = 4\,\text{k}$
$\therefore N_{CE} = -\dfrac{25}{20}(4\,\text{k}) = -5\,\text{k}$

Figure 8.31

392 Chapter 8 Approximate Analysis of Indeterminate Trusses and Braced Frames

Find the axial forces in the columns:

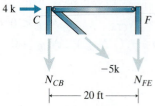

Figure 8.32

$$\circlearrowright \Sigma M_C = 0 = N_{FE}(20 \text{ ft})$$
$$N_{FE} = 0$$

$$+\downarrow \Sigma F_y = 0 = N_{CB} + \frac{15}{25}(-5 \text{ k})$$
$$N_{CB} = +3 \text{ k}$$

Cut the lower story:
Find the axial force in the brace:
Assume the shear is carried only by the brace.

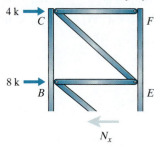

Figure 8.33

$$\pm \Sigma F_x = 0 = 4 \text{ k} + 8 \text{ k} - N_x$$
$$N_x = 12 \text{ k}$$

$$\therefore N_{BD} = -\frac{25}{20}(12 \text{ k}) = -15 \text{ k}$$

Find the axial forces in the columns:

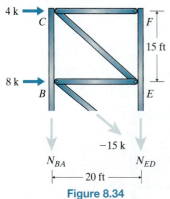

Figure 8.34

$$\circlearrowright \Sigma M_B = 0 = 4 \text{ k}(15 \text{ ft}) + N_{ED}(20 \text{ ft})$$
$$N_{ED} = -3 \text{ k}$$

$$+\downarrow \Sigma F_y = 0 = N_{BA} + \frac{15}{25}(-15 \text{ k}) + (-3 \text{ k})$$
$$N_{BA} = +12 \text{ k}$$

Summary:
If the wind reverses, the sense of the axial forces will reverse. Therefore, results are expressed as absolute values.

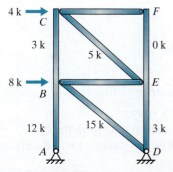

Figure 8.35

Evaluation of Results:
Observed Expected Features?
The results match the predicted senses from our qualitative analysis. ✓

Example 8.3

Our firm is designing a tall building. The architect would like us to use a cross braced frame and will expose the system visually. Although we are still early in the design process, the geotechnical engineers need to know preliminary forces on the foundations right away. We are tasked with providing the forces on the foundations due to wind load.

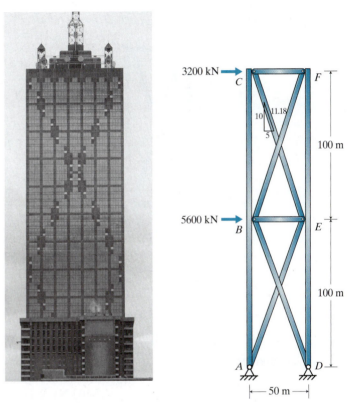

Figure 8.36

Approximate Analysis:

Since we only want to know the reactions, we need to cut the frame in just the lower panel.

Cut the lower frame:

Find the axial force in each brace: Assume the shear is carried only in the braces, and assume each brace carries an equal amount of shear.

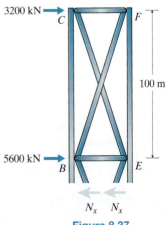

Figure 8.37

394 Chapter 8 Approximate Analysis of Indeterminate Trusses and Braced Frames

$$\pm \rightarrow \Sigma F_x = 0 = 3200 \text{ kN} + 5600 \text{ kN} - 2N_x$$
$$N_x = 4400 \text{ kN}$$
$$\therefore N_{CE} = -\frac{11.18}{5}(4400 \text{ kN}) = -9840 \text{ kN}$$
$$N_{FB} = +\frac{11.18}{5}(4400 \text{ kN}) = +9840 \text{ kN}$$

Find the axial forces in the columns:

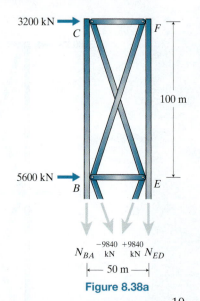

Figure 8.38a

$$+\circlearrowleft \Sigma M_B = 0 = 3200 \text{ kN}(100 \text{ m}) + N_{ED}(50 \text{ m}) + \frac{10}{11.18}(9840 \text{ kN})(50 \text{ m})$$
$$N_{ED} = -15{,}200 \text{ kN}$$
$$+\downarrow \Sigma F_y = 0 = N_{BA} + \frac{10}{11.18}(-9840 \text{ kN}) + \frac{10}{11.18}(9840 \text{ kN}) + (-15{,}200 \text{ kN})$$
$$N_{BA} = +15{,}200 \text{ kN}$$

FBD of joint A:

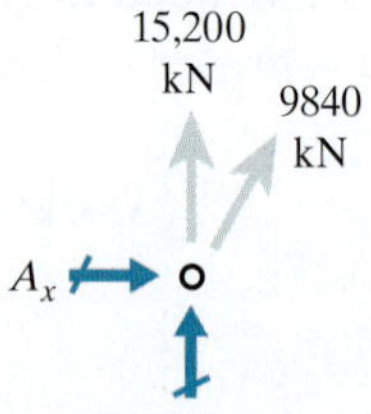

Figure 8.38b

$$\pm \rightarrow \Sigma F_x = 0 = A_x + \frac{5}{11.18}(9840 \text{ kN})$$
$$A_x = -4400 \text{ kN} \ (\pm \rightarrow)$$
$$+\uparrow \Sigma F_y = 0 = A_y + 15{,}200 \text{ kN} + \frac{10}{11.18}(9840 \text{ kN})$$
$$A_y = -24{,}000 \text{ kN} \ (+\uparrow)$$

FBD of joint D:

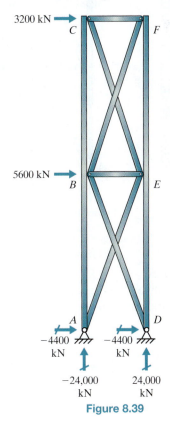

$$\xrightarrow{\pm} \Sigma F_x = 0 = D_x - \frac{5}{11.18}(-9840 \text{ kN})$$
$$D_x = -4400 \text{ kN } (\xrightarrow{\pm})$$

$$+\uparrow \Sigma F_y = 0 = D_y - 15{,}200 \text{ kN} + \frac{10}{11.18}(-9840 \text{ kN})$$
$$D_y = +24{,}000 \text{ kN}(+\uparrow)$$

Figure 8.38c

Summary:
The forces shown on the structure are provided by the foundations; the forces on the foundations due to the wind will have reversed directions.

Note also that if the wind reverses direction, the signs will reverse.

Figure 8.39

Evaluation of Results:

Satisfied Fundamental Principles?
Verify overall equilibrium:

$$\xrightarrow{\pm} \Sigma F_x = 3200 \text{ kN} + 5600 \text{ kN} - 4400 \text{ kN} - 4400 \text{ kN} = 0 \checkmark$$

$$+\uparrow \Sigma F_y = 24{,}000 \text{ kN} + (-24{,}000 \text{ kN}) = 0 \checkmark$$

$$\curvearrowright \Sigma M_A = 3200 \text{ kN}(200 \text{ m}) + 5600 \text{ kN}(100 \text{ m}) - 24{,}000 \text{ kN}(50 \text{ m}) = 0 \checkmark$$

Conclusion:
Based on this check, these preliminary results appear reasonable.

EXAMPLE 8.4

Our team is designing a large recreation facility. The senior engineer has chosen chevron braces for the lateral load-resisting system. Based on previous experience, the engineer anticipates we will need to use two bays of braces. Architectural constraints dictate that the two bays be different sizes and the two stories be different heights.

Our job is to estimate the axial force in each brace and column so that we can perform preliminary design of the members.

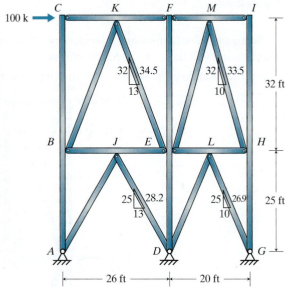

Figure 8.40

Approximate Analysis:

Cut the top level:
Find the axial force in each brace:
Assume the shear is carried only by the braces, and assume each brace carries an equal amount of shear as its horizontal component.

Figure 8.41a

$\pm \Sigma F_x = 0 = 100 \text{ k} - 4N_x \quad N_x = 25 \text{ k}$

$\therefore N_{KB} = +\dfrac{34.5}{13}(25 \text{ k}) = +66.3 \text{ k}$

$N_{KE} = -\dfrac{34.5}{13}(25 \text{ k}) = -66.3 \text{ k}$

$N_{ME} = +\dfrac{33.5}{10}(25 \text{ k}) = +83.8 \text{ k}$

$N_{MH} = -\dfrac{33.5}{10}(25 \text{ k}) = -83.8 \text{ k}$

Section 8.2 Braced Frames with Lateral Loads **397**

FBD of joint C:

$+\downarrow \Sigma F_y = 0 = N_{CB}$

Figure 8.41b

FBD of joint F:

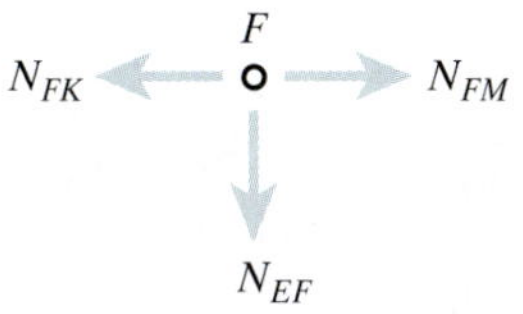

$+\downarrow \Sigma F_y = 0 = N_{EF}$

Figure 8.41c

FBD of joint I:

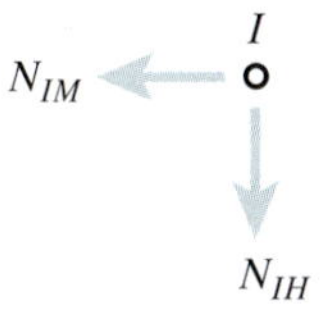

$N_{IM} \leftarrow \stackrel{I}{\circ}$

$\downarrow N_{IH}$

$+\downarrow \Sigma F_y = 0 = N_{IH}$

Figure 8.41d

Cut the bottom level:

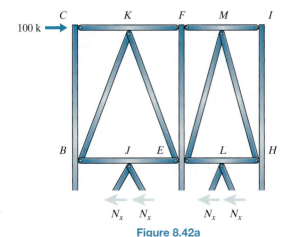

Figure 8.42a

$\xrightarrow{+} \Sigma F_x = 0 = 100\text{ k} - 4N_x \quad N_x = 25\text{ k}$

$\therefore N_{JA} = +\dfrac{28.2}{13}(25\text{ k}) = +54.2\text{ k}$

$N_{JD} = -\dfrac{28.2}{13}(25\text{ k}) = -54.2\text{ k}$

$N_{LD} = +\dfrac{26.9}{10}(25\text{ k}) = +67.2\text{ k}$

$N_{LG} = -\dfrac{26.9}{10}(25\text{ k}) = -67.2\text{ k}$

FBD of joint B:

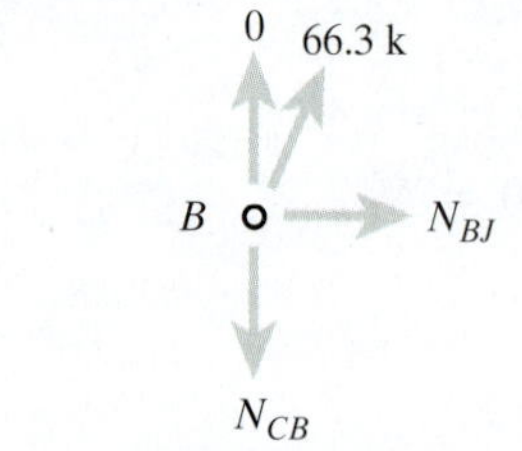

$$+\uparrow \Sigma F_y = 0 = \frac{32}{34.5}(66.3 \text{ k}) - N_{CB}$$

$$N_{CB} = +61.5 \text{ k}$$

Figure 8.42b

FBD of joint E:

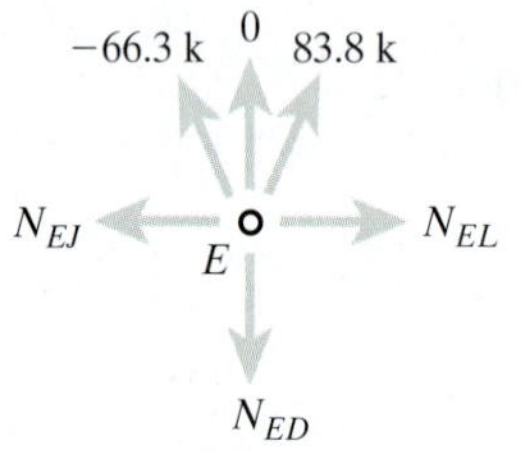

$$+\uparrow \Sigma F_y = 0 = \frac{32}{34.5}(-66.3 \text{ k}) + \frac{32}{33.5}(83.8 \text{ k}) - N_{ED}$$

$$N_{ED} = +18.6 \text{ k}$$

Figure 8.42c

FBD of joint H:

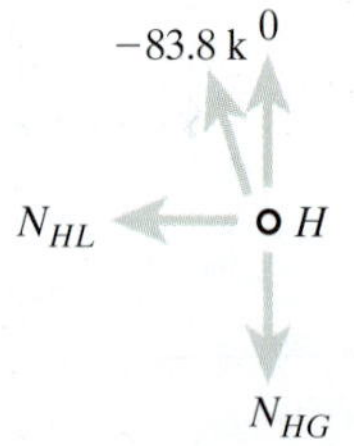

$$+\uparrow \Sigma F_y = 0 = \frac{32}{33.5}(-83.8 \text{ k}) - N_{HG}$$

$$N_{HG} = -80.0 \text{ k}$$

Figure 8.42d

Summary:
If the wind reverses, the sense of the axial forces will reverse. Therefore, results are expressed as absolute values.

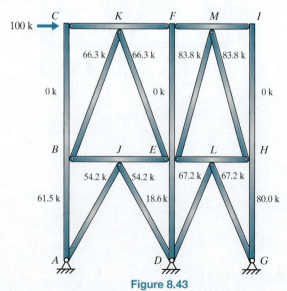

Figure 8.43

Observation:

Because the horizontal component of the brace force is set, a steeper brace carries a larger force than a shallow brace. So reducing the brace angle reduces the axial force requirement.

From mechanics of materials we know that buckling is governed largely by the length of the member. So shorter members are stronger in compression.

Therefore, braces at close to a 45° angle tend to be the most efficient.

Evaluation of Results:

Although not required, let's find the reactions in order to confirm that equilibrium is satisfied.

FBD of joint A:

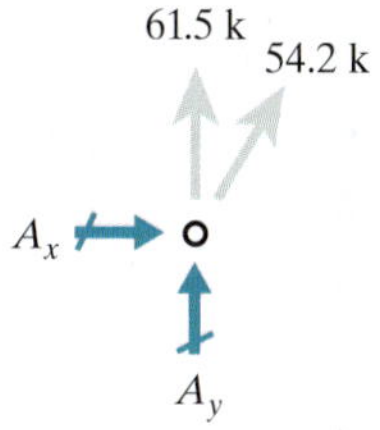

$$\xrightarrow{\pm} \Sigma F_x = 0 = A_x + \frac{13}{28.2}(54.2 \text{ k})$$

$$A_x = -25.0 \text{ k} \ (\xrightarrow{\pm})$$

$$+\uparrow \Sigma F_y = 0 = A_y + 61.5 \text{ k} + \frac{25}{28.2}(54.2 \text{ k})$$

$$A_y = -109.5 \text{ k} \ (+\uparrow)$$

Figure 8.44a

FBD of joint D:

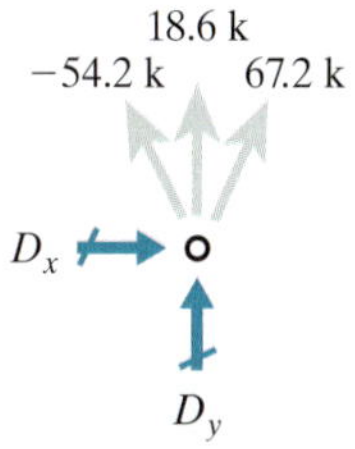

$$\xrightarrow{\pm} \Sigma F_x = 0 = D_x - \frac{13}{28.2}(-54.2 \text{ k}) + \frac{10}{26.9}(67.2 \text{ k})$$

$$D_x = -50.0 \text{ k} \ (\xrightarrow{\pm})$$

$$+\uparrow \Sigma F_y = 0 = D_y + \frac{25}{28.2}(-54.2 \text{ k}) + 18.6 \text{ k} + \frac{25}{26.9}(67.2 \text{ k})$$

$$D_y = -33.0 \text{ k} \ (+\uparrow)$$

Figure 8.44b

FBD of joint G:

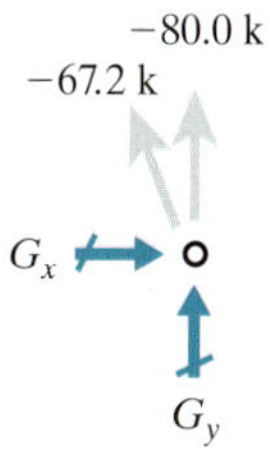

$$\xrightarrow{\pm} \Sigma F_x = 0 = G_x - \frac{10}{26.9}(-67.2 \text{ k})$$

$$G_x = -25.0 \text{ k} \ (\xrightarrow{\pm})$$

$$+\uparrow \Sigma F_y = 0 = G_y + \frac{25}{26.9}(-67.2 \text{ k}) + (-80.0 \text{ k})$$

$$G_y = +142.5 \text{ k} \ (+\uparrow)$$

Figure 8.44c

400 **Chapter 8** Approximate Analysis of Indeterminate Trusses and Braced Frames

FBD of the entire frame:

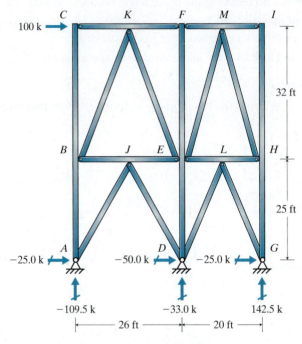

Figure 8.45

Evaluation of Results:

Satisfied Fundamental Principles?

Verify overall equilibrium:

$$\xrightarrow{+} \Sigma F_x = 100 \text{ k} - 25 \text{ k} - 50 \text{ k} - 25 \text{ k} = 0 \checkmark$$
$$+\uparrow \Sigma F_y = -109.5 \text{ k} - 33.0 \text{ k} + 142.5 \text{ k} = 0 \checkmark$$
$$\curvearrowleft_+ \Sigma M_A = 100 \text{ k}(57 \text{ ft}) - (-33.0 \text{ k})(26 \text{ ft}) - 142.5 \text{ k}(46 \text{ ft})$$
$$= 3 \text{ k} \cdot \text{ft} \approx 0 \checkmark$$

Conclusion:
Based on this check, these preliminary results appear reasonable.

8.3 Braced Frames with Gravity Loads

Introduction
Although the primary purpose of braced frames is to resist lateral loads, those frames might experience substantial gravity loads as well. Using a couple of assumptions, we can treat braced frames with gravity loads as trusses in order to perform approximate analysis.

How-To
In almost every structure, some parts have a dual purpose: support lateral loads and support gravity loads. When that happens to a braced frame, we need to use superposition (Section 2.2: Principle of Superposition) in order to perform approximate analysis. We have a good sense of what assumptions to make when there is only lateral load, but those assumptions become less accurate as we add more and more gravity load. Fortunately, we also have a good sense of what assumptions to make when there is only gravity load.

For a braced frame that experiences both types of loads, we analyze the frame with only the lateral load using the method in Section 8.2: Braced Frames with Lateral Loads. We also analyze the frame with only the gravity loads using the method described in this section. We then superimpose the results (Figure 8.46).

Truss Approximation
Because braced frames typically use pinned connections, they resemble trusses. The only difference between trusses and diagonally or cross braced frames is the continuity of the braced frame columns (Figure 8.47).

In order for us to treat diagonally and cross braced frames as trusses, we need to assume that the columns carry no shear. If the shear is zero

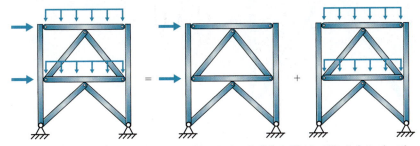

Figure 8.46 The principle of superposition states that the effects of both lateral and gravity loads on the frame are the same as the addition of the effects of lateral loads and the effects of gravity loads.

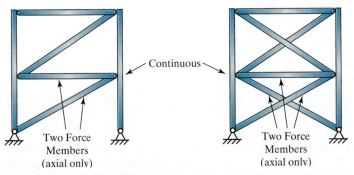

Figure 8.47 Diagonally and cross braced frames are made of two force members, except for columns in multistory frames.

everywhere in the columns, the moment is constant. Since the moment at the top of the columns must be zero, there is no moment in the columns either. That leaves only axial force in the columns.

For chevron-braced frames with gravity loads, we need an additional assumption. We see that need by looking at an approximate idealization of the beam; it is supported vertically at three locations (Figure 8.48). This idealization is not completely accurate because the vertical stiffnesses of the columns and the braces are not the same. A more accurate idealization is to model the middle support as a spring (Figure 8.49a) or the end supports as springs (Figure 8.49b), depending on the relative stiffnesses.

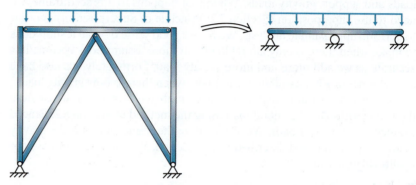

Figure 8.48 Beams in a chevron braced frame with gravity load can be idealized approximately as supported at three points.

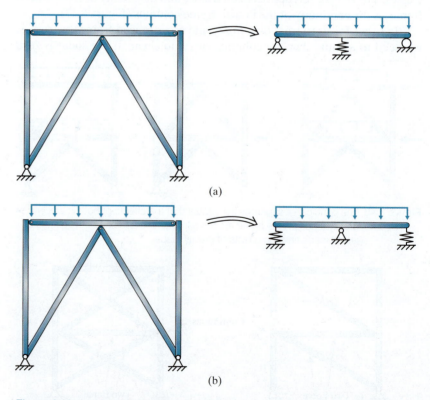

Figure 8.49 Accurate idealization of beam in a chevron braced frame subjected to gravity load: (a) If columns are stiffer than the braces, there is a spring support at the brace point; (b) If braces are stiffer than the columns, there are spring supports at the ends.

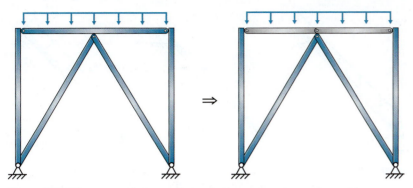

Figure 8.50 We assume that a gravity loaded beam in a chevron braced frame behaves as two members connected at the brace point.

No matter which of the three idealizations we use, the beam is indeterminate, so we need an assumption for analysis by hand.

The typical assumption we make is that there is zero moment at the brace point. Therefore, we are ignoring the continuity of the beam, and the frame acts as a truss (Figure 8.50).

With these assumptions we can treat braced frames as trusses, which means we can use the Method of Joints or Method of Sections to analyze the frames.

Distributed Loading

When we learned the Method of Joints and Method of Sections in an earlier course, we were taught that trusses are loaded only at the joints. That limitation results in only axial force in each member, but now we are dealing with distributed loading on members. In Chapter 4: Internal Force Diagrams, we saw that axial force and shear are uncoupled. That means one does not affect the other in a member, so we can calculate the effects independently.

Consider the beam in Figure 8.51a. The free body diagram of the beam has vertical and horizontal forces at each end (Figure 8.51b). That is the same as the beam being pin-pin supported (Figure 8.51c).

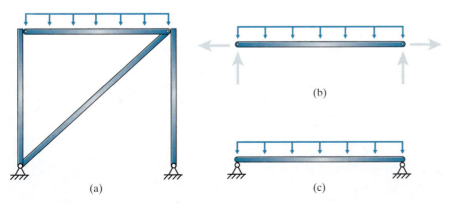

Figure 8.51 Idealizing a gravity loaded beam in a braced frame: (a) A beam experiencing distributed load; (b) FBD of the beam shows four forces at the ends; (c) The beam can be accurately idealized as pin-pin supported.

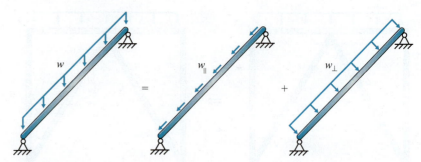

Figure 8.52 We use superposition to break down any distributed load acting on a member of the braced frame after idealizing the frame as a truss.

To begin analyzing any member of a brace frame that is experiencing a distributed load, we break the distributed load into components along the member, $w_\parallel$, and normal to the member, $w_\perp$ (Figure 8.52).

Although a pin-pin supported member is indeterminate, we can still find the peak shear and moment by analyzing it as simply supported with only the distributed load component normal to the member (Figure 8.53). That means for uniformly loaded members, the peak shear is $w_\perp l/2$, and the peak moment is $w_\perp l^2/8$.

Although we could find the changing axial force along the member, for approximate analysis we are typically content with finding the average axial force. To do that, we start by converting all distributed loads into equivalent point forces at the nodes. The equivalent point forces are the same magnitude as the reactions of a pin-pin supported member, but in the same direction as the applied loads (e.g., Figure 8.54).

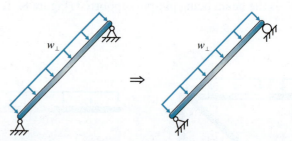

Figure 8.53 Although the idealized brace is pin-pin supported, we can find the peak shear and moment considering the brace to be simply supported.

Section 8.3 Braced Frames with Gravity Loads

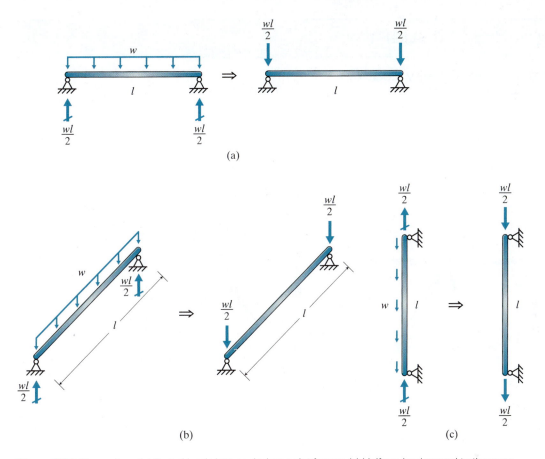

Figure 8.54 Converting distributed loads into equivalent point forces: (a) Uniform load normal to the member; (b) Uniform load on an inclined member; (c) Uniform load along the axis of the member.

Approximate Analysis Process

Once we have converted the distributed loads to point forces at the joints, we can analyze a braced frame as a truss. Together with our calculation of the peak shear and moment, we will then have all the internal force information that we typically need from an approximate analysis.

Step 1: Approximate the braced frame as a truss. If there are any chevron braces, approximate the beam as having an internal hinge at the brace point.

Step 2: If we desire the peak shear and moment in a member, find the normal component of the distributed load on that member. Analyze the member as simply supported with only the distributed load acting normal to the member.

Step 3: To find the average axial force in members, convert any distributed loads into equivalent point forces at the joints.

Step 4: With the frame approximated as a truss and any distributed loads converted to equivalent forces, use the Method of Joints or Method of Sections to find the axial forces and/or support reactions.

SECTION 8.3 HIGHLIGHTS

Braced Frames with Gravity Loads

Assumptions:
1. No shear develops in the columns.

For chevron braced frames only:
2. The beam can be reasonably approximated as having an internal hinge at the brace point.

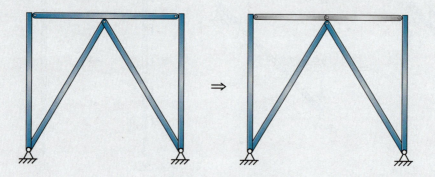

Analysis Steps:
1. Approximate the braced frame as a truss.
2. If we desire peak shear and moment in a member, analyze the member as simply supported with only the distributed load acting normal to the member.

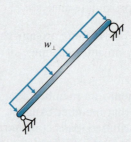

3. To find the average axial force in each member, convert any distributed loads into equivalent point forces at the joints.
4. Use the Method of Joints or Method of Sections to find the axial forces and/or support reactions.

Section 8.3 Braced Frames with Gravity Loads 407

EXAMPLE 8.5

Our firm is designing a two-story research facility. The senior engineer has chosen to use diagonal braces as the lateral load-resisting system. One-way slabs deliver the gravity load to the same frame. The geotechnical engineer needs to have an estimate of the force generated on each foundation due to dead load. For this preliminary analysis, we will guess the self-weight of each of the frame members to be 50 plf. Our job is to calculate the forces on the foundations under the diagonal braced frame; another team member is estimating the forces for the other supports.

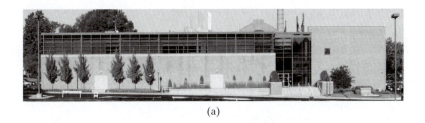

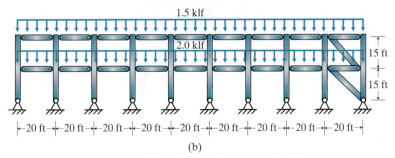

Figure 8.55a and b

Approximate Analysis:

Assumption:

- No shear is generated in the columns.

Equivalent point loads:
  Roof beams:

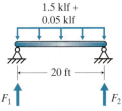

Figure 8.56

$$\curvearrowleft \Sigma M_2 = 0 = F_1(20 \text{ ft}) - 1.55 \text{ klf}(20 \text{ ft})(20 \text{ ft}/2)$$
$$F_1 = 15.5 \text{ k } (+\uparrow)$$
$$+\uparrow \Sigma F_y = 0 = 15.5 \text{ k} - 1.55 \text{ klf}(20 \text{ ft}) + F_2$$
$$F_2 = 15.5 \text{ k } (+\uparrow)$$

408   **Chapter 8**   Approximate Analysis of Indeterminate Trusses and Braced Frames

Floor beams:

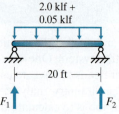

Figure 8.57

$$\circlearrowright \Sigma M_2 = 0 = F_1(20 \text{ ft}) - 2.05 \text{ klf}(20 \text{ ft})(20 \text{ ft}/2)$$
$$F_1 = 20.5 \text{ k} \ (+\uparrow)$$
$$+\uparrow \Sigma F_y = 0 = 20.5 \text{ k} - 2.05 \text{ klf}(20 \text{ ft}) + F_2$$
$$F_2 = 20.5 \text{ k} \ (+\uparrow)$$

Braces:

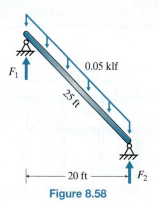

Figure 8.58

$$\circlearrowright \Sigma M_2 = 0 = F_1(20 \text{ ft}) - 0.05 \text{ klf}(25 \text{ ft})(20 \text{ ft}/2)$$
$$F_1 = 0.625 \text{ k} \ (+\uparrow)$$
$$+\uparrow \Sigma F_y = 0 = 0.625 \text{ k} - 0.05 \text{ klf}(25 \text{ ft}) + F_2$$
$$F_2 = 0.625 \text{ k} \ (+\uparrow)$$

Columns:

Figure 8.59

$$F_1 = \frac{wl}{2} = \frac{0.05 \text{ klf}(15 \text{ ft})}{2}$$
$$F_1 = 0.375 \text{ k} \ (+\uparrow)$$
$$F_2 = \frac{wl}{2} = \frac{0.05 \text{ klf}(15 \text{ ft})}{2}$$
$$F_2 = 0.375 \text{ k} \ (+\uparrow)$$

Section 8.3 Braced Frames with Gravity Loads 409

Equivalent forces on truss approximation:

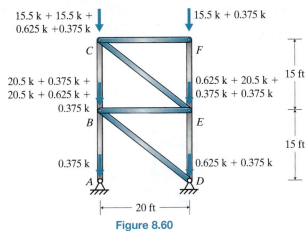

Figure 8.60

Truss analysis:

FBD of section cut at the bottom level:

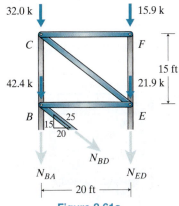

Figure 8.61a

$$\xrightarrow{+} \Sigma F_x = 0 = \frac{20}{25} N_{BD}$$

$$N_{BD} = 0$$

$$\curvearrowleft^+ \Sigma M_D = 0 = N_{BA}(20 \text{ ft}) + 42.4 \text{ k}(20 \text{ ft}) + 32.0 \text{ k}(20 \text{ ft})$$

$$N_{BA} = -74.4 \text{ k}$$

$$+\downarrow \Sigma F_y = 0 = -74.4 \text{ k} + 42.4 \text{ k} + 32.0 \text{ k} + 21.9 \text{ k} + 15.9 \text{ k} + N_{ED}$$

$$N_{ED} = -37.8 \text{ k}$$

FBD of joint A:

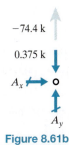

Figure 8.61b

$$\xrightarrow{+} \Sigma F_x = 0 = A_x$$
$$+\uparrow \Sigma F_y = 0 = -74.4 \text{ k} - 0.375 \text{ k} + A_y$$
$$A_y = 74.8 \text{ k } (+\uparrow)$$

FBD of joint D:

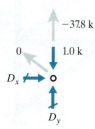

Figure 8.61c

$$\xrightarrow{+} \Sigma F_x = 0 = -0 + D_x$$
$$D_x = 0$$
$$+\uparrow \Sigma F_y = 0 = -37.8 \text{ k} - 1.0 \text{ k} + D_y$$
$$D_y = 38.8 \text{ k } (+\uparrow)$$

Summary:
The reaction directions shown are the foundations acting on the frame. Therefore, the frame acts down on the foundations.

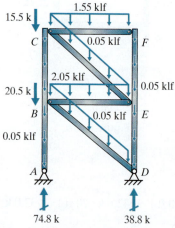

Figure 8.62

Evaluation of Results:

Satisfied Fundamental Principles?
Verify overall equilibrium:

$$+\uparrow \Sigma F_y = 74.8 \text{ k} + 38.8 \text{ k} - 15.5 \text{ k} - 20.5 \text{ k} - 2(0.05 \text{ klf})(30 \text{ ft})$$
$$-1.55 \text{ klf}(20 \text{ ft}) - 2.05 \text{ klf}(20 \text{ ft})$$
$$-2(0.05 \text{ klf})(25 \text{ ft}) = 0.1 \text{ k} \approx 0 \checkmark$$
$$\circlearrowleft \Sigma M_A = 1.55 \text{ klf}(20 \text{ ft})(20 \text{ ft}/2) + 2(0.05 \text{ klf})(25 \text{ ft})(20 \text{ ft}/2)$$
$$+ 2.05 \text{ klf}(20 \text{ ft})(20 \text{ ft}/2) + 0.05 \text{ klf}(30 \text{ ft})(20 \text{ ft})$$
$$- 38.8 \text{ k}(20 \text{ ft}) = -1 \text{ k} \cdot \text{ft} \ll \text{ each term } \therefore = 0 \checkmark$$

Conclusion:
Based on this check, these preliminary results appear reasonable.

Example 8.6

Our firm is designing the structure for a new supermarket. The project manager has chosen chevron braces as the lateral load-resisting system and has placed them in locations out of sight of the public. In order to perform detailed analysis of the structure, it would help to have a good idea of what the member properties will be. Therefore, we are tasked with performing approximate analysis of one of the braced frames under wind load. Our results will be combined with results from team members who considered the other relevant loads. Once the team knows the approximate internal forces, we can make a preliminary selection of member properties.

Another member of our team has determined the point forces that are carried to the frame from the adjacent beams.

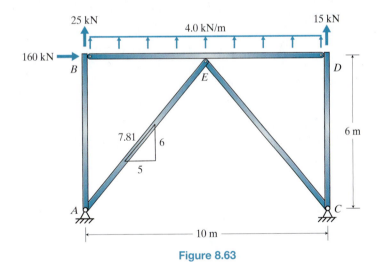

Figure 8.63

Approximate Analysis:

Superposition:
   Because the frame experiences both gravity and lateral loads, we must consider each type of load individually and then superimpose the results.

Assumption:
   • Material remains linear elastic.

412 **Chapter 8** Approximate Analysis of Indeterminate Trusses and Braced Frames

Gravity Load:

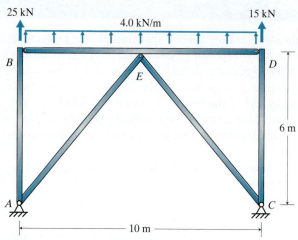

Figure 8.64

Assumptions:
- No shear is generated in the columns.
- A hinge in the beam is at the brace point.

Beam analysis:

Equivalent point loads:

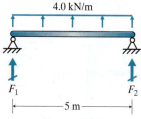

Figure 8.65

$$\circlearrowright \Sigma M_2 = 0 = F_1(5\,\text{m}) + (4.0\,\text{kN/m})(5\,\text{m})(5\,\text{m}/2)$$
$$F_1 = -10\,\text{kN} \ (+\uparrow)$$
$$+\uparrow \Sigma F_y = 0 = -10\,\text{kN} + (4.0\,\text{kN/m})(5\,\text{m}) + F_2$$
$$F_2 = -10\,\text{kN}(+\uparrow)$$

Peak internal forces for beam:

$$F_{\max} = \frac{(4.0\,\text{kN/m})(5\,\text{m})}{2} = 10.0\,\text{kN} \qquad M_{\max} = \frac{(4.0\,\text{kN/m})(5\,\text{m})^2}{8} = 12.5\,\text{kN}\cdot\text{m}$$

Truss approximation:

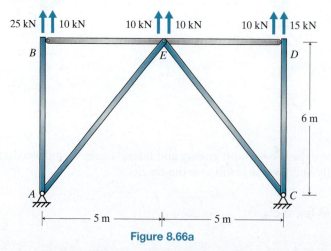

Figure 8.66a

Section 8.3 Braced Frames with Gravity Loads

FBD of joint B:

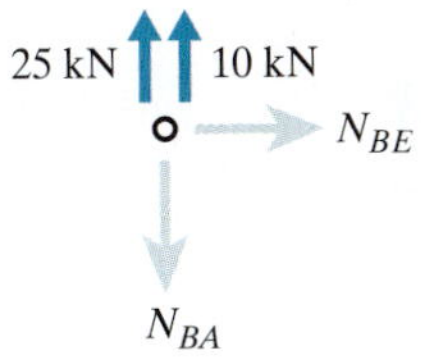

$\xrightarrow{+} \Sigma F_x = 0 = N_{BE}$

$+\uparrow \Sigma F_y = 0 = 25 \text{ kN} + 10 \text{ kN} - N_{BA}$

$N_{BA} = 35 \text{ kN}$

Figure 8.66b

FBD of joint D:

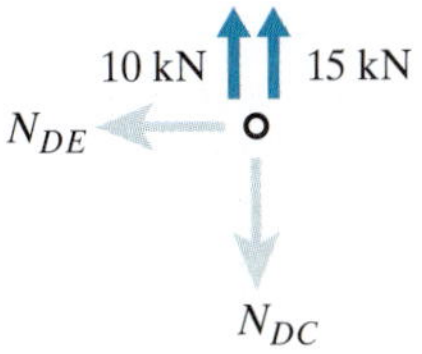

$\xrightarrow{+} \Sigma F_x = 0 = N_{DE}$

$+\uparrow \Sigma F_y = 0 = 10 \text{ kN} + 15 \text{ kN} - N_{DC}$

$N_{DC} = 25 \text{ kN}$

Figure 8.66c

FBD of joint E:

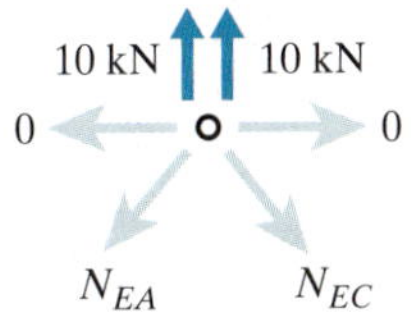

$\xrightarrow{+} \Sigma F_x = 0 = -\left(\dfrac{5}{7.81}\right)N_{EA} + \left(\dfrac{5}{7.81}\right)N_{EC}$

$N_{EA} = N_{EC}$

$+\uparrow \Sigma F_y = 0 = 10 \text{ kN} + 10 \text{ kN} - \left(\dfrac{6}{7.81}\right)N_{EA} - \left(\dfrac{6}{7.81}\right)N_{EC}$

Figure 8.66d

$0 = 10 \text{ kN} + 10 \text{ kN} - 2\left(\dfrac{6}{7.81}\right)N_{EA}$

$N_{EA} = 13.02 \text{ kN}$

$\therefore N_{EC} = 13.02 \text{ kN}$

FBD of joint A (so we can evaluate results):

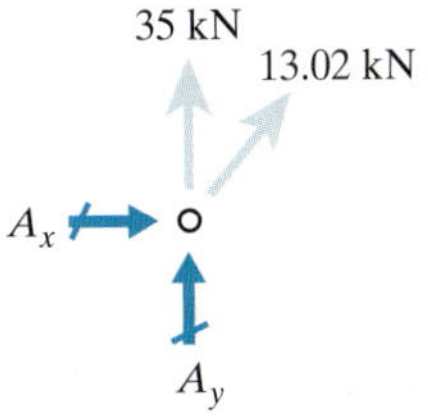

$+\uparrow \Sigma F_y = 0 = 35 \text{ kN} + \left(\dfrac{6}{7.81}\right)(13.02 \text{ kN}) + A_y$

$A_y = -45 \text{ kN} \ (+\uparrow)$

$\xrightarrow{+} \Sigma F_x = 0 = A_x + \left(\dfrac{5}{7.81}\right)(13.02 \text{ kN})$

$A_x = -8.34 \text{ kN} \ (\xrightarrow{+})$

Figure 8.66e

FBD of joint C (so we can evaluate results):

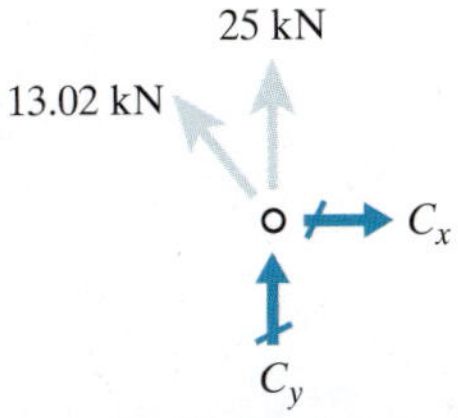

$+\uparrow \Sigma F_y = 0 = \left(\dfrac{6}{7.81}\right)(13.02 \text{ kN}) + 25 \text{ kN} + C_y$

$C_y = -35 \text{ kN} \ (+\uparrow)$

$\xrightarrow{+} \Sigma F_x = 0 = -\left(\dfrac{5}{7.81}\right)(13.02 \text{ kN}) + C_x$

$C_x = +8.34 \text{ kN} \ (\xrightarrow{+})$

Figure 8.66f

414 Chapter 8 Approximate Analysis of Indeterminate Trusses and Braced Frames

Summary:

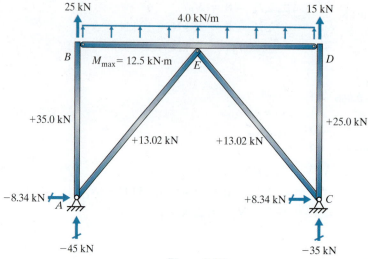

Figure 8.67

Evaluation of Results:

Satisfied Fundamental Principles?
Verify overall equilibrium:

$$\xrightarrow{\pm} \Sigma F_x = -8.34 \text{ kN} + 8.34 \text{ kN} = 0 \checkmark$$

$$+\uparrow \Sigma F_y = -45 \text{ kN} + 25 \text{ kN} + (4.0 \text{ kN/m})(10 \text{ m})$$
$$+15 \text{ kN} - 35 \text{ kN} = 0 \checkmark$$

$$+\circlearrowleft \Sigma M_C = -45 \text{ kN}(10 \text{ m}) + 25 \text{ kN}(10 \text{ m})$$
$$+ (4.0 \text{ kN/m})(20 \text{ m})(10 \text{ m}/2) = 0 \checkmark$$

Conclusion:
Based on this check, these approximate results appear reasonable.

Lateral Load:

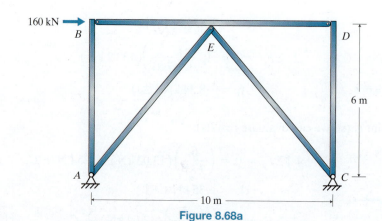

Figure 8.68a

Assumptions:
- No shear is generated in the columns.
- Shear is shared equally as horizontal components of brace forces.

Truss approximation:
FBD of top part of frame:

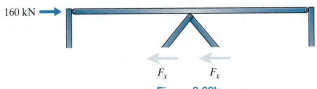

Figure 8.68b

$$\xrightarrow{+} \Sigma F_x = 0 = 160 \text{ kN} - 2F_x$$
$$F_x = 80 \text{ kN}$$
$$\therefore N_{EA} = +\left(\frac{7.81}{5}\right)(80 \text{ kN}) = +125.0 \text{ kN}$$

$$N_{EC} = -\left(\frac{7.81}{5}\right)(80 \text{ kN}) = -125.0 \text{ kN}$$

FBD of joint B:

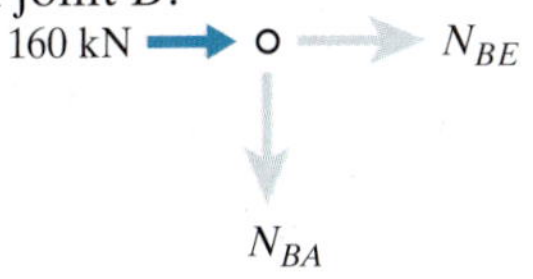

160 kN → ○ → N_{BE}
　　　　　↓
　　　　N_{BA}

$+\downarrow \Sigma F_y = 0 = N_{BA}$

Figure 8.68c

FBD of joint D:

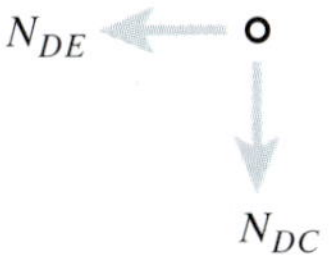

N_{DE} ← ○
　　　　　↓
　　　　N_{DC}

$+\downarrow \Sigma F_y = 0 = N_{DC}$

Figure 8.68d

FBD of joint A (so we can evaluate results):

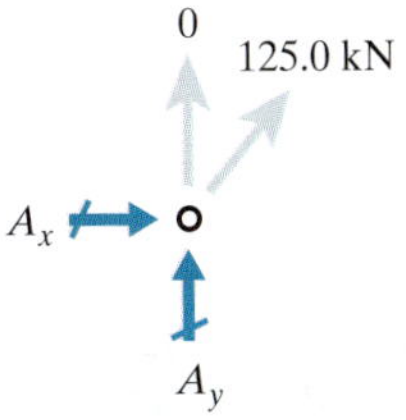

　　　　　0
　　　125.0 kN ↗
A_x ⇌ ○
　　　　↑
　　　A_y

Figure 8.68e

$+\uparrow \Sigma F_y = 0 = 0 + \left(\dfrac{6}{7.81}\right)(125.0 \text{ kN}) + A_y$
$A_y = -96.0 \text{ kN} \ (+\uparrow)$

$\xrightarrow{+} \Sigma F_x = 0 = A_x + \left(\dfrac{5}{7.81}\right)(125.0 \text{ kN})$
$A_x = -80.0 \text{ kN} \ (\xrightarrow{+})$

FBD of joint C:

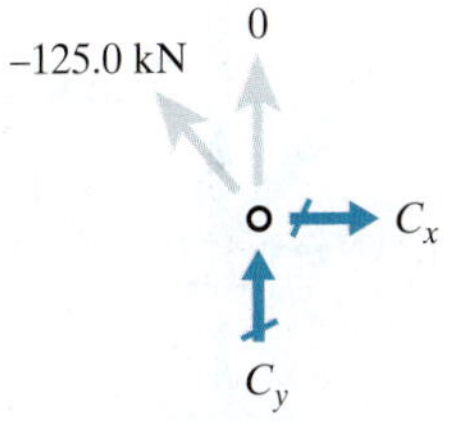

-125.0 kN ↖　0
　　　　　　↑
　　　　　　○ → C_x
　　　　　　↑
　　　　　C_y

Figure 8.68f

$+\uparrow \Sigma F_y = 0 = \left(\dfrac{6}{7.81}\right)(-125.0 \text{ kN}) + 0 + C_y$
$C_y = +96.0 \text{ kN} \ (+\uparrow)$

$\xrightarrow{+} \Sigma F_x = 0 = -\left(\dfrac{5}{7.81}\right)(-125.0 \text{ kN}) + C_x$
$C_x = -80.0 \text{ kN} \ (\xrightarrow{+})$

Summary:

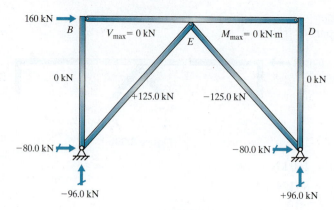

Figure 8.69

Evaluation of Results:

Satisfied Fundamental Principles?
Verify overall equilibrium:

$$\xrightarrow{\pm} \Sigma F_x = 160 \text{ kN} - 80 \text{ kN} - 80 \text{ kN} = 0 \checkmark$$
$$+\uparrow \Sigma F_y = -96.0 \text{ kN} + 96.0 \text{ kN} = 0 \checkmark$$
$$\underset{\curvearrowright}{+} \Sigma M_C = 160 \text{ kN}(6 \text{ m}) - 96.0 \text{ kN}(10 \text{ m}) = 0 \checkmark$$

Conclusion:
Based on this check, these approximate results appear reasonable.

Superimpose Results:

Add the internal forces and reactions:

$V_{max} = 10.0 \text{ kN} + 0 = 10.0 \text{ kN}$ $M_{max} = 12.5 \text{ kN} \cdot \text{m} + 0 = 12.5 \text{ kN} \cdot \text{m}$

$N_{AB} = +35 \text{ kN} + 0 = +35 \text{ kN} = 35 \text{ kN } (T)$
$N_{CD} = +25 \text{ kN} + 0 = +25 \text{ kN} = 25 \text{ kN } (T)$
$N_{AE} = +13.02 \text{ kN} + 125.0 \text{ kN} = +138.0 \text{ kN} = 138 \text{ kN } (T)$
$N_{CE} = +13.02 \text{ kN} - 125.0 \text{ kN} = -112.0 \text{ kN} = 112 \text{ kN } (C)$

$A_x = -8.34 \text{ kN} - 80.0 \text{ kN} = -88.3 \text{ kN } (\xrightarrow{\pm})$
$A_y = -45 \text{ kN} - 96.0 \text{ kN} = -141 \text{ kN } (+\uparrow)$
$C_x = +8.34 \text{ kN} - 80.0 \text{ kN} = -71.7 \text{ kN } (\xrightarrow{\pm})$
$C_y = -35 \text{ kN} + 96.0 \text{ kN} = +61 \text{ kN } (+\uparrow)$

Summary:

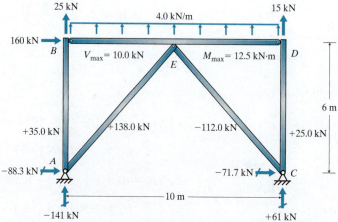

Figure 8.70

Homework Problems

8.1 Our firm is designing a rack for road maintenance equipment. To make a preliminary selection of member properties, the team needs to know the axial force generated in each member due to the various relevant loads. Our task is to calculate the axial forces due to the static equivalent seismic load.

a. Guess the directions of each of the four reactions.
b. List the assumptions you make in order to perform approximate analysis of the truss.
c. Use approximate analysis to find the magnitude and sense of the axial force in each member.
d. Determine the approximate reactions. Verify equilibrium of your results.
e. To demonstrate the effectiveness of approximate analysis, use structural analysis software to determine the reactions. Assume all members of the truss are made of Pipe 3 Std ($A = 2.07$ in^2, $I = 2.85$ in^4, $E = 29{,}000$ ksi).
f. Submit a printout of the displaced shape. Identify at least three features that suggest the deflected shape is reasonable.
g. Verify equilibrium of the computer aided analysis results.
h. Are the results of your computer aided analysis reasonable? Make a comprehensive argument to justify your answer.
i. Comment on why your guess in part (a) was or was not close to the solution in part (d).

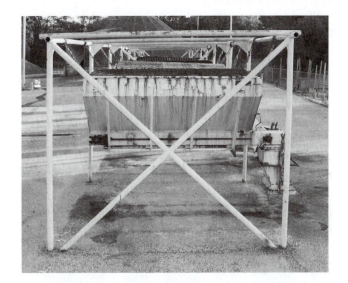

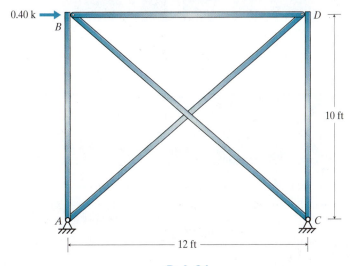

Prob. 8.1

8.2 Not only does a pedestrian bridge need to carry gravity loads, but it also must sustain wind load acting on the side of the bridge. The idealized unfactored wind loads on the truss at the bottom of this bridge have been calculated by a team member. For preliminary design, the team needs to know the peak axial forces due to the wind.

a. Guess which chord member will have the largest compression and which diagonal will have the largest compression.
b. Use qualitative truss analysis to predict the sense of the axial force (tension, compression, zero, or can't be determined) in each member.
c. List the assumptions you make in order to perform approximate analysis of the truss.
d. Use approximate analysis to find the magnitude and sense of the peak axial force in the top chord, in the bottom chord, and in the diagonals. Hint: If you construct the V and M diagrams as if the truss is a solid beam, the diagrams will guide you to where the axial forces will be maximum.
e. Use approximate analysis to find the magnitude and sense of the axial force in member JB.
f. To demonstrate the effectiveness of approximate analysis, use structural analysis software to determine the peak axial force in the top chord, in the bottom chord, and in the diagonals. Also find the axial force in member JB. Assume all members are round HSS sections ($E = 200$ GPa $= 200$ kN/mm^2). Assume the chords are HSS 152.4×6.4 ($A = 2.72 \times 10^3$ mm^2, $I = 7.33 \times 10^6$ mm^4) and all other members are HSS 48.3×3.7 ($A = 4.83 \times 10^2$ mm^2, $I = 1.22 \times 10^5$ mm^4).
g. Submit a printout of the displaced shape. Identify at least three features that suggest the deflected shape is reasonable.
h. Verify equilibrium of the computer aided analysis results.
i. Are the results of your computer aided analysis reasonable? Make a comprehensive argument to justify your answer.
j. Comment on why your guess in part (a) was or was not close to the solution in part (d).

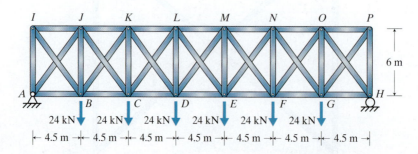

Prob. 8.2

8.3 Our firm is designing a pedestrian bridge to connect two office buildings. For aesthetic reasons, the architect has chosen a unique pattern for the diagonal members. Before designing the members, our team leader would like to know how the truss will behave with live load at different locations. We are tasked with evaluating the truss with a unit load at midspan.

a. Guess which diagonal member will have the largest tension.
b. List the assumptions you make in order to perform approximate analysis of the truss.
c. Use qualitative approximate truss analysis to predict the sense of the axial force (tension, compression, zero, or can't be determined) in each member. Note: Because the diagonals (not AL or TK) are thin members, they can support tension only; therefore, any diagonal that would be in compression under this load is actually a zero force member.
d. Use approximate analysis to find the magnitude of the axial force in each diagonal, recognizing that the diagonals can carry tension only.
e. Use approximate analysis to find the magnitude and sense of the axial force in member MC.
f. To demonstrate the effectiveness of approximate analysis, use structural analysis software to determine the tension in each diagonal. At this preliminary stage, make the following assumptions about the steel members of the truss ($E = 29{,}000$ ksi):

Top chord and AL and TK:	$A = 8 \text{ in}^2$
Bottom chord:	$A = 6 \text{ in}^2$
Verticals:	$A = 3 \text{ in}^2$
Diagonals:	$A = 2 \text{ in}^2$

g. Submit a printout of the displaced shape. Identify at least three features that suggest the deflected shape is reasonable.
h. Submit a printout of the axial force diagram for the truss. Identify at least three features that suggest the axial forces are reasonable.
i. Verify equilibrium of the computer aided analysis results.
j. Are the results of your computer aided analysis reasonable? Make a comprehensive argument to justify your answer.
k. Comment on why your guess in part (a) was or was not close to the solution in part (d).

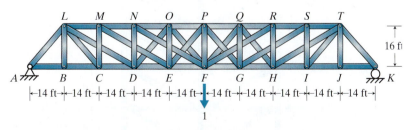

Prob. 8.3

8.4 Our team is designing a five-story building in a high seismic region. The architect would like to make the lateral load-resisting system visible on the outside, and our team leader has chosen to use two bays of diagonal braces. The geotechnical engineers need to know the approximate forces on the foundations due to static equivalent seismic loads right away. Therefore, our job is to perform approximate analysis of the frame.

a. Guess the directions of the six forces on the foundations.
b. List the assumptions you make in order to perform approximate analysis of the frame.
c. Use approximate analysis to find the magnitude and direction of each force acting on the foundations A, G, and M.
d. Verify equilibrium of your approximate analysis results.
e. To demonstrate the effectiveness of approximate analysis, use structural analysis software to determine the forces acting on the foundations. Initially assume all members are steel W12 × 65 sections ($E = 29{,}000$ ksi, $A = 19.1$ in^2, $I_y = 174$ in^4).
f. Submit a printout of the displaced shape. Identify at least three features that suggest the deflected shape is reasonable.
g. Verify equilibrium of the computer aided analysis results.
h. Are the results of your computer aided analysis reasonable? Make a comprehensive argument to justify your answer.
i. Comment on why your guess in part (a) was or was not close to the solution in part (c).

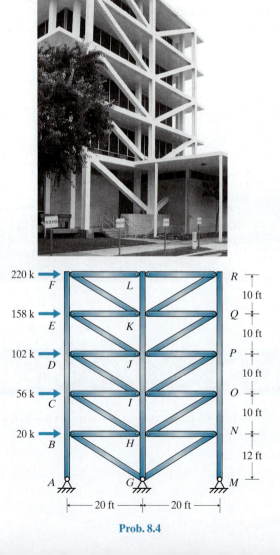

Prob. 8.4

8.5 and 8.6 Our team was selected to design a four-story apartment building. The team leader has chosen diagonally braced frames for the lateral load-resisting system. In order to make a preliminary selection of member properties, the team needs to know the axial forces in the braces and columns.

8.5 The design team divided up responsibility for the different loadings. Our task is to determine the unfactored forces due to wind load.

a. Guess the sense of the axial force (tension or compression) in each column and brace at the lowest level.
b. List the assumptions you make in order to perform approximate analysis of the frame.
c. Use approximate analysis to find the magnitude and sense of the axial force in each segment of the columns and in each of the braces.
d. To demonstrate the effectiveness of approximate analysis, use structural analysis software to determine the axial force in each segment of the columns and in each of the braces. For now, assume all members are the same steel section with the following properties: $E = 200$ GPa $= 200$ kN/mm^2, $A = 5 \times 10^3$ mm^2, $I = 3 \times 10^7$ mm^4.
e. Submit a printout of the displaced shape. Identify at least three features that suggest the deflected shape is reasonable.
f. Submit a printout of the axial force diagram for the frame. Identify at least three features that suggest the axial forces are reasonable.
g. Verify equilibrium of the computer aided analysis results.
h. Are the results of your computer aided analysis reasonable? Make a comprehensive argument to justify your answer.
i. Comment on why your guess in part (a) was or was not close to the solution in part (c).

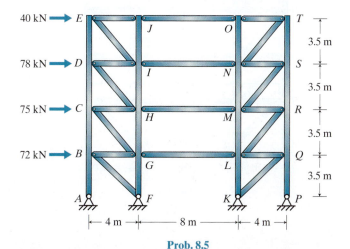

Prob. 8.5

8.6 The design team divided up responsibility for the different loadings. Our task is to determine the unfactored forces due to dead load.

a. Guess the sense of the axial force (tension or compression) in each column and brace at the lowest level.
b. List the assumptions you make in order to perform approximate analysis of the frame.
c. Use approximate analysis to find the magnitude and sense of the axial force in each segment of the columns and in each of the braces.
d. To demonstrate the effectiveness of approximate analysis, use structural analysis software to determine the axial force in each segment of the columns and in each of the braces. For now, assume all members are the same steel section with the following properties: $E = 200$ GPa $= 200$ kN/mm^2, $A = 5 \times 10^3$ mm^2, $I = 3 \times 10^7$ mm^4.
e. Submit a printout of the displaced shape. Identify at least three features that suggest the deflected shape is reasonable.
f. Submit a printout of the axial force diagram for the frame. Identify at least three features that suggest the axial forces are reasonable.
g. Verify equilibrium of the computer aided analysis results.
h. Are the results of your computer aided analysis reasonable? Make a comprehensive argument to justify your answer.
i. Comment on why your guess in part (a) was or was not close to the solution in part (c).

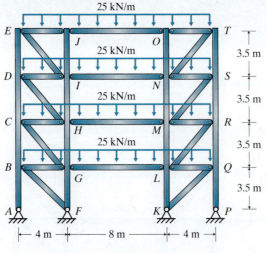

Prob. 8.6

8.7 Our firm has been selected to design the structural framing for new stadium bleachers. One of the design loads is live load that acts parallel to the seats. The result is a lateral load on the cross braced frame. In order to size the members, we need to know the axial force generated in each member due to the various unfactored loads. We have been tasked to analyze the frame subjected to the lateral live load.

a. Guess which member will have the largest axial compression.
b. List the assumptions you make in order to perform approximate analysis of the frame.
c. Use approximate analysis to find the axial force in each column and each brace (magnitude and sense).
d. To demonstrate the effectiveness of approximate analysis, use structural analysis software to determine the axial force in each column and brace. At this preliminary stage, make the following assumptions about the steel members of the frame ($E = 29{,}000$ ksi):

Beam:	$A = 6 \text{ in}^2$	$I = 150 \text{ in}^4$
Columns:	$A = 4 \text{ in}^2$	$I = 15 \text{ in}^4$
Braces:	$A = 2 \text{ in}^2$	$I = 3 \text{ in}^4$

e. Submit a printout of the displaced shape. Identify at least three features that suggest the deflected shape is reasonable.
f. Verify equilibrium of the computer aided analysis results.
g. Are the results of your computer aided analysis reasonable? Make a comprehensive argument to justify your answer.
h. Comment on why your guess in part (a) was or was not close to the solution in part (c).

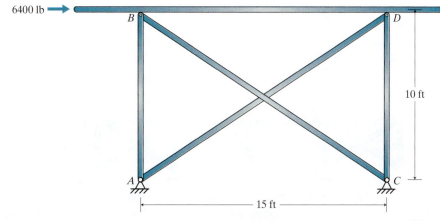

Prob. 8.7

8.8 Our team is designing a new sports arena in Europe. The team leader has chosen cross braced frames to provide lateral load resistance. In order to make important decisions about the type of foundations, the geotechnical engineers need to know the approximate forces on the foundations due to wind load. We have been tasked to perform approximate analysis of the frame to determine those values.

a. Guess the directions of the 10 forces on the foundations.
b. List the assumptions you make in order to perform approximate analysis of the frame.
c. Use approximate analysis to find the magnitude and direction of each force acting on the foundations: $A, E, I, M,$ and Q.
d. Verify equilibrium of your approximate analysis results.
e. To demonstrate the effectiveness of approximate analysis, use structural analysis software to determine the forces acting on the foundations. Initially assume the following sections for the steel members ($E = 200$ GPa $= 200$ kN/mm^2):

Columns: HEA 360
$$A = 142.8 \text{ cm}^2 \quad I_y = 7887 \text{ cm}^4$$

Girts (horizontal): HSS $250 \times 150 \times 10$
$$A = 74.9 \text{ cm}^2 \quad I_y = 2755 \text{ cm}^4$$

Braces: Round Pipe 139.7×8
$$A = 34.2 \text{ cm}^2 \quad I = 720 \text{ cm}^4$$

f. Submit a printout of the displaced shape. Identify at least three features that suggest the deflected shape is reasonable.
g. Verify equilibrium of the computer aided analysis results.
h. Are the results of your computer aided analysis reasonable? Make a comprehensive argument to justify your answer.
i. Comment on why your guess in part (a) was or was not close to the solution in part (c).

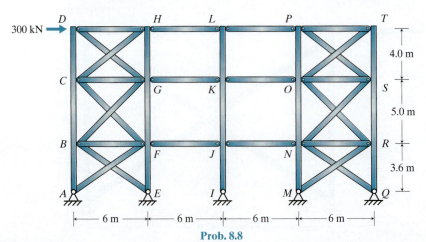

Prob. 8.8

8.9 Our firm has been selected to design the structural framing to support the press box at a stadium. In order to size the members, we need to know the axial force generated in each member due to the various unfactored loads. We have been tasked to analyze the frame subjected to wind load, which causes uplift and lateral force.

a. Guess the sense of the axial force in each brace and column segment.
b. List the assumptions you make in order to perform approximate analysis of the frame.
c. Use approximate analysis to find the axial force in each brace and each column segment (magnitude and sense). Hint: You need to use superposition because there are both lateral and gravity loads.
d. To demonstrate the effectiveness of approximate analysis, use structural analysis software to determine the axial force in each column segment and brace. At this preliminary stage, make the following assumptions about the steel members of the frame ($E = 29,000$ ksi):

Beams:	$A = 6$ in^2	$I = 150$ in^4
Columns:	$A = 4$ in^2	$I = 15$ in^4
Braces:	$A = 2$ in^2	$I = 3$ in^4

e. Submit a printout of the displaced shape. Identify at least three features that suggest the deflected shape is reasonable.
f. Submit a printout of the axial force diagram for the frame. Identify at least three features that suggest the axial forces are reasonable.
g. Verify equilibrium of the computer aided analysis results.
h. Are the results of your computer aided analysis reasonable? Make a comprehensive argument to justify your answer.
i. Comment on why your guess in part (a) was or was not close to the solution in part (c).

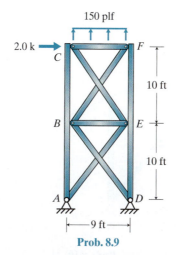

Prob. 8.9

8.10 Our firm has been selected to design new bleachers for sporting events. An announcer booth sits at the top of the structure. Wind on the booth is the largest lateral load on the system. In order for the geotechnical engineers to make preliminary designs for the foundations, they need to know the approximate forces on the foundations due to the various unfactored loads. We have been tasked with finding those forces due to wind on the front of the booth.

a. Guess the directions of the six forces on the foundations.
b. List the assumptions you make in order to perform approximate analysis of the frame.
c. Use approximate analysis to find the magnitude and direction of each force acting on the foundations: A, B, and C.
d. Verify equilibrium of your approximate analysis results.
e. To demonstrate the effectiveness of approximate analysis, use structural analysis software to determine the forces acting on the foundations. Initially assume the following sections for the steel members ($E = 200$ GPa $= 200$ kN/mm^2):

Columns: $A = 2 \times 10^3$ mm^2 $I = 4 \times 10^6$ mm^4
Braces: $A = 6 \times 10^2$ mm^2 $I = 6 \times 10^4$ mm^4
Girder: $A = 2 \times 10^3$ mm^2 $I = 2 \times 10^7$ mm^4

f. Submit a printout of the displaced shape. Identify at least three features that suggest the deflected shape is reasonable.
g. Verify equilibrium of the computer aided analysis results.
h. Are the results of your computer aided analysis reasonable? Make a comprehensive argument to justify your answer.
i. Comment on why your guess in part (a) was or was not close to the solution in part (c).

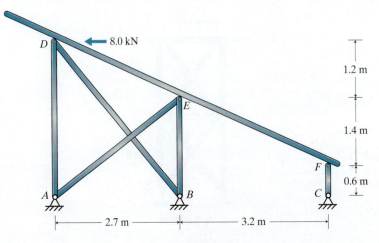

Prob. 8.10

8.11 The senior engineer on our team has chosen chevron braced frames as the lateral load-resisting system for a single-story structure we are designing. The bracing can be configured as a chevron or inverted chevron. We have been tasked with analyzing both configurations to determine the effects of changing the orientation of the braces.

a. Guess which system will have the largest vertical foundation reaction and for which support.
b. List the assumptions you make in order to perform approximate analysis of the frames.
c. Use approximate analysis to find the magnitude and sense of the axial force in the columns and braces for each of the configurations.
d. Determine the approximate reactions for each of the configurations. Verify equilibrium of your results for each configuration.
e. To demonstrate the effectiveness of approximate analysis, use structural analysis software to determine the axial forces in the columns and braces as well as the reactions for each configuration. Assume all members of the frames are made of the same material and section.
f. Submit a printout of the displaced shape for each configuration. On each, identify at least three features that suggest the deflected shape is reasonable.
g. Verify equilibrium of the computer aided analysis results for each configuration.
h. Are the results of your computer aided analysis reasonable? Make a comprehensive argument to justify your answer.
i. Comment on why your guess in part (a) was or was not close to the solution in part (d).

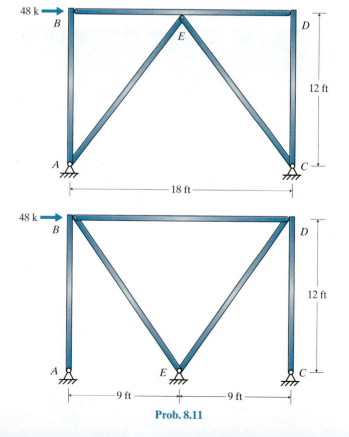

Prob. 8.11

428 **Chapter 8** Approximate Analysis of Indeterminate Trusses and Braced Frames

8.12 Our team is designing a new sports arena. The team leader has chosen chevron braced frames to provide lateral load resistance. In order to make important decisions about the type of foundations, the geotechnical engineers need to know the approximate forces on the foundations due to each of the unfactored loads. We have been tasked to perform approximate analysis of the frame in order to provide those values due to wind load.

a. Guess the directions of the 18 forces on the foundations.
b. List the assumptions you make in order to perform approximate analysis of the frame.
c. Use approximate analysis to find the magnitude and direction of each force acting on the foundations: A through I.
d. Verify equilibrium of your approximate analysis results.
e. To demonstrate the effectiveness of approximate analysis, use structural analysis software to determine the forces acting on the foundations. Based on previous work, an experienced member of the team guesses that the steel frame ($E = 29,000$ ksi) will have members close to the following when fully designed:

Columns: W14 × 90 $A = 26.5$ in^2 $I_y = 362$ in^4
Beams: W10 × 33 $A = 9.71$ in^2 $I_x = 171$ in^4
Braces: HSS 8 × 8 × ¼ $A = 7.10$ in^2 $I = 70.7$ in^4

f. Submit a printout of the displaced shape. Identify at least three features that suggest the deflected shape is reasonable.
g. Verify equilibrium of the computer aided analysis results.
h. Are the results of your computer aided analysis reasonable? Make a comprehensive argument to justify your answer.
i. Comment on why your guess in part (a) was or was not close to the solution in part (c).

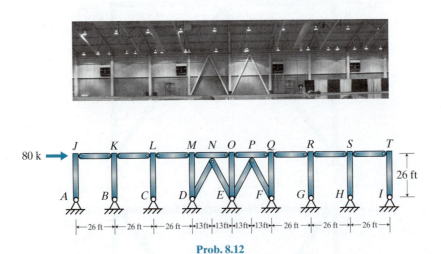

Prob. 8.12

8.13 and 8.14 Our firm is in the preliminary stages of designing a new high-rise building. The architectural vision calls for only three columns across the narrow side of the building at the bottom level but five columns at higher levels. Therefore, we need to transfer load from two of the columns to the three continuing columns. The senior member of the team in coordination with the architect has chosen chevron braces to provide the transfer. To size the diagonal members, the team needs to know the axial forces generated due to the various loads.

8.13 The chevron braces will not only transfer gravity loads but also provide lateral load resistance. We have been tasked with estimating the axial forces generated in the braces due to wind load.

a. Guess which brace will have the largest axial compression.
b. List the assumptions you make in order to perform approximate analysis of this segment of the frame.
c. Use approximate analysis to find the axial force in each brace (magnitude and sense).
d. Comment on why your guess in part (a) was or was not close to the solution in part (c).

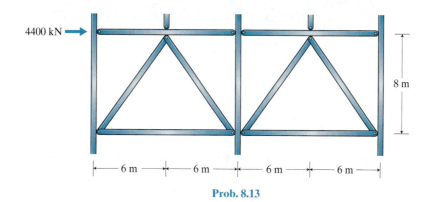

Prob. 8.13

8.14 We have been tasked with estimating the axial forces generated in the braces to transfer the unfactored live load.

a. Guess which brace will have the largest axial compression.
b. List the assumptions you make in order to perform approximate analysis of this segment of the frame.
c. Use approximate analysis to find the axial force in each brace (magnitude and sense).
d. Comment on why your guess in part (a) was or was not close to the solution in part (c).

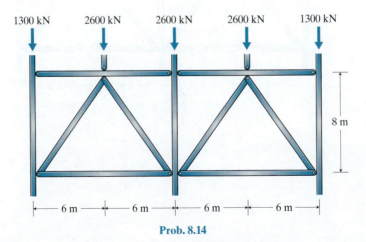

Prob. 8.14

8.15 and 8.16 Our team is designing a four-story office building. The principal in charge has experience with structures like this and has chosen to use chevron braces. Seismic requirements make it most economical to alternate chevrons and inverted chevrons, so the lateral load-resisting system looks similar to cross braces.

While we are working on the structural design, the geotechnical engineers are working on the foundation design. In order for them to begin, they need to know preliminary values for the unfactored forces on the foundations.

8.15 We have been tasked with finding approximate forces on the foundations due to unfactored wind load.

a. Guess the directions of the eight forces on the foundations.
b. List the assumptions you make in order to perform approximate analysis of the frame.
c. Use approximate analysis to find the magnitude and direction of each force acting on the foundations: $A, F, M,$ and R.
d. Verify equilibrium of your approximate analysis results.

e. To demonstrate the effectiveness of approximate analysis, use structural analysis software to determine the forces acting on the foundations. At this preliminary stage, an experienced member of the team guesses that the steel members (E = 29,000 ksi) of the frame will be close to the following:

Columns: W12 × 120 A = 35.2 in^2 I_x = 1070 in^4
Beams: W24 × 62 A = 18.2 in^2 I_x = 1550 in^4
Braces: HSS 8 × 8 × 3/8 A = 10.4 in^2 I = 100 in^4

f. Submit a printout of the displaced shape. Identify at least three features that suggest the deflected shape is reasonable.
g. Verify equilibrium of the computer aided analysis results.
h. Are the results of your computer aided analysis reasonable? Make a comprehensive argument to justify your answer.
i. Comment on why your guess in part (a) was or was not close to the solution in part (c).

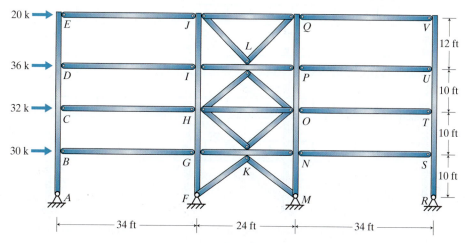

Prob. 8.15

8.16 We have been tasked with finding the approximate forces on the foundations due to unfactored dead load, ignoring the self-weight of the members.

a. Guess the directions of the eight forces on the foundations.
b. List the assumptions you make in order to perform approximate analysis of the frame.
c. Use approximate analysis to find the magnitude and direction of each force acting on the foundations: $A, F, M,$ and R.
d. Verify equilibrium of your approximate analysis results.
e. To demonstrate the effectiveness of approximate analysis, use structural analysis software to determine the forces acting on the foundations. At this preliminary stage, an experienced member of the team guesses that the steel members ($E = 29{,}000$ ksi) of the frame will be close to the following:

Columns:	W12 × 120	$A = 35.2 \text{ in}^2$	$I_x = 1070 \text{ in}^4$
Beams:	W24 × 62	$A = 18.2 \text{ in}^2$	$I_x = 1550 \text{ in}^4$
Braces:	HSS 8 × 8 × 3/8	$A = 10.4 \text{ in}^2$	$I = 100 \text{ in}^4$

f. Submit a printout of the displaced shape. Identify at least three features that suggest the deflected shape is reasonable.
g. Verify equilibrium of the computer aided analysis results.
h. Are the results of your computer aided analysis reasonable? Make a comprehensive argument to justify your answer.
i. Comment on why your guess in part (a) was or was not close to the solution in part (c).

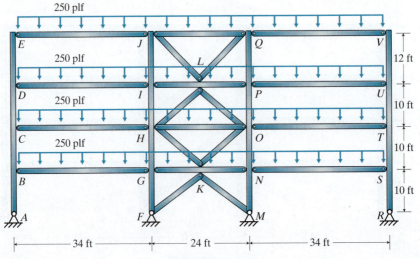

Prob. 8.16

8.17 Our firm has been selected to design the structural framing to support grandstands at a race track. In order to size the members, we need to know the axial force generated in each member due to the various unfactored loads. We have been tasked to analyze the frame subjected to wind load.

a. Guess the sense of the axial force in each brace and column segment.
b. List the assumptions you make in order to perform approximate analysis of the frame.
c. Use approximate analysis to find the axial force in each brace and each column segment (magnitude and sense).
d. To demonstrate the effectiveness of approximate analysis, use structural analysis software to determine the axial force in each column segment and brace. At this preliminary stage, an experienced team member suggests making the following assumptions about the steel members of the frame ($E = 29{,}000$ ksi):

Columns: $A = 50$ in^2 $I = 1000$ in^4
Beams: $A = 10$ in^2 $I = 100$ in^4
Braces: $A = 10$ in^2 $I = 20$ in^4

e. Submit a printout of the displaced shape. Identify at least three features that suggest the deflected shape is reasonable.
f. Submit a printout of the axial force diagram for the frame. Identify at least three features that suggest the axial forces are reasonable.
g. Verify equilibrium of the computer aided analysis results.
h. Are the results of your computer aided analysis reasonable? Make a comprehensive argument to justify your answer.
i. Comment on why your guess in part (a) was or was not close to the solution in part (c).

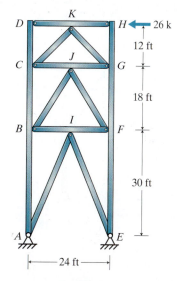

Prob. 8.17

8.18 Our team is designing a new airport terminal. The architect and structural team leader have decided to use a combination of chevron and inverted chevron braces to provide lateral load resistance while permitting the desired access. To make important decisions about the type of foundations, the geotechnical engineers need to know the approximate forces on the foundations due to each of the unfactored loads. We have been tasked to perform approximate analysis of the frame in order to provide those values due to static equivalent seismic load.

a. Guess the directions of the 14 forces on the foundations.
b. List the assumptions you make in order to perform approximate analysis of the frame.
c. Use approximate analysis to find the magnitude and direction of each force acting on the foundations: $A, D, E, I, M, P,$ and Q.
d. Verify equilibrium of your approximate analysis results.
e. To demonstrate the effectiveness of approximate analysis, use structural analysis software to determine the forces acting on the foundations. Based on previous work, an experienced member of the team guesses that the steel frame ($E = 200$ GPa $= 200$ kN/mm^2) will have members close to the following when fully designed:

Columns: HSS 457.2 × 12.7
$A = 1.65 \times 10^4$ mm^2 $I = 4.10 \times 10^8$ mm^4

Braces: HSS 323.9 × 12.7
$A = 1.15 \times 10^4$ mm^2 $I = 1.41 \times 10^8$ mm^4

Beams: W250 × 101
$A = 1.28 \times 10^4$ mm^2 $I = 1.64 \times 10^8$ mm^4

f. Submit a printout of the displaced shape. Identify at least three features that suggest the deflected shape is reasonable.
g. Verify equilibrium of the computer aided analysis results.
h. Are the results of your computer aided analysis reasonable? Make a comprehensive argument to justify your answer.
i. Comment on why your guess in part (a) was or was not close to the solution in part (c).

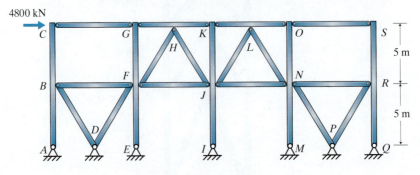

Prob. 8.18

CHAPTER 9

APPROXIMATE ANALYSIS OF RIGID FRAMES

Rigid frames provide several benefits including redundancy, but that also means they are indeterminate.

MOTIVATION

Rigid frames are a type of lateral load-resisting system. They are typically used instead of other systems when open space is important or required. One of the benefits of rigid frames is that they are highly indeterminate. This means that the structures have redundancy; if some part becomes overloaded, the load can redistribute without collapsing the structure.

A consequence of this redundancy is that there are more unknowns than equations of equilibrium. Therefore, we cannot analyze a rigid frame using only equilibrium. We need to either have additional equations (e.g., Chapter 12: Force Method) or change our approach (e.g., Chapter 13: Moment Distribution Method) in order to perform detailed analysis of the frame. Or we can make enough informed assumptions to reduce the number of unknowns down to the number of equations; we call this *approximate analysis*.

We use approximate analysis of rigid frames instead of detailed analysis for a variety of reasons:

1. Approximate analysis can quickly give us a sense of how different rigid frames will respond to the types of loads we expect. For example, we can compare two bays with three bays for the same overall structure width. With this information, we can make an informed choice of the frame layout.

2. Detailed analysis requires that we know member properties such as the cross-sectional area and moment of inertia, but in design we typically don't know those properties initially. Approximate analysis allows us to obtain the internal forces we can use to make the initial selection of member properties.

3. Detailed analysis can be a complex process prone to errors. Therefore, it is a good idea to validate our detailed analysis results. Approximate analysis is an efficient tool to help evaluate the results of our detailed analysis.

The key to quality approximate analysis results is quality assumptions. The closer the assumptions are to reality, the better the approximation. Therefore, we have developed different sets of assumptions for different situations. We call each set of assumptions an approximate analysis method.

Chapter 9 Approximate Analysis of Rigid Frames

9.1 Gravity Load Method

Introduction

Every rigid frame experiences gravity load, even if just its own self-weight. Therefore, we need a set of assumptions that help us analyze rigid frames subjected to the distribution of a typical gravity load. The key assumption we make is the locations of the inflection points, locations where the moment in the beam is zero.

How-To

Approximate analysis uses assumptions to close the gap between the number of unknowns in a structural system and the number of equations of equilibrium. Before we identify the assumptions for approximate analysis of rigid frames, let's identify how many assumptions we need. Consider the simple portal frame in Figure 9.1, which has six unknown reactions and three equations of equilibrium. The frame is three degrees indeterminate, so we need three assumptions. If the frame is two stories, we need to cut the frame in half in order to "open" the closed loop that the frame makes in the top story (Figure 9.2). Now there are six reactions and six internal forces at the cuts, for a total of 12 unknowns. There are two free body diagrams with three equations of equilibrium each, so we have six equations. Therefore, we need six assumptions for the two-story frame. Fortunately, we can make the same assumptions multiple times.

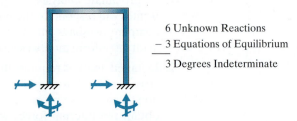

Figure 9.1 Degree of indeterminacy for a single-story portal frame with fixed supports.

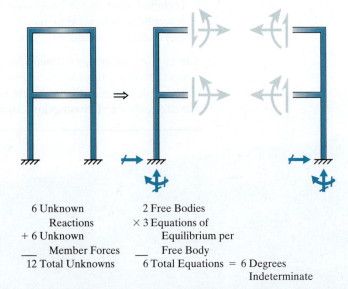

Figure 9.2 Degree of indeterminacy for a two-story portal frame with fixed supports.

First Assumption

To understand where our first assumption comes from, we need to study the displaced shape of a simple portal frame that has a uniform gravity load along the beam (Figure 9.3). Near the columns the deformed beam is concave down, or frowning. We learned in Chapter 4: Internal Force Diagrams that negative moment in the beam causes the beam to frown. Notice also that the middle section of the beam is concave up, or smiling. There the moment must be positive.

In Section 4.2: Constructing Diagrams by Deduction, we observed that a uniform distributed load results in a parabolic moment diagram, and we observed that a downward distributed load results in a concave down moment diagram. When we combine these observations with the knowledge that the moment diagram for this beam is negative at the ends and positive in the middle, we get the moment diagram shown in Figure 9.4. The importance of the moment diagram lies in the locations where the moment is zero. We call these locations *inflection points* because they are where the concavity of the deformed beam changes.

In general, when we cut a frame, we get three more equations of equilibrium, but there are three unknown internal forces at the cut. However, if we cut the frame at an inflection point, we still get three more equations of equilibrium but only two unknown internal forces. Because the moment is zero there, we make progress!

The challenge is knowing where the inflection points are. Since we don't know exactly where they are before we start the analysis, their locations will be our first assumptions. To help us narrow down the possible locations, let's consider two beams that have very different support conditions: pin-pin and fix-fix. For a pin-pin supported beam, the moment is zero at the supports (Figure 9.5); therefore, the inflection points are at the ends. For a fix-fix supported beam, we need tools from later chapters in order to perform detailed analysis to find the inflection points. Fortunately, that has already been done for typical loadings and can be found inside the back cover of this text. We see that for a uniform distributed load, the inflection points are 21% of the span length in from the supports (Figure 9.6).

The ends of a beam in a rigid frame do not rotate freely (i.e., pinned) nor do they remain horizontal (i.e., fixed). They behave somewhere in between. Therefore, the inflection points in a uniformly loaded beam in a rigid frame are somewhere between 0% and 21% of the span from the ends. The assumption we commonly make for a uniform load is right between the two—at 10% of the span.

The actual locations of the inflection points depend on the relative stiffness of the beam compared with the columns. If the beam is very stiff

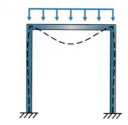

Figure 9.3 Displaced shape of a single-story portal frame subjected to a uniform gravity load along the beam.

Figure 9.4 Moment diagram for a uniformly loaded beam from a portal frame.

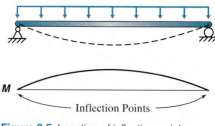

Figure 9.5 Location of inflection points for a uniformly loaded, pin-pin supported beam.

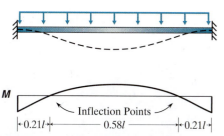

Figure 9.6 Location of inflection points for a uniformly loaded, fix-fix supported beam.

440 **Chapter 9** Approximate Analysis of Rigid Frames

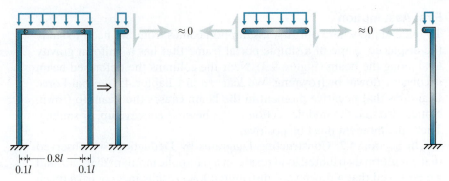

Figure 9.7 Free body diagrams of simple portal frame cut at the assumed inflection points.

compared with the columns, the beam behaves more like it is pin-pin supported and the inflection points are closer to the ends. If the columns are very stiff compared with the beam, the beam behaves more like it is fix-fix supported and the inflection points are closer to 21% in from the ends. Since the typical purpose of approximate analysis is to obtain an estimate of the internal forces or reactions, we usually use the middle value of 10% for uniformly loaded beams.

Second Assumption

Since there are two inflection points in each beam, these count as two assumptions per beam. For our frames with fixed supports in Figures 9.1 and 9.2, we still need another assumption. To visualize where our second assumption comes from, consider free body diagrams of the three pieces of the simple portal frame cut at the inflection points (Figure 9.7). Horizontal equilibrium of the middle beam piece dictates that the axial force in the beam is constant. Looking at the left column piece, we see that the majority of the load is downward: applied load and shear at the inflection point. Detailed analysis confirms that the axial force in the beam tends to be relatively small. Therefore, our second assumption is that the axial force in the beam is zero.

Pin Supports

If our simple portal frame had pin supports, we would need only one assumption (Figure 9.8). We assume the location of one inflection point and make *no* assumption about the axial force in the beam. There is still a second inflection point, but we don't assume its location. This one assumption leads us to six equations and six unknowns (Figure 9.9). The problem we encounter is that we must solve all six simultaneously; we have to use matrix methods.

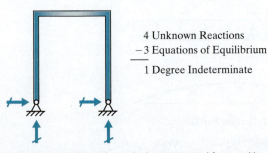

Figure 9.8 Degree of indeterminacy for a single-story portal frame with pinned supports.

Section 9.1 Gravity Load Method   441

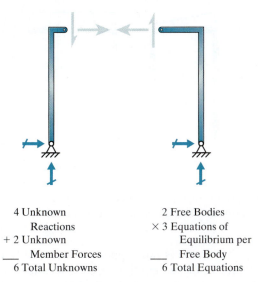

Figure 9.9 A single-story portal frame with pinned supports needs only one assumption in order to have the same number equations and unknowns.

If the loading and structure are symmetric, like our example portal frame, the second inflection point will be at the same distance from the column as the first (Figure 9.10). Therefore, we know the location of both inflection points while making only one assumption. With this approach we avoid needing to solve the system of equations simultaneously.

However, if the loading (Figure 9.11a) or the structure (Figure 9.11b) are not symmetric, we don't know the location of the second inflection point. It won't be at the same distance from the column as the first inflection point. In those cases, we can either solve the system of equations simultaneously, or we can make an additional assumption: the location of the second inflection point. If we make that additional assumption, we need to recognize that we have created an overdetermined system of equations. Mathematically, an overdetermined system means there is no unique solution. Structurally, the result is that equilibrium is not satisfied for every free body. If equilibrium is not satisfied, that is pulling us further away from the reality we are trying to approximate. Fortunately, for the typical purposes we have for approximate analysis, the results are still often useful even with the extra assumption.

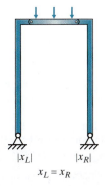

Figure 9.10 For a single-story portal frame with pinned supports, if the loading and structure are symmetric the second inflection point is at the same distance from the column as the first.

442 Chapter 9 Approximate Analysis of Rigid Frames

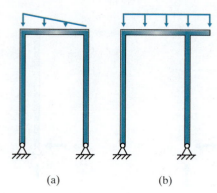

Figure 9.11 Single-story portal frames where the inflection points are not symmetrically located: (a) Asymmetric loading; (b) Asymmetric structure.

If our two-story frame is pin supported, we need four assumptions (Figure 9.12). Three of the assumptions are the locations of the inflection points: two for the top beam and one for the bottom beam (Figure 9.13a). If we assume the location of the fourth inflection point, we will not have a solvable system of equations. Therefore, the most reasonable fourth assumption is that the axial force in each beam is the same but not zero (Figure 9.13b).

The result is a system of nine equations and nine unknowns. Again, we must solve all nine simultaneously. If the loading and structure are symmetric, the last inflection point will be symmetrically located. With that knowledge, we won't need to solve the system of equations simultaneously. If the loading or the structure are not symmetric, we can assume the location of the final inflection point and still typically have useful results. We just need to recognize that equilibrium will not be satisfied.

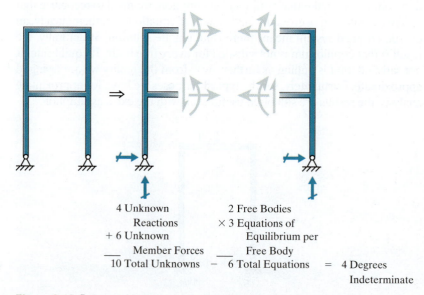

Figure 9.12 Degree of indeterminacy for a two-story portal frame with pinned supports.

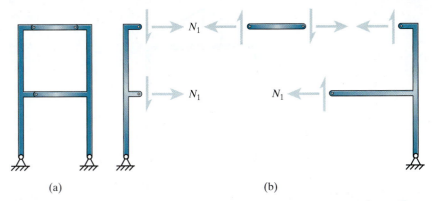

Figure 9.13 Assumptions made for a pin-supported, multistory, single-bay rigid frame with gravity loads: (a) Inflection points in beams; (b) Same beam axial force at each level.

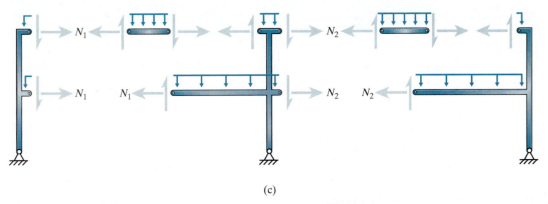

Figure 9.14 Assumptions made for a pin-supported, multistory, multibay rigid frame with gravity loads are made for each bay independently.

Remember that for a multibay frame, we remake the assumptions in each bay. Therefore, the axial forces in beams at different levels are the same in a bay, but they are not necessarily the same as the beams in the next bay (Figure 9.14).

Nonuniform Load

The most common gravity loading on rigid frames is uniform. However, other loadings also occur, such as the triangular loading associated with the two-way action of a floor (Section 1.4: Application of Gravity Loads). Different loadings result in different locations for the two inflection points in the beam. Fortunately, the process we used to bound the location of inflection points under a uniform load applies for all other loadings. Inside the back cover we have listed the inflection point locations for fix-fix beams with a variety of loadings. The actual inflection point locations are somewhere between the ends and those locations. A good initial guess is roughly halfway between the two (Figure 9.15).

444 Chapter 9 Approximate Analysis of Rigid Frames

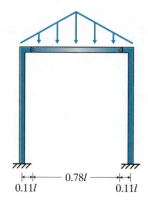

Figure 9.15 Assumed inflection point locations for a simple portal frame with triangular loading.

Section 9.1 Highlights

Gravity Load Method

Applicability:

- Rigid frame structures subjected to gravity loads.
- Beams rigidly connected to the columns; no pin connections.

Process:

1. Estimate the locations of inflection points in the beams.
    - If *uniformly* loaded, $0.1l$ from the columns.
    - For *other* loads, typically half the distance to the inflection points in a fix-fix beam (see the inside back cover).
    - If pinned at base, assume only one inflection point location for bottom-story beam only. If symmetric structure and loading, second inflection point will be symmetrically located. If asymmetric structure or loading, can solve system of equations simultaneously or can assume location of second inflection point.
    - Note: for cantilevered beams, the inflection point is at the free end; there is no assumption for these beams.
2. Draw a FBD of the beam between the inflection points.
    - Apply moment equilibrium to find the shear at one of the ends.
    - Apply vertical equilibrium to find the shear at the other end.
    - Repeat for all beams as needed.
    - Construct internal force diagrams for beams if desired.
3. Draw a FBD of the column and beams out to the inflection points.
    - Label the shears calculated at the inflection points.
    - If *fixed* at base, assume the axial force in each beam is zero.
    - If *pinned* at base, assume the axial force is the same for all beams in a bay.
    - Apply equilibrium to find the unknowns.
    - Repeat for all columns as needed.
    - Construct internal force diagrams for columns if desired.

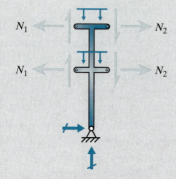

Example 9.1

Our team is designing a high-rise building. The architect and principal-in-charge have determined that the primary structural system will be a rigid frame. A team member has begun to model the system and would like our help verifying the results. Our job is to use approximate analysis to construct the moment diagram for a typical beam in the frame under live load.

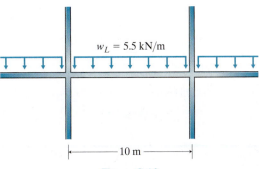

Figure 9.16

446   **Chapter 9**   Approximate Analysis of Rigid Frames

Approximate Analysis:

Estimate the inflection points at 10% of the girder length from the columns.

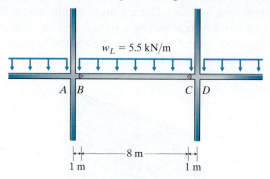

Figure 9.17

Cut at the inflection points:

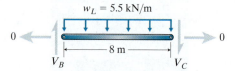

Figure 9.18a

$$\stackrel{+}{\curvearrowright}\Sigma M_C = 0 = V_B(8\text{ m}) - (5.5\text{ kN/m})(8\text{ m})(4\text{ m})$$
$$V_B = 22\text{ kN } (\downarrow + \uparrow)$$
$$M_{\text{max}+} = \frac{wl^2}{8} = \frac{(5.5\text{ kN/m})(8\text{ m})^2}{8} = +44\text{ kN} \cdot \text{m } (\smile + \frown)$$

FBD of piece AC:

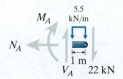

Figure 9.18b

$$\stackrel{+}{\curvearrowright}\Sigma M_A = 0 = M_A + (5.5\text{ kN/m})(1\text{ m})(0.5\text{ m}) + 22\text{ kN}(1\text{ m})$$
$$M_{\text{max}-} = M_A = -24.75\text{ kN} \cdot \text{m } (\smile + \frown)$$

Moment diagram for the girder:

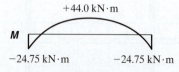

Figure 9.19

Example 9.2

Our firm is designing a new condominium building. The senior designer has decided to use a reinforced concrete rigid frame. For architectural reasons, the two bays have unequal widths.

Figure 9.20

In order for the geotechnical engineers to begin designing the foundations, they need to know the preliminary reactions. We have been tasked with estimating the foundation reactions due to the dead load. For now, we can ignore the self-weight of the columns.

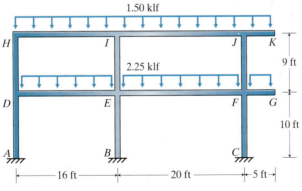

Figure 9.21

Approximate Analysis:

Estimate the inflection points at 10% of the beam spans from the columns. Assume that the axial force is zero in each of the beams.

448 Chapter 9 Approximate Analysis of Rigid Frames

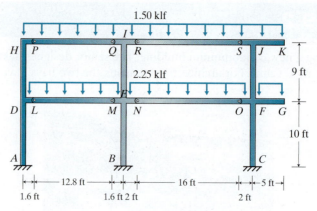

Figure 9.22a

FBD of beam PQ:

Figure 9.22b

$\curvearrowleft \Sigma M_Q = 0 = V_P(12.8 \text{ ft}) - 1.50 \text{ klf}(12.8 \text{ ft})(6.4 \text{ ft})$

$V_P = 9.6 \text{ k} (\downarrow + \uparrow)$

$+\uparrow \Sigma F_y = 0 = 9.6 \text{ k} - 1.50 \text{ klf}(12.8 \text{ ft}) - V_Q$

$V_Q = -9.6 \text{ k} (\downarrow + \uparrow)$

FBD of beam LM:

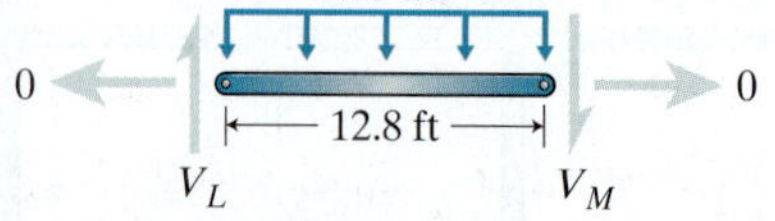

Figure 9.22c

$\curvearrowleft \Sigma M_M = 0 = V_L(12.8 \text{ ft}) - 2.25 \text{ klf}(12.8 \text{ ft})(6.4 \text{ ft})$

$V_L = 14.4 \text{ k} (\downarrow + \uparrow)$

$+\uparrow \Sigma F_y = 0 = 14.4 \text{ k} - 2.25 \text{ klf}(12.8 \text{ ft}) - V_M$

$V_M = -14.4 \text{ k} (\downarrow + \uparrow)$

Section 9.1 Gravity Load Method 449

FBD of beam RS:

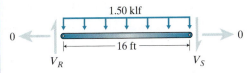

Figure 9.22d

$\curvearrowright \Sigma M_S = 0 = V_R(16\,\text{ft}) - 1.50\,\text{klf}(16\,\text{ft})(8\,\text{ft})$
$\qquad V_R = 12.0\,\text{k}\ (\downarrow + \uparrow)$

$+\uparrow \Sigma F_y = 0 = 12\,\text{k} - 1.50\,\text{klf}(16\,\text{ft}) - V_S$
$\qquad V_S = -12.0\,\text{k}\ (\downarrow + \uparrow)$

FBD of beam NO:

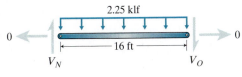

Figure 9.22e

$\curvearrowright \Sigma M_O = 0 = V_N(16\,\text{ft}) - 2.25\,\text{klf}(16\,\text{ft})(8\,\text{ft})$
$\qquad V_N = 18.0\,\text{k}\ (\downarrow + \uparrow)$

$+\uparrow \Sigma F_y = 0 = 18\,\text{k} - 2.25\,\text{klf}(16\,\text{ft}) - V_O$
$\qquad V_O = -18.0\,\text{k}\ (\downarrow + \uparrow)$

FBD of column ALP:

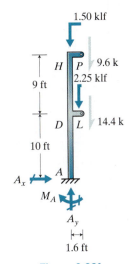

Figure 9.22f

$\curvearrowright \Sigma M_A = 0 = M_A + 9.6\,\text{k}(1.6\,\text{ft}) + 14.4\,\text{k}(1.6\,\text{ft}) + 1.50\,\text{klf}(1.6\,\text{ft})(0.8\,\text{ft})$
$\qquad\qquad + 2.25\,\text{klf}(1.6\,\text{ft})(0.8\,\text{ft})$
$\qquad M_A = -43.2\,\text{k}\cdot\text{ft}\ (\curvearrowright)$

$\xrightarrow{+} \Sigma F_x = 0 = A_x$
$\qquad A_x = 0$

$+\uparrow \Sigma F_y = 0 = A_y - 9.6\,\text{k} - 14.4\,\text{k} - 1.50\,\text{klf}(1.6\,\text{ft}) - 2.25\,\text{klf}(1.6\,\text{ft})$
$\qquad A_y = 30\,\text{k}\ (+\uparrow)$

450   **Chapter 9**   Approximate Analysis of Rigid Frames

FBD of column BMNQR:

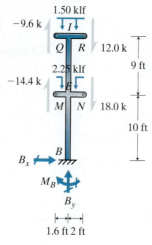

Figure 9.22g

$$\curvearrowleft \Sigma M_B = 0 = M_B + (-9.6\text{ k} - 14.4\text{ k})(1.6\text{ ft}) + (12\text{ k} + 18\text{ k})(2\text{ ft})$$
$$- (1.50\text{ klf} + 2.25\text{ klf})(1.6\text{ ft})(0.8\text{ ft})$$
$$+ (1.50\text{ klf} + 2.25\text{ klf})(2\text{ ft})(1\text{ ft})$$
$$M_B = -24.3\text{ k}\cdot\text{ft}\ (\curvearrowright)$$

$$\xrightarrow{+} \Sigma F_x = 0 = B_x$$
$$B_x = 0$$

$$+\uparrow \Sigma F_y = 0 = B_y + (-9.6\text{ k} - 14.4\text{ k}) - (1.50\text{ klf} + 2.25\text{ klf})(1.6\text{ ft} + 2\text{ ft})$$
$$- (12\text{ k} + 18\text{ k})$$
$$B_y = 67.5\text{ k}\ (+\uparrow)$$

FBD of column COS:

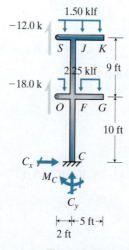

Figure 9.22h

$$\stackrel{+}{\curvearrowleft}\Sigma M_C = 0 = M_C + (-12\,\text{k} - 18\,\text{k})(2\,\text{ft}) - (1.50\,\text{klf} + 2.25\,\text{klf})(2\,\text{ft})(1\,\text{ft})$$
$$+ (1.50\,\text{klf} + 2.25\,\text{klf})(5\,\text{ft})(2.5\,\text{ft})$$
$$M_C = 20.6\,\text{k}\cdot\text{ft}\ (\stackrel{+}{\curvearrowleft})$$

$$\stackrel{+}{\rightarrow}\Sigma F_x = 0 = C_x$$
$$C_x = 0$$

$$+\uparrow\Sigma F_y = 0 = C_y + (-12\,\text{k} - 18\,\text{k}) - (1.50\,\text{klf} + 2.25\,\text{klf})(2\,\text{ft} + 5\,\text{ft})$$
$$C_y = 56.2\,\text{k}\ (+\uparrow)$$

FBD of the entire frame:

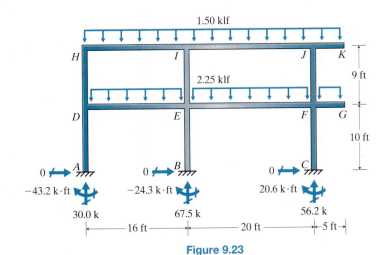

Figure 9.23

Evaluation of Results:
Satisfied Fundamental Principles?

$$\stackrel{+}{\rightarrow}\Sigma F_x = 0 + 0 + 0 = 0\ \checkmark$$

$$+\uparrow\Sigma F_y = 30\,\text{k} + 67.5\,\text{k} + 56.2\,\text{k}$$
$$- (1.5\,\text{klf} + 2.25\,\text{klf})(16\,\text{ft} + 20\,\text{ft} + 5\,\text{ft})$$
$$= -0.05\,\text{k}\ \approx 0\ \checkmark$$

$$\stackrel{+}{\curvearrowleft}\Sigma M_A = -43.2\,\text{k}\cdot\text{ft} - 24.3\,\text{k}\cdot\text{ft} + 20.6\,\text{k}\cdot\text{ft}$$
$$- 67.5\,\text{k}(16\,\text{ft}) - 56.2\,\text{k}(36\,\text{ft})$$
$$+ (1.50\,\text{klf} + 2.25\,\text{klf})(41\,\text{ft})(20.5\,\text{ft})$$
$$= 1.8\,\text{k}\cdot\text{ft} \approx 0\ \checkmark$$

EXAMPLE 9.3

Our team is designing the frame for a power substation. The dominant load is the self-weight of the frame. To perform preliminary design of the members, we would like to know the peak positive and negative moments in the beams and the peak axial force in the columns. In addition, the geotechnical engineers want to know the preliminary support reactions in order to begin designing the foundations. An experienced engineer on the team guesses that each member will weigh about 1 kN/m.

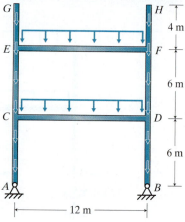

Figure 9.24

Approximate Analysis:

Estimate the inflection points at 10% of the beam spans from the columns. Even though we don't assume the locations of both inflection points for the bottom beam, we know that both inflection points will be at 10%, since both the loading and structure are symmetric. Because the columns are pinned at their bases, we assume that the axial force in each beam is the same.

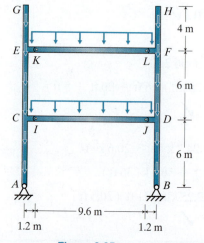

Figure 9.25a

FBD of beam KL (same as IJ):

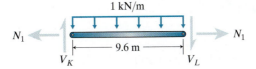

Figure 9.25b

$$\circlearrowright\!\!+\ \Sigma M_L = 0 = V_K(9.6\,\text{m}) - (1\,\text{kN/m})(9.6\,\text{m})(4.8\,\text{m})$$

$$V_K = 4.8\,\text{kN}\ (\downarrow + \uparrow)$$

$$M_{\text{max}+} = \frac{wl^2}{8} = \frac{(1\,\text{kN/m})(9.6\,\text{m})^2}{8} = +11.52\,\text{kN}\cdot\text{m}$$

FBD of piece EK (same as CI):

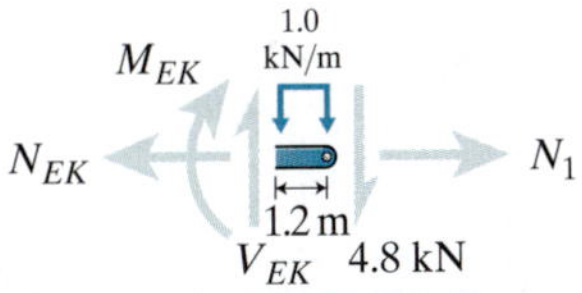

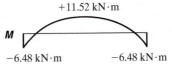

Figure 9.25c

$$\circlearrowright\!\!+\ \Sigma M_E = 0 = M_{EK} + (1\,\text{kN/m})(1.2\,\text{m})(0.6\,\text{m}) + 4.8\,\text{kN}(1.2\,\text{m})$$

$$M_{\text{max}-} = M_{EK} = -6.48\,\text{kN}\cdot\text{m}$$

Moment diagram for the beams:

Figure 9.26a

FBD of column AIK:

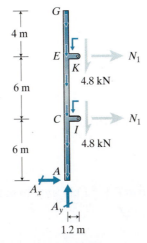

Figure 9.26b

454   Chapter 9   Approximate Analysis of Rigid Frames

$+\uparrow \Sigma F_y = 0 = A_y - (1\,\text{kN/m})(16\,\text{m}) - 2(1\,\text{kN/m})(1.2\,\text{m}) - 2(4.8\,\text{kN})$

$\quad A_y = 28\,\text{kN}\,(+\uparrow)$

$\curvearrowleft \Sigma M_A = 0 = 2(1\,\text{kN/m})(1.2\,\text{m})(0.6\,\text{m}) + 2(4.8\,\text{kN})(1.2\,\text{m})$

$\quad\quad\quad + N_1(6\,\text{m}) + N_1(12\,\text{m})$

$\quad N_1 = -0.72\,\text{kN}\,(\rightarrow + \leftarrow)$

$\xrightarrow{+} \Sigma F_x = 0 = A_x - 0.72\,\text{kN} - 0.72\,\text{kN}$

$\quad A_x = 1.44\,\text{kN}\,(\xrightarrow{\pm})$

FBD of the entire frame:

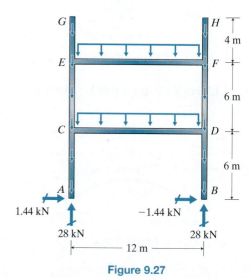

Figure 9.27

Evaluation of Results:

Satisfied Fundamental Principles?

$\xrightarrow{+} \Sigma F_x = 1.44\,\text{kN} - 1.44\,\text{kN} = 0$ ✓

$+\uparrow \Sigma F_y = 2(28\,\text{kN}) - 2(1\,\text{kN/m})(16\,\text{m})$

$\quad\quad - 2(1\,\text{kN/m})(12\,\text{m}) = 0$ ✓

$\curvearrowleft \Sigma M_A = 2(1\,\text{kN/m})(12\,\text{m})(6\,\text{m}) + (1\,\text{kN/m})(16\,\text{m})(12\,\text{m})$

$\quad\quad - 28\,\text{kN}(12\,\text{m}) = 0$ ✓

Example 9.4

Our team is designing a small framed structure to protect patients from the weather when they are dropped off at a doctor's office. The roof diaphragm will behave with two-way action.

Because this is part of a fast-track project, the foundations need to be designed and built right away. Therefore, the geotechnical engineers need to know the approximate foundation reactions as soon as possible. Our design team has divided up the various loadings to perform approximate analysis. We have been assigned snow load.

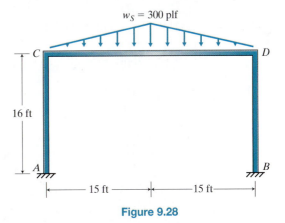

Figure 9.28

Approximate Analysis:

From the inside back cover, we see that the inflection points in a fixed-fixed beam that has this loading are at 22% of the span. Therefore, the estimated inflection points are at 11% of the beam span from the columns for this frame. Also, we assume that no axial force is generated in the beam.

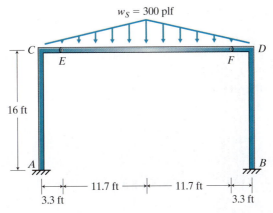

Figure 9.29a

456 Chapter 9 Approximate Analysis of Rigid Frames

FBD of beam EF:

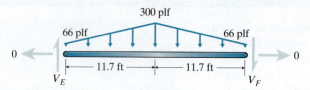

Figure 9.29b

$$\curvearrowleft \Sigma M_F = 0 = V_E(23.4 \text{ ft}) - 66 \text{ plf}(23.4 \text{ ft})(11.7 \text{ ft})$$
$$- \frac{1}{2}(300 \text{ plf} - 66 \text{ plf})(23.4 \text{ ft})(11.7 \text{ ft})$$

$$V_E = 2.14 \text{ k } (\downarrow + \uparrow)$$

$$+\uparrow \Sigma F_y = 0 = 2140 \text{ lb} - 66 \text{ plf}(23.4 \text{ ft})$$
$$- \frac{1}{2}(300 \text{ plf} - 66 \text{ plf})(23.4 \text{ ft}) - V_F$$

$$V_F = -2.14 \text{ k } (\downarrow + \uparrow)$$

FBD of column AE:

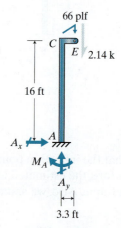

Figure 9.29c

$$\curvearrowleft \Sigma M_A = 0 = M_A + 2.14 \text{ k}(3.3 \text{ ft}) + \frac{1}{2}(0.066 \text{ klf})(3.3 \text{ ft})(2.2 \text{ ft})$$

$$M_A = -7.30 \text{ k} \cdot \text{ft } (\curvearrowleft)$$

$$\xrightarrow{+} \Sigma F_x = 0 = A_x$$

$$A_x = 0$$

$$+\uparrow \Sigma F_y = 0 = A_y - 2.14 \text{ k} - \frac{1}{2}(0.066 \text{ klf})(3.3 \text{ ft})$$

$$A_y = 2.25 \text{ k } (+\uparrow)$$

FBD of column BF:

$\curvearrowleft \Sigma M_B = 0 = M_B + (-2.14 \text{ k})(3.3 \text{ ft}) - \frac{1}{2}(0.066 \text{ klf})(3.3 \text{ ft})(2.2 \text{ ft})$

$M_B = 7.30 \text{ k} \cdot \text{ft} \ (\curvearrowleft)$

$\xrightarrow{+} \Sigma F_x = 0 = B_x$

$B_x = 0$

$+\uparrow \Sigma F_y = 0 = B_y + (-2.14 \text{ k}) - \frac{1}{2}(0.066 \text{ klf})(3.3 \text{ ft})$

$B_y = 2.25 \text{ k} \ (+\uparrow)$

Figure 9.29d

FBD of the entire frame:

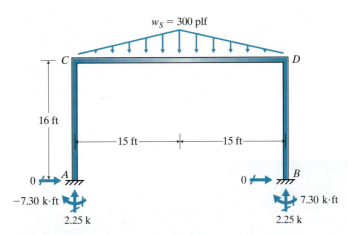

Figure 9.30

Evaluation of Results:
Satisfied Fundamental Principles?

$\curvearrowleft \Sigma M_A = -7.30 \text{ k} \cdot \text{ft} + \frac{1}{2}(0.300 \text{ klf})(30 \text{ ft})(15 \text{ ft})$
$+ 7.30 \text{ k} \cdot \text{ft} - 2.25 \text{ k}(30 \text{ ft}) = 0$ ✓

$\xrightarrow{+} \Sigma F_x = 0 + 0 = 0$ ✓

$+\uparrow \Sigma F_y = 2.25 \text{ k} - \frac{1}{2}(0.300 \text{ klf})(30 \text{ ft}) + 2.25 \text{ k} = 0$ ✓

9.2 Portal Method for Lateral Loads

Introduction

When a rigid frame is subjected to lateral loads, the inflection points are in completely different locations than when there are gravity loads. As we saw with the Gravity Load Method, assuming the locations of the inflection points is not enough, though; we need an additional assumption. When there is a lateral load, our final assumption depends on the overall shape of the rigid frame. For relatively short and wide frames, the Portal Method assumption tends to provide reasonable results.

How-To

Approximate analysis of rigid frames starts with our best estimate of the locations of inflection points. To develop a good sense of their locations, let's look at the displaced shape of a rigid frame (Figure 9.31).

From the photo we see that the beams change concavity at roughly midspan. The exact location is dependent on the relative stiffnesses of the members that frame into the two ends of the beam, but for approximate analysis we typically assume the inflection points are at midspan of the beams. The one exception is cantilevered beams (Figure 9.32). They have zero stiffness at the free end, so they do not help to resist lateral load. Another way to think about cantilevered beams is to note that the location of zero moment, the inflection point, is at the free end.

Looking back at the displaced shape of the rigid frame in Figure 9.31, we see that the columns change concavity at roughly midheight. Again, the exact location depends on the relative stiffnesses of the members that

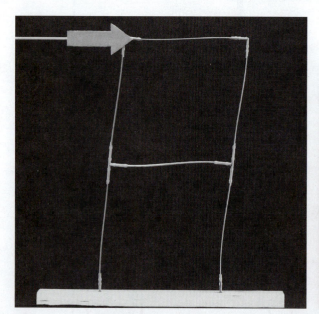

Figure 9.31 Two-story rigid frame with fixed supports experiencing lateral load. Note the locations where concavity changes in the members.

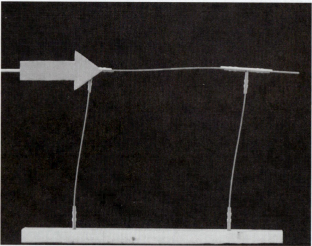

Figure 9.32 Rigid frame with cantilevered beam experiencing lateral load.

Section 9.2 Portal Method for Lateral Loads   459

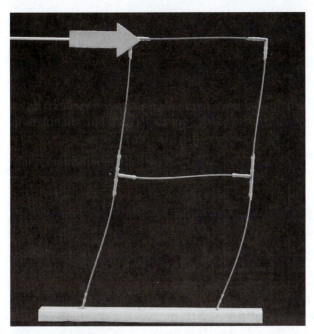

Figure 9.33 Rigid frame with pinned supports experiencing lateral load.

frame into the two ends of the column. For approximate analysis, we typically assume the inflection points are at midheight of each column. The one possible exception is the first level.

The frame in Figure 9.31 has fixed supports; therefore, the inflection points in the first level appear to be at midheight. But if the supports are pinned, the bottom-story columns don't change concavity (Figure 9.33). The location of zero moment in this situation is at the pins. The inflection points in the upper level are still at about midheight though.

Therefore, our first two assumptions for approximate analysis of rigid frames subjected to lateral loads are:

1. Inflection points are at midspan of each beam. Exception: Cantilevered beams have inflection points at their free ends.
2. Inflection points are at midheight of each column. Exception: The lowest level columns that are pin supported have inflection points at the supports.

These assumptions still leave us with too many unknowns. The third assumption we make is dependent upon the geometry of the rigid frame. For frames that are relatively short and wide, their deformation is dominated by shear racking. The Portal Method was developed for approximate analysis of these frames.

The interior columns of a rigid frame have two beams rigidly attached (Figure 9.34a), but the exterior columns have beams on only one side (Figure 9.34b). As a result, the interior columns are stiffer with respect to lateral deformation. Since stiffness attracts load, the interior columns carry more of the shear at a level than the exterior columns. The Portal Method assumption is that all interior columns have the same stiffness, so they carry the same shear. Further, we assume that the two exterior

460 Chapter 9 Approximate Analysis of Rigid Frames

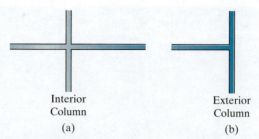

Figure 9.34 Types of columns in a rigid frame: (a) Interior column has beams rigidly attached on both sides; (b) Exterior column has a beam rigidly attached on only one side.

columns are half as stiff and, therefore, each exterior column carries half as much shear as each interior column.

If there are cantilevered beams, remember that they do not contribute stiffness to the columns. Therefore, a cantilevered beam does not change our assumption about the distribution of shear to an exterior column.

So, how do we use these three assumptions to perform approximate analysis? We start by applying the first two assumptions: locate all of the inflection points (Figure 9.35). Then we cut the frame at each level of the column inflection points and apply horizontal equilibrium (Figure 9.36). This lets us find the shear in each column. Now we have enough information to move piece by piece through the structure, finding the internal forces at the inflection points. To do so, we cut the structure into pieces by separating it at all of the inflection points. We must

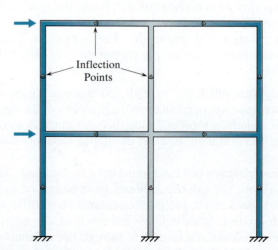

Figure 9.35 Assumed locations of inflection points for a rigid frame subjected to lateral load.

Section 9.2 Portal Method for Lateral Loads 461

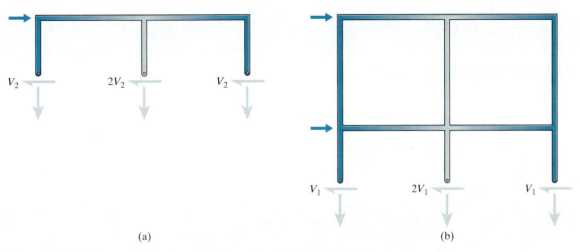

Figure 9.36 Rigid frame cut at column inflection points: (a) Top level; (b) Bottom level.

start applying equilibrium to a corner piece (Figure 9.37a); otherwise, there are too many unknowns (Figure 9.37b–d). We can then move to any piece next to that corner piece, and we keep moving to adjacent pieces until we have found all the desired forces at the inflection points.

Once we know the forces at the inflection points, we can calculate the reactions, create internal force diagrams, and even predict deflections (see Chapter 10: Approximate Lateral Displacements).

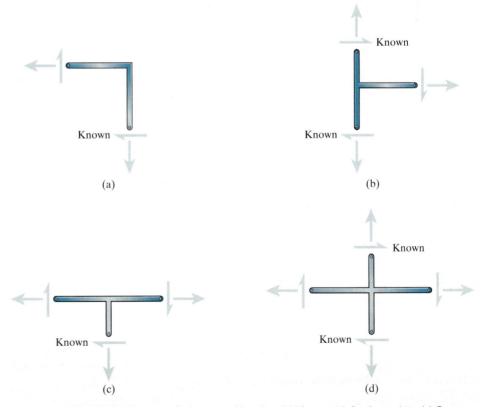

Figure 9.37 Free body diagrams of pieces cut from the rigid frame at inflection points: (a) Corner piece has three unknowns; (b) Interior piece with four unknowns; (c) Interior piece with five unknowns; (d) Interior piece with six unknowns.

Section 9.2 Highlights

Portal Method for Lateral Loads

Applicability:
- Rigid frame structures subjected to lateral loads.
- Beams rigidly connected to the columns; no pin connections.
- Best for short and wide frames but can be used with any size.

Process:
1. Assume the inflection points are at midspan of each beam.
   - Exception: For cantilevered beams, the inflection point is at the free end.
2. Assume the inflection points are at midheight of each column segment.
   - Exception: For the bottom level, if a column is pin supported, the inflection point is at the support.
3. Find the column shears.
   - Cut the frame at a row of column inflection points; draw a FBD of everything above the cut.

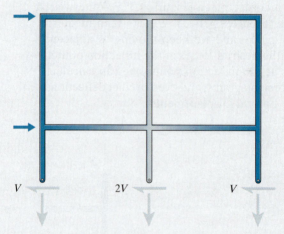

   - Assume the distribution of shear to the columns at the inflection points: all interior columns carry the same amount, which is twice what each exterior column carries.
   - Apply horizontal equilibrium to find the column shears.
   - Repeat at each level of column inflection points.
4. Find all other desired information.
   - Separate the frame into pieces by cutting it at all inflection points.
   - Draw a FBD of a top corner piece; label the known column shear.

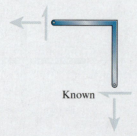

   - Apply equilibrium to find the other internal forces at the inflection points.
   - Move to an adjacent FBD and repeat the process.
   - Continue until the desired information is found.
   - Construct internal force diagrams if desired.

EXAMPLE 9.5

Our firm is designing a new condominium building. The senior designer has decided to use a reinforced concrete rigid frame. For architectural reasons, the two bays have unequal widths.

Figure 9.38

In order for the geotechnical engineers to begin designing the foundations, they need to know the preliminary reactions. We have been tasked with estimating the foundation reactions due to lateral wind load. Another member of the team will determine the effects of wind uplift on the roof.

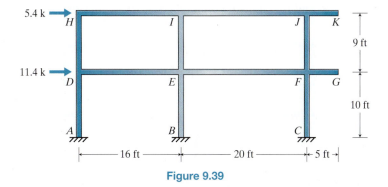

Figure 9.39

Approximate Analysis:

Estimate the inflection points at midspan of the beams, except the cantilevers. Estimate the inflection points at midheight of each column including the first level, since the foundations are idealized as fixed supports.

464 **Chapter 9** Approximate Analysis of Rigid Frames

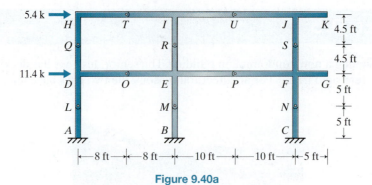

Figure 9.40a

FBD of the frame cut at the second level:

Cut the frame at the column inflection points: Q, R, S. Since the frame is wider than it is tall, use the Portal Method: assume the exterior columns carry the same shear, and the interior column carries twice that shear.

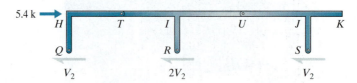

Figure 9.40b

$$\xrightarrow{+} \Sigma F_x = 0 = 5.4 \text{ k} - 4V_2$$
$$V_2 = 1.35 \text{ k} \;(\xrightarrow{+})$$

FBD of the frame cut at the first level:

Cut the frame at the column inflection points: L, M, N. Assume the exterior columns carry the same shear, and the interior column carries twice that shear.

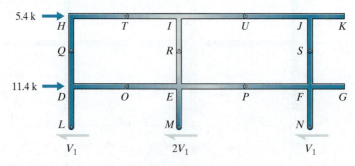

Figure 9.40c

$$\xrightarrow{+} \Sigma F_x = 0 = 5.4 \text{ k} + 11.4 \text{ k} - 4V_1$$
$$V_1 = 4.20 \text{ k} \;(\xrightarrow{+})$$

FBD of corner QHT:

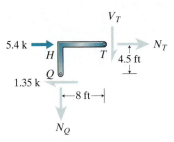

Figure 9.40d

$\circlearrowleft \Sigma M_H = 0 = 1.35 \text{ k}(4.5 \text{ ft}) + V_T(8 \text{ ft})$

$V_T = -0.76 \text{ k } (\downarrow + \uparrow)$

$+\downarrow \Sigma F_y = 0 = N_Q - 0.76 \text{ k}$

$N_Q = 0.76 \text{ k } (\updownarrow)$

FBD of piece TUR:

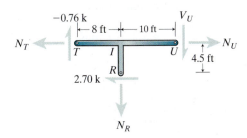

Figure 9.40e

$\circlearrowleft \Sigma M_I = 0 = -0.76 \text{ k}(8 \text{ ft}) + V_U(10 \text{ ft}) + 2.70 \text{ k}(4.5 \text{ ft})$

$V_U = -0.61 \text{ k } (\downarrow + \uparrow)$

$+\uparrow \Sigma F_y = 0 = -0.76 \text{ k} - N_R - (-0.61 \text{ k})$

$N_R = -0.15 \text{ k } (\updownarrow)$

FBD of corner UKS:

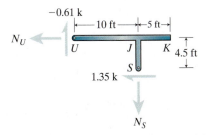

Figure 9.40f

We can use moment equilibrium on this piece as a quick check for errors.
Satisfied Fundamental Principles?

$\circlearrowleft \Sigma M_J = -0.61 \text{ k}(10 \text{ ft}) + 1.35 \text{ k}(4.5 \text{ ft}) = -0.02 \text{ k}\cdot\text{ft} \approx 0$ ✓

466 Chapter 9 Approximate Analysis of Rigid Frames

Continue:

$+\downarrow \Sigma F_y = 0 = -(-0.61 \text{ k}) + N_S$

$N_S = -0.61 \text{ k} (\updownarrow)$

FBD of piece LQO:

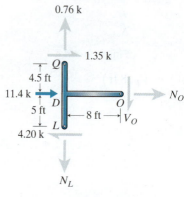

Figure 9.40g

$+\circlearrowleft \Sigma M_D = 0 = 1.35 \text{ k} (4.5 \text{ ft}) + V_O (8 \text{ ft}) + 4.20 \text{ k} (5 \text{ ft})$

$V_O = -3.38 \text{ k} (\downarrow + \uparrow)$

$+\uparrow \Sigma F_y = 0 = 0.76 \text{ k} - (-3.38 \text{ k}) - N_L$

$N_L = 4.14 \text{ k} (\updownarrow)$

FBD of piece ORPM:

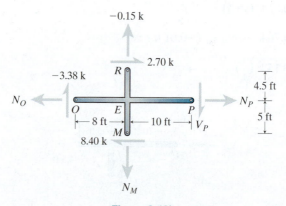

Figure 9.40h

$+\circlearrowleft \Sigma M_E = 0 = (-3.38 \text{ k})(8 \text{ ft}) + 2.70 \text{ k} (4.5 \text{ ft}) + V_P (10 \text{ ft}) + 8.40 \text{ k} (5 \text{ ft})$

$V_P = -2.71 \text{ k} (\downarrow + \uparrow)$

$+\uparrow \Sigma F_y = 0 = (-3.38 \text{ k}) + (-0.15 \text{ k}) - (-2.71 \text{ k}) - N_M$

$N_M = -0.82 \text{ k} (\updownarrow)$

FBD of piece PSGN:

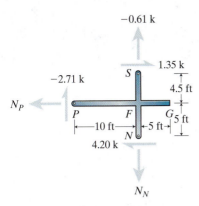

Figure 9.40i

We can use moment equilibrium on this piece as a quick check for errors.

Satisfied Fundamental Principles?

$$\curvearrowleft \Sigma M_F = (-2.71 \text{ k})(10 \text{ ft}) + 1.35 \text{ k}(4.5 \text{ ft}) + 4.20 \text{ k}(5 \text{ ft})$$
$$= -0.02 \text{ k} \cdot \text{ft} \approx 0 \checkmark$$

Continue:

$$+\uparrow \Sigma F_y = 0 = (-2.71 \text{ k}) + (-0.61 \text{ k}) - N_N$$
$$N_N = -3.32 \text{ k} \; (\updownarrow)$$

FBD of piece AL:

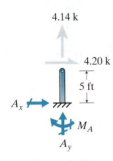

Figure 9.40j

$$\xrightarrow{+} \Sigma F_x = 0 = A_x + 4.20 \text{ k}$$
$$A_x = -4.20 \text{ k} \; (\xrightarrow{\pm})$$

$$+\uparrow \Sigma F_y = 0 = A_y + 4.14 \text{ k}$$
$$A_y = -4.14 \text{ k} \; (+\uparrow)$$

$$\curvearrowleft \Sigma M_A = 0 = M_A + 4.20 \text{ k}(5 \text{ ft})$$
$$M_A = -21.0 \text{ k} \cdot \text{ft} \; (\curvearrowleft)$$

468   **Chapter 9**   Approximate Analysis of Rigid Frames

FBD of piece BM:

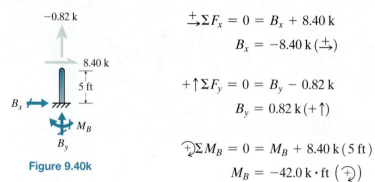

$$\xrightarrow{+} \Sigma F_x = 0 = B_x + 8.40 \text{ k}$$
$$B_x = -8.40 \text{ k} (\xrightarrow{\pm})$$

$$+\uparrow \Sigma F_y = 0 = B_y - 0.82 \text{ k}$$
$$B_y = 0.82 \text{ k} (+\uparrow)$$

$$\circlearrowleft \Sigma M_B = 0 = M_B + 8.40 \text{ k} (5 \text{ ft})$$
$$M_B = -42.0 \text{ k} \cdot \text{ft} (\circlearrowleft)$$

Figure 9.40k

FBD of piece CN:

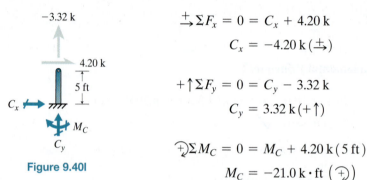

$$\xrightarrow{+} \Sigma F_x = 0 = C_x + 4.20 \text{ k}$$
$$C_x = -4.20 \text{ k} (\xrightarrow{\pm})$$

$$+\uparrow \Sigma F_y = 0 = C_y - 3.32 \text{ k}$$
$$C_y = 3.32 \text{ k} (+\uparrow)$$

$$\circlearrowleft \Sigma M_C = 0 = M_C + 4.20 \text{ k} (5 \text{ ft})$$
$$M_C = -21.0 \text{ k} \cdot \text{ft} (\circlearrowleft)$$

Figure 9.40l

FBD of the entire frame:

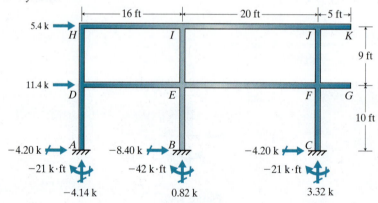

Figure 9.41

Evaluation of Results:
Satisfied Fundamental Principles?

$$\xrightarrow{+} \Sigma F_x = 5.4 \text{ k} + 11.4 \text{ k} - 4.2 \text{ k} - 8.4 \text{ k} - 4.2 \text{ k} = 0 \quad \checkmark$$
$$+\uparrow \Sigma F_y = -4.14 \text{ k} + 0.82 \text{ k} + 3.32 \text{ k} = 0 \quad \checkmark$$
$$\circlearrowleft \Sigma M_A = 5.4 \text{ k} (19 \text{ ft}) + 11.4 \text{ k} (10 \text{ ft}) - 0.82 \text{ k} (16 \text{ ft})$$
$$\quad - 3.32 \text{ k} (36 \text{ ft}) - 21 \text{ k} \cdot \text{ft} - 42 \text{ k} \cdot \text{ft}$$
$$\quad - 21 \text{ k} \cdot \text{ft} = -0.04 \text{ k} \cdot \text{ft} \approx 0 \quad \checkmark$$

Section 9.2 Portal Method for Lateral Loads   469

EXAMPLE 9.6

Our team is designing the rigging over a racetrack. The rigging must be able to carry equipment suspended all along its length. If an earthquake occurs, the rigging will experience lateral load proportional to the weight of the rigging and the suspended equipment.

Figure 9.42

In order to design the rigging, we want to create the moment diagram for the rigging subjected to a static equivalent seismic load.

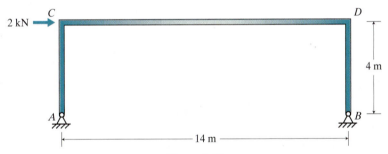

Figure 9.43

Approximate Analysis:

Estimate an inflection point at midspan of the beam. Estimate the inflection points at the base of each column, since there is only one level and the foundations are idealized as pinned supports.

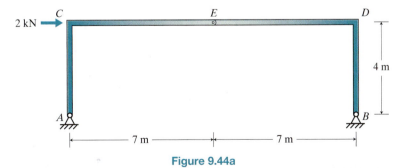

Figure 9.44a

470 Chapter 9 Approximate Analysis of Rigid Frames

FBD of the frame cut at the column inflection points:

Cut the frame at the column inflection points: A, B. Since the frame is wider than it is tall, use the Portal Method: assume the two columns carry the same shear.

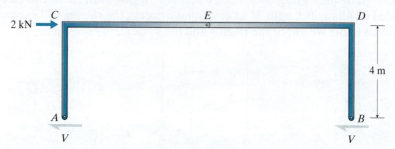

Figure 9.44b

$$\xrightarrow{+} \Sigma F_x = 0 = 2 \text{ kN} - 2V$$
$$V = 1 \text{ kN} (\xrightarrow{+})$$

FBD of corner EDB:

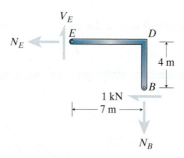

Figure 9.44c

$$\stackrel{+}{\curvearrowright} \Sigma M_D = 0 = V_E(7 \text{ m}) + 1 \text{ kN}(4 \text{ m})$$
$$V_E = -0.571 \text{ kN} (\downarrow + \uparrow)$$

FBD of BD:

Use to determine the internal moment at the top of the column.

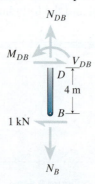

Figure 9.44d

$$\stackrel{+}{\curvearrowright} \Sigma M_D = 0 = 1 \text{ kN}(4 \text{ m}) - M_{DB}$$
$$M_{DB} = 4 \text{ kN} \cdot \text{m} (\stackrel{+}{\curvearrowright})$$

FBD of ED:

Use to determine the internal moment at the end of the beam.

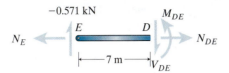

Figure 9.44e

$$\curvearrowleft \Sigma M_D = 0 = (-0.571 \text{ kN})(7 \text{ m}) - M_{DE}$$
$$M_{DE} = -4 \text{ kN} \cdot \text{m} \; (\curvearrowleft+)$$

FBD of corner CEA:

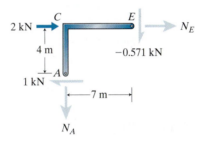

Figure 9.44f

We can use moment equilibrium on this piece as a quick check for errors.

Satisfied Fundamental Principles?

$$\curvearrowleft \Sigma M_C = 1 \text{ kN}(4 \text{ m}) + (-0.571 \text{ kN})(7 \text{ m}) = 0.003 \text{ k} \cdot \text{ft} \approx 0 \; \checkmark$$

FBD of piece AC:

Use to determine the internal moment at the top of the column.

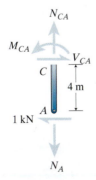

Figure 9.44g

472   **Chapter 9**   Approximate Analysis of Rigid Frames

$$\circlearrowleft \Sigma M_C = 0 = 1 \text{ kN}(4 \text{ m}) - M_{CA}$$
$$M_{CA} = 4 \text{ kN} \cdot \text{m} \, (\curvearrowdown)$$

FBD of CE:

Use to determine the internal moment at the end of the beam.

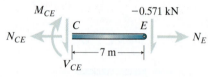

Figure 9.44h

$$\circlearrowleft \Sigma M_C = 0 = M_{CE} + (-0.571 \text{ kN})(7 \text{ m})$$
$$M_{CE} = 4 \text{ kN} \cdot \text{m} \, (\curvearrowright + \curvearrowleft)$$

Moment diagram:

The shear is constant along all members, so the moment must be linear along all members, and the moment is zero at the inflection points.

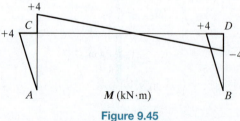

Figure 9.45

9.3 Cantilever Method for Lateral Loads

Introduction

When a rigid frame is relatively tall and narrow, it behaves differently under a lateral load than does a short and wide frame. Therefore, structural engineers have developed a different final assumption that tends to provide more accurate results for tall, narrow frames. We call this the Cantilever Method.

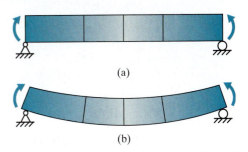

Figure 9.46 A simply supported beam subjected to pure bending: (a) Three planes through the cross-section identified before the beam deforms; (b) The three cross-sections are still planar after the beam deforms.

How-To

The lateral deformation of relatively tall, narrow rigid frames is dominated by flexural behavior. In mechanics of materials, we learned that with pure flexure, plane sections remain plane (Figure 9.46). For a cantilevered member, that is still true (Figure 9.47).

If plane sections remain plane, the distribution of axial strain through the cross-section must be linear (Figure 9.48a). *If* the material remains linear elastic, the distribution of axial stress must be linear because $\sigma = E\epsilon$ (Figure 9.48b).

If we assume that a tall, narrow rigid frame behaves essentially the same as a cantilevered beam, the distribution of axial stress through the frame cross-section will be linear (Figure 9.49). This is the third assumption of the Cantilever Method. The first two assumptions are about the locations of the inflection points and are the same as for the Portal Method:

1. Inflection points are at midspan of each beam. Exception: Cantilevered beams have inflection points at their free ends.
2. Inflection points are at midheight of each column. Exception: The lowest level columns that are pin supported have inflection points at the supports.

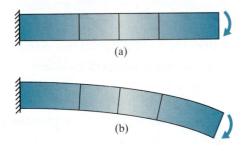

Figure 9.47 A cantilevered beam subjected to pure bending: (a) Three planes through the cross-section identified before the beam deforms; (b) The three cross-sections are still planar after the beam deforms.

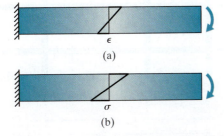

Figure 9.48 A cantilevered beam subjected to pure bending: (a) Axial strain distribution is linear; (b) Axial stress distribution is also linear *if* the material remains linear elastic.

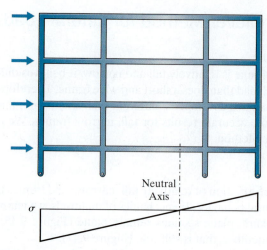

Figure 9.49 Distribution of axial stress across a plane in a rigid frame.

But how can we use this assumption that the distribution of axial stress to the columns at a level is linear? The key is to convert the stresses to axial forces in each column and to express each of the axial forces in terms of one unknown.

To start this conversion, we need to know the location of zero stress; that is called the neutral axis, and it runs vertically. In order to have vertical equilibrium, the neutral axis must be at the same location as the centroid of the column areas. As a reminder, we find the centroid using the following expression:

$$\bar{x} = \frac{\Sigma x_i A_i}{\Sigma A_i}$$

where

x_i = Distance from the reference point to a specific column
A_i = Cross-sectional area of the specific column
$\bar{x}$ = Distance from the reference point to the neutral axis (Figure 9.50)

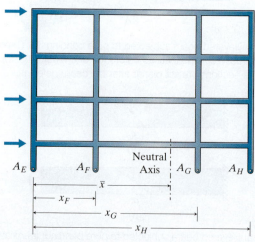

Figure 9.50 Calculating the location of the neutral axis at a level in a rigid frame. The reference location in this case is column E.

Figure 9.51 Change in column size at a level. Loads are smaller in the upper levels, so smaller cross-sectional areas can be used higher in the structure.

We need to be aware that the assumption of a linear distribution of stress does not mean that the neutral axis is in the same location all the way up the structure. It is common to reduce the size of the columns periodically because the axial demand typically decreases at each higher level (Figure 9.51). Since the column areas probably do not all decrease by the same percentage, the location of the neutral axis periodically shifts. Therefore, whenever the column arrangement or column size changes, we need to recalculate the location of the neutral axis starting at that level.

Once we know where the neutral axis is for a level, we can use similar triangles to relate all of the column axial stresses to one value. For example, from Figure 9.52 we get the following relationships:

$$\frac{\sigma_E}{d_E} = \frac{\sigma_F}{d_F} = \frac{-\sigma_G}{d_G} = \frac{-\sigma_H}{d_H}$$

where

σ = Axial stress in the column
d = Distance from the neutral axis to the column

Notice that on one side of the neutral axis the stresses are negative, compression. It doesn't matter which side we pick. If we guess wrong, the column axial force we calculate will be negative.

Chapter 9 Approximate Analysis of Rigid Frames

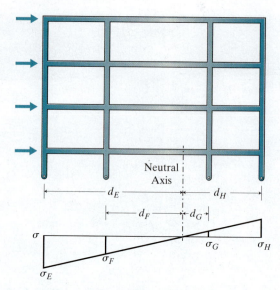

Figure 9.52 We can use similar triangles to relate all of the column axial stresses to one value.

If we choose the left column, σ_E, to be the value we use, then the other columns have the following axial stresses:

$$\sigma_F = \sigma_E \left(\frac{d_F}{d_E}\right) \qquad \sigma_G = -\sigma_E \left(\frac{d_G}{d_E}\right) \qquad \sigma_H = -\sigma_E \left(\frac{d_H}{d_E}\right)$$

The axial force in a column is the axial stress times the cross-sectional area. Therefore, we can express each of the column axial stresses in terms of the column axial forces:

$$\sigma_E = \frac{N_E}{A_E} \qquad \sigma_F = \frac{N_F}{A_F}$$

$$\sigma_G = \frac{N_G}{A_G} \qquad \sigma_H = \frac{N_H}{A_H}$$

When we substitute these equations into the previous expressions, we get the relationships between the column axial forces at a level:

$$\frac{N_F}{A_F} = \frac{N_E}{A_E}\left(\frac{d_F}{d_E}\right) \quad \Rightarrow \quad N_F = N_E \left(\frac{A_F d_F}{A_E d_E}\right)$$

$$\frac{N_G}{A_G} = \frac{-N_E}{A_E}\left(\frac{d_G}{d_E}\right) \quad \Rightarrow \quad N_G = -N_E \left(\frac{A_G d_G}{A_E d_E}\right)$$

$$\frac{N_H}{A_H} = \frac{-N_E}{A_E}\left(\frac{d_H}{d_E}\right) \quad \Rightarrow \quad N_H = -N_E \left(\frac{A_H d_H}{A_E d_E}\right)$$

In general, the column axial forces at a level are related in the following way (*if* the axial stress is distributed linearly):

$$N_i = N_R \left(\frac{A_i d_i}{A_R d_R} \right)$$

where

i = Column of interest
R = Reference column
A = Cross-sectional area of the column
d = Distance from the neutral axis to the column

Now we can apply moment equilibrium to the frame above a row of column inflection points. If we choose one of the inflection points at the cut for our summation, we have only one unknown: the axial force in the reference column (Figure 9.53).

We can repeat this process at each level of inflection points in the rigid frame. To find all the other desired internal forces and/or reactions, we follow the same process as the Portal Method:

- Cut the frame at all the inflection points.
- Apply equilibrium to a corner piece.
- Work across or down the structure until we find what we are looking for.

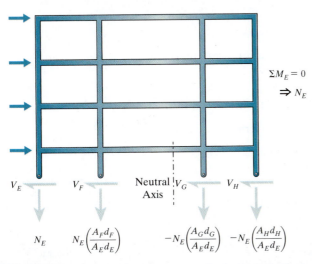

Figure 9.53 Free body diagram of a rigid frame cut at a level of inflection points. Each of the column axial forces is expressed in terms of the reference column.

Section 9.3 Highlights

Cantilever Method for Lateral Loads

Applicability:
- Rigid frame structures subjected to lateral loads.
- Beams rigidly connected to the columns; no pin connections.
- Best for tall and narrow frames but can be used with any size.

Process:
1. Assume the inflection points are at midspan of each beam.
   - Exception: For cantilevered beams, the inflection point is at the free end.
2. Assume the inflection points are at midheight of each column segment.
   - Exception: For the bottom level, if a column is pin supported, the inflection point is at the support.
3. Find the column axial forces.
   - Cut the frame at a row of column inflection points; draw a FBD of everything above the cut.

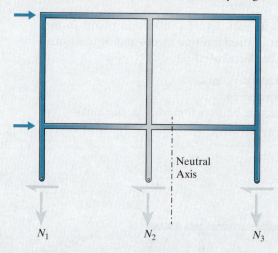

- Assume the axial stress in the columns is distributed linearly across the frame.
- Find the neutral axis given the column areas, $\bar{x} = \dfrac{\Sigma x_i A_i}{\Sigma A_i}$.
- Calculate the distance from each column to the neutral axis, d_i.
- Express the unknown axial force in each column, N_i, in terms of one column, N_R: $N_i = N_R \dfrac{A_i d_i}{A_R d_R}$
- Apply moment equilibrium to find the column axial forces.
- Repeat at each level of column inflection points.
4. Find all other desired information.
   - Separate the frame into pieces by cutting it at all inflection points.
   - Draw a FBD of a top corner piece; label the known column axial force.
   - Apply equilibrium to find the other internal forces at the inflection points.
   - Move to an adjacent FBD and repeat the process.
   - Continue until the desired information is found.
   - Construct internal force diagrams if desired.

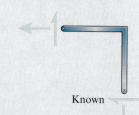

Known

Section 9.3 Cantilever Method for Lateral Loads   479

EXAMPLE 9.7

Knowing that the reactions we obtained using the Portal Method are approximate, our team leader for the condominium project recommended that we also use the Cantilever Method to estimate the reactions from the foundations. The second set of values will help us understand the variability of our estimates.

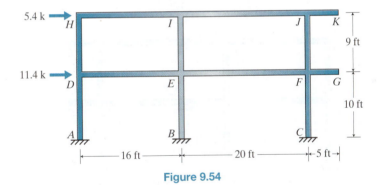

Figure 9.54

For now, we assume that each of the columns has the same cross-sectional area.

Approximate Analysis:

The assumed locations of the inflection points do not change. Estimate the inflection points at midspan of the beams, except the cantilevers. Estimate the inflection points at midheight of each column including the first level, since the foundations are idealized as fixed supports.

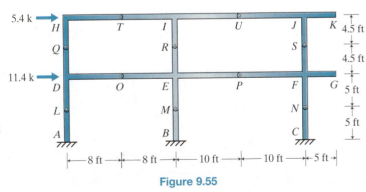

Figure 9.55

480 Chapter 9 Approximate Analysis of Rigid Frames

Locate the neutral axis:

Since the column areas and placement do not change from the first floor to the second, the neutral axis is at the same location for both floors. Determine the location of the neutral axis relative to the left column.

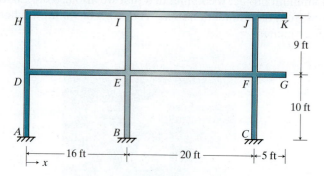

Figure 9.56a

$$\bar{x} = \frac{A(16 \text{ ft}) + A(36 \text{ ft})}{3A}$$

$$\bar{x} = 17.33 \text{ ft}$$

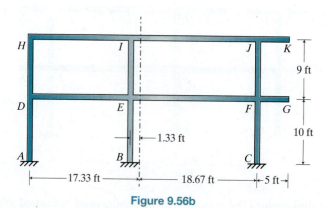

Figure 9.56b

FBD of the frame cut at the second level:

Cut the frame at the column inflection points: Q, R, S.

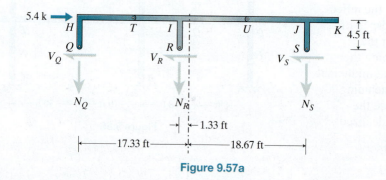

Figure 9.57a

Relate the column axial forces to N_Q:

$$N_R = N_Q \frac{A_R d_R}{A_Q d_Q} = N_Q \frac{A(1.33 \text{ ft})}{A(17.33 \text{ ft})} = 0.077 N_Q$$

$$N_S = N_Q \frac{A_S d_S}{A_Q d_Q} = N_Q \frac{A(-18.67 \text{ ft})}{A(17.33 \text{ ft})} = -1.077 N_Q$$

Apply moment equilibrium to find the axial forces:

$$+\!\!\upcurvearrowright \Sigma M_Q = 0 = 5.4 \text{ k}(4.5 \text{ ft}) + N_R(16 \text{ ft}) + N_S(36 \text{ ft})$$
$$-5.4 \text{ k}(4.5 \text{ ft}) = (0.077 N_Q)(16 \text{ ft}) + (-1.077 N_Q)(36 \text{ ft})$$
$$-24.3 \text{ k} \cdot \text{ft} = (-37.5 \text{ ft}) N_Q$$

$$N_Q = +0.65 \text{ k } (\updownarrow)$$

$$N_R = 0.077(0.65 \text{ k}) = +0.05 \text{ k } (\updownarrow)$$

$$N_S = -1.077(0.65 \text{ k}) = -0.70 \text{ k } (\updownarrow)$$

FBD of the frame cut at the first level:

Cut the frame at the column inflection points: L, M, N.

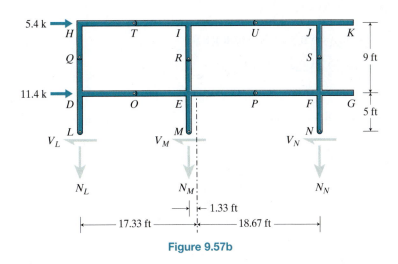

Figure 9.57b

Relate the column axial forces to N_L:

$$N_M = N_L \frac{A_M d_M}{A_L d_L} = N_L \frac{A(1.33 \text{ ft})}{A(17.33 \text{ ft})} = 0.077 N_L$$

$$N_N = N_L \frac{A_N d_N}{A_L d_L} = N_L \frac{A(-18.67 \text{ ft})}{A(17.33 \text{ ft})} = -1.077 N_L$$

482 **Chapter 9** Approximate Analysis of Rigid Frames

Apply moment equilibrium to find the axial forces:

$$\circlearrowright \Sigma M_L = 0 = 5.4\,\text{k}\,(14\,\text{ft}) + 11.4\,\text{k}\,(5\,\text{ft}) + N_M(16\,\text{ft}) + N_N(36\,\text{ft})$$

$$-5.4\,\text{k}\,(14\,\text{ft}) - 11.4\,\text{k}\,(5\,\text{ft}) = (0.077 N_L)(16\,\text{ft}) + (-1.077 N_L)(36\,\text{ft})$$

$$-132.6\,\text{k}\cdot\text{ft} = (-37.5\,\text{ft}) N_L$$

$$N_L = +3.54\,\text{k}\,(\uparrow\!\downarrow)$$

$$N_M = 0.077\,(3.54\,\text{k}) = +0.27\,\text{k}\,(\uparrow\!\downarrow)$$

$$N_N = -1.077\,(3.54\,\text{k}) = -3.81\,\text{k}\,(\uparrow\!\downarrow)$$

FBD of corner QHT:

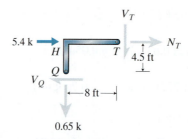

Figure 9.57c

$$+\downarrow \Sigma F_y = 0 = 0.65\,\text{k} + V_T$$

$$V_T = -0.65\,\text{k}\,(\downarrow+\uparrow)$$

$$\circlearrowright \Sigma M_H = 0 = V_Q(4.5\,\text{ft}) + (-0.65\,\text{k})(8\,\text{ft})$$

$$V_Q = 1.16\,\text{k}\,(\overset{\leftarrow}{+})$$

FBD of piece TUR:

Note that we could have worked down the left column and then moved to this piece.

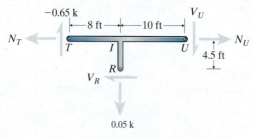

Figure 9.57d

$$+\uparrow \Sigma F_y = 0 = -0.65\,\text{k} - 0.05\,\text{k} - V_U$$

$$V_U = -0.70\,\text{k}\,(\downarrow+\uparrow)$$

$$\circlearrowright \Sigma M_I = 0 = -0.65\,\text{k}\,(8\,\text{ft}) + (-0.70\,\text{k})(10\,\text{ft}) + V_R(4.5\,\text{ft})$$

$$V_R = 2.71\,\text{k}\,(\overset{\leftarrow}{+})$$

FBD of corner UKS:

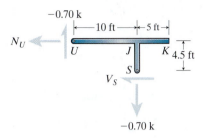

Figure 9.57e

We can use vertical equilibrium on this piece as a quick check for errors.
Satisfied Fundamental Principles?

$+\uparrow \Sigma F_y = -0.70 \text{ k} - (-0.70 \text{ k}) = 0$ ✓

Continue:

$+\circlearrowleft \Sigma M_J = 0 = -0.70 \text{ k}(10 \text{ ft}) + V_S(4.5 \text{ ft})$

$V_S = 1.56 \text{ k} \;(\overset{-}{\underset{+}{}})$

FBD of piece LQO:

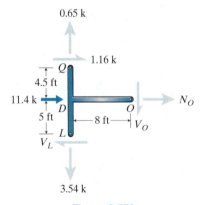

Figure 9.57f

$+\uparrow \Sigma F_y = 0 = 0.65 \text{ k} - 3.54 \text{ k} - V_O$

$V_O = -2.89 \text{ k} \;(\downarrow + \uparrow)$

$+\circlearrowleft \Sigma M_D = 0 = 1.16 \text{ k}(4.5 \text{ ft}) + (-2.89 \text{ k})(8 \text{ ft}) + V_L(5 \text{ ft})$

$V_L = 3.58 \text{ k} \;(\overset{-}{\underset{+}{}})$

484   **Chapter 9**   Approximate Analysis of Rigid Frames

FBD of piece ORPM:

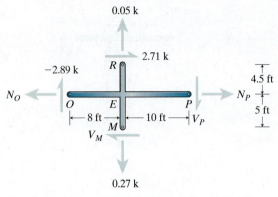

Figure 9.57g

$$+\uparrow \Sigma F_y = 0 = (-2.89 \text{ k}) + 0.05 \text{ k} - 0.27 \text{ k} - V_P$$
$$V_P = -3.11 \text{ k } (\downarrow + \uparrow)$$

$$\curvearrowright \Sigma M_E = 0 = (-2.89 \text{ k})(8 \text{ ft}) + 2.71 \text{ k}(4.5 \text{ ft}) + (-3.11 \text{ k})(10 \text{ ft}) + V_M(5 \text{ ft})$$
$$V_M = 8.40 \text{ k } (\overset{\rightarrow}{\leftarrow})$$

FBD of piece PSGN:

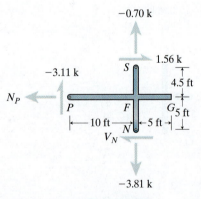

Figure 9.57h

We can use vertical equilibrium on this piece as a quick check for errors.

Satisfied Fundamental Principles?

$$+\uparrow \Sigma F_y = (-3.11 \text{ k}) + (-0.70 \text{ k}) - (-3.81 \text{ k}) = 0 \checkmark$$

Continue:

$$\curvearrowright \Sigma M_F = 0 = (-3.11 \text{ k})(10 \text{ ft}) + 1.56 \text{ k}(4.5 \text{ ft}) + V_N(5 \text{ ft})$$
$$V_N = 4.82 \text{ k } (\overset{\rightarrow}{\leftarrow})$$

Section 9.3 Cantilever Method for Lateral Loads **485**

FBD of piece AL:

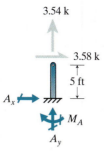

Figure 9.57i

$\xrightarrow{+} \Sigma F_x = 0 = A_x + 3.58 \text{ k}$
$A_x = -3.58 \text{ k} (\xrightarrow{+})$

$+\uparrow \Sigma F_y = 0 = A_y + 3.54 \text{ k}$
$A_y = -3.54 \text{ k} (+\uparrow)$

$\curvearrowleft^+ \Sigma M_A = 0 = M_A + 3.58 \text{ k} (5 \text{ ft})$
$M_A = -17.9 \text{ k} \cdot \text{ft} (\curvearrowleft^+)$

FBD of piece BM:

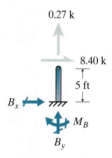

Figure 9.57j

$\xrightarrow{+} \Sigma F_x = 0 = B_x + 8.40 \text{ k}$
$B_x = -8.40 \text{ k} (\xrightarrow{+})$

$+\uparrow \Sigma F_y = 0 = B_y + 0.27 \text{ k}$
$B_y = -0.27 \text{ k} (+\uparrow)$

$\curvearrowleft^+ \Sigma M_B = 0 = M_B + 8.40 \text{ k} (5 \text{ ft})$
$M_B = -42.0 \text{ k} \cdot \text{ft} (\curvearrowleft^+)$

FBD of piece CN:

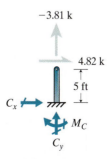

Figure 9.57k

$\xrightarrow{+} \Sigma F_x = 0 = C_x + 4.82 \text{ k}$
$C_x = -4.82 \text{ k} (\xrightarrow{+})$

$+\uparrow \Sigma F_y = 0 = C_y - 3.81 \text{ k}$
$C_y = 3.81 \text{ k} (+\uparrow)$

$\curvearrowleft^+ \Sigma M_C = 0 = M_C + 4.82 \text{ k} (5 \text{ ft})$
$M_C = -24.1 \text{ k} \cdot \text{ft} (\curvearrowleft^+)$

FBD of the entire frame:

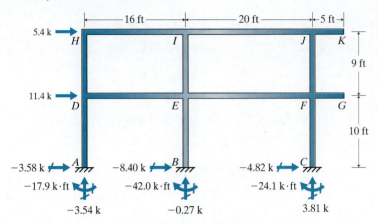

Figure 9.58

Evaluation of Results:

Satisfied Fundamental Principles?

$\xrightarrow{+} \Sigma F_x = 5.4\,k + 11.4\,k - 3.58\,k - 8.40\,k - 4.82\,k = 0$ ✓

$+\uparrow \Sigma F_y = -3.54\,k - 0.27\,k + 3.81\,k = 0$ ✓

$\circlearrowright \Sigma M_A = 5.4\,k\,(19\,\text{ft}) + 11.4\,k\,(10\,\text{ft}) + 0.27\,k\,(16\,\text{ft})$
$\quad - 3.81\,k\,(36\,\text{ft}) - 17.9\,k \cdot \text{ft} - 42.0\,k \cdot \text{ft}$
$\quad - 24.1\,k \cdot \text{ft} = -0.24\,k \cdot \text{ft} \approx 0$ ✓

Comparison of Results:

Reaction	Portal Method	Cantilever Method	Difference
A_x	−4.20 k	−3.58 k	17%
A_y	−4.14 k	−3.54 k	17%
M_A	−21.0 k·ft	−17.9 k·ft	17%
B_x	−8.40 k	−8.40 k	0%
B_y	0.82 k	−0.27 k	404%
M_B	−42.0 k·ft	−42.0 k·ft	0%
C_x	−4.20 k	−4.82 k	15%
C_y	3.32 k	3.81 k	15%
M_C	−21.0 k·ft	−24.1 k·ft	15%

The results are all very similar except for the smallest value, the vertical reaction at B. If we want to refine our prediction for that value, we can use analysis methods covered in later chapters.

Example 9.8

A development company wants to buy a historic building and renovate it to use for offices. The structural system is a rigid frame. Before finalizing the deal, the developer wants to know whether the existing framing is strong enough for the new use.

Figure 9.59

A member of the engineering team has modeled the frame using structural analysis software. To help evaluate the results of the computer aided analysis, we have been tasked with performing an approximate analysis to find the internal forces in a few key members.

Our first loading is the static equivalent seismic loading. Because the frame is as tall as it is wide, there is no strong preference for the approximate analysis method. We are going to find the axial force in the bottom-level columns, so the Cantilever Method will be most efficient.

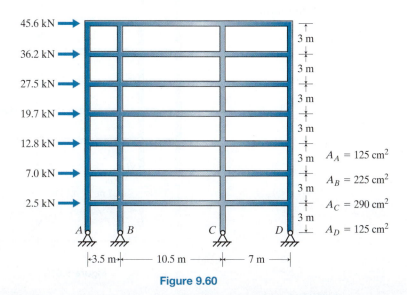

Figure 9.60

Chapter 9 Approximate Analysis of Rigid Frames

Approximate Analysis:

Estimate the inflection points at midspan of the beams. Estimate the inflection points at midheight of each column except the first level; the inflection points at the first level are at the supports, since the foundations are idealized as pins.

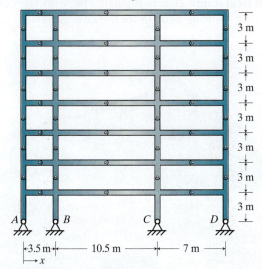

Figure 9.61a

Locate the neutral axis:

Since we are only finding the axial force in the bottom-level columns, we need to find the neutral axis only at that level. Let's determine the location of the neutral axis relative to the left column:

$$\bar{x} = \frac{(225 \text{ cm}^2)(3.5 \text{ m}) + (290 \text{ cm}^2)(14 \text{ m}) + (125 \text{ cm}^2)(21 \text{ m})}{125 \text{ cm}^2 + 225 \text{ cm}^2 + 290 \text{ cm}^2 + 125 \text{ cm}^2}$$

$\bar{x} = 9.77$ m

FBD of the frame cut at the bottom level:

Cut the frame at the column inflection points: A, B, C, D.

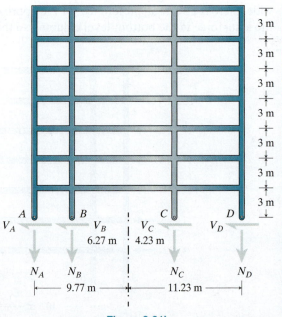

Figure 9.61b

Section 9.3 Cantilever Method for Lateral Loads

Relate the column axial forces to N_A:

$$N_B = N_A \frac{A_B d_B}{A_A d_A} = N_A \frac{(225 \text{ cm}^2)(6.27 \text{ m})}{(125 \text{ cm}^2)(9.77 \text{ m})} = 1.155 N_A$$

$$N_C = -N_A \frac{A_C d_C}{A_A d_A} = -N_A \frac{(290 \text{ cm}^2)(4.23 \text{ m})}{(125 \text{ cm}^2)(9.77 \text{ m})} = -1.004 N_A$$

$$N_D = -N_A \frac{A_D d_D}{A_A d_A} = -N_A \frac{(125 \text{ cm}^2)(11.23 \text{ m})}{(125 \text{ cm}^2)(9.77 \text{ m})} = -1.149 N_A$$

Apply moment equilibrium to find the axial forces:

$$\circlearrowright \Sigma M_A = 0 = 45.6 \text{ kN}(21 \text{ m}) + 36.2 \text{ kN}(18 \text{ m}) + 27.5 \text{ kN}(15 \text{ m})$$
$$+ 19.7 \text{ kN}(12 \text{ m}) + 12.8 \text{ kN}(9 \text{ m}) + 7.0 \text{ kN}(6 \text{ m})$$
$$+ 2.5 \text{ kN}(3 \text{ m}) + N_B(3.5 \text{ m}) + N_C(14 \text{ m}) + N_D(21 \text{ m})$$
$$- 2420 \text{ kN} \cdot \text{m} = 1.155 N_A (3.5 \text{ m}) + (-1.004 N_A)(14 \text{ m}) + (-1.149 N_A)(21 \text{ m})$$

$N_A = +71.0 \text{ kN} (\uparrow\downarrow)$

$N_B = +82.0 \text{ kN} (\uparrow\downarrow)$

$N_C = -71.2 \text{ kN} (\uparrow\downarrow)$

$N_D = -81.5 \text{ kN} (\uparrow\downarrow)$

FBD of the entire frame:

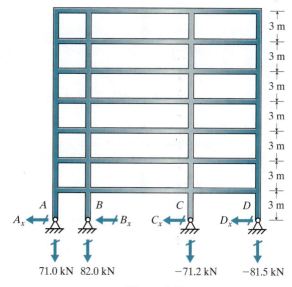

Figure 9.62

Evaluation of Results:
Verify vertical equilibrium:

Satisfied Fundamental Principles?

$+\downarrow \Sigma F_y = 71.0 \text{ kN} + 82.0 \text{ kN} - 71.2 \text{ kN} - 81.5 \text{ kN}$

$= +0.3 \text{ kN} \approx 0$ ✓

9.4 Combined Gravity and Lateral Loads

Introduction

In some circumstances, a single type of load results in both lateral load and gravity load on the rigid frame. In other circumstances, we want to find the reactions or internal forces due to both gravity and lateral loads. In these cases, we cannot reasonably predict where the inflection points will be. But, if we consider the lateral and gravity components separately, we have viable approximate analysis methods. We can then superimpose the results.

How-To

The approximate analysis methods from Sections 9.1–9.3 work only if there is just gravity load or just lateral load, but some loadings have both components. For example, wind causes lateral load as well as vertical load on the roof. Areas used for concerts or sporting events are designed for lateral live load in addition to vertical live load.

The challenge of combined loading is predicting the locations of the inflection points. Consider a single beam from a rigid frame. If the frame experienced only gravity load, we would know where to assume the inflection points are, and we would predict a parabolic moment diagram (Figure 9.63). If the frame experienced only lateral load, we would assume the inflection point is midspan of the beam, and the resulting moment diagram would be linear (Figure 9.64).

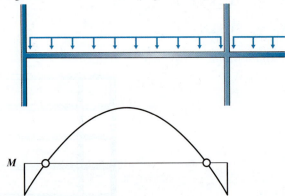

Figure 9.63 Moment diagram and inflection point locations for a beam in a rigid frame experiencing gravity load.

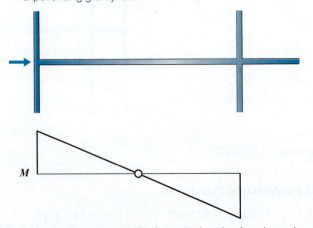

Figure 9.64 Moment diagram and inflection point location for a beam in a rigid frame experiencing lateral load.

Section 9.4 Combined Gravity and Lateral Loads

When the frame experiences both components of load (Figure 9.65), the moment diagram is a superposition of the individual diagrams (Figure 9.66a). The problem with predicting the locations of inflection points is that they shift based on the relative magnitudes of the two loads (Figure 9.66b–c). In fact, depending on the relative magnitudes of the loads, there might be one or two inflection points in the beam.

The solution is to analyze the frame by considering the gravity and lateral components separately, and then superimpose the results. This is valid *if* the material in the beams and columns remains linear elastic. We can superimpose individual internal forces, internal force diagrams, and reactions.

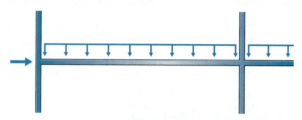

Figure 9.65 Beam in a rigid frame experiencing both gravity and lateral loads.

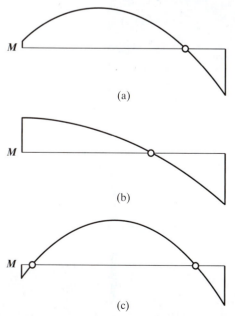

Figure 9.66 Moment diagrams for a beam in a rigid frame experiencing both gravity and lateral loads: (a) Balance of load magnitudes; (b) Larger lateral load; (c) Larger gravity load.

492   **Chapter 9**   Approximate Analysis of Rigid Frames

SECTION 9.4 HIGHLIGHTS

Combined Gravity and Lateral Loads

Process:

1. Assume the material remains linear elastic so that we can apply superposition.
2. Divide the loading into vertical (gravity) and lateral components.
3. Use the Gravity Load Method from Section 9.1 to analyze the rigid frame due to the vertical loading. Find the desired information (e.g., internal forces, reactions).
4. Use either the Portal Method or the Cantilever Method from Sections 9.2 and 9.3, respectively, to analyze the rigid frame due to the lateral loading. Find the desired information (e.g., internal forces, reactions).
5. Superimpose the desired information. Note: we must use the same sign convention for both Steps 3 and 4, because signs matter when we add the results.

EXAMPLE 9.9

Having completed an approximate analysis of the condominium for lateral wind load, our team leader has assigned us to also perform an approximate analysis for the wind uplift.

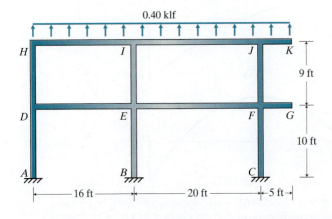

Figure 9.67

We are to combine our results with the results of the Portal Method in order to present preliminary requirements to the geotechnical engineers.

Approximate Analysis:

Estimate the inflection points at 10% of the beam spans from the columns, except for the cantilevered beam. Approximate the axial force in the beams as zero.

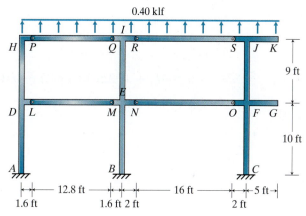

Figure 9.68a

FBD of beam PQ:

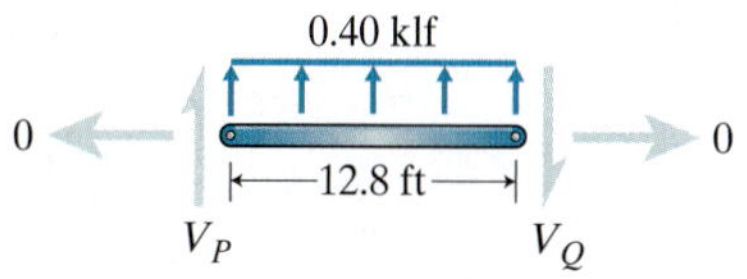

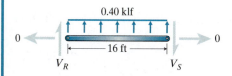

Figure 9.68b

$$\circlearrowright \Sigma M_Q = 0 = V_P(12.8 \text{ ft}) + 0.40 \text{ klf}(12.8 \text{ ft})(6.4 \text{ ft})$$
$$V_P = -2.56 \text{ k } (\downarrow + \uparrow)$$

$$+\uparrow \Sigma F_y = 0 = -2.56 \text{ k} + 0.40 \text{ klf}(12.8 \text{ ft}) - V_Q$$
$$V_Q = 2.56 \text{ k } (\downarrow + \uparrow)$$

FBD of beam LM:

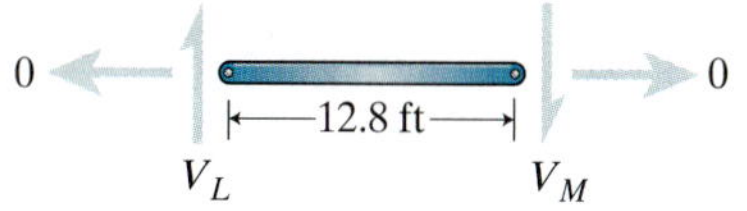

Figure 9.68c

$$\circlearrowright \Sigma M_M = 0 = V_L(12.8 \text{ ft})$$
$$V_L = 0$$

$$+\uparrow \Sigma F_y = 0 = 0 - V_M$$
$$V_M = 0$$

FBD of beam RS:

Figure 9.68d

$$\circlearrowright \Sigma M_S = 0 = V_R(16 \text{ ft}) + 0.40 \text{ klf}(16 \text{ ft})(8 \text{ ft})$$
$$V_R = -3.20 \text{ k } (\downarrow + \uparrow)$$

$$+\uparrow \Sigma F_y = 0 = -3.20 \text{ k} + 0.40 \text{ klf}(16 \text{ ft}) - V_S$$
$$V_S = 3.20 \text{ k } (\downarrow + \uparrow)$$

494   **Chapter 9**   Approximate Analysis of Rigid Frames

FBD of beam NO:

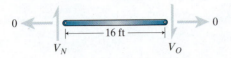

Figure 9.68e

$$+\circlearrowleft \Sigma M_O = 0 = V_N(16\text{ ft})$$
$$V_N = 0$$

$$+\uparrow \Sigma F_y = 0 = 0 - V_O$$
$$V_O = 0$$

FBD of column ALP:

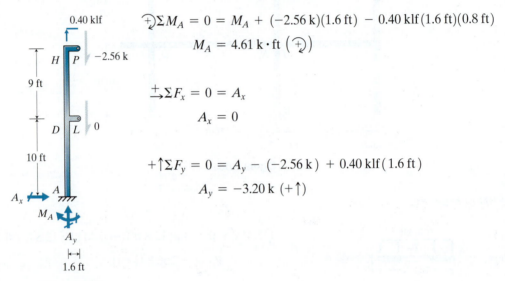

Figure 9.68f

$$+\circlearrowleft \Sigma M_A = 0 = M_A + (-2.56\text{ k})(1.6\text{ ft}) - 0.40\text{ klf}(1.6\text{ ft})(0.8\text{ ft})$$
$$M_A = 4.61\text{ k}\cdot\text{ft}\ (\circlearrowleft)$$

$$\xrightarrow{+}\Sigma F_x = 0 = A_x$$
$$A_x = 0$$

$$+\uparrow \Sigma F_y = 0 = A_y - (-2.56\text{ k}) + 0.40\text{ klf}(1.6\text{ ft})$$
$$A_y = -3.20\text{ k}\ (+\uparrow)$$

FBD of column BMNQR:

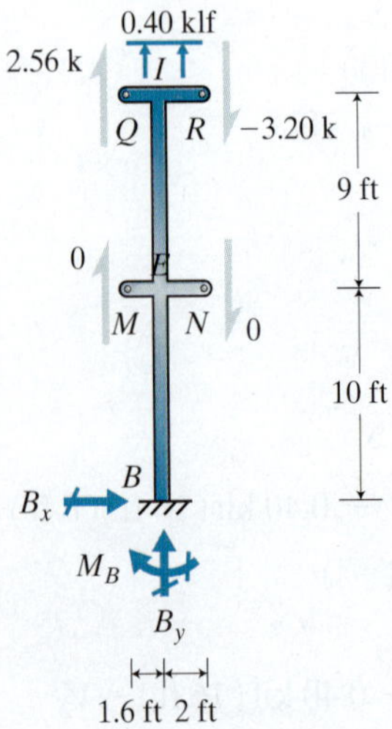

$$+\circlearrowleft \Sigma M_B = 0 = M_B + 2.56\text{ k}(1.6\text{ ft}) + 0.40\text{ klf}(1.6\text{ ft})(0.8\text{ ft})$$
$$- 0.40\text{ klf}(2\text{ ft})(1\text{ ft}) + (-3.20\text{ k})(2\text{ ft})$$
$$M_B = 2.59\text{ k}\cdot\text{ft}\ (\circlearrowleft)$$

$$\xrightarrow{+}\Sigma F_x = 0 = B_x$$
$$B_x = 0$$

$$+\uparrow \Sigma F_y = 0 = B_y + 2.56\text{ k} + 0.40\text{ klf}(1.6\text{ ft} + 2\text{ ft}) - (-3.20\text{ k})$$
$$B_y = -7.20\text{ k}\ (+\uparrow)$$

Figure 9.68g

Section 9.4 Combined Gravity and Lateral Loads

FBD of column COS:

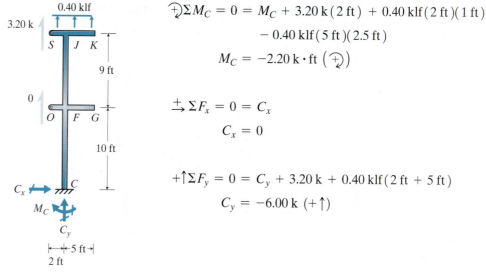

Figure 9.68h

$$+\circlearrowleft \Sigma M_C = 0 = M_C + 3.20\,\text{k}\,(2\,\text{ft}) + 0.40\,\text{klf}\,(2\,\text{ft})(1\,\text{ft})$$
$$- 0.40\,\text{klf}\,(5\,\text{ft})(2.5\,\text{ft})$$
$$M_C = -2.20\,\text{k}\cdot\text{ft}\ (\circlearrowleft)$$

$$\xrightarrow{+}\Sigma F_x = 0 = C_x$$
$$C_x = 0$$

$$+\uparrow \Sigma F_y = 0 = C_y + 3.20\,\text{k} + 0.40\,\text{klf}\,(2\,\text{ft} + 5\,\text{ft})$$
$$C_y = -6.00\,\text{k}\ (+\uparrow)$$

FBD of the entire frame with only gravity load:

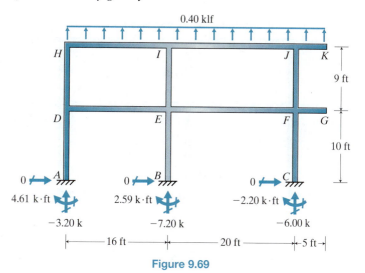

Figure 9.69

Evaluation of Results:

Verify overall equilibrium considering only gravity load:

Satisfied Fundamental Principles?

$$\xrightarrow{+}\Sigma F_x = 0 + 0 + 0 = 0\ \checkmark$$

$$+\uparrow \Sigma F_y = -3.20\,\text{k} - 7.20\,\text{k} - 6.00\,\text{k}$$
$$+ 0.40\,\text{klf}\,(16\,\text{ft} + 20\,\text{ft} + 5\,\text{ft}) = 0\ \checkmark$$

$$+\circlearrowleft \Sigma M_A = 4.61\,\text{k}\cdot\text{ft} + 2.59\,\text{k}\cdot\text{ft} - 2.20\,\text{k}\cdot\text{ft}$$
$$- (-7.20\,\text{k})(16\,\text{ft}) - (-6.00\,\text{k})(36\,\text{ft})$$
$$- 0.40\,\text{klf}\,(41\,\text{ft})(20.5\,\text{ft}) = 0\ \checkmark$$

496 **Chapter 9** Approximate Analysis of Rigid Frames

Superimpose Reactions:

Add the results from the Portal Method analysis (Example 9.5) to these results.

FBD of the entire frame with only lateral load:

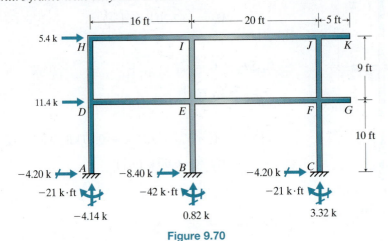

Figure 9.70

FBD of the entire frame with both components of wind load:

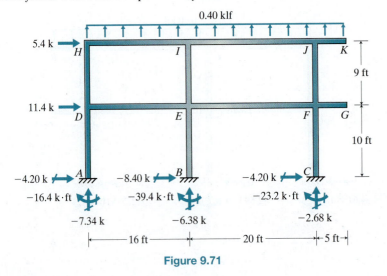

Figure 9.71

Evaluation of Results:
Satisfied Fundamental Principles?

$\xrightarrow{+} \Sigma F_x = 5.4 \text{ k} + 11.4 \text{ k} - 4.20 \text{ k} - 8.40 \text{ k} - 4.20 \text{ k} = 0$ ✓

$+\uparrow \Sigma F_y = -7.34 \text{ k} - 6.38 \text{ k} - 2.68 \text{ k}$
$\qquad + 0.40 \text{ klf}(16 \text{ ft} + 20 \text{ ft} + 5 \text{ ft}) = 0$ ✓

$\circlearrowleft \Sigma M_A = -16.4 \text{ k} \cdot \text{ft} - 39.4 \text{ k} \cdot \text{ft} - 23.2 \text{ k} \cdot \text{ft}$
$\qquad - (-6.38 \text{ k})(16 \text{ ft}) - (-2.68 \text{ k})(36 \text{ ft})$
$\qquad - 0.40 \text{ klf}(41 \text{ ft})(20.5 \text{ ft}) + 5.4 \text{ k}(19 \text{ ft})$
$\qquad + 11.4 \text{ k}(10 \text{ ft}) = -0.04 \text{ k} \cdot \text{ft} \approx 0$ ✓

Homework Problems

9.1 A highway overpass is supported by a two-bay rigid frame. In order to perform preliminary design of the members and foundation, we want to perform an approximate analysis. The foundations are idealized as fixed, but the geotechnical engineer estimates that the point of fixity (i.e., the depth where the column actually behaves as fixed) is several feet below the ground surface. The dead weight from the bridge can be approximated as a uniform distributed load.

a. Guess the direction of each of the reactions for each of the column bases.
b. List the assumptions you make in order to perform an approximate analysis of the frame.
c. Use approximate analysis to determine the peak positive and negative moments for the beam.
d. Use approximate analysis to find the reactions at the base of each column.
e. To demonstrate the effectiveness of approximate analysis, use structural analysis software to find the peak positive and negative moments for the beam. Also find the reactions at the base of each column. In order to provide better corrosion resistance and quicker strength gain, you anticipate that the design will be based on a concrete strength of 6000 psi; that results in an estimated modulus of elasticity, E, of 4400 ksi. A more experienced member of the team estimates that the beam will be 8 ft deep and 6 ft thick ($A = 48 \text{ ft}^2, I = 256 \text{ ft}^4$) and the columns will be 6 ft by 6 ft square ($A = 36 \text{ ft}^2, I = 108 \text{ ft}^4$).
f. Submit a printout of the displaced shape. Identify at least three features that suggest the displaced shape is reasonable.
g. Verify equilibrium of the computer aided analysis results.
h. Submit a printout of the moment diagram for the beam. Identify at least three features that suggest the moment diagram is reasonable.
i. Are the results of your computer aided analysis reasonable? Make a comprehensive argument to justify your answer.
j. Comment on why your guess in part (a) was or was not close to the solution in part (d).

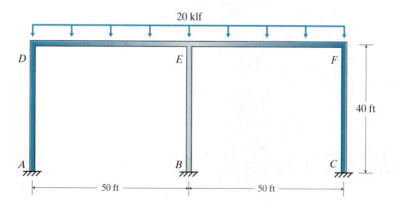

Prob. 9.1

Chapter 9 Approximate Analysis of Rigid Frames

9.2 Our firm is designing a single-story office building to span two existing buildings, creating a breezeway. Because of buried utilities, we cannot put both columns at the ends of the supporting beam. In order to begin preliminary selection of member properties and foundation design, we need to know the approximate internal forces and reactions due to each of the anticipated loads. The plan is to use shallow foundations that will act as pin supports.

We have been tasked with performing an approximate analysis of the frame considering dead load. Note that the roof and walls create point loads at the ends of the beam.

a. Guess the direction of each of the reactions for each of the column bases.
b. List the assumptions you make in order to perform an approximate analysis of the frame.
c. Use approximate analysis to construct the moment diagram for the beam.
d. Use approximate analysis to find the reactions at the base of each column.
e. To demonstrate the effectiveness of approximate analysis, use structural analysis software to find the peak positive and negative moments for the beam. Also find the reactions at the base of each column. In order to provide better corrosion resistance, you anticipate that the design will be based on a concrete strength of 30 MPa; that results in an estimated modulus of elasticity, E, of 26 GPa = 26 kN/mm². The team leader estimates that the beam and far column will be 20 cm wide and 40 cm deep ($A = 800 \text{ cm}^2, I = 1.067 \times 10^5 \text{ cm}^4$) and the interior column will be 20 cm wide by 60 cm deep ($A = 1200 \text{ cm}^2, I = 3.60 \times 10^5 \text{ cm}^4$).
f. Submit a printout of the displaced shape. Identify at least three features that suggest the displaced shape is reasonable.
g. Verify equilibrium of the computer aided analysis results.
h. Submit a printout of the moment diagram for the beam. Identify at least three features that suggest the moment diagram is reasonable.
i. Are the results of your computer aided analysis reasonable? Make a comprehensive argument to justify your answer.
j. Comment on why your guess in part (a) was or was not close to the solution in part (d).

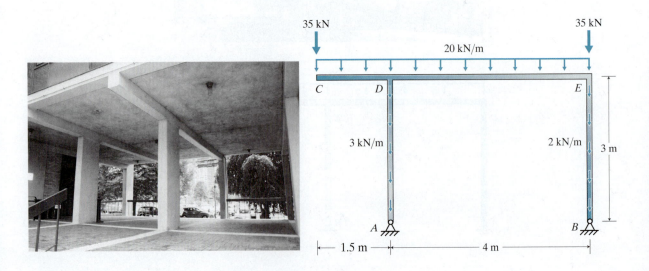

Prob. 9.2

9.3 Our team is designing an overpass structure that will be constructed in two phases. Because of the timing of the phases, the piers are separated into two parts. The geotechnical engineers have helped us estimate the point of fixity of the columns (i.e., the depth where the column actually behaves as fixed) well below the soil surface. Our job is to perform preliminary analysis of one of the piers under the loads expected when both phases are finished.

We have been tasked with performing an approximate analysis of the pier considering full live load.

a. Guess the direction of each of the reactions for each of the column bases.
b. List the assumptions you make in order to perform an approximate analysis of the frame.
c. Use approximate analysis to construct the moment diagram for the beam.
d. Use approximate analysis to find the reactions at the base of each column.
e. To demonstrate the effectiveness of approximate analysis, use structural analysis software to find the peak positive and negative moments for the beam. Also find the reactions at the base of each column. In order to provide better corrosion resistance and quicker strength gain, we guess that the design will be based on a concrete strength of 8000 psi; that results in an estimated modulus of elasticity, E, of 4600 ksi. An experienced member of the team estimates that the beam will be 3.5 ft deep and 3 ft thick ($A = 10.5 \text{ ft}^2, I = 10.7 \text{ ft}^4$) and the columns will be round with a 3-ft diameter ($A = 7.1 \text{ ft}^2, I = 4.0 \text{ ft}^4$).
f. Submit a printout of the displaced shape. Identify at least three features that suggest the displaced shape is reasonable.
g. Verify equilibrium of the computer aided analysis results.
h. Submit a printout of the moment diagram for the beam. Identify at least three features that suggest the moment diagram is reasonable.
i. Are the results of your computer aided analysis reasonable? Make a comprehensive argument to justify your answer.
j. Comment on why your guess in part (a) was or was not close to the solution in part (d).

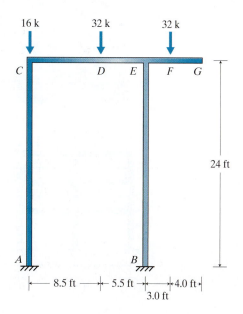

Prob. 9.3

9.4 The architect wants an open layout for a new medical facility. Therefore, the best lateral load-resisting system is a rigid frame. The floors and roof are concrete on metal decking, so they act one-way. The resulting dead load on an interior frame is a uniform distributed load. In order to make preliminary selection of member properties, the team leader would like an idea of the internal forces in the beams. At the same time, the geotechnical engineer would like to know the preliminary results for the requirements on the foundations.

a. Guess the direction of each of the reactions at the base of columns B1 and B2.
b. List the assumptions you make in order to perform an approximate analysis of the frame.
c. Use approximate analysis to determine the peak positive and negative moments for the roof beams and the other floor beams.
d. Use approximate analysis to find the reactions at the base of columns B1 and B2.
e. To demonstrate the effectiveness of approximate analysis, use structural analysis software to find the peak positive and negative moments for the roof beam and other floor beams for bay B1-B2. Also find the reactions at the base of columns B1 and B2. An experienced member of the team recommends we preliminarily estimate that all of the beams will be steel ($E = 200$ GPa $= 200$ kN/mm^2) W530 × 82s ($A = 1.045 \times 10^4$ mm^2, $I = 4.75 \times 10^8$ mm^4) and all of the columns will be steel W310 × 86s bending about their weak axes ($A = 1.097 \times 10^4$ mm^2, $I = 4.45 \times 10^7$ mm^4).
f. Submit a printout of the displaced shape. Identify at least four features that suggest the displaced shape is reasonable.
g. Verify equilibrium of the computer aided analysis results.
h. Are the results of your computer aided analysis reasonable? Make a comprehensive argument to justify your answer.
i. Comment on why your guess in part (a) was or was not close to the solution in part (d).

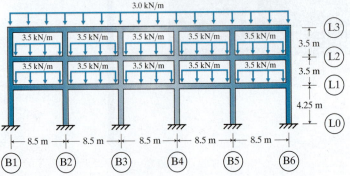

Prob. 9.4

9.5 Our firm has been selected to design a new mixed-use building. The lower two levels will be retail and offices, while the upper four floors will be residential. Our team leader has determined that flat-slab concrete construction will be the most cost effective. Based on previous experience, the team leader also wants us to design the columns to be the same size at each column line all the way up the structure; that will reduce the cost of formwork during construction. The geotechnical engineers think we will be able to use shallow foundations, thus saving even more expense. As a result, we should idealize the supports at the bottom of the columns as pins.

In order to begin designing the foundations, the geotechnical engineers need preliminary estimates of the support reactions that the foundations must provide. We have been tasked with finding those reactions for one column line due to anticipated dead load.

Although the flat slab will behave two-way, we can use a one-way approximation for this preliminary estimate. Therefore, we can approximate the dead load as uniform along the part of the floor and roof plates that act as beams. A more experienced team member's preliminary calculations suggest that the uniform dead load will be 840 plf along the beams and 270 plf along the height of the columns.

a. Guess which column will have the largest vertical reaction due to dead load and which will have the smallest.
b. List the assumptions you make in order to perform an approximate analysis of the frame.
c. Use approximate analysis to find the reactions (horizontal and vertical) at the base of each column.
d. To demonstrate the effectiveness of approximate analysis, use structural analysis software to find the reactions at the base of each column. The more experienced member's preliminary calculations were based on a slab thickness of 7.5 inches and an effective beam width of 5 ft ($A = 450$ in.^{2}, $I = 2110$ in.^{4}). That team member also anticipates we will use 16-inch-square columns ($A = 256$ in.^{2}, $I = 5460$ in.^{4}). The modulus of elasticity of the concrete will be about 3600 ksi.
e. Submit a printout of the displaced shape. Identify at least four features that suggest the displaced shape is reasonable.
f. Verify equilibrium of the computer aided analysis results.
g. Are the results of your computer aided analysis reasonable? Make a comprehensive argument to justify your answer.
h. Comment on why your guess in part (a) was or was not close to the solution in part (c).

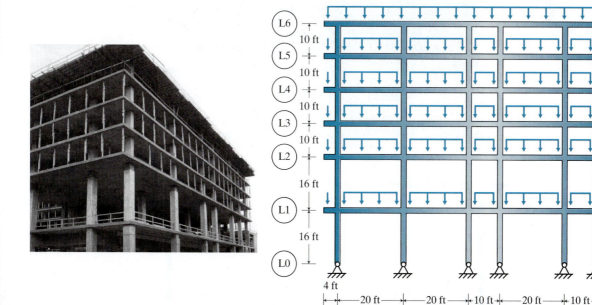

Prob. 9.5

9.6 A highway overpass is supported by a two-bay rigid frame. For the preliminary design of the members and foundation, we want to perform an approximate analysis. The foundations are idealized as fixed, but the geotechnical engineer estimates that the point of fixity (i.e., the depth where the column actually behaves as fixed) is several feet below the ground surface. A member of the team has calculated the earthquake loading as an equivalent static load.

a. Guess the direction of each of the reactions for each of the column bases.
b. What is the most appropriate method for approximate analysis of this frame? Why is that method most appropriate?
c. List the assumptions you make in order to perform an approximate analysis of the frame.
d. Use approximate analysis to determine the peak positive and negative moments for the beam.
e. Use approximate analysis to find the reactions at the base of each column.
f. To demonstrate the effectiveness of approximate analysis, use structural analysis software to find the peak positive and negative moments for the beam. Also find the reactions at the base of each column. In order to provide better corrosion resistance and quicker strength gain, we guess that the design will be based on a concrete strength of 6000 psi; that results in an estimated modulus of elasticity, E, of 4400 ksi. A more experienced member of the team estimates that the beam will be 8 ft deep and 6 ft thick ($A = 48 \text{ ft}^2, I = 256 \text{ ft}^4$) and the columns will be 6 ft by 6 ft square ($A = 36 \text{ ft}^2, I = 108 \text{ ft}^4$).
g. Submit a printout of the displaced shape. Identify at least three features that suggest the displaced shape is reasonable.
h. Verify equilibrium of the computer aided analysis results.
i. Are the results of your computer aided analysis reasonable? Make a comprehensive argument to justify your answer.
j. Comment on why your guess in part (a) was or was not close to the solution in part (e).

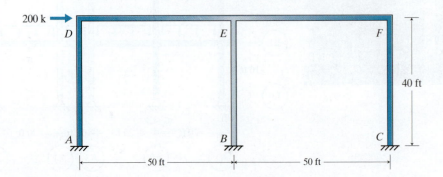

Prob. 9.6

9.7 Our firm is designing the bents that support a new highway overpass. The design team leader has chosen concrete for the bent. In order to begin designing the members, the team needs preliminary estimates of the peak moments in the beams and columns. We have been tasked with finding those moments due to lateral wind load.

Because of very soft soils, the geotechnical engineer has estimated the point of fixity of the columns well below the surface. The point of fixity is where the deep foundation behaves as a fixed support with essentially zero rotation.

a. Guess the direction of each of the reactions for each of the column bases.
b. What is the most appropriate method for approximate analysis of this frame? Why is that method most appropriate?
c. List the assumptions you make in order to perform an approximate analysis of the frame.
d. Use approximate analysis to determine the peak positive and negative moments for the cap beam and for each of the columns.
e. Use approximate analysis to find the reactions at the base of each column.
f. To demonstrate the effectiveness of approximate analysis, use structural analysis software to find the peak positive and negative moments for the cap beam and each column. Also find the reactions at the base of each column. In order to provide better corrosion resistance, we guess that the design will be based on a concrete strength of 30 MPa; that results in an estimated modulus of elasticity, E, of 26 GPa = 26 kN/mm². The team leader suggested trying square members 80 cm deep for both the cap beam and columns ($A = 6400$ cm², $I = 3.41 \times 10^6$ cm⁴).
g. Submit a printout of the displaced shape. Identify at least three features that suggest the displaced shape is reasonable.
h. Verify equilibrium of the computer aided analysis results.
i. Are the results of your computer aided analysis reasonable? Make a comprehensive argument to justify your answer.
j. Comment on why your guess in part (a) was or was not close to the solution in part (e).

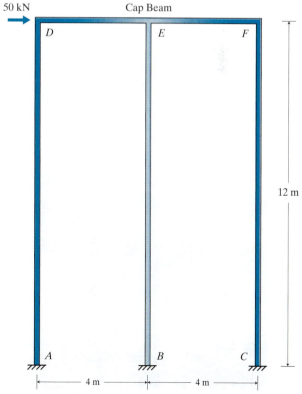

Prob. 9.7

9.8 As an alternative to the deep foundations, the team leader wants to consider using a grade beam, which would result in much shorter columns. Using a grade beam would allow us to use spread footings, which are less expensive.

a. Guess the direction of each of the reactions for each of the column bases.
b. What is the most appropriate method for approximate analysis of this frame? Why is that method most appropriate?
c. List the assumptions you make in order to perform an approximate analysis of the frame.
d. Use approximate analysis to determine the peak positive and negative moments for the cap beam, the grade beam, and each of the columns.
e. Use approximate analysis to find the reactions at the base of each column.
f. To demonstrate the effectiveness of approximate analysis, use structural analysis software to find the peak positive and negative moments for the cap beam, grade beam, and each column. Also, find the reactions at the base of each column. In order to provide better corrosion resistance, we guess that the design will be based on a concrete strength of 30 MPa; that results in an estimated modulus of elasticity, E, of 26 GPa = 26 kN/mm^2. The team leader suggested trying square members 80 cm deep for both the cap beam and columns ($A = 6400$ cm^2, $I = 3.41 \times 10^6$ cm^4) and a square member 120 cm deep for the grade beam ($A = 14{,}400$ cm^2, $I = 1.728 \times 10^7$ cm^4).
g. Submit a printout of the displaced shape. Identify at least three features that suggest the displaced shape is reasonable.
h. Verify equilibrium of the computer aided analysis results.
i. Are the results of your computer aided analysis reasonable? Make a comprehensive argument to justify your answer.
j. Comment on why your guess in part (a) was or was not close to the solution in part (e).

 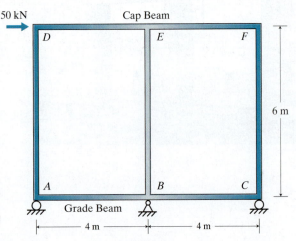

Prob. 9.8

9.9 The architect wants an open layout for a new medical facility. Therefore, the best lateral load-resisting system is a rigid frame. In order to make the preliminary selection of members, the team leader would like an idea of the internal forces in the beams. At the same time, the geotechnical engineer would like preliminary results for the requirements on the foundations. A team member has calculated the unfactored lateral wind load on each rigid frame.

Consider wind acting on the short face of the building. It is resisted by the longer rigid frame.

a. Guess the direction of each of the reactions at the base of columns B1 and B2.
b. What is the most appropriate method for approximate analysis of this frame? Why is that method most appropriate?
c. List the assumptions you make in order to perform an approximate analysis of the frame.
d. Use approximate analysis to determine the peak positive and negative moments for each of the beams in bay B1-B2.
e. Use approximate analysis to find the reactions at the base of columns B1 and B2.
f. To demonstrate the effectiveness of approximate analysis, use structural analysis software to find the peak positive and negative moments for each of the beams in bay B1-B2. Also, find the reactions at the base of columns B1 and B2. An experienced member of the team recommends we preliminarily estimate that all of the beams will be steel ($E = 200$ GPa $= 200$ kN/mm^2) W530 × 82s ($A = 1.045 \times 10^4$ mm^2, $I = 4.75 \times 10^8$ mm^4) and all of the columns will be steel W310 × 86s bending about their weak axes ($A = 1.097 \times 10^4$ mm^2, $I = 4.45 \times 10^7$ mm^4).
g. Submit a printout of the displaced shape. Identify at least four features that suggest the displaced shape is reasonable.
h. Verify equilibrium of the computer aided analysis results.
i. Are the results of your computer aided analysis reasonable? Make a comprehensive argument to justify your answer.
j. Comment on why your guess in part (a) was or was not close to the solution in part (e).

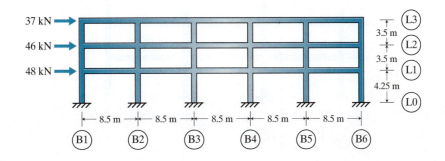

Prob. 9.9

9.10 Consider wind acting on the long face of the building. It is resisted by the shorter rigid frame.

a. Guess the direction of each of the reactions at the base of columns C2 and D2.
b. What is the most appropriate method for approximate analysis of this frame? Why is that method most appropriate?
c. List the assumptions you make in order to perform an approximate analysis of the frame.
d. Use approximate analysis to determine the peak positive and negative moments for each of the beams in bay C2-D2.
e. Use approximate analysis to find the reactions at the base of columns C2 and D2.
f. To demonstrate the effectiveness of approximate analysis, use structural analysis software to find the peak positive and negative moments for each of the beams in bay C2-D2. Also, find the reactions at the base of columns C2 and D2. An experienced member of the team recommends we preliminarily estimate that all of the beams will be steel ($E = 200$ GPa $= 200$ kN/mm^2) W530 × 82s ($A = 1.045 \times 10^4$ mm^2, $I = 4.75 \times 10^8$ mm^4) and all of the columns will be steel W310 × 86s bending about their strong axes ($A = 1.097 \times 10^4$ mm^2, $I = 1.977 \times 10^8$ mm^4).
g. Submit a printout of the displaced shape. Identify at least four features that suggest the displaced shape is reasonable.
h. Verify equilibrium of the computer aided analysis results.
i. Are the results of your computer aided analysis reasonable? Make a comprehensive argument to justify your answer.
j. Comment on why your guess in part (a) was or was not close to the solution in part (e).

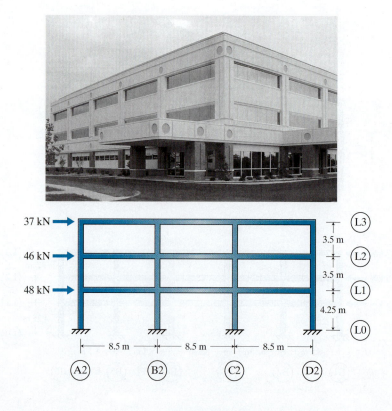

Prob. 9.10

9.11 We are designing a six-story office building to fit on a narrow site in a densely populated city. Based on the architectural requirements and local economics, our team leader has determined that steel will be the ideal structural material. In the short direction, the leader predicts that we need only one bay of rigid frame to carry the lateral load; therefore, the second bay uses pin connections and does not contribute to lateral load resistance.

In order to begin preliminary design of the frame members, the team needs to know the internal forces generated. We have been tasked with finding the internal forces for one column line due to lateral wind load.

a. Guess which rigid frame beam will have the largest moment due to lateral wind load.
b. What is the most appropriate method for approximate analysis of this frame? Why is that method most appropriate?
c. List the assumptions you make in order to perform an approximate analysis of the frame.
d. Use approximate analysis to find the peak moment in each rigid frame beam.
e. Use approximate analysis to find the peak moment in column A1.
f. To demonstrate the effectiveness of approximate analysis, use structural analysis software to find the peak moment in each rigid frame beam and the peak moment in column A1. An experienced member of the team recommends initially assuming that all of the columns are W10 × 45 ($A = 13.3$ in.^{2}, $I_x = 248$ in.^{4}, $I_y = 53.4$ in.^{4}) and all of the beams are W14 × 43 ($A = 12.6$ in.^{2}, $I_x = 428$ in.^{4}). The columns in the rigid frame bend about their primary axis, I_x, and the pin-connected column bends about its weak axis, I_y. This arrangement will add stiffness in the long direction.
g. Submit a printout of the displaced shape. Identify at least four features that suggest the displaced shape is reasonable.
h. Verify equilibrium of the computer aided analysis results.
i. Submit a printout of the moment diagram for the rigid frame. Identify at least three features that suggest the moment diagram is reasonable.
j. Are the results of your computer aided analysis reasonable? Make a comprehensive argument to justify your answer.
k. Comment on why your guess in part (a) was or was not close to the solution in part (d).

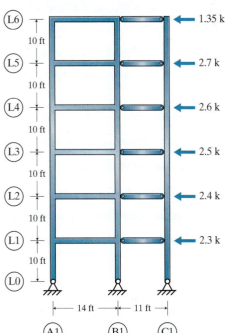

Prob. 9.11

9.12 Our firm is designing an eight-story office building in an urban location. The team leader has decided that the most cost-effective material to use for this situation is reinforced concrete. The primary lateral load-resisting system will be structural walls in the core; however, because we are using cast-in-place concrete, the columns and beams will act as rigid frames. Based on rigid diaphragm analysis (see Chapter 11), a team member has estimated the lateral wind load on one of the rigid frames.

In order to begin designing the foundations, the geotechnical engineers need preliminary estimates of only the vertical support reactions that the foundations must provide. We have been tasked with finding those reactions for one column line due to lateral wind load.

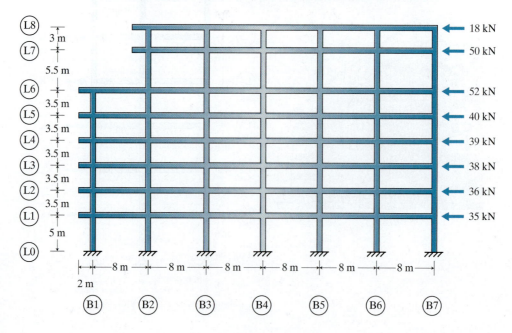

Prob. 9.12

An experienced member of the team estimates the following properties for the columns:

Column	Estimated Width (cm)	A (cm^2)	I (cm^4)
B1	30	900	6.75×10^4
B2	40	1600	2.13×10^5
B3	40	1600	2.13×10^5
B4	40	1600	2.13×10^5
B5	45	2025	3.42×10^5
B6	50	2500	5.21×10^5
B7	40	1600	2.13×10^5

a. Guess which column will have the largest vertical force due to lateral wind loading (can be up or down).
b. What is the most appropriate method for approximate analysis of this frame? Why is that method most appropriate?
c. For the desired results, what is the most efficient approximate analysis method? Use this method regardless of your answer to part (b).
d. List the assumptions you make in order to perform an approximate analysis of the frame.
e. Use approximate analysis to find the vertical reaction at the base of each column.
f. To demonstrate the effectiveness of approximate analysis, use structural analysis software to find the reactions at the base of each column. Preliminarily, we guess that the design will use concrete with a compressive strength of 35 MPa, which results in an estimated modulus of elasticity, E, of 28 GPa = 28 kN/mm^2. The experienced team member who estimated the column sizes also estimates that the beams will have these effective properties: $A = 3750$ cm^2, $I = 7.81 \times 10^5$ cm^4.
g. Submit a printout of the displaced shape. Identify at least three features that suggest the displaced shape is reasonable.
h. Verify equilibrium of the computer aided analysis results.
i. Are the results of your computer aided analysis reasonable? Make a comprehensive argument to justify your answer.
j. Comment on why your guess in part (a) was or was not close to the solution in part (e).

9.13 Our firm has been selected to design a new mixed-use building. The lower two levels will be retail and offices, while the upper four floors will be residential. Our team leader has determined that flat-slab concrete construction will be the most cost effective. Based on previous experience, the team leader also wants us to design the columns to be the same size at each column line all the way up the structure; that will reduce the cost of formwork during construction. The geotechnical engineers think we will be able to use shallow foundations, thus saving even more expense. As a result, we should idealize the supports at the bottom of the columns as pins.

In order to begin designing the foundations, the geotechnical engineers need preliminary estimates of only the vertical support reactions that the foundations must provide. We have been tasked with finding those reactions for one column line due to static equivalent seismic load.

a. Guess which column will have the largest vertical force due to seismic loading (can be up or down).
b. What is the most appropriate method for approximate analysis of this frame? Why is that method most appropriate?
c. For the desired results, vertical reactions, what is the most efficient approximate analysis method? Use this method regardless of your answer to part (b).
d. List the assumptions you make in order to perform an approximate analysis of the frame.
e. Use approximate analysis to find the vertical reaction at the base of each column.
f. To demonstrate the effectiveness of approximate analysis, use structural analysis software to find the reactions at the base of each column. The experienced member's preliminary calculations were based on a slab thickness of 7.5 inches and an effective beam width of 5 ft ($A = 450$ in.^{2}, $I = 2110$ in.^{4}). That team member also anticipates that we will use 16-inch-square columns ($A = 256$ in.^{2}, $I = 5460$ in.^{4}). The modulus of elasticity of the concrete will be about 3600 ksi.
g. Submit a printout of the displaced shape. Identify at least four features that suggest the displaced shape is reasonable.
h. Verify equilibrium of the computer aided analysis results.
i. Are the results of your computer aided analysis reasonable? Make a comprehensive argument to justify your answer.
j. Comment on why your guess in part (a) was or was not close to the solution in part (e).

Chapter 9 Approximate Analysis of Rigid Frames

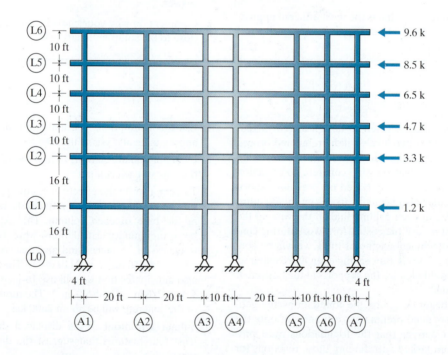

Prob. 9.13

9.14 Our team is designing an overpass structure that will be constructed in two phases. Because of the timing of the phases, the piers are separated into two parts. The geotechnical engineers have helped us estimate the point of fixity of the columns (i.e., the depth where the column actually behaves as fixed) well below the soil surface. Our job is to perform preliminary analysis of one of the piers under the loads expected before the second phase is added.

We have been tasked with performing an approximate analysis of the pier considering wind load, both vertical and lateral components.

a. Guess the direction of each of the reactions for each of the column bases.
b. List the assumptions you make in order to perform an approximate analysis of the frame.
c. Use approximate analysis to find the reactions at the base of each column.
d. To demonstrate the effectiveness of approximate analysis, use structural analysis software to find the peak positive and negative moments for the beam due to the combined vertical and lateral wind load. Also, find the reactions at the base of each column. In order to provide better corrosion resistance and quicker strength gain, we guess that the design will be based on a concrete strength of 8000 psi; that results in an estimated modulus of elasticity, E, of 4600 ksi. An experienced member of the team estimates that the beam will be 3.5 ft deep and 3 ft thick ($A = 10.5 \text{ ft}^2, I = 10.7 \text{ ft}^4$) and the columns will be round with a 3-ft diameter ($A = 7.1 \text{ ft}^2, I = 4.0 \text{ ft}^4$).
e. Submit a printout of the displaced shape. Identify at least three features that suggest the displaced shape is reasonable.
f. Verify equilibrium of the computer aided analysis results.
g. Are the results of your computer aided analysis reasonable? Make a comprehensive argument to justify your answer.
h. Comment on why your guess in part (a) was or was not close to the solution in part (c).

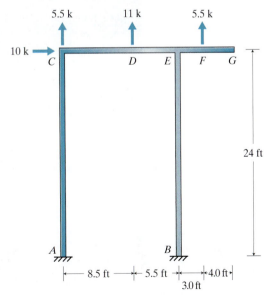

Prob. 9.14

9.15 Our firm is designing a small warehouse to be constructed in Europe. The senior designer has selected steel as the most economical structural material. Discussions with fabricators and erectors in the region where the warehouse will be built have led us to conclude that the most cost-effective option is to use pinned connections between the members and knee braces along the perimeter of the building. The knee braces cause the beams and columns to act as though they are rigidly connected.

In order to begin preliminary design of the steel members, the team needs preliminary estimates of the internal forces. We have been tasked with performing an approximate analysis of the end frame subjected to wind load, both vertical and lateral components.

a. Guess which column will have the largest moment.
b. List the assumptions you make in order to perform an approximate analysis of the frame.
c. Use approximate analysis to find the peak positive and negative moments in each beam and the peak positive and negative moments in each column.
d. To demonstrate the effectiveness of approximate analysis, use structural analysis software to find the peak positive and negative moments in each beam and each column due to the combined vertical and lateral wind load. The modulus of elasticity of steel, E, is 200 GPa = 200 kN/mm². Another member of the team has researched standard European sections and estimates that the columns will be HSS 150 × 150 × 6, square hollow tubes ($A = 34.2$ cm², $I = 1174$ cm⁴), and the beams will be IPE 360, I-shaped sections 360 mm deep ($A = 72.7$ cm², $I = 16{,}270$ cm⁴).
e. Submit a printout of the displaced shape. Identify at least three features that suggest the displaced shape is reasonable.
f. Verify equilibrium of the computer aided analysis results.
g. Submit a printout of the moment diagram for the entire frame. Identify at least three features that suggest the moment diagram is reasonable.
h. Are the results of your computer aided analysis reasonable? Make a comprehensive argument to justify your answer.
i. Comment on why your guess in part (a) was or was not close to the solution in part (c).

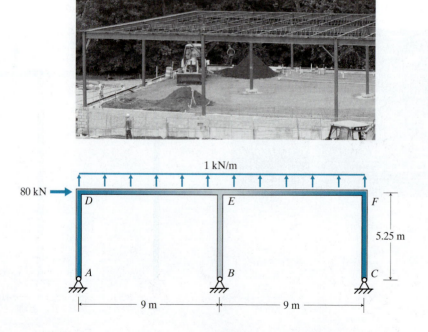

Prob. 9.15

9.16 Our team is designing a two-story office building. In order to provide an open layout and to maximize the views from the windows, the principal in charge of the project has chosen to use rigid frames. The senior engineer on the project has determined that steel will be the most cost-effective structural material for this situation.

In order to begin preliminary design of the frame members, the team needs estimates of the internal forces. In addition, the geotechnical engineers need estimates of the reactions that the foundations must apply in order to begin their design. We have been tasked with performing an approximate analysis of the middle column line subjected to wind load, both vertical and lateral components.

a. Guess which column support will have the largest vertical force, up or down.
b. List the assumptions you make in order to perform an approximate analysis of the frame.
c. Use approximate analysis to find the peak positive and negative moments in the top beam. Consider all three bays.
d. Use approximate analysis to find the reactions at the base of each column.
e. To demonstrate the effectiveness of approximate analysis, use structural analysis software to find the peak positive and negative moments in the top beam and all of the support reactions due to the combined vertical and lateral wind loads. The modulus of elasticity of steel, E, is 29,000 ksi. An experienced member of our team estimates that the columns will be W10 × 33s ($A = 9.41$ in.^{2}, $I = 171$ in.^{4}) and the beams will be W14 × 30s ($A = 8.85$ in.^{2}, $I = 291$ in.^{4}).
f. Submit a printout of the displaced shape. Identify at least four features that suggest the displaced shape is reasonable.
g. Verify equilibrium of the computer aided analysis results.
h. Submit a printout of the moment diagram for the top beam. Identify at least two features that suggest the moment diagram is reasonable.
i. Are the results of your computer aided analysis reasonable? Make a comprehensive argument to justify your answer.
j. Comment on why your guess in part (a) was or was not close to the solution in part (d).

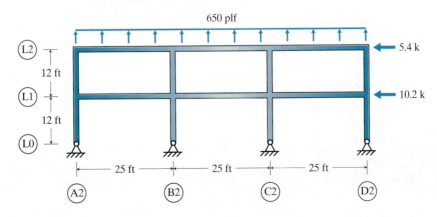

Prob. 9.16

CHAPTER 10

APPROXIMATE LATERAL DISPLACEMENTS

All lateral load-resisting systems displace, and knowing how much is important to proper design.

MOTIVATION

Lateral displacement of a structure is called *drift*. Drift is an important phenomenon to control in design because the things we attach to the structure are generally not very flexible. To protect attached materials from damage, our standard of practice (the generally accepted but not mandatory way of designing) is to keep the story drift between the values of height/400 and height/1000. In addition, we limit the drift of structures under seismic loads in order to limit damage to the lateral load-resisting elements. Therefore, we need to be able to predict the lateral displacement of the various types of lateral load-resisting elements: braced frames, rigid frames, and shear walls.

Most drift analysis is done using structural analysis software. However, the approximate methods covered in this chapter play several important roles for practicing engineers. For example, we use these approximate methods to help verify the results of computer aided analysis.

Even before we reach the detailed design stage, we use approximate analysis tools to help with preliminary design. The amount of drift depends on the magnitude of the applied load, the geometry of the frame, and the sizes of the members. However, we don't know the sizes of the members until they are designed. We can't design the members until we know the internal forces, and we can't accurately find the internal forces until we know the sizes of the members. Therefore, we can estimate the required member sizes to keep drift within allowable limits, thus starting the iterative design process with a good initial guess.

Approximate analysis methods also can help us quickly identify how many lateral load-resisting frames are needed in order to keep the total drift below the limit. Increasing the number of frames reduces the load on each frame and, therefore, reduces the drift. This process is often done faster by hand than by creating and running computer models.

It is important for us to note that the magnitudes of the lateral loads applied to a lateral load-resisting frame or wall can be significantly influenced by whether the load is transferred to the frame by a flexible diaphragm (Section 1.6: Distribution of Lateral Loads by Flexible Diaphragm) or a rigid diaphragm (Section 11.1: Distribution of Lateral Loads by Rigid Diaphragm). This chapter focuses on how to predict drift once we have determined the magnitudes of the loads.

10.1 Braced Frames—Story Drift Method

Introduction

The Story Drift Method of approximate analysis is the quickest for estimating the drift of braced frames. The derivation results in a single formula for drift; no direct calculation of internal forces is required. The Story Drift Method is not as accurate, however, as the Virtual Work Method discussed in Section 10.2.

This method predicts the drift of individual stories, which can be very helpful in preliminary design when we are choosing member sizes.

How-To

To understand the derivation of the Story Drift Method, we must first look at the displaced shape of a single-story, braced frame (Figure 10.1). Although this derivation uses a chevron braced frame, the derived theory still applies to cross braced and diagonally braced frames.

The goal is to find a relationship between the applied load and the lateral displacement. Since the columns and beams don't change shape, the lateral displacement must be due to a change in the length of the braces. Since the braces stay straight, the change in the length of the braces is due to axial force. So we need to link lateral force to axial force in the brace.

Consider the geometry change of the right brace in Figure 10.2. From trigonometry, we have

$$\cos\theta = \frac{\Delta_l}{\Delta} \quad \text{so} \quad \Delta = \frac{\Delta_l}{\cos\theta}$$

If Δ is very small, then l' and l'' are basically the same length, so Δ_l is the change in the length of the brace. From mechanics of materials, we know that the change in the length of the brace is given by

$$\Delta_l = \frac{Nl}{AE} \quad \text{so} \quad \Delta = \frac{Nl}{AE\cos\theta}$$

where

N = Axial force in the brace
l = Original length of the brace
A = Cross-sectional area of the brace
E = Modulus of elasticity of the brace

We saw in Section 8.2: Braced Frames with Lateral Loads that if the stiffness of each brace is the same, we can assume that the shear carried in each brace, V, is equal. Therefore, we divide the total shear carried by a braced frame by the total number of braces to find the shear carried in each brace (Figure 10.3). To find the force in one brace, N, we consider how the force vectors are related (Figure 10.4).

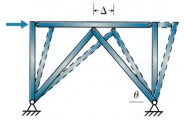

Figure 10.1 Displaced shape of a single-story, chevron braced frame subjected to lateral load.

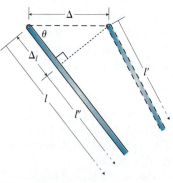

Figure 10.2 Close-up view of the end of the right brace deforming under lateral movement.

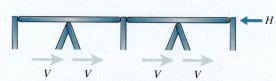

Figure 10.3 Lateral load is assumed to be carried only by the braces, and the horizontal component in each brace is equal.

From trigonometry, we have

$$\cos\theta = \frac{V}{N} \quad \text{so} \quad N = \frac{V}{\cos\theta}$$

We can substitute the expression for axial force, N, into the expression for drift, Δ. This gives the expression for the drift of one story of the braced frame, Δ_{story}:

$$\Delta_{story} = \frac{Vl}{AE\cos^2\theta}$$

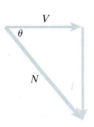

Figure 10.4 Force vector of axial force in brace broken into horizontal and vertical components.

To find the total drift of the frame, we sum the drifts calculated for each story:

$$\Delta_{total} = \Sigma\Delta_{story}$$

Bays with Different Brace Stiffnesses

Sometimes the braces at a level are not identical. They might be at different angles or made of different cross-sectional areas. Fortunately, this is not common, but when it happens the braces have different stiffnesses. Different stiffnesses means that the braces do not share the lateral load equally.

For most of the situations we will encounter, the effect of ignoring this difference in approximate analysis is small. Therefore, in Section 8.2: Braced Frames with Lateral Loads, we ignored any differences in brace stiffness. When predicting drift, however, we see the impact of the difference; we calculate different Δ_{story} values depending on which brace we consider. But now we have the tools we need to refine our approximation to account for these differences.

Deformations must be compatible, which means that no gaps or overlaps can occur in the displaced shape of the frame. *If* the axial deformation of the beams is negligible, we can have compatibility only if the story drift for each brace is the same (Figure 10.5). That means at any level of a braced frame, all of the bays at that level must drift the same amount.

To find the shear in each brace, we can use the story drift formula:

$$\Delta_a = \frac{V_a l_a}{A_a E_a \cos^2\theta_a}$$

$$\Delta_b = \frac{V_b l_b}{A_b E_b \cos^2\theta_b}$$

$$\Delta_c = \frac{V_c l_c}{A_c E_c \cos^2\theta_c}$$

where

V_i = Shear carried in one brace in the bay

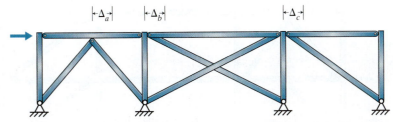

Figure 10.5 Drift of each braced bay must be the same for compatibility *if* the axial deformation of the beams is negligible.

Chapter 10 Approximate Lateral Displacements

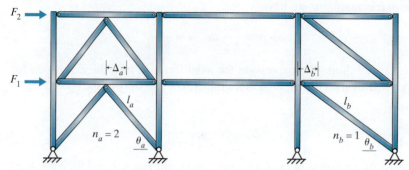

Figure 10.6 Two braced bays with different brace lengths and angles.

If there are more than two braced bays with unequal braces, the calculations for the different shears, V_i, can become lengthy. Let's focus on the most likely scenario: two bays with different braces (Figure 10.6).

We start by cutting the frame at one level and labeling the shear carried by each brace (Figure 10.7).

Using the Story Drift Method, we obtain the following predictions of drift for the two bays:

$$\Delta_a = \frac{V_a l_a}{A_a E_a \cos^2\theta_a}$$

$$\Delta_b = \frac{V_b l_b}{A_b E_b \cos^2\theta_b}$$

Since $\Delta_a = \Delta_b$, we have

$$\frac{V_a l_a}{A_a E_a \cos^2\theta_a} = \frac{V_b l_b}{A_b E_b \cos^2\theta_b}$$

So

$$V_b = V_a \frac{A_b E_b \cos^2\theta_b l_a}{A_a E_a \cos^2\theta_a l_b}$$

In order to keep the frame in Figure 10.7 in equilibrium, the applied lateral loads must equal the shears. To keep this general, let's define n as the number of braces in a single bay. Then equilibrium gives us the expression

$$\Sigma F_{\text{horiz}} = n_a V_a + n_b V_b$$

Substituting in our expression for V_b, we have

$$= n_a V_a + n_b V_a \frac{A_b E_b \cos^2\theta_b l_a}{A_a E_a \cos^2\theta_a l_b}$$

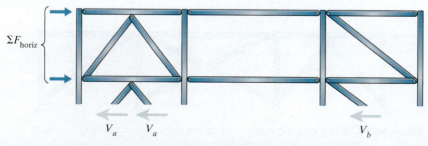

Figure 10.7 Shear distributed to each brace at a level in a frame.

We can use a common denominator to consolidate terms:

$$= n_a V_a \frac{A_a E_a \cos^2\theta_a l_b}{A_a E_a \cos^2\theta_a l_b} + n_b V_a \frac{A_b E_b \cos^2\theta_b l_a}{A_a E_a \cos^2\theta_a l_b}$$

$$= V_a \left(\frac{n_a A_a E_a \cos^2\theta_a l_b + n_b A_b E_b \cos^2\theta_b l_a}{A_a E_a \cos^2\theta_a l_b} \right)$$

Solving for the shear in a brace in the left bay, we obtain an expression for V_a and similarly an expression for V_b:

$$V_a = \frac{\Sigma F_{\text{horiz}} (A_a E_a \cos^2\theta_a l_b)}{n_a A_a E_a \cos^2\theta_a l_b + n_b A_b E_b \cos^2\theta_b l_a}$$

$$V_b = \frac{\Sigma F_{\text{horiz}} (A_b E_b \cos^2\theta_b l_a)}{n_a A_a E_a \cos^2\theta_a l_b + n_b A_b E_b \cos^2\theta_b l_a}$$

We can use one of these expressions to find the story drift.

$$\Delta_{\text{story}} = \Delta_a = \frac{V_a l_a}{A_a E_a \cos^2\theta_a}$$

$$= \frac{\Sigma F_{\text{horiz}} (A_a E_a \cos^2\theta_a l_b) l_a}{A_a E_a \cos^2\theta_a (n_a A_a E_a \cos^2\theta_a l_b + n_b A_b E_b \cos^2\theta_b l_a)}$$

$$\Delta_{\text{story}} = \frac{\Sigma F_{\text{horiz}} (l_a l_b)}{n_a A_a E_a \cos^2\theta_a l_b + n_b A_b E_b \cos^2\theta_b l_a}$$

Section 10.1 Highlights

Braced Frames—Story Drift Method

For any individual story of a braced frame (chevron, cross, or diagonally braced) when all braces are identical:

$$\Delta_{\text{story}} = \frac{Vl}{AE \cos^2\theta}$$

For any individual story with only two bays braced and the braces are not identical:

$$\Delta_{\text{story}} = \frac{\Sigma F_{\text{horiz}} (l_a l_b)}{n_a A_a E_a \cos^2\theta_a l_b + n_b A_b E_b \cos^2\theta_b l_a}$$

To estimate the total drift of a braced frame:

$$\Delta_{\text{total}} = \Sigma \Delta_{\text{story}}$$

Summary of Limitations and Assumptions:
- Assumes all shear is carried in the braces, none in the columns.
- Is not for eccentrically braced frames.
- Assumes lateral displacement is relatively small.
- Neglects any shortening of beams.
- Neglects the effect of change in the column length that will occur in multistory frames. [See Section 10.2: Virtual Work Method to account for this.]

Implication:
- Will always *underpredict* drift.

EXAMPLE 10.1

Our team is designing a supermarket in Canada. A team member has completed a preliminary design of the structural members. Now we need to confirm that the chevron braced frame does not drift too much with the sections chosen.

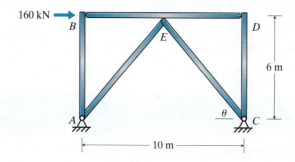

Figure 10.8a

Our job is to find the approximate drift and then use that value to verify the computer aided analysis results.

Preliminary selection of member properties:

Beam:	W360 × 64	$A = 81.4 \text{ cm}^2$	$I = 17{,}810 \text{ cm}^4$
Columns:	W310 × 60	$A = 75.3 \text{ cm}^2$	$I = 12{,}780 \text{ cm}^4$
Braces:	HSS 203.2 × 203.2 × 6.4	$A = 45.8 \text{ cm}^2$	$I = 2940 \text{ cm}^4$
Steel:	$E = 200 \text{ GPa} = 20{,}000 \text{ kN/cm}^2$		

Approximate Analysis:

Assume the shear is carried only in the braces. Assume the axial deformation of the beam is negligible. Since the braces are identical, the shear will be equal to each.

FBD of top part of frame:

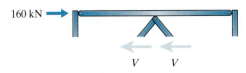

Figure 10.8b

$\overset{+}{\rightarrow} \Sigma F_x = 0 = 160 \text{ kN} - 2V$

$V = 80 \text{ kN}$

Use the Story Drift Method to predict the drift:

$$\Delta = \frac{Vl}{AE \cos^2\theta}$$

$l = \sqrt{(6 \text{ m})^2 + (5 \text{ m})^2} = 7.81 \text{ m}$

$\theta = \tan^{-1}\left(\frac{6 \text{ m}}{5 \text{ m}}\right) = 50.2°$

$\Delta = \dfrac{80 \text{ kN}(7.81 \text{ m})(1000 \text{ mm/m})}{(45.8 \text{ cm}^2)(20{,}000 \text{ kN/cm}^2)\cos^2(50.2°)}$

$= 1.66 \text{ mm}$

Evaluation of Results:

Approximation Predicted Outcomes?

Compare with the results of computer aided analysis (Direct Stiffness Method):

$\Delta = 1.66$ mm (the same to more than three significant figures!) ✓

Conclusion:

This comparison suggests that the computer aided analysis results are reasonable.

Use of Results:

Based on the materials used for the façade of the supermarket, we want to limit the drift to $h/500$.

$h/500 = 6000 \text{ mm}/500 = 12 \text{ mm}$

$\Delta \ll 12 \text{ mm}$

∴ Frame is adequately designed.

Chapter 10 Approximate Lateral Displacements

EXAMPLE 10.2

Our firm is designing a tall building. The architect would like us to use a cross braced frame and will expose the system visually. We are still in the schematic design phase, deciding which structural material will be best in this situation. We are on the team that is exploring the use of reinforced concrete as the primary structural material.

The team has made a preliminary selection of concrete properties and member sizes. Now we need a sense of the magnitude of the drift the building will experience with these member sizes. Since we need only a rough estimate, the Story Drift Method will be accurate enough.

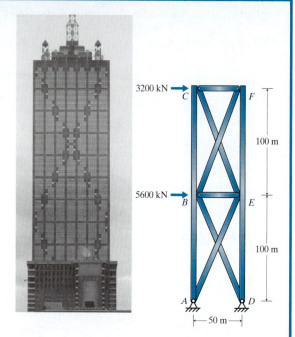

Figure 10.9a

Preliminary selection of member properties:

Beams:	60 cm × 100 cm	$A = 6000 \text{ cm}^2$	$I = 5.00 \times 10^6 \text{ cm}^4$
Columns:	60 cm × 60 cm	$A = 3600 \text{ cm}^2$	$I = 1.08 \times 10^6 \text{ cm}^4$
Braces:	60 cm × 60 cm	$A = 3600 \text{ cm}^2$	$I = 1.08 \times 10^6 \text{ cm}^4$
Concrete:	$E = 40 \text{ GPa} = 4000 \text{ kN/cm}^2$		

Approximate Analysis:

Assume the shear is carried only in the braces. Assume each brace carries an equal amount of shear. Assume the axial deformation of the beams is negligible.

FBD of frame cut at the top level:

$$\overset{+}{\rightarrow} \Sigma F_x = 0 = 3200 \text{ kN} - 2V_2$$

$$V_2 = 1600 \text{ kN}$$

Figure 10.9b

Story Drift Method:

$$l_2 = \sqrt{(100 \text{ m})^2 + (50 \text{ m})^2} = 111.8 \text{ m}$$

$$\theta_2 = \tan^{-1}\left(\frac{100 \text{ m}}{50 \text{ m}}\right) = 63.4°$$

$$\Delta_2 = \frac{V_2 l_2}{A_2 E_2 \cos^2 \theta_2} = \frac{1600 \text{ kN}(111.8 \text{ m})(1000 \text{ mm/m})}{(3600 \text{ cm}^2)(4000 \text{ kN/cm}^2)\cos^2(63.4°)}$$

$$= 62 \text{ mm}$$

FBD of frame cut at the bottom level:

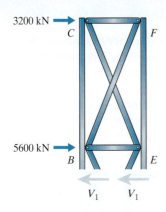

Figure 10.9c

$$\overset{+}{\rightarrow} \Sigma F_x = 0 = 3200 \text{ kN} + 5600 \text{ kN} - 2V_1$$
$$V_1 = 4400 \text{ kN}$$

Story Drift Method:

$$l_1 = \sqrt{(100 \text{ m})^2 + (50 \text{ m})^2} = 111.8 \text{ m}$$

$$\theta_1 = \tan^{-1}\left(\frac{100 \text{ m}}{50 \text{ m}}\right) = 63.4°$$

$$\Delta_1 = \frac{V_1 l_1}{A_1 E_1 \cos^2\theta_1}$$

$$= \frac{4400 \text{ kN}(111.8 \text{ m})(1000 \text{ mm/m})}{(3600 \text{ cm}^2)(4000 \text{ kN/cm}^2)\cos^2(63.4°)}$$

$$= 170 \text{ mm}$$

Total drift:

$$\Delta_{\text{total}} = \Delta_1 + \Delta_2$$
$$= 170 \text{ mm} + 62 \text{ mm} = 232 \text{ mm}$$

Use of Results:

Drift limits are typically for individual stories because they protect nonstructural elements attached to the structure (e.g., windows, partition walls). However, comparing the total drift to typical drift limits gives us a sense of whether this design will need to be revised.

$$\text{Factor} = \frac{h}{\Delta}$$

$$= \frac{200 \text{ m}}{232 \text{ mm}} = 862$$

$$\Delta_{\text{total}} = \frac{h_{\text{total}}}{862}$$

The typical drift limit factors are 500–1000, and the peak story drift will be larger than the average for the total height. Also, the Story Drift Method underpredicts drift, so we have overestimated the factor. Therefore, if reinforced concrete appears to be the best option for the structural system, we should carefully review the drift limits and refine our prediction of the drift.

EXAMPLE 10.3

Our team is designing a large recreation facility. The senior engineer has chosen chevron braces for the lateral load-resisting system. Based on previous experience, that engineer anticipates we will need to use two bays of braces. Architectural constraints dictate that the two bays will be different sizes and the two stories will be different heights.

The team has already made a preliminary selection of member properties. Before we create a computer model of the structure, we want to know whether the chosen members will be stiff enough to keep the drift well below the limit of $h/500$ for each story.

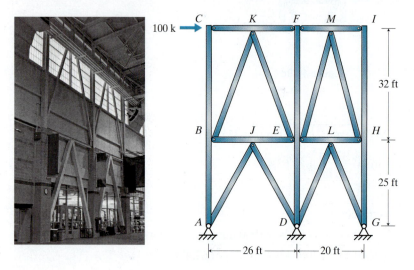

Figure 10.10a

Preliminary selection of member properties:

Beams:	W10 × 33	$A = 9.71$ in^2	$I_x = 171$ in^4
Columns:	W14 × 90	$A = 26.5$ in^2	$I_y = 362$ in^4
Braces:	HSS 8 × 8 × $\frac{1}{4}$	$A = 7.10$ in^2	$I = 70.7$ in^4
Steel:	$E = 29{,}000$ ksi		

Approximate Analysis:

Assume the shear is carried only in the braces. Assume each pair of braces in a bay shares the bay shear equally. Assume the axial deformation of the beams is negligible.

FBD of frame cut at the top level:

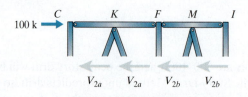

Figure 10.10b

Story Drift Method with unequal braces:

$$l_{2a} = \sqrt{(32\text{ ft})^2 + (13\text{ ft})^2} = 34.5\text{ ft} = 414\text{ in.}$$

$$\theta_{2a} = \tan^{-1}\left(\frac{32\text{ ft}}{13\text{ ft}}\right) = 67.9°$$

$$l_{2b} = \sqrt{(32\text{ ft})^2 + (10\text{ ft})^2} = 33.5\text{ ft} = 402\text{ in.}$$

$$\theta_{2b} = \tan^{-1}\left(\frac{32\text{ ft}}{10\text{ ft}}\right) = 72.6°$$

Since A and E are the same for all braces, and each bay has two braces, the drift formula becomes

$$\Delta_2 = \frac{\Sigma F_{\text{horiz}}(l_{2a}l_{2b})}{2AE\cos^2\theta_{2a}l_{2b} + 2AE\cos^2\theta_{2b}l_{2a}}$$

$$= \frac{\Sigma F_{\text{horiz}}(l_{2a}l_{2b})}{2AE[l_{2b}\cos^2\theta_{2a} + l_{2a}\cos^2\theta_{2b}]}$$

$$= \frac{100\text{ k}(414\text{ in.})(402\text{ in.})}{2(7.10\text{ in}^2)(29{,}000\text{ ksi})[(402\text{ in.})\cos^2(67.9°) + (414\text{ in.})\cos^2(72.6°)]}$$

$$= 0.43\text{ in.}$$

FBD of frame cut at the bottom level:

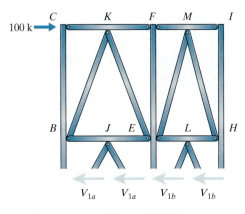

Figure 10.10c

Story Drift Method with unequal braces:

$$l_{1a} = \sqrt{(25\text{ ft})^2 + (13\text{ ft})^2} = 28.2\text{ ft} = 338\text{ in.}$$

$$\theta_{1a} = \tan^{-1}\left(\frac{25\text{ ft}}{13\text{ ft}}\right) = 62.5°$$

$$l_{1b} = \sqrt{(25\text{ ft})^2 + (10\text{ ft})^2} = 26.9\text{ ft} = 323\text{ in.}$$

$$\theta_{1b} = \tan^{-1}\left(\frac{25\text{ ft}}{10\text{ ft}}\right) = 68.2°$$

Since A and E are the same for all braces, and each bay has two braces, the drift formula becomes

$$\Delta_1 = \frac{\Sigma F_{\text{horiz}}(l_{1a}l_{1b})}{2AE[l_{1b}\cos^2\theta_{1a} + l_{1a}\cos^2\theta_{1b}]}$$

$$= \frac{100\text{ k}(338\text{ in.})(323\text{ in.})}{2(7.10\text{ in}^2)(29{,}000\text{ ksi})[(323\text{ in.})\cos^2(62.5°) + (338\text{ in.})\cos^2(68.2°)]}$$

$$= 0.23\text{ in.}$$

Evaluation of Results:

Approximation Predicted Outcomes?

Compare with the results of computer aided analysis (Direct Stiffness Method) at the middle column:

$\Delta_2 = 0.46$ in.

$\Delta_1 = 0.24$ in.

As expected, the Story Drift Method approximation underpredicts drift compared with the Direct Stiffness Method (approximately 4%–7% lower). ✓

Conclusion:

These comparisons suggest that the computer aided analysis results are reasonable.

Use of Results:

Compare each story's drift with its limit:

$$\text{Drift limit}_2 = \frac{h_2}{500} = \frac{32\text{ ft}(12\text{ in./ft})}{500} = 0.77\text{ in.}$$

$\Delta_2 = 0.43$ in.

$\Delta_2 \ll \text{Limit}_2$ $\therefore$ Upper-story design appears adequate.

$$\text{Drift limit}_1 = \frac{h_1}{500} = \frac{25\text{ ft}(12\text{ in./ft})}{500} = 0.60\text{ in.}$$

$\Delta_1 = 0.23$ in.

$\Delta_1 \ll \text{Limit}_1$ $\therefore$ Lower-story design appears adequate also.

10.2 Braced Frames—Virtual Work Method

Introduction

The Story Drift Method neglects the impact of changes in column length on the frame drift. Approximate analysis of the braced frame (Section 8.2: Braced Frames with Lateral Loads) can be combined with the Virtual Work Method of predicting deformations (Section 5.3: Virtual Work Method) to improve the estimate of drift for braced frames.

The Virtual Work Method provides displacement at only one point, so this method provides the total drift at a level. If we want to know the story drift, we need to find the total drift at the top and bottom of the level and take the difference. Therefore, we need to apply the Virtual Work Method twice in order to find one story drift.

How-To

For an approximate analysis of braced frames, we assume that all shear is carried as the horizontal component of braces at a level. We further assume that the horizontal component is the same in each brace. Knowing the force in each brace, we can use the equations of equilibrium to find the force in each column segment.

Displacement is predicted in the Virtual Work Method by the following expression:

$$\Delta = \Sigma \frac{nNl}{AE} + \Sigma \int \frac{mM}{EI} dx$$

If we neglect the effects of bending everywhere in the frame, the expression reduces to

$$\Delta = \Sigma \frac{nNl}{AE}$$

where
- n = Axial force in the member due to the virtual force
- N = Axial force in the member due to the real applied forces
- l = Length of the member
- A = Cross-sectional area of the member
- E = Modulus of elasticity of the member

Note that most of the energy stored in the structure will be in the braces and columns. Therefore, we can neglect the impact of deformation of the beams, since it has minimal impact.

Review of the formula in the Story Drift Method reveals that it is a special case of the Virtual Work Method. If we take the axial force in the columns to be zero, the Virtual Work Method reduces to the Story Drift Method. Since the axial force in columns at the top story of chevron braced frames is zero, the two methods will produce the exact same drift values for single-story chevron braced frames.

SECTION 10.2 HIGHLIGHTS

Braced Frames—Virtual Work Method

To estimate the total drift of a braced frame (considering braces and columns only):

$$\Delta_{total} = \Sigma \frac{nNl}{AE}$$

Summary of Limitations and Assumptions:
- Assumes all shear is carried in the braces, none in the columns.
- Neglects effects of bending everywhere in the frame.
- Neglects any elongation or shortening of the beams.
- Is not for eccentrically braced frames.

Implications:
- Will usually *underpredict* drift, but not always.
- Should be closer than Story Drift Method predictions.

Example 10.4

We are designing a steel parking garage in Europe. The team has completed preliminary design of the structural members. A coworker has performed a computer aided analysis of the chevron braced frame and determined a total drift of 27 mm. Our coworker is not convinced that the results of the computer aided analysis are reasonable and has come to you for help.

To help evaluate our coworker's results, let's use the Virtual Work Method to predict the total drift.

Figure 10.11a

Preliminary selection of member properties:

Beams:	IPE 450	$A = 98.8 \text{ cm}^2$	$I_x = 33{,}740 \text{ cm}^4$
Columns:	HEB 180	$A = 65.3 \text{ cm}^2$	$I_x = 3831 \text{ cm}^4$
Braces:	HSS 120 × 5	$A = 22.7 \text{ cm}^2$	$I = 98 \text{ cm}^4$
Steel:	$E = 200 \text{ GPa} = 20{,}000 \text{ kN/cm}^2$		

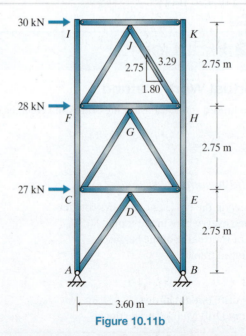

Figure 10.11b

Section 10.2 Braced Frames—Virtual Work Method

Approximate Analysis:
Assume the shear at a level is carried by only the braces. Assume the braces at a level share the shear equally. Assume the axial deformation of the beams is negligible.

Approximate analysis of the real system:
FBD of frame cut at the top level:
   Find the axial force in each brace.

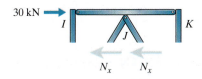

Figure 10.12a

$\xrightarrow{+} \Sigma F_x = 0 = 30 \text{ kN} - 2N_x$

$N_x = 15 \text{ kN}$

$\therefore N_{FJ} = +\dfrac{3.29}{1.80}(15 \text{ kN}) = +27.4 \text{ kN}$

$N_{HJ} = -\dfrac{3.29}{1.80}(15 \text{ kN}) = -27.4 \text{ kN}$

Find the axial forces in the columns.

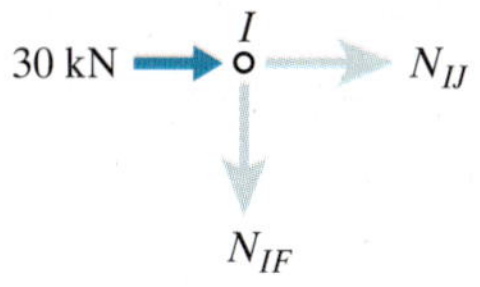

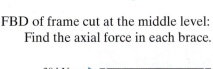

$+\downarrow \Sigma F_y = 0 = N_{IF}$

Figure 10.12b

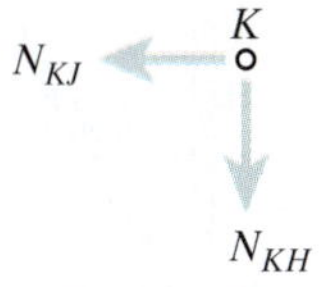

$+\downarrow \Sigma F_y = 0 = N_{KH}$

Figure 10.12c

FBD of frame cut at the middle level:
   Find the axial force in each brace.

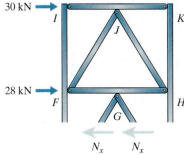

Figure 10.12d

$\xrightarrow{+} \Sigma F_x = 0 = 30 \text{ kN} + 28 \text{ kN} - 2N_x$

$N_x = 29 \text{ kN}$

$\therefore N_{CG} = +\dfrac{3.29}{1.80}(29 \text{ kN}) = +53.0 \text{ kN}$

$N_{EG} = -\dfrac{3.29}{1.80}(29 \text{ kN}) = -53.0 \text{ kN}$

Find the axial forces in the columns.

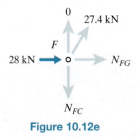

Figure 10.12e

$+\uparrow \Sigma F_y = 0 = \dfrac{2.75}{3.29}(27.4 \text{ kN}) - N_{FC}$

$N_{FC} = +22.9 \text{ kN}$

530   **Chapter 10**   Approximate Lateral Displacements

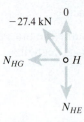

Figure 10.12f

$$+\uparrow \Sigma F_y = 0 = \frac{2.75}{3.29}(-27.4 \text{ kN}) - N_{HE}$$

$$N_{HE} = -22.9 \text{ kN}$$

FBD of frame cut at the bottom level:
Find the axial force in each brace.

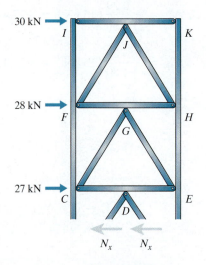

Figure 10.12g

$$\pm\!\!\!\to \Sigma F_x = 0 = 30 \text{ kN} + 28 \text{ kN} + 27 \text{ kN} - 2N_x$$

$$N_x = 42.5 \text{ kN}$$

$$\therefore N_{AD} = +\frac{3.29}{1.80}(42.5 \text{ kN}) = +77.7 \text{ kN}$$

$$N_{BD} = -\frac{3.29}{1.80}(42.5 \text{ kN}) = -77.7 \text{ kN}$$

Find the axial forces in the columns.

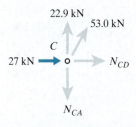

Figure 10.12h

$$+\uparrow \Sigma F_y = 0 = 22.9 \text{ kN} + \frac{2.75}{3.29}(53.0 \text{ kN}) - N_{CA}$$

$$N_{CA} = +67.2 \text{ kN}$$

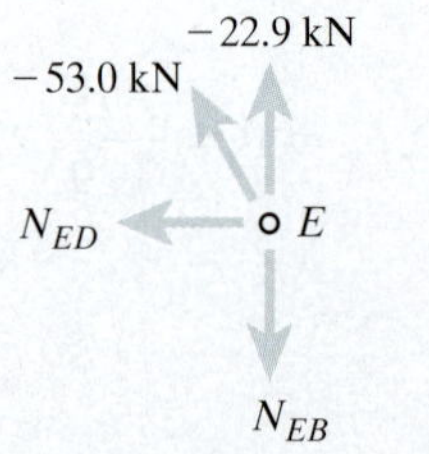

$$+\uparrow \Sigma F_y = 0 = -22.9 \text{ kN} + \frac{2.75}{3.29}(-53.0 \text{ kN}) - N_{EB}$$

$$N_{EB} = -67.2 \text{ kN}$$

Figure 10.12i

Section 10.2 Braced Frames—Virtual Work Method 531

Approximate analysis of the virtual system:

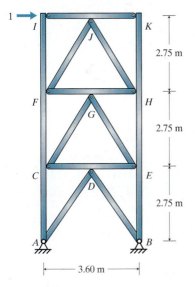

Figure 10.13a

FBD of frame cut at the top level:
Find the axial force in each brace.

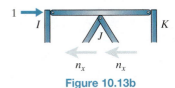

Figure 10.13b

$$\xrightarrow{+} \Sigma F_x = 0 = 1 - 2n_x$$

$$n_x = 0.5$$

$$\therefore n_{FJ} = +\frac{3.29}{1.80}(0.5) = +0.914$$

$$n_{HJ} = -\frac{3.29}{1.80}(0.5) = -0.914$$

Find the axial forces in the columns.

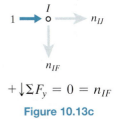

$+\downarrow \Sigma F_y = 0 = n_{IF}$

Figure 10.13c

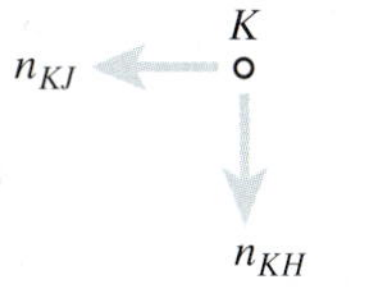

$+\downarrow \Sigma F_y = 0 = n_{KH}$

Figure 10.13d

FBD of frame cut at the middle level:
Find the axial force in each brace.

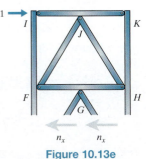

Figure 10.13e

$$\xrightarrow{+} \Sigma F_x = 0 = 1 - 2n_x$$

$$n_x = 0.5$$

$$\therefore n_{CG} = +\frac{3.29}{1.80}(0.5) = +0.914$$

$$n_{EG} = -\frac{3.29}{1.80}(0.5) = -0.914$$

532   Chapter 10   Approximate Lateral Displacements

Find the axial forces in the columns.

Figure 10.13f   Figure 10.13g

$$+\uparrow \Sigma F_y = 0 = \frac{2.75}{3.29}(0.914) - n_{FC}$$
$$n_{FC} = +0.764$$

$$+\uparrow \Sigma F_y = 0 = \frac{2.75}{3.29}(-0.914) - n_{HE}$$
$$n_{HE} = -0.764$$

FBD of frame cut at the bottom level:
Find the axial force in each brace.

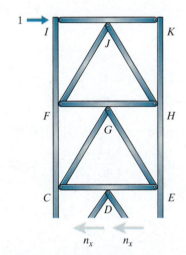

Figure 10.13h

$$\pm \Sigma F_x = 0 = 1 - 2n_x$$
$$n_x = 0.5$$
$$\therefore n_{AD} = +\frac{3.29}{1.80}(0.5) = +0.914$$
$$n_{BD} = -\frac{3.29}{1.80}(0.5) = -0.914$$

Find the axial forces in the columns.

Figure 10.13i   Figure 10.13j

$$+\uparrow \Sigma F_y = 0 = 0.764 + \frac{2.75}{3.29}(0.914) - n_{CA}$$
$$n_{CA} = +1.528$$

$$+\uparrow \Sigma F_y = 0 = -0.764 + \frac{2.75}{3.29}(-0.914) - n_{EB}$$
$$n_{EB} = -1.528$$

Summary of internal forces:

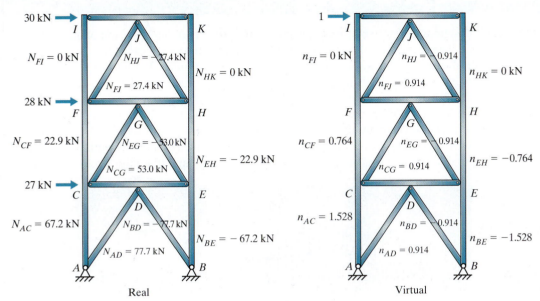

Figure 10.14

Virtual Work Method:

$$\Delta_I = \Sigma \frac{nNl}{AE}$$

$$= (2)\frac{(0.914)(27.4 \text{ kN})(3.29 \text{ m})}{(22.7 \text{ cm}^2)(20{,}000 \text{ kN/cm}^2)} + (2)\frac{(0.914)(53.0 \text{ kN})(3.29 \text{ m})}{(22.7 \text{ cm}^2)(20{,}000 \text{ kN/cm}^2)}$$

$$+ (2)\frac{(0.914)(77.7 \text{ kN})(3.29 \text{ m})}{(22.7 \text{ cm}^2)(20{,}000 \text{ kN/cm}^2)} + (2)\frac{(0.764)(22.9 \text{ kN})(2.75 \text{ m})}{(65.3 \text{ cm}^2)(20{,}000 \text{ kN/cm}^2)}$$

$$+ (2)\frac{(1.528)(67.2 \text{ kN})(2.75 \text{ m})}{(65.3 \text{ cm}^2)(20{,}000 \text{ kN/cm}^2)}$$

$$= 3.63 \times 10^{-4} \text{ m} + 7.02 \times 10^{-4} \text{ m} + 10.29 \times 10^{-4} \text{ m} + 0.74 \times 10^{-4} \text{ m} + 4.32 \times 10^{-4} \text{ m}$$

$$= 2.60 \times 10^{-3} \text{ m} = 2.60 \text{ mm}$$

Evaluation of Results:

Approximation Predicted Outcomes?

The Virtual Work Method always underpredicts drift, but the result should not be 1/10th of the result from the detailed analysis. Therefore, it appears that our coworker's result is *not* reasonable. ✗

Approximation Predicted Outcomes?

If we redo the computer aided analysis (Direct Stiffness Method), we obtain the following result:

$$\Delta = 2.69 \text{ mm}$$

As expected, the Virtual Work Method approximation underpredicts drift compared with the Direct Stiffness Method, but it is only about 3% lower. ✓ Therefore, this computer aided drift prediction appears reasonable.

EXAMPLE 10.5

Our firm is designing a new two-story department store to anchor an extension of an existing shopping mall. The architectural requirements are for maximum openness of the space; therefore, the lateral load-resisting system must be on the outer perimeter. For a store this size, an experienced engineer recommends we use two bays of chevron braces on each side of the building.

Members of the team have finished a preliminary design of the structural members. Before we finalize the selection of the members, we need to make sure that the frames keep the story drift sufficiently small. The manufacturer of the exterior precast panels requires that the story drift be below $h/1000$ in order to warrantee the panels. We have been tasked with predicting the story drift and determining whether it meets the precast panel manufacturer's requirements.

Figure 10.15a

Preliminary selection of member properties:

Beams:	W8 × 31	$A = 9.13$ in²	$I_x = 110$ in⁴
Columns:	W10 × 39	$A = 11.5$ in²	$I_x = 290$ in⁴
Braces:	HSS 5 × 5 × $\frac{1}{4}$	$A = 4.30$ in²	$I = 16.0$ in⁴
Steel:	$E = 29{,}000$ ksi		

Approximate Analysis:

Assume the shear at each level is carried by only the braces. Assume the braces at a level share the shear equally. Assume the axial deformation of the beams is negligible.

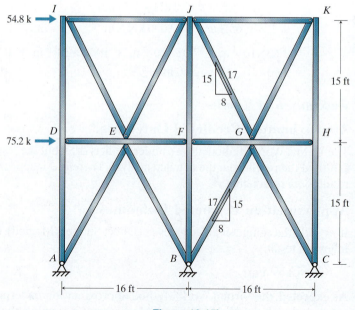

Figure 10.15b

Section 10.2 Braced Frames—Virtual Work Method

Approximate analysis of the real system:

FBD of frame cut at the top level:
    Find the axial force in each brace.

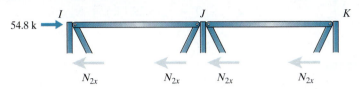

Figure 10.16a

$$\xrightarrow{+} \Sigma F_x = 0 = 54.8 \text{ k} - 4N_{2x}$$
$$N_{2x} = 13.7 \text{ k}$$
$$\therefore N_{2\text{brace}} = \frac{17}{8}(13.7 \text{ k}) = 29.1 \text{ k}$$

FBD of frame cut at the lower level:

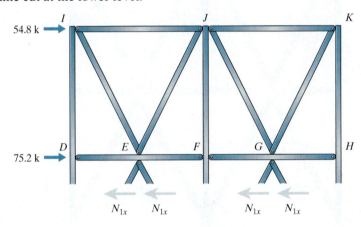

Figure 10.16b

$$\xrightarrow{+} \Sigma F_x = 0 = 54.8 \text{ k} + 75.2 \text{ k} - 4N_{1x}$$
$$N_{1x} = 32.5 \text{ k}$$
$$\therefore N_{2\text{brace}} = \frac{17}{8}(32.5 \text{ k}) = 69.1 \text{ k}$$

FBD of joint I:

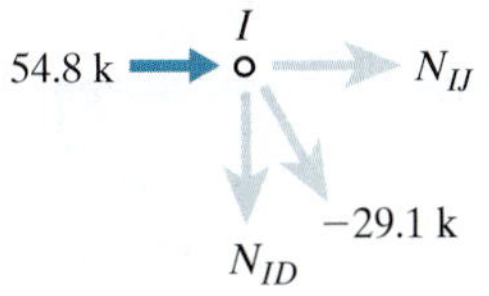

54.8 k → I → N_{IJ}
↓ N_{ID} ↘ −29.1 k

$$+\downarrow \Sigma F_y = 0 = N_{ID} + \frac{15}{17}(-29.1 \text{ k})$$
$$N_{ID} = +25.7 \text{ k}$$

Figure 10.16c

FBD of joint J:

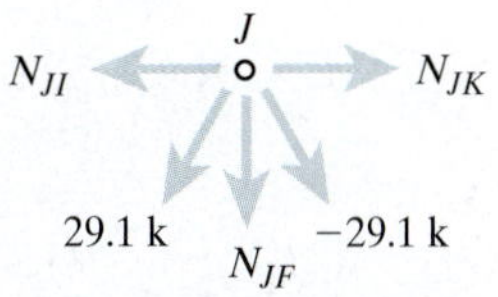

N_{JI} ← J → N_{JK}
29.1 k ↙ ↓ N_{JF} ↘ −29.1 k

$$+\downarrow \Sigma F_y = 0 = \frac{15}{17}(29.1 \text{ k}) + N_{JF} + \frac{15}{17}(-29.1 \text{ k})$$
$$N_{JF} = 0$$

Figure 10.16d

536 Chapter 10 Approximate Lateral Displacements

FBD of joint K:

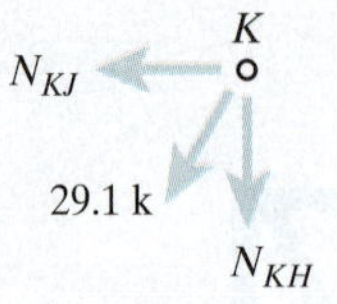

$N_{KJ} \leftarrow$

29.1 k $\downarrow$

$N_{KH} \downarrow$

$+\downarrow \Sigma F_y = 0 = \dfrac{15}{17}(+29.1 \text{ k}) + N_{KH}$

$N_{KH} = -25.7 \text{ k}$

Figure 10.16e

Approximate analysis of one virtual system:

Apply the virtual load to the top level to find the total drift of the top level.

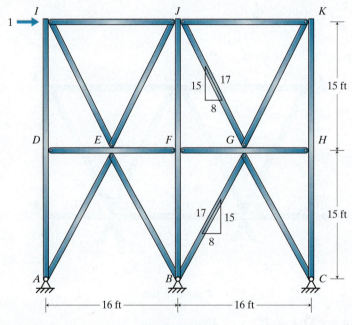

Figure 10.17a

FBD of frame cut at the top level:
   Find the axial force in each brace.

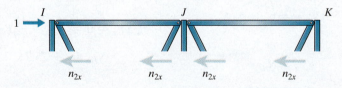

Figure 10.17b

$\xrightarrow{+} \Sigma F_x = 0 = 1 - 4n_{2x}$

$n_{2x} = 0.25$

$\therefore n_{2\text{brace}} = \dfrac{17}{8}(0.25) = 0.531$

Section 10.2 Braced Frames—Virtual Work Method 537

FBD of frame cut at the lower level:

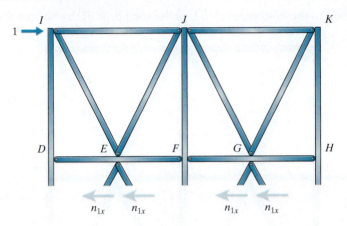

Figure 10.17c

$$\xrightarrow{+} \Sigma F_x = 0 = 1 - 4n_{1x}$$
$$n_{1x} = 0.25$$
$$\therefore n_{1\text{brace}} = \frac{17}{8}(0.25) = 0.531$$

FBD of joint I:

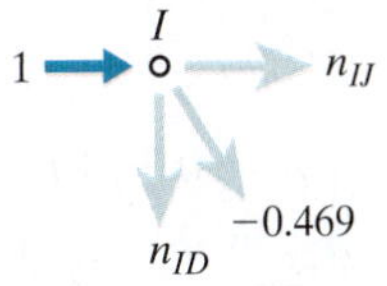

$$+\downarrow \Sigma F_y = 0 = n_{ID} + \frac{15}{17}(-0.531)$$
$$n_{ID} = +0.469$$

Figure 10.17d

FBD of joint J:

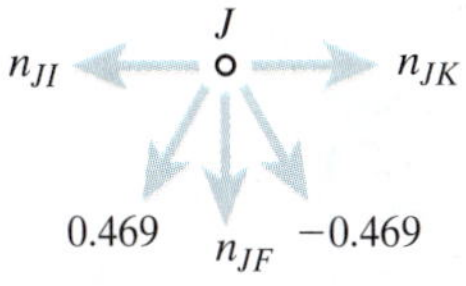

$$+\downarrow \Sigma F_y = 0 = \frac{15}{17}(0.531) + n_{JF} + \frac{15}{17}(-0.531)$$
$$n_{JF} = 0$$

Figure 10.17e

FBD of joint K:

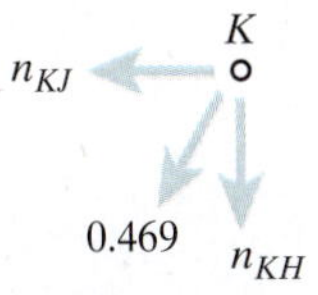

$$+\downarrow \Sigma F_y = 0 = \frac{15}{17}(+29.1 \text{ k}) + n_{KH}$$
$$n_{KH} = -0.469$$

Figure 10.17f

538   **Chapter 10**   Approximate Lateral Displacements

Approximate analysis of second virtual system:

Apply the virtual load to the bottom level to find the total drift of the bottom level.

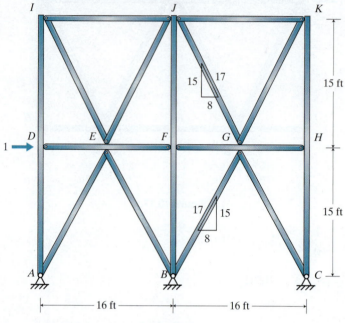

Figure 10.18a

FBD of frame cut at the top level:
    Find the axial force in each brace.

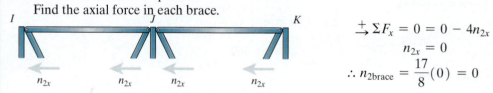

Figure 10.18b

$$\xrightarrow{+} \Sigma F_x = 0 = 0 - 4n_{2x}$$

$$n_{2x} = 0$$

$$\therefore n_{2\text{brace}} = \frac{17}{8}(0) = 0$$

FBD of frame cut at the lower level:

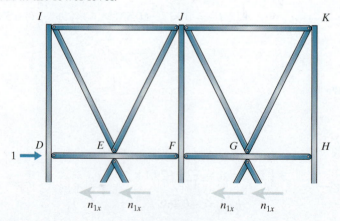

Figure 10.18c

$$\xrightarrow{+} \Sigma F_x = 0 = 1 - 4n_{1x}$$

$$n_{1x} = 0.25$$

$$\therefore n_{1\text{brace}} = \frac{17}{8}(0.25) = 0.531$$

Section 10.2 Braced Frames—Virtual Work Method   **539**

FBD of joint I:

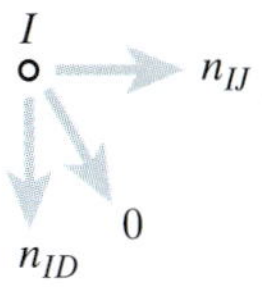

$$+\downarrow \Sigma F_y = 0 = n_{ID} + \frac{15}{17}(0)$$

$$n_{ID} = 0$$

Figure 10.18d

FBD of joint J:

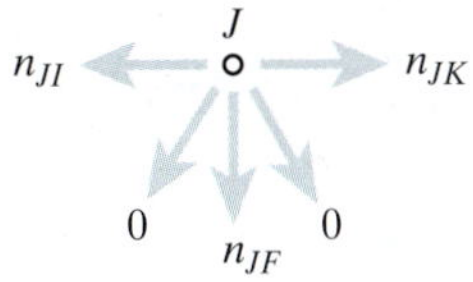

$$+\downarrow \Sigma F_y = 0 = \frac{15}{17}(0) + n_{JF} + \frac{15}{17}(0)$$

$$n_{JF} = 0$$

Figure 10.18e

FBD of joint K:

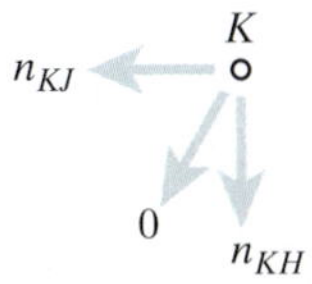

$$+\downarrow \Sigma F_y = 0 = \frac{15}{17}(0) + n_{KH}$$

$$n_{KH} = 0$$

Figure 10.18f

Summary of internal forces:

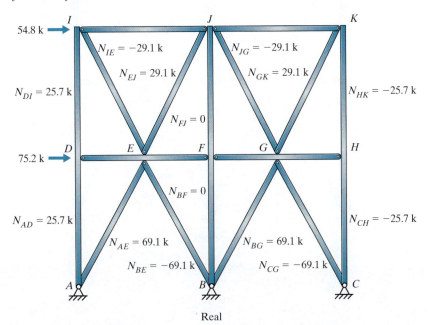

Real

Figure 10.19

540 Chapter 10 Approximate Lateral Displacements

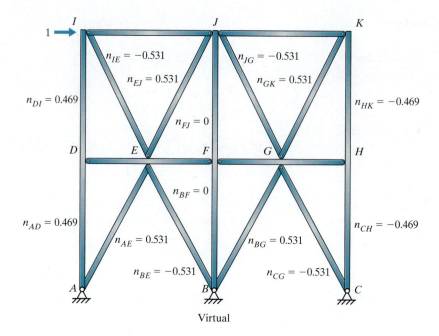

Figure 10.20

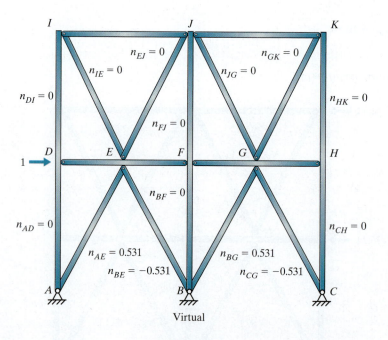

Figure 10.21

Virtual Work Method:

Find the total drift of the top level, Δ_I:

$$\Delta_I = (2)\frac{0.469(25.7\text{ k})(30\text{ ft}\cdot 12\text{ in./ft})}{(11.5\text{ in}^2)(29{,}000\text{ ksi})}$$

$$+ (4)\frac{0.531(29.1\text{ k})(17\text{ ft}\cdot 12\text{ in./ft})}{(11.5\text{ in}^2)(29{,}000\text{ ksi})}$$

$$+ (4)\frac{0.531(29.1\text{ k})(17\text{ ft}\cdot 12\text{ in./ft})}{(4.30\text{ in}^2)(29{,}000\text{ ksi})}$$

$$= 0.367\text{ in.}$$

Find the total drift of the bottom level, Δ_D:

$$\Delta_D = (2)\frac{0(29.1\text{ k})(30\text{ ft}\cdot 12\text{ in./ft})}{(11.5\text{ in}^2)(29{,}000\text{ ksi})}$$

$$+ (4)\frac{0(29.1\text{ k})(17\text{ ft}\cdot 12\text{ in./ft})}{(4.30\text{ in}^2)(29{,}000\text{ ksi})}$$

$$+ (4)\frac{0.531(69.1\text{ k})(17\text{ ft}\cdot 12\text{ in./ft})}{(4.30\text{ in}^2)(29{,}000\text{ ksi})}$$

$$= 0.240\text{ in.}$$

Calculate story drift:

Story drift of the top level:

$$\Delta_2 = \Delta_I - \Delta_D$$
$$\Delta_2 = 0.367\text{ in.} - 0.240\text{ in.} = 0.127\text{ in.}$$

Drift limit:

$$\Delta_2^{\text{Limit}} = \frac{h}{1000} = \frac{15\text{ ft}(12\text{ in./ft})}{1000} = 0.180\text{ in.}$$

Since $\Delta_2 < \Delta_2^{\text{Limit}}$, the section properties are probably adequate. This method typically underpredicts drift, but probably not this much.

Story drift of the bottom level:

$$\Delta_1 = \Delta_D$$
$$\Delta_1 = 0.240\text{ in.}$$

Drift limit:

$$\Delta_1^{\text{Limit}} = \frac{15\text{ ft}(12\text{ in./ft})}{1000} = 0.180\text{ in.}$$

Since $\Delta_1 > \Delta_1^{\text{Limit}}$ and this method typically underpredicts drift, the section properties are probably not adequate for this level.

Evaluation of Results:

Using the current member properties, a team member performed a computer aided analysis (Direct Stiffness Method), which resulted in these story drifts:

$\Delta_2 = 0.390$ in. $- 0.278$ in. $= 0.112$ in.
$\Delta_1 = 0.278$ in.

Approximation Predicted Outcomes?

Our results from the Story Drift Method are within 13%–14% of the results from the Direct Stiffness Method. ✓

Approximation Predicted Outcomes?

Our approximate total drift results underpredict the results from the Direct Stiffness Method, as expected. ✓

Conclusion:

Based on these observations, the computer aided analysis results could be reasonable. We would need to verify other aspects of the results to be more certain.

Use of Results:

We need to reduce the lower-story drift. Let's look at how much each member contributes to the total drift of each level.

Member	$\Delta_2^{Contrib}$	$\Delta_1^{Contrib}$
Top left column	0.007 in.	0.000 in.
Top middle column	0.000 in.	0.000 in.
Top right column	0.007 in.	0.000 in.
Bottom left column	0.007 in.	0.000 in.
Bottom middle column	0.000 in.	0.000 in.
Bottom right column	0.007 in.	0.000 in.
Top braces	0.101 in.	0.000 in.
Bottom braces	0.240 in.	0.240 in.
$\Sigma =$	0.367 in.	0.240 in.

Based on this information, the only way to reduce the lower-story drift is to increase the cross-sectional area of the bottom braces.

10.3 Rigid Frames—Stiff Beam Method

Introduction

The Stiff Beam Method of approximate analysis is the quickest for estimating the lateral displacement of rigid frames because the derivation results in a single formula. Similar to the Story Drift Method for braced frames, the Stiff Beam Method does not require the direct calculation of internal forces. This makes it a great choice for preliminary selection of member properties. Generally the Stiff Beam Method is not as accurate, however, as the Virtual Work Method discussed in Section 10.4.

Section 10.3 Rigid Frames—Stiff Beam Method

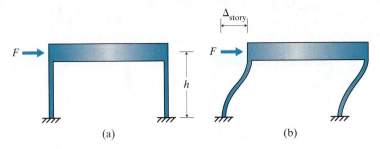

Figure 10.22 Single-story, single-bay rigid frame with an infinitely stiff beam and fixed supports: (a) Undeformed shape; (b) Displaced shape.

How-To

For an infinitely stiff beam, the deflected shape of a single-story, single-bay rigid frame is as shown in Figure 10.22. Because the beam is infinitely stiff, it does not bend. Therefore, the column is vertical at the fixed support and vertical where it meets the beam.

In this case, the inflection points in the columns are at midheight, and the displaced shape below and above the inflection points is identical (Figure 10.23).

At the inflection points, the moment is zero. Therefore, the free body diagrams of the bottom and top halves look like cantilevered beams with a point load at the end (Figure 10.24).

The formula for displacement at the end of a cantilevered beam can be found in the inside front cover:

$$\Delta' = \frac{Pl^3}{3EI}$$

where

P = Force applied at the tip of the cantilever—V in this case
l = Length of the cantilever—$h/2$ in this case
E = Modulus of elasticity of the cantilever
I = Moment of inertia of the cantilever—I_c in this case

When we make these substitutions, the displacement becomes

$$\Delta' = \frac{V(h/2)^3}{3EI_c} = \frac{Vh^3}{24EI_c}$$

Since Δ' is the displacement of only half the column, the total story drift is twice that value:

$$\Delta_{story} = 2\Delta'$$

Therefore, the drift for one story is predicted by the following:

$$\Delta_{story} = \frac{Vh^3}{12EI_c}$$

where

V = Shear in one column

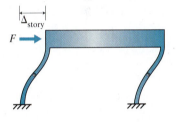

Figure 10.23 Location of inflection points if the rigid frame has an infinitely stiff beam.

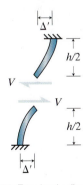

Figure 10.24 Free body diagrams of top and bottom halves of one column are identical cantilevers.

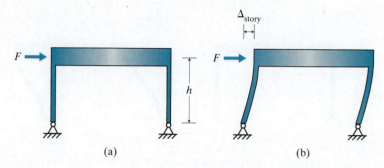

Figure 10.25 Single-story, single-bay rigid frame with an infinitely stiff beam and pinned supports: (a) Undeformed shape; (b) Displaced shape.

Pinned Base
That expression works for all stories of a multistory frame, except the bottom story if the columns are pinned at the base. When the column bases are pinned, the inflection points are at the base (Figure 10.25).

In this case, the story drift for the frame must be calculated with the full story height:

$$\Delta_{story} = \frac{Vh^3}{3EI_c}$$

This relationship applies for only the bottom level. Even if the columns are pinned at the base, the inflection points are still at midheight of the columns for all levels above the bottom level.

Finding V
If h and I_c are the same for all columns in the frame in that story, then

$$V = \frac{\Sigma F_{horiz}}{\text{\# columns}}$$

where
- V = Shear in each individual column in that story
- ΣF_{horiz} = Total horizontal force applied to the rigid frame from the top of the particular story to the top of the frame

If I_c is different for the columns in the frame in a particular story (but h and E are the same), then since Δ is the same for each column at that story, the formula for Δ' results in the following observation:

$$\frac{V_i}{I_i} = \frac{V_j}{I_j} = \frac{V_k}{I_k} = \text{constant}$$

where
i, j, k are different columns in the frame in that story

We can algebraically manipulate this observation to obtain the following expressions relating the shear in each column to one reference column that we call column i:

$$V_j = V_i \frac{I_j}{I_i}, \quad V_k = V_i \frac{I_k}{I_i}, \quad \text{etc.}$$

Equilibrium requires that the sum of all column shears in that frame at that story must equal the sum of all applied lateral loads from the top of the story to the top of the frame:

$$\Sigma F_{horiz} = V_i + V_j + V_k + \ldots$$

Substituting the expressions for the various column shears into this equilibrium equation and consolidating terms result in the following expression:

$$\Sigma F_{\text{horiz}} = V_i\left(\frac{I_i}{I_i} + \frac{I_j}{I_i} + \frac{I_k}{I_i} + \cdots\right)$$

If we solve this expression for the shear in our reference column, V_i, we obtain an equation for the shear in the reference column:

$$V_i = \frac{I_i \Sigma F_{\text{horiz}}}{(I_i + I_j + I_k + \cdots)}$$

We can use the shear in this reference column with its moment of inertia, I_i, to find the story drift.

Total Drift

To find the total drift of any story in the frame, we sum the story drift for that story and all the stories below it:

$$\Delta_{\text{total}} = \Sigma \Delta_{\text{story}}$$

SECTION 10.3 HIGHLIGHTS

Rigid Frames—Stiff Beam Method

For any individual story of a rigid frame:

$$\Delta_{\text{story}} = \frac{Vh^3}{12EI_c}$$

For the bottom story if the columns are pinned at the base:

$$\Delta_{\text{story}} = \frac{Vh^3}{3EI_c}$$

If all columns at that story have the same moment of inertia:

$$V = \frac{\Sigma F_{\text{horiz}}}{\# \text{ columns}}$$

If columns at that story have different moments of inertia:

$$V_i = \frac{I_i \Sigma F_{\text{horiz}}}{(I_i + I_j + I_k + \cdots)}$$

To estimate the total drift of a rigid frame:

$$\Delta_{\text{total}} = \Sigma \Delta_{\text{story}}$$

Summary of Limitations and Assumptions:

- Assumes the beam is infinitely more stiff than the columns so that the beams do not deform. See Section 10.4: Virtual Work Method to account for beam deformation.
- Assumes perfectly fixed or pinned supports at the base, not semifixed.
- Assumes changes in the axial length of the columns are negligible (so the beam remains horizontal).

Implication:

- Will always *underpredict* drift.

Example 10.6

Our firm is designing a new condominium building. The senior designer has decided to use a reinforced concrete rigid frame. For architectural reasons, the two bays have unequal widths. That same engineer has recommended using 5000-psi concrete, since some of the frame will be exposed to sea air. That strength of concrete will have a modulus of elasticity, E, of roughly 4000 ksi. Team members have completed a preliminary design of the members based on strength, but before we finalize the member sizes, we need to verify that the story drifts are less than the limit $h/500$.

Our task is to make a quick check of the story drifts. For this initial check, let's use the Stiff Beam Method.

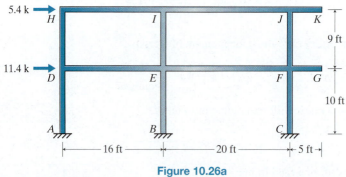

Figure 10.26a

Preliminary selection of member properties:

Beams:	$A = 96 \text{ in}^2$	$I = 512 \text{ in}^4$
Columns:	$A = 64 \text{ in}^2$	$I = 341 \text{ in}^4$

Section 10.3 Rigid Frames—Stiff Beam Method

Approximate Analysis:
Assume the beams do not bend.

FBD of frame cut at the top level:

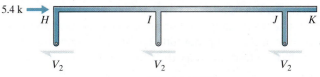

Figure 10.26b

$$\overset{+}{\rightarrow} \Sigma F_x = 0 = 5.4 \text{ k} - 3V_2$$
$$V_2 = 1.80 \text{ k} \, (\overset{\leftarrow}{+})$$

$$\Delta_2 = \frac{1.80 \text{ k} \, (9 \text{ ft} \cdot 12 \text{ in./ft})^3}{12 \, (4000 \text{ ksi})(341 \text{ in}^4)} = 0.139 \text{ in.}$$

$$\Delta_2^{\text{Limit}} = \frac{h_2}{500} = \frac{9 \text{ ft} \cdot 12 \text{ in./ft}}{500} = 0.216 \text{ in.}$$

Since $\Delta_2 > \Delta_2^{\text{Limit}}$, the member properties for this story appear okay, but this method always underpredicts the story drift.

FBD of frame cut at the bottom level:

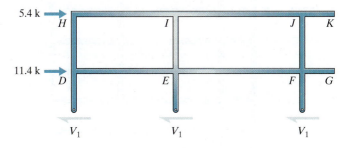

Figure 10.26c

$$\overset{+}{\rightarrow} \Sigma F_x = 0 = 5.4 \text{ k} + 11.4 \text{ k} - 3V_1$$
$$V_1 = 5.60 \text{ k} \, (\overset{\leftarrow}{+})$$

$$\Delta_1 = \frac{5.60 \text{ k} \, (10 \text{ ft} \cdot 12 \text{ in./ft})^3}{12 \, (4000 \text{ ksi})(341 \text{ in}^4)} = 0.591 \text{ in.}$$

$$\Delta_1^{\text{Limit}} = \frac{10 \text{ ft} \, (12 \text{ in./ft})}{500} = 0.240 \text{ in.}$$

Since $\Delta_1 > \Delta_1^{\text{Limit}}$ and we know that Δ_1 is an underprediction, the section properties for the lower level are not adequate.

548 Chapter 10 Approximate Lateral Displacements

EXAMPLE 10.7

A development company is looking to buy a historic building and renovate it to use for offices. The structural system is a rigid frame. Before finalizing the deal, the developer wants to know if the existing framing is strong enough for the new use.

A member of the team has modeled the frame using structural analysis software. In order to help evaluate the results of the computer aided analysis, we have been tasked with performing an approximate analysis to find the story drift for the bottom level.

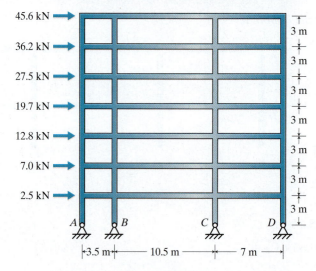

Figure 10.27a

Measured member properties:

Column A:	$A = 125 \text{ cm}^2$	$I = 22{,}200 \text{ cm}^4$
Column B:	$A = 225 \text{ cm}^2$	$I = 44{,}500 \text{ cm}^4$
Column C:	$A = 290 \text{ cm}^2$	$I = 59{,}500 \text{ cm}^4$
Column D:	$A = 125 \text{ cm}^2$	$I = 22{,}200 \text{ cm}^4$
Steel:	$E = 200 \text{ GPa} = 20{,}000 \text{ kN/cm}^2$	

Section 10.3 Rigid Frames—Stiff Beam Method

Approximate Analysis:

Assume the beams do not bend.

FBD of frame cut at the supports:

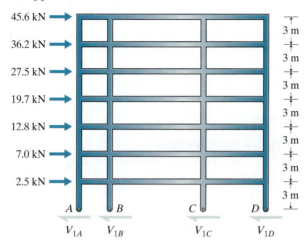

Figure 10.27b

Find the shear in column A:

Since the columns do not have identical moments of inertia, we must calculate the shear in one of the columns.

$$V_{1A} = \frac{I_A \Sigma F_{horiz}}{\Sigma I}$$

$$\Sigma F_{horiz} = 45.6 \text{ kN} + 36.2 \text{ kN} + 27.5 \text{ kN} + 19.7 \text{ kN} + 12.8 \text{ kN}$$
$$+ 7.0 \text{ kN} + 2.5 \text{ kN}$$
$$= 151.3 \text{ kN}$$

$$\Sigma I = 22{,}200 \text{ cm}^4 + 44{,}500 \text{ cm}^4 + 59{,}500 \text{ cm}^4 + 22{,}200 \text{ cm}^4$$
$$= 148{,}400 \text{ cm}^4$$

$$V_{1A} = \frac{22{,}200 \text{ cm}^4 (151.3 \text{ kN})}{(148{,}400 \text{ cm}^4)} = 22.6 \text{ kN}$$

Find the story drift:

Since we are finding the story drift of the lowest level with pinned supports, we must be careful to use the appropriate relationship.

$$\Delta_1 = \frac{V_{1A} h_1^3}{3 E I_A}$$

$$= \frac{22.6 \text{ kN} (300 \text{ cm})^3}{3 (20{,}000 \text{ kN/cm}^2)(22{,}200 \text{ cm}^4)} = 0.46 \text{ cm}$$

Use of Results:

This approximate analysis method will underpredict the story drift, so we expect the computer aided analysis results to be larger.

10.4 Rigid Frames—Virtual Work Method

Introduction
The Stiff Beam Method neglects the impact of beam bending on the frame drift. We can combine approximate analysis of a rigid frame (Chapter 9) with the Virtual Work Method of predicting deformations (Section 5.3) to improve the estimate of the drift of rigid frames. We can use the most appropriate approximate analysis method for the given structure: Portal Method or Cantilever Method.

How-To

Virtual Work Method
The Virtual Work Method predicts displacement by the following expression:

$$\Delta = \Sigma \frac{nNl}{AE} + \Sigma \int \frac{mM}{EI} dx$$

If we neglect the effects of axial deformation everywhere in the frame, the expression reduces to

$$\Delta = \Sigma \int \frac{mM}{EI} dx$$

where

m = Equation for the moment in the member due to the virtual force
M = Equation for the moment in the member due to the real applied forces
E = Modulus of elasticity of the member
I = Moment of inertia of the member

To obtain the equations for the moment due to the real applied forces and the virtual force, we can use either of the approximate analysis methods.

Approximate Analysis Derivation
As a reminder, the first two assumptions for performing an approximate analysis by either the Portal Method or Cantilever Method are as follows:

- Inflection points exist at midspan of beams.
- Inflection points exist at midheight of columns (columns with pinned bases are addressed at the end of this section).

Since the loading on the frame is idealized as acting only at the beam-column joints, the shear in each member is constant and, therefore, the moment varies linearly. The first and second assumptions tell us where the moment is zero. Figure 10.28 shows the resulting moment diagram for an example frame.

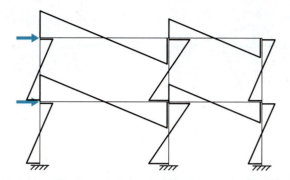

Figure 10.28 Moment diagram for a rigid frame based on the Portal and Cantilever Method assumptions.

If we cut a piece of the frame at the inflection points, we get the moment diagram in Figure 10.29a. The peak moment in each segment occurs at the joint and decreases to zero at the inflection point. Therefore, the moment in each beam and column segment can be expressed by the equations shown in Figure 10.29b.

The equations for moment do not change when we consider the virtual system. Only the magnitudes of v_c and v_b, the shears due to the virtual force, change. Therefore, the contribution to drift by one piece of beam can be calculated as follows:

$$\begin{aligned}
\int \frac{mM}{EI}dx &= \int \frac{[v_b(l/2) - v_b x_b][V_b(l/2) - V_b x_b]}{EI_b} dx_b \\
&= \int \frac{[v_b V_b(l/2)^2 - 2v_b V_b(l/2)x_b + v_b V_b x_b^2]}{EI_b} dx_b \\
&= \frac{v_b V_b}{EI_b}\left[\left(\frac{l}{2}\right)^2 x_b - \left(\frac{l}{2}\right)x_b^2 + \frac{1}{3}x_b^3\right]_0^{l/2} \\
&= \frac{v_b V_b}{EI_b}\left[\left(\frac{l}{2}\right)^3 - \left(\frac{l}{2}\right)^3 + \frac{1}{3}\left(\frac{l}{2}\right)^3\right] = \frac{v_b V_b l^3}{24EI_b}
\end{aligned}$$

Since each half of the beam has the same length and shear value, the contribution to drift from the entire beam is twice that value:

$$\frac{v_b V_b l^3}{12EI_b}$$

Similarly, the contribution to drift from a full story of column is given by

$$\frac{v_c V_c h^3}{12EI_c}$$

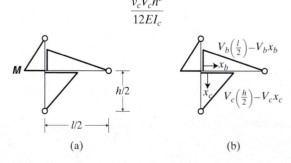

Figure 10.29 FBD of a piece of rigid frame cut at the inflection points: (a) Moment diagrams are linear; (b) Equations for moment as a function of distance from the joint.

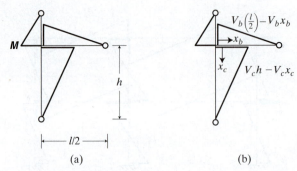

Figure 10.30 FBD of a piece of the bottom level of a rigid frame cut at the inflection points and pin supported at the base: (a) Moment diagram is zero at the base of the pinned column; (b) Equations for moment as a function of distance from the joint.

Columns Pinned at the Base

If a column in the bottom level of a frame is pinned at the base, the inflection point is at the base instead of at midheight. That changes the expression for the moment for just the bottom level of the column, but not for the beams or columns above the lowest level (Figure 10.30).

The contribution to drift by a column pinned at the base can be calculated as follows:

$$\int \frac{mM}{EI}dx = \int \frac{(v_c h - v_c x_c)(V_c h - V_c x_c)}{EI_c}dx_c$$

$$= \int \frac{(v_c V_c h^2 - 2v_c V_c h x_c + v_c V_c x_c^2)}{EI_c}dx_c$$

$$= \frac{v_c V_c}{EI_c}\left[h^2 x_c - h x_c^2 + \frac{1}{3}x_c^3\right]_0^h$$

$$= \frac{v_c V_c}{EI_c}\left[h^3 - h^3 + \frac{1}{3}h^3\right] = \frac{v_c V_c h^3}{3EI_c}$$

SECTION 10.4 HIGHLIGHTS

Rigid Frames—Virtual Work Method

To estimate the total drift of a rigid frame, sum the contributions from each beam and column:

$$\Delta_{total} = \sum_{\substack{\text{Pin}\\\text{Base}\\\text{Cols}}} \frac{v_c V_c h^3}{3EI_c} + \sum_{\substack{\text{All}\\\text{Other}\\\text{Cols}}} \frac{v_c V_c h^3}{12EI_c} + \sum \frac{v_b V_b l^3}{12EI_b}$$

Summary of Limitations and Assumptions:

- Assumes inflection points are at midspan of all beams.
- Assumes inflection points are at midheight or base of all columns; therefore, formula is not derived for columns semifixed at the base.
- Neglects axial deformation everywhere in the structure.

Implications:

- Because of the Portal and Cantilever Method assumptions, will *underpredict* or *overpredict* drift.
- Should be an improvement over the Stiff Beam Method, however, because it incorporates beam bending.

Example 10.8

Our firm is designing a new condominium building. A preliminary design of the members provided adequate strength, but the first-level story drift was larger than the limit of $h/500$. Therefore, a team member redesigned the first-level columns. The redesign did not affect the choice of concrete; the modulus of elasticity, E, is still roughly 4000 ksi.

Our task is to verify the story drift for both levels before the team creates a full computer model. The senior engineer wants us to use a more accurate prediction than the results of the Stiff Beam Method, so we will use the Virtual Work Method.

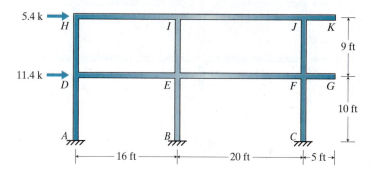

Figure 10.31

Revised member properties:

Beams:	$A = 96 \text{ in}^2$	$I = 512 \text{ in}^4$
Level 2 columns:	$A = 64 \text{ in}^2$	$I = 341 \text{ in}^4$
Level 1 columns:	$A = 96 \text{ in}^2$	$I = 1152 \text{ in}^4$

Approximate Analysis:

Since the frame is short and wide, use the Portal Method.

Assume the inflection points are at midspan of each beam, except the cantilevered beams. Assume the inflection points are at midheight of each column. Assume the shear at each level is the same for the outer columns and the inner column carries twice as much. Assume the effect of axial deformation of all members is negligible.

554 Chapter 10 Approximate Lateral Displacements

Approximate analysis of the real system:

Find the shear in each member.

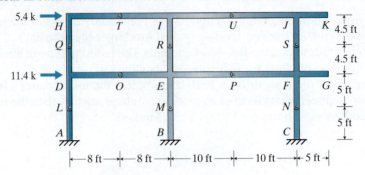

Figure 10.32a

FBD of frame cut at the top level:

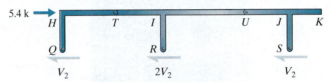

Figure 10.32b

$$\xrightarrow{+} \Sigma F_x = 0 = 5.4 \text{ k} - 4V_2$$
$$V_2 = 1.35 \text{ k} \; (\overset{+}{\rightarrow})$$

FBD of frame cut at the bottom level:

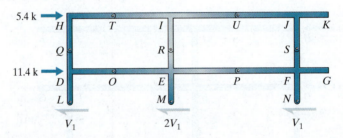

Figure 10.32c

$$\xrightarrow{+} \Sigma F_x = 0 = 5.4 \text{ k} + 11.4 \text{ k} - 4V_1$$
$$V_1 = 4.20 \text{ k} \; (\overset{+}{\rightarrow})$$

FBD of corner *QHT*:

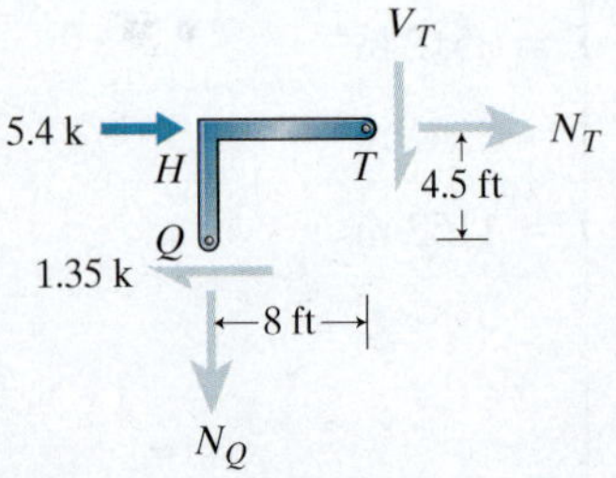

$$\overset{+}{\curvearrowright} \Sigma M_H = 0 = 1.35 \text{ k} (4.5 \text{ ft}) + V_T (8 \text{ ft})$$
$$V_T = -0.76 \text{ k} \; (\downarrow + \uparrow)$$

Figure 10.32d

FBD of piece *TUR*:

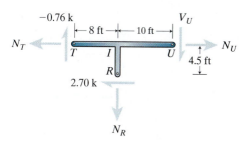

Figure 10.32e

$\overset{+}{\curvearrowleft}\Sigma M_I = 0 = -0.76 \text{ k} (8 \text{ ft}) + V_U (10 \text{ ft}) + 2.70 \text{ k} (4.5 \text{ ft})$
$V_U = -0.61 \text{ k} (\downarrow + \uparrow)$

FBD of piece *UKS*:

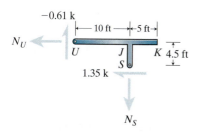

Figure 10.32f

We can use moment equilibrium on this piece as a quick check for errors.

Satisfied Fundamental Principles?

$\overset{+}{\curvearrowleft}\Sigma M_J = -0.61 \text{ k} (10 \text{ ft}) + 1.35 \text{ k} (4.5 \text{ ft}) = -0.02 \text{ k} \cdot \text{ft} \approx 0 \checkmark$

FBD of piece *LQO*:

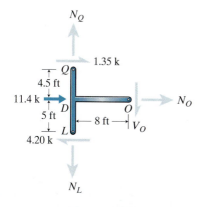

Figure 10.32g

$\overset{+}{\curvearrowleft}\Sigma M_D = 0 = 1.35 \text{ k} (4.5 \text{ ft}) + V_O (8 \text{ ft}) + 4.20 \text{ k} (5 \text{ ft})$
$V_O = -3.38 \text{ k} (\downarrow + \uparrow)$

FBD of piece *ORPM*:

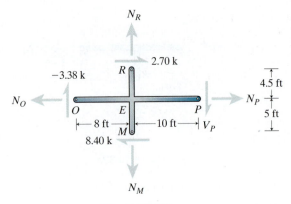

Figure 10.32h

$$+\circlearrowleft \Sigma M_E = 0 = (-3.38 \text{ k})(8 \text{ ft}) + 2.70 \text{ k}(4.5 \text{ ft}) + V_P(10 \text{ ft}) + 8.40 \text{ k}(5 \text{ ft})$$
$$V_P = -2.71 \text{ k} (\downarrow + \uparrow)$$

FBD of piece *PSGN*:

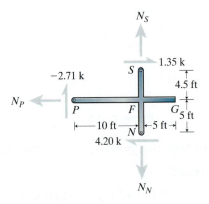

Figure 10.32i

We can use moment equilibrium on this piece as a quick check for errors.

Satisfied Fundamental Principles?

$$+\circlearrowleft \Sigma M_F = (-2.71 \text{ k})(10 \text{ ft}) + 1.35 \text{ k}(4.5 \text{ ft}) + 4.20 \text{ k}(5 \text{ ft})$$
$$= -0.02 \text{ k} \cdot \text{ft} \approx 0 \checkmark$$

Approximate analysis of one virtual system:

Apply the virtual load to the top level to find the total drift of the top level.

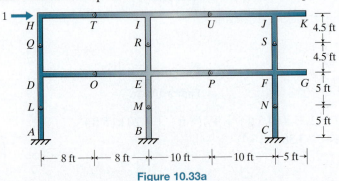

Figure 10.33a

FBD of frame cut at the top level:

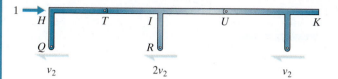

$$\pm \Sigma F_x = 0 = 1 - 4v_2$$
$$v_2 = 0.25\ (\overset{+}{\rightarrow})$$

Figure 10.33b

FBD of frame cut at the bottom level:

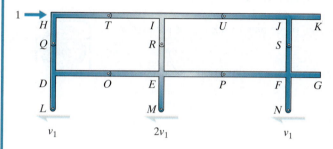

$$\pm \Sigma F_x = 0 = 1 - 4v_1$$
$$v_1 = 0.25\ (\overset{+}{\rightarrow})$$

Figure 10.33c

FBD of corner QHT:

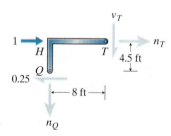

$$\curvearrowright \Sigma M_H = 0 = 0.25(4.5\text{ ft}) + v_T(8\text{ ft})$$
$$v_T = -0.141\ (\downarrow + \uparrow)$$

Figure 10.33d

FBD of piece TUR:

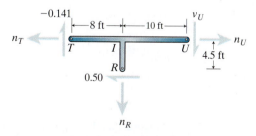

Figure 10.33e

$$\curvearrowright \Sigma M_I = 0 = -0.141(8\text{ ft}) + v_U(10\text{ ft}) + 0.50(4.5\text{ ft})$$
$$v_U = -0.112\ (\downarrow + \uparrow)$$

558 Chapter 10 Approximate Lateral Displacements

FBD of piece *UKS*:

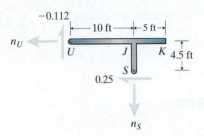

Figure 10.33f

We can use moment equilibrium on this piece as a quick check for errors.

Satisfied Fundamental Principles?

$$\underset{+}{\curvearrowleft}\Sigma M_J = -0.112(10\text{ ft}) + 0.25(4.5\text{ ft}) = 0.005 \approx 0 \checkmark$$

FBD of piece *LQO*:

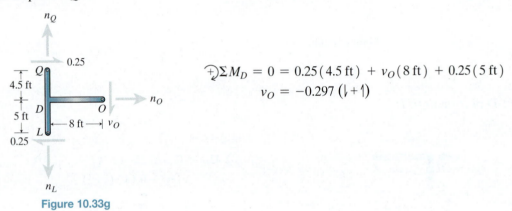

Figure 10.33g

$$\underset{+}{\curvearrowleft}\Sigma M_D = 0 = 0.25(4.5\text{ ft}) + v_O(8\text{ ft}) + 0.25(5\text{ ft})$$
$$v_O = -0.297\ (\downarrow+\uparrow)$$

FBD of piece *ORPM*:

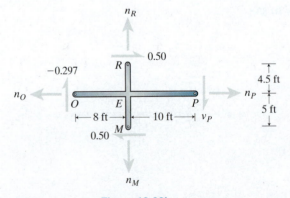

Figure 10.33h

$$\underset{+}{\curvearrowleft}\Sigma M_E = 0 = (-0.297)(8\text{ ft}) + 0.50(4.5\text{ ft}) + v_P(10\text{ ft}) + 0.50(5\text{ ft})$$
$$v_P = -0.237\ (\downarrow+\uparrow)$$

FBD of piece *PSGN*:

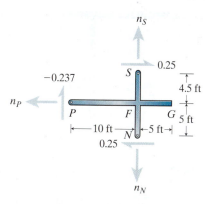

Figure 10.33i

We can use moment equilibrium on this piece as a quick check for errors.

Satisfied Fundamental Principles?

$$\circlearrowleft \Sigma M_F = (-0.237)(10 \text{ ft}) + 0.25(4.5 \text{ ft}) + 0.25(5 \text{ ft})$$
$$= 0.005 \approx 0 \checkmark$$

Approximate analysis of second virtual system:

Apply the virtual load to the middle level to find the total drift of the bottom level.

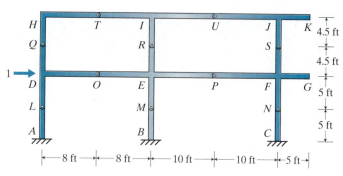

Figure 10.34a

FBD of frame cut at the top level:

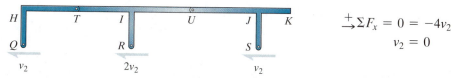

Figure 10.34b

$$\xrightarrow{+} \Sigma F_x = 0 = -4v_2$$
$$v_2 = 0$$

560 Chapter 10 Approximate Lateral Displacements

FBD of frame cut at the bottom level:

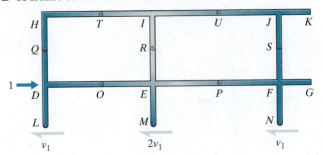

$$\xrightarrow{+}\Sigma F_x = 0 = 1 - 4v_1$$
$$v_1 = 0.25\ (\xrightarrow{+})$$

Figure 10.34c

FBD of corner QHT:

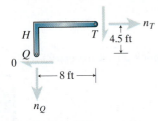

$$\curvearrowleft{+}\Sigma M_H = 0 = 0(4.5\text{ ft}) + v_T(8\text{ ft})$$
$$v_T = 0\ (\downarrow + \uparrow)$$

Figure 10.34d

FBD of piece TUR:

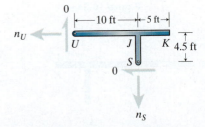

$$\curvearrowleft{+}\Sigma M_I = 0 = 0(8\text{ ft}) + v_U(10\text{ ft}) + 0(4.5\text{ ft})$$
$$v_U = 0\ (\downarrow + \uparrow)$$

Figure 10.34e

FBD of piece UKS:

Figure 10.34f

We can use moment equilibrium on this piece as a quick check for errors.

Satisfied Fundamental Principles?

$$\curvearrowleft{+}\Sigma M_J = 0(10\text{ ft}) + 0(4.5\text{ ft}) = 0\ \checkmark$$

FBD of piece *LQO*:

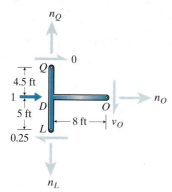

$$+\circlearrowleft \Sigma M_D = 0 = 0(4.5\text{ ft}) + v_O(8\text{ ft}) + 0.25(5\text{ ft})$$
$$v_O = -0.156 \ (\downarrow + \uparrow)$$

Figure 10.34g

FBD of piece *ORPM*:

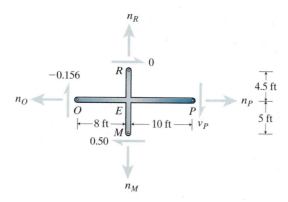

Figure 10.34h

$$+\circlearrowleft \Sigma M_E = 0 = (-0.156)(8\text{ ft}) + 0(4.5\text{ ft}) + v_P(10\text{ ft}) + 0.50(5\text{ ft})$$
$$v_P = -0.125 \ (\downarrow + \uparrow)$$

FBD of piece *PSGN*:

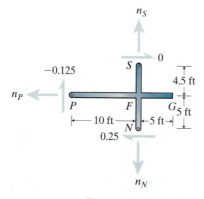

Figure 10.34i

We can use moment equilibrium on this piece as a quick check for errors.

Satisfied Fundamental Principles?

$$+\circlearrowleft \Sigma M_F = (-0.125)(10\text{ ft}) + 0(4.5\text{ ft}) + 0.25(5\text{ ft}) = 0 \ \checkmark$$

562 **Chapter 10** Approximate Lateral Displacements

Summary of internal forces:

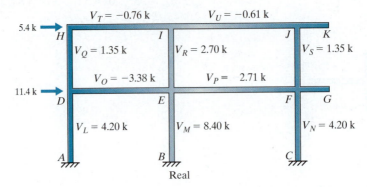

Figure 10.35

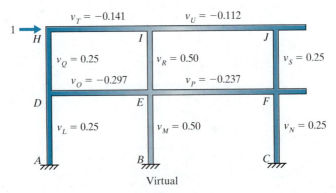

Figure 10.36

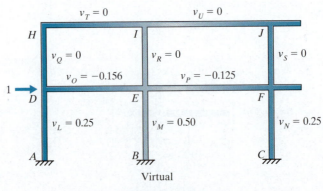

Figure 10.37

Virtual Work Method:
Find the total drift of the top level, Δ_H:

$$\Delta_H = (2)\frac{0.25(1.35\text{ k})(9\text{ ft}\cdot 12\text{ in./ft})^3}{12(4000\text{ ksi})(341\text{ in}^4)} + \frac{0.50(2.70\text{ k})(9\text{ ft}\cdot 12\text{ in./ft})^3}{12(4000\text{ ksi})(341\text{ in}^4)}$$

$$+ (2)\frac{0.25(4.20\text{ k})(9\text{ ft}\cdot 12\text{ in./ft})^3}{12(4000\text{ ksi})(1152\text{ in}^4)}$$

$$+ \frac{0.50(8.40\text{ k})(9\text{ ft}\cdot 12\text{ in./ft})^3}{12(4000\text{ ksi})(1152\text{ in}^4)}$$

$$+ \frac{(-0.141)(-0.76\text{ k})(16\text{ ft}\cdot 12\text{ in./ft})^3}{12(4000\text{ ksi})(512\text{ in}^4)}$$

$$+ \frac{(-0.112)(-0.61\text{ k})(20\text{ ft}\cdot 12\text{ in./ft})^3}{12(4000\text{ ksi})(512\text{ in}^4)}$$

$$+ \frac{(-0.297)(-3.38\text{ k})(16\text{ ft}\cdot 12\text{ in./ft})^3}{12(4000\text{ ksi})(512\text{ in}^4)}$$

$$+ \frac{(-0.237)(-2.71\text{ k})(20\text{ ft}\cdot 12\text{ in./ft})^3}{12(4000\text{ ksi})(512\text{ in}^4)}$$

$$= 1.072\text{ in.}$$

Find the total drift of the bottom level, Δ_D:

$$\Delta_D = (2)\frac{0(1.35\text{ k})(9\text{ ft}\cdot 12\text{ in./ft})^3}{12(4000\text{ ksi})(341\text{ in}^4)} + \frac{0(2.70\text{ k})(9\text{ ft}\cdot 12\text{ in./ft})^3}{12(4000\text{ ksi})(341\text{ in}^4)}$$

$$+ (2)\frac{0.25(4.20\text{ k})(9\text{ ft}\cdot 12\text{ in./ft})^3}{12(4000\text{ ksi})(1152\text{ in}^4)}$$

$$+ \frac{0.50(8.40\text{ k})(9\text{ ft}\cdot 12\text{ in./ft})^3}{12(4000\text{ ksi})(1152\text{ in}^4)}$$

$$+ \frac{0(-0.76\text{ k})(16\text{ ft}\cdot 12\text{ in./ft})^3}{12(4000\text{ ksi})(512\text{ in}^4)}$$

$$+ \frac{0(-0.61\text{ k})(20\text{ ft}\cdot 12\text{ in./ft})^3}{12(4000\text{ ksi})(512\text{ in}^4)}$$

$$+ \frac{(-0.156)(-3.38\text{ k})(16\text{ ft}\cdot 12\text{ in./ft})^3}{12(4000\text{ ksi})(512\text{ in}^4)}$$

$$+ \frac{(-0.125)(-2.71\text{ k})(20\text{ ft}\cdot 12\text{ in./ft})^3}{12(4000\text{ ksi})(512\text{ in}^4)}$$

$$= 0.539\text{ in.}$$

Story Drift Method:
Story drift of the top level:
$$\Delta_2 = \Delta_H - \Delta_D$$
$$\Delta_2 = 1.072\text{ in.} - 0.539\text{ in.} = 0.533\text{ in.}$$

Drift limit:
$$\Delta_2^{Limit} = \frac{h_2}{500} = \frac{9\text{ ft}(12\text{ in./ft})}{500} = 0.216\text{ in.}$$

Since $\Delta_2 > \Delta_2^{Limit}$ by a factor of 2, the section properties are probably not adequate. This method might overpredict drift, but probably not this much.

Story drift of the bottom level:
$$\Delta_1 = \Delta_D$$
$$\Delta_1 = 0.539\text{ in.}$$

Drift limit:
$$\Delta_1^{Limit} = \frac{10\text{ ft}(12\text{ in./ft})}{500} = 0.240\text{ in.}$$

Since $\Delta_1 > \Delta_1^{Limit}$ also by a factor of 2, the section properties are probably not adequate for this level either.

Evaluation of Results:

A team member has used the current member properties to perform a computer aided analysis (Direct Stiffness Method). That analysis resulted in the following story drifts:
$$\Delta_2 = 0.825\text{ in.} - 0.397\text{ in.} = 0.428\text{ in.}$$
$$\Delta_1 = 0.397\text{ in.}$$

Approximation Predicted Outcomes?

Our results are only 25%–36% higher than the results of the Direct Stiffness Method. ✓

Conclusion:

Based on this observation, the computer aided analysis results could be reasonable. We would need to verify other aspects of the results to be more certain.

Use of Results:

We need to reduce the story drifts. Let's look at how much each member contributes to the total drift of each level.

Member	$\Delta_2^{Contrib}$	$\Delta_1^{Contrib}$
Top left column	0.026 in.	0.000 in.
Top middle column	0.104 in.	0.000 in.
Top right column	0.026 in.	0.000 in.
Bottom left column	0.033 in.	0.033 in.
Bottom middle column	0.131 in.	0.131 in.
Bottom right column	0.033 in.	0.033 in.
Top left beam	0.031 in.	0.000 in.
Top right beam	0.038 in.	0.000 in.
Bottom left beam	0.289 in.	0.152 in.
Bottom right beam	0.361 in.	0.191 in.
$\Sigma =$	1.072 in.	0.539 in.

Based on this information, the most economical place to increase the member moment of inertia is the bottom beams. That reduces the drift for both levels. The next best place to focus in order to reduce the drift is the bottom middle column, BE.

10.5 Solid Walls—Single Story

Introduction

Walls are typically the stiffest option for resisting lateral loads and often the most economical. The economic advantage of walls for lateral load resistance is that they serve multiple purposes simultaneously. For example, structural walls can provide fire resistance, which is required by code. Structural walls can also be used as the building's façade. Structural walls, often called shear walls, are typically made of concrete, masonry, wood, or sometimes steel.

Structural walls tend to have small drifts, so they rarely exceed the drift limits for serviceability concerns. However, we still need to be able to predict the drift of structural walls. There are drift limits to ensure sufficient ductility in seismic events, and we need to check for interference during seismic events. Interference is when two structures deflect toward each other and collide. Single-story, separate buildings are rarely constructed close enough to each other to collide. However, large buildings or buildings with irregular shapes are often built with construction or expansion joints. That means the parts act as separate structures even though the users pass through as if they were one structure.

We frequently use the drift to find the rigidity of a wall. Rigidity is used to determine how much lateral load is distributed to each lateral load-resisting element when there is a rigid diaphragm (Chapter 11: Diaphragms).

How-To

The stiffness of a structural wall out of plane is very small. Therefore, we count on structural walls to carry load only in plane (Figure 10.38). That means we need to focus only on predicting the lateral displacement of walls in plane.

To make that prediction, we consider that a solid wall subjected to lateral load is essentially the same as a cantilevered beam (Figure 10.39). The key difference we want to account for is the effect of shear deformation. Beams are typically long and narrow; therefore, their displacement is dominated by flexural deformation. Our calculations in Chapter 5: Deformations ignored the contribution of shear to the total displacement.

Single-story walls can be very short and wide. In those cases, shear deformation dominates. To give the best possible predictions of drift, we use both components: shear and flexure.

From solid mechanics, we have the following expression for shear deformation of a solid beam or wall:

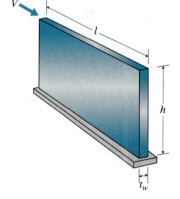

Figure 10.38 Dimensions of a structural wall relevant to predicting drift.

$$\Delta = \frac{1.2Vh}{AG}$$

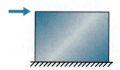 

Figure 10.39 A structural wall with lateral load has the same boundary conditions as a cantilevered beam with gravity load; we've just rotated it 90°.

where

A = Cross-sectional area of the wall being sheared (Figure 10.38)

$= t_w l$

t_w = Thickness of the wall

G = Shear modulus, a material property

$= \dfrac{E}{2(1+\nu)}$

ν = Poisson's ratio, the ratio of the strain perpendicular to the axis of a member to the axial strain of the member; it is a material property

If we substitute this expression for the shear modulus into the expression for drift, it gives us the following expression for the drift of a solid wall due to shear deformation:

$$\Delta_{sh} = \dfrac{2.4 V h (1+\nu)}{E t_w l}$$

To find the contribution from flexure, we can look inside the front cover of this text:

$$\Delta = \dfrac{V}{6EI}[-h^3 + 3h(h^2)] = \dfrac{Vh^3}{3EI}$$

where

I = In plane moment of inertia of the wall cross-section (Figure 10.38)

$= \dfrac{t_w l^3}{12}$

If we substitute the expression for moment of inertia into the expression for drift, we get the following expression for the drift of a solid wall due to just flexural deformation:

$$\Delta_{fl} = \dfrac{4 V h^3}{E t_w l^3}$$

We want to consider both contributions, so the resulting expression is the sum of both components: shear and flexure:

$$\Delta = \dfrac{2.4 V h (1+\nu)}{E t_w l} + \dfrac{4 V h^3}{E t_w l^3}$$

Now let's examine the different terms in this expression. The shear component of the drift increases linearly as the ratio of height to width, h/l, increases. But the flexural component of the drift increases as the cube of h/l. Therefore, shear deformation dominates for short and wide walls; flexure dominates for tall and narrow walls (Figure 10.40).

Coupled Walls

Sometimes we couple walls together to resist lateral load. Coupled walls are in the same plane and are connected by a link (Figure 10.41). If the walls are not identical, the shear carried by each is different.

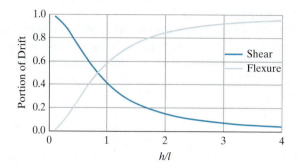

Figure 10.40 Relative contributions of shear and flexure to the total drift of a solid wall.

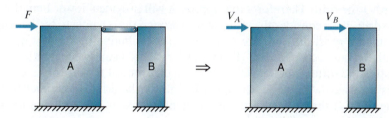

Figure 10.41 The total force applied to coupled walls is not necessarily shared equally.

We consider the link to stay the same length, so the walls drift the same amount (Figure 10.42). In these cases, we must use compatibility to determine how much shear is carried by each wall. Compatibility means that since we know the drifts are the same, compatible, we can back-calculate the shear carried by each.

We can use the drift formula to find the shear carried by each wall:

$$\Delta_A = \Delta_B = \cdots$$

$$\frac{2.4V_A h(1+\nu_A)}{E_A t_{wA} l_A} + \frac{4V_A h^3}{E_A t_{wA} l_A^3} = \frac{2.4V_B h(1+\nu_B)}{E_B t_{wB} l_B} + \frac{4V_B h^3}{E_B t_{wB} l_B^3} = \cdots$$

Solving each expression in terms of V_A gives

$$V_B = V_A \frac{\left[\dfrac{2.4h(1+\nu_A)}{E_A t_{wA} l_A} + \dfrac{4h^3}{E_A t_{wA} l_A^3}\right]}{\left[\dfrac{2.4h(1+\nu_B)}{E_B t_{wB} l_B} + \dfrac{4h^3}{E_B t_{wB} l_B^3}\right]}$$

$$V_C = V_A \cdots$$

From these relationships, we see that the walls must be made of the same material and have the same dimensions in order for the shears to be equal. Since equilibrium must be satisfied, we have a way of finding the shear carried by each wall:

$$F = V_A + V_B + \cdots$$

With the shear in each wall, we can calculate the drift.

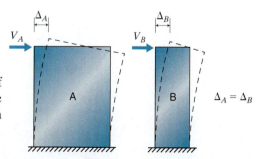

Figure 10.42 For coupled walls, the drift of each wall must be the same.

Walls with Openings

The drift formula is based on walls that have no openings. Openings for doors or windows reduce the stiffness, thereby increasing the drift. How much depends on the size, number, and locations of the openings. In all cases, the result is larger drift though. There are some by hand methods for predicting the drift of walls that have openings, but they are often not practical. The most common approach is to use the Finite Element Method (FEM) to model a wall with openings. The Finite Element Method, however, is outside the scope of this text.

Although our drift formula doesn't accurately predict the drift when structural walls have openings, we can still use the formula to bound the anticipated drift. Consider the structural wall in Figure 10.43a. If we ignore the opening and treat the wall as solid (Figure 10.43b), we will underpredict the drift. Therefore, this approach will provide a lower bound for the drift. If we consider the strips of wall around the opening to be the only parts that resist drift, we are ignoring the contribution of the part of the wall that connects the strips (Figure 10.43c). This means we are underestimating the stiffness of the wall, so we will overpredict the drift. Therefore, this approach will provide an upper bound for the drift. Remember from Figure 10.41 that if the strips are not identical, we must use compatibility of the strips to predict drift.

Rotation at the Base of a Wall

In certain situations, the base of the wall rotates. When that happens, the drift increases, so our drift predictions are lower bounds. Two of those situations are foundation movement and discontinuous walls.

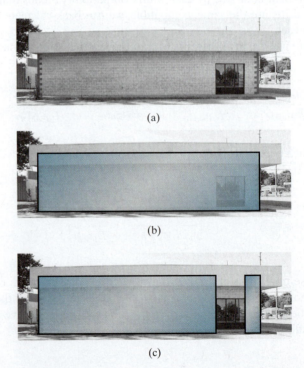

Figure 10.43 Idealizing a structural wall with an opening: (a) Actual wall; (b) Wall idealized as solid without openings; (c) Wall idealized as two coupled walls, ignoring the stiffness of the connector.

Section 10.5 Solid Walls—Single Story 569

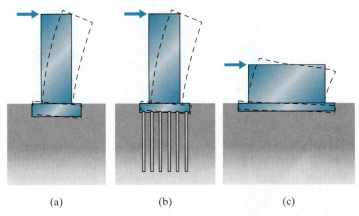

Figure 10.44 Vertical deformation patterns for foundations under laterally loaded walls: (a) Foundation remains straight as soil deforms elastically; (b) Deep foundation components deform axially; (c) Shear deformation under short, wide walls results in nonlinear distribution of vertical deformation.

We typically idealize the foundation support as fixed in the plane of the wall. That means the wall has no vertical displacement at the foundation. However, there is typically some vertical movement when the wall experiences lateral loading. The amount of movement depends on the soil conditions, type of foundation, and height/width ratio of the wall (Figure 10.44).

If we want to account for the soil-structure interaction in our drift prediction, the method of choice is usually the Finite Element Method, a computer aided analysis method. Entire books are dedicated to the Finite Element Method, so we leave the explanation of FEM to them.

Although not common, we are likely to encounter a structural wall that is not supported by a foundation at some point in our career. This happens in multi-story structures where the location of the wall has to be moved between levels (Figure 10.45). In these cases, the rotation at the bottom of the wall is resisted only by the torsional stiffness of the floor diaphragm and the flexural stiffness of any beam added directly along the bottom of the wall (Figure 10.46).

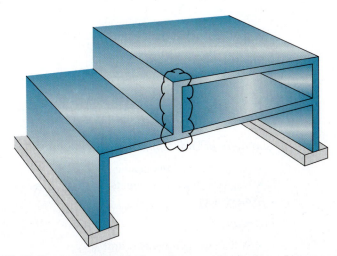

Figure 10.45 Structural wall that is discontinuous vertically; the wall is supported by a floor diaphragm that transfers load to the walls below.

570   **Chapter 10**   Approximate Lateral Displacements

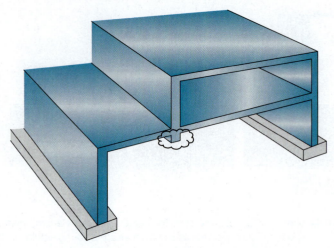

Figure 10.46 Beam under a structural wall to add stiffness.

Because the discontinuous wall *will* experience rotation in these cases, our by hand method for predicting drift will be a lower bound. To improve our prediction would require modeling with the Finite Element Method.

SECTION 10.5 HIGHLIGHTS

Solid Walls—Single Story

To estimate the total drift of a single-story wall:

$$\Delta = \frac{2.4Vh(1+\nu)}{Et_w l} + \frac{4Vh^3}{Et_w l^3}$$

where

V = Shear carried by the individual wall panel
ν = Poisson's ratio for the wall material
E = Modulus of elasticity for the wall material

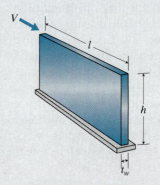

Coupled Walls:

If multiple walls are linked together in line, each drifts the same amount but carries different shear.

1. Set the drifts equal to each other:

$$\Delta_A = \Delta_B = \cdots$$

2. Rearrange the terms to express each shear as a function of one value:

$$V_i = V_A \frac{\left[\dfrac{2.4h(1+\nu_A)}{E_A t_{wA} l_A} + \dfrac{4h^3}{E_A t_{wA} l_A^3}\right]}{\left[\dfrac{2.4h(1+\nu_i)}{E_i t_{wi} l_i} + \dfrac{4h^3}{E_i t_{wi} l_i^3}\right]}$$

3. Apply horizontal equilibrium:

$$F = V_A + V_B + \cdots$$

4. Solve for the one shear value (e.g., V_A); then substitute in to find the other shear values if desired.
5. Use any of the shear values to calculate the drift if desired.

Summary of Limitations and Assumptions:

- Neglects effects of openings.
- Neglects effects of foundation deformation and movement.

Implication:

- Will always *underpredict* drift.

Example 10.9

Our firm is designing a convenience store that will have an awning over fuel pumps. In this region with moderate seismic activity, we need to make sure the store and awning don't slam into each other during a seismic event. To determine the minimum separation required between the two, we calculate the maximum inelastic response drift of each. The separation must be greater than the sum of the two drifts.

To find the maximum inelastic response drift, we multiply the static equivalent drift by an amplification factor; in this case, the amplification factor is 3.0.

The store will be made of precast panels connected at the top only; therefore, they act as linked walls. The panels have already been designed; the thickness of each panel is 14 cm. The concrete will have a modulus of elasticity of approximately 27 GPa and a Poisson's ratio of 0.18. Now our task is to predict the maximum inelastic response drift of the wall. Another team member is calculating that drift for the awning.

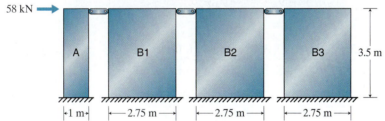

Figure 10.47

Approximate Analysis:

Neglect any foundation movement.

Find the shear in a panel:

Because they are coupled, all of the wall panels drift the same amount:

$$\Delta_A = \Delta_{B1} = \Delta_{B2} = \Delta_{B3}$$

Since three of the panels are identical, they all carry the same amount of shear:

$$V_{B1} = V_{B2} = V_{B3}$$

Express each panel shear in terms of the first panel, V_A:

$$V_B = V_A \frac{\left[\dfrac{2.4h(1+\nu_A)}{E_A t_{wA} l_A} + \dfrac{4h^3}{E_A t_{wA} l_A^3}\right]}{\left[\dfrac{2.4h(1+\nu_B)}{E_B t_{wB} l_B} + \dfrac{4h^3}{E_B t_{wB} l_B^3}\right]}$$

The modulus of elasticity and panel thickness are the same for all the panels, so the expression reduces to

$$V_B = V_A \frac{\left[\dfrac{2.4(1+\nu_A)}{l_A} + \dfrac{4h^2}{l_A^3}\right]}{\left[\dfrac{2.4(1+\nu_B)}{l_B} + \dfrac{4h^2}{l_B^3}\right]}$$

$$= V_A \frac{\left[\dfrac{2.4(1+0.18)}{(1\,\text{m})} + \dfrac{4(3.5\,\text{m})^2}{(1\,\text{m})^3}\right]}{\left[\dfrac{2.4(1+0.18)}{(2.75\,\text{m})} + \dfrac{4(3.5\,\text{m})^2}{(2.75\,\text{m})^3}\right]} = V_A \left[\dfrac{51.8}{3.39}\right]$$

$$= 15.3 V_A$$

Apply horizontal equilibrium:

$$\overset{+}{\rightarrow} \Sigma F_x = 0 = 58\,\text{kN} - V_A - 3(15.3V_A)$$

$$V_A = 1.237\,\text{kN}$$

Calculate the drift:

Static equivalent drift

$$\Delta = \frac{2.4 V_A h(1+\nu_A)}{E_A t_{wA} l_A} + \frac{4 V_A h^3}{E_A t_{wA} l_A^3}$$

$$= \frac{2.4(1.237\,\text{kN})(3500\,\text{mm})(1+0.18)}{(27\times 10^6\,\text{kN/m}^2)(0.14\,\text{m})(1\,\text{m})}$$

$$+ \frac{4(1.237\,\text{kN})(3.5\,\text{m})^3(1000\,\text{mm/m})}{(27\times 10^6\,\text{kN/m}^2)(0.14\,\text{m})(1\,\text{m})^3}$$

$$= 3.24\times 10^{-3}\,\text{mm} + 5.61\times 10^{-2}\,\text{mm}$$

$$\Delta_{\text{static equiv}} = 5.93\times 10^{-2}\,\text{mm}$$

Inelastic response drift

$$\Delta_{\text{max inelastic}} = 3(5.93\times 10^{-2}\,\text{mm}) = 0.18\,\text{mm}$$

Conclusion:

We can consider the convenience store as effectively stationary. Therefore, we only need to design for the awning drift.

Example 10.10

Our team is designing a new restaurant building. The senior engineer is considering a rigid diaphragm for the roof. If that is our final decision, we will need to calculate the rigidity of each wall to determine the distribution of lateral load. To prepare for those calculations, we have been tasked with finding the drift of one wall subject to a unit lateral load.

The preliminary design is for 8-in. concrete masonry unit (CMU) blocks with a net area compressive strength of 1500 psi. Therefore, the modulus of elasticity of the wall will be approximately 1350 ksi, and the Poisson's ratio will be about 0.25. For these blocks, the wall will have an equivalent thickness of 4.6 in.

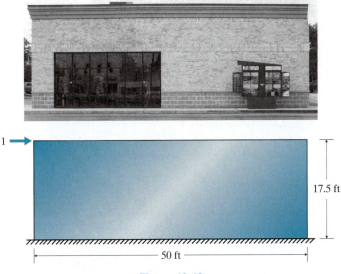

Figure 10.48

Approximate Analysis:

Neglect any foundation movement.

Solid wall:

Approximate the wall as solid. This will underpredict drift, so it is a lower bound estimate.

$$\Delta_{solid} = \frac{2.4Fh(1+\nu)}{Et_w l} + \frac{4Fh^3}{Et_w l^3}$$

$$= \frac{2.4(1)(17.5 \text{ ft})(1+0.25)}{(1350 \text{ ksi})(4.6 \text{ in.})(50 \text{ ft})} + \frac{4(1)(17.5 \text{ ft})^3}{(1350 \text{ ksi})(4.6 \text{ in.})(50 \text{ ft})^3}$$

$$= 1.691 \times 10^{-4} \text{ in./k} + 2.76 \times 10^{-5} \text{ in./k}$$
$$= 1.967 \times 10^{-4} \text{ in./k}$$

Coupled walls:

Consider the wall to be a pair of coupled walls on either side of the large window. This will overpredict drift, so it is an upper bound.

574　Chapter 10　Approximate Lateral Displacements

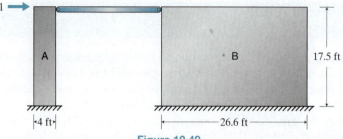

Figure 10.49

Find the shear in a panel:

Each drifts the same amount:

$$\Delta_A = \Delta_B$$

Express the shear in the right panel, V_B, in terms of the left, V_A:

$$V_B = V_A \frac{\left[\dfrac{2.4h(1+\nu_A)}{E_A t_{wA} l_A} + \dfrac{4h^3}{E_A t_{wA} l_A^3}\right]}{\left[\dfrac{2.4h(1+\nu_B)}{E_B t_{wB} l_B} + \dfrac{4h^3}{E_B t_{wB} l_B^3}\right]}$$

The modulus of elasticity and panel thickness are the same for all the panels, so the expression reduces to

$$V_B = V_A \frac{\left[\dfrac{2.4(1+\nu_A)}{l_A} + \dfrac{4h^2}{l_A^3}\right]}{\left[\dfrac{2.4(1+\nu_B)}{l_B} + \dfrac{4h^2}{l_B^3}\right]} = V_A \frac{\left[\dfrac{2.4(1+0.25)}{(4\text{ ft})} + \dfrac{4(17.5\text{ ft})^2}{(4\text{ ft})^3}\right]}{\left[\dfrac{2.4(1+0.25)}{(26.6\text{ ft})} + \dfrac{4(17.5\text{ ft})^2}{(26.6\text{ ft})^3}\right]} = V_A \left[\dfrac{19.89}{0.1779}\right] = 111.8 V_A$$

Apply horizontal equilibrium:

$$\xrightarrow{+} \Sigma F_x = 0 = 1 - V_A - (111.8) V_A$$

$$V_A = 8.87 \times 10^{-3}$$

Calculate the drift:

$$\Delta = \frac{2.4 V_A h (1+\nu_A)}{E_A t_{wA} l_A} + \frac{4 V_A h^3}{E_A t_{wA} l_A^3}$$

$$= \frac{2.4(8.87 \times 10^{-3})(17.5\text{ ft})(1+0.25)}{(1350\text{ ksi})(4.6\text{ in.})(4\text{ ft})} + \frac{4(8.87 \times 10^{-3})(17.5\text{ ft})^3}{(1350\text{ ksi})(4.6\text{ in.})(4\text{ ft})^3}$$

$$= 1.875 \times 10^{-5} \text{ in./k} + 4.78 \times 10^{-4} \text{ in./k}$$

$$= 4.97 \times 10^{-4} \text{ in./k}$$

Summary:

The anticipated drift, excluding any foundation movement, is bounded as follows:

$$1.967 \times 10^{-4} \text{ in./k} < \Delta < 4.97 \times 10^{-4} \text{ in./k}$$

10.6 Solid Walls—Multistory

Introduction

Structural walls are a popular choice for multistory buildings, but our formulas from the previous section are for a single load at the top of the wall. We want to be able to predict the total drift of a level when the levels above and below are receiving load as well. We need to be able to predict the total drift of different levels so that we can predict the story drift. Story drift is what we need to know to predict rigidity (see Section 11.1).

If we limit ourselves to solid walls, no openings, we can use solid mechanics and superposition to predict drift.

How-To

If the wall experiences force at multiple levels, we can use superposition, which lets us calculate drifts due to each individual force. We then add the individual results to obtain the effect of all the forces simultaneously (Figure 10.50). Remember that superposition is valid only *if* the material remains linear elastic.

We know how to predict the drift at the point of load application. Now we need to be able to predict the drift above and below the load.

Shear

Although not exactly correct, we make the following simplifications to predict the displaced shape due to just shear deformation:

1. Above the applied force, there is no shear deformation (Figure 10.51a on the next page). This is the same as saying that for a cantilevered beam, beyond the applied load there is no slope due to shear deformation (Figure 10.51b on the next page).
2. Below the applied force we assume a linear variation in drift (Figure 10.52 on the next page). This means we can use linear interpolation to find the drift at levels below the load.

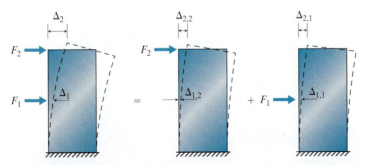

Figure 10.50 Effects of the two lateral forces can be predicted by adding the effects of each individual force.

576　Chapter 10　Approximate Lateral Displacements

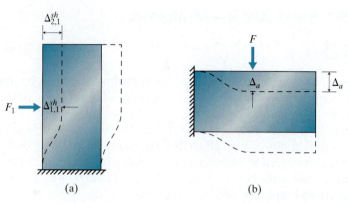

Figure 10.51 Shear deformation only for cantilevered members with a single point load: (a) Drift above the point of load application is the same as drift at the point of load application; (b) Displacement past the point of load application is the same as displacement at the point of load application.

These simplifications result in the following expressions for shear drift:

$$\Delta^{sh}_{\text{load}} = \frac{2.4(1+v)Vh_{\text{load}}}{Et_w l}$$

For $h_{\text{level}} \leq h_{\text{load}}$,

$$\Delta^{sh}_{\text{level}} = \frac{h_{\text{level}}}{h_{\text{load}}} \Delta^{sh}_{\text{load}}$$

For $h_{\text{level}} > h_{\text{load}}$,

$$\Delta^{sh}_{\text{level}} = \Delta^{sh}_{\text{load}}$$

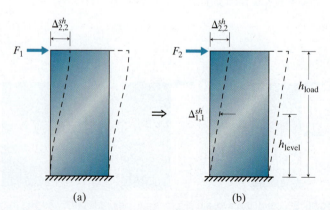

Figure 10.52 Drift of a solid wall due to just shear deformation: (a) Actual displaced shape; (b) Approximating displaced shape as linear.

Flexure

To find the drift anywhere up the wall, we can use the displacement formulas for a cantilevered beam from inside the front cover of this text (Figure 10.53a).

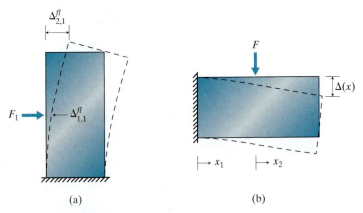

Figure 10.53 Displaced shape of cantilevered members considering only flexure: (a) Flexural drift of a solid wall due to a single load; (b) Flexural displacement of a cantilevered beam due to a single load.

For a wall (Figure 10.53b), these formulas become

$$\Delta^{fl}_{load} = \frac{4Vh^3_{load}}{Et_w l^3}$$

For $h_{level} \leq h_{load}$,

$$\Delta^{fl}_{level} = \frac{2V}{Et_w l^3}(3h^2_{level}h_{load} - h^3_{level})$$

For $h_{level} > h_{load}$,

$$\Delta^{fl}_{level} = \frac{2V}{Et_w l^3}[3h^2_{load}(h_{level} - h_{load}) + 2h^3_{load}]$$

Net Effect

The total drift of each level is the superposition of the shear and flexure drift predictions due to each load:

$$\Delta^{Total}_{level} = \Sigma(\Delta^{sh}_{level} + \Delta^{fl}_{level})$$

To find the story drift, we subtract the total drift at the bottom of the level from the total drift at the top of the level:

$$\Delta_{story\,i} = \Delta_{level\,i} - \Delta_{level\,i-1}$$

As indicated in Section 10.5, openings reduce stiffness and increase drift. Therefore, these formulas tend to underpredict the total and story drift for walls that have openings.

Section 10.6 Highlights

Solid Walls—Multistory

To estimate total drift at any level of a multistory wall:
1. Calculate the predicted drift at a level due to each load individually:

$$\Delta^{sh}_{load} = \frac{2.4(1+v)Vh_{load}}{Et_w l}$$

For $h_{level} \leq h_{load}$, $\Delta^{sh}_{level} = \frac{h_{level}}{h_{load}}\Delta^{sh}_{load}$

For $h_{level} > h_{load}$, $\Delta^{sh}_{level} = \Delta^{sh}_{load}$

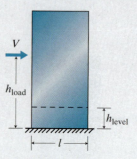

$$\Delta^{fl}_{load} = \frac{4Vh_{load}^3}{Et_w l^3}$$

For $h_{level} \leq h_{load}$, $\Delta^{fl}_{level} = \frac{2V}{Et_w l^3}(3h_{level}^2 h_{load} - h_{level}^3)$

For $h_{level} > h_{load}$, $\Delta^{fl}_{level} = \frac{2V}{Et_w l^3}[3h_{load}^2(h_{level} - h_{load}) + 2h_{load}^3]$

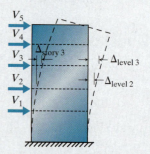

2. Superimpose the results to obtain the total drift at the level:

$$\Delta^{Total}_{level} = \Sigma(\Delta^{sh}_{level} + \Delta^{fl}_{level})$$

To estimate the story drift:
1. Calculate total drift at the top of the story, $\Delta_{level\ i}$.
2. Calculate total drift at the bottom of the story, $\Delta_{level\ i-1}$.
3. Find the difference:

$$\Delta_{story\ i} = \Delta_{level\ i} - \Delta_{level\ i-1}$$

Summary of Limitations and Assumptions:
- Neglects the effects of openings.
- Neglects the effects of foundation deformation and movement.
- Neglects any drift caused by shear above the force.
- Assumes linear distribution of shear deformation below the force.

Implication:
- Will almost always *underpredict* drift.

Section 10.6 Solid Walls—Multistory 579

EXAMPLE 10.11

Our firm has been hired to design a three-story apartment building in a region that often has high winds. An experienced engineer has determined that CMU structural walls will be the most cost-effective system in this situation.

The lateral load-resisting system in the short direction will be pairs of coupled CMU walls. Another member of our team has performed a preliminary design of the walls: 6-in. blocks with a modulus of elasticity of approximately 1500 ksi and a Poisson's ratio of approximately 0.25. Because the blocks will be filled with mortar, the effective thickness is the full 6 in.

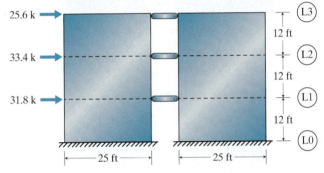

Figure 10.54

Our job is to verify that the wind drift of each story is below the limit of $h/500$.

Approximate Analysis:
Neglect any foundation movement. Since both walls are identical and coupled, they will drift the same amount at each level and each will carry half the shear at that level.

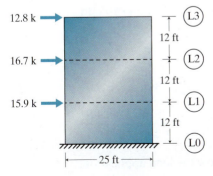

Figure 10.55

Use superposition to predict the total drift of each wall.

Drifts due to level 1 load:
    Shear:
        At the level of the load,

$$\Delta_{1,1}^{sh} = \frac{2.4(1+0.25)(15.9 \text{ k})(12 \text{ ft})}{(1500 \text{ ksi})(6 \text{ in.})(25 \text{ ft})} = 2.54\times10^{-3} \text{ in.}$$

        For levels above the load application,

$$\Delta_{2,1}^{sh} = \Delta_{1,1}^{sh} = 2.54\times10^{-3} \text{ in.}$$

$$\Delta_{3,1}^{sh} = \Delta_{1,1}^{sh} = 2.54\times10^{-3} \text{ in.}$$

    Flexure:
        At the level of the load,

$$\Delta_{1,1}^{fl} = \frac{4(15.9 \text{ k})(12 \text{ ft})^3}{(1500 \text{ ksi})(6 \text{ in.})(25 \text{ ft})^3} = 7.82\times10^{-4} \text{ in.}$$

        For levels above the load application,

$$\Delta_{2,1}^{fl} = \frac{2(15.9 \text{ k})}{(1500 \text{ ksi})(6 \text{ in.})(25 \text{ ft})^3}[3(12 \text{ ft})^2(24 \text{ ft} - 12 \text{ ft}) + 2(12 \text{ ft})^3]$$

$$= 1.954\times10^{-3} \text{ in.}$$

$$\Delta_{3,1}^{fl} = \frac{2(15.9 \text{ k})}{(1500 \text{ ksi})(6 \text{ in.})(25 \text{ ft})^3}[3(12 \text{ ft})^2(36 \text{ ft} - 12 \text{ ft}) + 2(12 \text{ ft})^3]$$

$$= 3.13\times10^{-3} \text{ in.}$$

    Total drifts due to level 1 load:

$$\Delta_{1,1}^{Total} = \Delta_{1,1}^{sh} + \Delta_{1,1}^{fl} = 2.54\times10^{-3} \text{ in.} + 7.82\times10^{-4} \text{ in.}$$

$$= 3.32\times10^{-3} \text{ in.}$$

$$\Delta_{2,1}^{Total} = \Delta_{2,1}^{sh} + \Delta_{2,1}^{fl} = 2.54\times10^{-3} \text{ in.} + 1.954\times10^{-3} \text{ in.}$$

$$= 4.49\times10^{-3} \text{ in.}$$

$$\Delta_{3,1}^{Total} = \Delta_{3,1}^{sh} + \Delta_{3,1}^{fl} = 2.54\times10^{-3} \text{ in.} + 3.13\times10^{-3} \text{ in.}$$

$$= 5.67\times10^{-3} \text{ in.}$$

Drifts due to level 2 load:
    Shear:
        At the level of the load,

$$\Delta_{2,2}^{sh} = \frac{2.4(1+0.25)(16.7 \text{ k})(24 \text{ ft})}{(1500 \text{ ksi})(6 \text{ in.})(25 \text{ ft})} = 5.34\times10^{-3} \text{ in.}$$

        For the level below the load application,

$$\Delta_{1,2}^{sh} = \left(\frac{h_{level}}{h_{load}}\right)\Delta_{2,2}^{sh} = \left(\frac{12 \text{ ft}}{24 \text{ ft}}\right)5.34\times10^{-3} \text{ in.} = 2.67\times10^{-3} \text{ in.}$$

For the level above the load application,
$$\Delta^{sh}_{3,2} = \Delta^{sh}_{2,2} = 5.34 \times 10^{-3} \text{ in.}$$

Flexure:
At the level of the load,
$$\Delta^{fl}_{2,2} = \frac{4(16.7 \text{ k})(24 \text{ ft})^3}{(1500 \text{ ksi})(6 \text{ in.})(25 \text{ ft})^3} = 6.57 \times 10^{-3} \text{ in.}$$

For the level below the load application,
$$\Delta^{fl}_{1,2} = \frac{2(16.7 \text{ k})}{(1500 \text{ ksi})(6 \text{ in.})(25 \text{ ft})^3}[3(12 \text{ ft})^2(24 \text{ ft}) - (12 \text{ ft})^3]$$
$$= 2.05 \times 10^{-3} \text{ in.}$$

For the level above the load application,
$$\Delta^{fl}_{3,2} = \frac{2(16.7 \text{ k})}{(1500 \text{ ksi})(6 \text{ in.})(25 \text{ ft})^3}[3(12 \text{ ft})^2(36 \text{ ft} - 24 \text{ ft}) + 2(24 \text{ ft})^3]$$
$$= 1.149 \times 10^{-2} \text{ in.}$$

Total drifts due to level 2 load:
$$\Delta^{Total}_{1,2} = \Delta^{sh}_{1,2} + \Delta^{fl}_{1,2} = 2.67 \times 10^{-3} \text{ in.} + 2.05 \times 10^{-3} \text{ in.}$$
$$= 4.72 \times 10^{-3} \text{ in.}$$
$$\Delta^{Total}_{2,2} = \Delta^{sh}_{2,2} + \Delta^{fl}_{2,2} = 5.34 \times 10^{-3} \text{ in.} + 6.57 \times 10^{-3} \text{ in.}$$
$$= 1.191 \times 10^{-2} \text{ in.}$$
$$\Delta^{Total}_{3,2} = \Delta^{sh}_{3,2} + \Delta^{fl}_{3,2} = 5.34 \times 10^{-3} \text{ in.} + 1.149 \times 10^{-2} \text{ in.}$$
$$= 1.683 \times 10^{-2} \text{ in.}$$

Drifts due to level 3 load:
Shear:
At the level of the load,
$$\Delta^{sh}_{3,3} = \frac{2.4(1 + 0.25)(12.8 \text{ k})(36 \text{ ft})}{(1500 \text{ ksi})(6 \text{ in.})(25 \text{ ft})} = 6.14 \times 10^{-3} \text{ in.}$$

For levels below the load application,
$$\Delta^{sh}_{1,3} = \left(\frac{h_{level}}{h_{load}}\right)\Delta^{sh}_{3,3} = \left(\frac{12 \text{ ft}}{36 \text{ ft}}\right)6.14 \times 10^{-3} \text{ in.} = 2.05 \times 10^{-3} \text{ in.}$$

$$\Delta^{sh}_{2,3} = \left(\frac{h_{level}}{h_{load}}\right)\Delta^{sh}_{3,3} = \left(\frac{24 \text{ ft}}{36 \text{ ft}}\right)6.14 \times 10^{-3} \text{ in.} = 4.09 \times 10^{-3} \text{ in.}$$

Flexure:
At the level of the load,
$$\Delta^{fl}_{3,3} = \frac{4(12.8 \text{ k})(36 \text{ ft})^3}{(1500 \text{ ksi})(6 \text{ in.})(25 \text{ ft})^3} = 1.699 \times 10^{-2} \text{ in.}$$

For levels below the load application,

$$\Delta^{fl}_{1,3} = \frac{2(12.8 \text{ k})}{(1500 \text{ ksi})(6 \text{ in.})(25 \text{ ft})^3}[3(12 \text{ ft})^2(36 \text{ ft}) - (12 \text{ ft})^3]$$

$$= 2.52 \times 10^{-3} \text{ in.}$$

$$\Delta^{fl}_{2,3} = \frac{2(12.8 \text{ k})}{(1500 \text{ ksi})(6 \text{ in.})(25 \text{ ft})^3}[3(24 \text{ ft})^2(36 \text{ ft}) - (24 \text{ ft})^3]$$

$$= 8.81 \times 10^{-3} \text{ in.}$$

Total drifts due to level 3 load:

$$\Delta^{\text{Total}}_{1,3} = \Delta^{sh}_{1,3} + \Delta^{fl}_{1,3} = 2.05 \times 10^{-3} \text{ in.} + 2.52 \times 10^{-3} \text{ in.} = 4.58 \times 10^{-3} \text{ in.}$$

$$\Delta^{\text{Total}}_{2,3} = \Delta^{sh}_{2,3} + \Delta^{fl}_{2,3} = 4.09 \times 10^{-3} \text{ in.} + 8.81 \times 10^{-3} \text{ in.} = 1.290 \times 10^{-2} \text{ in.}$$

$$\Delta^{\text{Total}}_{3,3} = \Delta^{sh}_{3,3} + \Delta^{fl}_{3,3} = 6.14 \times 10^{-3} \text{ in.} + 1.699 \times 10^{-2} \text{ in.} = 2.31 \times 10^{-2} \text{ in.}$$

Total drifts due to all loads:

$$\Delta^{\text{Total}}_1 = \Delta^{\text{Total}}_{1,1} + \Delta^{\text{Total}}_{1,2} + \Delta^{\text{Total}}_{1,3} = 3.32 \times 10^{-3} \text{ in.} + 4.72 \times 10^{-3} \text{ in.} + 4.58 \times 10^{-3} \text{ in.}$$

$$= 0.0126 \text{ in.}$$

$$\Delta^{\text{Total}}_2 = \Delta^{\text{Total}}_{2,1} + \Delta^{\text{Total}}_{2,2} + \Delta^{\text{Total}}_{2,3} = 4.49 \times 10^{-3} \text{ in.} + 1.191 \times 10^{-2} \text{ in.} + 1.290 \times 10^{-2} \text{ in.}$$

$$= 0.0293 \text{ in.}$$

$$\Delta^{\text{Total}}_3 = \Delta^{\text{Total}}_{3,1} + \Delta^{\text{Total}}_{3,2} + \Delta^{\text{Total}}_{3,3} = 5.67 \times 10^{-3} \text{ in.} + 1.683 \times 10^{-2} \text{ in.} + 2.31 \times 10^{-2} \text{ in.}$$

$$= 0.0456 \text{ in.}$$

Story drifts:

$$\Delta^{\text{Story}}_1 = \Delta^{\text{Total}}_1 = 0.0126 \text{ in.}$$

$$\Delta^{\text{Story}}_2 = \Delta^{\text{Total}}_2 - \Delta^{\text{Total}}_1 = 0.0293 \text{ in.} - 0.0126 \text{ in.} = 0.0167 \text{ in.}$$

$$\Delta^{\text{Story}}_3 = \Delta^{\text{Total}}_3 - \Delta^{\text{Total}}_2 = 0.0456 \text{ in.} - 0.0293 \text{ in.} = 0.0163 \text{ in.}$$

Use of Results:

Drift limits:

$$\Delta^{\text{Limit}}_1 = \frac{h_1}{500} = \frac{12 \text{ ft}(12 \text{ in./ft})}{500} = 0.29 \text{ in.}$$

$$\Delta^{\text{Limit}}_2 = \frac{h_2}{500} = \frac{12 \text{ ft}(12 \text{ in./ft})}{500} = 0.29 \text{ in.}$$

$$\Delta^{\text{Limit}}_3 = \frac{h_3}{500} = \frac{12 \text{ ft}(12 \text{ in./ft})}{500} = 0.29 \text{ in.}$$

Comparison:

$$\Delta^{\text{Story}}_1 < \Delta^{\text{Limit}}_1 \therefore \text{ Adequate design}$$

$$\Delta^{\text{Story}}_2 < \Delta^{\text{Limit}}_2 \therefore \text{ Adequate design}$$

$$\Delta^{\text{Story}}_3 < \Delta^{\text{Limit}}_3 \therefore \text{ Adequate design}$$

Homework Problems

10.1 Our team is designing a new sports arena. The team leader has chosen to use chevron braced frames to provide lateral load resistance. The team has performed a preliminary design of the frame. Before we model the structural system in a computer program, we want an idea of what the drift will be. To help us better understand the difference between the two approximate analysis methods, our supervisor wants us to use both.

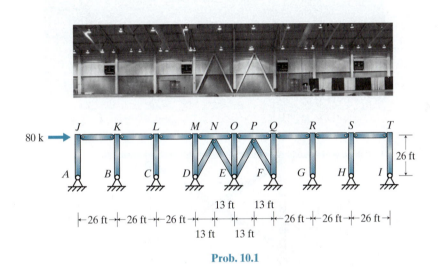

Prob. 10.1

Preliminary selection of member properties:

Columns:	W14 × 99	$A = 29.1$ in^2	$I_y = 402$ in^4
Beams:	W10 × 45	$A = 13.3$ in^2	$I_x = 248$ in^4
Braces:	HSS 8 × 8 × $\frac{3}{8}$	$A = 10.4$ in^2	$I = 100$ in^4
Steel:	$E = 29{,}000$ ksi		

a. Guess which approximate analysis method will result in the larger drift prediction.
b. List the assumptions you make in order to use the Story Drift Method to predict the drift of the frame.
c. Use the Story Drift Method to predict the total drift of the frame.
d. List the assumptions you make in order to use the Virtual Work Method to predict the drift of the frame.
e. Use the Virtual Work Method to predict the total drift of the frame.
f. Predict the total drift of the frame using computer aided analysis. Because of axial shortening of the beams, the drift value will be different at different joints; measure the drift at joint O.
g. Submit a printout of the displaced shape. Identify at least three features that suggest the displaced shape is reasonable.
h. Verify equilibrium of the computer aided analysis results.
i. Are the results of your computer aided analysis reasonable? Make a comprehensive argument to justify your answer.
j. Comment on why your guess in part (a) was or was not correct when compared with your results in parts (c) and (e).

10.2 Our firm has been selected to design the structural framing for new stadium bleachers. The team has performed a preliminary design of the frame and is now checking its seismic performance. The seismic drift limit for this frame is $h/67$; the limit exists to help ensure that the frame has enough ductility to provide the necessary seismic strength.

Before we model the frame in computer analysis software, we want to know whether the frame drifts more than the limit. To help us better understand the difference between the two approximate analysis methods, our supervisor wants us to use both.

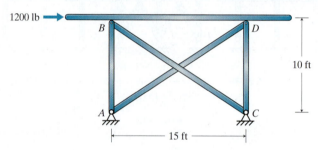

Prob. 10.2

Preliminary selection of member properties:

Columns:	W4 × 13	$A = 3.83$ in^2	$I_y = 3.86$ in^4
Beam:	W12 × 22	$A = 6.48$ in^2	$I_x = 1.56$ in^4
Braces:	L4 × 4 × $\frac{1}{4}$	$A = 1.93$ in^2	$I_z = 1.18$ in^4
Steel:	$E = 29{,}000$ ksi		

a. Guess which approximate analysis method will result in the larger drift prediction.
b. List the assumptions you make in order to use the Story Drift Method to predict the drift of the frame.
c. Use the Story Drift Method to predict the static equivalent total drift of the frame.
d. List the assumptions you make in order to use the Virtual Work Method to predict the drift of the frame.
e. Use the Virtual Work Method to predict the static equivalent total drift of the frame.
f. Compare the largest seismic drift prediction to the drift limit. Does the frame appear to be adequately designed for seismic response?
g. Predict the static equivalent total drift of the frame using computer aided analysis. Because of axial short-

ening of the beam, the drift value will be different at different joints; the best comparison will be joint D, since all of the beam shortening happens to the left of D. Compare this drift prediction to the limit. Does this change your conclusion about the adequacy of the current design?
h. Submit a printout of the displaced shape. Identify at least three features that suggest the displaced shape is reasonable.
i. Verify equilibrium of the computer aided analysis results.
j. Are the results of your computer aided analysis reasonable? Make a comprehensive argument to justify your answer.
k. Comment on why your guess in part (a) was or was not correct when compared with your results in parts (c) and (e).

10.3 Our team was selected to design a new four-story apartment building. The team leader has chosen to use diagonally braced frames for the lateral load-resisting system. Other members of the team have completed a preliminary design of the structural elements. Our job is to verify that the design meets the drift limits of the precast panels that will hang on the exterior of the frame. The manufacturer of the precast panels recommends a story drift limit of $h/1000$.

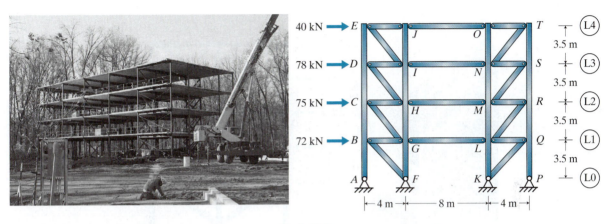

Prob. 10.3

Preliminary selection of member properties:

Beams: W360 × 44.6	$A = 57.1$ cm²	$I_x = 12{,}110$ cm⁴
Columns F and K: HSS 304.8 × 152.4 × 15.9	$A = 120.6$ cm²	$I_x = 13{,}360$ cm⁴
Columns A and P: HSS 203.2 × 152.4 × 15.9	$A = 90.3$ cm²	$I_x = 4745$ cm⁴
Braces L1 and L2: HSS 203.2 × 101.6 × 15.9	$A = 75.5$ cm²	$I_x = 3413$ cm⁴
Braces L3 and L4: HSS 203.2 × 101.6 × 7.9	$A = 41.5$ cm²	$I_x = 2123$ cm⁴
Steel:	$E = 200$ GPa $= 20{,}000$ kN/cm²	

a. Guess which story will drift the most.
b. List the assumptions you make in order to use the Story Drift Method to predict the drift of each level.
c. Use the Story Drift Method to predict the story drift for each level.
d. Compare the largest story drift prediction to the drift limit. Does the frame appear to be adequately designed?
e. Refine your drift predictions using computer aided analysis. Determine the story drift prediction for each level. Because of axial shortening of the beams, the drift value will be different at different joints; the best comparison will be along column line P, since all of the beam shortening happens to the left of that column line. Compare the largest story drift prediction to the limit. Does this change your conclusion about the adequacy of the current design?
f. Submit a printout of the displaced shape. Identify at least three features that suggest the displaced shape is reasonable.
g. Verify equilibrium of the computer aided analysis results.
h. Are the results of your computer aided analysis reasonable? Make a comprehensive argument to justify your answer.
i. Comment on why your guess in part (a) was or was not correct when compared with your results in part (c).

10.4 Our team is designing a five-story building in a region that has high seismic activity. The architect would like to make the lateral load-resisting system visible on the outside, and our team leader has chosen to use two bays of diagonal braces. The team has completed a preliminary design of the structural members. Now we are tasked with verifying that each story drift is below the allowable limit of $h/87$ for this structure.

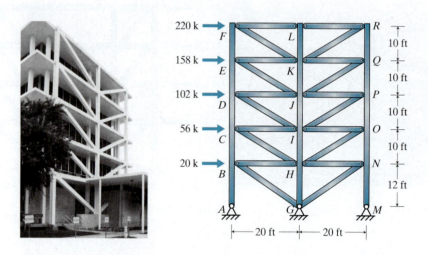

Prob. 10.4

Preliminary selection of member properties:

Columns A and M:	W12 × 72	$A = 21.1$ in^2	$I_y = 195$ in^4
Column G:	W12 × 65	$A = 19.1$ in^2	$I_y = 174$ in^4
Beams and braces:	W12 × 65	$A = 19.1$ in^2	$I_y = 174$ in^4
Steel:	$E = 29{,}000$ ksi		

a. Guess which story will drift the most.
b. List the assumptions you make in order to use the Story Drift Method to predict the drift of each level.
c. Use the Story Drift Method to predict the story drift for each level.
d. Compare each story drift prediction to the drift limit for that story. Does the frame appear to be adequately designed?
e. Refine your drift predictions using computer aided analysis with the static equivalent loads. Determine the story drift prediction for each level. Because of axial shortening of the beams, the drift value will be different at different joints; calculate the drift at column line G, since it is in the middle. Compare each story drift prediction to its limit. Does this change your conclusion about the adequacy of the current design?
f. Submit a printout of the displaced shape. Identify at least three features that suggest the displaced shape is reasonable.
g. Verify equilibrium of the computer aided analysis results.
h. Are the results of your computer aided analysis reasonable? Make a comprehensive argument to justify your answer.
i. Comment on why your guess in part (a) was or was not correct when compared with your results in part (c).

10.5 Our team is designing a new airport terminal in a region that has low seismic activity. The architect and structural team leader have decided to use a combination of chevron and inverted chevron braces to provide lateral load resistance while permitting the desired access. Other members of the team have completed a preliminary design of the structural members. Our job is to verify that each story drift is below the allowable limit of $h/50$ for this structure.

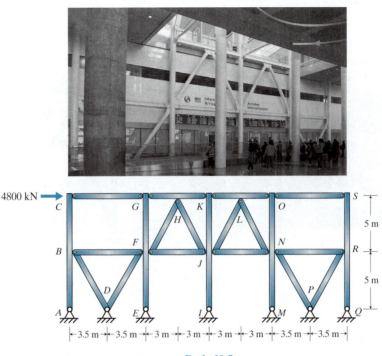

Prob. 10.5

Preliminary selection of member properties:

Columns:	HSS 457.2 × 9.5	$A = 1.25 \times 10^4$ mm^2	$I = 3.14 \times 10^8$ mm^4
Braces:	HSS 355.6 × 9.5	$A = 9.68 \times 10^3$ mm^2	$I = 1.45 \times 10^8$ mm^4
Beams:	W250 × 80	$A = 1.02 \times 10^4$ mm^2	$I = 1.26 \times 10^8$ mm^4
Steel:	$E = 200$ GPa $= 200$ kN/mm^2		

a. Guess which story will drift the most.
b. List the assumptions you make in order to use the Story Drift Method to predict the drift of each level.
c. Use the Story Drift Method to predict the story drift for each level.
d. Compare the largest story drift prediction to the drift limit. Does the frame appear to be adequately designed?
e. Refine your drift predictions using computer aided analysis with the static equivalent loads. Determine the story drift prediction for each level. Because of axial shortening of the beams, the drift value will be different at different joints; calculate the drift at column line I, since it is in the middle. Compare the largest story drift prediction to the drift limit. Does this change your conclusion about the adequacy of the current design?
f. Submit a printout of the displaced shape. Identify at least three features that suggest the displaced shape is reasonable.
g. Verify equilibrium of the computer aided analysis results.
h. Are the results of your computer aided analysis reasonable? Make a comprehensive argument to justify your answer.
i. Comment on why your guess in part (a) was or was not correct when compared with your results in part (c).

10.6 Our firm has been selected to design the structural framing to support the grandstands at a race track. A member of the team has completed a preliminary design of the structural members. The team leader wants to know if the total drift at the top of the frame is less than $h_{total}/500$; that is a good first indication of whether the side sway in a stiff wind will make spectators uncomfortable. If our design is close to this limit, we will need to perform detailed analysis of the motion and possibly redesign the braces.

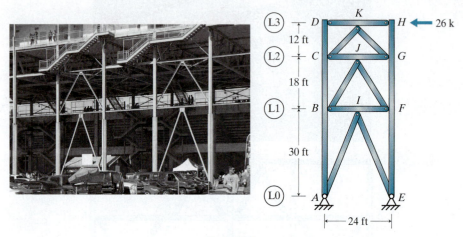

Prob. 10.6

Preliminary selection of member properties:

Columns:	W14 × 176	$A = 51.8 \text{ in}^2$	$I_y = 838 \text{ in}^4$
Beams:	W8 × 35	$A = 10.3 \text{ in}^2$	$I_x = 127 \text{ in}^4$
L1 braces:	HSS 7 × 0.375	$A = 7.29 \text{ in}^2$	$I = 40.4 \text{ in}^4$
L2 braces:	2L5 × 5 × $\frac{1}{2}$	$A = 9.58 \text{ in}^2$	$I_x = 22.4 \text{ in}^4$
L3 braces:	2L3 × 3 × $\frac{3}{8}$	$A = 4.22 \text{ in}^2$	$I_x = 3.49 \text{ in}^4$
Steel:	$E = 29{,}000 \text{ ksi}$		

a. Guess which level of braces contributes most to the total drift.
b. List the assumptions you make in order to use the Virtual Work Method to predict the total drift.
c. Use the Virtual Work Method to predict the total drift of the frame. Calculate the individual contributions of each pair of braces and each pair of column segments.
d. Compare the total drift prediction to the drift limit. Does the frame appear to be adequately designed?
e. Refine your drift predictions using computer aided analysis. Determine the total drift prediction. Because of axial shortening of the beams, the drift value will be different at different joints; the best comparison will be along column line *A*, since all of the beam shortening happens to the right of that column line. Compare the total drift prediction to the limit. Does this change your conclusion about the adequacy of the current design?
f. Submit a printout of the displaced shape. Identify at least three features that suggest the displaced shape is reasonable.
g. Verify equilibrium of the computer aided analysis results.
h. Are the results of your computer aided analysis reasonable? Make a comprehensive argument to justify your answer.
i. Comment on why your guess in part (a) was or was not correct when compared with your results in part (c).

10.7 Our firm has been selected to design the structural framing to support the press box at a stadium. The stadium is in a region with high seismic activity. A team member has completed a preliminary design of the frame. Our team leader points out that we planned a 1-inch gap between the bleachers and the press box, but we must confirm that the two will not strike each other during an earthquake. Therefore, we have been tasked with predicting the maximum inelastic drift of the frame during the design seismic event and comparing that value with the space available.

The maximum inelastic response drift is calculated by multiplying the static equivalent drift by an amplification factor; in this case, the amplification factor is 4.0. We use the maximum inelastic response drift to ensure sufficient structural separation. The maximum drift of the bleachers is 0.6 inch and we planned a gap of 1.0 inch, so the maximum inelastic drift that the press box can have is 0.4 inch.

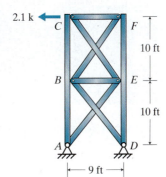

Prob. 10.7

Preliminary selection of member properties:

Columns:	W8 × 21	$A = 6.16 \text{ in}^2$	$I_y = 9.77 \text{ in}^4$
Beams:	W5 × 16	$A = 4.71 \text{ in}^2$	$I_x = 21.4 \text{ in}^4$
Braces:	L2 × 2 × $\frac{3}{8}$	$A = 1.37 \text{ in}^2$	$I_z = 0.204 \text{ in}^4$
Steel:	$E = 29{,}000 \text{ ksi}$		

a. Guess which element contributes most to the total drift: top-level columns, bottom-level columns, top-level braces, or bottom-level braces.
b. List the assumptions you make in order to use the Virtual Work Method to predict the total static equivalent drift.
c. Use the Virtual Work Method to predict the total static equivalent drift. Calculate the individual contributions of each pair of braces and each pair of column segments.
d. Convert the total static equivalent drift into the maximum inelastic response drift, and compare that value with the 0.4 inch available. Does the frame appear to be adequately designed?
e. Refine your prediction of the static equivalent drift using the static equivalent loads in a computer aided analysis. Determine the total static equivalent drift. Because of axial deformation of the beams, the drift value will be different at different joints; the best comparison will be along column line D, since all of the beam deformation happens to the left of that column line. Convert the total static equivalent drift into the maximum inelastic response drift, and compare that value with the 0.4 inch available. Does this change your conclusion about the adequacy of the current design?
f. Submit a printout of the displaced shape. Identify at least three features that suggest the displaced shape is reasonable.
g. Verify equilibrium of the computer aided analysis results.
h. Are the results of your computer aided analysis reasonable? Make a comprehensive argument to justify your answer.
i. Comment on why your guess in part (a) was or was not correct when compared with your results in part (c).

10.8 Our team is designing a new sports arena in Europe. The team leader has chosen cross braced frames to provide lateral load resistance. Members of the team have completed a preliminary design of the structure. Before we finalize the design, we need to know if the chosen members keep the drift below the limit. The manufacturer of the façade recommends that the total drift be kept below $h/800$; we need only consider total drift because the façade panels run the full height.

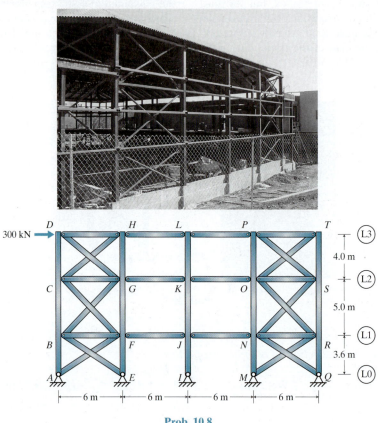

Prob. 10.8

Preliminary selection of member properties:

Columns:	HEA 360	$A = 142.8$ cm^2	$I_y = 7887$ cm^4
Girts (horiz):	HSS 250 × 150 × 10	$A = 74.9$ cm^2	$I_y = 2755$ cm^4
Braces:	Round Pipe 139.7 × 8	$A = 75.5$ cm^2	$I = 720$ cm^4
Steel:	$E = 200$ GPa $= 20{,}000$ kN/cm^2		

a. Guess which element contributes most to the total drift: column or brace, which level.
b. List the assumptions you make in order to use the Virtual Work Method to predict the total drift.
c. Use the Virtual Work Method to predict the total drift of the frame. Calculate the individual contributions of each level of braces and each level of column segments.
d. Compare the total drift prediction to the drift limit. Does the frame appear to be adequately designed?
e. Refine your drift prediction using computer aided analysis. Determine the total drift prediction. Because of axial shortening of the girts, the drift value will be different at different joints; measure at column line I, since it is in the middle. Compare the total drift prediction to the limit. Does this change your conclusion about the adequacy of the current design?
f. Submit a printout of the displaced shape. Identify at least three features that suggest the displaced shape is reasonable.
g. Verify equilibrium of the computer aided analysis results.
h. Are the results of your computer aided analysis reasonable? Make a comprehensive argument to justify your answer.
i. Comment on why your guess in part (a) was or was not correct when compared with your results in part (c).

10.9 Our team is designing a four-story office building. The principal in charge has experience with structures like this and has chosen to use chevron braces. Seismic requirements make it most economical to alternate chevrons and inverted chevrons, so the lateral load-resisting system looks similar to cross braces.

The team has completed a preliminary design considering all loads, including seismic and wind. Before we model the structure in structural analysis software and finalize the design, we want to know if the frame meets the drift limit requirements. Our job is to find the level that has the largest story drift and verify that the drift is below the limit of $h/500$.

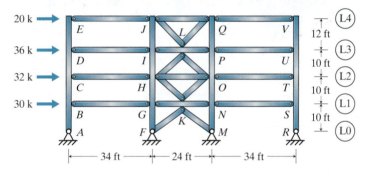

Prob. 10.9

Preliminary selection of member properties:

Columns:	W12 × 96	$A = 28.2$ in²	$I_x = 833$ in⁴
Beams:	W21 × 44	$A = 13.0$ in²	$I_x = 843$ in⁴
Braces:	HSS 5 × 5 × $\frac{1}{4}$	$A = 4.30$ in²	$I = 16.0$ in⁴
Steel:	$E = 29{,}000$ ksi		

a. Guess which level has the largest story drift index (Δ_{story}/h).
b. List the assumptions you make in order to use the Story Drift Method to predict the drift of each level.
c. Use the Story Drift Method to predict the story drift for each level, and convert those drifts to drift indexes.
d. List the assumptions you make in order to use the Virtual Work Method to predict the total drift of a level.
e. Use the Virtual Work Method to predict the story drift for the level with the largest drift index from part (c).
f. Compare the story drift prediction to the drift limit. Does the frame appear to be adequately designed?
g. Refine your drift predictions using computer aided analysis. Determine the story drift for each level. Because of axial shortening of the beams, the drift value will be different at different joints; measure at column line M, since most of the beam shortening will happen to the left of that column. Compare the story drifts to their limits. Does this change your conclusion about the adequacy of the current design?
h. Submit a printout of the displaced shape. Identify at least three features that suggest the displaced shape is reasonable.
i. Verify equilibrium of the computer aided analysis results.
j. Are the results of your computer aided analysis reasonable? Make a comprehensive argument to justify your answer.
k. Comment on why your guess in part (a) was or was not correct when compared with your results in part (c).

10.10 Our team is designing a large recreation facility. The senior engineer has chosen to use two bays of chevron braces for the lateral load-resisting system. After discussions with the architect, the engineer was able to get the middle column moved so that the bay widths are closer to equal.

The team updated the wind load on the frame and refined the selection of member properties for the new geometry and loading. However, the wall panel manufacturer sent revised requirements. The manufacturer recommends that the total drift be kept below $h/750$; we need only consider total drift because the wall panels run the full height.

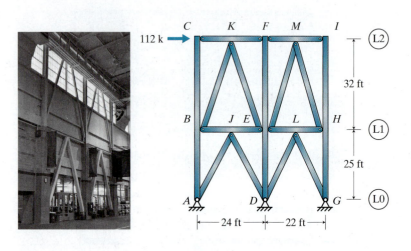

Prob. 10.10

Revised member properties:

Columns:	W14 × 109	$A = 32.0 \text{ in}^2$	$I_y = 447 \text{ in}^4$
Beams:	W10 × 33	$A = 9.71 \text{ in}^2$	$I_x = 171 \text{ in}^4$
Upper braces:	HSS 8 × 8 × $\frac{5}{8}$	$A = 16.4 \text{ in}^2$	$I = 146 \text{ in}^4$
Lower braces:	HSS 8 × 8 × $\frac{1}{4}$	$A = 7.10 \text{ in}^2$	$I = 70.7 \text{ in}^4$
Steel:	$E = 29{,}000 \text{ ksi}$		

a. Guess which element contributes most to the total drift: bottom-level columns, top-level columns, or a pair of braces.
b. List the assumptions you make in order to use the Virtual Work Method to predict the total drift.
c. Use the Virtual Work Method to predict the total drift of the frame. Calculate the individual contributions of each pair of braces and each level of column segments.
d. Compare the total drift prediction to the new drift limit. Does the frame appear to be adequately designed?
e. Refine your drift prediction using computer aided analysis. Determine the total drift prediction. Because of axial shortening of the beams, the drift value will be different at different joints; measure at column line D, since it is the middle column. Compare the total drift prediction to the limit. Does this change your conclusion about the adequacy of the current design?
f. Submit a printout of the displaced shape. Identify at least three features that suggest the displaced shape is reasonable.
g. Verify equilibrium of the computer aided analysis results.
h. Are the results of your computer aided analysis reasonable? Make a comprehensive argument to justify your answer.
i. Comment on why your guess in part (a) was or was not correct when compared with your results in part (c).

10.11 The superstructure of a bridge (everything from the girders up) is being modified, and as a result it will receive more wind load than it was originally designed for. The team leader wants us to calculate the lateral drift of the bridge under the new wind load. The leader wants to know if the drift exceeds $h/250$.

The barrier at ground level looks like a beam but is not connected in a way that would help resist lateral load. Therefore, the substructure and wind load can be idealized as shown.

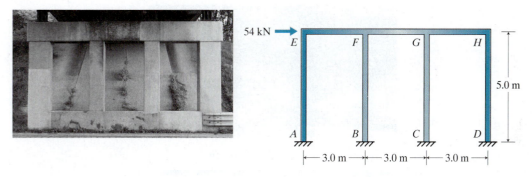

Prob. 10.11

Field measurements resulted in the following current member properties:

Beams:	$A = 4900 \text{ cm}^2$	$I = 2.0 \times 10^6 \text{ cm}^4$
Outer columns:	$A = 6500 \text{ cm}^2$	$I = 4.7 \times 10^6 \text{ cm}^4$
Inner columns:	$A = 6000 \text{ cm}^2$	$I = 3.6 \times 10^6 \text{ cm}^4$
Concrete:	$E = 28 \text{ GPa} = 28 \text{ kN/mm}^2$	

a. Guess the anticipated deflected shape for the frame.
b. Predict the drift using the Stiff Beam Method. Should the prediction overestimate or underestimate drift, or can't you be sure?
c. Predict the drift using the Virtual Work Method. Should the prediction overestimate or underestimate drift, or can't you be sure? Should this prediction be better or worse than the prediction in part (b)?
d. Using the largest result from parts (b) and (c), compare the predicted drift to the drift limit. Evaluate the adequacy of the existing frame to carry the new wind load.
e. Predict the drift using structural analysis software. Because of axial shortening of the beams, the drift value will be different at different joints; the best comparison will be at joint H, since all of the beam shortening happens to the left of that column line.
f. Submit a printout of the displaced shape. Identify at least three features that suggest the displaced shape is reasonable.
g. Verify equilibrium of the computer aided analysis results.
h. Are the results of your computer aided analysis reasonable? Make a comprehensive argument to justify your answer.
i. Comment on why your guess in part (a) was or was not close to the solution in part (f).

594 Chapter 10 Approximate Lateral Displacements

10.12 A highway overpass is supported by a two-bay rigid frame adjacent to another frame. The point of fixity for the columns is well below the ground surface. Our team has designed the frame to provide sufficient strength; however, we still need to confirm that the frame will not sway into the other frame during the design seismic event. There is a 1-inch gap between the frames. Our task is to predict the maximum inelastic response drift of the two-bay frame. If the maximum inelastic drift of this overpass stays below half the gap, the design is adequate. Another team member is finding the drift of the adjacent frame.

To find the maximum inelastic response drift, we multiply the static equivalent drift by an amplification factor. For this frame, the amplification factor is 4.5. To make a preliminary determination of whether drift will be an issue, the team leader recommends we use the Stiff Beam Method. We can use a more elaborate model if the drift might be an issue.

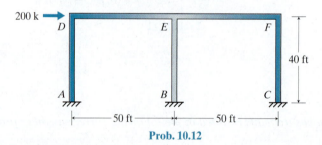

Prob. 10.12

Current member properties:

Columns:	$A = 36$ ft^2	$I = 108$ ft^4
Beams:	$A = 48$ ft^2	$I = 256$ ft^4
Concrete:	$E = 4400$ ksi	

a. Guess the displaced shape.
b. List the assumptions you make in order to use the Stiff Beam Method to predict the total drift.
c. Use the Stiff Beam Method to predict the static equivalent drift. Convert the static equivalent drift into the maximum inelastic response drift.
d. If the maximum inelastic response drift exceeds one-fourth of the gap, it is getting close to the limit; therefore, you should revise your drift predictions using the Virtual Work Method. Calculate the individual contributions of the beams and the columns. Does the frame appear adequate? Where would you recommend that the team increase the size of members in order to most effectively reduce the drift if it became necessary?
e. Refine your drift prediction using computer aided analysis. Use structural analysis software to predict the static equivalent drift; then calculate the maximum inelastic response drift by hand. Because of axial shortening of the beams, the drift value will be different at different joints; the best comparison will be at joint F, since all of the beam shortening happens to the left of that column line. Compare this inelastic response drift to the gap. Does this change your conclusion about the adequacy of the current design?
f. Submit a printout of the displaced shape. Identify at least three features that suggest the displaced shape is reasonable.
g. Verify equilibrium of the computer aided analysis results.
h. Are the results of your computer aided analysis reasonable? Make a comprehensive argument to justify your answer.
i. Comment on why your guess in part (a) was or was not correct when compared with your results in part (f).

10.13 Our firm is designing a six-story office building to fit on a narrow site in a densely populated city. The lateral load-resisting system has been designed for strength. Our design team leader wants to have a sense of whether drift will be a problem. Therefore, we have been tasked with estimating the story drift for each level. The team leader wants the story drift to stay below $h/500$. For this preliminary investigation, the Stiff Beam Method will be adequate.

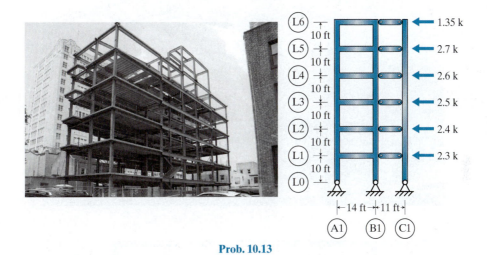

Prob. 10.13

Selected member properties:

Columns:	W10 × 39	$A = 11.5$ in²	$I_x = 209$ in⁴	$I_y = 45.0$ in⁴
Beams:	W14 × 53	$A = 15.6$ in²	$I_x = 541$ in⁴	
Steel:	$E = 29{,}000$ ksi			

Note that the columns in the rigid frame bend about their primary axis, I_x, and the pin-connected column bends about its weak axis, I_y.

a. Guess which level has the largest story drift.
b. List the assumptions you make in order to use the Stiff Beam Method to predict the total drift.
c. Use the Stiff Beam Method to predict the drift for each story.
d. Compare each story drift prediction to the drift limit for that level. Does the frame appear to be adequately designed?
e. Refine your drift prediction using computer aided analysis. Predict the story drift for each level. Because of axial shortening of the beams, the drift value will be different at different joints; the best comparison will be along column line A1, since all of the beam shortening happens to the right of that column line. Compare each story drift prediction to its limit. Does this change your conclusion about the adequacy of the current design?
f. Submit a printout of the displaced shape. Identify at least three features that suggest the displaced shape is reasonable.
g. Verify equilibrium of the computer aided analysis results.
h. Are the results of your computer aided analysis reasonable? Make a comprehensive argument to justify your answer.
i. Comment on why your guess in part (a) was or was not correct when compared with your results in part (c).

10.14 The architect wants an open layout for a new medical facility. Therefore, the best lateral load-resisting system is a rigid frame. Our team has finished designing the frames for strength, but sensitive test equipment requires that the story drift stay below $h/1000$. Our task is to verify the adequacy of the frames in the long direction. Note that the columns bend about their weak axis, I_y, in this direction. For this preliminary investigation, the Stiff Beam Method will be adequate.

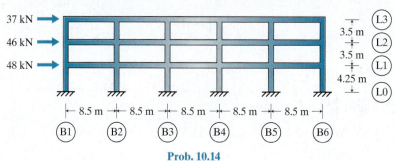

Prob. 10.14

Selected member properties:

Columns:	W310 × 79	$A = 1.00 \times 10^4$ mm²	$I_y = 3.99 \times 10^7$ mm⁴
Beams:	W530 × 66	$A = 8.39 \times 10^3$ mm²	$I_x = 3.51 \times 10^8$ mm⁴
Steel:	$E = 200$ GPa $= 200$ kN/mm²		

a. Guess which level has the largest story drift index (Δ_{story}/h).
b. List the assumptions you make in order to use the Stiff Beam Method to predict the total drift.
c. Use the Stiff Beam Method to predict the drift for each story.
d. Compare each story drift prediction to the drift limit for that level. Does the frame appear to be adequately designed?
e. Refine your drift prediction using computer aided analysis. Predict the story drift for each level. Because of axial shortening of the beams, the drift value will be different at different joints; the best comparison will be along column line B6, since all of the beam shortening happens to the left of that column line. Compare each story drift prediction to its limit. Does this change your conclusion about the adequacy of the current design?
f. Submit a printout of the displaced shape. Identify at least three features that suggest the displaced shape is reasonable.
g. Verify equilibrium of the computer aided analysis results.
h. Are the results of your computer aided analysis reasonable? Make a comprehensive argument to justify your answer.
i. Comment on why your guess in part (a) was or was not correct when compared with your results in part (c).

10.15 Our firm is designing an eight-story, reinforced concrete office building in an urban environment. The primary lateral load-resisting system will be structural walls in the core; however, because we are using cast-in-place concrete, the columns and beams will act as rigid frames. Coworkers have performed a rigid diaphragm analysis to determine the loads on the various frames and have designed the members to meet strength requirements.

Before we finalize the design, our team leader wants to know if any of the levels are susceptible to drift problems. The façade manufacturer requires that the story drift be below $h/500$. We have been asked to predict the story drift of each level of one of the frames. For this preliminary analysis, our team leader recommends using the Stiff Beam Method.

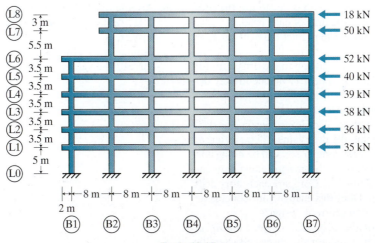

Prob. 10.15

Chapter 10 Approximate Lateral Displacements

Selected member properties:

Concrete:	$E = 38$ GPa $= 3800$ kN/cm^2	
Beams:	$A_{eff} = 3750$ cm^2	$I_{eff} = 7.81 \times 10^5$ cm^4

Column	Designed Width (cm)	A (cm^2)	I (cm^4)
B1	35	1225	1.25×10^5
B2	40	1600	2.13×10^5
B3	40	1600	2.13×10^5
B4	40	1600	2.13×10^5
B5	45	2025	3.42×10^5
B6	55	3025	7.63×10^5
B7	40	1600	2.13×10^5

a. Guess which level has the largest story drift index (Δ_{story}/h).
b. List the assumptions you make in order to use the Stiff Beam Method to predict the drift of each level.
c. Use the Stiff Beam Method to predict the drift for each level.
d. Compare each story drift prediction to the drift limit for that level. Does the frame appear to be adequately designed?
e. Refine your drift prediction using computer aided analysis. Predict the story drift for each level. Because of axial shortening of the beams, the drift value will be different at different joints; the best comparison will be along column line B2, since most of the beam shortening happens to the right of that column line. Compare each story drift prediction to its limit. Does this change your conclusion about the adequacy of the current design?
f. Submit a printout of the displaced shape. Identify at least three features that suggest the displaced shape is reasonable.
g. Verify equilibrium of the computer aided analysis results.
h. Are the results of your computer aided analysis reasonable? Make a comprehensive argument to justify your answer.
i. Comment on why your guess in part (a) was or was not correct when compared with your results in part (c).

10.16 Our firm is designing the bents that support a new highway overpass. Because of the soil conditions, our design team has performed preliminary design of two different structural systems: deep foundations and shallow foundations with a grade beam. The team has created computer models of both systems. Before we select the preferred system, the senior engineer wants to make sure the computer models are reasonable.

For the deep foundation system, the predicted drift is 1.37 mm. Our task is to evaluate that result.

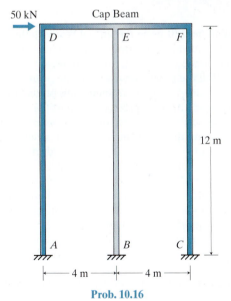

Prob. 10.16

a. Guess which component contributes most to the total drift: cap beams, exterior columns, or interior column.
b. Predict the drift using the Stiff Beam Method. Will the prediction overestimate or underestimate drift, or can't you be sure?
c. Predict the drift using the Virtual Work Method. Calculate the individual contributions of each component: cap beams, exterior columns, and interior column. Will the prediction overestimate or underestimate drift, or can't you be sure? Will this prediction be better or worse than the prediction in part (b)?
d. If we need to reduce the total drift, where would you recommend that the team increase the size of members in order to most effectively reduce the drift?
e. Predict the drift using structural analysis software. Because of axial shortening of the cap beam, the drift value will be different at different joints; the best comparison will be at joint F, since all of the beam shortening happens to the left of that column line.
f. Submit a printout of the displaced shape. Identify at least three features that suggest the displaced shape is reasonable.
g. Verify equilibrium of your computer aided analysis results.
h. Are the results of your computer aided analysis reasonable? Make a comprehensive argument to justify your answer.
i. Evaluate the reasonableness of your teammate's drift prediction.
j. Comment on why your guess in part (a) was or was not correct when compared to the solution in part (c).

Preliminary selection of member properties:

Cap beam:	$A = 5625 \text{ cm}^2$	$I = 2.64 \times 10^6 \text{ cm}^4$
Columns:	$A = 5625 \text{ cm}^2$	$I = 2.64 \times 10^6 \text{ cm}^4$
Concrete:	$E = 26 \text{ GPa} = 2600 \text{ kN/cm}^2$	

10.17 Our firm is designing the bents that support a new highway overpass. Because of the soil conditions, our design team has performed preliminary design of two different structural systems: deep foundations and shallow foundations with a grade beam. The team has created computer models of both systems. Before we select the preferred system, the senior engineer wants to make sure the computer models are reasonable.

For the shallow foundation system with a grade beam, the predicted drift is 0.98 mm. Our task is to evaluate that result.

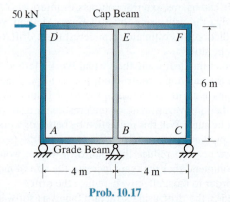

Prob. 10.17

Preliminary selection of member properties:

Cap beam:	$A = 5625 \text{ cm}^2$	$I = 2.64 \times 10^6 \text{ cm}^4$
Grade beam:	$A = 12{,}100 \text{ cm}^2$	$I = 1.22 \times 10^7 \text{ cm}^4$
Columns:	$A = 5625 \text{ cm}^2$	$I = 2.64 \times 10^6 \text{ cm}^4$
Concrete:	$E = 26 \text{ GPa} = 2600 \text{ kN/cm}^2$	

a. Guess which component contributes most to the total drift: cap beams, grade beams, exterior columns, or interior column.
b. Predict the drift using the Stiff Beam Method. Will the prediction overestimate or underestimate drift, or can't you be sure?
c. Predict the drift using the Virtual Work Method. Calculate the individual contributions of each component: cap beams, grade beams, exterior columns, and interior column. Will the prediction overestimate or underestimate drift, or can't you be sure? Should this prediction be better or worse than the prediction in part (b)?
d. If we need to reduce the total drift, where would you recommend that the team increase the size of members in order to most effectively reduce the drift?
e. Predict the drift using structural analysis software. Because of axial shortening of the cap beam, the drift value will be different at different joints; the best comparison will be at joint F, since all of the beam shortening happens to the left of that column line.
f. Submit a printout of the displaced shape. Identify at least three features that suggest the displaced shape is reasonable.
g. Verify equilibrium of your computer aided analysis results.
h. Are the results of your computer aided analysis reasonable? Make a comprehensive argument to justify your answer.
i. Evaluate the reasonableness of your teammate's drift prediction.
j. Comment on why your guess in part (a) was or was not correct when compared to the solution in part (c).

10.18 Our team is designing an overpass structure that will be constructed in two phases. Because of the timing of the phases, the piers are separated into two parts. The team has completed a preliminary design of the first pier to meet the strength requirements.

One of the team members has created a computer model of the frame. The predicted drift from that model is 0.030 in. Before we make any decisions based on the output from that model, the senior engineer wants a different team member to validate the results. We have been tasked with evaluating the computer aided drift prediction.

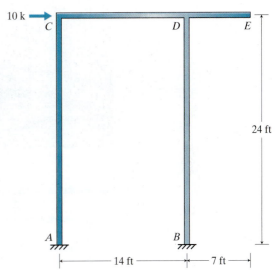

Prob. 10.18

Selected member properties:

Cap beam (3 ft × 3.25 ft):	$A = 9.75$ ft^2	$I = 8.58$ ft^4
Columns (3-ft diam):	$A = 7.07$ ft^2	$I = 3.98$ ft^4
Concrete:	$E = 4600$ ksi	

a. Guess which component contributes most to the total drift: cap beam or columns.
b. Predict the drift using the Stiff Beam Method. Will the prediction overestimate or underestimate drift, or can't you be sure?
c. Predict the drift using the Virtual Work Method. Calculate the individual contributions of each component: cap beam, columns. Will the prediction overestimate or underestimate drift, or can't you be sure? Should this prediction be better or worse than the prediction in part (b)?
d. If we need to reduce the total drift, where would you recommend that the team increase the size of members in order to most effectively reduce the drift?
e. Predict the drift using structural analysis software. Because of axial shortening of the cap beam, the drift value will be different along the beam; the best comparison will be at joint D, since all of the beam shortening happens to the left of that column line.
f. Submit a printout of the displaced shape. Identify at least three features that suggest the displaced shape is reasonable.
g. Verify equilibrium of your computer aided analysis results.
h. Are the results of your computer aided analysis reasonable? Make a comprehensive argument to justify your answer.
i. Evaluate the reasonableness of your teammate's drift prediction.
j. Comment on why your guess in part (a) was or was not correct when compared to the solution in part (c).

10.19 The architect wants an open layout for a new medical facility. Therefore, the best lateral load-resisting system is a rigid frame. As part of the design process, a coworker developed a computer model of the structure. To help validate the model, that coworker asked us to predict the total drift of the structure in the short direction. In order not to influence our analysis, the coworker is waiting to know what we predict before sharing their results.

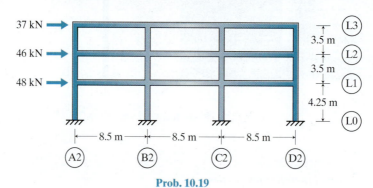

Prob. 10.19

Selected member properties:

Columns:	W310 × 79	$A = 1.00×10^4$ mm^2	$I_x = 1.77×10^8$ mm^4
Beams:	W530 × 66	$A = 8.39×10^3$ mm^2	$I_x = 3.51×10^8$ mm^4
Steel:	$E = 200$ GPa $= 200$ kN/mm^2		

a. Guess which single member contributes most to the total drift: level and bay.
b. List the assumptions you make in order to use the Virtual Work Method to predict the total drift.
c. Use the Virtual Work Method to predict the total drift for the frame. Calculate the contribution of each member.
d. Refine your drift prediction using computer aided analysis. Because of axial shortening of the beams, the drift value will be different at different joints; the best comparison will be along column line D2, since all of the beam shortening happens to the left of that column line.
e. Submit a printout of the displaced shape. Identify at least three features that suggest the displaced shape is reasonable.
f. Verify equilibrium of your computer aided analysis results.
g. Are the results of your computer aided analysis reasonable? Make a comprehensive argument to justify your answer.
h. Comment on why your guess in part (a) was or was not correct when compared with your results in part (c).

10.20 Our firm is designing a small warehouse to be constructed in Europe. Discussions with fabricators and erectors in the region where the warehouse will be built indicate that the most cost-effective option is to use pinned connections between the members and knee braces along the perimeter of the building. The knee braces cause the beams and columns to act as though they were rigidly connected.

Other members of the team have completed a preliminary design of the structural members. Our job is to verify that the story drift is below the allowable limit of $h/500$ for this structure. Another member of our team has predicted the total drift to be 760 mm, which is well in excess of the drift limit. Our task is to validate that result before we start redesigning the frame.

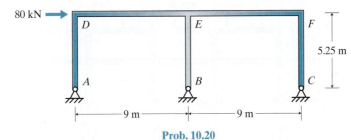

Prob. 10.20

Selected member properties:

Columns:	HSS 150 × 150 × 8	$A = 44.8$ cm²	$I = 1491$ cm⁴
Beams:	IPE 330	$A = 62.6$ cm²	$I_x = 11,770$ cm⁴
Steel:	$E = 200$ GPa $= 20,000$ kN/cm²		

a. Guess the displaced shape.
b. List the assumptions you make in order to use the Stiff Beam Method to predict the total drift.
c. Predict the total drift using the Stiff Beam Method.
d. List the assumptions you make in order to use the Virtual Work Method to predict the total drift.
e. Predict the total drift using the Virtual Work Method.
f. Using the largest result from parts (c) and (e), compare the predicted drift to the drift limit. Evaluate the adequacy of the frame as currently designed.
g. Refine your total drift prediction using computer aided analysis. Because of axial shortening of the beams, the drift value will be different at different joints; the best comparison will be at joint F, since all of the beam shortening happens to the left of that joint.
h. Submit a printout of the displaced shape. Identify at least three features that suggest the displaced shape is reasonable.
i. Verify equilibrium of your computer aided analysis results.
j. Are the results of your computer aided analysis reasonable? Make a comprehensive argument to justify your answer.
k. Evaluate the reasonableness of your teammate's drift prediction.
l. Comment on why your guess in part (a) was or was not correct when compared with your results in part (h).

10.21 Our team is designing a two-story office building. In order to provide an open layout and to maximize the views from the windows, the principal in charge of the project has chosen to use rigid frames. The project team leader has determined that steel will be the most cost-effective structural material for this situation.

Other members of the team have completed a preliminary design of the structural members. Our job is to verify that each story drift is below the allowable limit of $h/500$ for this structure when experiencing the design wind load.

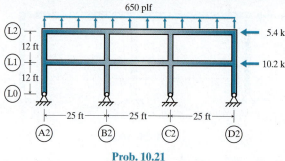

Prob. 10.21

Preliminary selection of member properties:

Columns:	W10 × 33	$A = 9.71$ in^2	$I_x = 171$ in^4
Beams:	W14 × 30	$A = 8.85$ in^2	$I_x = 291$ in^4
Steel:	$E = 29,000$ ksi		

a. Guess which level has the largest story drift index (Δ_{story}/h).
b. List the assumptions you make in order to use the Stiff Beam Method to predict the total drift.
c. Use the Stiff Beam Method to predict the drift for each story.
d. Compare each story drift prediction to the drift limit for that level. Does the frame appear to be adequately designed?
e. If one or more levels are close to or over the drift limit, revise your drift predictions using the Virtual Work Method. Calculate the individual contributions of each level of beams and each level of column segments. Where would you recommend that the team increase the size of members in order to most effectively reduce the drift if necessary?
f. Refine your drift prediction using computer aided analysis. Predict the story drift for each level. Because of axial shortening of the beams, the drift value will be different at different joints; the best comparison will be along column line A2, since all of the beam shortening happens to the right of that column line. Compare each story drift prediction to its limit. Does this change your conclusion about the adequacy of the current design?
g. Submit a printout of the displaced shape. Identify at least three features that suggest the displaced shape is reasonable.
h. Verify equilibrium of the computer aided analysis results.
i. Are the results of your computer aided analysis reasonable? Make a comprehensive argument to justify your answer.
j. Comment on why your guess in part (a) was or was not correct when compared with your results in part (c).

10.22 We are in the final stages of designing the walls of a new warehouse. The precast concrete panels have been designed to carry the design loads. One of the walls is made of 21 panels and must carry 126 k of in plane shear. Each panel is 8 feet wide, 26.75 feet tall, and 6 inches thick. We predict the concrete will have a modulus of elasticity of 4500 ksi and a Poisson's ratio of 0.2.

Prob. 10.22

The end of the wall has a large glass window. The window manufacturer requires that the maximum drift be below $h/1000$. We have a couple of detailing options that will affect the drift of the wall: (1) connect the panels so that they deform as one, or (2) let each panel deform individually.

a. Guess which option will have the smallest drift, or if the drifts will be identical.
b. List the assumptions you make in order to predict the drift for each option.
c. Predict the total drift of the wall if the 21 panels deform as one wall 168 ft wide (option 1). Determine whether this option keeps the total drift below the limit.
d. Predict the total drift of the wall if the 21 panels deform individually as coupled walls (option 2). Determine whether this option keeps the total drift below the limit.
e. Comment on why your guess in part (a) was or was not correct when compared with your results in parts (c) and (d).

10.23 Our team is designing a parking garage. The lateral load-resisting system includes several single-story concrete walls that support a rigid diaphragm. In order to determine how much lateral load is carried by each of the walls, the team needs to calculate the rigidity of each wall. To do that, we predict the story drift due to a unit lateral load.

The wall we are assigned has two large holes for mechanical equipment. For this preliminary design stage, we will ignore the effect of the holes. For now, assume the wall has these properties: $E = 30$ GPa, $\nu = 0.18$, $t_w = 0.35$ m.

a. Guess which effect will contribute the most to the drift: shear or flexure.
b. List the assumptions you make in order to predict the drift.
c. Predict the shear and flexure contributions to the drift. Combine them to predict the total drift.
d. We ignored the presence of the holes. Comment on the effect of that simplification on your prediction of the drift.
e. Comment on why your guess in part (a) was or was not correct when compared with your results in part (c).

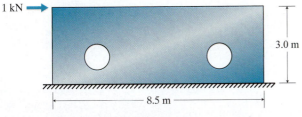

Prob. 10.23

10.24 Our firm is designing a single-story manufacturing facility that will utilize precast panels to make shear walls. The individual panels will be connected so that they behave as solid walls. However, one side will have a huge opening for windows to allow for natural lighting. We have been tasked with determining the shear carried in each part of that wall.

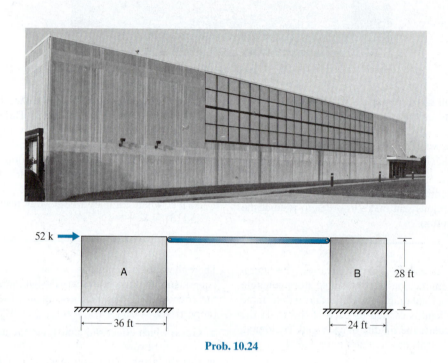

Prob. 10.24

As an approximation, we can ignore the short wall below the windows and consider the two wall segments as coupled walls. At this preliminary stage, let's assume the wall panels are 4 inches thick and made of concrete with these properties: $E = 4000$ ksi, $\nu = 0.18$.

a. Guess which wall segment will carry the largest force.
b. Determine the distribution of the lateral force to each wall segment.
c. Predict the total drift of the wall when subjected to the full lateral load.
d. We ignored the presence of the wall below the windows. Comment on the effect of that simplification on our prediction of forces in the wall segments.
e. Comment on why your guess in part (a) was or was not correct when compared with your results in part (b).

10.25 Our team has been designing a strip mall for a region that has moderate seismic activity. The lateral load-resisting system will be reinforced CMU walls. On one end, there are several openings in the wall. The team has already designed the larger part of the wall to have sufficient strength to carry the static equivalent seismic loading. The senior engineer wants to know how much shear each of the three CMU pillars will take during the design seismic event. This will let the team verify that the pillars have adequate shear strength as well.

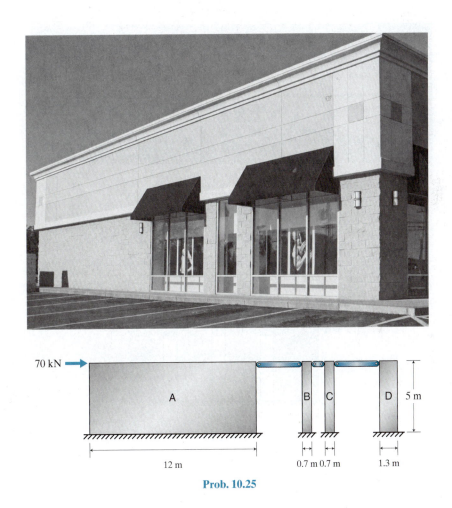

Prob. 10.25

The reinforced CMU wall will have these effective properties: $E = 15$ GPa $= 1500$ kN/cm^2, $\nu = 0.2$, $t_w = 15$ cm.

a. Guess which wall component will carry the smallest force.

b. Determine the distribution of the lateral force to each wall component.

c. Comment on why your guess in part (a) was or was not the same as the solution in part (b).

10.26 We are designing an office building with CMU exterior shear walls. The shear walls have been designed for strength, but our team leader wants to be sure that the walls do not drift too much. A team member has performed an analysis with the Finite Element Method and predicts the total drift will be 0.037 inch. Our job is to verify those results.

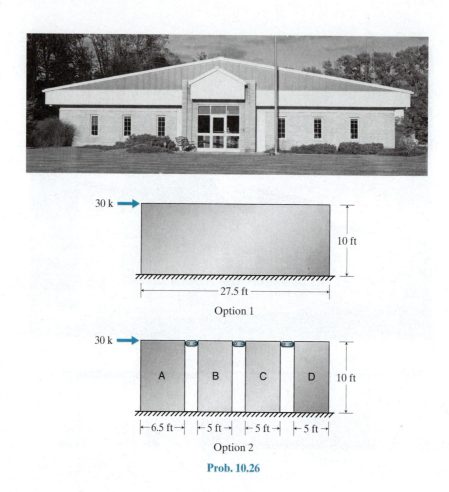

Prob. 10.26

At this end of the building, the wall is divided into two symmetric halves on either side of the doorway. To predict the total drift we can (1) ignore the presence of the windows, or (2) consider the wall half to be a set of coupled walls around the window strips.

The unreinforced CMU wall has these effective properties: $E = 1000$ ksi, $\nu = 0.2$, $t_w = 2.25$ in.

a. Guess how much the presence of window openings affects the drift of the wall. Express the guess as a percentage of the drift of the wall, ignoring the openings.
b. Predict the drift of the wall by ignoring the window openings.
c. Predict the drift of the wall by approximating the wall as four coupled wall segments around the windows.
d. Evaluate you teammate's results from the Finite Element Method.
e. Comment on the quality of your guess in part (a) based on your solutions to parts (b) and (c).

10.27 An old two-story warehouse that was converted to restaurants is for sale. An interested buyer wants to have a full structural evaluation of the building before purchasing it. One team member is verifying the strength of the CMU walls. Our task is to determine whether the story drift will approach $h/500$ for either story.

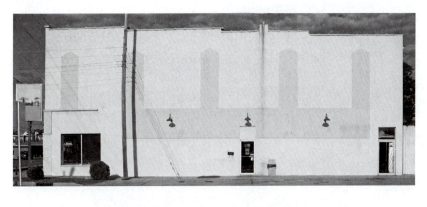

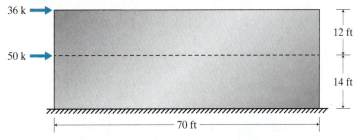

Prob. 10.27

The senior engineer recommends we use the following properties as conservative estimates: $t_w = 2.0$ in., $E = 1000$ ksi, $\nu = 0.25$. For this investigation, ignore the openings.

a. Guess which level will have the largest story drift ratio, Δ_{story}/h.
b. Predict the drift of each story. Convert those to story drift ratios.
c. We ignored the presence of the openings in the first level. Comment on the effect of that simplification on our prediction of the story drifts.
d. Evaluate the adequacy of the existing walls.
e. Comment on why your guess in part (a) was or was not the same as the solution in part (b).

10.28 Our firm has been selected to design a new two-story office building. We anticipate that both levels will use rigid diaphragms. Therefore, the team needs to calculate the story drift for each lateral load-resisting element at each level in order to be able to calculate rigidities. We have been tasked with predicting the story drift ratio for one of the CMU shear walls at the core.

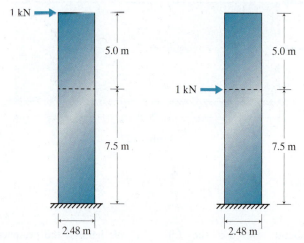

Prob. 10.28

In order to predict the story drift ratio at each level, we need to consider the wall to be loaded at one level at a time. To find the drift ratio for the top level, we consider the wall to have a unit load at the roof level. To find the drift ratio for the bottom level, we consider the wall to have a unit load at the middle-floor level.

An experienced member of the team recommends we use these effective properties for the CMU shear walls: $E = 12.5$ GPa $= 1250$ kN/cm², $\nu = 0.2$, $t_w = 12$ cm.

a. Guess which level will have the largest story drift ratio, Δ_{story}/h.
b. Predict the story drift of the top story when there is a unit load at only the roof level. Find the resulting story drift ratio.
c. Predict the story drift of the bottom story when there is a unit load at only the middle-floor level. Find the resulting story drift ratio.
d. Comment on why your guess in part (a) was or was not the same as the solution in part (b).

10.29 Our firm is designing a two-story apartment building. The senior engineer has determined that stick (wood) construction is the optimum choice. The client has had problems with serviceability issues in other buildings, so the principal-in-charge from our firm wants us to design the structure such that the story drift will be below $h/500$ for each level.

Prob. 10.29a

After a bit of online research, the senior engineer found research results that propose using an effective modulus of elasticity between 45 ksi and 55 ksi, and using the stud depth as the wall thickness. The effective Poisson's ratio will be between 0.2 and 0.3.

Our task is to predict the story drift for each level due to wind load. To make those predictions, the senior engineer recommends we ignore the window on the top level but consider the wall to stop at the edge of the doorways. The walls have been designed for strength, and 2 × 4 studs were used: $t_w = 3.5$ in.

a. Guess which level will have the largest story drift ratio, Δ_{story}/h.
b. Predict the drift of each story. Use the effective properties that will give the largest drift predictions. Convert the story drifts to story drift ratios.
c. Evaluate the adequacy of the system to meet the drift limit requirements.
d. Comment on why your guess in part (a) was or was not the same as the solution in part (b).

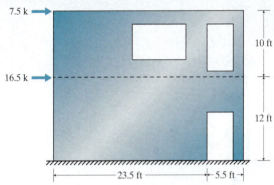

Prob. 10.29b

10.30 The owner of a historic building wants to add a large sign to the top of the building to match the look of the building when it was first constructed. Our firm has been hired to evaluate whether the building can support the large sign in its current condition.

The senior engineer on the project set the total drift of the building to $h/1000$ to protect the plaster interior of the structure. Field testing of the brick in the structural walls resulted in these estimated properties: $E = 7.5$ GPa $= 750$ kN/cm^2, $\nu = 0.25$, $t_w = 10$ cm (double wythe). For this evaluation, the senior engineer recommends we ignore the window and door openings. Although the wind load on the sign acts above the roof level, when predicting drift, we can consider it to act at the roof level.

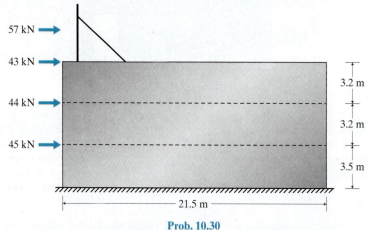

Prob. 10.30

a. Guess which story's loading will contribute the most to the total drift.
b. Predict the drift at the top of the structure due to each story's load. Combine those results to predict the total drift of the structure.
c. We ignored the presence of the openings. Comment on the effect of that simplification on our prediction of the total drift.
d. Evaluate the adequacy of the structure to carry the peak wind load before the sign is added. Evaluate the adequacy of the structure to carry the peak wind load when it includes the sign.
e. Comment on why your guess in part (a) was or was not the same as the solution in part (b).

10.31 Our firm is designing a three-story apartment complex in a region with moderate seismic activity. The structural system will be stick (wood) construction. The complex is three buildings that connect architecturally to form a U-shape. The buildings have been designed for strength. Now our design team needs to verify that the buildings do not contact each other during a seismic event. To make that determination, our team needs to predict the maximum inelastic response drift of each building. We have been tasked with finding that drift for the longer buildings.

To find the maximum inelastic response drift, we multiply the static equivalent drift by an amplification factor; for this type of construction, the amplification factor is 2.0.

A team member points out that predicting the drift of stick construction according to the building code involves the following components: bending, shear, nail slip, and hold down strap stretch. An experienced engineer on the project recommends we start with a simplified prediction that considers only flexure and shear. If the predicted drift is less than 10 mm, the additional drift from the other components will not likely push the total drift over the 25-mm limit we have set.

In order to use the code method for predicting drift, our drift formula becomes

$$\Delta = \frac{2.4Vh(1 + \nu_{sheathing})}{E_{sheathing} t_{sheathing} l} + \frac{4Vh^3}{E_{stud} A_{stud} l^2}$$

That means the shear contribution to drift is based on the sheathing properties, and the flexural contribution is based on the stud properties. We can use that knowledge to adjust our formulas for the drift of a multistory wall.

For this project, we anticipate the following wall properties:

$E_{sheathing} = 2.5$ GPa $= 2.5$ kN/mm^2
$\nu_{sheathing} = 0.3$
$t_{sheathing} = 11$ mm
$E_{stud} = 9.0$ GPa $= 9.0$ kN/mm^2
$A_{stud} = 2(38$ mm $\times 140$ mm$) = 10{,}640$ mm^2 (double studs at ends)

To make the drift prediction, the experienced engineer recommends that we consider only the strips of wall between the windows. That leaves us with 15 coupled wall strips of different sizes: 8 at 2 m wide and 7 at 3 m wide. Although not exactly correct, we can approximate the lateral load distribution as equal to all of the strips and predict the drift of the narrowest.

Prob. 10.31a

a. Guess which story's loading will contribute the most to the total drift.
b. Predict the drift at the top of the structure due to each story's static equivalent seismic load. Combine those results to predict the total static equivalent seismic drift of the structure. Convert that drift into the maximum inelastic response drift.
c. We ignored the presence of wall above and below the window openings. Comment on the effect of that simplification on our prediction of the total drift.
d. We assumed the coupled wall strips would each carry the same amount of shear and used the narrowest strip to predict drift. Comment on the effect of that simplification on our prediction of the total drift.
e. Evaluate whether the long building exceeds the currently set limit for inelastic response drift.
f. Comment on why your guess in part (a) was or was not the same as the solution in part (b).

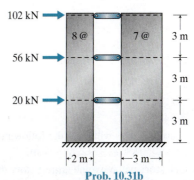

Prob. 10.31b

614 Chapter 10 Approximate Lateral Displacements

10.32 We are designing the structural walls for a four-story hotel. A member of the team has already performed a preliminary design of the walls based on gravity loads. The walls will be concrete masonry units. The center hallway means that the system will behave as two walls linked together; the result is that each wall carries one-half of the lateral load applied at that level. Our task is to verify that the story drift is below the limit of $h/500$.

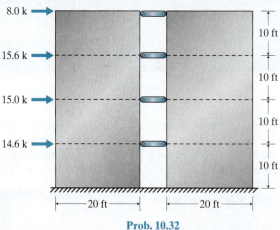

Prob. 10.32

The chosen masonry units will have the following effective properties: $E = 1350$ ksi, $\nu = 0.2$, $t_w = 2$ in.

a. Guess which story will have the largest story drift.
b. Predict the total drift at each level.

c. Calculate the story drift for each level. Evaluate whether or not each story drift is within the limit.
d. Comment on why your guess in part (a) was or was not close to the solution in part (c).

CHAPTER 11

Diaphragms

Not only do diaphragms carry gravity loads, but they also play an important role in the lateral load-resisting system.

MOTIVATION

We use the term *diaphragm* for floors and roofs because they transfer lateral loads to the lateral load-resisting elements. Other systems can also be diaphragms if they are designed to transfer lateral loads (e.g., ramps, horizontal braced frames, or trusses).

If the maximum in plane displacement of a diaphragm is more than two times the average story drift of the lateral load-resisting elements that support it, we call it a *flexible diaphragm*. We learned in Section 1.6: Distribution of Lateral Loads by Flexible Diaphragm that we can use the Equivalent Beam Model to determine the forces generated in each lateral element.

If the diaphragm is very stiff, however, it remains relatively undeformed as the lateral elements drift. Then we call it a *rigid diaphragm*. Most cast-in-place and many precast concrete diaphragms qualify as rigid. In some cases when the diaphragm is not very stiff, we add diagonal bracing in plane to make it behave rigidly. Therefore, many building floors and bridge decks are best idealized as rigid diaphragms. Because these diaphragms don't change shape, we must use compatibility to determine the forces generated in each of the lateral load-resisting elements that support the diaphragm.

When analyzing floor and roof diaphragms, we often think of the internal forces caused by gravity loads, such as self-weight and live load. These internal forces are due to loads that act normal to the surface. They cause the diaphragm to deflect out of plane; therefore, these internal forces are often called out of plane forces. But lateral loads typically transfer from the walls to the floor or roof diaphragms and act in the same plane as the diaphragms (e.g., Section 1.5: Application of Lateral Loads). Therefore, lateral loads cause in plane forces for flexible, rigid, and semirigid diaphragms.

The in plane shear is transferred from the diaphragm to a lateral load-resisting element along the length of the element. If the shear strength of the diaphragm is not sufficient to make the transfer over that length, we can enlist the help of beams along the axis of the lateral load-resisting element. We call those beams *collector beams* or *drag struts*.

The in plane moment causes tension along the outer edge of the diaphragm. That tension can cause deep cracks or tears. To strengthen the outer edge, we use diaphragm *chords*.

11.1 Distribution of Lateral Loads by Rigid Diaphragm

Introduction

Floor and roof diaphragms are typically characterized as either flexible or rigid. If the diaphragm is much more flexible in plane than the lateral load-resisting elements that support it, we call it a flexible diaphragm. But if the diaphragm is very stiff in plane compared with the lateral load-resisting elements, we call it a rigid diaphragm.

When lateral load acts on a rigid diaphragm, the diaphragm does not change shape. Therefore, it translates and rotates in response to lateral loading. Analysis of this type of diaphragm requires us to consider compatibility of displacements.

How-To

Most diaphragms are neither perfectly flexible nor perfectly rigid. Accurately modeling their behavior requires finite element modeling with careful consideration to connections and openings. Fortunately, the actual behavior is bounded by the perfectly flexible and perfectly rigid behaviors. That means in most cases we don't need to go to the trouble of creating and calibrating a finite element model in order to find a reasonable upper bound on the force developed in each lateral load-resisting element. We just need to calculate those forces considering the diaphragm to be flexible and to be rigid. The largest force generated in each element from the two analyses will be the upper bound for that element.

We covered how to analyze a flexible diaphragm in Section 1.6: Distribution of Lateral Loads by Flexible Diaphragm. A flexible diaphragm acts like a beam on supports, but a rigid diaphragm behaves much differently. A rigid diaphragm does not change shape, so it translates and rotates, which causes force to develop in each lateral load-resisting element.

We use a few key concepts to determine how much force develops in each lateral element. We start by considering the translation and rotation effects separately. We superimpose the results at the end. To find the forces due to translation or rotation, we use compatibility of displacements. Compatibility lets us relate all of the forces to one unknown force. Equilibrium allows us to find that one unknown force.

In order to begin the analysis process, we need to know the *rigidity* of each lateral load-resisting element. Rigidity is a measure of how stiff a lateral element is. It is the force required to cause a unit displacement of drift. We can calculate rigidity with the following formula:

$$R = \frac{V}{\Delta}$$

where
- R = Rigidity of a lateral load-resisting element (e.g., rigid frame, braced frame, structural wall)
- V = Force applied to the lateral element
- Δ = Story drift in the lateral element due to the force, V

Section 11.1 Distribution of Lateral Loads by Rigid Diaphragm

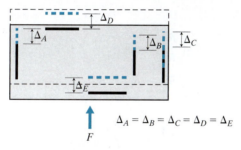

Figure 11.1 Each lateral load-resisting element displaces the same amount when we consider only translation of a rigid diaphragm.

We don't need to know the actual force generated in the element in order to calculate rigidity, since rigidity is the force per unit drift.

Translation
If we consider only the translation of the rigid diaphragm, every lateral load-resisting element displaces the same amount (Figure 11.1).

The lateral load-resisting elements have very little stiffness out of plane, so we neglect any force that might be generated by out of plane deformation. We calculate only the force developed due to in plane deformation. In Figure 11.1, that means only elements A, B, and C will generate force to resist the lateral loading. Elements D and E will provide negligible resistance to translation in the y-direction.

We can use rigidity to relate the story drift of each element to the force generated in each element:

$$\Delta_A = \frac{V_A}{R_A} \qquad \Delta_B = \frac{V_B}{R_B} \qquad \Delta_C = \frac{V_C}{R_C}$$

For a rigid diaphragm that experiences only translation, the drifts are all the same:

$$\Delta_A = \Delta_B = \Delta_C$$
$$\Rightarrow \frac{V_A}{R_A} = \frac{V_B}{R_B} = \frac{V_C}{R_C}$$

We can use this relationship to express each force in terms of one unknown—for example, V_A:

$$V_B = V_A\left(\frac{R_B}{R_A}\right) \qquad V_C = V_A\left(\frac{R_C}{R_A}\right) \qquad V_A = V_A\left(\frac{R_A}{R_A}\right)$$

The diaphragm must be in equilibrium:

$$+\uparrow \Sigma F_y = 0 = F - V_A - V_B - V_C$$

so

$$V_A\left(\frac{R_A}{R_A}\right) + V_A\left(\frac{R_B}{R_A}\right) + V_A\left(\frac{R_C}{R_A}\right) = F$$

$$V_A\left(\frac{R_A + R_B + R_C}{R_A}\right) = F$$

$$V_A = \left(\frac{R_A}{R_A + R_B + R_C}\right)F$$

In general,

$$V_i^F = \frac{R_i}{\Sigma R_j}F$$

where

$j =$ All lateral load-resisting elements whose plane is in the same direction as the applied lateral load

Since all of the lateral elements displace in the same direction, they all develop force in the same direction: the forces oppose the applied lateral load (Figure 11.2).

Center of Rigidity

Before we can relate all of the forces due to rotation, we need to know the location of the point about which the diaphragm rotates. That point is called the *center of rigidity*. As the diaphragm rotates, the new positions of the diaphragm and lateral elements are shown as dashed lines in Figure 11.3. The actual drift of an element is measured as the change in position, but *if* the rotation angle, θ, is relatively small, we can reasonably approximate the drift as the component along the axis of the element (Figures 11.4 and 11.5).

Because the diaphragm does not change shape, the magnitudes of the element drifts are linearly related. The elements oriented along the x-axis in Figure 11.5, D and E, have linearly related drifts, and they lead us to the y-coordinate location of the center of rigidity. The elements oriented along the y-axis, A, B, and C, also have linearly related drifts and

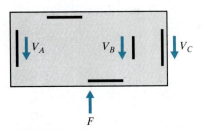

Figure 11.2 Forces generated in lateral load-resisting elements always oppose the applied load when we consider only translation.

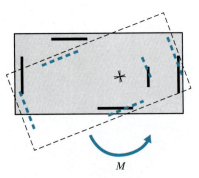

Figure 11.3 Deformed shape of a rigid diaphragm subjected to rotation. Note that the lateral load-resisting elements experience both translation and rotation even though the diaphragm only rotates.

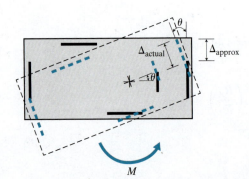

Figure 11.4 A relatively small angle of rotation allows us to approximate the deformation of the lateral elements as translation along the axis of the elements.

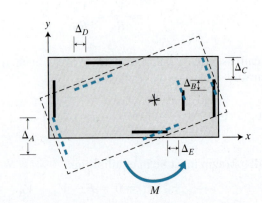

Figure 11.5 Approximate drifts for each lateral load-resisting element due to only rotation of a rigid diaphragm.

Section 11.1 Distribution of Lateral Loads by Rigid Diaphragm

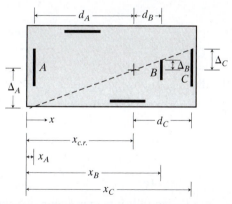

Figure 11.6 Approximate drifts for each lateral load-resisting element due to only rotation of a rigid diaphragm.

determine the x-coordinate location of the center of rigidity. To illustrate, let's focus on the elements in the y-direction (Figure 11.6).

Using similar triangles, we can relate the three drifts. Note that on one side of the center of rigidity, the drift, Δ, and distance to the element, d, are negative; on the other side, they are positive. Maintaining this sign convention will be important.

$$\frac{\Delta_A}{d_A} = \frac{\Delta_B}{d_B} = \frac{\Delta_C}{d_C}$$

$$\Rightarrow \Delta_B = \frac{d_B}{d_A}\Delta_A$$

$$\Delta_C = \frac{d_C}{d_A}\Delta_A$$

and

$$\Delta_A = \frac{d_A}{d_A}\Delta_A$$

From the definition of rigidity, we can express the force developed in each element in terms of the drift:

$$V_i = R_i \Delta_i$$

Even though we are considering only rotation of the diaphragm, the forces generated in the lateral elements must still satisfy equilibrium in the x- and y-directions:

$$+\uparrow \Sigma F_y = 0 = V_A + V_B + V_C$$
$$\Rightarrow R_A \Delta_A + R_B \Delta_B + R_C \Delta_C = 0$$

We can express each force in terms of one unknown drift, Δ_A:

$$R_A \frac{d_A}{d_A}\Delta_A + R_B \frac{d_B}{d_A}\Delta_A + R_C \frac{d_C}{d_A}\Delta_A = 0$$

$$(R_A d_A + R_B d_B + R_C d_C)\left(\frac{\Delta_A}{d_A}\right) = 0$$

Since $\left(\dfrac{\Delta_A}{d_A}\right) \ne 0,$

$$R_A d_A + R_B d_B + R_C d_C = 0$$

We can express the distances between the lateral elements and the center of rigidity, d_i, as the differences between distances from a reference location, x_i:

$$d_A = x_A - x_{c.r.}$$
$$d_B = x_B - x_{c.r.}$$
$$d_C = x_C - x_{c.r.}$$

Substituting these into the equilibrium expression, we get

$$R_A(x_A - x_{c.r.}) + R_B(x_B - x_{c.r.}) + R_C(x_C - x_{c.r.}) = 0$$
$$R_A x_A + R_B x_B + R_C x_C = (R_A + R_B + R_C)x_{c.r.}$$

For all elements with their axis in the *y*-direction,

$$\Rightarrow x_{c.r.} = \dfrac{\Sigma R_i x_i}{\Sigma R_i}$$

Similarly, for all elements with their axis in the *x*-direction,

$$y_{c.r.} = \dfrac{\Sigma R_i y_i}{\Sigma R_i}$$

Rotation

If we consider only rotation of the rigid diaphragm, then every lateral load-resisting element undergoes the same amount of rotation. We have shown that the displacement along the axis of the element, Δ_i, is a reasonable approximation of the actual displacement. We can draw the perpendicular distance from the center of rigidity to the line of action of each lateral element, d_i (Figure 11.7).

The perpendicular distance and element drift make two legs of a right triangle. One of the interior angles is θ, so we have

$$\tan(\theta) = \dfrac{\Delta_i}{d_i}$$

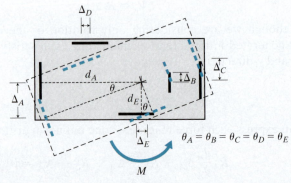

Figure 11.7 Each lateral load-resisting element rotates the same amount when we consider only rotation of a rigid diaphragm.

If the rotation is relatively small, we have

$$\tan(\theta) \approx \theta$$

so

$$\theta = \frac{\Delta_i}{d_i}$$

Using our definition of rigidity, we can express the rotation in terms of the force generated in each element:

$$\theta = \frac{V_i}{R_i d_i}$$

Since all the rotations are the same,

$$\frac{V_A}{R_A d_A} = \frac{V_B}{R_B d_B} = \frac{V_C}{R_C d_C} = \frac{V_D}{R_D d_D} = \frac{V_E}{R_E d_E}$$

We can use this equation to express each force in terms of one unknown force—for example, V_A:

$$V_B = V_A \left(\frac{R_B d_B}{R_A d_A} \right)$$

$$V_C = V_A \left(\frac{R_C d_C}{R_A d_A} \right)$$

$$V_D = V_A \left(\frac{R_D d_D}{R_A d_A} \right)$$

$$V_E = V_A \left(\frac{R_E d_E}{R_A d_A} \right)$$

and

$$V_A = V_A \left(\frac{R_A d_A}{R_A d_A} \right)$$

Based on the deformations in Figure 11.7, we see that each force generated opposes the applied moment (Figure 11.8).

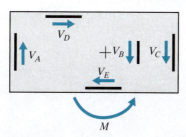

Figure 11.8 Each lateral load-resisting element develops shear in a direction to resist the applied moment.

Moment equilibrium must be satisfied for the diaphragm:

$$\circlearrowleft +\Sigma M_{c.r.} = M - V_A d_A - V_B d_B - V_C d_C - V_D d_D - V_E d_E$$

We can substitute in the expressions for the forces to obtain

$$M = V_A\left(\frac{R_A d_A^2}{R_A d_A}\right) + V_A\left(\frac{R_B d_B^2}{R_A d_A}\right) + V_A\left(\frac{R_C d_C^2}{R_A d_A}\right) + V_A\left(\frac{R_D d_D^2}{R_A d_A}\right) + V_A\left(\frac{R_E d_E^2}{R_A d_A}\right)$$

$$M = \frac{V_A}{R_A d_A}(\Sigma R_j d_j^2)$$

This equation has only one unknown, the shear developed in element A:

$$V_A = \left(\frac{R_A d_A}{\Sigma R_j d_j^2}\right) M$$

Because we can relate all of the shears back to this one, we can find any force generated in a lateral element with the following expression:

$$V_i^M = \left(\frac{R_i d_i}{\Sigma R_j d_j^2}\right) M$$

where

M = Applied moment, which includes the applied resultant force acting at a distance from the center of rigidity

j = All lateral load-resisting elements that directly support the diaphragm

Superposition

The total force generated in each lateral load-resisting element is the superposition of the translation and rotation effects:

$$V_i^T = V_i^F + V_i^M$$

Remember that the force generated in a lateral element might be in different directions due to the two effects. So, for some lateral elements the effects add, and for some lateral elements the effects oppose each other. Therefore, signs matter.

Section 11.1 Highlights

Distribution of Lateral Loads by Rigid Diaphragm

A rigid diaphragm is a floor or roof that is so stiff in plane that it does not deform when subjected to lateral loading.

Rigidity is the stiffness of a lateral load-resisting element:

$$R_i = \frac{V_i}{\Delta_i}$$

where

R_i = Rigidity
V_i = Force applied to the lateral load-resisting element
Δ_i = Resulting story drift of the lateral element

The center of rigidity is the point about which a rigid diaphragm rotates:

$$x_{c.r.} = \frac{\Sigma R_i x_i}{\Sigma R_i}$$

where
- R_i = Rigidity of the lateral elements whose plane is in the y-direction
- x_i = Perpendicular distance from a reference point to the plane of the lateral element; signs matter
- $x_{c.r.}$ = Distance from the reference point to the center of rigidity

$$y_{c.r.} = \frac{\Sigma R_i y_i}{\Sigma R_i}$$

where
- R_i = Rigidity of the lateral elements whose plane is in the x-direction
- y_i = Perpendicular distance from a reference point to the plane of the lateral element; signs matter
- $y_{c.r.}$ = Distance from the reference point to the center of rigidity

Translation:

$$V_i^F = \frac{R_i}{\Sigma R_j} F$$

where
- V_i^F = Force generated in element i due to translation; the direction opposes F
- F = Lateral force applied directly to the diaphragm
- j = All lateral elements whose plane is in the same direction as F

Rotation:

$$V_i^M = \left(\frac{R_i d_i}{\Sigma R_j d_j^2}\right) M$$

where
- V_i^M = Force generated in element i due to rotation; the direction opposes M
- M = Moment applied; can include direct application and effect of F being eccentric from the center of rigidity
- d = Perpendicular distance between the plane of the lateral element and the center of rigidity
- j = All lateral elements

Net force generated:

$$V_i^T = V_i^F + V_i^M$$

Signs matter; each effect might be in a different direction.

Assumptions:
- Linear elastic material behavior.
- Relatively small rotation.
- Perfectly rigid diaphragm.

EXAMPLE 11.1

Our firm has been selected to design a new manufacturing building in an industrial complex. Because of the sensitivity of the equipment that will be mounted on the roof, we anticipate using a rigid diaphragm. The arrangement of the manufacturing process has dictated the layout of the gravity system and the placement of the diagonal braces for lateral resistance.

Figure 11.9

Based on the geometry of the diagonal braces, we have been able to estimate the rigidities of the different braced frames. The next step in the process is to design the members of the lateral load-resisting system. To do that, we need to know the shear generated in each of the braced frames.

We are using the Directional Procedure to determine the effects of wind on the structure. The Directional Procedure has four different load cases: (1) maximum uniform wind in one direction at a time, (2) reduced uniform wind in one direction with torsion, (3) reduced uniform wind in two directions simultaneously, and (4) even further reduced uniform wind in two directions simultaneously with torsion. Our task is to determine the shear generated in each of the braced frames for case 2, considering wind from the north.

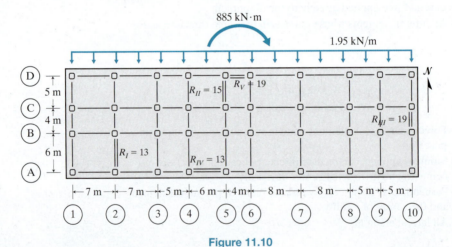

Figure 11.10

Center of Rigidity:

Choose corner A1 as the reference location.

    x-coordinate:

        Braced frames *I*, *II*, and *III* contribute because they are oriented in the *y*-direction.

$$x_{c.r.} = \frac{\Sigma R_i x_i}{\Sigma R_i} = \frac{13(7\text{ m}) + 15(25\text{ m}) + 19(55\text{ m})}{13 + 15 + 19}$$

$$= 32.1 \text{ m}$$

y-coordinate:
   Braced frames *IV* and *V* contribute because they are oriented in the *x*-direction.

$$y_{c.r.} = \frac{\Sigma R_i y_i}{\Sigma R_i} = \frac{13(0\text{ m}) + 19(15\text{ m})}{13 + 19}$$

$$= 8.9 \text{ m}$$

Translation:

Only braced frames *I*, *II*, and *III* resist translation caused by force in the *y*-direction.

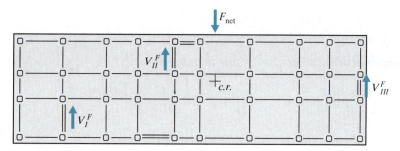

Figure 11.11

Net force:

$$F = 1.95 \text{ kN/m}(55 \text{ m}) = 107.2 \text{ kN } (+\downarrow)$$

Force generated in each lateral element due to translation:

$$V_i^F = \frac{R_i}{\Sigma R_j} F$$

$$V_I^F = \frac{13}{\Sigma(13 + 15 + 19)}(107.2 \text{ kN}) = 29.7 \text{ kN } (+\uparrow)$$

$$V_{II}^F = \frac{15}{\Sigma(13 + 15 + 19)}(107.2 \text{ kN}) = 34.2 \text{ kN } (+\uparrow)$$

$$V_{III}^F = \frac{19}{\Sigma(13 + 15 + 19)}(107.2 \text{ kN}) = 43.3 \text{ kN } (+\uparrow)$$

Rotation:

All braced frames resist the rotation.

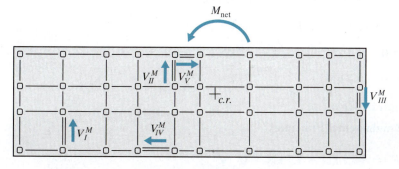

Figure 11.12

628 Chapter 11 Diaphragms

Element	R	d	Rd	Rd²
I	13	25.1 m	326 m	8190 m²
II	15	7.1 m	107 m	756 m²
III	19	22.9 m	435 m	9964 m²
IV	13	8.9 m	116 m	1030 m²
V	19	6.1 m	116 m	707 m²

$$\Sigma = 20{,}647 \text{ m}^2$$

Net moment:

$$M = 885 \text{ kN}\cdot\text{m} + 107.2 \text{ kN}\cdot(55 \text{ m}/2 - 32.1 \text{ m}) = 1378 \text{ kN}\cdot\text{m} \;(\circlearrowright)$$

Force generated in each lateral element due to rotation:

$$V_i^M = \left(\frac{R_i d_i}{\Sigma R_j d_j^2}\right) M$$

$$V_I^M = \frac{326 \text{ m}}{20{,}650 \text{ m}^2}(1378 \text{ kN}\cdot\text{m}) = 21.8 \text{ kN} \;(+\uparrow)$$

$$V_{II}^M = \frac{107 \text{ m}}{20{,}650 \text{ m}^2}(1378 \text{ kN}\cdot\text{m}) = 7.1 \text{ kN} \;(+\uparrow)$$

$$V_{III}^M = \frac{435 \text{ m}}{20{,}650 \text{ m}^2}(1378 \text{ kN}\cdot\text{m}) = 29.0 \text{ kN} \;(+\downarrow)$$

$$V_{IV}^M = \frac{116 \text{ m}}{20{,}650 \text{ m}^2}(1378 \text{ kN}\cdot\text{m}) = 7.7 \text{ kN} \;(\overset{+}{\leftarrow})$$

$$V_V^M = \frac{116 \text{ m}}{20{,}650 \text{ m}^2}(1378 \text{ kN}\cdot\text{m}) = 7.7 \text{ kN} \;(\overset{+}{\rightarrow})$$

Net Forces:

Let's choose +x and +y as positive directions for our sign convention.

$$V_i^T = V_i^F + V_i^M$$

$$V_I^T = 29.7 \text{ kN} + 21.8 \text{ kN} = 51.5 \text{ kN} \;(+\uparrow)$$

$$V_{II}^T = 34.2 \text{ kN} + 7.1 \text{ kN} = 41.3 \text{ kN} \;(+\uparrow)$$

$$V_{III}^T = 43.3 \text{ kN} - 29.0 \text{ kN} = 14.3 \text{ kN} \;(+\uparrow)$$

$$V_{IV}^T = 0 - 7.7 \text{ kN} = -7.7 \text{ kN} \;(\overset{+}{\rightarrow})$$

$$V_V^T = 0 + 7.7 \text{ kN} = 7.7 \text{ kN} \;(\overset{+}{\rightarrow})$$

Evaluation of Results:
Satisfied Fundamental Principles?

$$\overset{+}{\rightarrow}\Sigma F_x = -7.7 \text{ kN} + 7.7 \text{ kN} = 0 \;\checkmark$$

$$+\uparrow\Sigma F_y = -107.2 \text{ kN} + 51.5 \text{ kN} + 41.3 \text{ kN} + 14.3 \text{ kN} = -0.1 \text{ kN} \approx 0 \;\checkmark$$

Example 11.2

We are designing a new building that will be used as an office and shop. To meet the owner's tight construction schedule, the building will have precast wall panels and precast roof panels.

Figure 11.13

The entire perimeter is made of precast double-t panels that serve as shear walls. We have designed the connections such that the panels act coupled. Therefore, the rigidity of each element is proportional to the number of panels. A support beam runs from B2 to B3.

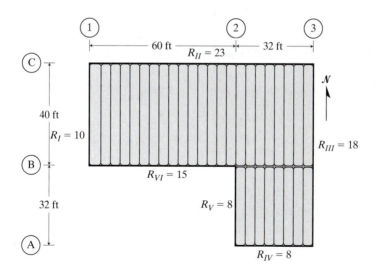

Figure 11.14

The building is in a zone with moderate seismic activity. A team member has calculated that the equivalent static seismic load will be 0.15 of the dead weight of the roof (seismic response coefficient $C_s = 0.15$). The resulting distributed lateral loads have been calculated.

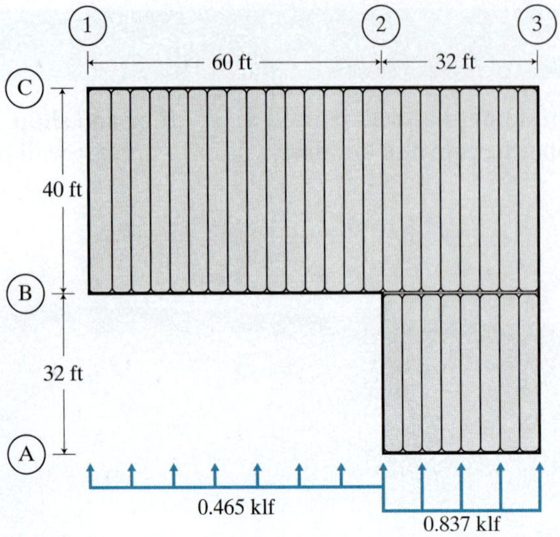

630   **Chapter 11**   Diaphragms

Figure 11.15

Our job is to determine the shear generated in each wall due to the design seismic event in the north-south direction.

Center of Rigidity:

Choose corner B1 as the reference location.

    x-coordinate:
        Walls *I*, *III*, and *V* contribute because they are oriented in the *y*-direction.

$$x_{c.r.} = \frac{\Sigma R_i x_i}{\Sigma R_i} = \frac{10(0\text{ ft}) + 18(92\text{ ft}) + 8(60\text{ ft})}{10 + 18 + 8}$$

$$= 59.3\text{ ft}$$

    y-coordinate:
        Walls *II*, *IV*, and *VI* contribute because they are oriented in the *x*-direction. Note that wall *IV* has a negative *y*-coordinate.

$$y_{c.r.} = \frac{\Sigma R_i y_i}{\Sigma R_i} = \frac{23(40\text{ ft}) + 8(-32\text{ ft}) + 15(0\text{ ft})}{23 + 8 + 15}$$

$$= 14.4\text{ ft}$$

Section 11.1 Distribution of Lateral Loads by Rigid Diaphragm

Translation:

Only walls *I*, *III*, and *V* resist translation caused by force in the *y*-direction.

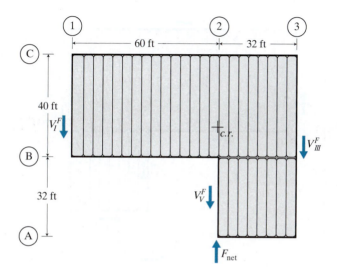

Figure 11.16

Net force:

$$F = 0.465 \text{k/ft}\,(60 \text{ ft}) + 0.837 \text{k/ft}\,(32 \text{ ft})$$
$$= 27.9 \text{ k} + 26.8 \text{ k} = 54.7 \text{ k} \ (+\uparrow)$$

Force generated in each lateral element due to translation:

$$V_i^F = \frac{R_i}{\Sigma R_j} F$$

$$V_I^F = \frac{10}{\Sigma(10 + 18 + 8)}(54.7 \text{ k}) = 15.2 \text{ k} \ (+\downarrow)$$

$$V_{III}^F = \frac{18}{\Sigma(10 + 18 + 8)}(54.7 \text{ k}) = 27.4 \text{ k} \ (+\downarrow)$$

$$V_V^F = \frac{8}{\Sigma(10 + 18 + 8)}(54.7 \text{ k}) = 12.2 \text{ k} \ (+\downarrow)$$

Rotation:

All walls resist the rotation.

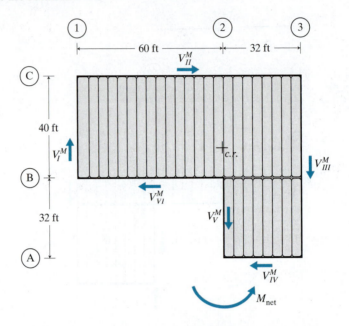

Figure 11.17

Element	R	d	Rd	Rd²
I	10	59.3 ft	593 ft	35,160 ft²
II	23	25.6 ft	589 ft	15,070 ft²
III	18	32.7 ft	589 ft	19,250 ft²
IV	8	46.4 ft	371 ft	17,220 ft²
V	8	0.7 ft	6 ft	4 ft²
VI	15	14.4 ft	216 ft	3,110 ft²
			$\Sigma =$	89,810 ft²

Net moment:

$$M = 27.9 \text{ k} (30 \text{ ft} - 59.3 \text{ ft}) + 26.8 \text{ k} (76 \text{ ft} - 59.3 \text{ ft})$$
$$= -817 \text{ k} \cdot \text{ft} + 448 \text{ k} \cdot \text{ft} = -369 \text{ k} \cdot \text{ft} \ (\curvearrowright{+})$$

Since M is negative, the shear in the walls is opposite the directions we drew. To avoid confusion, let's just carry the negative sign and leave the arrows the way we drew them.

Force generated in each lateral element due to rotation:

$$V_i^M = \left(\frac{R_i d_i}{\Sigma R_j d_j^2} \right) M$$

$$V_I^M = \frac{593 \text{ ft}}{89,810 \text{ ft}^2} (-369 \text{ k} \cdot \text{ft}) = -2.4 \text{ k} \ (+\uparrow)$$

$$V_{II}^M = \frac{589 \text{ ft}}{89,810 \text{ ft}^2} (-369 \text{ k} \cdot \text{ft}) = -2.4 \text{ k} \ (\overset{+}{\rightarrow})$$

$$V_{III}^M = \frac{589 \text{ ft}}{89{,}810 \text{ ft}^2}(-369 \text{ k}\cdot\text{ft}) = -2.4 \text{ k } (+\downarrow)$$

$$V_{IV}^M = \frac{371 \text{ ft}}{89{,}810 \text{ ft}^2}(-369 \text{ k}\cdot\text{ft}) = -1.5 \text{ k } (\underset{\leftarrow}{\pm})$$

$$V_{V}^M = \frac{6 \text{ ft}}{89{,}810 \text{ ft}^2}(-369 \text{ k}\cdot\text{ft}) = 0 \text{ k } (+\downarrow)$$

$$V_{VI}^M = \frac{216 \text{ ft}}{89{,}810 \text{ ft}^2}(-369 \text{ k}\cdot\text{ft}) = -0.9 \text{ k } (\underset{\leftarrow}{\pm})$$

Net Forces:

Let's choose $+x$ and $+y$ as positive directions for our sign convention.

$$V_i^T = V_i^F + V_i^M$$

$$V_I^T = -15.2 \text{ k} - 2.4 \text{ k} = -17.6 \text{ k } (\underset{\rightarrow}{\pm})$$

$$V_{II}^T = \phantom{-15.2 \text{ k}} 0 - 2.4 \text{ k} = -2.4 \text{ k } (\underset{\rightarrow}{\pm})$$

$$V_{III}^T = -27.4 \text{ k} + 2.4 \text{ k} = -25.0 \text{ k } (+\uparrow)$$

$$V_{IV}^T = \phantom{-15.2 \text{ k}} 0 + 1.5 \text{ k} = 1.5 \text{ k } (\underset{\rightarrow}{\pm})$$

$$V_V^T = -12.2 \text{ k} - 0 \phantom{.5 \text{ k}} = -12.2 \text{ k } (+\uparrow)$$

$$V_{VI}^T = \phantom{-15.2 \text{ k}} 0 + 0.9 \text{ k} = 0.9 \text{ k } (+\uparrow)$$

Evaluation of Results:

Satisfied Fundamental Principles?

$$\underset{\rightarrow}{\pm}\Sigma F_x = -2.4 \text{ k} + 1.5 \text{ k} + 0.9 \text{ k} = 0 \checkmark$$
$$+\uparrow\Sigma F_y = 54.7 \text{ k} - 17.6 \text{ k} - 25.0 \text{ k} - 12.2 \text{ k} = -0.1 \text{ kN} \approx 0 \checkmark$$

11.2 In Plane Shear: Collector Beams

Introduction

Both rigid and flexible diaphragms carry in plane loading to the lateral load-resisting elements as shear. The diaphragm is typically attached to the lateral element along the length of the element. That means the shear force is transferred over that length. If the shear is too great for that connection, the most economical option is typically to use collector beams.

How-To

Diaphragms experience in plane loading due to lateral loads on the structure. Those loads might come from pressures on the outer walls of the structure carried to the diaphragms (Section 1.5: Application of Lateral Loads). The other source is inertial forces due to earthquakes.

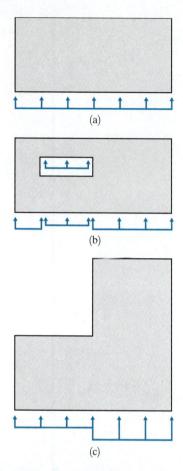

When the loading is caused by seismic events, earthquakes, the load is distributed where the mass is concentrated. For most structures, we consider that loading to be at the diaphragm levels. Within the diaphragm, the load is distributed based on the distribution of the mass. For example, a rectangular diaphragm without openings has a uniform load distribution (Figure 11.18a). If there is an opening, the loading is distributed to the various subareas of the diaphragm (Figure 11.18b). If the diaphragm is not rectangular, the load distribution is proportional to the depth of the subarea (Figure 11.18c).

The most accurate idealization of seismic loading on a diaphragm is to consider the diaphragm broken into many small subareas of roughly the same size. The loading can then be idealized as a point force at the centroid of the subarea (Figure 11.19). This level of refinement is usually only necessary for very irregularly shaped diaphragms.

In addition to distributed loading, the diaphragm might experience point loads from lateral load-resisting elements above the diaphragm. In most cases, the force in the lateral element follows the lateral element past the diaphragm. However, there are certain situations that will cause force to leave a lateral element and transfer across the diaphragm to a different lateral element. Sometimes constraints of the situation require that a lateral element stop at one diaphragm level; the result is a discontinuous load path (Figure 11.20a). Some structures change size at different levels. Where the plan of a structure changes, there is often a different number of lateral elements below the diaphragm than above the diaphragm (Figure 11.20b). The additional elements change the distribution of force, so some lateral elements are off loaded. If the diaphragm is rigid or semirigid (i.e., anything other than flexible), changes in the relative rigidity of the lateral elements cause a redistribution of force from above the diaphragm to below (Figure 11.20c).

Figure 11.18 Distribution of static equivalent seismic loading to diaphragms: (a) A rectangular diaphragm has uniform distributed loading; (b) Loading is distributed to the various subareas of the diaphragm around openings; (c) Loading is proportional to the subarea depth with a nonrectangular diaphragm.

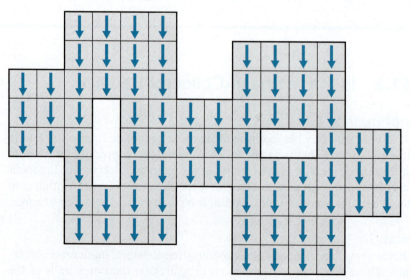

Figure 11.19 Distribution of static equivalent seismic loading by considering the diaphragm to be an assembly of many small areas.

Section 11.2 In Plane Shear: Collector Beams

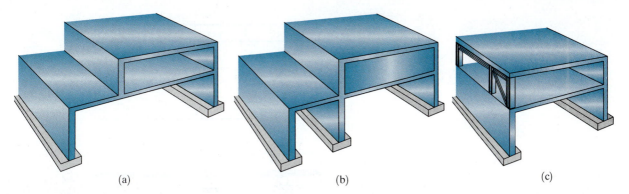

Figure 11.20 Situations that cause lateral force to transfer from lateral load-resisting elements above a diaphragm to different lateral element locations below the diaphragm: (a) Discontinuous load path for lateral elements; (b) Different number of lateral elements at different levels; (c) Large changes in relative rigidity between levels.

Although we did not show examples of these situations in Section 1.6: Distribution of Lateral Loads by Flexible Diaphragm or Section 11.1: Distribution of Lateral Loads by Rigid Diaphragm, those methods still apply. In addition to the distributed load at the diaphragm, we add in point forces from all of the lateral elements above the diaphragm (Figure 11.21).

Calculating In Plane Shear

The peak in plane shear always occurs at a lateral load-resisting element. Therefore, we won't need to calculate the shear all along the diaphragm until we get to Section 11.3: In Plane Moment: Diaphragm Chords. We do, however, need to know the shear at *each* lateral element.

In most cases, we want to know only the total force being transferred from the diaphragm to the lateral element. Those are the forces we calculate when analyzing the diaphragm as flexible or rigid.

In some situations, we want to know the shear on each side of the lateral element. The total force generated in the supporting lateral element is the sum of these two shears. To find the shears adjacent to the

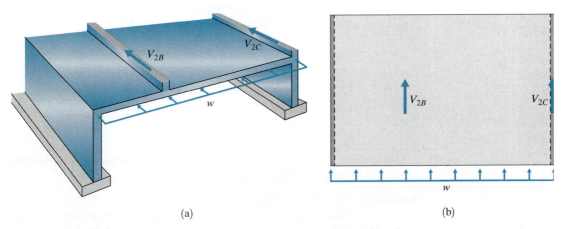

Figure 11.21 Modeling a diaphragm subjected to distributed in plane loading as well as shear forces from lateral load-resisting elements above the diaphragm: (a) Cutaway showing forces on diaphragm; (b) Plan view of diaphragm.

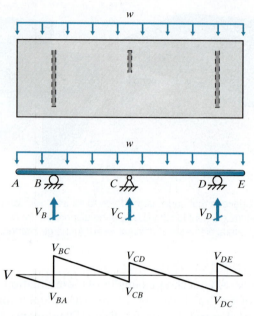

Figure 11.22 In plane shear generated in a diaphragm subjected to uniform distributed load.

lateral elements, we can use integration (Section 4.1: Internal Forces by Integration). As an example, consider the diaphragm in Figure 11.22.

From Section 4.1 we learned that the change in shear is the integral of the distributed load over a certain length:

$$\Delta V_{ab} = \int_a^b w(x)\,dx$$

Since we know that the shear at the free end is zero, we can calculate the shear to the left of wall B directly:

$$V_{BA} = 0 + \int_A^B w\,dx$$

At the wall, there is a jump in the shear diagram. We use that to find the shear at the right of wall B:

$$V_{BC} = V_{BA} + V_B$$

To find the shear at the left of wall C, we add the change in shear across the segment to the shear at the right of wall B:

$$V_{CB} = V_{BC} + \int_B^C w\,dx$$

Following this process, we can work our way across the diaphragm, finding V_{CD}, V_{DC}, and V_{DE} as well.

Evaluating Shear Capacity

Up to now throughout this text, we have idealized lateral force on a lateral load-resisting element as a point load (Figure 11.23a). In reality, the diaphragm is typically connected to the lateral element along the length of the element. That means the lateral force is most appropriately

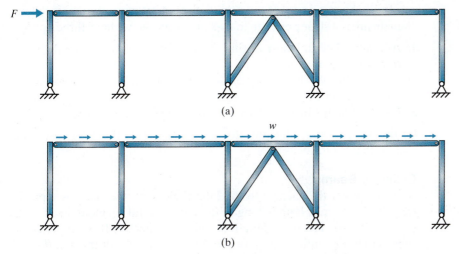

Figure 11.23 Idealizing lateral load on lateral elements: (a) Typical idealization of the load as a point force at each level; (b) More realistic idealization is a distributed shear along the connection between the diaphragm and lateral element.

idealized as a distributed shear load (Figure 11.23b). This more accurate idealization is necessary only when we are dealing with in plane shear and moment in the diaphragm.

The diaphragm and its connection to the lateral element have a shear capacity often expressed in force per unit length. Calculating that shear capacity depends on the materials used and the type of connection. Therefore, we leave that to other references.

The shear capacity is for a single failure plane. That plane might experience load from both sides (Figure 11.24a) or from just one side (Figure 11.24b). If the connection is such that loading is coming from both sides, we *must* compare the shear capacity with the total force transferred to the lateral element (e.g., V_B from Figure 11.22). If the connection is such that loading is calculated from only one side, we might choose to compare the shear capacity with the in plane shear on that side (e.g., V_{BC} from Figure 11.22).

Standard practice, however, is to ignore the possibility of two shearing planes. This is because it is often not clear whether the connection will actually behave such that both faces can generate adequate capacity simultaneously. For connections that do provide two shearing planes, our standard practice is conservative.

To check the in plane shear strength of the diaphragm, we assume that the shear is distributed uniformly along the connection. Compatibility of displacements implies that this is a reasonable assumption. The result is the following requirement:

$$\phi v_n \geq \frac{V_i}{l_i}$$

where
- ϕv_n = Usable shear capacity of the diaphragm, in units of force/length
- V_i = Shear being transferred to the lateral element i
- l_i = Connection length between the diaphragm and the lateral element i

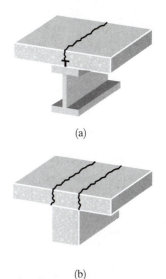

Figure 11.24 Examples of connections between diaphragms and lateral load-resisting elements: (a) Shear studs result in a single shearing plane to transfer force; (b) Cast-in-place concrete results in two shearing planes to transfer force.

If the shear capacity of the diaphragm and its connection to the lateral element are not strong enough to make the transfer, we have three options:

1. Add more lateral load-resisting elements in order to reduce the shear transferred to any one.
2. Increase the shear capacity of the diaphragm or its connection to the lateral element.
3. Use a collector beam to increase the transfer length.

Of the three options, the collector beam is typically the most cost effective, so that is our focus here.

Collector Beams

In order to get the shear into the lateral element, the shear is typically transferred over a length longer than the lateral element itself (e.g., Figure 11.23b). The extra length is called a collector. If we transfer the shear along the entire length of the diaphragm, we call this a *full-depth collector*. We use full-depth collectors most of the time because they have two key advantages: (1) the strength requirements on the collector are reduced, and (2) the depth of transfer has implications for the in plane moment capacity (Section 11.3: In Plane Moment: Diaphragm Chords). If we count on using only some of the available diaphragm length to transfer the shear, we call this a *partial-depth collector*.

Collectors are also called drag struts or drag beams. If a diaphragm is transferring load from a lateral element above the diaphragm to a different lateral element below the diaphragm (e.g., Figure 11.20), we often call the extension a distributor instead of a collector. The way we treat it is the same though.

The collector is physically a beam, a width of slab that acts like a beam, or both. The collector is oriented in the same direction as the lateral element and must connect to the lateral element. Because the shear force is being carried through the collector beam to the lateral element, the path must be continuous (Figure 11.25a). We cannot select a beam away from the lateral element to be a collector (Figure 11.25b). If we want or need to utilize additional length, we can extend the collector on one side of the lateral element (Figure 11.25c) or on both sides (Figure 11.25d). Most of the time we use full-depth collectors, which means we use the maximum length available along the line of the lateral element (Figure 11.25e).

We need to address three structural design issues for collector beams:

1. The axial force generated in collectors must be accounted for when designing the beam. Often this results in combined axial force and bending in the beam.
2. The axial force generated in collectors must be transferred through connections. The connections must be designed to handle these loads.
3. Wind and seismic loads can act in either direction; therefore, we must consider the axial force in the members and connections to be either tension or compression.

Analysis of Collector Forces

The first step in analysis of the collectors is to determine the distributed load along the collector and lateral element:

$$v_i = \frac{V_i}{\Sigma l}$$

Section 11.2 In Plane Shear: Collector Beams

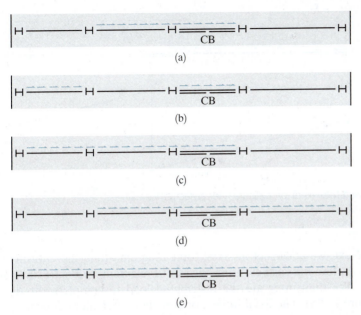

Figure 11.25 Plan view of options when collector beams are needed to transfer diaphragm shear to a lateral element: (a) Acceptable collector choice because of continuous load path; (b) Unacceptable collector choice because of discontinuous load path; (c) Two collector beam segments on one side of the lateral element; (d) Collector beam segments on opposite sides of the lateral element; (e) Full-depth collector uses all available segments.

where
- v_i = Shear distributed along the collector, in units of force per length
- V_i = Shear force transferred to the lateral element
- Σl = Total length over which the shear is transferred, collector and lateral element

If we are using full-depth collectors, Σl is the maximum length available. We then compare the distributed load to the diaphragm capacity. With full-depth collectors, if there is not enough capacity, we need to increase the diaphragm capacity or add more lateral elements.

If we are using partial-depth collectors, we probably don't know the transfer length in advance. In that case, we start by calculating the length needed given the shear capacity of the diaphragm:

$$l_{min} = \frac{V_i}{\phi v_n}$$

where
- ϕv_n = Usable shear capacity of the diaphragm, in units of force per length
- l_{min} = Minimum transfer length to keep the transfer stress below the limit

Once we know the required transfer length, we can select the necessary collector beam segments to provide at least the minimum length. With the selected collector length, we can use the formula above for v_i to obtain the distributed shear.

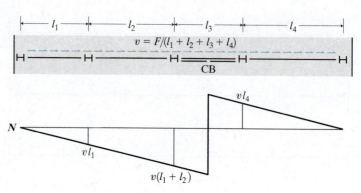

Figure 11.26 Axial force diagram for collector beams.

Next, we construct an axial force diagram along the entire column line. At the end farthest from the lateral element, the net axial force is zero. We then integrate the distributed load to complete the diagram (Figure 11.26). The axial force along the lateral element depends on the type of lateral element and the orientation of the element members. In Figure 11.26, the lateral element is a chevron brace, so all of the axial force is transferred at the apex of the chevron. For cross-braced frames, there are jumps in the axial force diagram at each end of the braced frame, since each diagonal takes some of the axial force. From the diagram, we can determine the axial force required at each connection and the axial force required in each collector segment.

SECTION 11.2 HIGHLIGHTS

In Plane Shear: Collector Beams

- Peak in plane shear is always at the connection to a lateral element.
- Standard of practice is to use the total shear force transferred from both sides of the element.
- Shear is transferred over a finite length.
- Design requires that the shear stress being transferred is less than the allowable:

$$\phi v_n \geq \frac{V}{l}$$

- For a partial-depth collector, use only some of the available diaphragm depth to transfer the shear.
- For a full-depth collector, use the full depth of the diaphragm to transfer the shear.
- If using a partial-depth collector, we can calculate the minimum required transfer length:

$$l_{min} = \frac{V_i}{\phi v_n}$$

- Construct an axial force diagram for the collector beam to determine the peak axial force in each collector segment and the axial force that each segment connection must carry.

Example 11.3

We are designing a new building that will be used as an office and shop. To meet the owner's tight construction schedule, the building will have precast wall panels and precast roof panels.

Figure 11.27

Treating the roof as a rigid diaphragm, we have determined the shear generated in each lateral load-resisting element if there is seismic loading in the north-south direction (Example 11.2).

This loading gives the peak shear for each element oriented in the north-south direction. To find the peak shear for elements oriented in the east-west direction, we need to use loading in the east-west direction.

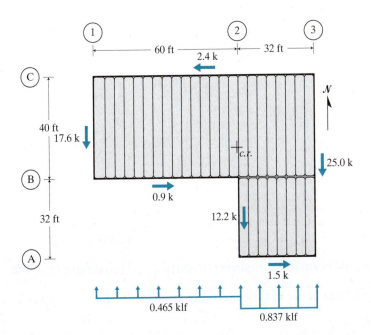

Figure 11.28

The plan is to use full-depth collectors for each of the lateral elements. The shear capacity is governed by the connection between the precast planks and the lateral element. Therefore, we need to know the required shear capacity for each connection. The wall along line 2 does not extend the full depth of the diaphragm, so we will use a collector beam. We need to determine the peak axial force generated in that collector beam and the peak axial force in the connection to the wall.

Shear along wall B1-C1:

$$v_{B1-C1} = \frac{17.6 \text{ k}}{40 \text{ ft}} = 0.440 \text{ klf}$$

∴ Design connection along B1-C1 for 0.440 klf

Shear along wall A3-C3:

$$v_{A3-C3} = \frac{25.0 \text{ k}}{72 \text{ ft}} = 0.347 \text{ klf}$$

∴ Design connection along A3-C3 for 0.347 klf

Shear along line 2:

$$v_{A2-C2} = \frac{12.2 \text{ k}}{72 \text{ ft}} = 0.169 \text{ klf}$$

∴ Design connection along A2-B2 for 0.169 klf

Axial force diagram:

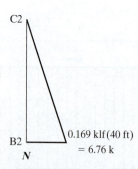

Note that once the axial force reaches the wall, it transforms into shear in the wall. Therefore, there is no axial force on the diagram between A2 and B2.

Design the collector beam (precast planks) to carry a peak axial force of 6.76 k.

Design the connection between the collector beam and the wall to carry a peak axial force of 6.76 k.

Example 11.4

Our firm has been selected to design a new manufacturing building in an industrial complex. Because of the sensitivity of the equipment that will be mounted on the roof, we anticipate using a rigid diaphragm. The arrangement of the manufacturing process has dictated the layout of the gravity system and the placement of the diagonal braces for lateral resistance.

Figure 11.29

Using the Directional Procedure for determining the effects of wind on the structure, we have calculated the shear generated in each lateral load-resisting element for case 2, considering wind from the north (Example 11.1). This wind direction causes the largest forces in the north-south–oriented lateral elements.

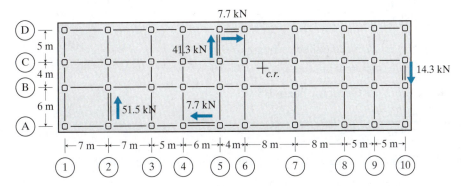

Figure 11.30

Client requirements have determined the direction of each diagonal brace.

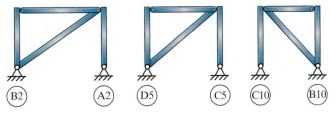

Figure 11.31

The shear capacity of the diaphragm is 4.75 kN/m. Our team leader wants us to determine whether the diaphragm is sufficiently strong to transfer the shear just along the braced frame for each column line. If any are not strong enough, we are to determine the fewest possible collector beam segments. For any collectors, we are to label the connection force on a plan view of the framing.

Shear along column line 2:

$$l_{min}^{Line\ 2} = \frac{51.5\ kN}{4.75\ kN/m} = 10.8\ m$$

10.8 m > 6 m ∴ Need collector beams

$$l_{A2-D2} = 6\ m + 4\ m + 5\ m = 15\ m$$

15 m > 10.8 m ∴ Collector beams B2-C2 and C2-D2

$$v_{A2-D2} = \frac{51.5\ kN}{15\ m} = 3.43\ kN/m$$

The shear starts gathering at D2.

$$F_{C2} = 3.43 kN/m(5\ m) = 17.2\ kN$$

$$F_{B2} = 3.43 kN/m(5\ m + 4\ m) = 30.9\ kN$$

Since the diagonal attaches at A2, the shear continues to gather.

$$F_{A2} = 3.43 kN/m(5\ m + 4\ m + 6\ m) = 51.4\ kN$$

Shear along column line 5:

$$l_{min}^{Line\ 5} = \frac{41.3\ kN}{4.75 kN/m} = 8.7\ m$$

8.7 m > 5 m ∴ Need collector beams

$$l_{B5-D5} = 5\ m + 4\ m = 9\ m$$

9 m > 8.7 m ∴ Collector beam B5-C5

$$v_{B5-D5} = \frac{41.3\ kN}{9\ m} = 4.59 kN/m$$

Since the diagonal attaches at C5, the shear gathers toward this point.

$$F_{C5-D5} = 4.59 kN/m(5\ m) = 23.0\ kN$$

$$F_{C5-B5} = 4.59 kN/m(4\ m) = 18.4\ kN$$

Shear along column line 10:

$$l_{min}^{Line\ 10} = \frac{14.3\ kN}{4.75\ kN/m} = 3.0\ m$$

3.0 m < 4 m ∴ Adequate without collector beams

$$v_{B10-C10} = \frac{14.3\ kN}{4\ m} = 3.58\ kN/m$$

The shear gathers toward the diagonal at C10.

$$F_{C10} = 3.58\ kN/m(4\ m) = 14.3\ kN$$

Summary:

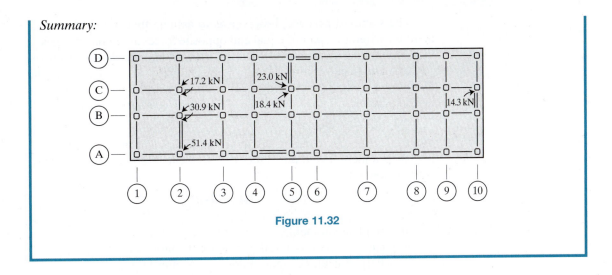

Figure 11.32

11.3 In Plane Moment: Diaphragm Chords

Introduction
When lateral load acts on a horizontal diaphragm, it causes in plane moment. We must ensure that the diaphragm can support the moment without tearing or experiencing large cracks. To carry the moment, we typically use a pair of beams on opposite sides of the diaphragm. The axial forces generated in the beams make a couple moment to counteract the internal moment. We call those beams chords, and we call the axial forces chord forces.

How-To
Lateral loads cause in plane moment in diaphragms. When we design the diaphragm for the in plane moment, it is accepted practice to assume that plane sections remain plane, which means that the flexure stresses are linear across the depth of the diaphragm (Figure 11.33a). This approach can require large tension strains in the diaphragm, which might result in unacceptably large tears or cracks at the edge of the diaphragm.

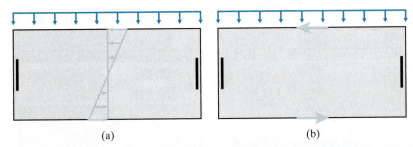

Figure 11.33 Diaphragms are typically designed to carry in plane moment in one of two ways: (a) Distributed stresses across the depth of the diaphragm; (b) Pair of equal and opposite forces.

The standard practice, however, is to assume that the in plane moment is carried by a pair of equal and opposite forces at the outer edges of the diaphragm (Figure 11.33b). These forces act like the chords of a truss, so we call them chord forces. To carry these forces, we can strengthen the diaphragm along the outer edge, count on the existing perimeter beams, or add perimeter beams. All of these solutions we refer to as *chords*.

Chord Forces

The forces in the two chords create a couple moment that is equal to the in plane moment. From our statics course, we learned how to calculate a couple moment:

$$M = F \cdot d$$

where
 M = In plane moment
 F = Chord forces required to balance the moment
 d = Perpendicular distance between the chord forces

Since we know the in plane moment, we can rearrange the relationship to find the required chord force:

$$C = T = \frac{M}{d}$$

where
 C = Compression chord force
 T = Tension chord force

Knowing the chord force is important in two aspects of design. First, the chord probably experiences moment due to the gravity load at the same time it experiences axial force from the lateral load. That means the strength of the chord must be determined considering the interaction of the two internal forces. Second, the chord is often made of segments connected together or connected to columns or walls. We need to consider the axial force when designing those connections.

The depth used to calculate the chord force is affected by our choice of collector length. If we use full-depth collectors to transfer shear, we can use the full depth to carry the moment (Figure 11.34a). However, if we use partial-depth collectors, only the depth used to transfer shear can be counted on to carry the moment (Figure 11.34b). Note that in these situations, large cracks might form in the diaphragm outside the depth of

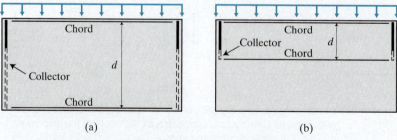

Figure 11.34 The length of shear transfer affects the usable depth for carrying in plane moment: (a) Full-depth collectors allow use of the full depth of the diaphragm to carry moment; (b) Partial-depth collectors limit the depth of the diaphragm for carrying moment.

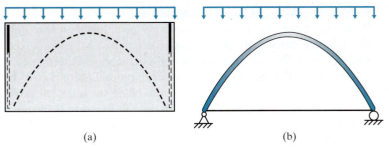

Figure 11.35 Many diaphragms distribute internal stresses more like an arch: (a) Compression struts carry loading out to the lateral elements, causing outward thrust on the lateral elements; (b) An arch tends to widen under load, but the tie keeps the ends in place.

the chords. For detailed information on dealing with partial-depth collectors and the resulting chord forces, see Moehle et al. (2010).

If we use the beam analogy for diaphragms, the in plane moment varies with position along the diaphragm. That implies that the required chord force changes along the length of the chord. In reality, many diaphragms behave more like tied arches (Figure 11.35). In a tied arch, the tension is constant along the tie. To adequately account for this range of behaviors, standard practice is to design each chord for the peak chord force between the lateral elements (Figure 11.36). The method for calculating the peak moment between each lateral element depends on the flexibility of the diaphragm.

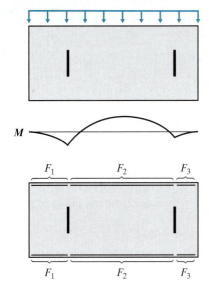

Figure 11.36 Lateral loading causes in plane moments that vary along the diaphragm; standard practice is to design each chord segment for the peak moment in that segment.

Calculating M_{max} in Flexible Diaphragms

If the maximum in plane displacement of a diaphragm is more than two times the average story drift of the lateral load-resisting elements that support it, we call the diaphragm flexible. We learned in Section 1.6: Distribution of Lateral Loads by Flexible Diaphragm that we can use the Equivalent Beam Model to determine the forces generated in each lateral element (Figure 11.37). If the equivalent beam is indeterminate, we can use the approximate analysis method in Section 1.6 or one of the detailed analysis methods in Chapter 12: Force Method or Chapter 13: Moment Distribution Method to find the reactions. The reactions for the equivalent beam are the forces generated in each lateral element.

We can use the Equivalent Beam Model to find the in plane moment as well. Once we have the equivalent beam reactions, we can use the method from Section 4.1: Internal Forces by Integration to find the expressions for shear and moment between each of the lateral elements. The peak moment will be either at the lateral element or where the shear is zero.

$$V(x_i) = \int w(x_i) \, dx_i$$

$$M(x_i) = \int V(x_i) \, dx_i$$

Take note that if we use the approximate analysis method from Section 1.6, we are essentially assuming that the moment is zero at the lateral elements, except cantilevered ends. That means we are assuming that each segment of the diaphragm between two lateral elements is acting as a simply supported beam (Figure 11.38). Sometimes that approximation will overpredict the peak moment, but sometimes it will underpredict.

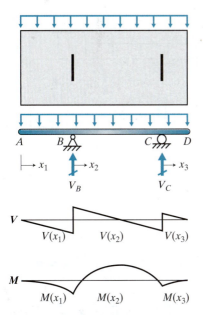

Figure 11.37 Analyzing a flexible diaphragm using the Equivalent Beam Model allows us to calculate the force generated in each lateral element, expressions for shear in each segment of the diaphragm, and expressions for moment.

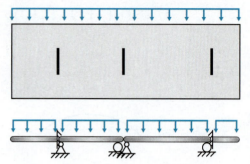

Figure 11.38 Approximate analysis of the Equivalent Beam Model is the same as considering each diaphragm segment independently.

Calculating M_{max} in Rigid Diaphragms

If the maximum in plane displacement is much less than two times the average story drift of the lateral elements, we call the diaphragm rigid. We covered how to find the force generated in each lateral element in Section 11.1: Distribution of Lateral Loads by Rigid Diaphragm. The distribution of in plane shear is not exactly the same as for a flexible diaphragm though. If we try to use the method from Section 4.1, the moment expression will not close to zero at the end of the diaphragm. This is because of the moment carried by lateral elements orthogonal to the loading (e.g., walls IV and V in Figure 11.39).

The standard practice in this situation is to use the Corrected Equivalent Beam Model. We still idealize the diaphragm as a beam and the lateral elements as supports, but we use an equivalent trapezoidal load (Figure 11.40). By *equivalent* we mean that the trapezoidal load and lateral element forces are in equilibrium. Since there are two relevant equations of equilibrium, we can find the two unknown distributed load magnitudes: w_1 and w_2.

Using the example equivalent beam in Figure 11.40 as a generalization for all rigid diaphragms, we can derive expressions for w_1 and w_2:

$$+\uparrow \Sigma F_y = 0 = \Sigma V_{yi} - \frac{1}{2}(w_1 + w_2)(l)$$

$$w_2 = \frac{2\Sigma V_{yi}}{l} - w_1$$

$$\circlearrowleft \Sigma M_O = 0 = w_1(l)\left(\frac{l}{2}\right) + \frac{1}{2}(w_2 - w_1)(l)\left(\frac{2l}{3}\right) - \Sigma(V_{yi} \cdot x_i)$$

$$w_1\frac{l^2}{2} - w_1\frac{l^2}{3} + w_2\frac{l^2}{3} = \Sigma(V_{yi} \cdot x_i)$$

$$w_1\frac{l^2}{6} + w_2\frac{l^2}{3} = \Sigma(V_{yi} \cdot x_i)$$

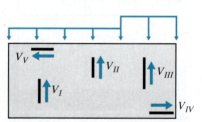

Figure 11.39 Rigid diaphragm results in forces generated in all lateral elements, not just the ones oriented with the loading.

If we substitute the expression for w_2 into this last expression, we get one equation with one unknown:

$$w_1\frac{l^2}{6} + \left(\frac{2\Sigma V_{yi}}{l} - w_1\right)\frac{l^2}{3} = \Sigma(V_{yi} \cdot x_i)$$

$$w_1\frac{l^2}{6} = \frac{2(\Sigma V_{yi})l}{3} - \Sigma(V_{yi} \cdot x_i)$$

$$w_1 = \frac{4(\Sigma V_{yi})}{l} - \frac{6\Sigma(V_{yi} \cdot x_i)}{l^2}$$

Substituting this into the expression for w_2 gives us

$$w_2 = \frac{2\Sigma V_{yi}}{l} - w_1 = \frac{2\Sigma V_{yi}}{l} - \frac{4(\Sigma V_{yi})}{l} + \frac{6\Sigma(V_{yi} \cdot x_i)}{l^2}$$

$$w_2 = -\frac{2(\Sigma V_{yi})}{l} + \frac{6\Sigma(V_{yi} \cdot x_i)}{l^2}$$

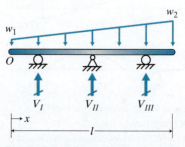

Figure 11.40 Actual loading on a rigid diaphragm is converted to an equivalent trapezoidal distributed load.

Since each segment of the corrected equivalent beam has essentially the same free body diagram (Figure 11.41), we can use the method from Section 4.1 to derive expressions for finding the peak moment between lateral elements. We use equilibrium from segment to segment to find the shear and moment at the ends of the segment: V_a, M_a, V_b, and M_b.

The distributed load is a function of position along the segment:

$$w(x) = \frac{(w_b - w_a)}{l} x + w_a$$

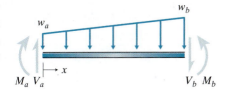

Figure 11.41 Free body diagram of a corrected equivalent beam segment between lateral elements a and b.

We integrate the equivalent distributed load to obtain a quadratic expression for shear along the segment:

$$V(x) = \int -\left[\frac{(w_b - w_a)}{l} x + w_a\right] dx$$

$$V(x) = -\frac{(w_b - w_a)}{2l} x^2 - w_a x + C_V$$

At $x = 0$, $V(x) = V_a$:

$$\therefore C_V = V_a$$

So,

$$V(x) = \frac{(w_a - w_b)}{2l} x^2 - w_a x + V_a$$

We integrate the shear expression to obtain a cubic expression for moment along the segment:

$$M(x) = \int \left[\frac{(w_a - w_b)}{2l} x^2 - w_a x + V_a\right] dx$$

$$M(x) = \frac{(w_a - w_b)}{6l} x^3 - \frac{w_a}{2} x^2 + V_a x + C_M$$

At $x = 0$, $M(x) = M_a$:

$$\therefore C_M = M_a$$

So,

$$M(x) = \frac{(w_a - w_b)}{6l} x^3 - \frac{w_a}{2} x^2 + V_a x + M_a$$

The peak moment will occur at one of the ends, M_a or M_b, or where the shear is zero. Using the quadratic formula, we can find where the shear is zero:

$$x = \frac{-(-w_a) \pm \sqrt{(-w_a)^2 - 4 \frac{(w_a - w_b)}{2l} V_a}}{2 \frac{(w_a - w_b)}{2l}}$$

$$x = \frac{w_a \pm \sqrt{w_a^2 + 2(w_b - w_a) V_a / l}}{(w_a - w_b)} (l)$$

At most, only one of the two roots will occur within the range of x. If no roots are within the range, the peak moment is guaranteed to be at one of the ends.

Calculating M_{max} in Semirigid Diaphragms

To find the required chord forces in a semirigid diaphragm, we typically use the Finite Element Method. However, if we analyze the diaphragm as flexible and analyze it as rigid, the results will bound the solution for a semirigid diaphragm. Therefore, we can use these by hand methods to bound the computer aided analysis results. In many cases, using the more conservative value has minimal impact on the design.

SECTION 11.3 HIGHLIGHTS

In Plane Moment: Diaphragm Chords

Chord Forces:

- Based on peak moment in the diaphragm segment between the lateral elements.
- Considered constant along the segment between the lateral elements.

$$C = T = \frac{M_{max}}{d}$$

Finding M_{max}:

Flexible Diaphragms: Equivalent Beam Model

1. Forces in the lateral elements are the reactions of a beam with the given loading.
2. Integrate to find $V(x)$'s and $M(x)$'s.
3. Peak moment is at a lateral element or where $V = 0$.

Rigid Diaphragms: Corrected Equivalent Beam Model

1. Find forces in the lateral elements using the method from Section 11.1.
2. Convert the lateral loading into an equivalent trapezoidal distributed load:

$$w_1 = \frac{4(\Sigma V_{yi})}{l} - \frac{6\Sigma(V_{yi} \cdot x_i)}{l^2}$$

$$w_2 = -\frac{2(\Sigma V_{yi})}{l} + \frac{6\Sigma(V_{yi} \cdot x_i)}{l^2}$$

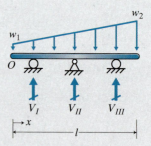

3. Linearly interpolate to find w at the ends of each segment.
4. Use equilibrium to find V and M at the left end of each segment.

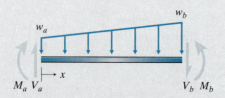

5. Generate expressions for the moment in each segment:

$$M(x) = \frac{(w_a - w_b)}{6l}x^3 - \frac{w_a}{2}x^2 + V_a x + M_a$$

6. Evaluate the moment expressions at the ends of the segments and where $V = 0$:

$$x = \frac{w_a \pm \sqrt{w_a^2 + 2(w_b - w_a)V_a/l}}{(w_a - w_b)}(l)$$

Example 11.5

Our team is designing a single-story office building that has a flexible roof. Architectural requirements have led the senior engineer to use rigid frames for lateral load resistance. We are at the point of the design where we are determining the chord forces that the perimeter beams and connections must carry. We are using full-depth collectors to transfer shear to each lateral element. Our task is to determine the chord forces for one of the wind cases: uniform wind from the south.

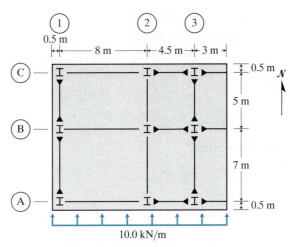

Figure 11.42

Find Shear in Each Lateral Element:

Equivalent Beam Model:

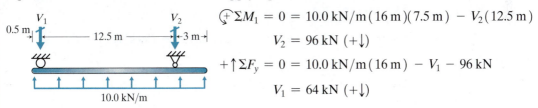

Figure 11.43

Apply equilibrium:

$$\circlearrowright \Sigma M_1 = 0 = 10.0 \text{ kN/m}(16 \text{ m})(7.5 \text{ m}) - V_2(12.5 \text{ m})$$

$$V_2 = 96 \text{ kN} \ (+\downarrow)$$

$$+\uparrow \Sigma F_y = 0 = 10.0 \text{ kN/m}(16 \text{ m}) - V_1 - 96 \text{ kN}$$

$$V_1 = 64 \text{ kN} \ (+\downarrow)$$

Expressions for Shear and Moment:

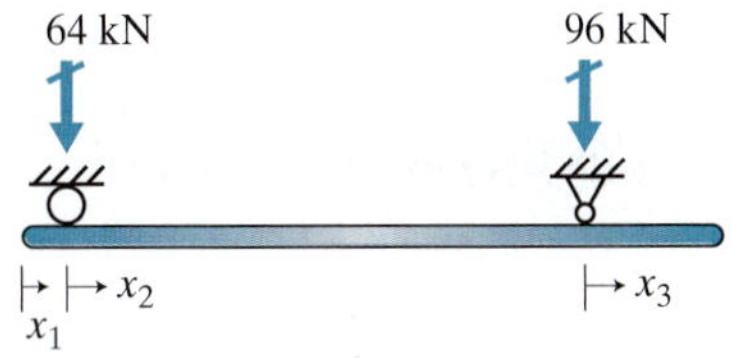

Figure 11.44

$$V(x_1) = \int(-10 \text{ kN/m}) dx_1 = (-10 \text{ kN/m}) x_1 + C_1$$

At $x_1 = 0$, $V = 0$:

$$\therefore C_1 = 0$$

So,

$$V(x_1) = (-10 \text{ kN/m}) x_1$$

$$V(x_2) = \int(-10 \text{ kN/m}) dx_2 = (-10 \text{ kN/m})x_2 + C_2$$
At $x_2 = 0, x_1 = 0.5$ m, $V(x_2) = V(x_1) + 64$ kN:
$$V(x_2) = -10 \text{ kN/m}(0.5 \text{ m}) + 64 \text{ kN} = 59 \text{ kN}$$
$$\therefore C_2 = 59 \text{ kN}$$

So,
$$V(x_2) = (-10 \text{ kN/m})x_2 + 59 \text{ kN}$$

$$V(x_3) = \int(-10 \text{ kN/m}) dx_3 = (-10 \text{ kN/m})x_3 + C_3$$
At $x_3 = 0, x_2 = 12.5$ m, $V(x_3) = V(x_2) + 96$ kN:
$$V(x_3) = -10 \text{ kN/m}(12.5 \text{ m}) + 59 \text{ kN} + 96 \text{ kN} = 30 \text{ kN}$$
$$\therefore C_3 = 30 \text{ kN}$$

So,
$$V(x_3) = (-10 \text{ kN/m})x_3 + 30 \text{ kN}$$

Satisfied Fundamental Principles?

Check that the shear closes to zero:

At $x_3 = 3$ m:
$$V(x_3) = -10 \text{ kN/m}(3 \text{ m}) + 30 \text{ kN} = 0 \checkmark$$

$$M(x_1) = \int(-10 \text{ kN/m})x_1 \cdot dx_1 = (-5 \text{ kN/m})x_1^2 + C_4$$
At $x_1 = 0, M = 0$:
$$\therefore C_4 = 0$$
So,
$$M(x_1) = (-5 \text{ kN/m})x_1^2$$

$$M(x_2) = \int(-10 \text{ kN/m} \cdot x_2 + 59 \text{ kN}) dx_2 = (-5 \text{ kN/m})x_2^2 + (59 \text{ kN})x_2 + C_5$$
At $x_2 = 0, x_1 = 0.5$ m, $M(x_2) = M(x_1)$:
$$M(x_2) = -5 \text{ kN/m}(0.5 \text{ m})^2 = -1.25 \text{ kN} \cdot \text{m}$$
$$\therefore C_5 = -1.25 \text{ kN} \cdot \text{m}$$

So,
$$M(x_2) = (-5 \text{ kN/m})x_2^2 + (59 \text{ kN})x_2 - 1.25 \text{ kN} \cdot \text{m}$$

$$M(x_3) = \int(-10 \text{ kN/m} \cdot x_3 + 30 \text{ kN}) dx_3 = (-5 \text{ kN/m})x_3^2 + (30 \text{ kN})x_3 + C_6$$
At $x_3 = 0, x_2 = 12.5$ m, $M(x_3) = M(x_2)$:
$$M(x_3) = -5 \text{ kN/m}(12.5 \text{ m})^2 + 59 \text{ kN}(12.5 \text{ m}) - 1.25 \text{ kN} \cdot \text{m} = -45 \text{ kN} \cdot \text{m}$$
$$\therefore C_6 = -45 \text{ kN} \cdot \text{m}$$

So,
$$M(x_3) = (-5 \text{ kN/m})x_3^2 + (30 \text{ kN})x_3 - 45 \text{ kN} \cdot \text{m}$$

Satisfied Fundamental Principles?

Check that the moment closes to zero:

At $x_3 = 3$ m:
$$M(x_3) = -5 \text{ kN/m}(3 \text{ m})^2 + 30 \text{ kN}(3 \text{ m}) - 45 \text{ kN} \cdot \text{m} = 0 \checkmark$$

Peak Moments:

Since the equivalent beam is cantilevered on the left end, M_{1max} is at the lateral element at $x_1 = 0.5$ m:

$$M(x_1) = -5 \text{ kN/m}(0.5 \text{ m})^2$$
$$M_{1max} = -1.25 \text{ kN} \cdot \text{m}$$

Since the equivalent beam is cantilevered on the other end, M_{3max} is at the lateral element at $x_3 = 0$ m:

$$M(x_3) = -5 \text{ kN/m}(0)^2 + 30 \text{ kN}(0) - 45 \text{ kN} \cdot \text{m}$$
$$M_{3max} = -45 \text{ kN} \cdot \text{m}$$

Between lateral elements, M_{2max} is where $V = 0$ or at a lateral element:

$$V(x_2) \equiv 0 = (-10 \text{ kN/m})x_2 + 59 \text{ kN}$$
$$x_2 = 5.9 \text{ m}$$
$$M(x_2) = -5 \text{ kN/m}(5.9 \text{ m})^2 + 59 \text{ kN}(5.9 \text{ m}) - 1.25 \text{ kN} \cdot \text{m} = 172.8 \text{ kN} \cdot \text{m}$$
$$M(x_2) = 172.8 \text{ kN} \cdot \text{m} > 1.25 \text{ kN} \cdot \text{m} \text{ and } 45 \text{ kN} \cdot \text{m}$$
$$M_{2max} = 172.8 \text{ kN} \cdot \text{m}$$

Chord Forces:

The depth between chords is $5 \text{ m} + 7 \text{ m} = 12 \text{ m}$.

In the range of x_1:

$$T_1 = \frac{1.25 \text{ kN} \cdot \text{m}}{12 \text{ m}} = 0.10 \text{ kN}$$

Note: since this is a very short span and small force, no beam will be added.

In the range of x_2:

$$T_2 = \frac{172.8 \text{ kN} \cdot \text{m}}{12 \text{ m}} = 14.40 \text{ kN}$$

In the range of x_3:

$$T_3 = \frac{45 \text{ kN} \cdot \text{m}}{12 \text{ m}} = 3.75 \text{ kN}$$

Summary:

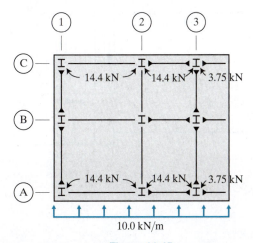

Figure 11.45

EXAMPLE 11.6

We are designing a new building that will be used as an office and shop. To meet the owner's tight construction schedule, the building will have precast wall panels and precast roof panels.

Figure 11.46

Treating the roof as a rigid diaphragm, we have determined the shear generated in each lateral load-resisting element if there is seismic loading in the north-south direction (Example 11.2).

Our assignment is to determine the chord forces that will be carried on the perimeter of the diaphragm due to this loading. The plan is to use full-depth collectors for each of the lateral elements.

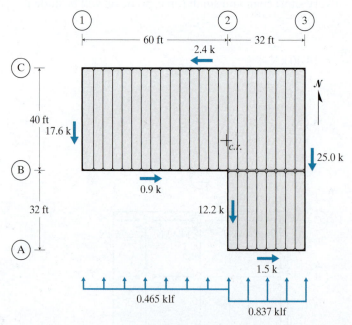

Figure 11.47

Approximate Analysis:

Treat each diaphragm segment between lateral elements as an independent diaphragm.

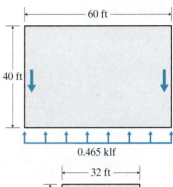

$$M_{1max} = \frac{0.465 \text{ klf}(60 \text{ ft})^2}{8} = 209 \text{ k}$$

$$T_1 = \frac{209 \text{ k}}{40 \text{ ft}} = 5.22 \text{ k}$$

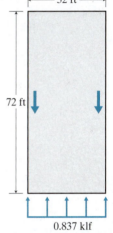

$$M_{2max} = \frac{0.837 \text{ klf}(32 \text{ ft})^2}{8} = 107 \text{ k}$$

$$T_2 = \frac{107 \text{ k}}{72 \text{ ft}} = 1.49 \text{ k}$$

Figure 11.48

Corrected Equivalent Beam:

Find the equivalent linear distributed load.

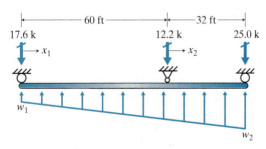

Figure 11.49

$+\downarrow \Sigma V_y = 17.6 \text{ k} + 12.2 \text{ k} + 25.0 \text{ k} = 54.8 \text{ k}$

$\circlearrowleft^+ \Sigma V_y \cdot x = 17.6 \text{ k}(0) + 12.2 \text{ k}(60 \text{ ft}) + 25.0 \text{ k}(92 \text{ ft}) = 3032 \text{ k} \cdot \text{ft}$

$$w_1 = \frac{4(\Sigma V_{yi})}{l} - \frac{6\Sigma(V_{yi} \cdot x_i)}{l^2}$$

$$w_1 = \frac{4(54.8 \text{ k})}{92 \text{ ft}} - \frac{6(3032 \text{ k} \cdot \text{ft})}{(92 \text{ ft})^2} = 2.383 \text{ klf} - 2.149 \text{ klf} = 0.234 \text{ klf}$$

$$w_2 = -\frac{2(\Sigma V_{yi})}{l} + \frac{6\Sigma(V_{yi} \cdot x_i)}{l^2}$$

$$w_2 = -\frac{2(54.8 \text{ k})}{92 \text{ ft}} + \frac{6(3032 \text{ k} \cdot \text{ft})}{(92 \text{ ft})^2} = -1.191 \text{ klf} + 2.149 \text{ klf} = 0.958 \text{ klf}$$

Values at End of Each Segment:
At $x_1 = 0$ ft:
  $w_{a1} = 0.234$ klf
  $V_{a1} = 17.6$ k
  $M_{a1} = 0$

At $x_2 = 0$ ft:
  Linearly interpolate to find w:
  $$w_{a2} = \frac{(0.958 \text{ klf} - 0.234 \text{ klf})}{92 \text{ ft}}(60 \text{ ft}) + 0.234 \text{ klf} = 0.706 \text{ klf}$$

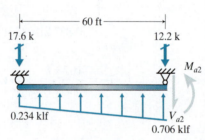

Figure 11.50

$$+\uparrow \Sigma F_y = 0 = -17.6 \text{ k} - 12.2 \text{ k} + \frac{1}{2}(0.234 \text{ klf} + 0.706 \text{ klf})(60 \text{ ft}) - V_{a2}$$

$$V_{a2} = -1.6 \text{ k}$$

$$\circlearrowleft \Sigma M_{a2} = 0 = 17.6 \text{ k}(60 \text{ ft}) - 0.234 \text{ klf}(60 \text{ ft})(30 \text{ ft})$$

$$- \frac{1}{2}(0.706 \text{ klf} - 0.234 \text{ klf})(60 \text{ ft})(20 \text{ ft}) + M_{a2}$$

$$M_{a2} = -352 \text{ k} \cdot \text{ft}$$

Note that because the free body diagram is upside down compared with our derived formulas, we must reverse the signs on V_{a2} and M_{a2} before using them to find $M(x)$ and $M_{\max}$.

  $V_{a2} = 1.6$ k
  $M_{a2} = 352$ k $\cdot$ ft

Expressions for Moment:

$$M(x) = \frac{(w_a - w_b)}{6l}x^3 - \frac{w_a}{2}x^2 + V_a x + M_a$$

$$M(x_1) = \frac{(0.234 \text{ klf} - 0.706 \text{ klf})}{6(60 \text{ ft})}x_1^3 - \frac{0.234 \text{ klf}}{2}x_1^2 + (17.6 \text{ k})x_1$$

$$M(x_2) = \frac{(0.706 \text{ klf} - 0.958 \text{ klf})}{6(32 \text{ ft})}x_2^3 - \frac{0.706 \text{ klf}}{2}x_2^2 + (1.6 \text{ k})x_2 + 352 \text{ k} \cdot \text{ft}$$

Section 11.3 In Plane Moment: Diaphragm Chords

Satisfied Fundamental Principles?

Check that the moment closes to zero:

At $x_2 = 32$ ft:

$$M(x_2) = \frac{(0.706 \text{ klf} - 0.958 \text{ klf})}{6(32 \text{ ft})}(32 \text{ ft})^3 - \frac{0.706 \text{ klf}}{2}(32 \text{ ft})^2$$
$$+ 1.6 \text{ k}(32 \text{ ft}) + 352 \text{ k} \cdot \text{ft}$$
$$= -1.3 \text{ k} \cdot \text{ft} \ll \text{each term} \therefore = 0 \checkmark$$

Peak Moments:

$$x = \frac{w_a \pm \sqrt{w_a^2 + 2(w_b - w_a)V_a/l}}{(w_a - w_b)}(l)$$

Left segment:

$$x_1 = \frac{0.234 \text{ klf} \pm \sqrt{(0.234 \text{ klf})^2 + 2(0.706 \text{ klf} - 0.234 \text{ klf})(17.6 \text{ k})/(60 \text{ ft})}}{(0.234 \text{ klf} - 0.706 \text{ klf})}(60 \text{ ft})$$

$$= \frac{0.234 \text{ klf} \pm 0.576 \text{ klf}}{-0.472 \text{ klf}}(60 \text{ ft})$$

$$= -103.0 \text{ ft} \text{ or } 43.5 \text{ ft}$$

Only one value is within the range of x_1: 43.5 ft.

$$M_{\max}(x_1) = \frac{(0.234 \text{ klf} - 0.706 \text{ klf})}{6(60 \text{ ft})}(43.5 \text{ ft})^3 - \frac{0.234 \text{ klf}}{2}(43.5 \text{ ft})^2 + 17.6 \text{ k}(43.5 \text{ ft})$$

$$M_{\max}(x_1) = 436 \text{ k} \cdot \text{ft}$$

Check the value at the end of the segment, $x_1 = 60$ ft:

$$M(x_1) = \frac{(0.234 \text{ klf} - 0.706 \text{ klf})}{6(60 \text{ ft})}(60 \text{ ft})^3 - \frac{0.234 \text{ klf}}{2}(60 \text{ ft})^2 + 17.6 \text{ k}(60 \text{ ft})$$

$$M_{b1} = 352 \text{ k} \cdot \text{ft} < 436 \text{ k} \cdot \text{ft} \therefore \text{ not the maximum}$$

Right segment:

$$x_2 = \frac{0.706 \text{ klf} \pm \sqrt{(0.706 \text{ klf})^2 + 2(0.958 \text{ klf} - 0.706 \text{ klf})(1.6 \text{ k})/(32 \text{ ft})}}{(0.706 \text{ klf} - 0.958 \text{ klf})}(32 \text{ ft})$$

$$= \frac{0.706 \text{ klf} \pm 0.724 \text{ klf}}{-0.252 \text{ klf}}(32 \text{ ft})$$

$$= -181.6 \text{ ft} \text{ or } 2.3 \text{ ft}$$

Only one value is within the range of x_2: 2.3 ft.

$$M_{\max}(x_2) = \frac{(0.706 \text{ klf} - 0.958 \text{ klf})}{6(32 \text{ ft})}(2.3 \text{ ft})^3 - \frac{0.706 \text{ klf}}{2}(2.3 \text{ ft})^2$$
$$+ 1.6 \text{ k}(2.3 \text{ ft}) + 352 \text{ k} \cdot \text{ft}$$

$$M_{\max}(x_2) = 354 \text{ k} \cdot \text{ft}$$

Chord Forces:

Left segment:

$$M_{max}(x_1) = 436 \text{ k} \cdot \text{ft}$$

$$T(x_1) = \frac{436 \text{ k} \cdot \text{ft}}{40 \text{ ft}} = 10.90 \text{ k}$$

Right segment:

$$M_{max}(x_2) = 354 \text{ k} \cdot \text{ft}$$

$$T(x_2) = \frac{354 \text{ k} \cdot \text{ft}}{72 \text{ ft}} = 4.92 \text{ k}$$

Evaluation of Results:

Approximation Predicted Outcomes?
Left segment approximate result: $T_1 = 5.22$ k (52% low)
Right segment approximate result: $T_2 = 1.49$ k (70% low)

We expect these results to be low because the interior lateral element carries much less shear than the outer lateral elements; therefore, the moment stays positive throughout the diaphragm and acts more like a simply supported member without the interior support. ✓

Summary:

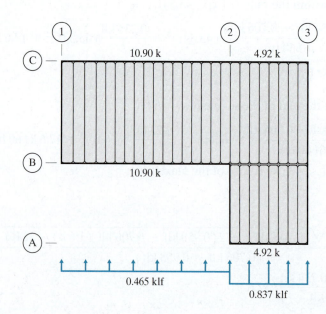

Figure 11.51

Example 11.7

Our team is designing a small reinforced concrete office. The senior engineer has decided that the lateral load-resisting system will be structural walls and that we will use partial-depth collectors if needed.

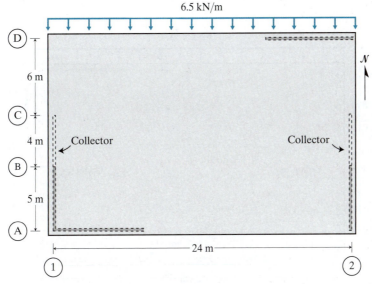

Figure 11.52

The team has determined the requirements for the partial-depth collectors due to wind in the north-south direction. Our job is to determine the chord forces and locations to carry the in plane moment due to wind from the north.

The lateral elements in the north-south direction are identical, so they have the same rigidity. A more experienced member of the team points out that because of the uniform loading and the symmetrical rigidity, the load carried by each of the lateral elements is the same whether we treat the diaphragm as flexible or rigid.

Equivalent beam:

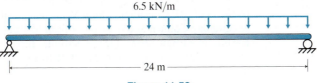

Figure 11.53

Peak Moment:

$$M_{max} = \frac{wl^2}{8} = \frac{6.5 \text{ kN/m}(24 \text{ m})^2}{8}$$
$$= 468 \text{ kN} \cdot \text{m}$$

Chapter 11 Diaphragms

Chord Forces:

Partial-depth collectors mean partial depth to carry the moment.

$$T = \frac{468 \text{ kN} \cdot \text{m}}{(5 \text{ m} + 4 \text{ m})} = 52 \text{ kN}$$

Summary:

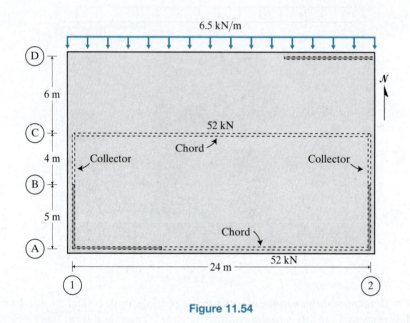

Figure 11.54

Reference Cited

Moehle, J. P., J. D. Hooper, D. J. Kelly, and T. R. Meyer. 2010. "Seismic design of cast-in-place concrete diaphragms, chords, and collectors: A guide for practicing engineers," *NEHRP Seismic Design Technical Brief No. 3*, produced by the NEHRP Consultants Joint Venture, a partnership of the Applied Technology Council and the Consortium of Universities for Research in Earthquake Engineering, for the National Institute of Standards and Technology, Gaithersburg, MD, NIST GCR 10-917-4.

Homework Problems

11.1 The architect wants an open layout for a new medical facility, so the best lateral load-resisting system is a rigid frame. In order to make preliminary selection of member properties, the team leader would like to have an idea of the forces generated in each of the members due to seismic loading. To perform that analysis, we first need to find the shear generated in each rigid frame due to seismic loading.

One of the seismic load cases is an earthquake in the north-south direction with accidental torsion. Seismic load acts at the center of mass, which is at the centroid of this diaphragm. Each rigid frame in the short direction is the same, so they all have the same rigidity. Each rigid frame in the long direction is the same, so they all have the same rigidity. Another team member has estimated that the rigidity of each long frame is 1.6 times that of each short frame.

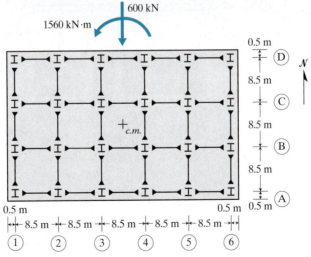

Prob. 11.1

a. Guess which rigid frame will experience the largest shear force due to the seismic loading.
b. Find the center of rigidity for the system based on the preliminary rigidity values.
c. Determine the force developed in each rigid frame (magnitude and direction) due to an earthquake in the north-south direction.
d. Comment on why your guess in part (a) was or was not correct when compared with your results in part (c).

11.2 A tall building will be made with rigid diaphragms at each floor level. Our team has calculated the net shear and moment at the lowest diaphragm due to wind load. We have been tasked with determining how much shear will develop in each of the chevron braced frames. The frames are identical; therefore, they all have the same rigidity. A teammate points out that the braced frames are orthogonal, so it might be easiest to break the wind load into components.

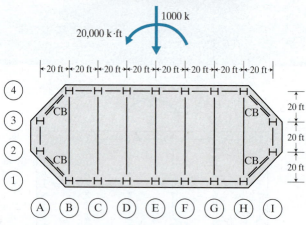

Prob. 11.2

a. Guess which braced frame will experience the largest shear force due to the wind loading.
b. Find the center of rigidity for the system.
c. Determine the force developed in each braced frame (magnitude and direction) due to the wind loading.
d. Comment on why your guess in part (a) was or was not correct when compared with your results in part (c).

11.3 Our team is designing an office building that will be enclosed with glass. Working with the architect, the senior engineer has determined that CMU structural walls in certain locations will be the best lateral load-resisting system. We anticipate that the roof will act as a rigid diaphragm. The senior engineer estimated the rigidity of each wall based on its length. Now our job is to predict the shear developed in each wall due to wind from the north.

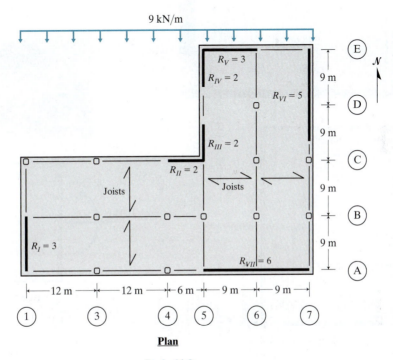

Plan

Prob. 11.3

a. Guess which structural wall will experience the largest shear force due to the wind loading.
b. Find the center of rigidity for the system.
c. Determine the force developed in each structural wall (magnitude and direction) due to the wind loading.
d. Comment on why your guess in part (a) was or was not correct when compared with your results in part (c).

11.4 An old two-story warehouse converted to restaurants is for sale. An interested buyer wants to have a full structural evaluation of the building before purchasing it. A team member has calculated the design wind loads using current codes, and another team member has estimated the rigidity of each wall based on length. Our job is to find the effect of one of the wind cases on the second floor, assuming it acts as a rigid diaphragm.

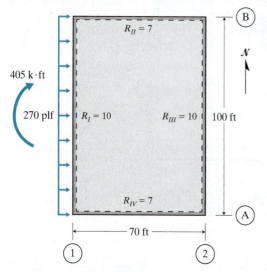

Prob. 11.4

11.5 The owner of a historic building wants to add a large sign to the top of the building to match the look of the building when it was first constructed. Our firm has been hired to evaluate whether the building can support the large sign in its current condition. Part of that evaluation is to consider the seismic loading on the structure with the added weight of the sign. A team member has calculated the new location of the center of mass and the resulting seismic loading. That team member has also estimated the rigidity of the long walls to be 1200 and the short walls to be 900.

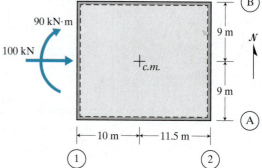

Prob. 11.5

a. Guess which structural wall will experience the largest shear force due to this wind load case.
b. Find the center of rigidity for the system.
c. Determine the force developed in each structural wall (magnitude and direction) due to this load case.
d. If we consider the diaphragm connected along the full length of each of the walls, what is the peak shear strength (force/length) that the diaphragm must have for this load case?
e. Using the Corrected Equivalent Beam Model, what are the peak chord forces due to this load case?
f. Comment on why your guess in part (a) was or was not correct when compared with your results in part (c).

a. Guess which structural wall will experience the largest shear force due to this seismic loading.
b. Find the center of rigidity for the system.
c. Determine the force developed in each structural wall (magnitude and direction) due to this load case.
d. If we consider the diaphragm connected along the full length of each of the walls, what is the peak shear strength (force/length) that the diaphragm must have for this load case?
e. Using the Corrected Equivalent Beam Model, what are the peak chord forces due to this load case?
f. Comment on why your guess in part (a) was or was not correct when compared with your results in part (c).

11.6 and 11.7 Our team is designing a new two-story department store. The senior engineer has chosen steel braced frames for the lateral load-resisting elements ($E = 29,000$ ksi). The type and placement of the braced frames have been decided based on architectural requirements. The opening in the middle is for escalators. We are still in the preliminary design phase, but the team anticipates that the upper floor will behave as a rigid diaphragm and the roof will likely behave as a flexible diaphragm.

A more experienced team member recommends that we preliminarily consider the braces to be square tubes: HSS $6 \times 6 \times \frac{1}{4}$ ($A = 5.24$ in^2, $I = 28.6$ in^4).

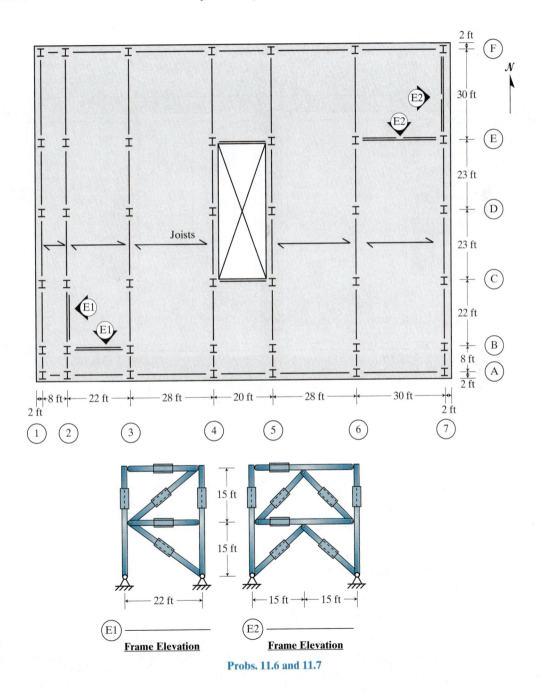

Probs. 11.6 and 11.7

666   **Chapter 11**   Diaphragms

11.6 Before finalizing the design, our team leader wants to know the shear force generated in each lateral element if the upper floor behaves as a rigid diaphragm. Our job is to consider wind from the north. We must consider two cases: uniform load and wind that also causes moment in either direction.

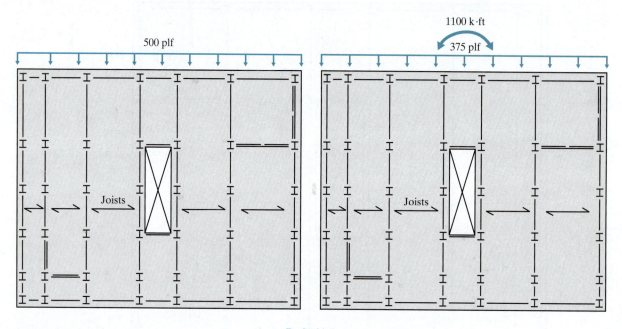

Prob. 11.6

a. Guess which loading will cause the largest shear in any lateral element.
b. Using the information on the previous page and the Story Drift Method, predict the story drift of the lower level of each braced frame due to a unit lateral load. Convert each story drift into rigidity.
c. Find the center of rigidity for the system based on the preliminary section properties.
d. Determine the force developed in each braced frame (magnitude and direction) due to wind from the north. Perform the analysis using the uniform wind load. Repeat the analysis using the wind and moment. Use the moment direction that generates the largest forces in the lateral elements. Identify the largest force generated in each lateral element.
e. Comment on why your guess in part (a) was or was not correct when compared with your results in part (d).

11.7 While some team members are analyzing the upper floor, our team leader wants us to analyze the roof level, which will behave as a flexible diaphragm. Let's start by considering uniform wind from the north. Note from page 665 that at the roof level the diagonal braces attach at columns C2 and B3.

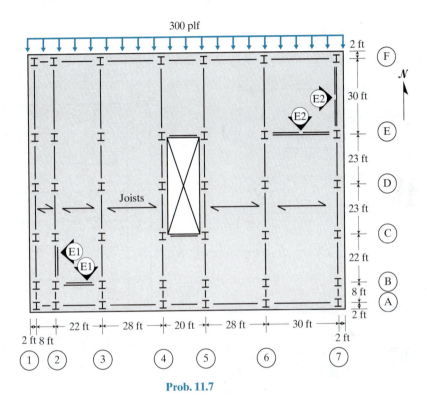

Prob. 11.7

a. Guess which will experience the largest axial connection force: full-depth collectors or chords.
b. Using the Equivalent Beam Model, predict the shear generated at each lateral load-resisting element.
c. Based on full-depth collectors, generate the axial force diagram for each column line with a lateral element. Label the axial force that each collector connection must carry on a plan view of the roof level.
d. Using the Equivalent Beam Model, predict the peak moment generated in each segment of the diaphragm.
e. Based on full-depth collectors, calculate the chord forces needed to carry the in plane moment. Label the axial force that each chord connection must carry on a plan view of the roof level.
f. Comment on why your guess in part (a) was or was not correct when compared with your results in parts (c) and (e).

668 Chapter 11 Diaphragms

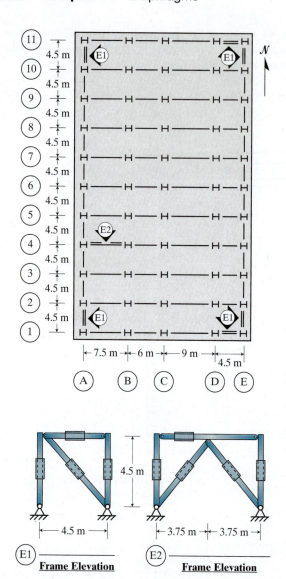

Probs. 11.8 and 11.9

11.8 and 11.9 Our firm has been selected to design the structure for a new grocery store in a location that experiences lots of snow. Because of the potential snow loads, the senior engineer has decided to use a concrete roof and steel framing. That roof will likely behave as a rigid diaphragm. Architectural requirements have determined the placement of the columns and braced frames. For each of the diagonally braced frames, the diagonal attaches to the column at the corner of the building.

11.8 In order to begin designing the members in the braced frames, our team needs to know how much shear is carried by each lateral element. A more experienced team member suggests that we assume the columns and braces are all HSS $127 \times 127 \times 6.4$ ($E = 200$ GPa $= 200$ kN/mm^2, $A = 2780$ mm^2, $I = 6.67 \times 10^6$ mm^4). Our task is to analyze the structure considering wind from the east.

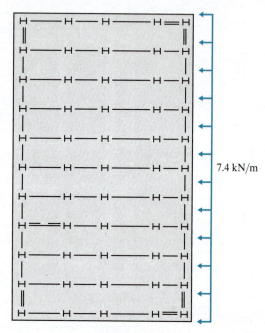

Prob. 11.8

a. Guess which braced frame will experience the largest shear due to wind from the east.
b. Using the Story Drift Method, predict the story drift of each braced frame due to a unit lateral load. Convert each story drift into rigidity.
c. Find the center of rigidity for the system based on the preliminary section properties.
d. Determine the force developed in each braced frame (magnitude and direction) due to wind from the east.
e. Comment on why your guess in part (a) was or was not correct when compared with your results in part (d).

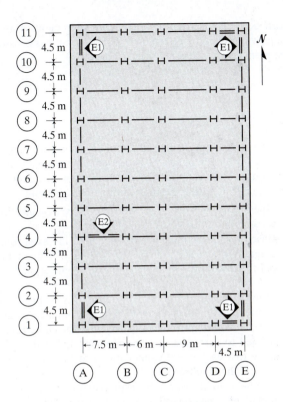

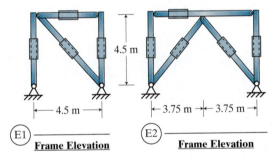

Probs. 11.8 and 11.9 (Repeated)

11.9 A team member has determined the shear that will develop in each of the lateral elements if wind comes from the south. Our task is to find the effects of the in plane forces. The team leader wants us to use full-depth collectors.

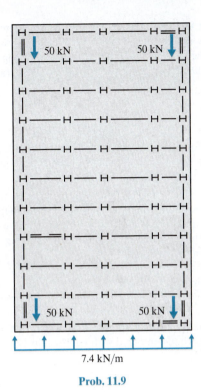

Prob. 11.9

a. Guess which will experience the largest axial connection force: full-depth collectors or chords.
b. Based on full-depth collectors, generate the axial force diagram for each column line with a lateral element. Label the axial force that each collector connection must carry on a plan view of the diaphragm.
c. Using the Corrected Equivalent Beam Model, predict the peak moment generated in the diaphragm.
d. Based on full-depth collectors, calculate the chord forces needed to carry the in plane moment. Label the axial force that each chord connection must carry on a plan view of the diaphragm.
e. Comment on why your guess in part (a) was or was not correct when compared with your results in parts (b) and (d).

Chapter 11 Diaphragms

11.10 and 11.11 Our firm is designing the structural system for a new natatorium. The team leader has chosen to use steel braced frames as the lateral load-resisting elements ($E = 29{,}000$ ksi). The locations of the braced frames have been decided based on architectural requirements. Based on previous experience, a member of the team has guessed the member sizes for the braced frames.

11.10 Before our team performs a computer aided analysis, our job is to estimate the shear developed in each of the braced frames due to wind from the east.

a. Guess which braced frame will experience the largest shear assuming the roof behaves as a rigid diaphragm.
b. Using the Story Drift Method, predict the total drift of each braced frame due to a unit lateral load. Convert each total drift into rigidity.
c. Find the center of rigidity for the system based on the preliminary section properties.
d. Determine the force developed in each braced frame (magnitude and direction) due to wind from the east.
e. Comment on why your guess in part (a) was or was not correct when compared with your results in part (d).

11.11 We have not finalized the roof design yet, so it might act as a flexible diaphragm. If we assume it is a flexible diaphragm, our task is to find the effects of the in plane forces. The team leader wants us to use full-depth collectors.

a. Guess which will experience the largest axial connection force: full-depth collectors or chords.
b. Based on full-depth collectors, generate the axial force diagram for each column line with a lateral element. Label the axial force that each collector connection must carry on a plan view of the diaphragm.
c. Using the Equivalent Beam Model, predict the peak moment generated in each segment of the roof diaphragm.
d. Based on full-depth collectors, calculate the chord forces needed to carry the in plane moment. Label the axial force that each chord connection must carry on a plan view of the diaphragm.
e. Comment on why your guess in part (a) was or was not correct when compared with your results in parts (b) and (d).

Homework Problems **671**

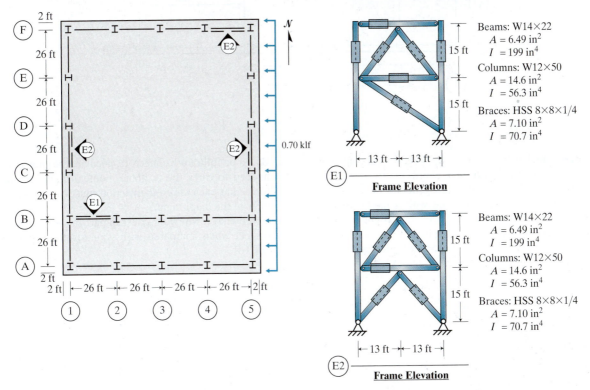

Probs. 11.10 and 11.11

11.12 Our firm is designing a two-story office building. The facility has a cast-in-place concrete floor supported by steel beams. The box with an × represents an opening in the floor where the elevators are. Other team members are designing the roof system for gravity loads. Our task is to determine the in plane requirements assuming the roof acts as a flexible diaphragm.

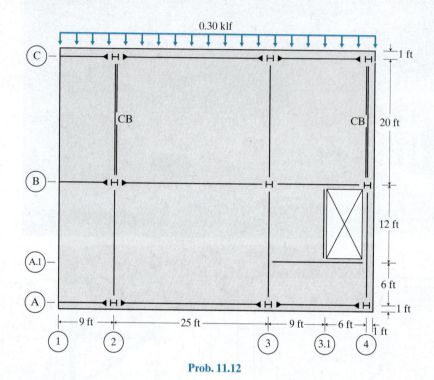

Prob. 11.12

a. Guess which will experience the largest axial connection force: full-depth collectors or chords.
b. Based on full-depth collectors, generate the axial force diagram for each column line with a lateral element. Label the axial force that each collector connection must carry on a plan view of the diaphragm.
c. Using the Equivalent Beam Model, predict the peak moment generated in each segment of the diaphragm.
d. Based on full-depth collectors, calculate the chord forces needed to carry the in plane moment. Label the axial force that each chord connection must carry on a plan view of the diaphragm.
e. Comment on why your guess in part (a) was or was not correct when compared with your results in parts (b) and (d).

11.13 A local car dealer wants to build a new showroom in a location where potential customers can see all four sides of the building. Our firm is providing the structural design services. In order to show vehicles on all four sides, the only support for the roof will be a few structural walls inside the building. The concrete roof will likely behave as a semirigid diaphragm. To bound the effects of wind load, another team member is analyzing the roof as a rigid diaphragm. Our job is to analyze the roof as a flexible diaphragm.

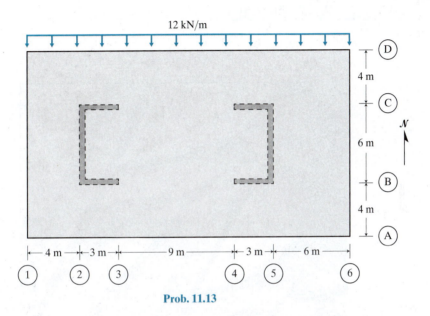

Prob. 11.13

a. Guess which segment of the diaphragm will have the largest chord forces.
b. Collectors for shear will be created by putting extra reinforcement in the concrete roof. If the shear capacity of the connection between the flexible diaphragm and the wall is 20 kN/m, what is the shortest length of partial-depth collector we need? Show the collectors on a plan view of the diaphragm.
c. Using the Equivalent Beam Model, predict the peak moment generated in each segment of the diaphragm.
d. Chords will be created by putting extra reinforcement in the concrete. Based on the shortest partial-depth collectors, calculate the chord forces needed to carry the in plane moment in each segment.
e. Based on full-depth collectors, calculate the chord forces needed to carry the in plane moment in each segment.
f. Comment on why your guess in part (a) was or was not correct when compared with your results in part (c).

CHAPTER 12

Force Method

Indeterminate structures have redundancy, which tends to make them more resilient and tends to reduce internal forces.

Motivation

Most structures are intentionally made to be indeterminate because such structures have important advantages:

- Indeterminate structures have redundancy, which means there are alternative load paths through the structure; therefore, they are more resistant to collapse.
- The peak internal forces are typically reduced, thus reducing the size requirements, self-weight, and cost for individual members.

Indeterminate structures present a challenge for analysis though: they have more unknowns than equations of equilibrium. In previous chapters, we added assumptions to make up the difference between unknowns and equations; that is called approximate analysis. Approximate analysis can be very helpful for evaluating structural schemes and making preliminary designs of members. However, we typically want more accurate analysis results when we finalize our designs. Analysis methods that do not make assumptions about the location of inflection points or the distribution of internal forces are called *detailed* analysis methods. We cover three detailed analysis methods in this text. The first is the Force Method. The other two are the Moment Distribution Method (Chapter 13) and the Direct Stiffness Method (Chapters 14 and 15).

The Force Method supplements the equations of equilibrium with equations of compatibility. This means the Force Method requires us to set up and solve a system of simultaneous linear equations.

Although we could use the Force Method to analyze almost any indeterminate structure, it turns out to be a very efficient analysis tool for structures that are only one or two degrees indeterminate. In fact, the Force Method is often faster than modeling the structure with computer software when the structure is only one degree indeterminate. The Force Method can be used with any type of structure (i.e., beam, frame, truss) but is typically used with beams and trusses.

12.1 One Degree Indeterminate Beams

Introduction
The Force Method requires us to convert an indeterminate beam into a determinate one; then we calculate the deformations of the determinate version. If the deformations are easy to calculate (e.g., we can apply formulas from the inside front cover of this text), this method can be a very efficient analysis tool.

How-To

Overview of the Force Method
The basic idea of the Force Method is to turn an indeterminate structure into a determinate structure that we can analyze using just the equations of equilibrium. We then use superposition to reproduce the original behavior. Figure 12.1 illustrates the process.

The top beam in Figure 12.1 is 1° indeterminate, which means we need to remove one unknown in order to make the beam determinate. In this case, we chose the vertical reaction at B. The result is the middle beam. Because that beam is determinate, we can calculate the internal forces using the equations of equilibrium and we can use any of the methods from Chapter 5: Deformations to predict Δ_{By}.

To reproduce the behavior of the original beam, we need to apply a vertical force at B to push the beam back up (the third beam in Figure 12.1). Again, we can calculate the vertical displacement Δ'_{By} using the tools we already have. The vertical force, F_B, will be an unknown in the calculation of Δ'_{By}. The original beam had zero vertical displacement at B; therefore, the following must be true:

$$0 = \Delta_{By} + \Delta'_{By}$$

We call this an *equation of compatibility*. In order to have compatibility, the two vertical displacements must be equal and opposite. The only unknown in the equation is the vertical force, F_B, so we can solve for it. That force is actually the vertical reaction at B, B_y, in the original structure. Once we know B_y, only three unknown reactions are left along with three

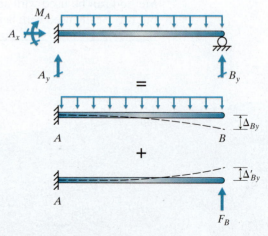

Figure 12.1 Illustration of the Force Method process.

Section 12.1 One Degree Indeterminate Beams

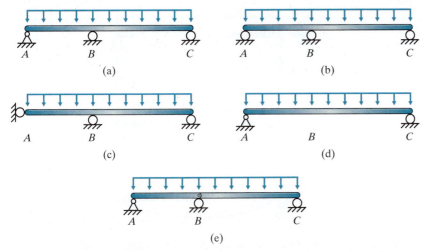

Figure 12.2 Choosing the redundant for an indeterminate beam: (a) Original 1° indeterminate beam; (b) Choice of A_x as the redundant is not valid because it results in an unstable beam; (c) Choice of the vertical reaction A_y is valid; (d) Choice of the vertical reaction B_y is valid; (e) Choice of the internal moment M_B is valid.

equations of equilibrium. We now have a structure we can analyze with the skills we already learned.

Redundants

The degree of indeterminacy is the difference between the number of unknowns and the number of equations of equilibrium for a structure. To convert an indeterminate structure into a determinate structure, we need to remove or "release" the extra unknowns equal to the degree of indeterminacy. We call those extra unknowns *redundants*.

Any reaction or internal force can be chosen as a redundant as long as we obey one rule: the released structure cannot be unstable. Figure 12.2a shows a 1° indeterminate beam. If we choose the horizontal reaction at A as the redundant, the released structure is unstable because the beam can move freely in the horizontal direction (Figure 12.2b); therefore, A_x is not an acceptable redundant. There are many valid redundants to choose from though. For example, we could release the vertical reaction A_y (Figure 12.2c) or B_y (Figure 12.2d). We could even release the internal moment at B, M_B (Figure 12.2e).

Although there are typically several valid choices of redundants for a structure, some options make the problem easier to solve. The key is how we predict the deformations of the released structure. Predicting Δ_{Ay} in Figure 12.2c requires us to use one of the methods from Chapter 5: Deformations. However, we can predict Δ_{By} by using the formulas inside the front cover of this text. Therefore, choosing B_y as the redundant (Figure 12.2d) will be faster and less vulnerable to mistakes.

Compatibility

Compatibility means that the superposition of effects must result in the original condition. When used with the Force Method, compatibility means that the net deformation at the redundant in the released structure must be equal to the deformation at that same location in the original structure. In the example in Figure 12.1, the vertical displacement at B in

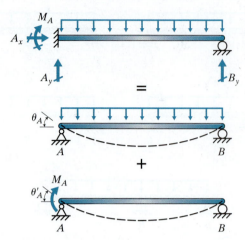

Figure 12.3 Illustration of the Force Method process with a moment reaction redundant.

the original structure was zero, so the net vertical displacement at B for the released structure also had to be zero.

The choice of redundant affects the type of compatibility. For example, if we choose a force as the redundant (reaction or internal), the compatibility must be of the displacements (Figure 12.1). If we choose the moment as the redundant (reaction or internal), the compatibility must be of the rotations/slopes (Figure 12.3).

The equation of compatibility for Figure 12.3 is

$$0 = \theta_A + \theta'_A$$

We are solving the same problem as in Figure 12.1, but we chose a different redundant. That does not affect our results. Using the M_A we calculate from this compatibility equation and applying equilibrium to the beam result in the same B_y we would calculate from Figure 12.1.

If we choose an internal moment as the redundant (e.g., Figure 12.2e), it is important to note that the compatibility is the rotation between the two ends (Figure 12.4a). To calculate the slope between the ends, we calculate the slope of each end and add them (Figure 12.4b). We use the same process to calculate the slope between the ends due to the redundant, M_B.

We call the Force Method a "detailed" analysis method because it does not make assumptions about the location of inflection points or the distribution of internal forces as we do in approximate analysis methods. The Force Method does have underlying assumptions, however. We are counting on predictions of deformation in this method. Therefore, the assumptions inherent in the deformation prediction method influence our results. We learned in Chapter 5: Deformations that the assumptions in those predictions are the following:

- Plane sections remain plane, or the shear deformation is negligible.
- Deformations or slopes are relatively small.
- Material remains linear elastic.

Because the Force Method makes the same assumptions as the method embedded in structural analysis software (Direct Stiffness Method), our

Section 12.1 One Degree Indeterminate Beams

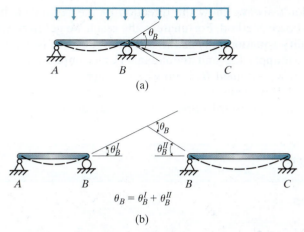

Figure 12.4 Choosing an internal moment as the redundant: (a) Released beam has a hinge at B, and we calculate the slope between the ends; (b) We calculate the slope between the ends by calculating the slope generated at each end and adding them.

results from the Force Method should match the results of computer aided analysis to within roundoff error.

Flexibility Coefficients

When we apply the redundant to the released structure, we don't actually know the magnitude of the redundant. To deal with this, we apply a unit redundant and then measure the deformation. This deformation is called a flexibility coefficient, f (Figure 12.5).

To get the deformation due to the actual redundant, we scale the flexibility coefficient by the redundant. The equation of compatibility for the beam in Figure 12.5 becomes

$$0 = \Delta_{By} + B_y \cdot f_{By}$$

In this form, we see the unknown reaction B_y and we can solve for it. In order to obtain the correct units for the redundant (e.g., B_y, M_A), we need to apply the unit redundant without dimension units (e.g., no kN, k·ft).

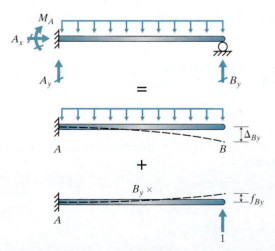

Figure 12.5 Scaling the effect of a unit redundant in the Force Method process.

We don't always know the direction of the redundant before we apply the Force Method. Fortunately, the result we get from solving the compatibility equation tells us the direction. Positive is defined by the direction we apply the unit redundant. For example, for the beam in Figure 12.5, we assumed B_y is up when we applied the unit redundant at B upward. If the value we calculate for B_y is negative, it means the actual reaction is downward, opposite the way we drew the unit redundant.

Section 12.1 Highlights

One Degree Indeterminate Beams

Force Method Steps:

1. Choose the redundant and identify the associated deformation.

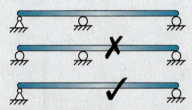

- Cannot choose a redundant that leaves the released structure unstable.
- Choose the redundant that makes the deformation prediction as easy as possible.
- Can be a force reaction, moment reaction, or internal moment.
  - If the redundant is a force, measure the displacement.
  - If the redundant is a moment, measure the slope.
  - If the redundant is an internal moment, measure the slope change between the ends.

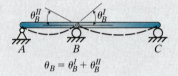

$$\theta_B = \theta_B^I + \theta_B^{II}$$

2. Release the structure from the redundant.
3. Predict the relevant deformation at the redundant for the released structure due to the applied loads.

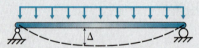

4. Predict the relevant deformation at the redundant for the released structure due to a unit redundant (flexibility coefficient).

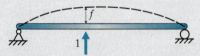

5. Write the equation of compatibility.
6. Solve the compatibility equation for the unknown redundant.
7. Use equations of equilibrium to find all other reactions for the original indeterminate structure. Then calculate any desired quantity (e.g., internal forces, deformations).

Assumptions:

Same as for predicting deformations.

- Plane sections remain plane, or the shear deformation is negligible.
- Deformations or slopes are relatively small.
- Material remains linear elastic.

EXAMPLE 12.1

We are tasked with designing a beam to support a moving hoist. The track for the hoist attaches to the beam at midspan. Since the hoist will only load the beam sometimes, it is a live load. In order to allow the hoist to move smoothly along the track, we have tight limits on the live load deflection. One option is to make the beam simply supported and use a sufficiently large moment of inertia, I. Another option is to make the end against the wall a fixed support. The fixed support will probably cost more than a pinned support, but we might be able to save even more money on a smaller beam. To help choose between the options, we want to know how much of an impact adding a fixed support has compared to using simple supports.

Figure 12.6

Before we can predict the peak displacement of the beam with a fixed support, we need to calculate the reactions.

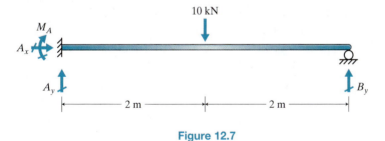

Figure 12.7

Estimated Solution:

Degree of indeterminacy:
    4 unknowns
    −3 equations
      1° indeterminate

682   **Chapter 12**   Force Method

Assumption:
Assume the inflection point is at $0.25l$ from the fixed end, A.

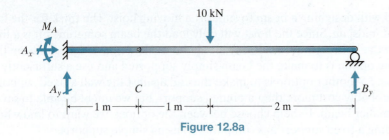

Figure 12.8a

Approximate analysis:
FBD of CB:

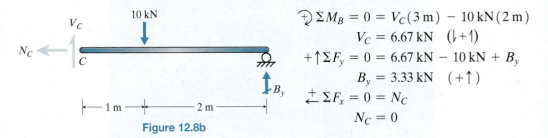

Figure 12.8b

$$\circlearrowleft^+ \Sigma M_B = 0 = V_C(3\text{ m}) - 10\text{ kN}(2\text{ m})$$
$$V_C = 6.67 \text{ kN} \quad (\downarrow + \uparrow)$$
$$+\uparrow \Sigma F_y = 0 = 6.67\text{ kN} - 10\text{ kN} + B_y$$
$$B_y = 3.33 \text{ kN} \quad (+\uparrow)$$
$$\xleftarrow{+} \Sigma F_x = 0 = N_C$$
$$N_C = 0$$

FBD of AC:

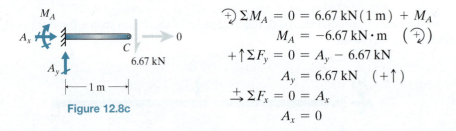

Figure 12.8c

$$\circlearrowleft^+ \Sigma M_A = 0 = 6.67\text{ kN}(1\text{ m}) + M_A$$
$$M_A = -6.67 \text{ kN} \cdot \text{m} \quad (\circlearrowleft^+)$$
$$+\uparrow \Sigma F_y = 0 = A_y - 6.67\text{ kN}$$
$$A_y = 6.67 \text{ kN} \quad (+\uparrow)$$
$$\xrightarrow{+} \Sigma F_x = 0 = A_x$$
$$A_x = 0$$

Summary:

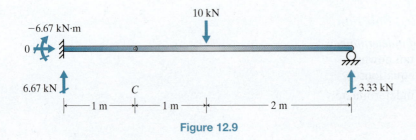

Figure 12.9

Detailed Analysis:

Since the beam is 1° indeterminate, we need to choose only one redundant. Let's choose M_A.

Released structure with applied load:

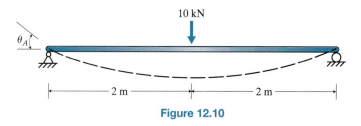

Figure 12.10

From the inside front cover, we have

$$\theta_A = \frac{Pl^2}{16EI} = \frac{10\,\text{kN}(4\,\text{m})^2}{16EI} = \frac{10\,\text{kN}\cdot\text{m}^2}{EI} \quad (\curvearrowleft +)$$

Note that this is positive because we switched the sign convention from the front cover in order to match our sign convention for positive internal moment.

Released structure with unit redundant:

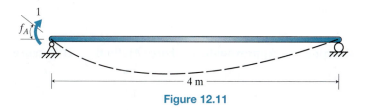

Figure 12.11

From the inside front cover, we have

$$f_A = \frac{M_o l}{3EI} = \frac{1(4\,\text{m})}{3EI} = \frac{1.333\,\text{m}}{EI} \quad (\curvearrowleft +)$$

Again, this is positive because we used the opposite sign convention from the one inside the front cover.

Compatibility:

We have one equation, since there is one redundant.

$$0 = \theta_A + M_A \cdot f_A$$

$$0 = \frac{10\,\text{kN}\cdot\text{m}^2}{EI} + M_A \cdot \frac{1.333\,\text{m}}{EI}$$

$$M_A = -7.50\,\text{kN}\cdot\text{m} \quad (\curvearrowright)$$

Find the remaining reactions:

$$+\circlearrowleft \Sigma M_A = 0 = -7.50 \text{ kN} \cdot \text{m} + 10 \text{ kN}(2 \text{ m}) - B_y(4 \text{ m})$$
$$B_y = 3.12 \text{ kN} \quad (+\uparrow)$$
$$+\uparrow \Sigma F_y = 0 = A_y - 10 \text{ kN} + 3.12 \text{ kN}$$
$$A_y = 6.88 \text{ kN} \quad (+\uparrow)$$
$$\xrightarrow{+} \Sigma F_x = 0 = A_x$$
$$A_x = 0$$

Summary:

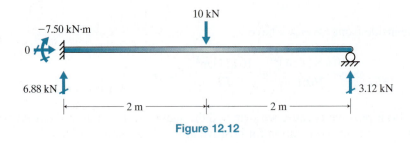

Figure 12.12

Evaluation of Results:
Approximation Predicted Outcomes?

Comparison with the approximate analysis results:

Reaction	Approximate	Force Method	Difference
A_x	0 kN	0 kN	0%
A_y	6.67 kN	6.88 kN	−3%
M_A	−6.67 kN·m	−7.50 kN·m	−11%
B_y	3.33 kN	3.12 kN	7%

All of the approximate reactions are within 11% of the results of the detailed analysis. ✓

Satisfied Fundamental Principles?

The reactions match the computer aided analysis results to more than three significant figures, as expected. ✓

Reaction	Force Method	Computer Aided	Difference
A_x	0 kN	0 kN	0%
A_y	6.88 kN	6.88 kN	0%
M_A	−7.50 kN·m	−7.50 kN·m	0%
B_y	3.12 kN	3.12 kN	0%

Conclusion:

These observations suggest we have reasonable detailed analysis results.

EXAMPLE 12.2

As additional verification of our results in Example 12.1, we can choose a different redundant and reanalyze.

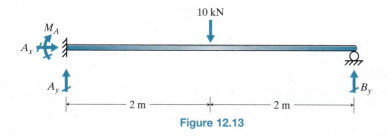

Figure 12.13

Detailed Analysis:

Let's choose B_y as the redundant this time.

Released structure with applied load:

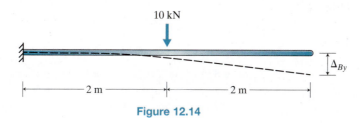

Figure 12.14

From the inside front cover, we have

$$\Delta_{By} = -\frac{P}{6EI}(3a^2b + 2a^3) = -\frac{10 \text{ kN}}{6EI}[3(2\text{ m})^2(2\text{ m}) + 2(2\text{ m})^3]$$

$$= -\frac{66.67 \text{ kN} \cdot \text{m}^3}{EI} \quad (+\uparrow)$$

Released structure with unit redundant:

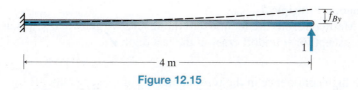

Figure 12.15

From the inside front cover, we have

$$f_{By} = -\frac{2Pa^3}{6EI} = -\frac{2(-1)(4\text{ m})^3}{6EI} = \frac{21.33\text{ m}^3}{EI} \quad (+\uparrow)$$

The applied force is negative in the formula because inside the front cover the force is down, but this force is up.

Compatibility:
We have one equation, since there is only one redundant:

$$0 = \Delta_B + B_y \cdot f_{By}$$

$$0 = -\frac{66.67\text{ kN}\cdot\text{m}^3}{EI} + B_y \cdot \frac{21.33\text{ m}^3}{EI}$$

$$B_y = 3.13\text{ kN} \quad (+\uparrow)$$

Find the remaining reactions:

$$\circlearrowleft \Sigma M_A = 0 = M_A + 10\text{ kN}(2\text{ m}) - 3.13\text{ kN}(4\text{ m})$$
$$M_A = -7.48\text{ kN}\cdot\text{m} \quad (\circlearrowright)$$

$$+\uparrow \Sigma F_y = 0 = A_y - 10\text{ kN} + 3.13\text{ kN}$$
$$A_y = 6.87\text{ kN} \quad (+\uparrow)$$

$$\xrightarrow{+} \Sigma F_x = 0 = A_x$$
$$A_x = 0$$

Summary:

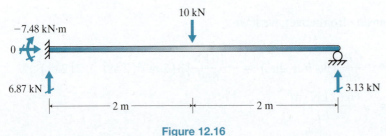

Figure 12.16

Evaluation of Results:
Satisfied Fundamental Principles?
These Force Method results are identical to the results we get when M_A is used as the redundant, except for roundoff error in the last digit. ✓

Conclusion:
Now we can have high confidence in the reasonableness of these results.

Example 12.3

The Department of Transportation is planning some repairs for an existing overpass. The contractor wants to stage materials on the short span. Before okaying the plan, the DOT wants to know the potential effect on the bridge girders. Our task is to find the peak moment generated by the proposed loading.

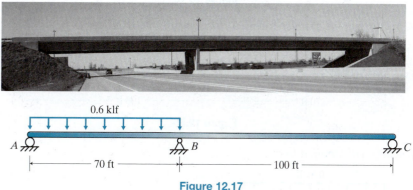

Figure 12.17

The lead engineer wants to be sure our answer is reasonable; therefore, one team member will be doing a computer aided analysis and we will be performing a detailed analysis by hand.

Estimated Solution:

Treat the continuous girder as two simply supported girders. This should provide an upper bound on the peak moment, since continuity tends to reduce the peak internal forces.

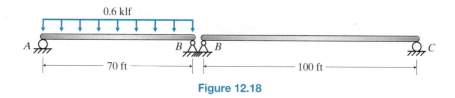

Figure 12.18

The peak moment is in the left girder:

$$M_{max} = \frac{wl^2}{8} = \frac{0.6 \text{ klf}(70 \text{ ft})^2}{8} = 367 \text{ k} \cdot \text{ft} \quad (\circlearrowright + \circlearrowleft)$$

Detailed Analysis:

Redundant:

  Degree of indeterminacy:
  4 unknowns
  $\underline{-3}$ equations
  $1°$ indeterminate

Choosing B_y is valid, but we would need to use one of the methods from Chapter 5: Deformations to predict Δ_{By}. Notice that since there is no axial force in the beam, we can show removing B_y as moving the pinned support to one of the other ends.

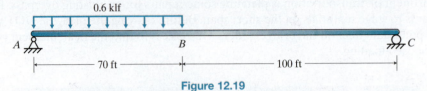

Figure 12.19

Choose M_B so that we can use formulas from inside the front cover.

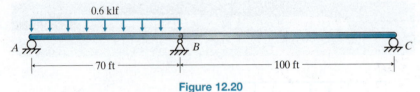

Figure 12.20

Released structure with applied loads:

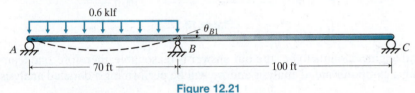

Figure 12.21

From the inside front cover, we have

$$\theta_{B1} = \frac{wl^3}{24EI} = \frac{0.6 \text{ klf}(70 \text{ ft})^3}{24EI} = \frac{8575 \text{ k} \cdot \text{ft}^2}{EI} \quad (\measuredangle +)$$

$$\theta_{B2} = 0$$

Add to find the angle between the segments:

$$\theta_B = \theta_{B1} + \theta_{B2}$$
$$= \frac{8575 \text{ k} \cdot \text{ft}^2}{EI} + 0 = \frac{8575 \text{ k} \cdot \text{ft}^2}{EI}$$

Released structure with unit redundant:

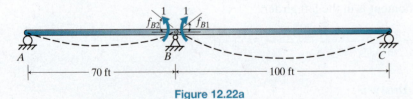

Figure 12.22a

From the inside front cover, we have

$$f_{B1} = -\frac{M_o l}{3EI} = -\frac{(-1)(70 \text{ ft})}{3EI} = \frac{23.33 \text{ ft}}{EI} \quad (\measuredangle +)$$

$$f_{B2} = -\frac{M_o l}{3EI} = -\frac{(-1)(100 \text{ ft})}{3EI} = \frac{33.33 \text{ ft}}{EI} \quad (+\measuredangle)$$

Add to find the angle between the segments:

$$f_B = \frac{23.33 \text{ ft}}{EI} + \frac{33.33 \text{ ft}}{EI} = \frac{56.67 \text{ ft}}{EI}$$

Compatibility:
We have one equation, since there is only one redundant:

$$0 = \theta_B + M_B \cdot f_B$$

$$0 = \frac{8575 \text{ k} \cdot \text{ft}^2}{EI} + M_B \frac{56.67 \text{ ft}}{EI}$$

$$M_B = -151.3 \text{ k} \cdot \text{ft} \quad (\circlearrowright + \circlearrowleft)$$

Find the reactions:
FBD of AB:

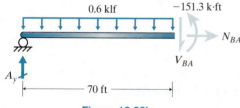

Figure 12.22b

$$\curvearrowleft \Sigma M_B = 0 = A_y(70 \text{ ft})$$
$$- 0.6 \text{ klf}(70 \text{ ft})(70 \text{ ft}/2)$$
$$- (-151.3 \text{ k} \cdot \text{ft})$$
$$A_y = 18.8 \text{ k} \quad (+\uparrow)$$
$$+\uparrow \Sigma F_y = 0 = 18.8 \text{ k} - 0.6 \text{ klf}(70 \text{ ft}) - V_{BA}$$
$$V_{BA} = -23.2 \text{ k} \quad (\downarrow + \uparrow)$$
$$\xrightarrow{+} \Sigma F_x = 0 = N_{BA}$$
$$N_{BA} = 0$$

FBD of BC:

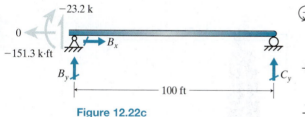

Figure 12.22c

$$\curvearrowleft \Sigma M_B = 0 = -(-151.3 \text{ k} \cdot \text{ft})$$
$$+ C_y(100 \text{ ft})$$
$$C_y = -1.5 \text{ k} \quad (+\uparrow)$$
$$+\uparrow \Sigma F_y = 0 = -23.2 \text{ k} + B_y - 1.5 \text{ k}$$
$$B_y = 24.7 \text{ k} \quad (+\uparrow)$$
$$\xrightarrow{+} \Sigma F_x = 0 = 0 + B_x$$
$$B_x = 0$$

Find the peak moment:
Expressions for the shear:

$$V(x_1) = -\int w \, dx_1 = -0.6 \text{ klf}(x_1) + C_1$$

At $x_1 = 0$, $V = 18.8$ k:

$$\therefore C_1 = 18.8 \text{ k}$$

So,

$$V(x_1) = -0.6 \text{ klf}(x_1) + 18.8 \text{ k}$$

$$V(x_2) = -\int w \, dx_2 = C_2$$

At $x_2 = 0$, $V = V(x_1 = 70 \text{ ft}) + 24.7$ k:

$$\therefore C_2 = -0.6 \text{ klf}(70 \text{ k}) + 18.8 \text{ k} + 24.7 \text{ k} = 1.5 \text{ k}$$

So,

$$V(x_2) = 1.5 \text{ k}$$

Find the locations of the peak moment:
For the range of x_1,
$$-0.6 \text{ klf}(x_1) + 18.8 \text{ k} \equiv 0$$
$$x_1 = 31.3 \text{ ft}$$

For the range of x_2, the shear is never zero, so there is no peak moment.

Expression for the moment:
$$M(x_1) = \int V(x_1) dx_1 = -0.3 \text{ klf}(x_1^2) + 18.8 \text{ k}(x_1) + C_3$$
At $x_1 = 0, M = 0$:
$$\therefore C_3 = 0$$

So,
$$M(x_1) = -0.3 \text{ klf}(x_1^2) + 18.8 \text{ k}(x_1)$$

Peak moments:
Peak positive moment at $x_1 = 31.3$ ft:
$$M(x_1) = -0.3 \text{ klf}(31.3 \text{ ft})^2 + 18.8 \text{ k}(31.3 \text{ ft}) = 295 \text{ k} \cdot \text{ft}$$

Peak negative moment at $x_1 = 70$ ft, if negative at all:
$$M(x_1) = -0.3 \text{ klf}(70 \text{ ft})^2 + 18.8 \text{ k}(70 \text{ ft}) = -154 \text{ k} \cdot \text{ft}$$

Summary:

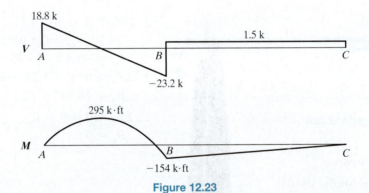

Figure 12.23

Evaluation of Results:

Approximation Predicted Outcomes?
The peak moment is smaller than the result from the approximate analysis, as expected. ✓
The approximate analysis results are 24% larger than the detailed analysis results for the peak positive moment.

Conclusion:
These observations suggest we have a reasonable result.

Note that the computer aided analysis result for the peak moment is 296 k·ft, so the Force Method result essentially matches. The difference is due to roundoff error in our integration. ✓

12.2 Multi-Degree Indeterminate Beams

Introduction

The Force Method is also valid for multi-degree indeterminate structures. Because we have precalculated deformation formulas only for beams, the method is most efficient for beams. We can use the Virtual Work Method (Section 5.3) to predict deformations for frames, but the Moment Distribution Method (Chapter 13) will typically be more efficient for detailed analysis of frames.

The Force Method requires a compatibility equation for each degree of indeterminacy. Therefore, a multi-degree indeterminate beam requires the setup and solution of a linear system of simultaneous equations. For beams that are beyond second degree indeterminate, the Moment Distribution Method is typically faster.

How-To

When we have structures that are more than one degree indeterminate, the Force Method process is essentially the same as we covered in Section 12.1. There are some important adaptations though.

Redundants and Compatibility

The Force Method uses equations of compatibility to supplement the equations of equilibrium. Therefore, we need to have the same number of compatibility equations as the degree of indeterminacy. Since each compatibility equation is based on a redundant, we need to select the same number of redundants as the degree of indeterminacy.

For example, the beam in Figure 12.24 is two degrees indeterminate, so we need to choose two redundants. We can choose any two as long as

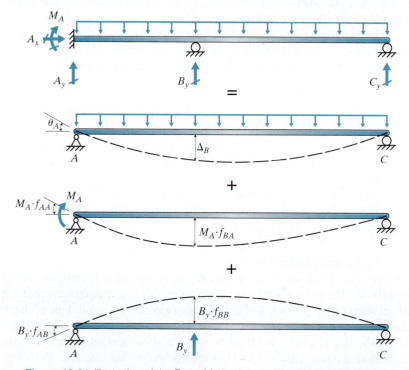

Figure 12.24 Illustration of the Force Method process with two redundants.

the released structure is stable. In the case of Figure 12.24, we chose M_A and B_y. The resulting compatibility equations are for the slope at A and the vertical displacement at B:

$$0 = \theta_A + M_A \cdot f_{AA} + B_y \cdot f_{AB}$$
$$0 = \Delta_B + M_A \cdot f_{BA} + B_y \cdot f_{BB}$$

Notice that to have compatibility, we must consider the deformation caused by each redundant at each redundant location. So, for the beam in Figure 12.24, we need to calculate two deformations for the released structure due to each load, and we must consider each redundant as a separate load.

One of the implications of this requirement is that as the degree of indeterminacy increases, the number of deformations we need to predict increases exponentially. For the 1° indeterminate beam in Figure 12.5, we needed to calculate only two deformations: Δ_{By} and f_{By}. For the 2° indeterminate beam in Figure 12.24, we needed to calculate six deformations: $\theta_A, \Delta_B, f_{AA}, f_{BA}, f_{AB}$, and f_{BB}. For a 3° indeterminate structure, we would need to calculate 12 deformations. That is why the Force Method is typically used only for structures up to 2° indeterminate.

Maxwell-Betti's Reciprocal Theorem

In the late 1800s, Enrico Betti discovered that the work done by a set of forces P through the deformations caused by a different set of forces Q must be the same as the work done by the set of forces Q through the deformations caused by set P. We demonstrated that in our derivation of the Virtual Work Method (Section 5.3). So what is the relevance to the Force Method? Maxwell showed that Betti's theorem has valuable implications for flexibility coefficients.

The Maxwell-Betti Reciprocal Theorem states that the deformation at redundant i due to the application of a redundant at j is the same as the deformation at redundant j due to the application of a redundant at i:

$$f_{ij} = f_{ji} \quad \text{for} \quad i \neq j$$

The benefit to the Force Method is that we don't need to calculate all of the flexibility coefficients for multi-degree indeterminate structures. For example, if we take advantage of Maxwell-Betti's Reciprocal Theorem, we only need to calculate a total of five deformations for a 2° indeterminate structure. For the beam in Figure 12.24, $f_{AB} = f_{BA}$ even though one is a slope and one is a displacement. Note, however, that the theorem does not apply to the diagonal terms of the flexibility coefficient matrix. So, for the beam in Figure 12.24, $f_{AA} \neq f_{BB}$.

Solving for the Redundants

To find the unknown redundants, we need to solve the compatibility equations. They make a system of simultaneous linear equations that are often susceptible to roundoff errors. We typically lose one digit of accuracy in the solution process for these equations. Therefore, when calculating the deformations for the Force Method, we should carry one more digit than we normally carry. That means we should calculate deformations with four or five digits of precision.

Section 12.2 Highlights

Multi-Degree Indeterminate Beams

Force Method Steps:

1. Calculate the degree of indeterminacy; the degree is the number of redundants.
2. Choose the redundants and identify the associated deformation for each.
   - Cannot choose redundants that leave the released structure unstable.
   - Choose redundants that make the deformation prediction as easy as possible.
   - Can be a combination of force reactions, moment reactions, and/or internal moments.

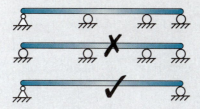

   If the redundant is a force, measure the displacement.
   If the redundant is a moment, measure the slope.
3. Release the structure from all the redundants.
4. Predict the relevant deformations at all of the redundants for the released structure due to the applied loads.

5. Predict the relevant deformations at all of the redundants for the released structure due to a single unit redundant (flexibility coefficients). Repeat the process with a single unit load for each redundant, one at a time.

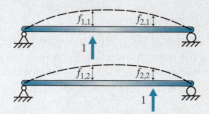

6. Write the equations of compatibility, one for each redundant.
7. Solve the system of compatibility equations for the unknown redundants.
8. Use equations of equilibrium to find all other reactions for the original indeterminate structure; then calculate any desired quantity (e.g., internal forces, deformations).

Maxwell-Betti's Reciprocal Theorem:

Certain flexibility coefficients are equal:

$$f_{ij} = f_{ji} \quad \text{for} \quad i \neq j$$

Assumptions:

Same as for predicting deformations.

- Plane sections remain plane, or the shear deformation is negligible.
- Deformations or slopes are relatively small.
- Material remains linear elastic.

EXAMPLE 12.4

A large piece of equipment is being delivered via special transport to its final destination. Our firm has been hired to evaluate the adequacy of the bridges the equipment must cross. We have been assigned to evaluate a three-span overpass. One of the concerns is overloading the splice that holds pieces of the girder together. A splice transfers axial force, shear, and moment. Using the concept of influence lines, we have determined where the transport must be in order to cause the largest shear on the splice: just to the right of the splice. Now we need to find the actual shear generated by the transport and equipment.

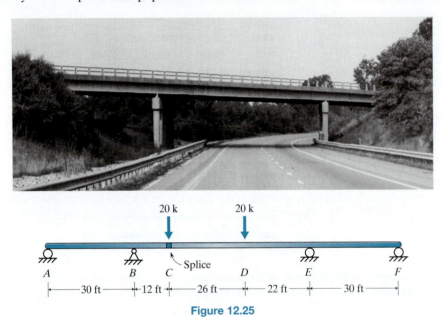

Figure 12.25

The bridge deck will distribute the point forces out to the various bridge girders across the width. A common method for determining the effect on each girder is the Distribution Factor Method. Another team member will perform those calculations, but first we need to determine the net shear at the splice due to the equipment and transport.

Estimated Solution:

Consider the middle segment to be simply supported. This should provide an upper bound on the shear, since continuity tends to reduce the peak internal forces.

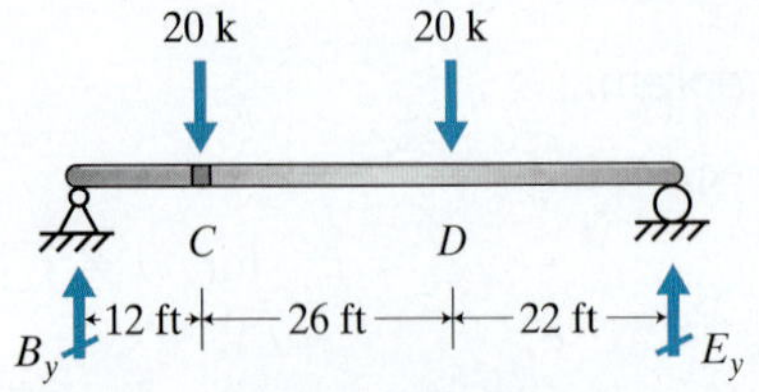

Figure 12.26a

$$+\circlearrowleft \Sigma M_E = 0 = B_y(60 \text{ ft}) - 20 \text{ k}(48 \text{ ft}) - 20 \text{ k}(22 \text{ ft})$$
$$B_y = 23.3 \text{ k} \quad (+\uparrow)$$
$$+\uparrow \Sigma F_y = 0 = 23.3 \text{ k} - 20 \text{ k} - 20 \text{ k} + E_y$$
$$E_y = 16.7 \text{ k} \quad (+\uparrow)$$

Section 12.2 Multi-Degree Indeterminate Beams

FBD of BC:

Since the vehicle is just to the right of C, it does not show up on this FBD.

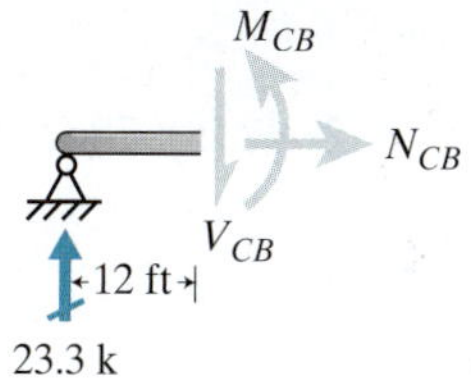

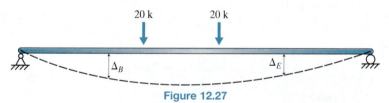

$$+\uparrow \Sigma F_y = 0 = 23.3 \text{ k} - V_{CB}$$
$$V_{CB} = 23.3 \text{ k} \quad (\downarrow + \uparrow)$$

The peak approximate shear is 23.3 k.

Figure 12.26b

Detailed Analysis:

Redundants:

Degree of indeterminacy

5 unknowns
$\underline{-3}$ equations
2° indeterminate

Let's choose B_y and E_y as redundants because we can use fomulas from the inside front cover of this text to predict the deformations. Notice that since there is no axial force in the beam, we can show removing B_y as moving the pinned support to one of the other ends.

Released structure with applied loads:

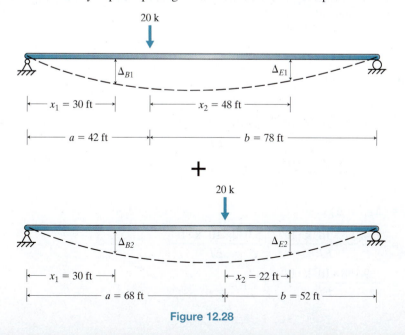

Figure 12.27

This loading is not covered directly by one of the situations inside the front cover. However, we can create it by superimposing the effects of each of the point loads:

Figure 12.28

Chapter 12 Force Method

From inside the front cover, we have

$$\Delta_{B1} = -\frac{Pb}{6EIl}[-x_1^3 + (a^2 + 2ab)x_1]$$

$$= -\frac{20\,\text{k}(78\,\text{ft})}{6EI(120\,\text{ft})}\left[-(30\,\text{ft})^3 + [(42\,\text{ft})^2 + 2(42\,\text{ft})(78\,\text{ft})](30\,\text{ft})\right]$$

$$= -\frac{4.820 \times 10^5\,\text{k}\cdot\text{ft}^3}{EI}$$

$$\Delta_{E1} = -\frac{Pa}{6EIl}\left[x_2^3 - 3bx_2^2 + 2(b^2 - ab)x_2 + 2ab^2\right]$$

$$= -\frac{20\,\text{k}(42\,\text{ft})}{6EI(120\,\text{ft})}\Big[(48\,\text{ft})^3 - 3(78\,\text{ft})(48\,\text{ft})^2$$
$$+ 2[(78\,\text{ft})^2 - (42\,\text{ft})(78\,\text{ft})](48\,\text{ft}) + 2(42\,\text{ft})(78\,\text{ft})^2\Big]$$

$$= -\frac{4.108 \times 10^5\,\text{k}\cdot\text{ft}^3}{EI}$$

$$\Delta_{B2} = -\frac{Pb}{6EIl}[-x_1^3 + (a^2 + 2ab)x_1]$$

$$= -\frac{20\,\text{k}(52\,\text{ft})}{6EI(120\,\text{ft})}\left[-(30\,\text{ft})^3 + [(68\,\text{ft})^2 + 2(68\,\text{ft})(52\,\text{ft})](30\,\text{ft})\right]$$

$$= -\frac{4.678 \times 10^5\,\text{k}\cdot\text{ft}^3}{EI}$$

$$\Delta_{E2} = -\frac{Pa}{6EIl}\left[x_2^3 - 3bx_2^2 + 2(b^2 - ab)x_2 + 2ab^2\right]$$

$$= -\frac{20\,\text{k}(68\,\text{ft})}{6EI(120\,\text{ft})}\Big[(22\,\text{ft})^3 - 3(52\,\text{ft})(22\,\text{ft})^2$$
$$+ 2[(52\,\text{ft})^2 - (68\,\text{ft})(52\,\text{ft})](22\,\text{ft}) + 2(68\,\text{ft})(52\,\text{ft})^2\Big]$$

$$= -\frac{5.030 \times 10^5\,\text{k}\cdot\text{ft}^3}{EI}$$

Superimpose the results:

$$\Delta_B = \Delta_{B1} + \Delta_{B2}$$

$$= -\frac{4.820 \times 10^5\,\text{k}\cdot\text{ft}^3}{EI} - \frac{4.678 \times 10^5\,\text{k}\cdot\text{ft}^3}{EI}$$

$$= -\frac{9.498 \times 10^5\,\text{k}\cdot\text{ft}^3}{EI} \quad (+\uparrow)$$

$$\Delta_E = \Delta_{E1} + \Delta_{E2}$$

$$= -\frac{4.108 \times 10^5\,\text{k}\cdot\text{ft}^3}{EI} - \frac{5.030 \times 10^5\,\text{k}\cdot\text{ft}^3}{EI}$$

$$= -\frac{9.138 \times 10^5\,\text{k}\cdot\text{ft}^3}{EI} \quad (+\uparrow)$$

Released structure with unit redundant at B:

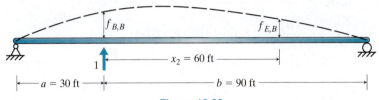

Figure 12.29

From the inside front cover, we have

$$f_{B,B} = -\frac{Pa}{6EIl}\left[x_2^3 - 3bx_2^2 + 2(b^2 - ab)x_2 + 2ab^2\right]$$

$$= -\frac{(-1)(30\text{ ft})}{6EI(120\text{ ft})}\Big[(0\text{ ft})^3 - 3(90\text{ ft})(0\text{ ft})^2$$
$$+ 2\left[(90\text{ ft})^2 - (30\text{ ft})(90\text{ ft})\right](0\text{ ft}) + 2(30\text{ ft})(90\text{ ft})^2\Big]$$

$$= \frac{2.025\times 10^4\text{ ft}^3}{EI}\quad (+\uparrow)$$

$$f_{E,B} = -\frac{Pa}{6EIl}\left[x_2^3 - 3bx_2^2 + 2(b^2 - ab)x_2 + 2ab^2\right]$$

$$= -\frac{(-1)(30\text{ ft})}{6EI(120\text{ ft})}\Big[(60\text{ ft})^3 - 3(90\text{ ft})(60\text{ ft})^2$$
$$+ 2\left[(90\text{ ft})^2 - (30\text{ ft})(90\text{ ft})\right](60\text{ ft}) + 2(30\text{ ft})(90\text{ ft})^2\Big]$$

$$= \frac{1.575\times 10^4\text{ ft}^3}{EI}\quad (+\uparrow)$$

Released structure with unit redundant at E:

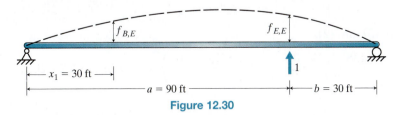

Figure 12.30

From the inside front cover, we have

$$f_{B,E} = -\frac{Pb}{6EIl}\left[-x_1^3 + (a^2 + 2ab)x_1\right]$$

$$= -\frac{(-1)(30\text{ ft})}{6EI(120\text{ ft})}\Big[-(30\text{ ft})^3 + \left[(90\text{ ft})^2 + 2(90\text{ ft})(30\text{ ft})\right](30\text{ ft})\Big]$$

$$= \frac{1.575\times 10^4\text{ ft}^3}{EI}\quad (+\uparrow)$$

Evaluation of Results:

Satisfied Fundamental Principles?

Maxwell-Betti's Reciprocal Theorem says $f_{E,B}$ should equal $f_{B,E}$; they are equal. ✓

$$f_{E,E} = -\frac{Pa}{6EIl}\left[x_2^3 - 3bx_2^2 + 2(b^2 - ab)x_2 + 2ab^2\right]$$

$$= -\frac{(-1)(90\,\text{ft})}{6EI(120\,\text{ft})}\Big[(0\,\text{ft})^3 - 3(30\,\text{ft})(0\,\text{ft})^2$$
$$+ 2\big[(30\,\text{ft})^2 - (90\,\text{ft})(30\,\text{ft})\big](0\,\text{ft}) + 2(90\,\text{ft})(30\,\text{ft})^2\Big]$$

$$= \frac{2.025 \times 10^4\,\text{ft}^3}{EI} \quad (+\uparrow)$$

In general, we should expect $f_{B,B} \neq f_{E,E}$, but in this case they are equal due to the symmetry of the two released structures with unit loads.

Compatibility:

We have two equations, since there are two redundants:

$$0 = \Delta_B + B_y \cdot f_{B,B} + E_y \cdot f_{B,E}$$

$$0 = \Delta_E + B_y \cdot f_{E,B} + E_y \cdot f_{E,E}$$

$$0 = -\frac{9.498 \times 10^5\,\text{k}\cdot\text{ft}^3}{EI} + B_y\left(\frac{2.025 \times 10^4\,\text{ft}^3}{EI}\right) + E_y\left(\frac{1.575 \times 10^4\,\text{ft}^3}{EI}\right) \quad (1)$$

$$0 = -\frac{9.138 \times 10^5\,\text{k}\cdot\text{ft}^3}{EI} + B_y\left(\frac{1.575 \times 10^4\,\text{ft}^3}{EI}\right) + E_y\left(\frac{2.025 \times 10^4\,\text{ft}^3}{EI}\right) \quad (2)$$

Solve (1) for E_y:

$$E_y\left(\frac{1.575 \times 10^4\,\text{ft}^3}{EI}\right) = \frac{9.498 \times 10^5\,\text{k}\cdot\text{ft}^3}{EI} - B_y\left(\frac{2.025 \times 10^4\,\text{ft}^3}{EI}\right)$$

$$E_y = 60.30\,\text{k} - 1.286 B_y \quad (3)$$

Substitute (3) into (2):

$$0 = -\frac{9.138 \times 10^5\,\text{k}\cdot\text{ft}^3}{EI} + B_y\left(\frac{1.575 \times 10^4\,\text{ft}^3}{EI}\right) + (60.30\,\text{k} - 1.286 B_y)\left(\frac{2.025 \times 10^4\,\text{ft}^3}{EI}\right)$$

$$0 = -\frac{9.138 \times 10^5\,\text{k}\cdot\text{ft}^3}{EI} + B_y\left(\frac{1.575 \times 10^4\,\text{ft}^3}{EI}\right) + \frac{12.211 \times 10^5\,\text{k}\cdot\text{ft}^3}{EI}$$

$$- B_y\left(\frac{2.604 \times 10^4\,\text{ft}^3}{EI}\right)$$

$$B_y\left(\frac{1.029 \times 10^4\,\text{ft}^3}{EI}\right) = \frac{3.073 \times 10^5\,\text{k}\cdot\text{ft}^3}{EI}$$

$$B_y = 29.9\,\text{k} \quad (+\uparrow)$$

Substitute the result back into (3):

$$E_y = 60.30\,\text{k} - 1.286(29.9\,\text{k}) = 21.8\,\text{k} \quad (+\uparrow)$$

Find the shear at the splice:

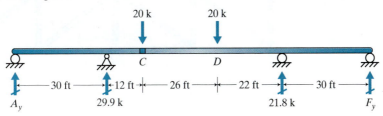

Figure 12.31

Apply equilibrium to find the left reaction:

$$\circlearrowright \Sigma M_F = 0 = A_y(120 \text{ ft}) + 29.9 \text{ k}(90 \text{ ft}) - 20 \text{ k}(78 \text{ ft}) - 20 \text{ k}(52 \text{ ft}) + 21.8 \text{ k}(30 \text{ ft})$$
$$A_y = -6.2 \text{ k} \quad (+\uparrow)$$

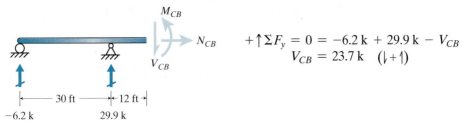

$+\uparrow \Sigma F_y = 0 = -6.2 \text{ k} + 29.9 \text{ k} - V_{CB}$
$V_{CB} = 23.7 \text{ k} \quad (\downarrow + \uparrow)$

Figure 12.32

Evaluation of Results:

Approximation Predicted Outcomes?
The detailed analysis prediction of the shear at C is only 2% larger than the result from our approximate analysis.

Conclusion:
This suggests we have a reasonable result.

Note that our detailed analysis result matches the computer aided analysis result to three significant figures, as expected. ✓

12.3 Indeterminate Trusses

Introduction
The Force Method applies well to indeterminate trusses that are only one or two degrees indeterminate. We have to use the Virtual Work Method (Section 5.3) to calculate each of the deformations. Therefore, the Force Method can become laborious when we are analyzing trusses that are more than two degrees indeterminate.

How-To
Before applying the Force Method to analyze an indeterminate truss, we need to know both the degree of indeterminacy and the cause. To find the degree of indeterminacy, we use the concepts from the Method

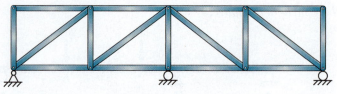

Figure 12.33 Truss made indeterminate by extra support reactions.

of Joints. There is one unknown internal force, the axial force, for each member plus the unknown support reactions:

r = Unknown reactions

n = Unknown axial forces

There are two equations of equilibrium for each joint if we are analyzing a two-dimensional truss:

j = Number of joints

The degree of indeterminacy is the difference between the number of unknowns and the number of equations of equilibrium:

Degree = $r + n - 2j$

When we analyze an indeterminate truss, the cause of indeterminacy affects our approach. There are two possibilities. First, trusses can be indeterminate due to extra supports. For example, the truss in Figure 12.33 is 1° indeterminate: $4 + 17 - 2(10) = 1$. It has four unknown reactions, and the truss as a free body has three equations of equilibrium. Since the difference between the number of reactions and equations of equilibrium matches the degree of indeterminacy, all of the redundants are reactions. To solve this type of problem, we use the same Force Method approach as described for beams.

The second cause of indeterminacy for trusses is extra members. The truss in Figure 12.34 is 4° indeterminate: $3 + 21 - 2(10) = 4$. It has only three support reactions, so those do not contribute to the degree of indeterminacy.

If a truss is indeterminate due to extra members, we must choose the axial forces in members to be the redundants. Releasing the axial force in a member is the same as cutting the member.

We still need to choose the same number of redundants as the degree of indeterminacy, and our rule still applies: the released truss cannot be unstable. For example, the truss in Figure 12.35a is 1° indeterminate due to an extra member. Cutting member AE would leave the truss unstable (Figure 12.35b), but we can cut member BF and still have a stable truss (Figure 12.35c).

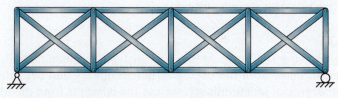

Figure 12.34 Truss made indeterminate by extra members.

Section 12.3 Indeterminate Trusses   **701**

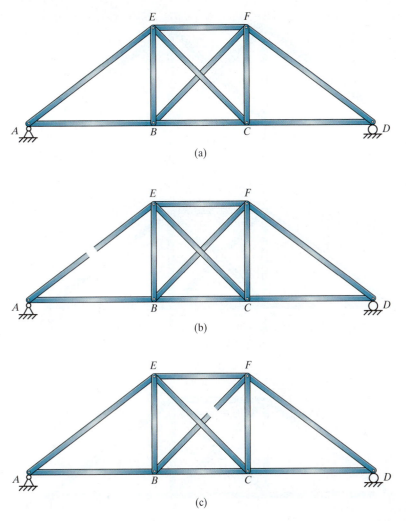

Figure 12.35 Example of a truss made indeterminate by an extra member: (a) Original truss; (b) Invalid choice of redundant because it makes the truss unstable; (c) Valid choice of redundant.

Once we choose the redundants, the next step in the Force Method process is to predict the relevant deformations of the released structure. For a truss that has extra members, the deformations are gaps or overlaps at the cuts. To find those deformations, we need to use the Virtual Work Method from Section 5.3.

Predicting Deformations for 1° Indeterminate Trusses

To predict the gap or overlap at the cut member, we must use the Virtual Work Method (Section 5.3). That means we must analyze the truss twice for each displacement: once with the real loads applied and once with the virtual force applied.

As an example, for the truss in Figure 12.36a, we can choose member BF as the redundant. One of the displacements we need to measure is the overlap in BF caused by the applied loads, Δ_{BF} (Figure 12.36b). The other is the overlap in BF caused by the unit redundant, f_{BF} (Figure 12.36c).

Chapter 12 Force Method

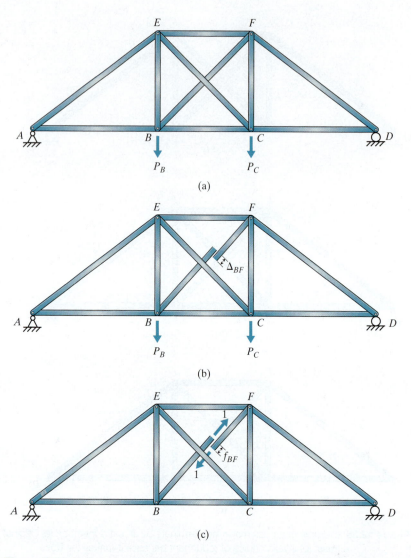

Figure 12.36 Deformations needed in order to analyze a 1° indeterminate truss: (a) Original truss; (b) Released truss with applied loads; (c) Released truss with unit redundant.

To calculate Δ_{BF}, we analyze the released truss with the real loads (Figure 12.37a). Remember that the cut member, BF, does not carry any axial force. Our second analysis is of the released structure with the virtual force. Since we want to measure the overlap that member BF experiences, we apply a unit tension on both ends of the cut member (Figure 12.37b). When we choose tension, a positive result means overlap.

We also need to find the flexibility coefficient, f_{BF}. That is the overlap of the cut member due to a unit redundant, which is the tension in the cut member. For our example truss, the real system is shown in Figure 12.38a and the virtual system in Figure 12.38b.

Figure 12.38 is not a misprint. For a 1° indeterminate truss, the real and virtual systems for predicting the flexibility coefficient will always be the same. In fact, we already performed this analysis when we analyzed the virtual system for Δ_{BF} (Figure 12.37b). That means that to calculate Δ and f for a 1° indeterminate truss, we actually only need to analyze the released truss twice.

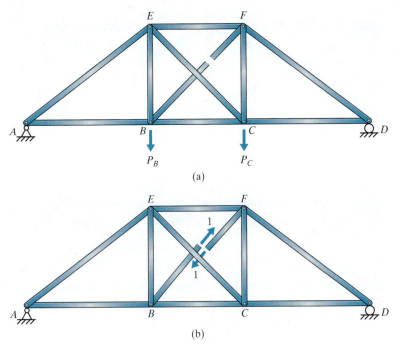

Figure 12.37 Analyses needed to use the Virtual Work Method to predict Δ_{BF}: (a) Released truss with applied loads; (b) Released truss with virtual force.

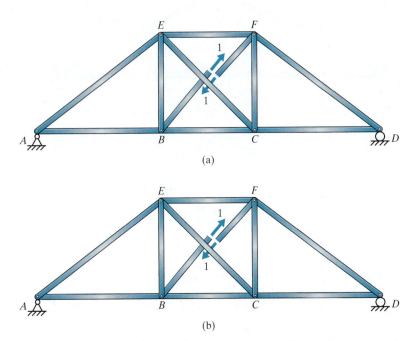

Figure 12.38 Analyses needed to use the Virtual Work Method to predict f_{BF}: (a) Released truss with applied redundant; (b) Released truss with virtual force.

Predicting Deformations for a Multi-Degree Indeterminate Truss

We can use the Force Method to analyze multi-degree indeterminate trusses as well. Beyond 2° indeterminate, though, the method can become very time consuming.

704　Chapter 12　Force Method

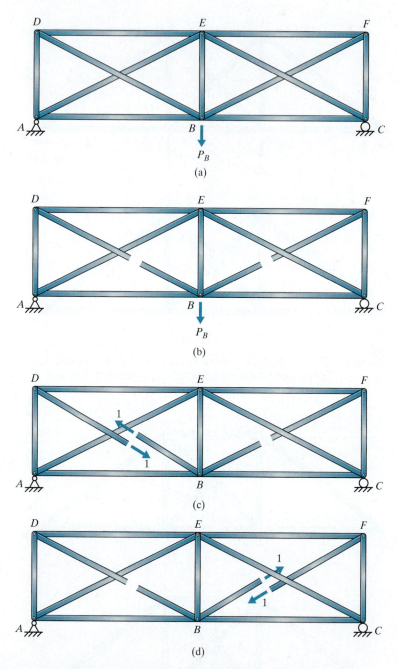

Figure 12.39 Analyses needed for 2° indeterminate truss: (a) Original truss; (b) Released truss with applied load; (c) Released truss with one unit redundant; (d) Released truss with second unit redundant.

The process for multi-degree indeterminate trusses is the same as for beams. The challenge is the bookkeeping needed to calculate the necessary deformations. Consider, for example, the 2° indeterminate truss in Figure 12.39a. We will need to choose two redundants (e.g., N_{BD} and N_{BF}) and predict six displacements (e.g., Δ_{BD}, Δ_{BF}, $f_{BD,BD}$, $f_{BD,BF}$, $f_{BF,BD}$, and $f_{BF,BF}$). Fortunately, to calculate these six displacements using the Virtual Work Method, we only need to analyze the released truss

Table 12.1 Combinations of Truss Analysis Results Used to Calculate the Various Displacements for a 2° Indeterminate Truss

Quantity	Real System	Virtual System
Δ_{BD}	12.39(b)	12.39(c)
Δ_{BF}	12.39(b)	12.39(d)
$f_{BD,BD}$	12.39(c)	12.39(c)
$f_{BD,BF}$	12.39(d)	12.39(c)
$f_{BF,BD}$	12.39(c)	12.39(d)
$f_{BF,BF}$	12.39(d)	12.39(d)

three times (Figures 12.39b–d). Then we can reuse those three results multiple times to calculate the six displacements. The combinations are shown in Table 12.1.

Maxwell-Betti's Theorem still applies, so we actually need to calculate only five unique displacements, but we still need all three truss analyses. The time required to calculate the sixth displacement given the truss analysis results is relatively short. Therefore, it makes an inexpensive check for errors.

SECTION 12.3 HIGHLIGHTS

Indeterminate Trusses

Redundants:

- If the truss has extra supports, choose support redundants.
- If the truss has extra members, choose member axial forces as redundants.
- For both cases, use the Virtual Work Method to predict the displacements.

Support Redundants:

Follow the same steps as for indeterminate beams.

Member Redundants:

Follow the same steps as for indeterminate beams with the following clarifications:
- To release the truss from a member axial force, it is as if we cut the member.
- Tip: Choose to measure overlap rather than the gap created at the cut in the member; therefore, a positive result means tension in the member.

Assumptions:

Same as for predicting deformations.
- Deformations are relatively small.
- Material remains linear elastic.

EXAMPLE 12.5

We work for a steel fabricator as a connection designer. Our company just got a job from a local gym to design and fabricate a pull-up bar that can be mounted on the wall.

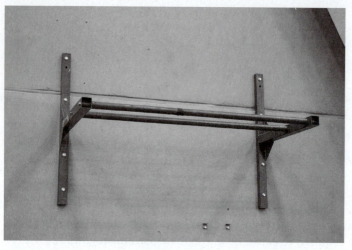

Figure 12.40

The gym has tried a rigid frame version but wants something stiffer, like a truss. Because of constraints on where the bar can be put, the members cannot be symmetric. We have been tasked with analyzing the truss to find the design force in each member.

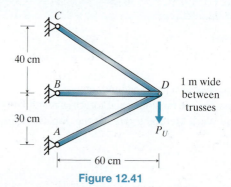

Figure 12.41

We start by researching the unfactored loads.

- Initially let's assume the self-weight of the members is 0.4 N/cm.
- We should design for a 1.33-kN person hanging from the apparatus.
- A person jumping onto the apparatus will have a larger effect; the most reasonable impact factor is 1.5.

Design Loading:

Dead load:

Find the length of each member:

$$l_{CD} = \sqrt{(60 \text{ cm})^2 + (40 \text{ cm})^2} = 72.1 \text{ cm}$$
$$l_{AD} = \sqrt{(60 \text{ cm})^2 + (30 \text{ cm})^2} = 67.1 \text{ cm}$$

Total length:
$$l_{Total} = 2\,(72.1\text{ cm} + 60\text{ cm} + 67.1\text{ cm}) + 100\text{ cm} = 498\text{ cm}$$

Conservatively consider all of the weight acting at point D. Half of the weight is carried by each truss.

$$P_D = \frac{1}{2}(0.4\text{ N/cm})(498\text{ cm}) = 99.6\text{ N} = 0.0996\text{ kN}$$

Live load:

Live load is multiplied by the impact factor. Half the load is carried by each truss.

$$P_L = \frac{1}{2}(1.5)(1.33\text{ kN}) = 1.00\text{ kN}$$

Design load:

Load combination 2 from Section 1.2: Load Combinations.

$$P_U = 1.2\,(0.0996\text{ kN}) + 1.6\,(1.00\text{ kN}) = 1.720\text{ kN}$$

Estimated Solution:

Assume the vertical force on the truss is shared equally as the vertical components of the axial forces in the diagonal members, AD and CD.

FBD of D:

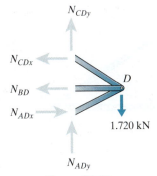

Figure 12.42a

$$N_{ADy} = N_{CDy} = \frac{1}{2}(1.720\text{ kN}) = 0.860\text{ kN}$$

Convert the vertical components to the axial forces:

$$N_{AD} = \frac{67.1}{30}(0.860\text{ kN}) = 1.924\text{ kN }(C)$$

$$N_{CD} = \frac{72.1}{40}(0.860\text{ kN}) = 1.550\text{ kN }(T)$$

Find the third axial force:

$$\xrightarrow{+}\Sigma F_x = 0 = -\frac{60}{30}(0.860\text{ kN}) + N_{BD} + \frac{60}{40}(0.860\text{ kN})$$

$$N_{BD} = 0.430\text{ kN }(T)$$

Summary:

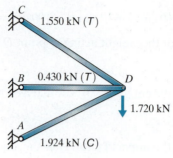

Figure 12.42b

Detailed Analysis:

Redundants:

Degree of indeterminacy:

6 Unknown Reactions	4 Joints
+3 Unknown Member Forces	×2 Equations of Equilibrium per Joint
9 Total Unknowns	8 Total Equations = 1 Degree Indeterminate

The truss has an extra member. For this truss, we can choose any member as the redundant and still have a stable truss. Let's choose the axial force in member AD as the redundant.

Released truss with applied load:

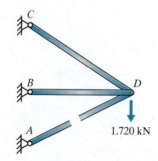

Figure 12.43a

FBD of D:

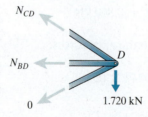

Figure 12.43b

$$+\uparrow \Sigma F_y = 0 = \frac{40}{72.1} N_{CD} - 1.720 \text{ kN}$$

$$N_{CD} = 3.10 \text{ kN}$$

$$\xleftarrow{+} \Sigma F_x = 0 = \frac{60}{72.1}(3.10 \text{ kN}) + N_{BD}$$

$$N_{BD} = -2.58 \text{ kN}$$

Section 12.3 Indeterminate Trusses

Released truss with unit redundant:

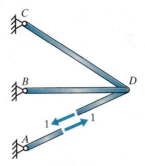

Figure 12.44a

FBD of D:

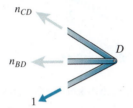

Figure 12.44b

$$+\uparrow \Sigma F_y = 0 = \frac{40}{72.1} n_{CD} - \frac{30}{67.1}(1)$$

$$n_{CD} = 0.806$$

$$\xleftarrow{+} \Sigma F_x = 0 = \frac{60}{72.1}(0.806) + n_{BD} + \frac{60}{67.1}(1)$$

$$n_{BD} = -1.565$$

Predict the displacements:

Find Δ_{AD}:

$$\Delta_{AD} = \sum_{j=1}^{3} \frac{N_j n_j l_j}{AE}$$

$$= \sum \frac{(0)(1)(67.1 \text{ cm})}{AE} + \frac{(-2.58 \text{ kN})(-1.565)(60 \text{ cm})}{AE}$$

$$+ \frac{(3.10 \text{ kN})(0.806)(72.1 \text{ cm})}{AE}$$

$$= \frac{422.4 \text{ kN} \cdot \text{cm}}{AE}$$

Find f_{AD}:

$$f_{AD} = \sum_{j=1}^{3} \frac{n_j^2 l_j}{AE}$$

$$= \sum \frac{(1)^2(67.1 \text{ cm})}{AE} + \frac{(-1.565)^2(60 \text{ cm})}{AE} + \frac{(0.806)^2(72.1 \text{ cm})}{AE}$$

$$= \frac{260.9 \text{ cm}}{AE}$$

710 Chapter 12 Force Method

Compatibility:

$$0 = \Delta_{AD} + N_{AD} \cdot f_{AD}$$

$$0 = \frac{422.4 \text{ kN} \cdot \text{cm}}{AE} + N_{AD}\frac{260.9 \text{ cm}}{AE}$$

$$N_{AD} = -1.619 \text{ kN} = 1.619 \text{ kN } (C)$$

Other member forces:
  FBD of D:

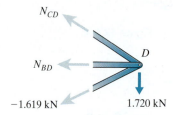

Figure 12.45

$$+\uparrow \Sigma F_y = 0 = \frac{40}{72.1}N_{CD} - \frac{30}{67.1}(-1.619 \text{ kN}) - 1.720 \text{ kN}$$

$$N_{CD} = 1.796 \text{ kN } (T)$$

$$\pm \Sigma F_x = 0 = \frac{60}{72.1}(1.796 \text{ kN}) + N_{BD} + \frac{60}{67.1}(-1.619 \text{ kN})$$

$$N_{BD} = -0.047 \text{ kN} = 0.047 \text{ kN } (C)$$

Summary:

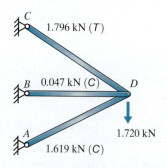

Figure 12.46

Evaluation of Results:

Approximation Predicted Outcomes?
  The detailed analysis predictions of the axial forces in AD and CD match the approximate analysis results to within 20%. The results for BD reverse the sign, but they are smaller numbers.

Conclusion:
This suggests we have a reasonable result.

Note that our detailed analysis results match the computer aided analysis results to three significant figures, as expected. ✓

Homework Problems

12.1 Our team is tasked to design the supports for a beam for large air conditioner units. The beam will carry only one unit initially, but the facility may expand in the future. According to the manufacturer's literature, the air conditioner unit will apply a uniformly distributed load on the beam.

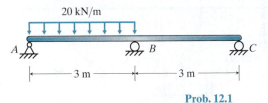

20 kN/m

W250×32.7
$A = 4190$ mm^2
$I = 4.91 \times 10^7$ mm^4
$E = 200$ GPa $= 200$ kN/mm^2

3 m — 3 m

Prob. 12.1

a. Guess the displaced shape of the beam under this loading.
b. Using approximate analysis, find the reactions. You can do so by assuming the location of the inflection point.
c. Use the Force Method to determine the reactions for the beam subjected to one air conditioner unit.
d. Using structural analysis software, determine the reactions for the beam.
e. Submit a printout of the displaced shape. Identify at least three features that suggest the displaced shape is reasonable.
f. Verify equilibrium of the computer aided analysis results.
g. Are the results of your computer aided analysis reasonable? Make a comprehensive argument to justify your answer.
h. Comment on why your guess in part (a) was or was not close to the solution in part (e).

712 Chapter 12 Force Method

12.2 and 12.3 Our firm has won the contract to design a new two-span overpass. The principal in charge believes the most cost-effective solution is to provide continuity at the middle support. Therefore, the girders will be indeterminate. The team is checking many different live load situations to find the largest internal forces generated.

The design truck we are most concerned with is a fully loaded, five-axle semi with a relatively short span. Although there would actually be five point loads from the truck, for our preliminary calculations we can combine the tandem loads.

The bridge deck will distribute the point forces to the various bridge girders across the width. Another team member will use the Distribution Factor Method to calculate that distribution after we finish our analyses.

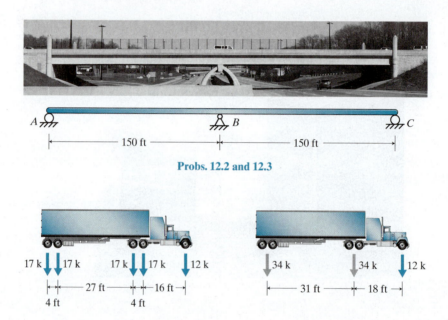

Probs. 12.2 and 12.3

12.2 To find the peak shear generated in the bridge near B, we consider the truck to be just past B (e.g., 1 inch past B).

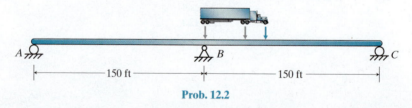

Prob. 12.2

a. Guess the shape of the shear diagram under this loading.
b. Using approximate analysis, find the shear just to the right of B. You can do so by assuming the two spans are each simply supported; this is the same as assuming the inflection point is at B.
c. Use the Force Method to determine the vertical reactions for the bridge.
d. Construct and label the shear diagram for the bridge based on your results from part (c).
e. Using structural analysis software, determine the reactions for the bridge.
f. Submit a printout of the displaced shape. Identify at least three features that suggest the displaced shape is reasonable.
g. Submit a printout of the shear diagram. Identify at least three features that suggest the shear diagram is reasonable.
h. Verify equilibrium of the computer aided analysis results.
i. Are the results of your computer aided analysis reasonable? Make a comprehensive argument to justify your answer.
j. Comment on why your guess in part (a) was or was not close to the solution in part (d).

12.3 To find the peak positive moment generated in the bridge, we must consider the truck to be in several different locations. For a good first guess, let's put the middle tandem at midspan.

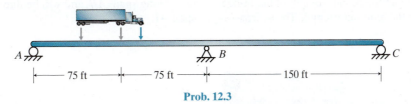

Prob. 12.3

a. Guess the shape of the moment diagram under this loading.
b. Using approximate analysis, find the moment at midspan. You can do so by assuming the two spans are each simply supported; this is the same as assuming the inflection point is at B.
c. Use the Force Method to determine the vertical reactions for the bridge.
d. Construct and label the moment diagram for the bridge based on your results from part (c).
e. Using structural analysis software, determine the reactions for the bridge.
f. Submit a printout of the displaced shape. Identify at least three features that suggest the displaced shape is reasonable.
g. Submit a printout of the moment diagram. Identify at least three features that suggest the moment diagram is reasonable.
h. Verify equilibrium of the computer aided analysis results.
i. Are the results of your computer aided analysis reasonable? Make a comprehensive argument to justify your answer.
j. Comment on why your guess in part (a) was or was not close to the solution in part (d).

12.4 Maintenance is scheduled for an overpass, and the contractor wants to store materials and equipment on one part of the overpass. Before the Department of Transportation can give approval, our team must verify that the girders are adequate. In order to make that determination, we need to find the resulting peak moment in the girders. We have been tasked with the analysis of one girder.

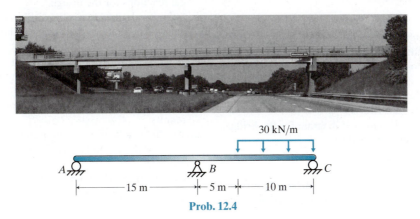

Prob. 12.4

a. Guess the shape of the moment diagram.
b. Using approximate analysis, find the vertical reactions. You can do so by assuming the location of the inflection point.
c. Use the Force Method to determine the vertical reactions for the girder.
d. Construct and label the moment diagram for the girder based on your results from part (c).
e. Using structural analysis software, determine the vertical reactions for the girder.
f. Submit a printout of the displaced shape. Identify at least three features that suggest the displaced shape is reasonable.
g. Submit a printout of the moment diagram. Identify at least three features that suggest the moment diagram is reasonable.
h. Verify equilibrium of the computer aided analysis results.
i. Are the results of your computer aided analysis reasonable? Make a comprehensive argument to justify your answer.
j. Comment on why your guess in part (a) was or was not close to the solution in part (d).

714 Chapter 12 Force Method

12.5 and 12.6 An existing overpass in Canada is showing signs of distress. While a team is in the field obtaining information in order to predict the actual strength of the bridge, our job is to calculate the required strength. The overpass is two lanes wide.

Probs. 12.5 and 12.6

12.5 A team member has used the concepts for influence lines to determine that the peak negative moment at B will be due to full live load along the entire bridge.

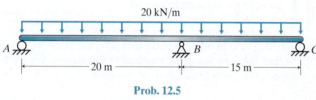

Prob. 12.5

a. Guess the shape of the moment diagram.
b. Using approximate analysis, find the vertical reactions. You can do so by assuming the location of an inflection point.
c. Use the Force Method to determine the vertical reactions for the bridge.
d. Construct and label the moment diagram for the bridge based on your results from part (c).
e. Using structural analysis software, determine the vertical reactions for the bridge.
f. Submit a printout of the displaced shape. Identify at least three features that suggest the displaced shape is reasonable.
g. Submit a printout of the moment diagram. Identify at least three features that suggest the moment diagram is reasonable.
h. Verify equilibrium of the computer aided analysis results.
i. Are the results of your computer aided analysis reasonable? Make a comprehensive argument to justify your answer.
j. Comment on why your guess in part (a) was or was not close to the solution in part (d).

12.6 A team member used the concepts for influence lines to determine that the peak positive moment will be in the long span, AB, and will be due to live load only on span AB.

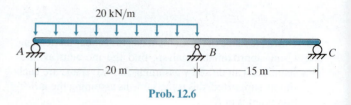

Prob. 12.6

a. Guess the shape of the moment diagram.
b. Using approximate analysis, find the vertical reactions. You can do so by assuming the location of the inflection point.
c. Use the Force Method to determine the vertical reactions for the bridge.
d. Construct and label the moment diagram for the bridge based on your results from part (c).
e. Using structural analysis software, determine the vertical reactions for the bridge.
f. Submit a printout of the displaced shape. Identify at least three features that suggest the displaced shape is reasonable.
g. Submit a printout of the moment diagram. Identify at least three features that suggest the moment diagram is reasonable.
h. Verify equilibrium of the computer aided analysis results.
i. Are the results of your computer aided analysis reasonable? Make a comprehensive argument to justify your answer.
j. Comment on why your guess in part (a) was or was not close to the solution in part (d).

12.7 A pedestrian walkway must be suspended over a maintenance corridor. To do that, our firm has chosen to support the walkway on a frame fixed to the adjacent building and pinned to a column at the other end. We have been tasked with analyzing the girder subjected to full live load.

12.8 Our team is designing the structure to support a corridor. A team member has performed a preliminary design if the beams are simply supported. Fixing the beams at one end should reduce the peak internal forces, thus reducing the size of the members. To find out if that is a worthwhile change, we have been asked to analyze the beam fixed at one end and carrying the full live load.

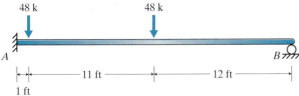

Prob. 12.7

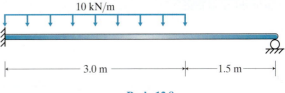

Prob. 12.8

a. Guess the shape of the moment diagram.
b. Using approximate analysis, find the reactions. You can do so by assuming the location of the inflection point.
c. Use the Force Method to determine the reactions for the girder subjected to live load.
d. Construct and label the moment diagram for the girder based on your results from part (c).
e. Using structural analysis software, determine the reactions for the girder.
f. Submit a printout of the displaced shape. Identify at least three features that suggest the displaced shape is reasonable.
g. Submit a printout of the moment diagram. Identify at least three features that suggest the moment diagram is reasonable.
h. Verify equilibrium of the computer aided analysis results.
i. Are the results of your computer aided analysis reasonable? Make a comprehensive argument to justify your answer.
j. Comment on why your guess in part (a) was or was not close to the solution in part (d).

a. Guess the shape of the moment diagram.
b. Using approximate analysis, find the reactions. You can do so by assuming the location of the inflection point.
c. Use the Force Method to determine the reactions for the beam subjected to live load.
d. Construct and label the moment diagram for the beam based on your results from part (c).
e. Using structural analysis software, determine the reactions for the beam.
f. Submit a printout of the displaced shape. Identify at least three features that suggest the displaced shape is reasonable.
g. Submit a printout of the moment diagram. Identify at least three features that suggest the moment diagram is reasonable.
h. Verify equilibrium of the computer aided analysis results.
i. Are the results of your computer aided analysis reasonable? Make a comprehensive argument to justify your answer.
j. Comment on why your guess in part (a) was or was not close to the solution in part (d).

12.9 and 12.10 Our firm has been contracted by a state Department of Transportation to design a three-span overpass. To design the bridge, we must find the peak positive and negative moments on each girder due to the various load combinations. Before we can combine the effects of the various loads, we need to know the unfactored effects.

Probs. 12.9 and 12.10

12.9 Our task is to analyze one girder considering only the unfactored dead load.

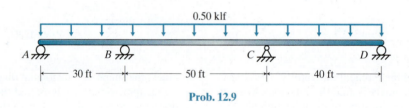

Prob. 12.9

a. Guess the displaced shape.
b. Using approximate analysis, find the vertical reactions. You can do so by assuming the locations of the inflection points.
c. Use the Force Method to determine the vertical reactions for the girder subjected to dead load.
d. Construct and label the moment diagram for the girder based on your results from part (c).
e. Using structural analysis software, determine the vertical reactions for the girder.
f. Submit a printout of the displaced shape. Identify at least three features that suggest the displaced shape is reasonable.
g. Submit a printout of the moment diagram. Identify at least three features that suggest the moment diagram is reasonable.
h. Verify equilibrium of the computer aided analysis results.
i. Are the results of your computer aided analysis reasonable? Make a comprehensive argument to justify your answer.
j. Comment on why your guess in part (a) was or was not close to the solution in part (f).

12.10 Using the concepts for influence lines, a team member has determined that the peak positive moment in span BC will be due to load only on span BC. Our task is to analyze the bridge subjected to unfactored live load only on span BC.

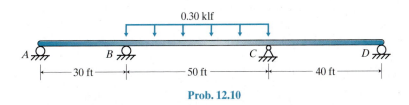

Prob. 12.10

a. Guess the displaced shape.
b. Using approximate analysis, find the vertical reactions. You can do so by assuming the locations of the inflection points.
c. Use the Force Method to determine the vertical reactions for the girder subjected to this live load.
d. Construct and label the moment diagram for the girder based on your results from part (c).
e. Using structural analysis software, determine the vertical reactions for the girder.
f. Submit a printout of the displaced shape. Identify at least three features that suggest the displaced shape is reasonable.
g. Submit a printout of the moment diagram. Identify at least three features that suggest the moment diagram is reasonable.
h. Verify equilibrium of the computer aided analysis results.
i. Are the results of your computer aided analysis reasonable? Make a comprehensive argument to justify your answer.
j. Comment on why your guess in part (a) was or was not close to the solution in part (f).

12.11 Our firm has been selected to design an overpass as part of a highway expansion project. The elevation of the roadway is set to match the grade of the other lanes. To provide adequate height clearance below the overpass, we need to use relatively shallow members. An experienced member of the team recommends we use deeper members for the outside spans to draw moment away from the shallower interior spans. We are in the preliminary design phase, so that experienced engineer suggests we start by assuming that the exterior spans have moments of inertia double those of the interior spans. The cross-sectional areas for the different spans will be essentially equal. Our task is to find the peak moment in each span due to dead load.

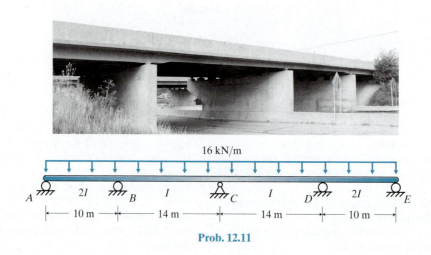

Prob. 12.11

a. Guess the shape of the moment diagram.
b. Using approximate analysis, find the vertical reactions. You can do so by assuming each span is a simply supported member; that is the same as assuming the inflection points are over the supports.
c. Use the Force Method to determine the vertical reactions for the girder. You may use computer software to solve the system of equations, but provide a printout of the inputs and results.
d. Construct and label the moment diagram for the girder based on your results from part (c).
e. Using structural analysis software, determine the vertical reactions for the girder.
f. Submit a printout of the displaced shape. Identify at least three features that suggest the displaced shape is reasonable.
g. Submit a printout of the moment diagram. Identify at least three features that suggest the moment diagram is reasonable.
h. Verify equilibrium of the computer aided analysis results.
i. Are the results of your computer aided analysis reasonable? Make a comprehensive argument to justify your answer.
j. Comment on why your guess in part (a) was or was not close to the solution in part (d).

12.12 A cantilevered balcony is showing signs of distress. Until it can be repaired, we need to reduce the peak moment. The principal in charge recommends adding two rows of post shores. Our role is to determine how much that would reduce the peak moment in the balcony.

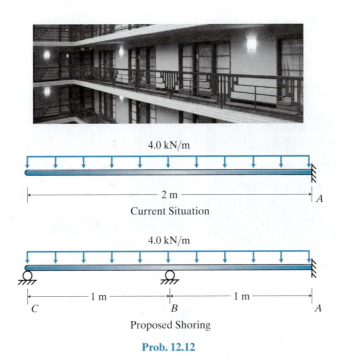

Prob. 12.12

a. Guess the shape of the moment diagram of the balcony with two rows of shores (i.e., the proposed shoring).
b. Calculate the peak moment for a strip of balcony without added supports subjected to only its self-weight (i.e., the current situation).
c. Using approximate analysis, find the reactions for a strip of balcony with two rows of shores. You can do so by assuming the locations of the inflection points.
d. Use the Force Method to determine the reactions for a strip of balcony with two rows of shores.
e. Construct and label the moment diagram for the balcony based on your results from part (d). By how much is the peak moment reduced when we add the shores?
f. Using structural analysis software, determine the reactions for a strip of balcony with two rows of shores.
g. Submit a printout of the displaced shape. Identify at least three features that suggest the displaced shape is reasonable.
h. Submit a printout of the moment diagram. Identify at least three features that suggest the moment diagram is reasonable.
i. Verify equilibrium of the computer aided analysis results.
j. Are the results of your computer aided analysis reasonable? Make a comprehensive argument to justify your answer.
k. Comment on why your guess in part (a) was or was not close to the solution in part (e).

12.13 The contractor for a new high-rise building must temporarily brace the large excavation for the basement. Our firm has been selected to design the bracing.

Using information from the geotechnical report and anticipated surcharge, another team member has calculated the resulting loading on a 1-meter-wide strip of sheet piling.

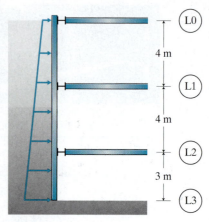

Prob. 12.13a

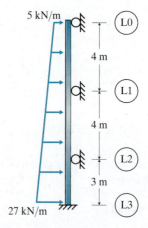

Prob. 12.13b

a. Guess the displaced shape of the strip of sheet piling.
b. Using approximate analysis, find the reactions for the strip of sheet piling. You can do so by assuming the locations of the inflection points.
c. Use the Force Method to determine the reactions for the strip of sheet piling.
d. Using structural analysis software, determine the reactions for the strip of sheet piling.
e. Submit a printout of the displaced shape. Identify at least three features that suggest the displaced shape is reasonable.
f. Verify equilibrium of the computer aided analysis results.
g. Are the results of your computer aided analysis reasonable? Make a comprehensive argument to justify your answer.
h. Comment on why your guess in part (a) was or was not close to the solution in part (e).

12.14 We are interviewing for a job with a structural engineering firm. The senior engineer conducting the interview has given us a problem to analyze. The engineer wants to test the breadth of our skills, so there are multiple parts to the problem.

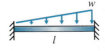

Prob. 12.14

a. Guess the displaced shape of the beam.
b. Using approximate analysis, find an upper bound for the peak moment.
c. Use the Force Method to determine the reactions for the beam. The interviewer gives a suggestion: since there is no horizontal load, we can consider one end free to move in the horizontal direction. That effectively reduces the degree of indeterminacy by one.
d. Develop an expression for vertical displacement along the entire beam.
e. Using structural analysis software and assumed values for w, l, I, and E, determine the reactions for the beam. Indicate the assumed values.
f. Submit a printout of the displaced shape. Identify at least three features that suggest the displaced shape is reasonable.
g. Submit a printout of the moment diagram. Identify at least three features that suggest the moment diagram is reasonable.
h. Verify equilibrium of the computer aided analysis results.
i. Make a comprehensive argument that your computer aided analysis results are reasonable.
j. Comment on why your guess in part (a) was or was not close to the solution in part (f).

12.15 We work for a steel fabricator as a connection designer. Our company just got a job from a local gym to design and fabricate a pull-up bar that can be mounted on the wall.

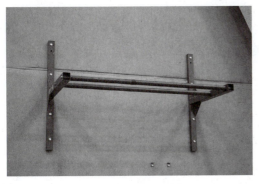

Prob. 12.15a

The gym has tried a rigid frame version but wants something stiffer, like a truss. For the safety of the users, there cannot be any members below the bar. We have been tasked with analyzing the truss to find the force in each member.

a. Guess the sense of the axial force in each member.
b. Using approximate analysis, find the axial force in each member (magnitude and sense).

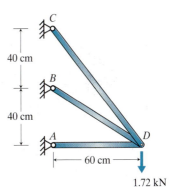

Prob. 12.15b

c. Use the Force Method to determine the axial force in each member (magnitude and sense).
d. Using structural analysis software, determine the axial force in each member (magnitude and sense).
e. Submit a printout of the displaced shape. Identify at least three features that suggest the displaced shape is reasonable.
f. Verify equilibrium of the computer aided analysis results.
g. Are the results of your computer aided analysis reasonable? Make a comprehensive argument to justify your answer.
h. Comment on why your guess in part (a) was or was not close to the solution in part (c).

12.16 During warm weather, salt spreaders are stored on the rack shown in the figure. The ends of the rack are indeterminate trusses to provide lateral stability. If there is an earthquake, the effect will be lateral load at the top of the trusses. We have been hired to determine the equivalent static load developed in each of the truss members if there is an earthquake. Another engineer has determined the equivalent static load on one truss. The truss is made of standard pipes with these properties: $E = 29{,}000$ ksi, $A = 1.02$ in^2, $I = 0.627$ in^4.

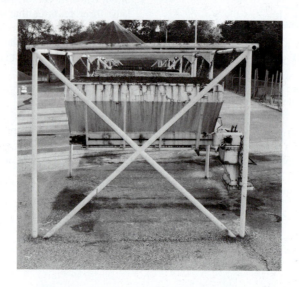

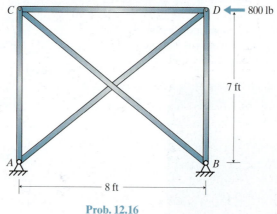

Prob. 12.16

a. Guess the sense of the axial force in each member.
b. Using approximate analysis, find the axial force in each member (magnitude and sense).
c. Use the Force Method to determine the axial force in each member (magnitude and sense).
d. Using structural analysis software, determine the axial force in each member (magnitude and sense).
e. Submit a printout of the displaced shape. Identify at least three features that suggest the displaced shape is reasonable.
f. Verify equilibrium of the computer aided analysis results.
g. Are the results of your computer aided analysis reasonable? Make a comprehensive argument to justify your answer.
h. Comment on why your guess in part (a) was or was not close to the solution in part (c).

12.17 Our firm has been selected to design a high-rise building that presents a unique challenge. Half of the building will hang over active railroad tracks. Therefore, there can be no foundations on one side. The principal in charge and the architect have decided to use a roof truss to transfer the load from one side of the building to the other.

The team is in the preliminary design phase, so the members have not been designed yet. In preparation for member design, the team leader wants to have a sense of how the truss will behave.

The truss will be made of steel ($E = 200$ GPa $= 200$ kN/mm^2). An experienced engineer estimates that the members will have the following cross-sectional areas:

① $A_1 = 600$ cm^2
② $A_2 = 800$ cm^2
③ $A_3 = 500$ cm^2
④ $A_4 = 200$ cm^2
⑤ $A_5 = 100$ cm^2

Prob. 12.17a

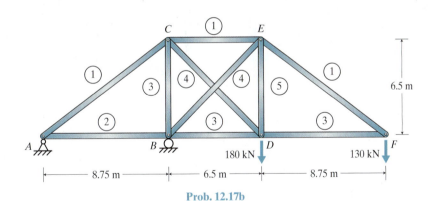

Prob. 12.17b

a. Guess the sense of the axial force in each member.
b. Using approximate analysis, find the axial force in each member (magnitude and sense).
c. Use the Force Method to determine the axial force in each member (magnitude and sense).
d. Using structural analysis software, determine the axial force in each member (magnitude and sense).
e. Submit a printout of the displaced shape. Identify at least three features that suggest the displaced shape is reasonable.
f. Verify equilibrium of the computer aided analysis results.
g. Are the results of your computer aided analysis reasonable? Make a comprehensive argument to justify your answer.
h. Comment on why your guess in part (a) was or was not close to the solution in part (c).

Chapter 12 Force Method

12.18 and 12.19 Our firm is designing a new warehouse. The emergency exit will require stairs at one end. The senior engineer has chosen a steel truss to support the stairs and landing. Our task is to analyze the truss due to various loadings.

Based on earlier designs, we estimate using the following sections:

Horizontal:	C6 × 13	$A = 3.82 \text{ in}^2$
Verticals:	L4 × 4 × 3/8	$A = 2.86 \text{ in}^2$
Diagonals:	L2 × 2 × 1/4	$A = 0.944 \text{ in}^2$
Steel:		$E = 29{,}000 \text{ ksi}$

Probs. 12.18 and 12.19

12.18 Because of the stairs, the dead load on the truss is not uniform.

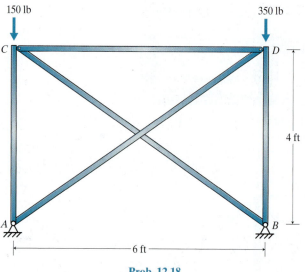

Prob. 12.18

a. Guess the sense of the axial force in each member.
b. Using approximate analysis, find the axial force in each member (magnitude and sense).
c. Use the Force Method to determine the axial force in each member (magnitude and sense).
d. Using structural analysis software, determine the axial force in each member (magnitude and sense).
e. Submit a printout of the displaced shape. Identify at least three features that suggest the displaced shape is reasonable.
f. Verify equilibrium of the computer aided analysis results.
g. Are the results of your computer aided analysis reasonable? Make a comprehensive argument to justify your answer.
h. Comment on why your guess in part (a) was or was not close to the solution in part (c).

12.19 By code, the live load on stairs is only vertical. However, bleachers and stadium seating are also designed for horizontal live load. The senior engineer on our team wants to make sure this truss can handle the lateral live load as well.

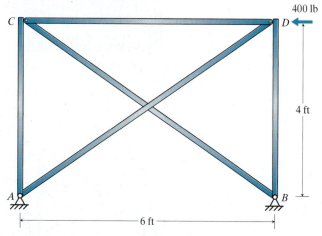

Prob. 12.19

a. Guess the sense of the axial force in each member.
b. Using approximate analysis, find the axial force in each member (magnitude and sense).
c. Use the Force Method to determine the axial force in each member (magnitude and sense).
d. Using structural analysis software, determine the axial force in each member (magnitude and sense).
e. Submit a printout of the displaced shape. Identify at least three features that suggest the displaced shape is reasonable.
f. Verify equilibrium of the computer aided analysis results.
g. Are the results of your computer aided analysis reasonable? Make a comprehensive argument to justify your answer.
h. Comment on why your guess in part (a) was or was not close to the solution in part (c).

12.20 We are designing a garage to store maintenance vehicles. Our team leader recommends that we use a cross braced frame over the back third of the building to provide lateral load resistance. The configuration of the braced frame means it will behave like a truss. At this preliminary stage, let's assume all the members have the same cross-sectional area. Another engineer on the team has calculated the loads on the truss due to wind. Our job is to analyze the truss.

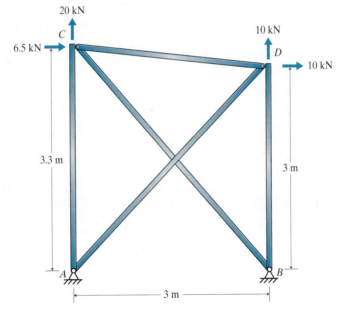

Prob. 12.20

a. Guess the sense of the axial force in each member.
b. Using approximate analysis, find the axial force in each member (magnitude and sense).
c. Use the Force Method to determine the axial force in each member (magnitude and sense).
d. Using structural analysis software, determine the axial force in each member (magnitude and sense). Indicate the assumed values for cross-sectional area and modulus of elasticity.
e. Submit a printout of the displaced shape. Identify at least three features that suggest the displaced shape is reasonable.
f. Verify equilibrium of the computer aided analysis results.
g. Are the results of your computer aided analysis reasonable? Make a comprehensive argument to justify your answer.
h. Comment on why your guess in part (a) was or was not close to the solution in part (c).

CHAPTER 13

MOMENT DISTRIBUTION METHOD

Rigid frames can be very effective structural systems.

Motivation

The Force Method (Chapter 12) can be a very efficient analysis method for structures that are up to a couple degrees indeterminate. However, many bridge girders and most rigid frames are more highly indeterminate. For many years pre-computers, the analysis of a large rigid frame had to be done by approximate analysis (Chapter 9). Seeking an efficient, detailed analysis method, Hardy Cross (1930) developed the Moment Distribution Method.

The Moment Distribution Method is a tool that is applicable only to rigid frames and continuous beams, but it can be very efficient even for highly indeterminate structures.

The method has several benefits beyond providing detailed analysis results for the designed structure. When we are in the preliminary design phase, we can obtain detailed analysis results without knowing the beam and column cross-sectional properties. This is helpful because we need the analysis results in order to choose the cross-sectional properties. For the Moment Distribution Method, we only need to guess the relative magnitudes of the properties. In addition, the Moment Distribution Method shows us how relative stiffness affects load flow through a structure. This allows us to make smarter structural layout choices in the schematic design phase.

Chapter 13 Moment Distribution Method

13.1 Overview of Method

Overview

The Moment Distribution Method was developed by Hardy Cross (1930) as a way to analyze rigid frames without making the assumptions of approximate analysis. At that time, which predated computers by decades, engineers were looking for an efficient way to analyze a highly indeterminate structure by hand.

Before we describe how the method works, we need to define the term *node*. This term is used with both the Moment Distribution Method and the Direct Stiffness Method (Chapters 14 and 15). A node is a place where multiple members meet or where a member ends (Figure 13.1). The Moment Distribution Method is limited to continuous beams and rigid frames, so the connection between members at a node is rigid.

The Moment Distribution Method is based on our ability to make two predictions accurately:

1. How much moment applied to a node will distribute to each member.
2. If moment is applied to the pinned end of a pin-fixed member, the amount of moment that is generated at the fixed end.

Fortunately, we can make both of these predictions without approximate analysis. We cover how in Sections 13.2 and 13.3 of this chapter.

When we know that we can calculate how much moment at a node distributes to each member and how much carries over to the other end of the member, the method basically has three steps:

1. Initially imagine that all nodes are clamped against rotation and then calculate the moment carried by the clamp. The moment required to hold one member in place when it is loaded is called the *fixed-end moment*; these moments have been precalculated for a variety of loadings inside the back cover of this text. The net moment at the node from all the fixed-end moments is called the *unbalanced moment*.
2. One at a time, release each node that should be able to rotate.

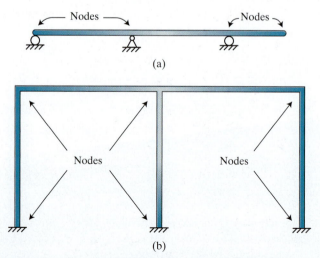

Figure 13.1 Locations of nodes: (a) Continuous beam; (b) Rigid frame.

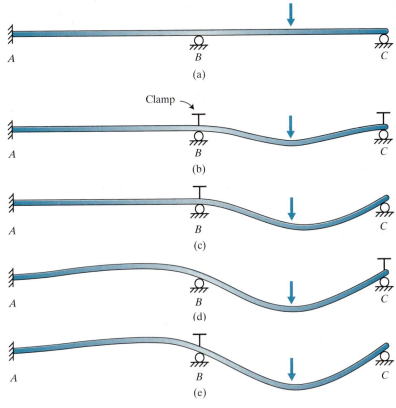

Figure 13.2 The Moment Distribution Method process: (a) Indeterminate beam; (b) Clamp all nodes and calculate the fixed-end moments; (c) Release node C and distribute the unbalanced moment; (d) Reclamp node C and release node B; (e) Reclamp node B and release node C.

3. When a node is released to rotate, distribute the unbalanced moment to the members and carry over the appropriate moment to the clamped ends. Reclamp the node before moving to the next.

We cover the implementation of this method in Sections 13.4 and 13.5. The actual moment at the end of each member is the superposition of all the iterations.

As an example, consider the beam in Figure 13.2a. We clamp the nodes so that they cannot rotate (Figure 13.2b). Because of the loading on member BC, fixed-end moments are generated in the clamps at B and C. These fixed-end moments keep the ends of the member from moving. We can choose any node to release first, but we need to choose nodes that should be able to rotate: B and C. Releasing node A is not wrong, but it won't do anything. Let's choose node C to start. When we release it, it can rotate (Figure 13.2c). The net moment at the clamp, the unbalanced moment, goes into the member, causing it to rotate into the shape we see. Next we reclamp node C and release node B. Node B is able to rotate, and its net moment is zero (Figure 13.2d). Node C, however, is locked into its previous deformed shape, so unbalanced moment is accumulating. Therefore, we need to repeat the process. We go back to node C and release it (Figure 13.2e). This time the resulting increase in rotation is smaller. We continue iterations until the unbalanced moment at each node is sufficiently small.

13.2 Fixed-End Moments and Distribution Factors

Introduction

In order to use the Moment Distribution Method to analyze a structure, we need to prepare by calculating several quantities. We begin the method by imagining that each node is unable to rotate. Any loads applied to the member will cause moment to build up at the clamped ends. Those moments are called fixed-end moments, and we need to calculate them before starting the distribution process.

The other information we need is how much moment at a node distributes to each member attached at that node. The proportion shared by each member is based on the stiffness of each member and is called the distribution factor.

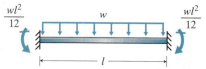

Figure 13.3 The moments generated at the ends of a fixed-fixed beam subjected to uniform load.

How-To

Fixed-End Moments

The Moment Distribution Method begins with all nodes clamped, fixed against rotation, so that any load between the nodes causes moments in the clamps. Those moments are the same as if we were analyzing a beam fixed supported at each end (Figure 13.3). To calculate the fixed-end moments (*FEM*), we can use the Force Method (Chapter 12). Over time, engineers have made those calculations for a variety of loadings, and the fixed-end moments for the most common loadings are provided inside the back cover of this text. The moment in the clamp due to loading on a member is the magnitude of the fixed-end reaction in the direction of the fixed-end reaction.

Cantilevered members are different. We don't clamp the free end of a cantilever, so we get a fixed-end moment only at the attached end (Figure 13.4). We can use equilibrium to calculate that moment because it is as if the member is fixed at one end and free at the other.

Sometimes the load occurs at the node rather than in between. One possibility is a couple moment. In this case, we treat the applied moment as if there was a cantilever with a fixed-end moment of the same magnitude but opposite direction (Figure 13.5). The fixed-end moment for the imaginary cantilever has the opposite direction because the effect of the cantilever on the other members is in the opposite direction of the fixed-end moment.

Force applied at a node, however, does not generate any moment in the clamp. Consider the simply supported beam in Figure 13.6. The point

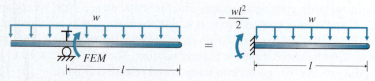

Figure 13.4 The fixed-end moment for a cantilevered member is the same as the reaction of a member fixed at one end.

Section 13.2 Fixed-End Moments and Distribution Factors

Figure 13.5 A couple moment applied at a node is treated like the fixed-end moment for a cantilever but reversed.

force does not cause deformation in the beam, and no deformation means no fixed-end moment.

When multiple members meet at a node, we add the fixed-end moments, and we call the net moment the *unbalanced moment, UM* (Figure 13.7). To avoid confusion, let's choose a sign convention: clockwise moment is positive. We use this sign convention for every moment in the Moment Distribution Method.

Figure 13.6 Force applied at a node causes no fixed-end moments.

Moment-Rotation Relationship

When we release a clamp, the end of each member at the clamp rotates because of the moment it receives. The relationship between the amount of rotation and the amount of moment is called *stiffness, K*.

For the Moment Distribution Method, we are concerned with only one stiffness: the moment required to generate a unit rotation at the pinned end of a pin-fixed member (Figure 13.8). This one stiffness is enough because we imagine every node clamped (i.e., fixed) except one. That one can rotate, so it is like being pin supported.

Using the Force Method (Chapter 12) and a deformation prediction method (Chapter 5), we get the following relationship between the applied moment, M, and the resulting rotation, θ:

$$M = \frac{4EI}{l}\theta$$

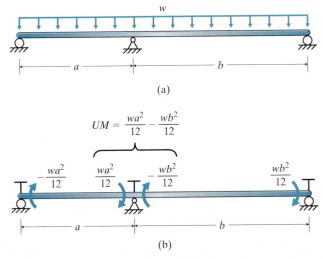

Figure 13.7 Fixed-end moments: (a) Beam with load on the two members; (b) Fixed-end moments for each member and the net unbalanced moment at the middle support.

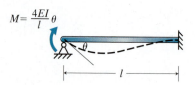

Figure 13.8 Relationship between end moment and rotation for a pin-fixed member.

Figure 13.9 Relationship between end moment and moment generated at opposite end for a pin-fixed member.

So now we know that stiffness for the Moment Distribution Method depends on only the length of the member, l, Young's modulus, E, and the moment of inertia, I:

$$K = \frac{4EI}{l}$$

Looking at the displaced shape of the beam in Figure 13.8, we can infer that there must be moment generated at the fixed support in order to hold the beam in that shape. This means that applying a moment at the pinned end of the pin-fixed member results in moment generated at the other end as well. We call that moment the *carryover moment* because it "carries over" from the pin end. Our Force Method analysis gives us the carryover moment (Figure 13.9). The moment generated at the other end is exactly half of the moment applied at the pinned end, and the moment generated is in the same direction as the applied moment.

Moment Distribution

We saw at the beginning of this section that to start the Moment Distribution Method, we calculate the unbalanced moment at each node. The unbalanced moment is the net moment in the clamp, and it resists rotation. When we release a clamp, the members at the node rotate and the net moment becomes zero. For the net moment to be zero, the internal moments generated must be in the opposite direction of the unbalanced moment in the clamp. We can see this happening for the simple one-member structure in Figure 13.10.

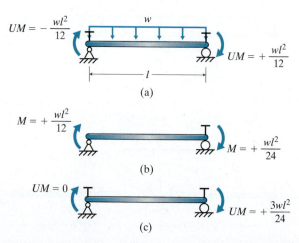

Figure 13.10 Distributing unbalanced moment when there is only one member at a node: (a) Unbalanced moment generated; (b) Moment distributed into the member when the clamp is released, and the carryover moment generated at the opposite end; (c) Unbalanced moment at each node after superimposing effects.

Section 13.2 Fixed-End Moments and Distribution Factors

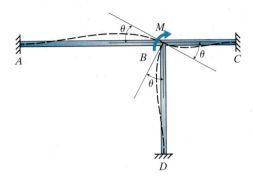

Figure 13.11 Moment applied to an unclamped node in a rigid frame causes rotation of the attached members.

When we superimpose the effects, the net moment at the left end is zero. At the right end, however, more moment has accumulated at the clamp.

When more than one member is rigidly connected at a node, the members share the negative unbalanced moment. The percentage that each member receives is called the *distribution factor, DF*. To understand how we calculate distribution factors, let's start by looking at a node that is free to rotate in a rigid frame (Figure 13.11). Note that in the Moment Distribution Method the applied moment, M, in Figure 13.11 would be $-UM$.

If the connection between the members is perfectly rigid, the members must all rotate the same amount. Remember from Section 1.3: Structure Idealization that most connections actually behave as semirigid, but we idealize them as perfectly pinned or perfectly rigid. The Moment Distribution Method is for rigid frames only, so each of the members at node B in Figure 13.11 will rotate through the same angle:

$$\theta_{BA} = \theta_{BC} = \theta_{BD} = \theta$$

We have learned how moment and rotation at the end of a pin-fixed member are related:

$$M = \frac{4EI}{l}\theta = K\theta$$

If we rearrange the terms, we have a way to predict the rotation due to the moment at the end of the member:

$$\theta = \frac{M}{K}$$

Knowing that each of the members that meet at node B in Figure 13.11 rotates through the same angle, we can relate the moments generated in each member:

$$\frac{M_{BA}}{K_{BA}} = \frac{M_{BC}}{K_{BC}} = \frac{M_{BD}}{K_{BD}}$$

We can solve this expression so that each moment is a function of one unknown moment, M_{BA}:

$$M_{BC} = \frac{K_{BC}}{K_{BA}} M_{BA}$$

$$M_{BD} = \frac{K_{BD}}{K_{BA}} M_{BA}$$

To have equilibrium, the moments generated at the ends of the members must add to equal the applied moment:

$$M = M_{BA} + M_{BC} + M_{BD}$$

$$= M_{BA}\left(\frac{K_{BA}}{K_{BA}}\right) + M_{BA}\left(\frac{K_{BC}}{K_{BA}}\right) + M_{BA}\left(\frac{K_{BD}}{K_{BA}}\right)$$

$$= M_{BA}\left(\frac{K_{BA} + K_{BC} + K_{BD}}{K_{BA}}\right)$$

We can solve this expression for the unknown moment, M_{BA}:

$$M_{BA} = M\left(\frac{K_{BA}}{K_{BA} + K_{BC} + K_{BD}}\right)$$

In general, we have the following relationship between the applied moment and the moment distributed to a specific member:

$$M_i = M\left(\frac{K_i}{\Sigma K_j}\right)$$

So the percentage of the moment carried by each member is the same as the ratio of the member's stiffness, K_i, to the total stiffness of all members that meet rigidly at that node, ΣK_j. We call this percentage the distribution factor:

$$DF_i = \frac{K_i}{\Sigma K_j}$$

Special Cases

Some distribution factors occur repeatedly, and we can precalculate them. When multiple members meet at a node that is supported by a pin or roller (Figure 13.12a), we calculate the distribution factors using the relative stiffness formula. But when a single member ends at a pin (Figure 13.12b) or roller (Figure 13.12c), all of the moment goes to the member, so its distribution factor is 1. If a single member (Figure 13.12d) or multiple members (Figure 13.12e) end at a fixed support, all of the moment goes to the support, so the distribution factors to the members are all zero. A member that cantilevers out from the frame or continuous beam has no stiffness. If we go back to the moment-rotation relationship, we see that the stiffness is the moment required to cause a unit rotation:

$$K = \frac{M}{\theta} = \frac{\text{moment}}{\text{unit rotation}}$$

The connected end of a cantilevered member can rotate without any moment, so its stiffness is zero. Therefore, the distribution factor for a cantilevered member is zero (Figure 13.12f).

13.3 Beams and Sidesway Inhibited Frames

Introduction
This is where we put all the pieces together in order to use the Moment Distribution Method to analyze beams and sidesway inhibited frames. If one joint can displace relative to the joint at the other end of a member,

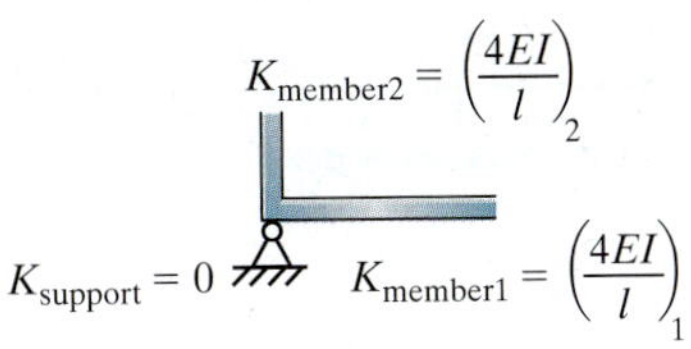

Section 13.3 Beams and Sidesway Inhibited Frames

(a) $K_{support} = 0$, $K_{member1} = \left(\dfrac{4EI}{l}\right)_1$, $K_{member2} = \left(\dfrac{4EI}{l}\right)_2$

$$DF_{member2} = \dfrac{K_2}{\Sigma K_j} = \dfrac{(4EI/l)_2}{0 + (4EI/l)_1 + (4EI/l)_2}$$

$$DF_{member1} = \dfrac{K_1}{\Sigma K_j} = \dfrac{(4EI/l)_1}{0 + (4EI/l)_1 + (4EI/l)_2}$$

(b) $K_{support} = 0$, $K_{member} = \dfrac{4EI}{l}$

$$DF_{member} = \dfrac{4EI/l}{0 + 4EI/l} = 1$$

(c) $K_{support} = 0$, $K_{member} = \dfrac{4EI}{l}$

$$DF_{member} = \dfrac{4EI/l}{0 + 4EI/l} = 1$$

(d) $K_{support} = \infty$, $K_{member} = \dfrac{4EI}{l}$

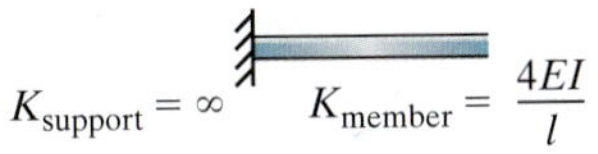

$$DF_{member} = \dfrac{4EI/l}{\infty + 4EI/l} = 0$$

(e) $K_{support} = \infty$, $K_{member1} = \left(\dfrac{4EI}{l}\right)_1$, $K_{member2} = \left(\dfrac{4EI}{l}\right)_2$

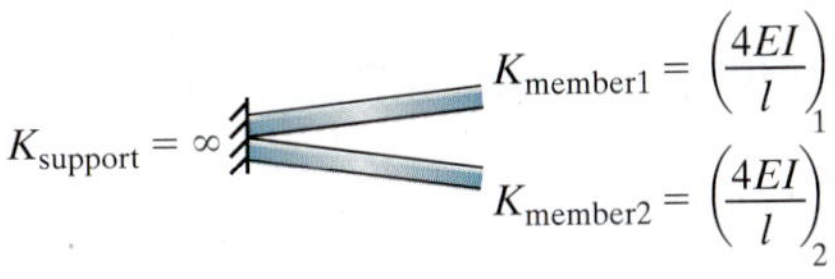

$$DF_{member1} = \dfrac{(4EI/l)_1}{\infty + (4EI/l)_1 + (4EI/l)_2} = 0$$

$$DF_{member2} = \dfrac{(4EI/l)_2}{\infty + (4EI/l)_1 + (4EI/l)_2} = 0$$

(f) $K_{cantilever} = 0$, $K_{member1} = \left(\dfrac{4EI}{l}\right)_1$, $K_{member2} = \left(\dfrac{4EI}{l}\right)_2$

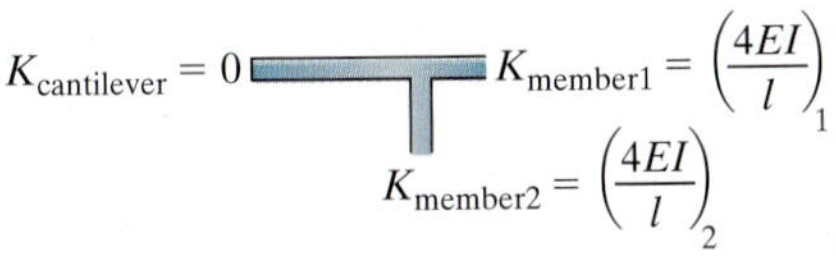

$$DF_{cantilever} = \dfrac{0}{0 + (4EI/l)_1 + (4EI/l)_2} = 0$$

Figure 13.12 Distribution factors for special cases: (a) Multiple members meet at a pin or roller support; (b) Single member ends at a pin support; (c) Single member ends at a roller support; (d) Single member ends at a fixed support; (e) Multiple members end at a fixed support; (f) Cantilevered member.

we call the frame a *sidesway* frame. Applying the Moment Distribution Method to sidesway frames requires additional steps that we cover in Section 13.4. Fortunately, beams typically do not have relative displacement at their joints. In addition, some rigid frames are supported such that they cannot move laterally at the joints.

How-To
Section 13.1 gave an overview of how the Moment Distribution Method works. Now we are ready to describe the steps in detail. To do so, let's use the indeterminate beam in Figure 13.13 as a demonstration.

1. *Determine the Stiffness of Each Member*

We covered this in Section 13.2. The stiffness is the moment generated for a unit rotation at that end. For cantilevers, the stiffness is zero. For all other members, the stiffness, K, is given by

$$K = \dfrac{4EI}{l}$$

736 Chapter 13 Moment Distribution Method

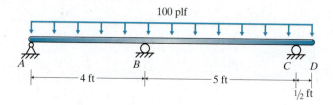

Figure 13.13 Indeterminate beam used to demonstrate the Moment Distribution Method.

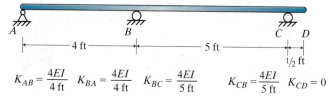

Figure 13.14 Stiffness of each member of the indeterminate beam.

For our example beam, we get the stiffnesses shown in Figure 13.14.

2. *Determine the Distribution Factors*

We covered the distribution factors in Section 13.2 as well. The distribution factor, DF, is the relative stiffness of a member compared with all the members that rigidly attach at that node. The distribution factors for our example beam are shown in Figure 13.15. Notice that the distribution factor at A is one of the special cases described in Section 13.2.

3. *Calculate the Fixed-End Moments*

The Moment Distribution Method starts with us imagining that every node is clamped so that it cannot rotate. Any loads applied between the nodes will cause moments in the clamps. In Section 13.2, we defined those moments as fixed-end moments, and we pointed out that for many loadings, the precalculated moments are listed inside the back cover of this text. Figure 13.16 shows the fixed-end moments for the example beam. Notice that even though the cantilevered end has zero stiffness, it does contribute moment.

4. *Iteratively Distribute the Moments*

With the fixed-end moments we are ready to begin releasing and reapplying the imaginary clamps one at a time. To facilitate tracking all of the moments, we typically use a table (Table 13.1).

It does not matter what node we start with or what sequence we use to release the nodes. Typically it is most efficient to release each node in some sequence and then repeat that sequence over and over. The only

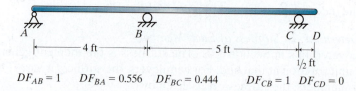

Figure 13.15 Distribution factor for each member at each node of the indeterminate beam.

Section 13.3 Beams and Sidesway Inhibited Frames 737

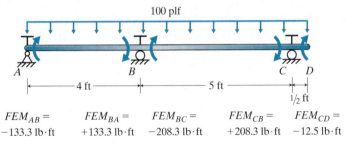

$FEM_{AB} =$ $FEM_{BA} =$ $FEM_{BC} =$ $FEM_{CB} =$ $FEM_{CD} =$
-133.3 lb·ft $+133.3$ lb·ft -208.3 lb·ft $+208.3$ lb·ft -12.5 lb·ft

Figure 13.16 Fixed-end moments at the end of each member of the indeterminate beam.

Table 13.1 Table of Preparatory Information and Fixed-End Moments for the Indeterminate Beam

NODE	A	B		C		
MEMBER	AB	BA	BC	CB	CD	
K	4EI/4	4EI/4	4EI/5	4EI/5	0	
DF	1	0.556	0.444	1	0	−UM
FEM (lb·ft)	−133.3	133.3	−208.3	208.3	−12.5	

nodes we do not need to include are the ones originally fix supported; they have a distribution factor of zero, so no moment gets distributed into the member. The key to the process is continuing the pattern until the unbalanced moment is sufficiently small.

The unbalanced moment, UM, at a node is the sum of all the moments collecting at that node due to fixed-end moments and carryover moments. When we release a clamp, the opposite of the unbalanced moment, $-UM$, is distributed to the attached members using the distribution factors. For our example beam, we could choose any node to start with; let's arbitrarily choose node B. The unbalanced moment is -75.0 lb·ft, so we distribute $+75.0$ lb·ft. Table 13.2 shows how the moment distributes.

When a node is released, the net moment after distribution is zero. To show that in the table, we typically draw a line under the distributed moments at that node (Table 13.3). Everything above the line at that node adds to zero.

Table 13.2 Moment Distribution Method Table Showing the Moments Distributing into the Members at Released Node B

NODE	A	B		C		
MEMBER	AB	BA	BC	CB	CD	
K	4EI/4	4EI/4	4EI/5	4EI/5	0	
DF	1	0.556	0.444	1	0	−UM
FEM (lb·ft)	−133.3	133.3	−208.3	208.3	−12.5	
Release B		41.7	33.3	$UM = -75.0$		75.0
		0.556(75.0)	0.444(75.0)			

738 Chapter 13 Moment Distribution Method

Table 13.3 Moment Distribution Method Table Showing that the Net Moment at a Node is Zero After Distributing into the Members

NODE	A	B		C		
MEMBER	AB	BA	BC	CB	CD	
K	4EI/4	4EI/4	4EI/5	4EI/5	0	
DF	1	0.556	0.444	1	0	−UM
FEM (lb·ft)	−133.3	133.3	−208.3	208.3	−12.5	
Release B		41.7	33.3 $\Sigma M = 0$			75.0

We learned in Section 13.2 that the moment distributed into the member causes a carryover moment at the other end of the member. Let's record that on the next line in our table (Table 13.4).

From here we can choose to release either node *A* or *C* for our example beam. Let's arbitrarily choose node *C*. The unbalanced moment includes the fixed-end moments from member *BC*, the cantilever, and the carryover moment from *B*. None of −UM distributes back into the cantilever, so all of it goes into member *CB* (Table 13.5).

We calculate the carryover moment to the node *B* end of member *BC*. Since no moment distributed to the cantilever, there is no moment to carry over there (Table 13.6).

Next we release node *A* and calculate the carryover moments. We then repeat the sequence again and again. Now our question is when do we stop?

The more iterations we use, the more accurate our answers will be. In most cases, convergence to within 1% approximate error is plenty of accuracy. To calculate the approximate error, we can use the ratio of the peak unbalanced moment at any node to the typical starting fixed-end moment. We could choose to use the smallest fixed-end moment, but that might lead to many more iterations than we need if the smallest fixed-end moment is much smaller than the others. If we want more accuracy, we

Table 13.4 Moment Distribution Method Table Showing the Carryover Moments After Distributing Moment at Node *B*

NODE	A	B		C		
MEMBER	AB	BA	BC	CB	CD	
K	4EI/4	4EI/4	4EI/5	4EI/5	0	
DF	1	0.556	0.444	1	0	−UM
FEM (lb·ft)	−133.3	133.3	−208.3	208.3	−12.5	
Release B		41.7	33.3			75.0
Carryover	20.8			16.7		

Table 13.5 Moment Distribution Method Table Showing the Moments Distributing into the Members at Released Node C

NODE	A	B		C		
MEMBER	AB	BA	BC	CB	CD	
K	4EI/4	4EI/4	4EI/5	4EI/5	0	
DF	1	0.556	0.444	1	0	−UM
FEM (lb·ft)	−133.3	133.3	−208.3	208.3	−12.5	
Release B		41.7	33.3			75.0
Carryover	20.8			16.7		$UM_C = 212.5$
Release C				−212.5	0.0	−212.5
			1(−212.5)		0(−212.5)	

Table 13.6 Moment Distribution Method Table Showing the Carryover Moment After Distributing Moment at Node C.

NODE	A	B		C		
MEMBER	AB	BA	BC	CB	CD	
K	4EI/4	4EI/4	4EI/5	4EI/5	0	
DF	1	0.556	0.444	1	0	−UM
FEM (lb·ft)	−133.3	133.3	−208.3	208.3	−12.5	
Release B		41.7	33.3			75.0
Carryover	20.8			16.7		
Release C				−212.5	0.0	−212.5
Carryover			−106.3			

can choose a smaller percentage threshold. However, don't forget that we have roundoff error. If we carry only three digits in our calculations, iterating to 0.01% error exceeds the accuracy of our calculations.

For our example beam (Figure 13.16), the fixed-end moment from the cantilever is much smaller than the other fixed-end moments, so let's use the other fixed-end moments. If we use 133 lb·ft as our typical value for the example beam, then we should continue iterating until the unbalanced moment at each node is at or below about 1.3 lb·ft. In Table 13.7, the unbalanced moment at A and C is zero, but at B the unbalanced moment is −3.1 lb·ft. Therefore, we should continue iterating.

5. *Find the Net Moments*

The Moment Distribution Method is basically a superposition of smaller and smaller effects. Therefore, once the iteration process is complete, we need to superimpose all of the moments distributed to each end of each member. To do that, we sum the moments in each column of the table (Table 13.8).

Table 13.7 Moment Distribution Method Table Showing the Unbalanced Moment at Each Node After Several Iterations.

NODE	A	B		C		
MEMBER	AB	BA	BC	CB	CD	
K	4EI/4	4EI/4	4EI/5	4EI/5	0	
DF	1	0.556	0.444	1	0	−UM
FEM (lb·ft)	−133.3	133.3	−208.3	208.3	−12.5	
Release B		41.7	33.3			75.0
Carryover	20.8			16.7		
Release C				−212.5	0.0	−212.5
Carryover			−106.3			
Release A	112.5					112.5
Carryover		56.3				
Release B		27.8	22.2			50.0
Carryover	13.9			11.1		
Release C				−11.1	0.0	−11.1
Carryover			−5.6			
Release A	−13.9					−13.9
Carryover		−6.9				
Release B		6.9	5.6			12.5
Carryover	3.5			2.8		
Release C				−2.8	0.0	−2.8
Carryover			−1.4	$UM_C = 0$		
Release A	−3.5					−3.5
Carryover	$UM_A = 0$	−1.7				
		$UM_B = -3.1$				

Table 13.8 Moment Distribution Method Table Showing the Net Moments After Sufficient Iterations.

Release A	−3.5					−3.5
Carryover			−1.7			
Release B			1.7	1.4		3.1
Carryover	0.9				0.7	
Total M	Σ = 0.9	259.0	−259.0	13.2	−12.5	

Section 13.3 Beamsand Sidesway Inhibited Frames 741

Table 13.9 Moment Distribution Method Table Showing Net Moment at Each Node as a Check for Errors.

Release A	−3.5						−3.5
Carryover		−1.7					
Release B		1.7	1.4				3.1
Carryover	0.9				0.7		
Total M	0.9	259.0	−259.0		13.2	−12.5	

$M_A = 0.9$ $M_B = 259.0 - 259.0$ $M_C = 13.2 - 12.5$ **Satisfied Fundamental Principles?**
$\approx 0 \checkmark$ $= 0 \checkmark$ $= 0.7 \approx 0 \checkmark$

Before we go any further, now is a great time to check our results for basic math or procedural errors. We can do that easily by considering that each node must be in equilibrium. Therefore, the net force of all the members at a node must equal zero. For our example beam (Table 13.9), there is only one member at node A, so the net moment is the moment in that member: 0.9 lb·ft. The moment should be zero, but this is within the 1% threshold of 1.3 lb·ft we set. Node B is in equilibrium because the net moment is zero. The net moment at node C is 0.7 lb·ft, which is also within the threshold. So all the nodes are effectively in equilibrium.

6. *Complete the Analysis*

Once we know the moment at each end of each member, we can find the reactions, internal forces, and deformations. To do so, we use the free body diagrams (FBDs) of individual members to find the shears and some reactions (e.g., Figure 13.17). We find other reactions by combining

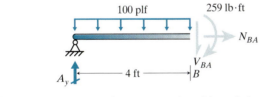

$\circlearrowleft + \Sigma M_B = 0 = A_y(4 \text{ ft}) - 100 \text{ plf}(4 \text{ ft})(4 \text{ ft}/2) + 259 \text{ lb} \cdot \text{ft}$
$\quad A_y = 135 \text{ lb} \quad (+\uparrow)$

$\circlearrowleft + \Sigma M_B = 0 = -259 \text{ lb} \cdot \text{ft} + 100 \text{ plf}(5 \text{ ft})(5 \text{ ft}/2) + V_{CB}(5 \text{ ft}) + 13.2 \text{ lb} \cdot \text{ft}$
$\quad V_{CB} = -201 \text{ lb} \quad (\downarrow +\uparrow)$

Figure 13.17 Applying equilibrium to members of the demonstration beam: (a) FBD of AB used to find A_y; (b) FBD of BC used to find V_{CB}.

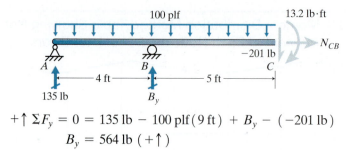

Figure 13.18 Applying equilibrium to an assembly of members from the demonstration beam to find reaction B_y.

members once we know information from the individual members (e.g., Figure 13.18). After we find all of the reactions, we can use the tools from Chapter 4: Internal Force Diagrams to construct axial force, shear, and moment diagrams for the beam or rigid frame. With the internal forces, we can use the tools from Chapter 5: Deformations to predict the slopes and displacements anywhere in the structure.

Sections 13.1 to 13.3 Highlights

Beams and Sidesway Inhibited Frames

Moment Distribution Steps:

1. Determine the stiffness of each member.
   - Cantilevers have no stiffness.
   - All other members: $K = \dfrac{4EI}{l}$
2. Determine the distribution factors.
   - End of single-member pin or roller support:

     $DF = 1$ $DF = 1$
   - End of member fixed support:

     $DF = 0$
   - End of cantilever connected to structure: $DF = 0$
   - All other members: $DF_i = \dfrac{K_i}{\Sigma K_j}$
3. Calculate the fixed-end moments.
   - Imagine every node, except the free end of any cantilevered members, clamped so that it cannot rotate.
   - Calculate the reactions required to keep the ends from rotating when loads are applied to the member (fixed-fixed condition unless cantilevered).
4. Iteratively distribute the moments.
   a. Choose one node and calculate the total moment carried by the clamp (unbalanced moment).
   b. Release the clamp; distribute the opposite of the unbalanced moment to the members rigidly connected at the node. The node is now in equilibrium.
   c. Consider the node clamped again after the distribution.
   d. For each member meeting at that node, carry over half of the moment distributed to the member to the other end.
   e. Repeat the process node by node multiple times until the unbalanced moment at all nodes is less than or equal to 1% of the typical fixed-end moment in the problem.
5. Find the net moment at each end of each member by summing all the moments down the table column.
6. Use the internal moments and equations of equilibrium to find the reactions. Then calculate any desired quantity (e.g., internal forces, deformations).

Assumptions:

Same as for idealization.
- Connections are perfectly rigid, so they maintain their angles.

Same as for predicting deformations.
- Plane sections remain plane, or the shear deformation is negligible.
- Displacements or slopes are relatively small.
- Material remains linear elastic.

Section 13.3 Beams and Sidesway Inhibited Frames   743

EXAMPLE 13.1

A large piece of equipment is being delivered via special transport to its final destination. Our firm has been hired to evaluate the adequacy of the bridges the equipment must cross. We have been assigned to evaluate a three-span overpass. One of the concerns is overloading the splice that holds pieces of the girder together. Using the concept of influence lines, we have determined where the transport must be in order to cause the largest shear on the splice: just to the right of the splice. Now we need to find the actual shear generated by the transport and equipment.

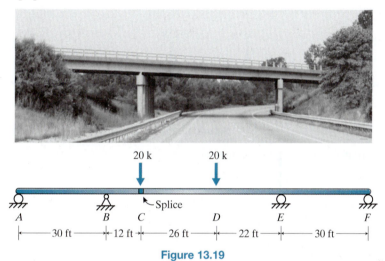

Figure 13.19

The bridge deck will distribute the point forces out to the various bridge girders across the width. A common method for determining the effect on each girder is the Distribution Factor Method. Another team member will perform those calculations, but first we need to determine the net shear at the splice due to the equipment and transport.

The senior engineer on the team wants to be sure we have reasonable results, so one engineer is assigned to use the Force Method to find the shear at C. Our job is to use the Moment Distribution Method. We are also tasked with finding the four vertical reactions.

Detailed Analysis:

Determine the distribution factors:

Stiffness of each member:

$$K_{AB} = K_{BA} = \frac{4EI}{30 \text{ ft}}$$

$$K_{BE} = K_{EB} = \frac{4EI}{60 \text{ ft}}$$

$$K_{EF} = K_{FE} = \frac{4EI}{30 \text{ ft}}$$

Distribution factors:

$$DF_{AB} = 1 \quad \text{(single member at end with roller support)}$$

$$DF_{BA} = \frac{4EI/30 \text{ ft}}{4EI/30 \text{ ft} + 4EI/60 \text{ ft}} = 0.667$$

$$DF_{BE} = \frac{4EI/60 \text{ ft}}{4EI/30 \text{ ft} + 4EI/60 \text{ ft}} = 0.333$$

$$DF_{EB} = \frac{4EI/60 \text{ ft}}{4EI/30 \text{ ft} + 4EI/60 \text{ ft}} = 0.333$$

$$DF_{EF} = \frac{4EI/30 \text{ ft}}{4EI/30 \text{ ft} + 4EI/60 \text{ ft}} = 0.667$$

$DF_{FE} = 1$ (single member at end with roller support)

Calculate the fixed-end moments:

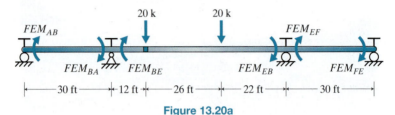

Figure 13.20a

For the members with no load between the nodes:

$$FEM_{AB} = FEM_{BA} = FEM_{EF} = FEM_{FE} = 0$$

For member BE, use formulas from inside the back cover and superposition:

$$FEM_{BE} = -\frac{20 \text{ k}(12 \text{ ft})(48 \text{ ft})^2}{(60 \text{ ft})^2} - \frac{20 \text{ k}(38 \text{ ft})(22 \text{ ft})^2}{(60 \text{ ft})^2} = -255.8 \text{ k} \cdot \text{ft}$$

$$FEM_{EB} = +\frac{20 \text{ k}(12 \text{ ft})^2(48 \text{ ft})}{(60 \text{ ft})^2} + \frac{20 \text{ k}(38 \text{ ft})^2(22 \text{ ft})}{(60 \text{ ft})^2} = +214.9 \text{ k} \cdot \text{ft}$$

Distribute the moments:

Stop when the peak UM is no more than 2.1 k·ft.
Moment distribution is shown in Table 13.10

Resulting unbalanced moments:

$UM_A = 2.1 \text{ k} \cdot \text{ft}$
$UM_B = -0.5 \text{ k} \cdot \text{ft}$
$UM_E = 0.0 \text{ k} \cdot \text{ft}$
$UM_F = -1.1 \text{ k} \cdot \text{ft}$

All are at or below our stopping threshold of 2.1 k·ft.
∴ Okay to stop

Evaluation of Results:

Satisfied Fundamental Principles?

Check equilibrium of the net moments:

$\Sigma M_A = 2.1 \text{ k} \cdot \text{ft} \approx 0$ ✓
$\Sigma M_B = 186.6 \text{ k} \cdot \text{ft} - 187.1 \text{ k} \cdot \text{ft} = -0.5 \text{ k} \cdot \text{ft} \approx 0$ ✓
$\Sigma M_E = 166.3 \text{ k} \cdot \text{ft} - 166.4 \text{ k} \cdot \text{ft} = -0.1 \text{ k} \cdot \text{ft} \approx 0$ ✓
$\Sigma M_F = -1.1 \text{ k} \cdot \text{ft} \approx 0$ ✓

Table 13.10

NODE	A	B		E		F	
MEMBER	AB	BA	BE	EB	EF	FE	
K	4EI/30	4EI/30	4EI/60	4EI/60	4EI/30	4EI/30	
DF	1	0.667	0.333	0.333	0.667	1	−UM
FEM (k·ft)			−255.8	214.9			
Release B		170.5	85.3				255.8
Carryover	85.3			42.6			
Release E				−85.8	−171.7		−257.5
Carryover			−42.9			−85.8	
Release F						85.8	85.8
Carryover					42.9		
Release A	−85.3						−85.3
Carryover		−42.6					
Release B		57.0	28.5				85.6
Carryover	28.5			14.3			
Release E				−19.1	−38.1		−57.2
Carryover			−9.5			−19.1	
Release F						19.1	19.1
Carryover					9.5		
Release A	−28.5						−28.5
Carryover		−14.3					
Release B		15.9	7.9				23.8
Carryover	7.9			4.0			
Release E				−4.5	−9.0		−13.5
Carryover			−2.2			−4.5	
Release F						4.5	4.5
Carryover					2.2		
Release A	−7.9						−7.9
Carryover		−4.0					
Release B		4.1	2.1				6.2
Carryover	2.1			1.0			
Release E				−1.1	−2.2		−3.3
Carryover			−0.5			−1.1	
Total M	2.1	186.6	−187.1	166.3	−166.4	−1.1	

746   **Chapter 13**   Moment Distribution Method

Find the shear at C:
   FBD of *BE*:

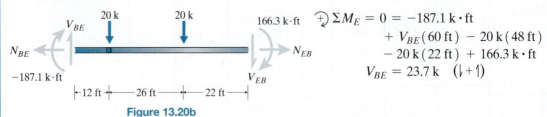

$$+\circlearrowleft \Sigma M_E = 0 = -187.1 \text{ k}\cdot\text{ft} + V_{BE}(60 \text{ ft}) - 20 \text{ k}(48 \text{ ft}) - 20 \text{ k}(22 \text{ ft}) + 166.3 \text{ k}\cdot\text{ft}$$
$$V_{BE} = 23.7 \text{ k} \quad (\downarrow + \uparrow)$$

Figure 13.20b

FBD of *BC*:

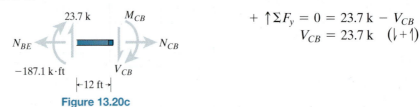

$$+\uparrow \Sigma F_y = 0 = 23.7 \text{ k} - V_{CB}$$
$$V_{CB} = 23.7 \text{ k} \quad (\downarrow + \uparrow)$$

Figure 13.20c

Find the vertical reactions:
   FBD of *AB*:

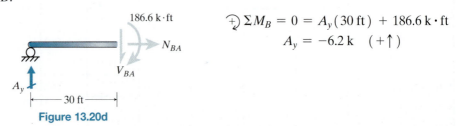

$$+\circlearrowleft \Sigma M_B = 0 = A_y(30 \text{ ft}) + 186.6 \text{ k}\cdot\text{ft}$$
$$A_y = -6.2 \text{ k} \quad (+\uparrow)$$

Figure 13.20d

FBD of *AE*:

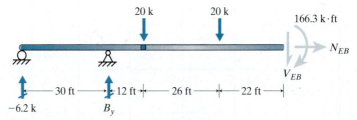

Figure 13.20e

$$+\circlearrowleft \Sigma M_E = 0 = (-6.2 \text{ k})(90 \text{ ft}) + B_y(60 \text{ ft}) - 20 \text{ k}(48 \text{ ft}) - 20 \text{ k}(22 \text{ ft}) + 166.3 \text{ k}\cdot\text{ft}$$
$$B_y = 29.9 \text{ k} \quad (+\uparrow)$$

FBD of *EF*:

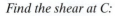

$$+\circlearrowleft \Sigma M_E = 0 = -166.4 \text{ k}\cdot\text{ft} - F_y(30 \text{ ft})$$
$$F_y = -5.5 \text{ k} \quad (+\uparrow)$$

Figure 13.20f

FBD of BF:

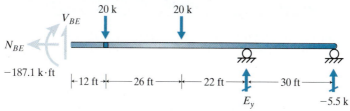

Figure 13.20g

$$+\circlearrowleft \Sigma M_B = 0 = -187.1 \text{ k} \cdot \text{ft} + 20 \text{ k}(12 \text{ ft}) + 20 \text{ k}(38 \text{ ft})$$
$$-E_y(60 \text{ ft}) - (-5.5 \text{ k})(90 \text{ ft})$$
$$E_y = 21.8 \text{ k} \quad (+\uparrow)$$

Summary of reactions:

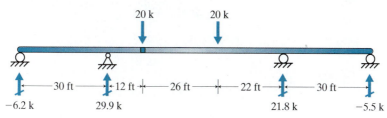

Figure 13.21

Evaluation of Results:
Satisfied Fundamental Principles?

Check equilibrium:

$$+\uparrow \Sigma F_y = -6.2 \text{ k} + 29.9 \text{ k} - 20 \text{ k} - 20 \text{ k} + 21.8 \text{ k} - 5.5 \text{ k} = 0 \checkmark$$
$$+\circlearrowleft \Sigma M_A = -29.9 \text{ k}(30 \text{ ft}) + 20 \text{ k}(42 \text{ ft}) + 20 \text{ k}(68 \text{ ft})$$
$$-21.8 \text{ k}(90 \text{ ft}) - (-5.5 \text{ k})(120 \text{ ft})$$
$$= 1 \text{ k} \cdot \text{ft} \approx 0 \checkmark$$

Satisfied Fundamental Principles?

Our predictions of the shear at C and the vertical reactions match the Force Method results from Example 12.4 exactly. ✓

Conclusion:

These observations are strong evidence that our results from the Moment Distrubution Method are reasonable.

Comparison with computer aided analysis results:

The reactions match the results of the computer aided analysis to within ½%, which is within roundoff error. ✓

Reaction	Moment Distribution	Computer Aided	Difference
A_y	−6.2 k	−6.2 k	0%
B_x	0 k	0 k	0%
B_y	29.9 k	29.9 k	0%
E_y	21.8 k	21.9 k	0.5%
F_y	−5.5 k	−5.5 k	0%

EXAMPLE 13.2

Our team is designing the awning for a new office building. The team leader wants to explore a couple different structural schemes. One of them is a rigid frame. Our task is to analyze the rigid frame option carrying snow load.

Figure 13.22

In order to compare the different structural schemes, the team leader wants to know the peak moment in each member.

Estimated Solution:

Approximate member BC as a simply supported beam. This should provide an upper bound to the peak moment in the beam.

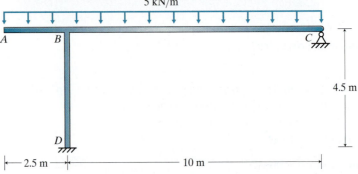

Figure 13.23

$$M_{max} = \frac{wl^2}{8} = \frac{5 \text{ kN/m}(10 \text{ m})^2}{8} = 62.5 \text{ kN} \cdot \text{m}$$

Section 13.3 Beams and Sidesway Inhibited Frames

Detailed Analysis:

Determine the distribution factors:

Stiffness of each member:

$$K_{BA} = 0 \quad (\text{cantilevered member})$$

$$K_{BC} = K_{CB} = \frac{4EI}{10 \text{ m}}$$

$$K_{BD} = K_{DB} = \frac{4EI}{4.5 \text{ m}}$$

Distribution factors:

$$DF_{BA} = 0 \quad (\text{cantilevered member})$$

$$DF_{BC} = \frac{4EI/10 \text{ m}}{0 + 4EI/10 \text{ m} + 4EI/4.5 \text{ m}} = 0.310$$

$$DF_{BD} = \frac{4EI/4.5 \text{ m}}{0 + 4EI/10 \text{ m} + 4EI/4.5 \text{ m}} = 0.690$$

$$DF_{CB} = 1 \quad (\text{single member at end with roller support})$$

$$DF_{DB} = 0 \quad (\text{end of member at fixed support})$$

Calculate the fixed-end moments:

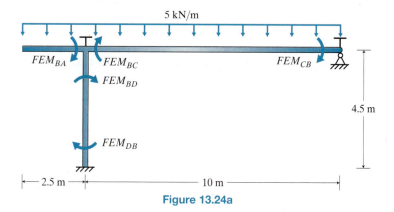

Figure 13.24a

For the cantilever, use statics:

$$FEM_{BA} = 5 \text{ kN/m} (2.5 \text{ m})(2.5 \text{ m}/2) = 15.6 \text{ kN} \cdot \text{m}$$

For member BC, use formulas from inside the back cover:

$$FEM_{BC} = -\frac{5 \text{ kN/m} (10 \text{ m})^2}{12} = -41.7 \text{ kN} \cdot \text{m}$$

$$FEM_{CB} = +\frac{5 \text{ kN/m} (10 \text{ m})^2}{12} = +41.7 \text{ kN} \cdot \text{m}$$

For the column with no load between the nodes:

$$FEM_{BD} = FEM_{DB} = 0$$

Distribute the moments:
Stop when the peak *UM* is no more than 1.6 kN·m. There is no need to release node *D* because it is a fixed support.

Table 13.11

NODE		B		C	D	
MEMBER	BA	BC	BD	CB	DB	
K	0	4EI/10	4EI/4.5	4EI/10	4EI/4.5	
DF	0	0.310	0.690	1	0	−UM
FEM (kN·m)	15.6	−41.7		41.7		
Release C				−41.7		−41.7
Carryover		−20.9				
Release B	0.0	14.6	32.4			47.0
Carryover				7.3	16.2	
Release C				−7.3		−7.3
Carryover		−3.6				
Release B	0.0	1.1	2.5			3.6
Carryover				0.6	1.3	
Total M	15.6	−50.5	34.9	0.6	17.5	

Unbalanced moments:
$$UM_B = 0.0 \text{ kN·m}$$
$$UM_C = 0.6 \text{ kN·m}$$
All are at or below our stopping threshold of 1.6 kN·m.
∴ Okay to stop

Evaluation of Results:

Satisfied Fundamental Principles?

Check equilibrium of the net moments:
$$\Sigma M_B = 15.6 \text{ kN·m} - 50.5 \text{ kN·m} + 34.9 \text{ kN·m} = 0 \checkmark$$
$$\Sigma M_C = 0.6 \text{ kN·m} \approx 0 \checkmark$$

Find the reactions to evaluate the results:
FBD of *BC*:

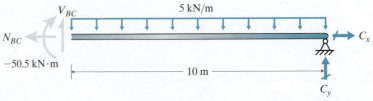

Figure 13.24b

Section 13.3 Beams and Sidesway Inhibited Frames

$$+\circlearrowleft \Sigma M_B = 0 = -50.5 \text{ kN} \cdot \text{m} + 5 \text{ kN/m}(10 \text{ m})(10 \text{ m}/2) - C_y(10 \text{ m})$$
$$C_y = 20.0 \text{ kN} \quad (+\uparrow)$$
$$+\uparrow \Sigma F_y = 0 = V_{BC} - 5 \text{ kN/m }(10 \text{ m}) + 20.0 \text{ kN}$$
$$V_{BC} = 30.0 \text{ kN} \quad (\downarrow + \uparrow)$$

FBD of BD:

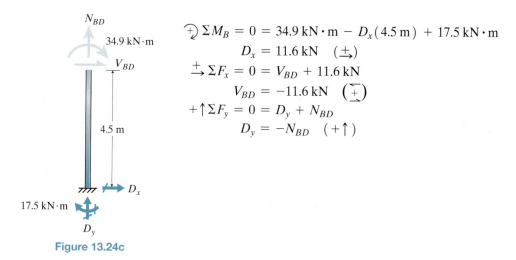

$$+\circlearrowleft \Sigma M_B = 0 = 34.9 \text{ kN} \cdot \text{m} - D_x(4.5 \text{ m}) + 17.5 \text{ kN} \cdot \text{m}$$
$$D_x = 11.6 \text{ kN} \quad (\xrightarrow{+})$$
$$\xrightarrow{+} \Sigma F_x = 0 = V_{BD} + 11.6 \text{ kN}$$
$$V_{BD} = -11.6 \text{ kN} \quad (\xrightarrow{+})$$
$$+\uparrow \Sigma F_y = 0 = D_y + N_{BD}$$
$$D_y = -N_{BD} \quad (+\uparrow)$$

Figure 13.24c

FBD of AC:
The moment is in the opposite direction of the internal force from column BD, not a moment obtained from distribution.

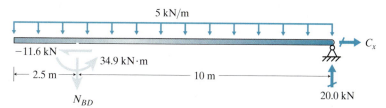

Figure 13.24d

$$+\uparrow \Sigma F_y = 0 = -5 \text{ kN/m}(12.5 \text{ m}) - N_{BD} + 20.0 \text{ kN}$$
$$N_{BD} = -42.5 \text{ kN} \quad (\updownarrow)$$
$$\therefore D_y = 42.5 \text{ kN} \quad (+\uparrow)$$
$$\xrightarrow{+} \Sigma F_x = 0 = -(-11.6 \text{ kN}) + C_x$$
$$C_x = -11.6 \text{ kN} \quad (\xrightarrow{+})$$

752 Chapter 13 Moment Distribution Method

Summary of reactions:

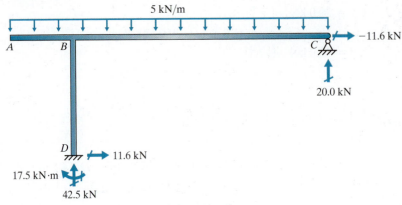

Figure 13.25

Evaluation of Results:

Satisfied Fundamental Principles?

Check equilibrium:

$$\xrightarrow{+} \Sigma F_x = 11.6 \text{ kN} + (-11.6 \text{ kN}) = 0 \checkmark$$
$$+\uparrow \Sigma F_y = -5 \text{ kN/m}(12.5 \text{ m}) + 42.5 \text{ kN} + 20 \text{ kN} = 0 \checkmark$$
$$\curvearrowleft_+ \Sigma M_C = -5 \text{ kN/m}(12.5 \text{ m})(12.5 \text{ m}/2) + 17.5 \text{ kN} \cdot \text{m}$$
$$+ 42.5 \text{ kN}(10 \text{ m}) - 11.6 \text{ kN}(4.5 \text{ m})$$
$$= -0.3 \text{ kN} \cdot \text{m} \approx 0 \checkmark$$

Find the peak moment in each member:
   Cantilever AB:
       Draw the moment as positive for the diagram (not clockwise).

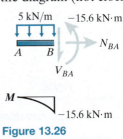

Figure 13.26

Beam BC:

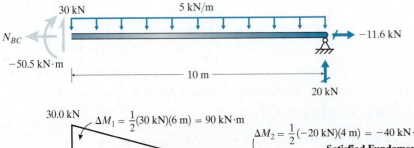

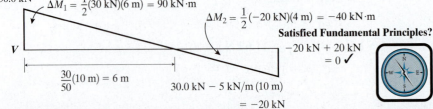

Satisfied Fundamental Principles?
$$-20 \text{ kN} + 20 \text{ kN}$$
$$= 0 \checkmark$$

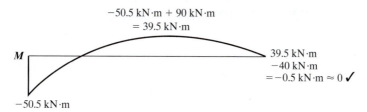

Figure 13.27

Column DB:
  Draw the internal moment as positive for the diagram (not clockwise).

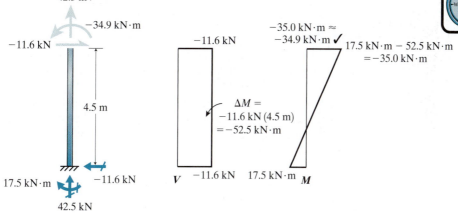

Figure 13.28

Summary of peak moments:
$M_{AB\,max} = -15.6$ kN·m $(\circlearrowright+\circlearrowleft)$
$M_{BC\,max} = -50.5$ kN·m $(\circlearrowright+\circlearrowleft)$
$M_{DB\,max} = -34.9$ kN·m $(\stackrel{+}{\leftarrow})$

Evaluation of Results:

Approximation Predicted Outcomes?
  Our approximate prediction of the peak moment in beam BC is larger than the actual, as expected. Our actual peak moment is 19% smaller than the approximate result. ✓

Conclusion:
  Our reactions satisfy equilibrium.
  Our moment diagrams close to zero or within roundoff error of zero.
  Our detailed analysis result for beam BC is below the upper bound.
  This evidence suggests that we have reasonable results.

Comparison with computer aided analysis results:
  Most of the reactions match the results of the the computer aided analysis to within 1%, which is within roundoff error. ✓

Reaction	Moment Distribution	Computer Aided	Difference
D_x	11.6 kN	11.5 kN	1%
D_y	42.5 kN	42.5 kN	0%
M_D	17.5 kN·m	17.0 kN·m	3%
C_x	−11.6 kN	−11.5 kN	1%
C_y	20.0 kN	20.0 kN	0%

One reaction, M_D, is not small and has a 3% difference; this difference is not all due to roundoff error and is explained in Section 13.4: Sidesway Frames.

754 Chapter 13 Moment Distribution Method

13.4 Sidesway Frames

Introduction

Most rigid frames are sidesway frames because we tend to use the rigid connections as the lateral load-resisting system. Sidesway means that one joint of a member displaces relative to the other due to flexural deformation.

Not only are fixed-end moments generated by loads between the joints, but they are also generated by the relative displacement of the ends of the member. We can calculate the magnitude if we know the actual displacements. Since we typically don't know the displacements before we perform the analysis, however, we need to take a different approach.

How-To

If a rigid frame can move laterally, drift, it will under most loadings. Of course, lateral load will cause nodes to drift, but a frame will also drift if it is unsymmetric or if the loading is unsymmetric (Figure 13.29). Looking at the fixed-end moments inside the back cover of this text, we see that moments are generated when one end of a member displaces relative to the other end (Figure 13.30). We must account for those when using the Moment Distribution Method.

The challenge is that we need to know the magnitude of the drift in order to calculate the fixed-end moments. To calculate the magnitude of the drift, we need the results of the Moment Distribution Method, which depends on the fixed-end moments. To solve this conundrum, Cross used superposition.

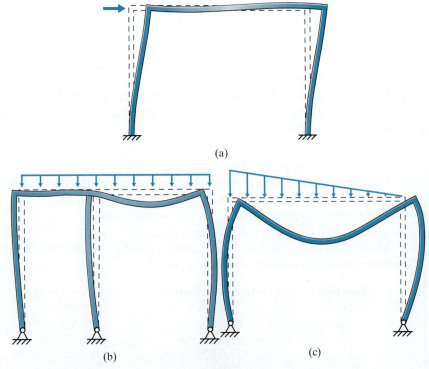

Figure 13.29 Lateral deformation of a rigid frame for various reasons: (a) Lateral load; (b) Unsymmetric frame subjected to uniform gravity load; (c) Unsymmetric gravity load.

Superposition Process for Single-Story Frames

The idea is to separate the effects of the loading from the effects of the lateral displacement, sideway. That means we need to perform two analyses for the same structure: one with the applied loads and one with the sideway. For this approach to be valid, the two load cases must add to exactly match the original loading (Figure 13.31).

To capture the effects of just the loading, we need to prevent the frame from moving laterally. To do that, we add a horizontal roller to the original structure. This gives us the middle load case in Figure 13.31. We can analyze this frame using the procedure we outlined in Section 13.3: Beams and Sideway Inhibited Frames, and from our results we can calculate the reaction at the horizontal roller, R.

To capture the effects of the sideway, we apply a horizontal force. In order for this load case to cancel out the horizontal reaction from the sideway inhibited load case, we must apply the reaction magnitude, R, in the opposite direction. That looks like the load case on the right in Figure 13.31. We still have a problem, though: we need to know the drift, Δ, caused by the force, R, in order to calculate the fixed-end moments.

Instead of trying to figure out that complicated problem, we can reverse the process: find the force, R', that causes a chosen drift, Δ'. Of course, we have no idea whether the drift we choose will cause the force we want. In fact, it probably won't. But *if* the material remains linear elastic, the relationship between the lateral force and the drift is linear. That means we can scale our results to match the reaction from the sideway inhibited load case, R. The process is shown in Figure 13.32.

If we choose the drift, Δ', the fixed-end moment is set by the formula in Figure 13.30, so we actually choose the magnitude of the fixed-end moment most of the time. The structure is exactly the same for both load cases; therefore, the member stiffnesses and distribution factors do not change. That means we can begin the moment distribution process as soon as we select the fixed-end moments. Remember that we begin by clamping each node so that it cannot rotate. As a result, the columns will experience fixed-end moments but the beams will not (Figure 13.33).

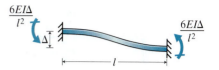

Figure 13.30 The moments generated at the ends of a fixed-fixed beam subjected to relative displacement of one end.

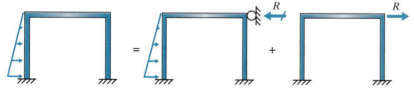

Figure 13.31 The Moment Distribution Method for a sideway frame is the superposition of loading effects and sideway effects.

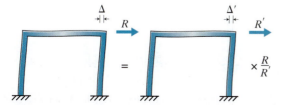

Figure 13.32 The results of analysis are proportional to the applied load, so they can be scaled to match a different applied load magnitude.

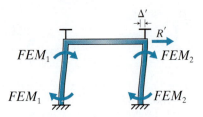

Figure 13.33 For the sidesway load case, only the columns experience fixed-end moments.

After the moment distribution converges sufficiently, we can calculate the force associated with our chosen fixed-end moments, R'. To scale the results, we multiply the sidesway load case results by the *scaling factor, SF*:

$$SF = \frac{R}{R'}$$

All of the results scale, so we can use the scaling factor on the net moments, other internal forces, reactions, and deformations.

Assumptions

We now see how the lateral movement of one end of a member relative to the other end causes additional moments. We have completely ignored one source of lateral movement in both sidesway and sidesway inhibited frames: member length change.

As axial force is transferred through the beams, the beams change length. That means the drift of each column will be different (Figure 13.34a). Even if the frame has no net drift, there can be moments introduced due to length change of the beam (Figure 13.34b).

Don't forget that the columns also change length based on their axial force and cross-sectional area. If two columns carry the same axial force but have different cross-sectional areas, they will shorten by different amounts, which causes moments in the beam (Figure 13.35a). Even if the columns are all the same size, they might carry different axial forces (Figure 13.35b).

Fortunately, in Section 5.3: Virtual Work Method, we saw that the length change normally contributes much less to the overall behavior of a rigid frame than bending. So ignoring the effects of member length change in the Moment Distribution Method is typically reasonable. The only place we might find the effect of length change is when comparing Moment Distribution Method results with computer aided analysis results. Even if we carry lots of digits and use a very small stopping threshold, our by hand calculation results might not match the computer results because computer aided analysis, which uses the Direct Stiffness Method (Chapter 15), considers member length change.

Superposition Process for Multistory Frames

Applying the Moment Distribution Method to multistory frames has an additional challenge: the story drift is not the same at each level. Remember that our approach for calculating the sidesway effect is to choose the drift and calculate the force that caused it. We can't correctly choose all the story drifts in the correct ratios because each frame will behave differently (Figure 13.36).

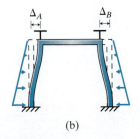

Figure 13.34 Drift of columns due to length change of the beam: (a) Deformation causes different drift magnitudes for the columns; (b) Deformation causes columns for a no-sidesway situation to have drift.

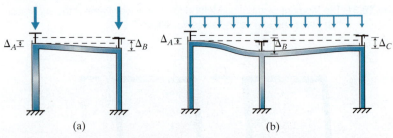

Figure 13.35 Differential displacement of beams due to length change of the columns: (a) Different cross-sectional areas result in different axial shortening of columns; (b) Different axial forces result in different axial shortening of columns.

Section 13.4 Sidesway Frames

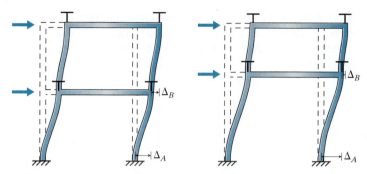

Figure 13.36 Each rigid frame will experience a different combination of story drifts.

The key is to get the sidesway load case back to one story drift at a time. So, for a two-story rigid frame, we would need to perform moment distribution for three load cases and superimpose the results (Figure 13.37).

The original structure was free to sidesway, so the sum of the forces at each level must equal zero. We call this compatibility. For our example frame in Figure 13.37, we have the following equations of compatibility:

$$\Sigma R_{1i} = 0 = R_1 + R_{11} + R_{12}$$
$$\Sigma R_{2i} = 0 = R_2 + R_{21} + R_{22}$$

But we don't obtain the actual forces R_{11}, R_{12}, and so on from analysis of our load cases. We choose the drift at each level, so we obtain forces R'_{11}, R'_{12}, and so on. We need to scale each set of forces by its own scaling factor. For our example frame, the compatibility equations become

$$0 = R_1 + SF_1(R'_{11}) + SF_2(R'_{12})$$
$$0 = R_2 + SF_1(R'_{21}) + SF_2(R'_{22})$$

There is one scaling factor for each sidesway load case, and there was one sidesway load case for each story of the frame. There is one compatibility equation for each story of the frame, so we will always have a linear

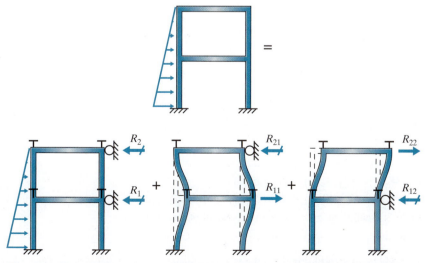

Figure 13.37 Different load cases that must be superimposed to apply the Moment Distribution Method to analyze a two-story rigid frame.

758   Chapter 13   Moment Distribution Method

system of equations to solve for the scaling factors. For example, for a 10-story frame, we will need to analyze 11 load cases and solve a system of 10 equations and 10 unknowns. Clearly, as the number of stories increases, the Moment Distribution Method becomes less and less efficient for analysis of rigid frames that will sidesway.

Once we know the scaling factors for the frame, we can superimpose the moment results. With the moments, we can find any internal forces, reactions, or displacements we desire.

SECTION 13.4 HIGHLIGHTS

Sidesway Frames
Sidesway frames drift due to lateral loads, unsymmetric gravity loading, or an unsymmetric frame.

Single-Story Frame:
Use the same analysis process as for beams and sidesway inhibited frames, except multiple load cases result in different fixed-end moments:

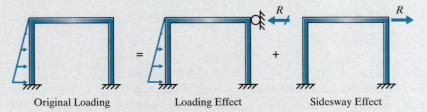

Original Loading   Loading Effect   Sidesway Effect

1. After performing moment distribution on the loading effect load case, find the reaction R.
2. Choose the fixed-end moments for the sidesway effect load case, and find the associated force R'.
3. Modify the results of the sidesway effect load case using the scaling factor so we have results as if R was applied:

$$SF = \frac{R}{R'}$$

4. Superimpose the results of the loading effect load case and the scaled sidesway effect load case to get the results for the original loading.

Multistory Frame:
Use the same analysis process as for a single story, but there is a load case for each story:

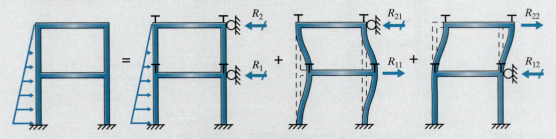

1. Solve the system of compatibility equations for the scaling factors; there will be one per story:

$$0 = R_1 + SF_1(R'_{11}) + SF_2(R'_{12})$$
$$0 = R_2 + SF_1(R'_{21}) + SF_2(R'_{22})$$

2. Modify each sidesway load case result by its scaling factor, and superimpose all of the results.

Additional Assumption:
There is negligible length change in all members, so no additional fixed-end moments.

Example 13.3

A school installed a lightweight frame to mark the start and finish line at the track. For a special event coming up, the organizers want to hang a single heavy piece of equipment from the frame. They contacted our firm to evaluate the frame to determine whether it is safe to hang the equipment from it. After inspecting the frame in the field, our team has determined that the weakest parts are the moment connections at the corners and the connections to the foundations.

One of our concerns is a possible seismic event while the equipment is hanging from the frame. Another team member has calculated the static equivalent seismic load.

Figure 13.38

Our task is to predict the resulting moment at each of the beam-column connections and the reactions at the foundations.

Estimated Solution:

Approximate member BC as a fix-fix supported beam. If there was only vertical load, this would provide an upper bound to the moments at the ends. However, since there is also lateral load on the frame, this is only an approximation.

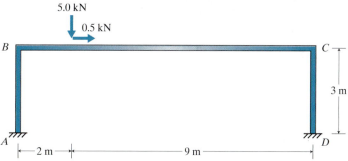

Figure 13.39

Use the formulas from inside the back cover:

$$M_{BC} = -\frac{5 \text{ kN}(2 \text{ m})(9 \text{ m})^2}{(11 \text{ m})^2} = -6.69 \text{ kN} \cdot \text{m}$$

760 Chapter 13 Moment Distribution Method

$$M_{CB} = +\frac{5\,\text{kN}\,(9\,\text{m})(2\,\text{m})^2}{(11\,\text{m})^2} = +1.49\,\text{kN}\cdot\text{m}$$

Detailed Analysis:

Assumption:
  The effect of the length change of members is negligible.

Determine the distribution factors:
  Stiffness of each member:

$$K_{AB} = K_{BA} = \frac{4EI}{3\,\text{m}}$$

$$K_{BC} = K_{CB} = \frac{4EI}{11\,\text{m}}$$

$$K_{CD} = K_{DC} = \frac{4EI}{3\,\text{m}}$$

Distribution factors:

$$DF_{AB} = 0 \quad (\text{end of member at fixed support})$$

$$DF_{BA} = \frac{4EI/3\,\text{m}}{4EI/3\,\text{m} + 4EI/11\,\text{m}} = 0.786$$

$$DF_{BC} = \frac{4EI/11\,\text{m}}{4EI/3\,\text{m} + 4EI/11\,\text{m}} = 0.214$$

$$DF_{CB} = \frac{4EI/11\,\text{m}}{4EI/11\,\text{m} + 4EI/3\,\text{m}} = 0.214$$

$$DF_{CD} = \frac{4EI/3\,\text{m}}{4EI/11\,\text{m} + 4EI/3\,\text{m}} = 0.786$$

$$DF_{DC} = 0 \quad (\text{end of member at fixed support})$$

Calculate the load effects (no sidesway):
  Put a horizontal roller at C to stop sidesway.
  Fixed-end moments:

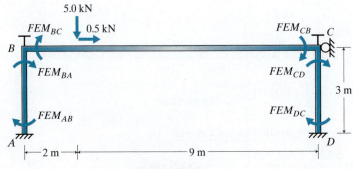

Figure 13.40a

For the columns with no load between the nodes:
$$FEM_{AB} = FEM_{BA} = FEM_{CD} = FEM_{DC} = 0$$

For the beam, we calculated the moments for approximate analysis:
$$FEM_{BC} = -6.69 \text{ kN} \cdot \text{m}$$
$$FEM_{CB} = +1.49 \text{ kN} \cdot \text{m}$$

Distribute the moments:
Stop when the peak UM is no more than 0.015 kN·m. There is no need to release node A or D because they are fixed supports.

Table 13.12

NODE	A	B		C		D	
MEMBER	AB	BA	BC	CB	CD	DC	
K	4EI/3	4EI/3	4EI/11	4EI/11	4EI/3	4EI/3	
DF	0	0.786	0.214	0.214	0.786	0	−UM
FEM (kN·m)			−6.690	1.490			
Release B		5.256	1.434				6.690
Carryover	2.628			0.717			
Release C				−0.473	−1.734		−2.207
Carryover			−0.236			−0.867	
Release B		0.186	0.051				0.236
Carryover	0.093			0.025			
Release C				−0.005	−0.020		−0.025
Carryover			−0.003			−0.010	
Total M_{NS}	2.721	5.442	−5.444	1.754	−1.754	−0.877	

Unbalanced moments:
$$UM_B = -0.002 \text{ kN} \cdot \text{m}$$
$$UM_C = 0.000 \text{ kN} \cdot \text{m}$$

All are at or below our stopping threshold of 0.015 kN·m.
$\therefore$ Okay to stop

Evaluation of Results:

Satisfied Fundamental Principles?

Check equilibrium of the net moments:
$$\Sigma M_B = 5.442 \text{ kN} \cdot \text{m} - 5.444 \text{ kN} \cdot \text{m}$$
$$= -0.002 \text{ kN} \cdot \text{m} \approx 0 \checkmark$$
$$\Sigma M_C = 1.754 \text{ kN} \cdot \text{m} - 1.754 \text{ kN} \cdot \text{m} = 0 \checkmark$$

762   **Chapter 13**   Moment Distribution Method

Find the lateral reaction:
  FBD of *AB*:

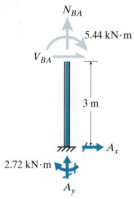
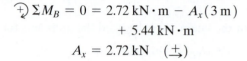

$$\circlearrowleft^{+} \Sigma M_B = 0 = 2.72 \text{ kN} \cdot \text{m} - A_x(3 \text{ m})$$
$$+ 5.44 \text{ kN} \cdot \text{m}$$
$$A_x = 2.72 \text{ kN} \quad (\xrightarrow{+})$$

Figure 13.40b

FBD of *CD*:

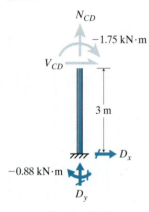
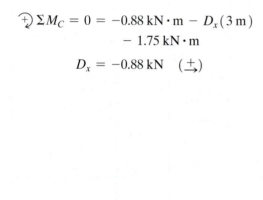

$$\circlearrowleft^{+} \Sigma M_C = 0 = -0.88 \text{ kN} \cdot \text{m} - D_x(3 \text{ m})$$
$$- 1.75 \text{ kN} \cdot \text{m}$$
$$D_x = -0.88 \text{ kN} \quad (\xrightarrow{+})$$

Figure 13.40c

FBD of entire frame:

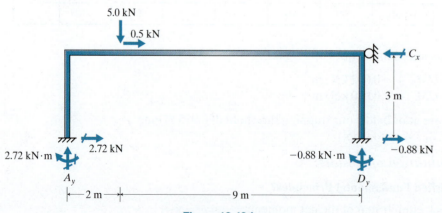

Figure 13.40d

$$\xrightarrow{+} \Sigma F_x = 0 = 2.72 \text{ kN} + 0.5 \text{ kN} - C_x + (-0.88 \text{ kN})$$
$$C_x = 2.34 \text{ kN} \quad (\xleftarrow{+})$$

Calculate the sidesway effects:
Apply an unknown lateral load at C.

Fixed-end moments:

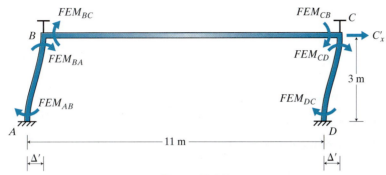

Figure 13.41a

For the columns, use formulas from inside the back cover:

$$FEM_{AB} = -\frac{6EI\Delta'}{l^2}$$

Since we choose Δ', we can choose it to have a certain fixed-end moment:

$$FEM_{AB} \equiv -100 \text{ kN} \cdot \text{m}$$

Since both columns have the same E, I, and l:

$$FEM_{AB} = FEM_{BA} = FEM_{CD} = FEM_{DC} = -100 \text{ kN} \cdot \text{m}$$

For the beam, there is no load or differential movement:

$$FEM_{BC} = FEM_{CB} = 0$$

Distribute the moments:

Stop when the peak *UM* is no more than 1 kN·m. There is still no need to release node *A* or *D* because they are fixed supports.

Table 13.13

NODE	A	B		C		D	
MEMBER	AB	BA	BC	CB	CD	DC	
K	4EI/3	4EI/3	4EI/11	4EI/11	4EI/3	4EI/3	
DF	0	0.786	0.214	0.214	0.786	0	−UM
FEM (kN·m)	−100.0	−100.0			−100.0	−100.0	
Release B		78.6	21.4				100.0
Carryover	39.3			10.7			
Release C				19.1	70.2		89.3
Carryover			9.6			35.1	
Release B		−7.5	−2.0				−9.6
Carryover	−3.8			−1.0			
Total M'ₛ	−64.5	−28.9	29.0	28.8	−29.8	−64.9	

Unbalanced moments:

$UM_B = 0.0 \text{ kN} \cdot \text{m}$

$UM_C = -1.0 \text{ kN} \cdot \text{m}$

All are at or below our stopping threshold of 1 kN·m.

∴ Okay to stop

Evaluation of Results:

Satisfied Fundamental Principles?

Check equilibrium of the net moments:

$\Sigma M_B = -28.9 \text{ kN} \cdot \text{m} + 29.0 \text{ kN} \cdot \text{m} = 0.1 \text{ kN} \cdot \text{m} \approx 0$ ✓

$\Sigma M_C = 28.8 \text{ kN} \cdot \text{m} - 29.8 \text{ kN} \cdot \text{m} = -1.0 \text{ kN} \cdot \text{m} \approx 0$ ✓

Find the lateral force:

FBD of AB:

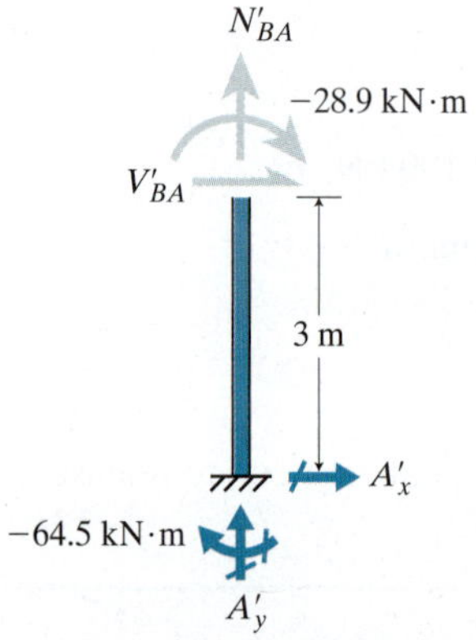

$\circlearrowleft \Sigma M_B = 0 = -64.5 \text{ kN} \cdot \text{m} - A'_x(3 \text{ m})$
$ - 28.9 \text{ kN} \cdot \text{m}$

$A'_x = -31.1 \text{ kN} \quad (\overset{+}{\to})$

Figure 13.41b

FBD of CD:

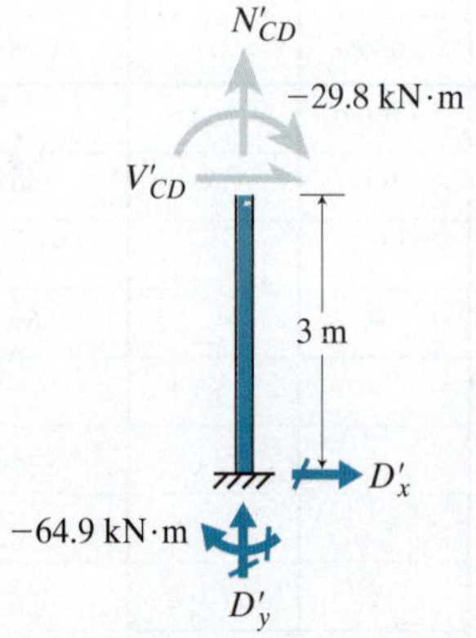

$\circlearrowleft \Sigma M_C = 0 = -64.9 \text{ kN} \cdot \text{m} - D'_x(3 \text{ m})$
$ - 29.8 \text{ kN} \cdot \text{m}$

$D'_x = -31.6 \text{ kN} \quad (\overset{+}{\to})$

Figure 13.41c

FBD of entire frame:

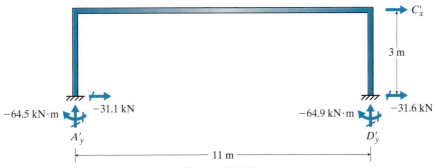

Figure 13.41d

$$\overset{+}{\rightarrow}\Sigma F_x = 0 = (-31.1 \text{ kN}) + (-31.6 \text{ kN}) + C'_x$$
$$C'_x = 62.7 \text{ kN} \ (\overset{+}{\rightarrow})$$

Scale results:

Scale factor:

$$SF = \frac{2.34 \text{ kN}}{62.7 \text{ kN}} = 0.0373$$

Scale moments:

$$M_{S\,AB} = 0.0373(-64.5 \text{ kN} \cdot \text{m}) = -2.41 \text{ kN} \cdot \text{m}$$
$$M_{S\,BA} = 0.0373(-28.9 \text{ kN} \cdot \text{m}) = -1.08 \text{ kN} \cdot \text{m}$$
$$M_{S\,BC} = 0.0373(+29.0 \text{ kN} \cdot \text{m}) = +1.08 \text{ kN} \cdot \text{m}$$
$$M_{S\,CB} = 0.0373(+28.8 \text{ kN} \cdot \text{m}) = +1.07 \text{ kN} \cdot \text{m}$$
$$M_{S\,CD} = 0.0373(-29.8 \text{ kN} \cdot \text{m}) = -1.11 \text{ kN} \cdot \text{m}$$
$$M_{S\,DC} = 0.0373(-64.9 \text{ kN} \cdot \text{m}) = -2.42 \text{ kN} \cdot \text{m}$$

Superimpose the effects:

Superimpose the moments:

$$M_{AB} = +2.72 \text{ kN} \cdot \text{m} - 2.41 \text{ kN} \cdot \text{m} = +0.31 \text{ kN} \cdot \text{m} \ (\curvearrowright)$$
$$M_{BA} = +5.44 \text{ kN} \cdot \text{m} - 1.08 \text{ kN} \cdot \text{m} = +4.36 \text{ kN} \cdot \text{m} \ (\curvearrowright)$$
$$M_{BC} = -5.44 \text{ kN} \cdot \text{m} + 1.08 \text{ kN} \cdot \text{m} = -4.36 \text{ kN} \cdot \text{m} \ (\curvearrowleft)$$
$$M_{CB} = +1.75 \text{ kN} \cdot \text{m} + 1.07 \text{ kN} \cdot \text{m} = +2.82 \text{ kN} \cdot \text{m} \ (\curvearrowleft)$$
$$M_{CD} = -1.75 \text{ kN} \cdot \text{m} - 1.11 \text{ kN} \cdot \text{m} = -2.86 \text{ kN} \cdot \text{m} \ (\curvearrowright)$$
$$M_{DC} = -0.88 \text{ kN} \cdot \text{m} - 2.42 \text{ kN} \cdot \text{m} = -3.30 \text{ kN} \cdot \text{m} \ (\curvearrowright)$$

Superimpose the horizontal reactions:

We need to scale the sidesway values:

$$A_x = +2.72 \text{ kN} + 0.0373(-31.1 \text{ kN}) = +1.56 \text{ kN} \ (\overset{+}{\rightarrow})$$
$$D_x = -0.88 \text{ kN} + 0.0373(-31.6 \text{ kN}) = -2.06 \text{ kN} \ (\overset{+}{\rightarrow})$$

Other reactions:

We found the moment and horizontal reactions by superimposing the moment distribution results. Let's use equilibrium to find the vertical reactions.

FBD of BC for the original frame:
The moments are net values from distribution, not the opposite values from the column FBDs.

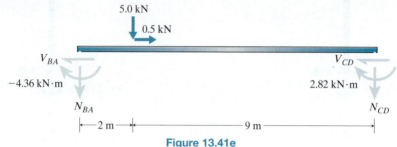
Figure 13.41e

$$\stackrel{+}{\curvearrowleft} \Sigma M_C = 0 = -4.36 \text{ kN} \cdot \text{m} - N_{BA}(11 \text{ m}) - 5 \text{ kN}(9 \text{ m}) + 2.86 \text{ kN} \cdot \text{m}$$
$$N_{BA} = -4.22 \text{ kN} \quad (\uparrow\downarrow)$$
$$+\uparrow \Sigma F_y = 0 = -(-4.22 \text{ kN}) - 5.0 \text{ kN} - N_{CD}$$
$$N_{CD} = -0.78 \text{ kN} \quad (\uparrow\downarrow)$$

FBD of AB for the original frame:
The moment is the net value from distribution, not the opposite value from the beam FBD.

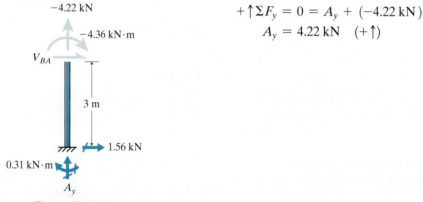
Figure 13.41f

$$+\uparrow \Sigma F_y = 0 = A_y + (-4.22 \text{ kN})$$
$$A_y = 4.22 \text{ kN} \quad (+\uparrow)$$

FBD of CD for the original frame:
The moment is the net value from distribution, not the opposite value from the beam FBD.

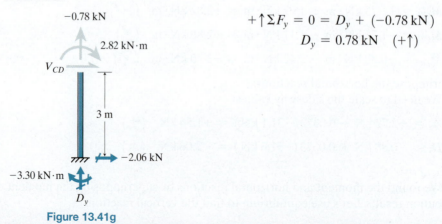
Figure 13.41g

$$+\uparrow \Sigma F_y = 0 = D_y + (-0.78 \text{ kN})$$
$$D_y = 0.78 \text{ kN} \quad (+\uparrow)$$

Summary:

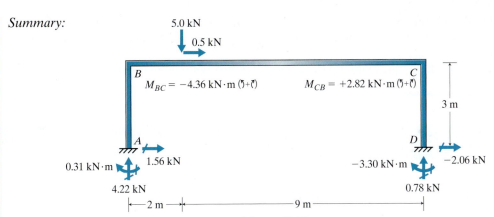

Figure 13.42

Evaluation of Results:

Satisfied Fundamental Principles?

Check equilibrium:
Note that the internal moments at the beam-column connections are shown but do not factor into the equilibrium calculations of the whole frame.

$$\xrightarrow{+} \Sigma F_x = 1.56 \text{ kN} + 0.5 \text{ kN} - 2.06 \text{ kN} = 0 \checkmark$$
$$+\uparrow \Sigma F_y = 4.22 \text{ kN} - 5.0 \text{ kN} + 0.78 \text{ kN} = 0 \checkmark$$
$$\circlearrowleft_+ \Sigma M_A = 0.31 \text{ kN} \cdot \text{m} + 5.0 \text{ kN}(2 \text{ m}) + 0.5 \text{ kN}(3 \text{ m})$$
$$\qquad -3.30 \text{ kN} \cdot \text{m} - 0.78 \text{ kN}(11 \text{ m})$$
$$\qquad = -0.07 \text{ kN} \cdot \text{m} \approx 0 \checkmark$$

Approximation Predicted Outcomes?

Our approximate prediction of the moment at BC is about 50% larger than our detailed result, but our approximate prediction of the moment at CB is about 50% of our detailed result. This suggests that our results are the right order of magnitude. ✓

Conclusion:
Our reactions satisfy equilibrium.
Since our reactions are based on the calculation of the internal moments, this suggests we have reasonable results.

Comparison with computer aided analysis results:
If we assume the size of the pipes, we can use their individual areas to calculate the cross-sectional properties for the frame. If we use the following net properties for the frame members, we get the computer aided analysis results in the table:

$A = 20 \text{ cm}^2 \quad I = 6300 \text{ cm}^4 \quad E = 200 \text{ GPa} = 20{,}000 \text{ kN/cm}^2$

Reaction	Moment Distribution	Computer Aided	Difference
A_x	1.56 kN	1.50 kN	4%
A_y	4.22 kN	4.23 kN	0.2%
M_A	0.31 kN·m	0.18 kN·m	72%
D_x	−2.06 kN	−2.00 kN	3%
D_y	0.78 kN	0.77 kN	1%
M_D	−3.03 kN·m	−3.18 kN·m	5%

768 Chapter 13 Moment Distribution Method

Most of the reactions match the results of the computer aided analysis to within 5%. ✓
Roundoff error accounts for about 1% of the difference; the rest is due to length change of the members.
The reaction with a much larger percentage difference is a relatively small magnitude.

Example 13.4

Our client is renovating an existing building and adding new mechanical equipment on the roof. To house the new equipment, we are designing a small rigid frame. The frame will attach to very stiff beams, so the frame will be effectively a single story with fixed supports.

The members have been designed to carry the gravity load. Now the team needs to verify that the members are adequate with regard to lateral load. To help in that process, our role is to analyze the frame subjected to lateral wind load. We need to find the axial force and peak moment generated in each column due to that wind load.

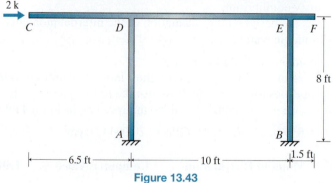

Figure 13.43

Preliminary member properties:

Beams:	W10 × 22	$A = 6.49 \text{ in}^2$	$I = 118 \text{ in}^4$
Column AD:	HSS5 × 5 × 1/2	$A = 7.88 \text{ in}^2$	$I = 26.0 \text{ in}^4$
Column BE:	HSS5 × 5 × 1/4	$A = 4.30 \text{ in}^2$	$I = 16.0 \text{ in}^4$
Steel:	$E = 29{,}000 \text{ ksi}$		

Section 13.4 Sidesway Frames

Estimated Solution:

Estimate an inflection point at midspan of the middle beam; there is no inflection point in the cantilevers. Estimate an inflection point at midheight of each column.

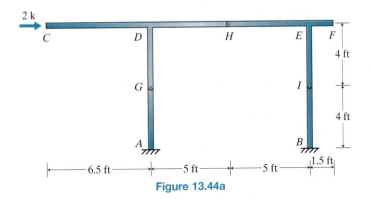

Figure 13.44a

FBD of frame cut at column inflection points:
Since the frame is wider (10 ft) than it is tall (8 ft), use the Portal Method: assume the shear to each column is the same.

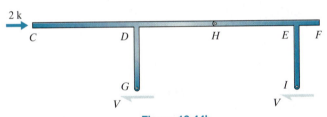

Figure 13.44b

$$\xrightarrow{+} \Sigma F_x = 0 = 2\,\text{k} - 2V$$
$$V = 1\,\text{k} \ (\downarrow{+})$$

FBD of AG:
Because of the Portal Method assumptions, the moments at the top and bottom of each column are the same magnitude and the peak moment in the two columns will be the same.

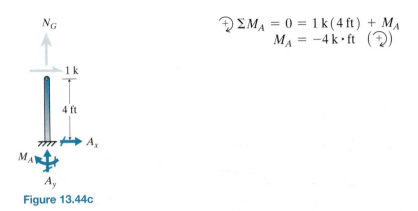

$$\curvearrowright{+} \Sigma M_A = 0 = 1\,\text{k}(4\,\text{ft}) + M_A$$
$$M_A = -4\,\text{k} \cdot \text{ft} \ (\curvearrowright{+})$$

Figure 13.44c

FBD of *GCH*:
Because there are only two columns, equilibrium requires that the axial force in each will be equal and opposite.

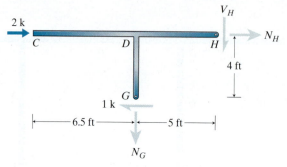

Figure 13.44d

$$+\circlearrowleft \Sigma M_H = 0 = 1\,\text{k}\,(4\,\text{ft}) - N_G(5\,\text{ft})$$
$$N_G = 0.8\,\text{k} \;(\uparrow\downarrow)$$

Summary:
   Axial force in columns: 0.8 k
   Peak moment in columns: 4 k·ft

Detailed Analysis:

Assumption:
   The effect of the length change of members is negligible.

Determine the distribution factors:
   Stiffness of each member:

$$K_{DC} = 0 \quad (\text{cantilevered member})$$

$$K_{AD} = K_{DA} = \frac{4(29{,}000\,\text{ksi})(26\,\text{in}^4)}{8\,\text{ft}} = 3.77 \times 10^5 \,\frac{\text{k} \cdot \text{in}^2}{\text{ft}}$$

$$K_{BE} = K_{EB} = \frac{4(29{,}000\,\text{ksi})(16\,\text{in}^4)}{8\,\text{ft}} = 2.32 \times 10^5 \,\frac{\text{k} \cdot \text{in}^2}{\text{ft}}$$

$$K_{DE} = K_{ED} = \frac{4(29{,}000\,\text{ksi})(118\,\text{in}^4)}{10\,\text{ft}} = 13.69 \times 10^5 \,\frac{\text{k} \cdot \text{in}^2}{\text{ft}}$$

$$K_{EF} = 0 \quad (\text{cantilevered member})$$

Distribution factors:

$$DF_{AD} = 0 \quad (\text{end of member at fixed support})$$
$$DF_{DC} = 0 \quad (\text{cantilevered member})$$
$$DF_{DA} = \frac{3.77}{0 + 3.77 + 13.69} = 0.216$$
$$DF_{DE} = \frac{13.69}{0 + 3.77 + 13.69} = 0.784$$

$$DF_{ED} = \frac{13.69}{13.69 + 0 + 2.32} = 0.855$$

$$DF_{EB} = \frac{2.32}{13.69 + 0 + 2.32} = 0.145$$

$DF_{EF} = 0$ (cantilevered member)

$DF_{BE} = 0$ (end of member at fixed support)

Calculate the load effects (no sidesway):
Put a horizontal roller at F to stop sidesway.

Fixed-end moments:

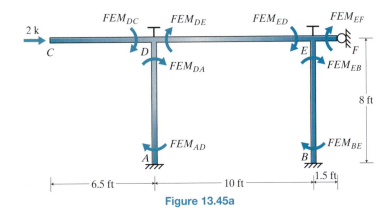

Figure 13.45a

Load along the axis of the cantilevers causes no moment:
$$FEM_{DC} = FEM_{EF} = 0$$

For the columns with no load between the nodes:
$$FEM_{AD} = FEM_{DA} = FEM_{BE} = FEM_{EB} = 0$$

For beam DE, no load between the nodes:
$$FEM_{DE} = FEM_{ED} = 0$$

Distribute the moments:
There are no moments to distribute, so there are no net moments anywhere in the frame.

Table 13.14

NODE	A	D			E			B	
MEMBER	AD	DC	DA	DE	ED	EB	EF	BE	
K	3.77	0	3.77	13.69	13.69	2.32	0	2.32	
DF	0	0	0.216	0.784	0.855	0.145	0	0	−UM
FEM (k·ft)									
Total M_{NS}	0	0	0	0	0	0	0	0	

Find the lateral reaction:
FBD of AD:

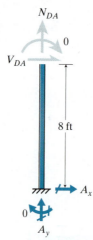

$$\circlearrowleft^+ \Sigma M_D = 0 = A_x(8\text{ ft})$$
$$A_x = 0$$

Figure 13.45b

FBD of BE:

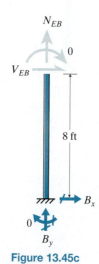

$$\circlearrowleft^+ \Sigma M_E = 0 = B_x(8\text{ ft})$$
$$B_x = 0$$

Figure 13.45c

FBD of entire frame:

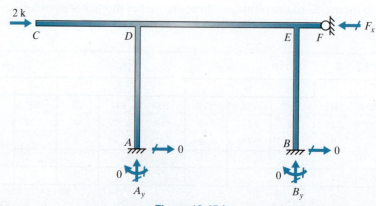

Figure 13.45d

$\xrightarrow{+} \Sigma F_x = 0 = 2\text{ k} - F_x$

$F_x = 2\text{ k} \quad (\xrightarrow{\pm})$

Observation: If there are no fixed-end moments, the lateral load is carried directly to the horizontal roller.

Calculate the sidesway effects:
Apply an unknown lateral load at F.

Fixed-end moments:

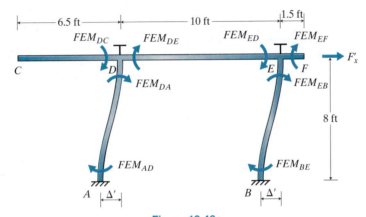

Figure 13.46a

Loading along the axis of the cantilevers causes no moment:

$$FEM_{DC} = FEM_{EF} = 0$$

For the beam, there is no load or differential movement:

$$FEM_{DE} = FEM_{ED} = 0$$

For the columns, use formulas from inside the back cover:

$$FEM = -\frac{6EI\Delta'}{l^2}$$

Since we choose Δ', we can choose it to have a certain fixed-end moment:

$$FEM_{BE} = FEM_{BE} = -\frac{6EI_{BE}\Delta'}{l^2_{BE}} \equiv -100\text{ k}\cdot\text{ft}$$

Since the columns have different I values but the same Δ':

$$\Delta' = \frac{100\text{ k}\cdot\text{ft}(l^2_{BE})}{6EI_{BE}} = \frac{100\text{ k}\cdot\text{ft}(8\text{ ft})^2}{6E(16.0\text{ in}^4)}$$

$$FEM_{AD} = FEM_{DA} = -\frac{6EI_{AD}\Delta'}{l^2_{AD}} = -\left[\frac{6E(26.0\text{ in}^4)}{(8\text{ ft})^2}\right]\left[\frac{100\text{ k}\cdot\text{ft}(8\text{ ft})^2}{6E(16.0\text{ in}^4)}\right]$$

$$= -162.5\text{ k}\cdot\text{ft}$$

Distribute the moments:
Stop when the peak UM is no more than 1 k·ft. There is no need to release node A or B because they are fixed supports.

Table 13.15

NODE	A	D			E			B	
MEMBER	AD	DC	DA	DE	ED	EB	EF	BE	
K	3.77	0	3.77	13.69	13.69	2.32	0	2.32	
DF	0	0	0.216	0.784	0.855	0.145	0	0	−UM
FEM (k·ft)	−162.5		−162.5			−100.0		−100.0	
Release D		0.0	35.1	127.4					162.5
Carryover	17.5				63.7				
Release E					31.0	5.3	0.0		36.3
Carryover				15.5				2.6	
Release D		0.0	−3.4	−12.2					−15.5
Carryover	−1.7				−6.1				
Release E					5.2	0.9	0.0		6.1
Carryover				2.6				0.4	
Release D		0.0	−0.6	−2.0					−2.6
Carryover	−0.3				−1.0				
Total M'_S	−147.0	0.0	−131.4	131.3	92.8	−93.8	0.0	−97.0	

Unbalanced moments:

$UM_D = 0.0$ k·ft
$UM_E = -1.0$ k·ft

All are at or below our stopping threshold of 1 k·ft.
∴ Okay to stop

Evaluation of Results:

Satisfied Fundamental Principles?

Check equilibrium of the net moments:
$\Sigma M_B = -131.4$ k·ft $+ 131.3$ k·ft $= -0.1$ k·ft ≈ 0 ✓
$\Sigma M_C = 92.8$ k·ft $- 93.8$ k·ft $= -1.0$ k·ft ≈ 0 ✓

Find the lateral force:
FBD of AD:

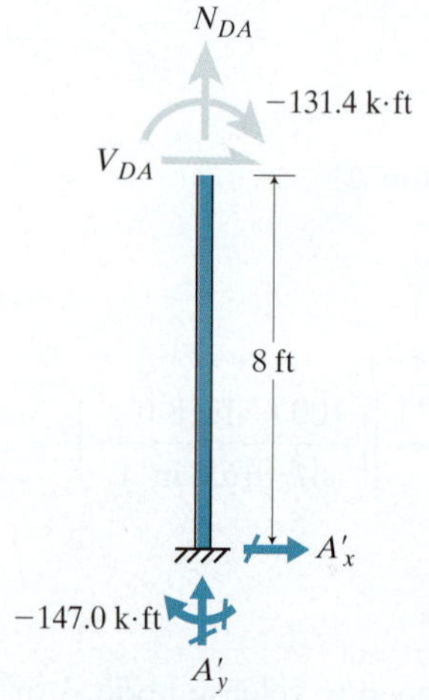

$\circlearrowright \Sigma M_D = 0 = -147.0$ k·ft $- A'_x(8\text{ ft}) + (-131.4$ k·ft$)$
$A'_x = -34.8$ k $(\xrightarrow{\pm})$

Figure 13.46b

FBD of BE:

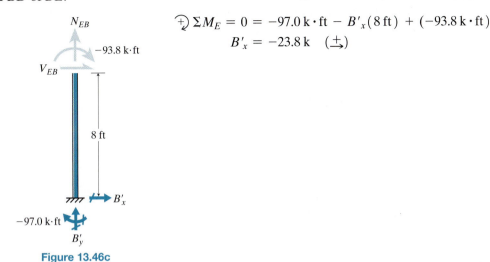

$$\circlearrowleft^{+} \Sigma M_E = 0 = -97.0 \text{ k·ft} - B'_x(8 \text{ ft}) + (-93.8 \text{ k·ft})$$
$$B'_x = -23.8 \text{ k} \quad (\overset{+}{\rightarrow})$$

Figure 13.46c

FBD of entire frame:

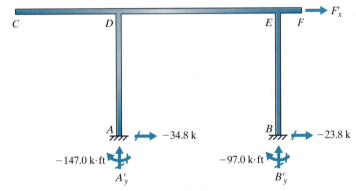

Figure 13.46d

$$\overset{+}{\rightarrow} \Sigma F_x = 0 = -34.8 \text{ k} - 23.8 \text{ k} + F'_x$$
$$F'_x = 58.6 \text{ k} \quad (\overset{+}{\rightarrow})$$

Scale results:
Scale factor:

$$SF = \frac{2 \text{ k}}{58.6 \text{ k}} = 0.0341$$

Scale moments:

$$M_{AD} = 0 + 0.0341(-147.0 \text{ k·ft}) = -5.01 \text{ k·ft} \quad (\circlearrowleft^{+})$$
$$M_{DA} = 0 + 0.0341(-131.4 \text{ k·ft}) = -4.48 \text{ k·ft} \quad (\circlearrowleft^{+})$$
$$M_{AD} = 0 + 0.0341(-93.8 \text{ k·ft}) = -3.20 \text{ k·ft} \quad (\circlearrowleft^{+})$$
$$M_{AD} = 0 + 0.0341(-97.0 \text{ k·ft}) = -3.31 \text{ k·ft} \quad (\circlearrowleft^{+})$$

Column axial forces:
We found the peak column moments by scaling the moment distribution results. Let's use equilibrium to find the column axial forces.

FBD of CF for the original frame:
The moments are the opposite of values from the column FBDs (scaled results).

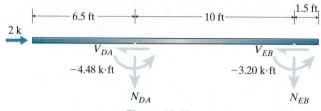

Figure 13.46e

$$\circlearrowleft^+ \Sigma M_E = 0 = -4.48 \text{ k·ft} + N_{DA}(10 \text{ ft}) + (-3.20 \text{ k·ft})$$
$$N_{DA} = 0.77 \text{ k} \quad (\uparrow\downarrow)$$
$$+\downarrow \Sigma F_y = 0 = 0.77 \text{ k} + N_{EB}$$
$$N_{EB} = -0.77 \text{ k} \quad (\uparrow\downarrow)$$

Summary:

Axial force in columns:	0.77 k
Peak moment in AD:	5.01 k·ft
Peak moment in BE:	3.31 k·ft

Evaluation of Results:

Approximation Predicted Outcomes?
  Our approximate prediction of the column axial force is about 4% larger than our detailed result, and our approximate prediction of the peak moment in each column is within 21% of our detailed results. ✓

Conclusion:
  These observations suggest we have reasonable results.

Comparison with computer aided analysis results:
  All of the reactions match the computer aided analysis results to within 1%, which is within roundoff error. ✓
  Therefore, the effect of length change was negligible.

Reaction	Moment Distribution	Computer Aided	Difference
A_x	−1.19 k	−1.19 k	0%
A_y	−0.77 k	−0.77 k	0%
M_A	−5.01 k·ft	−5.03 k·ft	0.4%
B_x	−0.81 kN	−0.81 kN	0%
B_y	0.77 k	0.77 k	0%
M_B	−3.31 k·ft	−3.30 k·ft	0.3%

Reference Cited

Cross, H. 1930. "Analysis of Continuous Frames by Distributing Fixed-End Moments." *Proceedings of the American Society of Civil Engineers* 56: 919–28.

Homework Problems

13.1 Our team is tasked to design the support beam for large air conditioner units. The beam will carry only one unit initially, but the facility may expand in the future. According to the manufacturer's literature, the air conditioner unit will apply a uniformly distributed load on the beam. For preliminary design, let's assume the beam is a steel W250 × 32.7 ($E = 200$ GPa $= 200$ kN/mm², $A = 4190$ mm², $I = 4.91 \times 10^7$ mm⁴).

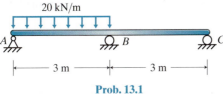

Prob. 13.1

a. Guess the shape of the moment diagram under this loading.
b. Using approximate analysis, find the peak moment. You can do so by assuming the location of the inflection point.
c. Use the Moment Distribution Method to determine the moment at B.
d. Construct and label the moment diagram for the beam based on your results from part (c).
e. Using structural analysis software, determine the reactions for the beam.
f. Submit a printout of the displaced shape. Identify at least three features that suggest the displaced shape is reasonable.
g. Submit a printout of the moment diagram. Identify at least three features that suggest the moment diagram is reasonable.
h. Verify equilibrium of the computer aided analysis results.
i. Are the results of your computer aided analysis reasonable? Make a comprehensive argument to justify your answer.
j. Comment on why your guess in part (a) was or was not close to the solution in part (d).

13.2 Cantilevered beams support the walkway for a high-rise hotel. For upcoming renovations, the contractor will be using a live load larger than the design value. The plan is to provide a line of supplemental supports. Our firm has been hired to design the temporary support system.

Prob. 13.2a

In order to select the support poles, we need to know how much load they will carry. Our task is to analyze one of the beams with support and carrying just the proposed live load.

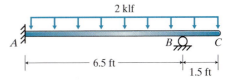

Prob. 13.2b

a. Guess the displaced shape.
b. Using approximate analysis, find the reactions. You can do so by assuming the location of the inflection point.
c. Use the Moment Distribution Method to determine the moments at A and B.
d. Find the other reactions based on your results from part (c).
e. Using structural analysis software, determine the reactions for the beam.
f. Submit a printout of the displaced shape. Identify at least three features that suggest the displaced shape is reasonable.
g. Verify equilibrium of the computer aided analysis results.
h. Are the results of your computer aided analysis reasonable? Make a comprehensive argument to justify your answer.
i. Comment on why your guess in part (a) was or was not close to the solution in part (f).

778　Chapter 13　Moment Distribution Method

13.3 An existing overpass in Canada is showing some signs of distress. While a team is in the field obtaining information to use to predict the actual strength of the bridge, our job is to calculate the required strength. The overpass is two lanes wide. One of the key load cases is full live load.

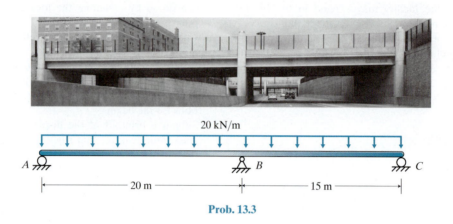

Prob. 13.3

a. Guess the displaced shape.
b. Using approximate analysis, find the vertical reactions. You can do so by assuming the location of the inflection point.
c. Use the Moment Distribution Method to determine the moment at B.
d. Find the vertical reactions based on your results from part (c).
e. Using structural analysis software, determine the reactions for the beam.
f. Submit a printout of the displaced shape. Identify at least three features that suggest the displaced shape is reasonable.
g. Verify equilibrium of the computer aided analysis results.
h. Are the results of your computer aided analysis reasonable? Make a comprehensive argument to justify your answer.
i. Comment on why your guess in part (a) was or was not close to the solution in part (f).

13.4 and 13.5 Our firm has been contracted by a state Department of Transportation to design a three-span overpass. To design the bridge, we must find the peak positive and negative moments on each girder due to the various load combinations. Before we can combine the effects of the various loads, we need to know the unfactored effects.

Probs. 13.4 and 13.5

13.4 Our task is to analyze one girder considering full live load.

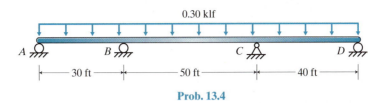

Prob. 13.4

a. Guess the shape of the moment diagram under this loading.
b. Using approximate analysis, find the vertical reactions. You can do so by assuming the locations of the inflection points.
c. Use the Moment Distribution Method to determine the moments at B and C.
d. Find the vertical reactions based on your results from part (c).
e. Using structural analysis software, determine the reactions for the girder.
f. Submit a printout of the displaced shape. Identify at least three features that suggest the displaced shape is reasonable.
g. Submit a printout of the moment diagram. Identify at least three features that suggest the moment diagram is reasonable.
h. Verify equilibrium of the computer aided analysis results.
i. Are the results of your computer aided analysis reasonable? Make a comprehensive argument to justify your answer.
j. Comment on why your guess in part (a) was or was not close to the solution in part (g).

13.5 An important load case is live load on only two adjacent spans.

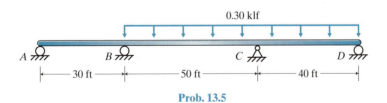

Prob. 13.5

a. Guess the shape of the moment diagram under this loading.
b. Using approximate analysis, find the vertical reactions. You can do so by assuming the locations of the inflection points.
c. Use the Moment Distribution Method to determine the moments at B and C.
d. Construct and label the moment diagram for the girder based on your results from part (c).
e. Using structural analysis software, determine the reactions for the girder.
f. Submit a printout of the displaced shape. Identify at least three features that suggest the displaced shape is reasonable.
g. Submit a printout of the moment diagram. Identify at least three features that suggest the moment diagram is reasonable.
h. Verify equilibrium of the computer aided analysis results.
i. Are the results of your computer aided analysis reasonable? Make a comprehensive argument to justify your answer.
j. Comment on why your guess in part (a) was or was not close to the solution in part (d).

13.6 and 13.7 Our firm has been selected to design an overpass as part of a highway expansion project. The elevation of the roadway is set to match the grade of the other lanes. To provide adequate height clearance below the overpass, we need to use relatively shallow members. The team leader would like us to explore deeper members for the outside spans to draw moment away from the shallower interior spans. The cross-sectional areas for the different spans will be essentially equal.

Probs. 13.6 and 13.7

13.6 To know whether this idea will work, the team leader would like us to analyze the overpass using the same shallow depth all the way across. We are to compare our results with the results of analysis of the overpass with deeper members for the outer spans.

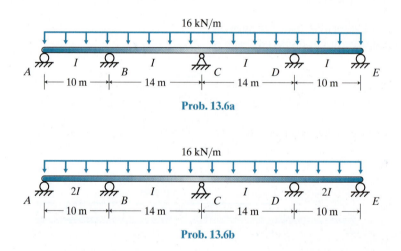

a. Guess which version will have the larger moment at point B and which will have the larger moment at point C.
b. Using approximate analysis, find the moment at B and at C. You can do so by assuming the locations of the inflection points.
c. Use the Moment Distribution Method to determine the moments generated at B, C, and D in the girder with uniform section properties.
d. Use the Moment Distribution Method to determine the moments generated at B, C, and D in the girder with deeper outer spans.
e. Provide feedback to the team leader about the effect of making the outer spans deeper.
f. Using structural analysis software, analyze the version with the smaller peak moment over a support. Determine the moments at B, C, and D.
g. Submit a printout of the displaced shape. Identify at least three features that suggest the displaced shape is reasonable.
h. Determine the reactions from the computer aided analysis. Verify equilibrium of the computer results.
i. Are the results of your computer aided analysis reasonable? Make a comprehensive argument to justify your answer.
j. Comment on why your guess in part (a) was or was not close to your observations in parts (c) and (d).

13.7 Assuming we use deeper members for the outside spans, the team leader also wants to know the effect of a single truck on one of the longer spans.

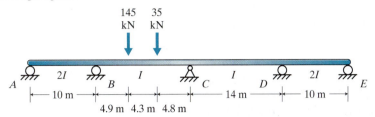

Prob. 13.7

a. Guess the displaced shape.
b. Using approximate analysis, find the peak moment in span BC. You can do so by assuming the locations of the inflection points.
c. Use the Moment Distribution Method to determine the moments generated at B, C, and D in the girder.
d. Construct and label the moment diagram for the girder based on your results from part (c).
e. Using structural analysis software, determine the moments at B, C, and D.
f. Submit a printout of the displaced shape. Identify at least three features that suggest the displaced shape is reasonable.
g. Submit a printout of the moment diagram. Identify at least three features that suggest the moment diagram is reasonable.
h. Determine the reactions from the computer aided analysis. Verify equilibrium of the computer results.
i. Are the results of your computer aided analysis reasonable? Make a comprehensive argument to justify your answer.
j. Comment on why your guess in part (a) was or was not close to the solution in part (f).

13.8 A pedestrian walkway must be suspended over a maintenance corridor. To do that, our firm has chosen to support the walkway on a frame fixed to the adjacent building and on a column at the other end. One of our team members has completed a preliminary design of the members. Now the team needs to reanalyze the frame with the chosen section properties to find the revised internal forces.

Prob. 13.8

a. Guess the shape of the moment diagram.
b. Using approximate analysis, find the reactions. You can do so by assuming the locations of the inflection points.
c. Use the Moment Distribution Method to determine the moments generated at the nodes.
d. Construct and label the moment diagram for the frame based on your results from part (c).
e. Using structural analysis software, determine the reactions for the frame.
f. Submit a printout of the displaced shape. Identify at least three features that suggest the displaced shape is reasonable.
g. Submit a printout of the moment diagram. Identify at least three features that suggest the moment diagram is reasonable.
h. Verify equilibrium of the computer results.
i. Are the results of your computer aided analysis reasonable? Make a comprehensive argument to justify your answer.
j. Comment on why your guess in part (a) was or was not close to the solution in part (d).

782 Chapter 13 Moment Distribution Method

13.9 and 13.10 Owners of a building in Europe with a below-grade window well would like to put on a roof in order to turn the space into a greenhouse. Originally, the design team considered making the roof removable, but the potential for wind uplift made that option impractical. The design team leader wants to consider two alternatives: pinned supports on the two ends, and fixed supports on the two ends.

The team leader anticipates that we will use 45×145 wood members ($E = 10$ GPa, $A = 65.2$ cm^2, $I = 1143$ cm^4) as a rigid frame.

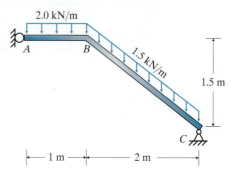

Probs. 13.9 and 13.10

13.9 An engineer on our team has calculated the dead load on one frame. Our task is to analyze the pinned support alternative.

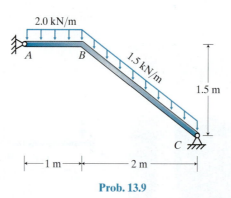

Prob. 13.9

a. Guess the shape of the moment diagram.
b. Using approximate analysis, find the reactions. You can do so by assuming the location of the inflection point.
c. Use the Moment Distribution Method to determine the moment generated at B.
d. Construct and label the moment diagram for the frame based on your results from part (c).
e. Using structural analysis software, determine the reactions for the frame.
f. Submit a printout of the displaced shape. Identify at least three features that suggest the displaced shape is reasonable.
g. Submit a printout of the moment diagram. Identify at least three features that suggest the moment diagram is reasonable.
h. Verify equilibrium of the computer results.
i. Are the results of your computer aided analysis reasonable? Make a comprehensive argument to justify your answer.
j. Comment on why your guess in part (a) was or was not close to the solution in part (d).

13.10 A team member has calculated the dead load on one frame. Our task is to analyze the fixed support alternative.

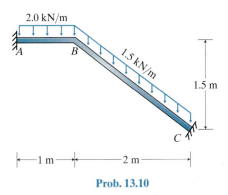

Prob. 13.10

a. Guess the displaced shape.
b. Using approximate analysis, find the reactions. You can do so by assuming the locations of the inflection points.
c. Use the Moment Distribution Method to determine the moments generated at the nodes.
d. Calculate the reactions for the frame based on your results from part (c).
e. Using structural analysis software, determine the reactions for the frame.
f. Submit a printout of the displaced shape. Identify at least three features that suggest the displaced shape is reasonable.
g. Submit a printout of the moment diagram. Identify at least three features that suggest the moment diagram is reasonable.
h. Verify equilibrium of the computer results.
i. Are the results of your computer aided analysis reasonable? Make a comprehensive argument to justify your answer.
j. Comment on why your guess in part (a) was or was not close to the solution in part (f).

13.11 The architect's vision of the entrance to a building is an asymmetric triangle. Our design team has explored the possibility of using a pin-roller combination to support the rigid frame, but the deformations were too large. Therefore, we are exploring the effect of using pin supports on both ends. Our task is to analyze the frame considering only dead load. The architect wants to use HSS tubes for the frame to complete the aesthetic. An experienced engineer on the team predicts that we will use HSS8 × 3 × 3/8 sections ($E = 29{,}000$ ksi, $A = 6.88$ in^2, $I = 48.5$ in^4).

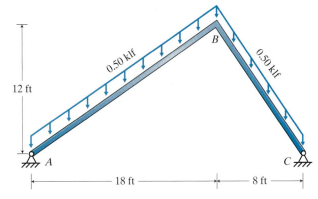

Prob. 13.11

a. Guess the displaced shape.
b. Using approximate analysis, find the reactions. You can do so by assuming the location of an inflection point.
c. Use the Moment Distribution Method to determine the moment generated at B.
d. Calculate the reactions for the frame based on your results from part (c).
e. Using structural analysis software, determine the reactions for the frame.
f. Submit a printout of the displaced shape. Identify at least three features that suggest the displaced shape is reasonable.
g. Submit a printout of the moment diagram. Identify at least three features that suggest the moment diagram is reasonable.
h. Verify equilibrium of the computer results.
i. Are the results of your computer aided analysis reasonable? Make a comprehensive argument to justify your answer.
j. Comment on why your guess in part (a) was or was not close to the solution in part (f).

13.12 and 13.13 Our team is designing a small rigid framed structure to protect patients from the weather when they are dropped off at a doctor's office. To design the members, we need to know the internal forces. Our design team has divided up the various loadings for analysis.

To ensure consistency, everyone has agreed to use the following preliminary section properties. We are using estimates of the cracked moment of inertia, since it depends on the moments in the frame.

Beam:	8 in. × 12 in.	$A = 96$ in^2	$I_{cr} = 800$ in^4
Columns:	8 in. × 8 in.	$A = 64$ in^2	$I_{cr} = 200$ in^4
Concrete:	$E = 3600$ ksi		

Probs. 13.12 and 13.13

13.12 We have been assigned the live roof load. The roof diaphragm will behave with two-way action, so the loading on the rigid frame is triangular. An experienced engineer on the team points out that although the frame can sidesway, the structure and this loading are symmetric, so it will not move laterally.

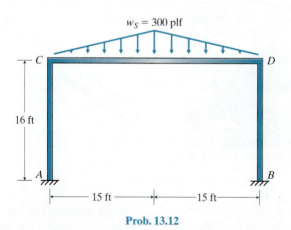

Prob. 13.12

a. Guess the shape of the moment diagram.
b. Using approximate analysis, find the internal moments at C and D. You can do so by assuming the locations of the inflection points.
c. Use the Moment Distribution Method to determine the moments generated at the nodes.
d. Calculate the reactions for the frame based on your results from part (c).
e. Using structural analysis software, determine the reactions for the frame.
f. Submit a printout of the displaced shape. Identify at least three features that suggest the displaced shape is reasonable.
g. Submit a printout of the moment diagram. Identify at least three features that suggest the moment diagram is reasonable.
h. Verify equilibrium of the computer results.
i. Are the results of your computer aided analysis reasonable? Make a comprehensive argument to justify your answer.
j. Comment on why your guess in part (a) was or was not close to the solution in part (g).

Homework Problems

13.13 We have been assigned the seismic load. Another engineer on the team has calculated the static equivalent seismic load based on an estimate of the self-weight of the roof.

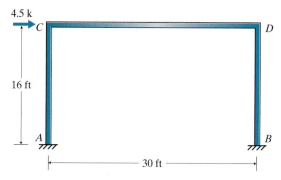

Prob. 13.13

a. Guess the displaced shape.
b. Using approximate analysis, find the internal moments at C and D.
c. Use the Moment Distribution Method to determine the moments generated at the nodes.
d. Calculate the reactions for the frame based on your results from part (c).
e. Using structural analysis software, determine the reactions for the frame.
f. Submit a printout of the displaced shape. Identify at least three features that suggest the displaced shape is reasonable.
g. Submit a printout of the moment diagram. Identify at least three features that suggest the moment diagram is reasonable.
h. Verify equilibrium of the computer results.
i. Are the results of your computer aided analysis reasonable? Make a comprehensive argument to justify your answer.
j. Comment on why your guess in part (a) was or was not close to the solution in part (f).

13.14 and 13.15 Our firm is designing the bents that support a new highway overpass. The design team leader has chosen concrete for the bents. There are two competing structural designs: deep foundations that act as fixed supports, and a grade beam on foundations that act as pin and roller supports. To decide between the two, our team leader wants more information about each.

For our preliminary analysis, we can use the uncracked moment of inertia for the members. Based on initial design calculations, the cap beam and columns are 1 m × 1 m: $A = 1 \text{ m}^2$, $I = 0.0833 \text{ m}^4$, $E = 30 \text{ GPa} = 30 \text{ kN/mm}^2$.

Probs. 13.14 and 13.15

13.14 We have been assigned the design that uses deep foundations as fixed supports. The geotechnical engineers have updated their calculations to predict that the point of fixity will be only a couple meters below the soil surface. We are to analyze the rigid frame with lateral wind load.

a. Guess the shape of the moment diagram.
b. Using approximate analysis, find the peak internal moment in each column.
c. Use the Moment Distribution Method to determine the moments generated at the nodes.
d. Calculate the reactions for the frame based on your results from part (c).
e. Using structural analysis software, determine the reactions for the frame.
f. Submit a printout of the displaced shape. Identify at least three features that suggest the displaced shape is reasonable.
g. Submit a printout of the moment diagram. Identify at least three features that suggest the moment diagram is reasonable.
h. Verify equilibrium of the computer results.
i. Are the results of your computer aided analysis reasonable? Make a comprehensive argument to justify your answer.
j. Comment on why your guess in part (a) was or was not close to the solution in part (g).

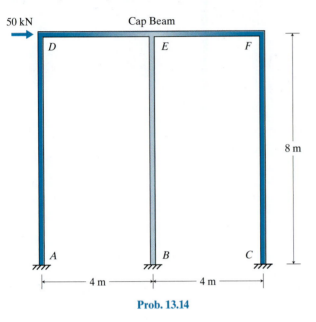

Prob. 13.14

13.15 We have been assigned the design that uses a grade beam. The grade beam allows us to have a frame that is stiff laterally but does not require deep foundations. We are to analyze the rigid frame with lateral wind load. Based on preliminary design, the grade beam will have these properties: $A = 2.25 \text{ m}^2$, $I = 0.422 \text{ m}^4$, $E = 30 \text{ GPa} = 30 \text{ kN/mm}^2$.

a. Guess the displaced shape.
b. Using approximate analysis, find the peak internal moment in each column.
c. Use the Moment Distribution Method to determine the moments generated at the nodes.
d. Calculate the reactions for the frame based on your results from part (c).
e. Using structural analysis software, determine the reactions for the frame.
f. Submit a printout of the displaced shape. Identify at least three features that suggest the displaced shape is reasonable.
g. Submit a printout of the moment diagram. Identify at least three features that suggest the moment diagram is reasonable.
h. Verify equilibrium of the computer results.
i. Are the results of your computer aided analysis reasonable? Make a comprehensive argument to justify your answer.
j. Comment on why your guess in part (a) was or was not close to the solution in part (f).

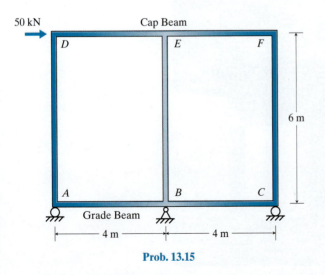

Prob. 13.15

13.16 Our firm has been selected to design a new highway overpass. The principal in charge has chosen to use an integral abutment design. That means the entire structure acts as a rigid frame. Lateral earth pressures might not be the same on both ends of the overpass, so we have been asked to determine the impact if that happens.

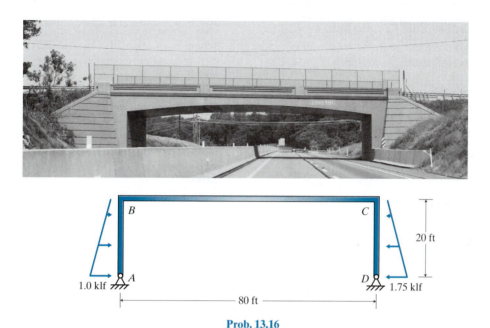

Prob. 13.16

The frame has been designed for gravity loads. Each 1-ft-wide strip has the following uncracked properties:

Deck:	$t = 18$ in.	$A = 216$ in^2	$I = 5830$ in^4
Ends:	$t = 8$ in.	$A = 96$ in^2	$I = 512$ in^4
Concrete:	$E = 4400$ ksi		

a. Guess the displaced shape.
b. Using approximate analysis, find the reactions. It helps to think in terms of the net applied force when choosing the assumptions.
c. Use the Moment Distribution Method to determine the moments generated at the nodes.
d. Calculate the reactions for the frame based on your results from part (c).
e. Using structural analysis software, determine the reactions for the frame.
f. Submit a printout of the displaced shape. Identify at least three features that suggest the displaced shape is reasonable.
g. Submit a printout of the moment diagram. Identify at least three features that suggest the moment diagram is reasonable.
h. Verify equilibrium of the computer results.
i. Are the results of your computer aided analysis reasonable? Make a comprehensive argument to justify your answer.
j. Comment on why your guess in part (a) was or was not close to the solution in part (f).

788 Chapter 13 Moment Distribution Method

13.17 Our team is designing an overpass structure that will be built in two phases. Because of the timing of the phases, the piers are separated into two parts. The geotechnical engineers have helped us estimate the point of fixity of the columns (i.e., the depth where the column actually behaves as fixed) well below the soil surface. The team is performing a preliminary analysis of one of the piers under the loads expected before the second phase is added. We are responsible for wind loading, lateral and uplift.

To provide consistent results, the team has agreed to use the following preliminary section properties:

Cap beam:	3 ft × 3.5 ft	$A = 10.5$ ft^2	$I = 10.7$ ft^4
Columns:	3 ft diameter	$A = 7.1$ ft^2	$I = 4.0$ ft^4
Concrete:	$E = 4600$ ksi		

a. Guess the displaced shape.
b. Using approximate analysis, find the reactions. It helps to use superposition.
c. Use the Moment Distribution Method to determine the moments generated at the nodes.
d. Calculate the reactions for the frame based on your results from part (c).
e. Using structural analysis software, determine the reactions for the frame.
f. Submit a printout of the displaced shape. Identify at least three features that suggest the displaced shape is reasonable.
g. Submit a printout of the moment diagram. Identify at least three features that suggest the moment diagram is reasonable.
h. Verify equilibrium of the computer results.
i. Are the results of your computer aided analysis reasonable? Make a comprehensive argument to justify your answer.
j. Comment on why your guess in part (a) was or was not close to the solution in part (f).

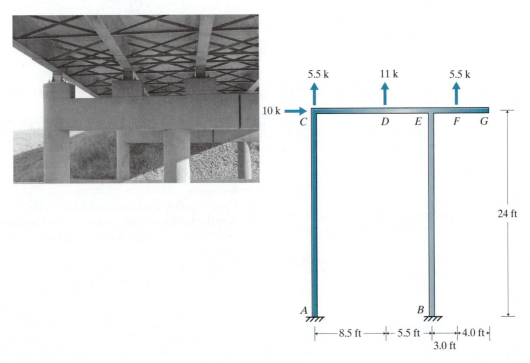

Prob. 13.17

13.18 Our firm is designing a single-story office building to span between two existing buildings, creating a breezeway. Because of buried utilities, we cannot put both columns at the ends of the supporting beam. The plan is to use shallow foundations that will act as pin supports. An engineer on our team has performed computer aided analysis of the frame using section properties developed from preliminary design. The engineer is concerned that the results do not seem reasonable and has asked us to redo the analysis.

Our teammate used these section properties:

Beams:	20 cm×40 cm	$A = 800 \text{ cm}^2$	$I = 1.067 \times 10^5 \text{ cm}^4$
Left column:	20 cm×60 cm	$A = 1200 \text{ cm}^2$	$I = 3.60 \times 10^5 \text{ cm}^4$
Right column:	20 cm×40 cm	$A = 800 \text{ cm}^2$	$I = 1.067 \times 10^5 \text{ cm}^4$
Concrete:	$E = 28 \text{ GPa} = 28 \text{ kN/mm}^2$		

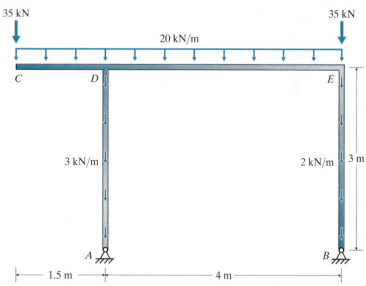

Prob. 13.18

a. Guess the displaced shape.
b. Using approximate analysis, find the reactions.
c. Use the Moment Distribution Method to determine the moments generated at the nodes. Note from the pictures inside the back cover of this text that the distributed axial loads on the columns do not cause fixed-end moments.
d. Calculate the reactions for the frame based on your results from part (c).
e. Using structural analysis software, determine the reactions for the frame.
f. Submit a printout of the displaced shape. Identify at least three features that suggest the displaced shape is reasonable.
g. Submit a printout of the moment diagram. Identify at least three features that suggest the moment diagram is reasonable.
h. Verify equilibrium of the computer results.
i. Are the results of your computer aided analysis reasonable? Make a comprehensive argument to justify your answer.
j. Comment on why your guess in part (a) was or was not close to the solution in part (f).

790   Chapter 13   Moment Distribution Method

13.19 Building a house on a rocky hillside allows for amazing views of the valley below. One new home is to rest on a rigid frame with pinned supports. In order to begin design of the frame, we would like to know the peak axial force and moment in each member. For this preliminary stage of design, a team member suggests we guess that all members are steel W12 × 40 ($E = 29{,}000$ ksi, $A = 11.7$ in^2, $I = 307$ in^4).

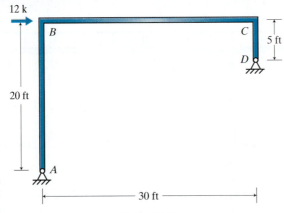

Prob. 13.19

a. Guess the shape of the moment diagram.
b. Using approximate analysis, find the peak internal moment in each column.
c. Use the Moment Distribution Method to determine the moments generated at the nodes.
d. Calculate the reactions for the frame based on your results from part (c).
e. Using structural analysis software, determine the reactions for the frame.
f. Submit a printout of the displaced shape. Identify at least three features that suggest the displaced shape is reasonable.
g. Submit a printout of the moment diagram. Identify at least three features that suggest the moment diagram is reasonable.
h. Verify equilibrium of the computer results.
i. Are the results of your computer aided analysis reasonable? Make a comprehensive argument to justify your answer.
j. Comment on why your guess in part (a) was or was not close to the solution in part (g).

13.20 Our client is a manufacturer in Europe. The manufacturer is replacing the cooling equipment for one of its facilities, but the new equipment is heavier than the old. The facility manager is concerned that the support structure will not be strong enough for the new equipment. Therefore, our firm has been hired to evaluate the structure under the new loads. Before incorporating the self-weight of the structure, our team leader wants to know the effect of just the new cooling equipment. The team leader has assigned one frame to us.

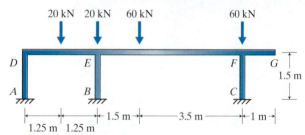

Prob. 13.20

The facility manager sent drawings for the support structure. The frame members have the following properties:

Column AD:	HSS 160 × 5	A = 30.7 cm²	I = 1225 cm⁴
Column BE:	HSS 160 × 12	A = 69.5 cm²	I = 2502 cm⁴
Column CF:	HSS 160 × 10	A = 58.9 cm²	I = 2186 cm⁴
Beam DE:	IPE 300	A = 53.8 cm²	I = 8356 cm⁴
Beam EF:	IPE 400	A = 84.5 cm²	I = 23,130 cm⁴
Beam FG:	IPE 400	A = 84.5 cm²	I = 23,130 cm⁴
Steel:	E = 200 GPa = 200 kN/mm²		

a. Guess the displaced shape.
b. Using approximate analysis, find the peak internal moment in each beam.
c. Use the Moment Distribution Method to determine the moments generated at the nodes.
d. Construct and label the moment diagram for the beams based on your results from part (c).
e. Using structural analysis software, determine the reactions for the frame.
f. Submit a printout of the displaced shape. Identify at least three features that suggest the displaced shape is reasonable.
g. Submit a printout of the moment diagram for the entire frame. Identify at least three features that suggest the moment diagram is reasonable.
h. Verify equilibrium of the computer results.
i. Are the results of your computer aided analysis reasonable? Make a comprehensive argument to justify your answer.
j. Comment on why your guess in part (a) was or was not close to the solution in part (f).

CHAPTER 14

DIRECT STIFFNESS METHOD FOR TRUSSES

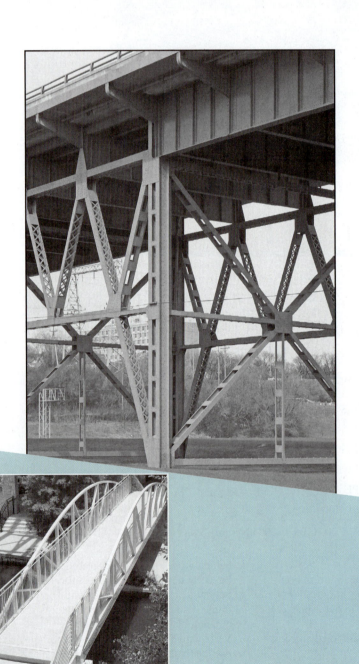

Trusses can make aesthetic statements while being structurally efficient.

MOTIVATION

The Direct Stiffness Method is the detailed analysis method used in almost all structural software. Although the method was developed before computers existed, now we use the method almost exclusively in computer aided analysis.

So, why should we learn how to perform this method by hand?

1. Understanding how the method works increases the likelihood that we will create an idealized model that reasonably represents the actual behavior.
2. Understanding how the method works can help us reproduce loadings that we want to model even if the software does not have that loading programmed.
3. The method shows us how to calculate stiffness, so we can anticipate which members will carry more load and anticipate what to change to shift where load is carried in a structure.

Let's start our discussion with trusses because truss members carry only axial load. We add the complexity of frame members in Chapter 15: Direct Stiffness Method for Frames.

Chapter 14 Direct Stiffness Method for Trusses

14.1 Overview of Method

Introduction
Rather than start by deriving all the pieces of the Direct Stiffness Method, let's begin with a simple truss and deduce how the method works. In later sections, we formally derive all the formulas we need.

How-To
The Direct Stiffness Method works the same for determinate and indeterminate trusses. Since we can analyze a determinate truss using only equations of equilibrium (e.g., Method of Joints, Method of Sections), let's use an indeterminate truss to demonstrate how the Direct Stiffness Method works (Figure 14.1). The method can be used on both two- and three-dimensional trusses; the only difference is that the bookkeeping more than doubles for three dimensions. Because our goal is to understand the method, a two-dimensional truss makes a great example.

In the Direct Stiffness Method, we refer to the members of the structure as *elements*. Each element spans between points called *nodes*. For a truss, nodes are the same as joints. Let's begin by applying equilibrium to a free body diagram of node D (Figure 14.2).

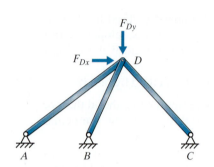

Figure 14.1 Indeterminate 2D truss.

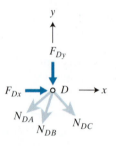

Figure 14.2 FBD of node D.

$$\xrightarrow{+} \Sigma F_x = 0 = F_{Dx} - N_{DAx} - N_{DBx} + N_{DCx}$$
$$+\downarrow \Sigma F_y = 0 = F_{Dy} + N_{DAy} + N_{DBy} + N_{DCy}$$

These two equations have three unknowns: N_{DA}, N_{DB}, and N_{DC}. Therefore, we can't find the unknown axial forces in this format. We need to express the axial forces in terms of just two unknowns. The key to the Direct Stiffness Method is recognizing that the number of unknown deformations at a node is the same as the number of equations of equilibrium. *If we can express internal force in terms of deformations at the nodes, we will have the same number of equations as unknowns.* We can! The relationship between internal force at a node and deformation at a node is called *element stiffness, k*.

For our node in Figure 14.2, the two unknown deformations are Δ_{Dx} and Δ_{Dy}. Each component of internal force in a member depends on both deformations:

$$N_{DAx} = f(\Delta_{Dx}, \Delta_{Dy}) = k^{DA}_{x,x} \Delta_{Dx} + k^{DA}_{x,y} \Delta_{Dy}$$
$$N_{DAy} = f(\Delta_{Dx}, \Delta_{Dy}) = k^{DA}_{y,x} \Delta_{Dx} + k^{DA}_{y,y} \Delta_{Dy}$$
$$N_{DBx} = f(\Delta_{Dx}, \Delta_{Dy}) = k^{DB}_{x,x} \Delta_{Dx} + k^{DB}_{x,y} \Delta_{Dy}$$
$$N_{DBy} = f(\Delta_{Dx}, \Delta_{Dy}) = k^{DB}_{y,x} \Delta_{Dx} + k^{DB}_{y,y} \Delta_{Dy}$$
$$N_{DCx} = f(\Delta_{Dx}, \Delta_{Dy}) = k^{DC}_{x,x} \Delta_{Dx} + k^{DC}_{x,y} \Delta_{Dy}$$
$$N_{DCy} = f(\Delta_{Dx}, \Delta_{Dy}) = k^{DC}_{y,x} \Delta_{Dx} + k^{DC}_{y,y} \Delta_{Dy}$$

The stiffness is different for each deformation, each component of internal force, and each member. Therefore, we use subscripts and superscripts on k to distinguish each:

$$k^{\text{member}}_{\text{direction of internal force component, direction of deformation}}$$

If we put these expressions for the internal force components into the equations of equilibrium for node D, we get

$$0 = F_{Dx} - k_{x,x}^{DA}\Delta_{Dx} - k_{x,y}^{DA}\Delta_{Dy} - k_{x,x}^{DB}\Delta_{Dx} - k_{x,y}^{DB}\Delta_{Dy} + k_{x,x}^{DC}\Delta_{Dx} + k_{x,y}^{DC}\Delta_{Dy}$$

$$0 = F_{Dy} + k_{y,x}^{DA}\Delta_{Dx} + k_{y,y}^{DA}\Delta_{Dy} + k_{y,x}^{DB}\Delta_{Dx} + k_{y,y}^{DB}\Delta_{Dy} + k_{y,x}^{DC}\Delta_{Dx} + k_{y,y}^{DC}\Delta_{Dy}$$

If we move the applied forces to the other side of the equation and consolidate terms, it becomes easier to see that we have two equations with two unknowns:

$$-F_{Dx} = (-k_{x,x}^{DA} - k_{x,x}^{DB} + k_{x,x}^{DC})\Delta_{Dx} + (-k_{x,y}^{DA} - k_{x,y}^{DB} + k_{x,y}^{DC})\Delta_{Dy}$$

$$-F_{Dy} = (k_{y,x}^{DA} + k_{y,x}^{DB} + k_{y,x}^{DC})\Delta_{Dx} + (k_{y,y}^{DA} + k_{y,y}^{DB} + k_{y,y}^{DC})\Delta_{Dy}$$

This leads to an interesting observation: all of the members that meet at a node contribute stiffness to the equilibrium equation. We call the summation of contributions the *structural stiffness* and give it the symbol K. For node D, the equilibrium equations then become

$$-F_{Dx} = K_{x,x}\Delta_{Dx} + K_{x,y}\Delta_{Dy}$$

$$-F_{Dy} = K_{y,x}\Delta_{Dx} + K_{y,y}\Delta_{Dy}$$

Solving the two equations gives us the deformations. With those deformations, we can use the element stiffnesses to find the internal force components.

This is the essence of the Direct Stiffness Method for both trusses and frames:

1. Write equations of equilibrium for the nodes.
2. Express the unknown internal forces as products of deformations and element stiffnesses.
3. Solve the system of equations for the deformations.
4. Use the deformations to calculate reactions and internal forces.

The formulas we used in this example are unique to this example, so we still need to derive the specifics of the method.

14.2 Transformation and Element Stiffness Matrices

Introduction

We begin the Direct Stiffness Method by calculating the stiffness for each element. If we derive a formula for a generic truss element, we can use it to calculate the components of element stiffness for each member of each truss. For the purposes of this text, we will stick to two-dimensional truss elements, but the process can be extrapolated to three-dimensional elements.

How-To

To begin the process of formulating the element stiffness for a truss member, let's consider the coordinate system to be along the member. We call this the *local coordinate system* because it might point in a different direction for each element. The industry standard is to line up the local x-axis

796 Chapter 14 Direct Stiffness Method for Trusses

Figure 14.3 The local x-axis follows along the axis of the element, but we still need to know in which direction.

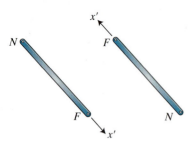

Figure 14.4 The local x-axis always runs from the near node to the far node of an element.

along the element. To distinguish between a local axis and the global axis, we use a prime on the local axis (e.g., x').

So, we line up the x'-axis with the element, but there are still two options (Figure 14.3). The solution is to identify one node of each element as the near node and the other as the far node. The x'-axis will always run from the near node to the far node (Figure 14.4). We can choose which is the near node, but once picked for that member, it should not be changed.

Now we are ready to determine the stiffness of a truss element in local coordinates.

Element Stiffness in Local Coordinates

A truss element experiences only one type of internal force, axial, so the stiffness of a truss element is the relationship between the axial displacement and the axial force. Let's use the symbol d' to represent the deformation at a node of a single element in local coordinates and the symbol q' to represent the internal force at a node of a single element in local coordinates.

In general, both nodes of a truss element might displace different amounts: d'_N and d'_F (Figure 14.5). To find the components of stiffness, we consider the deformations one at a time.

From mechanics of materials, we know that the relationship between axial displacement, d', and axial force, q', is

$$q' = \frac{AE}{l}d'$$

where

A = Cross-sectional area
E = Modulus of elasticity
l = Length of the element

If we pin the far end and displace the near end, we get axial forces at each end of the member (Figure 14.6). In order for equilibrium to be satisfied, the axial force at the far end is equal and opposite to the axial force at the near end. If we pin the near end and displace the far end, we get different axial forces (Figure 14.7).

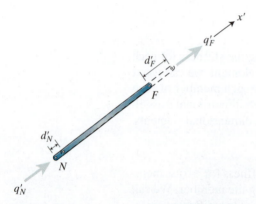

Figure 14.5 A truss element can have different axial displacements at the two nodes.

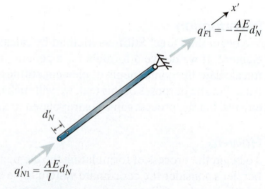

Figure 14.6 Axial forces generated when a truss element is displaced at just the near end.

Section 14.2 Transformation and Element Stiffness Matrices

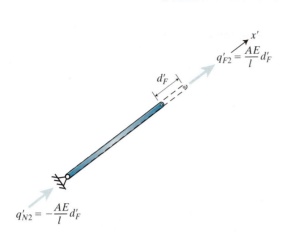

Figure 14.7 Axial forces generated when a truss element is displaced at just the far end.

Superimposing the two situations results in the following net internal forces for a truss element that has different displacements at the two ends:

$$q'_N = +\frac{AE}{l}d'_N - \frac{AE}{l}d'_F$$

$$q'_F = -\frac{AE}{l}d'_N + \frac{AE}{l}d'_F$$

If we convert this system of equations to matrix form, we have

$$[\mathbf{q'}] = [\mathbf{k'}][\mathbf{d'}]$$

where

$$[\mathbf{q'}] = \begin{bmatrix} q'_N \\ q'_F \end{bmatrix} = \text{Internal forces for an element in local coordinates}$$

$$[\mathbf{k'}] = \begin{bmatrix} \frac{AE}{l} & -\frac{AE}{l} \\ -\frac{AE}{l} & \frac{AE}{l} \end{bmatrix} = \text{Element stiffness matrix in local coordinates}$$

$$[\mathbf{d'}] = \begin{bmatrix} d'_N \\ d'_F \end{bmatrix} = \text{Displacements for an element in local coordinates}$$

So why use q? Internal forces for a truss element are axial, N. But internal forces for a frame element can also be shear, V, and moment, M. Therefore, we use q to generically refer to all internal forces.

Transformation Matrix

We saw in Section 14.1 that we need to express element stiffness in a common coordinate system in order to add the contributions from each element. We call that common system the global coordinate system, and it is something we get to choose.

To change truss stiffness from local to global coordinates, we need something called the *transformation matrix*, **T**. From mathematics, a transformation matrix is used to convert, or transform, one coordinate system into another. In the Direct Stiffness Method, we are transforming local into global coordinates or global into local coordinates. We need to be able to do both.

Let's start with the transformation from global into local coordinates. To demonstrate how that is done, we can look at the displacement of a node. The near node of a single truss element can displace in the global x-direction, d_{Nx} (Figure 14.8a), and the global y-direction, d_{Ny} (Figure 14.8b). Each of those displacements has a component along the axis of the element: d'_{N1} and d'_{N2}. We repeat the process for the far node (Figure 14.9).

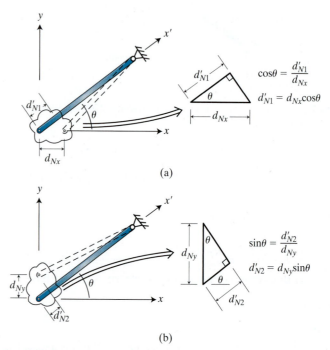

Figure 14.8 Displacements at near node divided into component along the axis of the element: (a) Global x-displacement; (b) Global y-displacement.

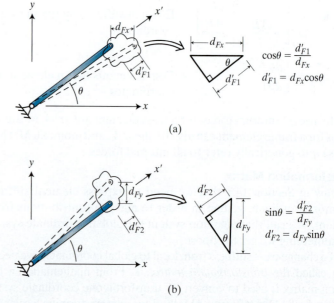

Figure 14.9 Displacements at far node divided into component along the axis of the element: (a) Global x-displacement; (b) Global y-displacement.

Section 14.2 Transformation and Element Stiffness Matrices

The net local displacement is the sum of the contributions from the global displacements:

$$d'_N = d'_{N1} + d'_{N2} = d_{Nx}\cos\theta + d_{Ny}\sin\theta$$
$$d'_F = d'_{F1} + d'_{F2} = d_{Fx}\cos\theta + d_{Fy}\sin\theta$$

If we convert this system of equations to matrix form, we have

$$[\mathbf{d'}] = [\mathbf{T}][\mathbf{d}]$$

where

$$[\mathbf{d'}] = \begin{bmatrix} d'_N \\ d'_F \end{bmatrix} \quad = \text{Displacements for an element in local coordinates}$$

$$[\mathbf{T}] = \begin{bmatrix} \cos\theta & \sin\theta & 0 & 0 \\ 0 & 0 & \cos\theta & \sin\theta \end{bmatrix} = \text{Transformation matrix}$$

$$[\mathbf{d}] = \begin{bmatrix} d_{Nx} \\ d_{Ny} \\ d_{Fx} \\ d_{Fy} \end{bmatrix} \quad = \text{Displacements for an element in global coordinates}$$

To see how to transform from local into global, let's again look at displacement at a node. Axial displacement of the near node has both x- and y-components in global coordinates (Figure 14.10a) as does axial displacement of the far node (Figure 14.10b).

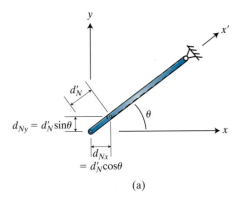

(a)

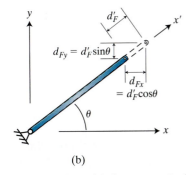

(b)

Figure 14.10 Axial displacements divided into global components: (a) Near node; (b) Far node.

800 Chapter 14 Direct Stiffness Method for Trusses

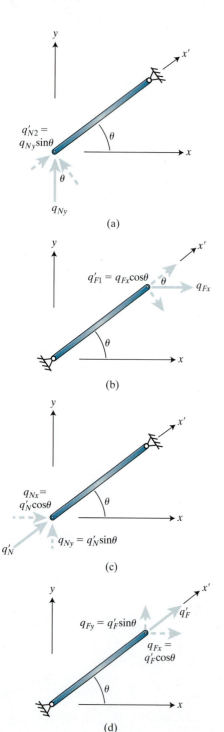

Figure 14.11 Transforming internal forces: (a) Near node force in the global y-direction; (b) Far node force in the global x-direction; (c) Near node force in the local x-direction; (d) Far node force in the local x-direction.

This gives us a set of four expressions for the global components of displacement:

$$d_{Nx} = d'_N \cos\theta \qquad d_{Fx} = d'_F \cos\theta$$
$$d_{Ny} = d'_N \sin\theta \qquad d_{Fy} = d'_F \sin\theta$$

We can convert this system of equations into matrix form:

$$[\mathbf{d}] = [\mathbf{T}]^T [\mathbf{d}']$$

where

$$[\mathbf{d}] = \begin{bmatrix} d_{Nx} \\ d_{Ny} \\ d_{Fx} \\ d_{Fy} \end{bmatrix} = \text{Displacements for an element in global coordinates}$$

$$[\mathbf{T}]^T = \begin{bmatrix} \cos\theta & 0 \\ \sin\theta & 0 \\ 0 & \cos\theta \\ 0 & \sin\theta \end{bmatrix} = \text{Transpose of the transformation matrix}$$

$$[\mathbf{d}'] = \begin{bmatrix} d'_N \\ d'_F \end{bmatrix} = \text{Displacements for an element in local coordinates}$$

So it turns out that we use the same transformation matrix to go from global to local and from local to global coordinates. The only difference is that we transpose the matrix.

We can repeat the derivation process for internal forces (Figure 14.11). The result is the same transformation matrix:

$$[\mathbf{q}'] = [\mathbf{T}][\mathbf{q}]$$
$$[\mathbf{q}] = [\mathbf{T}]^T [\mathbf{q}']$$

where

$$[\mathbf{q}'] = \begin{bmatrix} q'_N \\ q'_F \end{bmatrix} = \text{Internal forces for an element in local coordinates}$$

$$[\mathbf{T}] = \begin{bmatrix} \cos\theta & \sin\theta & 0 & 0 \\ 0 & 0 & \cos\theta & \sin\theta \end{bmatrix} = \text{Transformation matrix}$$

$$[\mathbf{q}] = \begin{bmatrix} q_{Nx} \\ q_{Ny} \\ q_{Fx} \\ q_{Fy} \end{bmatrix} = \text{Internal forces for an element in global coordinates}$$

Typically we know the global coordinates of the nodes but not the angle between the global and local axes for a member, θ. We could calculate the angle for each member, but it is typically easier to use the coordinates to calculate $\cos\theta$ and $\sin\theta$ directly:

$$\cos\theta = \frac{\text{width in } x\text{-direction}}{\text{length}} = \frac{x_F - x_N}{\sqrt{(x_F - x_N)^2 + (y_F - y_N)^2}}$$

$$\sin\theta = \frac{\text{height in } y\text{-direction}}{\text{length}} = \frac{y_F - y_N}{\sqrt{(x_F - x_N)^2 + (y_F - y_N)^2}}$$

Note that the sign matters when we calculate $\cos\theta$ and $\sin\theta$. Therefore, we must calculate the numerator from the far coordinate to the near coordinate.

Element Stiffness in Global Coordinates

We still need to convert the element stiffness from local coordinates to global coordinates in order to be able to add the stiffnesses from multiple members at a node. To see how to make that conversion, we start with the transformation of internal forces from local to global coordinates:

$$[\mathbf{q}] = [\mathbf{T}]^T[\mathbf{q}']$$

But we know that $[\mathbf{q}']$ can be expressed as a function of the displacements in local coordinates:

$$[\mathbf{q}'] = [\mathbf{k}'][\mathbf{d}']$$

so

$$[\mathbf{q}] = [\mathbf{T}]^T[\mathbf{k}'][\mathbf{d}']$$

We can also transform the displacements from local into global coordinates:

$$[\mathbf{d}'] = [\mathbf{T}][\mathbf{d}]$$

so

$$[\mathbf{q}] = [\mathbf{T}]^T[\mathbf{k}'][\mathbf{T}][\mathbf{d}]$$

By definition, stiffness is the force generated by a unit deformation. So the force in global coordinates is related to the deformations in global coordinates by the element stiffness in global coordinates, $[\mathbf{k}]$:

$$[\mathbf{q}] = [\mathbf{k}][\mathbf{d}]$$

The two expressions must be equal, so the following must be true:

$$[\mathbf{k}] = [\mathbf{T}]^T[\mathbf{k}'][\mathbf{T}]$$

When we substitute in the matrices for each term, we get the following product:

$$[\mathbf{k}] = \begin{bmatrix} \cos\theta & 0 \\ \sin\theta & 0 \\ 0 & \cos\theta \\ 0 & \sin\theta \end{bmatrix} \begin{bmatrix} \dfrac{AE}{l} & -\dfrac{AE}{l} \\ -\dfrac{AE}{l} & \dfrac{AE}{l} \end{bmatrix} \begin{bmatrix} \cos\theta & \sin\theta & 0 & 0 \\ 0 & 0 & \cos\theta & \sin\theta \end{bmatrix}$$

Chapter 14 Direct Stiffness Method for Trusses

Evaluating the expression, we get the formula for element stiffness in global coordinates for a two-dimensional truss element:

$$[k] = \begin{matrix} q_{Nx} \\ q_{Ny} \\ q_{Fx} \\ q_{Fy} \end{matrix} \begin{bmatrix} \overset{d_{Nx}}{\dfrac{AE}{l}\cos^2\theta} & \overset{d_{Ny}}{\dfrac{AE}{l}\cos\theta\sin\theta} & \overset{d_{Fx}}{-\dfrac{AE}{l}\cos^2\theta} & \overset{d_{Fy}}{-\dfrac{AE}{l}\cos\theta\sin\theta} \\ \dfrac{AE}{l}\cos\theta\sin\theta & \dfrac{AE}{l}\sin^2\theta & -\dfrac{AE}{l}\cos\theta\sin\theta & -\dfrac{AE}{l}\sin^2\theta \\ -\dfrac{AE}{l}\cos^2\theta & -\dfrac{AE}{l}\cos\theta\sin\theta & \dfrac{AE}{l}\cos^2\theta & \dfrac{AE}{l}\cos\theta\sin\theta \\ -\dfrac{AE}{l}\cos\theta\sin\theta & -\dfrac{AE}{l}\sin^2\theta\sin\theta & \dfrac{AE}{l}\cos\theta\sin\theta & \dfrac{AE}{l}\sin^2\theta \end{bmatrix}$$

Each row in [**k**] refers to an internal force, and each column is associated with a deformation. The columns are labeled above so we can see how each combination of deformation and internal force are related. Each term in the matrix is called a *stiffness coefficient, k*.

Notice that the units we select for each of the terms in the stiffness coefficient (i.e., A, E, l) are important. We are converting from a displacement, d, to an internal force, q, so the units on the stiffness coefficients need to make the proper change. As an example, consider what happens if we measure displacement in inches, area in square inches, modulus of elasticity in ksi, and length in feet:

$$q = \frac{(\text{in}^2)(\text{k/in}^2)}{\text{ft}}(\text{in.}) = \frac{\text{k}\cdot\text{in.}}{\text{ft}}$$

Although technically valid, these are not units that we use, so we have no sense of whether the results are reasonable. Instead, we should use consistent units—for example, displacement in inches, area in square inches, modulus of elasticity in ksi, and length in inches:

$$q = \frac{(\text{in}^2)(\text{k/in}^2)}{\text{in.}}(\text{in.}) = \text{k}$$

Or we could use the following units, which are still consistent: displacement in feet, area in square inches, modulus of elasticity in ksi, and length in feet:

$$q = \frac{(\text{in}^2)(\text{k/in}^2)}{\text{ft}}(\text{ft}) = \text{k}$$

Typically we need to carry only three significant figures in structural analysis calculations in order to have adequate accuracy. However, solving a system of equations requires more significant figures. Therefore, we will carry four significant figures until the final results.

Section 14.2 Highlights

Transformation and Element Stiffness Matrices

Terms:

- A = Cross-sectional area of an element
- d = Displacement at a node for an element (global coordinates)
- d' = Displacement at a node for an element (local coordinates)
- E = Modulus of elasticity for an element
- k = Stiffness coefficient for an element (global coordinates)
- k' = Stiffness coefficient for an element (local coordinates)
- l = Length of an element
- q = Internal force for an element (global coordinates)
- q' = Internal force for an element (local coordinates)
- $\mathbf{T}$ = Transformation matrix (global into local coordinates)
- $\mathbf{T}^T$ = Transpose of transformation matrix (local into global coordinates)

Use of Transformation Matrix:

$$[\mathbf{d'}] = [\mathbf{T}][\mathbf{d}] \qquad [\mathbf{d}] = [\mathbf{T}]^T[\mathbf{d'}]$$

$$[\mathbf{q'}] = [\mathbf{T}][\mathbf{q}] \qquad [\mathbf{q}] = [\mathbf{T}]^T[\mathbf{q'}]$$

$$[\mathbf{T}] = \begin{bmatrix} \cos\theta & \sin\theta & 0 & 0 \\ 0 & 0 & \cos\theta & \sin\theta \end{bmatrix} \qquad [\mathbf{T}]^T = \begin{bmatrix} \cos\theta & 0 \\ \sin\theta & 0 \\ 0 & \cos\theta \\ 0 & \sin\theta \end{bmatrix}$$

Element Stiffness in Local Coordinates:

$$[\mathbf{k'}] = \begin{bmatrix} \dfrac{AE}{l} & -\dfrac{AE}{l} \\ -\dfrac{AE}{l} & \dfrac{AE}{l} \end{bmatrix}$$

Element Stiffness in Global Coordinates:

$$[\mathbf{k}] = \begin{array}{c} q_{Nx} \\ q_{Ny} \\ q_{Fx} \\ q_{Fy} \end{array} \begin{bmatrix} \dfrac{AE}{l}\cos^2\theta & \dfrac{AE}{l}\cos\theta\sin\theta & -\dfrac{AE}{l}\cos^2\theta & -\dfrac{AE}{l}\cos\theta\sin\theta \\ \dfrac{AE}{l}\cos\theta\sin\theta & \dfrac{AE}{l}\sin^2\theta & -\dfrac{AE}{l}\cos\theta\sin\theta & -\dfrac{AE}{l}\sin^2\theta \\ -\dfrac{AE}{l}\cos^2\theta & -\dfrac{AE}{l}\cos\theta\sin\theta & \dfrac{AE}{l}\cos^2\theta & \dfrac{AE}{l}\cos\theta\sin\theta \\ -\dfrac{AE}{l}\cos\theta\sin\theta & -\dfrac{AE}{l}\sin^2\theta & \dfrac{AE}{l}\cos\theta\sin\theta & \dfrac{AE}{l}\sin^2\theta \end{bmatrix}$$

with column labels d_{Nx}, d_{Ny}, d_{Fx}, d_{Fy}.

$$\cos\theta = \dfrac{x_F - x_N}{\sqrt{(x_F - x_N)^2 + (y_F - y_N)^2}}$$

$$\sin\theta = \dfrac{y_F - y_N}{\sqrt{(x_F - x_N)^2 + (y_F - y_N)^2}}$$

Special Considerations:

- Make sure the stiffness coefficients have units that are consistent with the units for the displacements and internal forces we are calculating.
- Carry four significant figures in order to have sufficient accuracy.

Chapter 14 Direct Stiffness Method for Trusses

> ### EXAMPLE 14.1
>
> Our firm has designed a large pipeline along the edge of a canal in Singapore. Our team has been tasked with designing a structure to support the pipe. Other engineers on the team have designed the truss members for strength.
>
>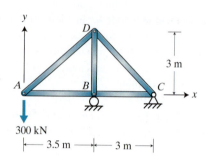
>
> **Figure 14.12**
>
> Because we are new to the firm, our supervisor wants to verify our knowledge of the Direct Stiffness Method. A determinate truss is a great way to do that. We start by finding the element stiffness matrix for each member of the truss.
>
> *Member properties:*
>
> | Diagonals: | Double angle | 120 × 120 × 8 | $A = 37.6 \text{ cm}^2$ |
> |---|---|---|---|
> | All others: | Universal columns | 250 × 250 × 66.5 | $A = 84.7 \text{ cm}^2$ |
> | Steel: | $E = 200 \text{ GPa} = 20{,}000 \text{ kN/cm}^2$ | | |
>
> ### *Element Stiffnesses in Global Coordinates:*
>
> *Define the near and far nodes:*
> The selection is our choice. For simplicity, let's work alphabetically across each element. The node whose letter comes earlier in the alphabet will be the near node.
>
> *Element AB:*
>
> $l_{AB} = 350 \text{ cm}$
>
> $\left(\dfrac{AE}{l}\right)_{AB} = \dfrac{(84.7 \text{ cm}^2)(20{,}000 \text{ kN/cm}^2)}{350 \text{ cm}} = 4840 \text{ kN/cm}$
>
> $\cos\theta = \dfrac{350 \text{ cm} - 0 \text{ cm}}{\sqrt{(350 \text{ cm} - 0 \text{ cm})^2 + (0 \text{ cm} - 0 \text{ cm})^2}} = 1.0000$
>
> $\sin\theta = \dfrac{0 \text{ cm} - 0 \text{ cm}}{\sqrt{(350 \text{ cm} - 0 \text{ cm})^2 + (0 \text{ cm} - 0 \text{ cm})^2}} = 0.0000$
>
> $\dfrac{AE}{l}\cos^2\theta = (4840 \text{ kN/cm})(1.0000)^2 = 4840 \text{ kN/cm}$
>
> $\dfrac{AE}{l}\cos\theta\sin\theta = (4840 \text{ kN/cm})(1.0000)(0.0000) = 0$
>
> $\dfrac{AE}{l}\sin^2\theta = (4840 \text{ kN/cm})(0.0000)^2 = 0$

$$[\mathbf{k}_{AB}] = \begin{matrix} q_{Ax} \\ q_{Ay} \\ q_{Bx} \\ q_{By} \end{matrix} \begin{bmatrix} \overset{d_{Ax}}{4840 \text{ kN/cm}} & \overset{d_{Ay}}{0} & \overset{d_{Bx}}{-4840 \text{ kN/cm}} & \overset{d_{By}}{0} \\ 0 & 0 & 0 & 0 \\ -4840 \text{ kN/cm} & 0 & 4840 \text{ kN/cm} & 0 \\ 0 & 0 & 0 & 0 \end{bmatrix}$$

Element AD:

$$l_{AD} = \sqrt{(350 \text{ cm})^2 + (300 \text{ cm})^2} = 461.0 \text{ cm}$$

$$\left(\frac{AE}{l}\right)_{AD} = \frac{(37.6 \text{ cm}^2)(20{,}000 \text{ kN/cm}^2)}{461.0 \text{ cm}} = 1631.2 \text{ kN/cm}$$

$$\cos\theta = \frac{350 \text{ cm} - 0 \text{ cm}}{\sqrt{(350 \text{ cm} - 0 \text{ cm})^2 + (300 \text{ cm} - 0 \text{ cm})^2}} = 0.7593$$

$$\sin\theta = \frac{300 \text{ cm} - 0 \text{ cm}}{\sqrt{(350 \text{ cm} - 0 \text{ cm})^2 + (300 \text{ cm} - 0 \text{ cm})^2}} = 0.6508$$

$$\frac{AE}{l}\cos^2\theta = (1631.2 \text{ kN/cm})(0.7593)^2 = 940.4 \text{ kN/cm}$$

$$\frac{AE}{l}\cos\theta\sin\theta = (1631.2 \text{ kN/cm})(0.7593)(0.6508) = 806.1 \text{ kN/cm}$$

$$\frac{AE}{l}\sin^2\theta = (1631.2 \text{ kN/cm})(0.6508)^2 = 690.9 \text{ kN/cm}$$

$$[\mathbf{k}_{AD}] = \begin{matrix} q_{Ax} \\ q_{Ay} \\ q_{Dx} \\ q_{Dy} \end{matrix} \begin{bmatrix} \overset{d_{Ax}}{940.4 \text{ kN/cm}} & \overset{d_{Ay}}{806.1 \text{ kN/cm}} & \overset{d_{Dx}}{-940.4 \text{ kN/cm}} & \overset{d_{Dy}}{-806.1 \text{ kN/cm}} \\ 806.1 \text{ kN/cm} & 690.9 \text{ kN/cm} & -806.1 \text{ kN/cm} & -690.9 \text{ kN/cm} \\ -940.4 \text{ kN/cm} & -806.1 \text{ kN/cm} & 940.4 \text{ kN/cm} & 806.1 \text{ kN/cm} \\ -806.1 \text{ kN/cm} & -690.9 \text{ kN/cm} & 806.1 \text{ kN/cm} & 690.9 \text{ kN/cm} \end{bmatrix}$$

Element BC:

$$l_{BC} = 300 \text{ cm}$$

$$\left(\frac{AE}{l}\right)_{BC} = \frac{(84.7 \text{ cm}^2)(20{,}000 \text{ kN/cm}^2)}{300 \text{ cm}} = 5647 \text{ kN/cm}$$

$$\cos\theta = \frac{650 \text{ cm} - 350 \text{ cm}}{\sqrt{(650 \text{ cm} - 300 \text{ cm})^2 + (0 \text{ cm} - 0 \text{ cm})^2}} = 1.0000$$

$$\sin\theta = \frac{0 \text{ cm} - 0 \text{ cm}}{\sqrt{(650 \text{ cm} - 300 \text{ cm})^2 + (0 \text{ cm} - 0 \text{ cm})^2}} = 0.0000$$

$$\frac{AE}{l}\cos^2\theta = (5647 \text{ kN/cm})(1.0000)^2 = 5647 \text{ kN/cm}$$

$$\frac{AE}{l}\cos\theta\sin\theta = (5647 \text{ kN/cm})(1.0000)(0.0000) = 0$$

$$\frac{AE}{l}\sin^2\theta = (5647 \text{ kN/cm})(0.0000)^2 = 0$$

$$[\mathbf{k}_{BC}] = \begin{matrix} q_{Bx} \\ q_{By} \\ q_{Cx} \\ q_{Cy} \end{matrix} \begin{bmatrix} \overset{d_{Bx}}{5647 \text{ kN/cm}} & \overset{d_{By}}{0} & \overset{d_{Cx}}{-5647 \text{ kN/cm}} & \overset{d_{Cy}}{0} \\ 0 & 0 & 0 & 0 \\ -5647 \text{ kN/cm} & 0 & 5647 \text{ kN/cm} & 0 \\ 0 & 0 & 0 & 0 \end{bmatrix}$$

Element BD:

$l_{BD} = 300 \text{ cm}$

$\left(\dfrac{AE}{l}\right)_{BD} = \dfrac{(84.7 \text{ cm}^2)(20{,}000 \text{ kN/cm}^2)}{300 \text{ cm}} = 5647 \text{ kN/cm}$

$\cos\theta = \dfrac{350 \text{ cm} - 350 \text{ cm}}{\sqrt{(350 \text{ cm} - 350 \text{ cm})^2 + (300 \text{ cm} - 0 \text{ cm})^2}} = 0.0000$

$\sin\theta = \dfrac{300 \text{ cm} - 0 \text{ cm}}{\sqrt{(350 \text{ cm} - 350 \text{ cm})^2 + (300 \text{ cm} - 0 \text{ cm})^2}} = 1.0000$

$\dfrac{AE}{l}\cos^2\theta = (5647 \text{ kN/cm})(0.0000)^2 = 0$

$\dfrac{AE}{l}\cos\theta\sin\theta = (5647 \text{ kN/cm})(0.0000)(1.0000) = 0$

$\dfrac{AE}{l}\sin^2\theta = (5647 \text{ kN/cm})(1.0000)^2 = 5647 \text{ kN/cm}$

$$[\mathbf{k}_{BD}] = \begin{array}{c} q_{Bx} \\ q_{By} \\ q_{Dx} \\ q_{Dy} \end{array} \begin{bmatrix} \overset{d_{Bx}}{0} & \overset{d_{By}}{0} & \overset{d_{Dx}}{0} & \overset{d_{Dy}}{0} \\ 0 & 5647 \text{ kN/cm} & 0 & -5647 \text{ kN/cm} \\ 0 & 0 & 0 & 0 \\ 0 & -5647 \text{ kN/cm} & 0 & 5647 \text{ kN/cm} \end{bmatrix}$$

Element CD:

$l_{CD} = \sqrt{(300 \text{ cm})^2 + (300 \text{ cm})^2} = 424.3 \text{ cm}$

$\left(\dfrac{AE}{l}\right)_{CD} = \dfrac{(37.6 \text{ cm}^2)(20{,}000 \text{ kN/cm}^2)}{424.3 \text{ cm}} = 1772.3 \text{ kN/cm}$

$\cos\theta = \dfrac{350 \text{ cm} - 650 \text{ cm}}{\sqrt{(350 \text{ cm} - 650 \text{ cm})^2 + (300 \text{ cm} - 0 \text{ cm})^2}} = -0.7071$

$\sin\theta = \dfrac{300 \text{ cm} - 0 \text{ cm}}{\sqrt{(350 \text{ cm} - 650 \text{ cm})^2 + (300 \text{ cm} - 0 \text{ cm})^2}} = 0.7071$

$\dfrac{AE}{l}\cos^2\theta = (1772.3 \text{ kN/cm})(-0.7071)^2 = 886.1 \text{ kN/cm}$

$\dfrac{AE}{l}\cos\theta\sin\theta = (1772.3 \text{ kN/cm})(-0.7071)(0.7071) = -886.1 \text{ kN/cm}$

$\dfrac{AE}{l}\sin^2\theta = (1772.3 \text{ kN/cm})(0.7071)^2 = 886.1 \text{ kN/cm}$

$$[\mathbf{k}_{CD}] = \begin{array}{c} q_{Cx} \\ q_{Cy} \\ q_{Dx} \\ q_{Dy} \end{array} \begin{bmatrix} \overset{d_{Cx}}{886.1 \text{ kN/cm}} & \overset{d_{Cy}}{-886.1 \text{ kN/cm}} & \overset{d_{Dx}}{-886.1 \text{ kN/cm}} & \overset{d_{Dy}}{886.1 \text{ kN/cm}} \\ -886.1 \text{ kN/cm} & 886.1 \text{ kN/cm} & 886.1 \text{ kN/cm} & -886.1 \text{ kN/cm} \\ -886.1 \text{ kN/cm} & 886.1 \text{ kN/cm} & 886.1 \text{ kN/cm} & -886.1 \text{ kN/cm} \\ 886.1 \text{ kN/cm} & -886.1 \text{ kN/cm} & -886.1 \text{ kN/cm} & 886.1 \text{ kN/cm} \end{bmatrix}$$

14.3 Compiling the System of Equations

Introduction
The Direct Stiffness Method ultimately is a system of simultaneous linear equations based on equilibrium at the nodes. The system of equations is made up of the structural stiffness matrix, the load vector, and the deformation vector.

How-To
In Section 14.2, we used the prime symbol (e.g., d', q') to denote local coordinates and no prime (e.g., d, q) to denote global coordinates for element information. The system of equations for a structure is made up of information that describes the whole structure rather than individual elements, so now we need different symbols. The industry standard is to use capital letters to refer to the structure:

- $[D]$ = Deformations at the nodes for the whole structure (global coordinates)
- $[K]$ = Structural stiffness matrix (global coordinates)
- $[Q]$ = Net loads that act on the structure at the nodes (global coordinates)

Degrees of Freedom
In order to assemble the information for the whole structure, we need a bookkeeping system. The system we use is *degrees of freedom* (DOF). A degree of freedom is a deformation associated with an internal force at one end of an element. For a truss element in local coordinates, each end has only one DOF: axial displacement. When we consider a two-dimensional truss in global coordinates, each end has two DOFs: d_x and d_y.

At a node, all the elements must move together, so the elements share the same degrees of freedom. That means the number of DOFs for a structure is set by the number of nodes. For example, a truss with four nodes has eight degrees of freedom (Figure 14.13a). We can number the DOFs in any sequence we want (Figure 14.13b, c). However, to make the bookkeeping easier, we often start with the x-direction and then the y-direction, node by node (Figure 14.13d).

Structural Stiffness Matrix
We saw in Section 14.1 that the structural stiffness is the combination of stiffnesses from all the elements. The key is to recognize that stiffness is unique to direction and location in the structure. Therefore, we add element stiffnesses only in the same global direction at the same node. Fortunately, that corresponds to degrees of freedom.

The structural stiffness matrix, $[K]$, has as many rows and columns as the structure has DOFs. As an illustration, the example truss in Figure 14.13 has eight DOFs, so $[K]$ is an 8×8 matrix. The matrix can be assembled in any DOF order, but since it is part of a system of equations, the other terms have to be in the same order. To reduce the

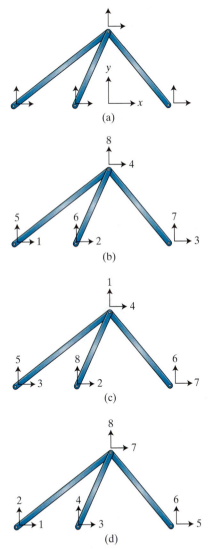

Figure 14.13 Degrees of freedom for a 2D truss with four nodes: (a) Each node has two DOFs, which line up with the global axes; (b) Numbering all x-direction DOFs and then y-direction DOFs is valid; (c) Randomly numbering the DOFs is valid; (d) The typical numbering system is x-direction and then y-direction, one node at a time.

likelihood of confusion or error, we will create the matrix in increasing DOF sequence. For our example truss of Figure 14.13, the matrix looks like the following:

$$[\mathbf{K}] = \begin{array}{c} \\ 1 \\ 2 \\ 3 \\ 4 \\ 5 \\ 6 \\ 7 \\ 8 \end{array} \begin{bmatrix} 1 & 2 & 3 & 4 & 5 & 6 & 7 & 8 \\ K_{11} & K_{12} & K_{13} & K_{14} & K_{15} & K_{16} & K_{17} & K_{18} \\ K_{21} & K_{22} & K_{23} & K_{24} & K_{25} & K_{26} & K_{27} & K_{28} \\ K_{31} & K_{32} & K_{33} & K_{34} & K_{35} & K_{36} & K_{37} & K_{38} \\ K_{41} & K_{42} & K_{43} & K_{44} & K_{45} & K_{46} & K_{47} & K_{48} \\ K_{51} & K_{52} & K_{53} & K_{54} & K_{55} & K_{56} & K_{57} & K_{58} \\ K_{61} & K_{62} & K_{63} & K_{64} & K_{65} & K_{66} & K_{67} & K_{68} \\ K_{71} & K_{72} & K_{73} & K_{74} & K_{75} & K_{76} & K_{77} & K_{78} \\ K_{81} & K_{82} & K_{83} & K_{84} & K_{85} & K_{86} & K_{87} & K_{88} \end{bmatrix}$$

Each term in the matrix, K_{ij}, is called a structural stiffness coefficient. They are always in global coordinates. We calculate each term by summing the contributions from all the elements that meet at that node:

$$K_{ij} = \sum_{Elem} k_{ij}^{Elem}$$

where

$Elem$ = All of the elements that have coefficients associated with the DOF combination ij

As an illustration, let's use the DOF numbering in Figure 14.13d for our example truss. If we consider node D to be the far node for each element (Figure 14.14), we get the following element stiffnesses:

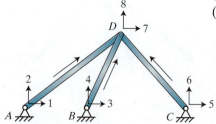

Figure 14.14 Degrees of freedom and local axes chosen for a 2D truss.

$$[\mathbf{k}_{AD}] = \begin{array}{c} \\ 1 \\ 2 \\ 7 \\ 8 \end{array} \begin{bmatrix} 1 & 2 & 7 & 8 \\ k_{11}^{AD} & k_{12}^{AD} & k_{17}^{AD} & k_{18}^{AD} \\ k_{21}^{AD} & k_{22}^{AD} & k_{27}^{AD} & k_{28}^{AD} \\ k_{71}^{AD} & k_{72}^{AD} & k_{77}^{AD} & k_{78}^{AD} \\ k_{81}^{AD} & k_{82}^{AD} & k_{87}^{AD} & k_{88}^{AD} \end{bmatrix}$$

$$[\mathbf{k}_{BD}] = \begin{array}{c} \\ 3 \\ 4 \\ 7 \\ 8 \end{array} \begin{bmatrix} 3 & 4 & 7 & 8 \\ k_{33}^{BD} & k_{34}^{BD} & k_{37}^{BD} & k_{38}^{BD} \\ k_{43}^{BD} & k_{44}^{BD} & k_{47}^{BD} & k_{48}^{BD} \\ k_{73}^{BD} & k_{74}^{BD} & k_{77}^{BD} & k_{78}^{BD} \\ k_{83}^{BD} & k_{84}^{BD} & k_{87}^{BD} & k_{88}^{BD} \end{bmatrix}$$

$$[\mathbf{k}_{CD}] = \begin{array}{c} \\ 5 \\ 6 \\ 7 \\ 8 \end{array} \begin{bmatrix} 5 & 6 & 7 & 8 \\ k_{55}^{CD} & k_{56}^{CD} & k_{57}^{CD} & k_{58}^{CD} \\ k_{65}^{CD} & k_{66}^{CD} & k_{67}^{CD} & k_{68}^{CD} \\ k_{75}^{CD} & k_{76}^{CD} & k_{77}^{CD} & k_{78}^{CD} \\ k_{85}^{CD} & k_{86}^{CD} & k_{87}^{CD} & k_{88}^{CD} \end{bmatrix}$$

Only element AD has a coefficient associated with DOFs 1 1, so the structural stiffness coefficient comes from only one element for our example truss:

$$K_{11} = k_{11}^{AD}$$

However, all three elements have coefficients associated with DOFs 8 7, so the structural stiffness coefficient is the sum of all three element stiffnesses:

$$K_{87} = k_{87}^{AD} + k_{87}^{BD} + k_{87}^{CD}$$

We follow this process for every combination of DOFs for the structure.

Some important observations result from this process. Not all combinations of DOF numbers are represented in the element stiffness matrices. As an illustration, our example truss does not have the combinations 1 3 and 6 2 as well as others. Remember the physical meaning of a stiffness coefficient: the load generated by a unit deformation. If the DOF combination does not exist, it means that the deformation does not cause any load at that particular DOF, so the stiffness is zero.

We can find these cases by elimination. There are element stiffness coefficients only when the DOF numbers are at the same node or connected by at least one element. Since no element in our example truss connects nodes B and C, all combinations of DOFs 3 and 4 to 5 and 6 have zero stiffness:

$$K_{35} = 0 \qquad K_{36} = 0$$
$$K_{45} = 0 \qquad K_{46} = 0$$
$$K_{53} = 0 \qquad K_{54} = 0$$
$$K_{63} = 0 \qquad K_{64} = 0$$

Another important observation comes from our formula for element stiffness in global coordinates:

$$[\mathbf{k}] = \begin{bmatrix} \frac{AE}{l}\cos^2\theta & \frac{AE}{l}\cos\theta\sin\theta & -\frac{AE}{l}\cos^2\theta & -\frac{AE}{l}\cos\theta\sin\theta \\ \frac{AE}{l}\cos\theta\sin\theta & \frac{AE}{l}\sin^2\theta & -\frac{AE}{l}\cos\theta\sin\theta & -\frac{AE}{l}\sin^2\theta \\ -\frac{AE}{l}\cos^2\theta & -\frac{AE}{l}\cos\theta\sin\theta & \frac{AE}{l}\cos^2\theta & \frac{AE}{l}\cos\theta\sin\theta \\ -\frac{AE}{l}\cos\theta\sin\theta & -\frac{AE}{l}\sin^2\theta & \frac{AE}{l}\cos\theta\sin\theta & \frac{AE}{l}\sin^2\theta \end{bmatrix}$$

The matrix is symmetric about the diagonal:

$$k_{ij} = k_{ji}, \quad i \neq j$$

As we add the contributions from each of the elements, this continues to be true. Therefore, the structural stiffness matrix $[\mathbf{K}]$ is also symmetric about the diagonal:

$$K_{ij} = K_{ji}, \quad i \neq j$$

We should have expected this based on the Maxwell-Betti Reciprocal Theorem we described in Section 12.2: Multi-Degree Indeterminate Beams.

The theorem shows that the matrix of flexibility coefficients is symmetric about the diagonal. If we look at the physical meaning of a flexibility coefficient and the physical meaning of a stiffness coefficient, we discover that one is the inverse of the other. So, the Maxwell-Betti Reciprocal Theorem applies to element stiffness as well as structural stiffness coefficients.

Load Vector

We explained in Section 14.1 that the Direct Stiffness Method is a set of equilibrium equations. Part of equilibrium is the net load applied to the node. Since we are dealing with load on the structure, we use the capital letter: Q. In general, the load might be a force or a moment, but for a truss, the load is only a force.

The load vector, $[\mathbf{Q}]$, is made up of the net load that acts on the structure at each degree of freedom. Consistent with how we assemble the structural stiffness matrix, we assemble the load vector in increasing DOF number sequence:

$$[\mathbf{Q}] = \begin{bmatrix} Q_1 \\ Q_2 \\ \vdots \\ Q_{n-1} \\ Q_n \end{bmatrix}$$

The load might be applied, a reaction, or a combination of both. Applied loads are known values. Even if a degree of freedom does not have an applied load, we still know the value: zero. Reactions, however, are unknown values. Consider our example truss (Figure 14.15).

The load vector for our example truss has six unknown loads ($A_x, A_y, B_x, B_y, C_x,$ and C_y) and two known loads (F_{Dx} and 0):

$$[\mathbf{Q}] = \begin{matrix} 1 \\ 2 \\ 3 \\ 4 \\ 5 \\ 6 \\ 7 \\ 8 \end{matrix} \begin{bmatrix} A_x \\ A_y \\ B_x \\ B_y \\ C_x \\ C_y \\ F_{Dx} \\ 0 \end{bmatrix}$$

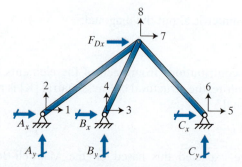

Figure 14.15 Example truss with six reactions and two applied loads.

For now, we will limit ourselves to a DOF that has either an applied load or a reaction. From statics, we know that the load applied at a support goes directly into the support, so we can deal with the situation outside of the Direct Stiffness Method if we want. We cover how the Direct Stiffness Method can deal with this situation in Section 14.5: Additional Loadings.

Deformation Vector

The Direct Stiffness Method also has a vector of deformations. In general, those deformations might be displacements or rotations, but for a truss, they are strictly displacements. Again, since we are dealing with deformations for the structure, we use the capital letter: D.

The deformation vector, $[D]$, contains the deformation of every DOF for the structure. As with the load vector and structural stiffness matrix, we assemble the deformation vector in increasing DOF number sequence:

$$[D] = \begin{bmatrix} D_1 \\ D_2 \\ \vdots \\ D_{n-1} \\ D_n \end{bmatrix}$$

Many of the deformations are unknown, but we know some when we assemble the vector. We know the deformations for all the DOFs that are supported by reactions. In most cases, supported deformations are zero; however, that is not always the reality we want to model. Differential foundation settlement and deformation of column baseplates are examples of situations where a DOF can be supported but has a nonzero deformation. As an illustration, consider our example truss (Figure 14.16).

If the left support moves Δ_{Ay} up relative to the other supports, we get the following deformation vector:

$$[D] = \begin{matrix} 1 \\ 2 \\ 3 \\ 4 \\ 5 \\ 6 \\ 7 \\ 8 \end{matrix} \begin{bmatrix} 0 \\ \Delta_{Ay} \\ 0 \\ 0 \\ 0 \\ 0 \\ \Delta_{Dx} \\ \Delta_{Dy} \end{bmatrix}$$

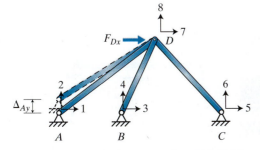

Figure 14.16 Example truss with displacement of a supported degree of freedom.

The value of Δ_{Ay} is positive because it is in the direction of the DOF. The only unknowns are Δ_{Dx} and Δ_{Dy} because only those DOFs are not supported.

System of Equations

The governing system of equations is summarized as

$$[Q] = [K][D]$$

We defined the terms earlier in this section.

SECTION 14.3 HIGHLIGHTS

Compiling the System of Equations

Additional Terms:

D_i = Deformation of a structure at DOF i
K_{ij} = Structural stiffness coefficient for DOFs $i\,j$
Q_i = Net load on a structure at DOF i
$\mathbf{D}$ = Structural deformation vector (global coordinates)
$\mathbf{K}$ = Structural stiffness matrix (global coordinates)
$\mathbf{Q}$ = Structural load vector (global coordinates)

Structural Stiffness Matrix: $[K]$

$$K_{ij} = \sum_{Elem} k_{ij}^{Elem}$$

- If no element has DOF combination $i\,j$, then $K_{ij} = 0$.
- $[K]$ is built in order of increasing DOF number.

Structural Load Vector: $[Q]$

- Support reactions are unknown Q_i.
- No applied load at DOF i and no support reaction means $Q_i = 0$.

Structural Deformation Vector: $[D]$

- No support at DOF i means D_i is unknown.
- Support at DOF i means D_i is known.
  – For most supported DOFs, $D_i = 0$.
  – If a specific support deformation is modeled, that deformation becomes the D_i.

EXAMPLE 14.2

Our firm has designed a large pipeline along the edge of a canal in Singapore. Our team has been tasked with designing a structure to support the pipe. Other engineers on the team have designed the truss members for strength.

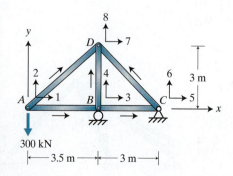

Figure 14.17

Because we are new to the firm, our supervisor wants to verify our knowledge of the Direct Stiffness Method. A determinate truss is a great way to do that.

We are at the point where we can compile the system of equations for the truss. Following our office's standard convention, we numbered the DOFs moving alphabetically through the nodes. In Example 14.1, we chose the near node for each element; the local x'-direction is shown near the middle of each element.

Structural Stiffness Matrix:

Element stiffness matrices with DOFs noted and partitioned into pairs of DOFs:

$$[\mathbf{k}_{AB}] = \begin{array}{c} \\ 1 \\ 2 \\ 3 \\ 4 \end{array} \begin{bmatrix} \overset{1}{4840 \text{ kN/cm}} & \overset{2}{0} & \overset{3}{-4840 \text{ kN/cm}} & \overset{4}{0} \\ 0 & 0 & 0 & 0 \\ -4840 \text{ kN/cm} & 0 & 4840 \text{ kN/cm} & 0 \\ 0 & 0 & 0 & 0 \end{bmatrix}$$

$$[\mathbf{k}_{AD}] = \begin{array}{c} \\ 1 \\ 2 \\ 7 \\ 8 \end{array} \begin{bmatrix} \overset{1}{940.4 \text{ kN/cm}} & \overset{2}{806.1 \text{ kN/cm}} & \overset{7}{-940.4 \text{ kN/cm}} & \overset{8}{-806.1 \text{ kN/cm}} \\ 806.1 \text{ kN/cm} & 690.9 \text{ kN/cm} & -806.1 \text{ kN/cm} & -690.9 \text{ kN/cm} \\ -940.4 \text{ kN/cm} & -806.1 \text{ kN/cm} & 940.4 \text{ kN/cm} & 806.1 \text{ kN/cm} \\ -806.1 \text{ kN/cm} & -690.9 \text{ kN/cm} & 806.1 \text{ kN/cm} & 690.9 \text{ kN/cm} \end{bmatrix}$$

$$[\mathbf{k}_{BC}] = \begin{array}{c} \\ 3 \\ 4 \\ 5 \\ 6 \end{array} \begin{bmatrix} \overset{3}{5647 \text{ kN/cm}} & \overset{4}{0} & \overset{5}{-5647 \text{ kN/cm}} & \overset{6}{0} \\ 0 & 0 & 0 & 0 \\ -5647 \text{ kN/cm} & 0 & 5647 \text{ kN/cm} & 0 \\ 0 & 0 & 0 & 0 \end{bmatrix}$$

$$[\mathbf{k}_{BD}] = \begin{array}{c} \\ 3 \\ 4 \\ 7 \\ 8 \end{array} \begin{bmatrix} \overset{3}{0} & \overset{4}{0} & \overset{7}{0} & \overset{8}{0} \\ 0 & 5647 \text{ kN/cm} & 0 & -5647 \text{ kN/cm} \\ 0 & 0 & 0 & 0 \\ 0 & -5647 \text{ kN/cm} & 0 & 5647 \text{ kN/cm} \end{bmatrix}$$

$$[\mathbf{k}_{CD}] = \begin{array}{c} \\ 5 \\ 6 \\ 7 \\ 8 \end{array} \begin{bmatrix} \overset{5}{886.1 \text{ kN/cm}} & \overset{6}{-886.1 \text{ kN/cm}} & \overset{7}{-886.1 \text{ kN/cm}} & \overset{8}{886.1 \text{ kN/cm}} \\ -886.1 \text{ kN/cm} & 886.1 \text{ kN/cm} & 886.1 \text{ kN/cm} & -886.1 \text{ kN/cm} \\ -886.1 \text{ kN/cm} & 886.1 \text{ kN/cm} & 886.1 \text{ kN/cm} & -886.1 \text{ kN/cm} \\ 886.1 \text{ kN/cm} & -886.1 \text{ kN/cm} & -886.1 \text{ kN/cm} & 886.1 \text{ kN/cm} \end{bmatrix}$$

Compile the terms to form the structural stiffness:
  Work in 2 × 2 chunks.
    Elements AB and AD share DOFs 1 and 2.

$$K_{11} = k_{11}^{AB} + k_{11}^{AD} = +4840 \text{ kN/cm} + 940.4 \text{ kN/cm} = 5780 \text{ kN/cm}$$

$$K_{12} = k_{12}^{AB} + k_{12}^{AD} = 0 + 806.1 \text{ kN/cm} = 806.1 \text{ kN/cm}$$

$$K_{21} = k_{21}^{AB} + k_{21}^{AD} = 0 + 806.1 \text{ kN/cm} = 806.1 \text{ kN/cm}$$

$$K_{22} = k_{22}^{AB} + k_{22}^{AD} = 0 + 690.9 \text{ kN/cm} = 690.9 \text{ kN/cm}$$

814   **Chapter 14**   Direct Stiffness Method for Trusses

Only element AB has DOF pairs 1 and 2 with 3 and 4.

$K_{13} = k_{13}^{AB} = -4840 \text{ kN/cm}$

$K_{14} = k_{14}^{AB} = 0$

$K_{23} = k_{23}^{AB} = 0$

$K_{24} = k_{24}^{AB} = 0$

No element has DOF pairs 1 and 2 with 5 and 6.

$K_{15} = K_{16} = K_{25} = K_{26} = 0$

Only element AD has DOF pairs 1 and 2 with 7 and 8.

$K_{17} = k_{17}^{AD} = -940.4 \text{ kN/cm}$

$K_{18} = k_{18}^{AD} = -806.1 \text{ kN/cm}$

$K_{27} = k_{27}^{AD} = -806.1 \text{ kN/cm}$

$K_{28} = k_{28}^{AD} = -690.9 \text{ kN/cm}$

All three elements share DOFs 3 and 4.

$K_{33} = k_{33}^{AB} + k_{33}^{BC} + k_{33}^{BD} = 4840 \text{ kN/cm} + 5647 \text{ kN/cm} + 0 = 10{,}487 \text{ kN/cm}$

$K_{34} = k_{34}^{AB} + k_{34}^{BC} + k_{34}^{BD} = 0 + 0 + 0 = 0$

$K_{43} = k_{43}^{AB} + k_{43}^{BC} + k_{43}^{BD} = 0 + 0 + 0 = 0$

$K_{44} = k_{44}^{AB} + k_{44}^{BC} + k_{44}^{BD} = 0 + 0 + 5647 \text{ kN/cm} = 5647 \text{ kN/cm}$

Continue the process for all 64 entries in $[\mathbf{K}]$.

$$[\mathbf{K}] = \begin{array}{c} 1 \\ 2 \\ 3 \\ 4 \\ 5 \\ 6 \\ 7 \\ 8 \end{array} \begin{bmatrix} 5780 \tfrac{kN}{cm} & 806.1 \tfrac{kN}{cm} & -4840 \tfrac{kN}{cm} & 0 & 0 & 0 & -940.4 \tfrac{kN}{cm} & -806.1 \tfrac{kN}{cm} \\ 806.1 \tfrac{kN}{cm} & 690.9 \tfrac{kN}{cm} & 0 & 0 & 0 & 0 & -806.1 \tfrac{kN}{cm} & -690.9 \tfrac{kN}{cm} \\ -4840 \tfrac{kN}{cm} & 0 & 10487 \tfrac{kN}{cm} & 0 & -5647 \tfrac{kN}{cm} & 0 & 0 & 0 \\ 0 & 0 & 0 & 5647 \tfrac{kN}{cm} & 0 & 0 & 0 & -5647 \tfrac{kN}{cm} \\ 0 & 0 & -5647 \tfrac{kN}{cm} & 0 & 6533 \tfrac{kN}{cm} & -886.1 \tfrac{kN}{cm} & -886.1 \tfrac{kN}{cm} & 886.1 \tfrac{kN}{cm} \\ 0 & 0 & 0 & 0 & -886.1 \tfrac{kN}{cm} & 886.1 \tfrac{kN}{cm} & 886.1 \tfrac{kN}{cm} & -886.1 \tfrac{kN}{cm} \\ -940.4 \tfrac{kN}{cm} & -806.1 \tfrac{kN}{cm} & 0 & 0 & -886.1 \tfrac{kN}{cm} & 886.1 \tfrac{kN}{cm} & 1826.5 \tfrac{kN}{cm} & -80.0 \tfrac{kN}{cm} \\ -806.1 \tfrac{kN}{cm} & -690.9 \tfrac{kN}{cm} & 0 & -5647 \tfrac{kN}{cm} & 886.1 \tfrac{kN}{cm} & -886.1 \tfrac{kN}{cm} & -80.0 \tfrac{kN}{cm} & 7224 \tfrac{kN}{cm} \end{bmatrix}$$

Evaluation of Results:

Satisfied Fundamental Principles?

The matrix is symmetric about the diagonal, as expected. ✓

Load Vector:

Three unknown reactions:

$$[\mathbf{Q}] = \begin{array}{c} 1 \\ 2 \\ 3 \\ 4 \\ 5 \\ 6 \\ 7 \\ 8 \end{array} \begin{bmatrix} 0 \\ -300 \text{ kN} \\ 0 \\ B_y \\ C_x \\ C_y \\ 0 \\ 0 \end{bmatrix}$$

Section 14.4 Finding Deformations, Reactions, and Internal Forces **815**

Deformation Vector:

Five unknown displacements:

$$[\mathbf{D}] = \begin{matrix} 1 \\ 2 \\ 3 \\ 4 \\ 5 \\ 6 \\ 7 \\ 8 \end{matrix} \begin{bmatrix} \Delta_{Ax} \\ \Delta_{Ay} \\ \Delta_{Bx} \\ 0 \\ 0 \\ 0 \\ \Delta_{Dx} \\ \Delta_{Dy} \end{bmatrix}$$

System of Equations:

$$\begin{matrix} 1 \\ 2 \\ 3 \\ 4 \\ 5 \\ 6 \\ 7 \\ 8 \end{matrix} \begin{bmatrix} 0 \\ -300 \text{ kN} \\ 0 \\ B_y \\ C_x \\ C_y \\ 0 \\ 0 \end{bmatrix} = \begin{bmatrix} 0 & 5780\tfrac{\text{kN}}{\text{cm}} & 806.1\tfrac{\text{kN}}{\text{cm}} & -4840\tfrac{\text{kN}}{\text{cm}} & 0 & 0 & 0 & -940.4\tfrac{\text{kN}}{\text{cm}} & -806.1\tfrac{\text{kN}}{\text{cm}} \\ & 806.1\tfrac{\text{kN}}{\text{cm}} & 690.9\tfrac{\text{kN}}{\text{cm}} & 0 & 0 & 0 & 0 & -806.1\tfrac{\text{kN}}{\text{cm}} & -690.9\tfrac{\text{kN}}{\text{cm}} \\ & -4840\tfrac{\text{kN}}{\text{cm}} & 0 & 10487\tfrac{\text{kN}}{\text{cm}} & 0 & -5647\tfrac{\text{kN}}{\text{cm}} & 0 & 0 & 0 \\ & 0 & 0 & 0 & 5647\tfrac{\text{kN}}{\text{cm}} & 0 & 0 & 0 & -5647\tfrac{\text{kN}}{\text{cm}} \\ & 0 & 0 & -5647\tfrac{\text{kN}}{\text{cm}} & 0 & 6533\tfrac{\text{kN}}{\text{cm}} & -886.1\tfrac{\text{kN}}{\text{cm}} & -886.1\tfrac{\text{kN}}{\text{cm}} & 886.1\tfrac{\text{kN}}{\text{cm}} \\ & 0 & 0 & 0 & 0 & -886.1\tfrac{\text{kN}}{\text{cm}} & 886.1\tfrac{\text{kN}}{\text{cm}} & 886.1\tfrac{\text{kN}}{\text{cm}} & -886.1\tfrac{\text{kN}}{\text{cm}} \\ & -940.4\tfrac{\text{kN}}{\text{cm}} & -806.1\tfrac{\text{kN}}{\text{cm}} & 0 & 0 & -886.1\tfrac{\text{kN}}{\text{cm}} & 886.1\tfrac{\text{kN}}{\text{cm}} & 1826.5\tfrac{\text{kN}}{\text{cm}} & -80.0\tfrac{\text{kN}}{\text{cm}} \\ & -806.1\tfrac{\text{kN}}{\text{cm}} & -690.9\tfrac{\text{kN}}{\text{cm}} & 0 & -5647\tfrac{\text{kN}}{\text{cm}} & 886.1\tfrac{\text{kN}}{\text{cm}} & -886.1\tfrac{\text{kN}}{\text{cm}} & -80.0\tfrac{\text{kN}}{\text{cm}} & 7224\tfrac{\text{kN}}{\text{cm}} \end{bmatrix} \begin{bmatrix} \Delta_{Ax} \\ \Delta_{Ay} \\ \Delta_{Bx} \\ 0 \\ 0 \\ 0 \\ \Delta_{Dx} \\ \Delta_{Dy} \end{bmatrix}$$

14.4 Finding Deformations, Reactions, and Internal Forces

Introduction
Solving the system of simultaneous linear equations requires some special steps. Once we find the deformations, though, we can find the reactions and internal forces.

How-To
The governing system of equations has the same number of unknowns as the number of equations, but for the following reasons we can't solve the system in the format used in Section 14.3.

1. There are unknowns on both sides of the equation. Some of the Q's are unknown and some of the D's are unknown.
2. Even if all the unknowns are on one side, the structural stiffness matrix $[\mathbf{K}]$ is singular. To solve the equation, we need to invert $[\mathbf{K}]$: $[\mathbf{K}]^{-1}[\mathbf{Q}] = [\mathbf{D}]$. But we cannot invert a singular matrix.

Partitioning the System of Equations
In order to solve the system of equations, we need to break down the system into two subsystems of equations. The criterion we will use is the unknowns. One subsystem will have all of the deformation unknowns; the other will have all of the reaction unknowns.

All of the deformation unknowns occur at DOFs that do not have supports, so we call them *free* and identify the subsystem with the subscript *f*. All of the reaction unknowns occur at DOFs that have supports, so we call them *supported* and identify the subsystem with the subscript *s*.

Because this is a system of simultaneous equations, we can put the equations in any order. We want to gather all of the equations associated with free DOFs at the top and all the equations associated with supported DOFs at the bottom:

$$\begin{bmatrix} \mathbf{Q}_f \\ \mathbf{Q}_s \end{bmatrix} = \begin{bmatrix} \mathbf{K}_f \\ \mathbf{K}_s \end{bmatrix} [\mathbf{D}]$$

This format is still not solvable because the unknowns in [**D**] are still commingled with the knowns. We want all of the unknowns at the top, so we want all of the deformations associated with free DOFs at the top. When we shift the rows of [**D**], we must also shift the columns of [**K**]. The result is the following partitioned system of equations:

$$\begin{bmatrix} \mathbf{Q}_f \\ \mathbf{Q}_s \end{bmatrix} = \begin{bmatrix} \mathbf{K}_{ff} & \mathbf{K}_{fs} \\ \mathbf{K}_{sf} & \mathbf{K}_{ss} \end{bmatrix} \begin{bmatrix} \mathbf{D}_f \\ \mathbf{D}_s \end{bmatrix}$$

The two subsets of equations we can solve are

$$[\mathbf{Q}_f] = [\mathbf{K}_{ff}][\mathbf{D}_f] + [\mathbf{K}_{fs}][\mathbf{D}_s]$$
$$[\mathbf{Q}_s] = [\mathbf{K}_{sf}][\mathbf{D}_f] + [\mathbf{K}_{ss}][\mathbf{D}_s]$$

Solve for Deformations

To find the unknown deformations, we solve the first subset of equations:

$$[\mathbf{Q}_f] = [\mathbf{K}_{ff}][\mathbf{D}_f] + [\mathbf{K}_{fs}][\mathbf{D}_s]$$

We do that by moving all of the known terms to one side, leaving the unknown deformations multiplied by a subset of the structural stiffness matrix:

$$[\mathbf{K}_{ff}][\mathbf{D}_f] = [\mathbf{Q}_f] - [\mathbf{K}_{fs}][\mathbf{D}_s]$$

We solve this expression by symbolically inverting [$\mathbf{K}_{ff}$]. This subset of the structural stiffness matrix is typically not singular, so we can invert it and there is a solution:

$$[\mathbf{D}_f] = [\mathbf{K}_{ff}]^{-1}([\mathbf{Q}_f] - [\mathbf{K}_{fs}][\mathbf{D}_s])$$

Mathematically, we don't actually invert the matrix; that is very inefficient. Instead, we use a numerical method like Gaussian elimination or the Gauss-Seidel Iterative Method. All of the solution methods typically result in the loss of at least one digit of accuracy; that is why we carry at least four digits until we have our final results (e.g., reactions, internal forces).

As we look at the first subset of equations, we note that any initial support deformations, [$\mathbf{D}_s$], contribute to the deformations at the free DOFs. So, even if there are no applied loads (i.e., [$\mathbf{Q}_f$] = **0**), there will

Section 14.4 Finding Deformations, Reactions, and Internal Forces

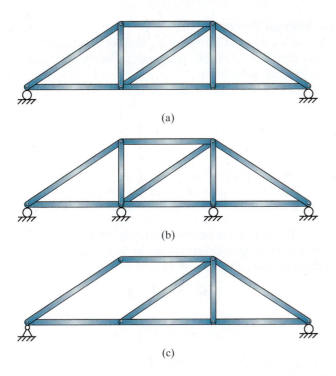

Figure 14.18 Different structural situations that result in a singular [K_{ff}]: (a) Insufficient constraints—a 2D structure needs three constrained DOFs to be stable; (b) Improper constraints—a 2D structure needs to be constrained against three rigid body motions: translation in the x-direction, translation in the y-direction, and rotation; (c) Internally unstable structure—a truss must be made of triangular regions to be stable.

be deformations at the free DOFs if we have initial support deformations. If there are no initial support deformations, the term [K_{fs}][D_s] is zero.

Notice that we said [K_{ff}] is typically not singular. When it is singular, that means we have an unstable structure. That instability might be due to insufficient constraints (e.g., Figure 14.18a), improper constraints (e.g., Figure 14.18b), or internal instability (e.g., Figure 14.18c). Structural analysis software programs typically use this check to generate a warning to the user before attempting to solve the system of equations.

Solve for Reactions

We use the second subset of equations to find the reactions:

$$[Q_s] = [K_{sf}][D_f] + [K_{ss}][D_s]$$

We need to know the deformations before we can make this calculation, but then the unknowns are all on the left side of the equation. That means we can find the reactions with matrix multiplication; we don't need to invert any matrices.

Again, if there are any initial deformations of supported DOFs, they are incorporated into this calculation.

Solve for Internal Forces

When calculating internal forces at the ends of an element, we typically want to know the forces in local coordinates. For a truss element, the only internal force is the axial force.

In Section 14.2, we learned that the internal forces are related to the nodal deformations:

$$[\mathbf{q'}] = [\mathbf{k'}][\mathbf{d'}]$$

where

$[\mathbf{q'}]$ = Internal forces at the ends of an element (local coordinates)
$[\mathbf{k'}]$ = Element stiffness matrix (local coordinates)
$[\mathbf{d'}]$ = Deformations at the ends of an element (local coordinates)

Calculating $[\mathbf{k'}]$ for an element is relatively easy, but the results of our previous calculations give deformations in global coordinates. Therefore, let's use the transformation matrix to convert $[\mathbf{d'}]$ to $[\mathbf{d}]$:

$$[\mathbf{d'}] = [\mathbf{T}][\mathbf{d}]$$

so

$$[\mathbf{q'}] = [\mathbf{k'}][\mathbf{T}][\mathbf{d}]$$

Our definition of $[\mathbf{d}]$ is the deformations at nodes in global coordinates for a specific element. Those deformations are just a subset of the structural deformation vector, $[\mathbf{D}]$. No transformation is required. We only need to be sure that we are pulling the entries associated with the DOFs of the particular element. Remember that the complete structural deformation vector includes deformations at supported DOFs that might or might not be zero.

We can substitute formulas for each of the terms in this expression to obtain the following matrix:

$$\begin{bmatrix} q'_N \\ q'_F \end{bmatrix} = \begin{bmatrix} \dfrac{AE}{l} & -\dfrac{AE}{l} \\ -\dfrac{AE}{l} & \dfrac{AE}{l} \end{bmatrix} \begin{bmatrix} \cos\theta & \sin\theta & 0 & 0 \\ 0 & 0 & \cos\theta & \sin\theta \end{bmatrix} \begin{bmatrix} d_{Nx} \\ d_{Ny} \\ d_{Fx} \\ d_{Fy} \end{bmatrix}$$

If no force is applied along the length of a truss element, equilibrium requires that the internal force at one end is equal and opposite to the internal force at the other end. Therefore, we really only need to use part of this expression. In Figure 14.5, we saw that the near node internal force is drawn in the compression direction, and the far node internal force is drawn in the tension direction. So, a positive value for the far node means the element is in tension. Let's focus on just that calculation:

$$q'_F = \dfrac{AE}{l} [-\cos\theta \quad -\sin\theta \quad \cos\theta \quad \sin\theta] \begin{bmatrix} d_{Nx} \\ d_{Ny} \\ d_{Fx} \\ d_{Fy} \end{bmatrix}$$

Section 14.4 Highlights

Finding Deformations, Reactions, and Internal Forces

Additional Terms:

- f = Free degree of freedom, not supported
- s = Supported degree of freedom, restrained against movement
- $\mathbf{D}_f$ = Deformations at free DOFs, initially unknown
- $\mathbf{D}_s$ = Deformations at supported DOFs, known, might not be zero
- $\mathbf{Q}_f$ = Net loads at free DOFs, known, might be zero
- $\mathbf{Q}_s$ = Net loads at supported DOFs, initially unknown, reactions

Finding Deformations:

Reorder the terms in the system of equations to group the free DOFs and the supported DOFs:

$$\begin{bmatrix} \mathbf{Q}_f \\ \mathbf{Q}_s \end{bmatrix} = \begin{bmatrix} \mathbf{K}_{ff} & \mathbf{K}_{fs} \\ \mathbf{K}_{sf} & \mathbf{K}_{ss} \end{bmatrix} \begin{bmatrix} \mathbf{D}_f \\ \mathbf{D}_s \end{bmatrix}$$

Solve a subset of the equations for deformations before finding anything else:

$$[\mathbf{Q}_f] = [\mathbf{K}_{ff}][\mathbf{D}_f] + [\mathbf{K}_{fs}][\mathbf{D}_s]$$

$$[\mathbf{D}_f] = [\mathbf{K}_{ff}]^{-1}([\mathbf{Q}_f] - [\mathbf{K}_{fs}][\mathbf{D}_s])$$

Use a numerical method to symbolically invert $[\mathbf{K}_{ff}]$. Can be done by hand, many calculators, or a variety of software packages.

Finding Reactions:

Use the deformations and the second subset of equations:

$$[\mathbf{Q}_s] = [\mathbf{K}_{sf}][\mathbf{D}_f] + [\mathbf{K}_{ss}][\mathbf{D}_s]$$

Finding Internal Forces:

Use the deformations for the DOFs specifically at the ends of the desired element. Reactions are not needed for this calculation.

$$q'_F = \frac{AE}{l}[-\cos\theta \quad -\sin\theta \quad \cos\theta \quad \sin\theta]\begin{bmatrix} d_{Nx} \\ d_{Ny} \\ d_{Fx} \\ d_{Fy} \end{bmatrix}$$

A positive result means the element is in tension; a negative result means compression.

EXAMPLE 14.3

Our firm has designed a large pipeline along the edge of a canal in Singapore. Our team has been tasked with designing a structure to support the pipe. Other engineers on the team have designed the truss members for strength.

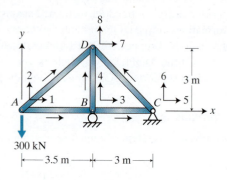

Figure 14.19

Because we are new to the firm, our supervisor wants to verify our knowledge of the Direct Stiffness Method. A determinate truss is a great way to do that.

We are at the point where we can calculate the deformations, reactions, and internal forces. In Example 14.2, we assembled the system of equations.

Estimated Solution:

We did this in Example 5.5.

Summary of results:

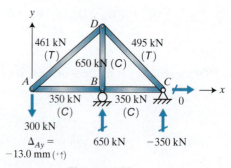

Figure 14.20

Direct Stiffness Method:

Partition the system of equations:

    Free DOFs: 1, 2, 3, 7, 8
    Supported DOFs: 4, 5, 6

Section 14.4 Finding Deformations, Reactions, and Internal Forces

$$[\mathbf{Q}_f] = [\mathbf{K}_{ff}][\mathbf{D}_f] + [\mathbf{K}_{fs}][\mathbf{D}_s]$$

$$\begin{matrix}1\\2\\3\\7\\8\end{matrix}\begin{bmatrix}0\\-300\text{ kN}\\0\\0\\0\end{bmatrix} = \begin{matrix}1\\2\\3\\7\\8\end{matrix}\begin{bmatrix}\overset{1}{5780\tfrac{\text{kN}}{\text{cm}}} & \overset{2}{806.1\tfrac{\text{kN}}{\text{cm}}} & \overset{3}{-4840\tfrac{\text{kN}}{\text{cm}}} & \overset{7}{-940.9\tfrac{\text{kN}}{\text{cm}}} & \overset{8}{-806.1\tfrac{\text{kN}}{\text{cm}}}\\806.1\tfrac{\text{kN}}{\text{cm}} & 690.9\tfrac{\text{kN}}{\text{cm}} & 0 & -806.1\tfrac{\text{kN}}{\text{cm}} & -690.9\tfrac{\text{kN}}{\text{cm}}\\-4840\tfrac{\text{kN}}{\text{cm}} & 0 & 10487\tfrac{\text{kN}}{\text{cm}} & 0 & 0\\-940.9\tfrac{\text{kN}}{\text{cm}} & -806.1\tfrac{\text{kN}}{\text{cm}} & 0 & 1826.5\tfrac{\text{kN}}{\text{cm}} & -80.0\tfrac{\text{kN}}{\text{cm}}\\-806.1\tfrac{\text{kN}}{\text{cm}} & -690.9\tfrac{\text{kN}}{\text{cm}} & 0 & -80.0\tfrac{\text{kN}}{\text{cm}} & 7224\tfrac{\text{kN}}{\text{cm}}\end{bmatrix}\begin{bmatrix}\Delta_{Ax}\\ \Delta_{Ay}\\ \Delta_{Bx}\\ \Delta_{Dx}\\ \Delta_{Dy}\end{bmatrix} + \begin{matrix}1\\2\\3\\7\\8\end{matrix}\begin{bmatrix}\overset{4}{0} & \overset{5}{0} & \overset{6}{0}\\0 & 0 & 0\\0 & -5647\tfrac{\text{kN}}{\text{cm}} & 0\\0 & -886.1\tfrac{\text{kN}}{\text{cm}} & 886.1\tfrac{\text{kN}}{\text{cm}}\\-5647\tfrac{\text{kN}}{\text{cm}} & 886.1\tfrac{\text{kN}}{\text{cm}} & -886.1\tfrac{\text{kN}}{\text{cm}}\end{bmatrix}\begin{bmatrix}0\\0\\0\end{bmatrix}\begin{matrix}4\\5\\6\end{matrix}$$

$$[\mathbf{Q}_s] = [\mathbf{K}_{sf}][\mathbf{D}_f] + [\mathbf{K}_{ss}][\mathbf{D}_s]$$

$$\begin{matrix}4\\5\\6\end{matrix}\begin{bmatrix}B_y\\C_x\\C_y\end{bmatrix} = \begin{matrix}4\\5\\6\end{matrix}\begin{bmatrix}\overset{1}{0} & \overset{2}{0} & \overset{3}{0} & \overset{7}{0} & \overset{8}{-5647\tfrac{\text{kN}}{\text{cm}}}\\0 & 0 & -5647\tfrac{\text{kN}}{\text{cm}} & -886.1\tfrac{\text{kN}}{\text{cm}} & 886.1\tfrac{\text{kN}}{\text{cm}}\\0 & 0 & 0 & 886.1\tfrac{\text{kN}}{\text{cm}} & -886.1\tfrac{\text{kN}}{\text{cm}}\end{bmatrix}\begin{bmatrix}\Delta_{Ax}\\ \Delta_{Ay}\\ \Delta_{Bx}\\ \Delta_{Dx}\\ \Delta_{Dy}\end{bmatrix} + \begin{matrix}4\\5\\6\end{matrix}\begin{bmatrix}\overset{4}{5647\tfrac{\text{kN}}{\text{cm}}} & \overset{5}{0} & \overset{6}{0}\\0 & 6533\tfrac{\text{kN}}{\text{cm}} & -886.1\tfrac{\text{kN}}{\text{cm}}\\0 & -886.1\tfrac{\text{kN}}{\text{cm}} & 886.1\tfrac{\text{kN}}{\text{cm}}\end{bmatrix}\begin{bmatrix}0\\0\\0\end{bmatrix}$$

Find the deformations:

$$[\mathbf{D}_f] = [\mathbf{K}_{ff}]^{-1}([\mathbf{Q}_f] - [\mathbf{K}_{fs}][\mathbf{D}_s])$$

$$\begin{bmatrix}\Delta_{Ax}\\ \Delta_{Ay}\\ \Delta_{Bx}\\ \Delta_{Dx}\\ \Delta_{Dy}\end{bmatrix} = \begin{matrix}1\\2\\3\\7\\8\end{matrix}\begin{bmatrix}\overset{1}{5780\tfrac{\text{kN}}{\text{cm}}} & \overset{2}{806.1\tfrac{\text{kN}}{\text{cm}}} & \overset{3}{-4840\tfrac{\text{kN}}{\text{cm}}} & \overset{7}{-940.9\tfrac{\text{kN}}{\text{cm}}} & \overset{8}{-806.1\tfrac{\text{kN}}{\text{cm}}}\\806.1\tfrac{\text{kN}}{\text{cm}} & 690.9\tfrac{\text{kN}}{\text{cm}} & 0 & -806.1\tfrac{\text{kN}}{\text{cm}} & -690.9\tfrac{\text{kN}}{\text{cm}}\\-4840\tfrac{\text{kN}}{\text{cm}} & 0 & 10487\tfrac{\text{kN}}{\text{cm}} & 0 & 0\\-940.9\tfrac{\text{kN}}{\text{cm}} & -806.1\tfrac{\text{kN}}{\text{cm}} & 0 & 1826.5\tfrac{\text{kN}}{\text{cm}} & -80.0\tfrac{\text{kN}}{\text{cm}}\\-806.1\tfrac{\text{kN}}{\text{cm}} & -690.9\tfrac{\text{kN}}{\text{cm}} & 0 & -80.0\tfrac{\text{kN}}{\text{cm}} & 7224\tfrac{\text{kN}}{\text{cm}}\end{bmatrix}^{-1}\left(\begin{bmatrix}0\\-300\text{ kN}\\0\\0\\0\end{bmatrix} - \begin{bmatrix}0\\0\\0\\0\\0\end{bmatrix}\right)$$

Solve the system of equations:

$$\begin{bmatrix}\Delta_{Ax}\\ \Delta_{Ay}\\ \Delta_{Bx}\\ \Delta_{Dx}\\ \Delta_{Dy}\end{bmatrix} = \begin{bmatrix}0.1343\text{ cm}\\-1.3011\text{ cm}\\0.0620\text{ cm}\\-0.5101\text{ cm}\\-0.1151\text{ cm}\end{bmatrix} = \begin{bmatrix}1.3\text{ mm}\\-13.0\text{ mm}\\0.6\text{ mm}\\-5.1\text{ mm}\\-1.2\text{ mm}\end{bmatrix}\begin{matrix}1\\2\\3\\7\\8\end{matrix}$$

Find the reactions:

$$[\mathbf{Q}_s] = [\mathbf{K}_{sf}][\mathbf{D}_f] + [\mathbf{K}_{ss}][\mathbf{D}_s]$$

$$\begin{bmatrix}B_y\\C_x\\C_y\end{bmatrix} = \begin{matrix}4\\5\\6\end{matrix}\begin{bmatrix}\overset{1}{0} & \overset{2}{0} & \overset{3}{0} & \overset{7}{0} & \overset{8}{-5647\tfrac{\text{kN}}{\text{cm}}}\\0 & 0 & -5647\tfrac{\text{kN}}{\text{cm}} & -886.1\tfrac{\text{kN}}{\text{cm}} & 886.1\tfrac{\text{kN}}{\text{cm}}\\0 & 0 & 0 & 886.1\tfrac{\text{kN}}{\text{cm}} & -886.1\tfrac{\text{kN}}{\text{cm}}\end{bmatrix}\begin{bmatrix}0.1343\text{ cm}\\-1.3011\text{ cm}\\0.0620\text{ cm}\\-0.5101\text{ cm}\\-0.1151\text{ cm}\end{bmatrix} + \begin{bmatrix}0\\0\\0\end{bmatrix}$$

Multiply the matrices:

$$\begin{bmatrix}B_y\\C_x\\C_y\end{bmatrix} = \begin{bmatrix}650\text{ kN}\\-0.1\text{ kN}\\-350\text{ kN}\end{bmatrix}\begin{matrix}4\\5\\6\end{matrix}$$

Find the internal forces:

$$q'_F = \frac{AE}{l}[-\cos\theta \quad -\sin\theta \quad \cos\theta \quad \sin\theta]\begin{bmatrix}d_{Nx}\\d_{Ny}\\d_{Fx}\\d_{Fy}\end{bmatrix}$$

Member AB:

$$q'^{AB}_F = 4840 \frac{kN}{cm}[-1 \quad 0 \quad 1 \quad 0]\begin{bmatrix} 0.1343 \text{ cm} \\ -1.3011 \text{ cm} \\ 0.0620 \text{ cm} \\ 0 \end{bmatrix} = -349.9 \text{ kN}$$

$$N_{AB} = 350 \text{ kN } (C)$$

Member AD:

$$q'^{AD}_F = 1631.2 \frac{kN}{cm}[-0.7593 \quad -0.6508 \quad 0.7593 \quad 0.6508]\begin{bmatrix} 0.1343 \text{ cm} \\ -1.3011 \text{ cm} \\ -0.5101 \text{ cm} \\ -0.1151 \text{ cm} \end{bmatrix} = 460.9 \text{ kN}$$

$$N_{AD} = 461 \text{ kN } (T)$$

Member BC:

$$q'^{BC}_F = 5647 \frac{kN}{cm}[-1 \quad 0 \quad 1 \quad 0]\begin{bmatrix} 0.0620 \text{ cm} \\ 0 \\ 0 \\ 0 \end{bmatrix} = -350.1 \text{ kN}$$

$$N_{BC} = 350 \text{ kN } (C)$$

Member BD:

$$q'^{BD}_F = 5647 \frac{kN}{cm}[0 \quad -1 \quad 0 \quad 1]\begin{bmatrix} 0.0620 \text{ cm} \\ 0 \\ -0.5101 \text{ cm} \\ -0.1151 \text{ cm} \end{bmatrix} = -650.0 \text{ kN}$$

$$N_{BD} = 650 \text{ kN } (C)$$

Member CD:

$$q'^{CD}_F = 1772.3 \frac{kN}{cm}[0.7071 \quad -0.7071 \quad -0.7071 \quad 0.7071]\begin{bmatrix} 0 \\ 0 \\ -0.5101 \text{ cm} \\ -0.1151 \text{ cm} \end{bmatrix} = 494.9 \text{ kN}$$

$$N_{CD} = 495 \text{ kN } (T)$$

Summary:

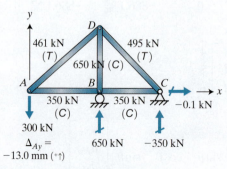

Figure 14.21

Evaluation of Results:

Satisfied Fundamental Principles?
Verify equilibrium of the Direct Stiffness Method results:
$$+\circlearrowleft \Sigma M_C = -300 \text{ kN}(6.5 \text{ m}) + 650 \text{ kN}(3 \text{ m}) = 0 \checkmark$$
$$+\uparrow \Sigma F_y = -300 \text{ kN} + 650 \text{ kN} - 350 \text{ kN} = 0 \checkmark$$
$$\xrightarrow{+} \Sigma F_x = -0.1 \text{ kN} \ll \text{other terms} \therefore \approx 0 \checkmark$$

Satisfied Fundamental Principles?
Compare the results from the Method of Joints and the Virtual Work Method with the results from the Direct Stiffness Method:

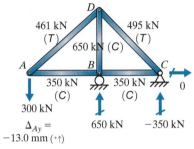
Method of Joints & Virtual Work

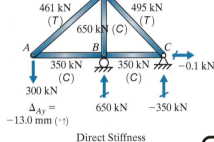
Direct Stiffness

Figure 14.22

The results are identical to at least three significant figures. ✓
    The horizontal reaction at C is within three significant figures compared with the magnitudes of the other reactions.

EXAMPLE 14.4

Our firm has designed a large pipeline along the edge of a canal in Singapore. Our team has been tasked with designing a structure to support the pipe. Other engineers on the team have designed the truss members for strength.

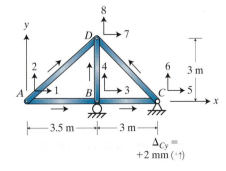

Figure 14.23

824 Chapter 14 Direct Stiffness Method for Trusses

Because we are new to the firm, our supervisor wants to verify our knowledge of the Direct Stiffness Method. A determinate truss is a great way to do that.

The connection to the concrete support at *C* will actually deform as the truss tries to lift up due to the live load. An engineer who is experienced in connection design predicts that the lift under the full live load will be 2 mm. Our job is to determine the effect of that lift on the vertical displacement of *A*. We found the vertical displacement due to the live load in Example 14.3. Our supervisor wants us to find the effect of just the lift at *C*. We can superimpose the results.

Estimated Solution:

Because the reactions are determinate, the truss will not change shape as the support at *C* rises. Therefore, we can use similar triangles to predict the vertical displacement at *A*.

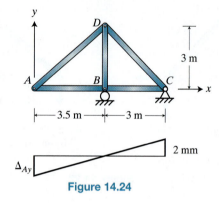

Figure 14.24

$$\frac{\Delta_{Ay}}{3.5 \text{ m}} = \frac{2 \text{ mm}}{3 \text{ m}}$$

$$\Delta_{Ay} = 2.33 \text{ mm} \quad (+\downarrow)$$

Since the truss does not change shape, the elements do not get longer or shorter. That means there are no internal forces.

Since the reactions are determinate and there is no applied load, the reactions are all zero.

Direct Stiffness Method:

Assemble Q_f and D_s:

Since we are considering only the support deformation, there are no applied loads:

$$[\mathbf{Q}_f] = \begin{matrix} 1 \\ 2 \\ 3 \\ 7 \\ 8 \end{matrix} \begin{bmatrix} 0 \\ 0 \\ 0 \\ 0 \\ 0 \end{bmatrix}$$

The only support deformation is at DOF 6:

$$[\mathbf{D}_s] = \begin{matrix} 4 \\ 5 \\ 6 \end{matrix} \begin{bmatrix} 0 \\ 0 \\ 2 \text{ mm} \end{bmatrix}$$

Section 14.4 Finding Deformations, Reactions, and Internal Forces

Find the deformations:

$$[\mathbf{Q}_f] = [\mathbf{K}_{ff}][\mathbf{D}_f] + [\mathbf{K}_{fs}][\mathbf{D}_s]$$

$$\begin{matrix} & & 1 & 2 & 3 & 7 & 8 \\ 1 \\ 2 \\ 3 \\ 7 \\ 8 \end{matrix} \begin{bmatrix} 0 \\ 0 \\ 0 \\ 0 \\ 0 \end{bmatrix} = \begin{matrix} 1 \\ 2 \\ 3 \\ 7 \\ 8 \end{matrix} \begin{bmatrix} 5780\,\tfrac{kN}{cm} & 806.1\,\tfrac{kN}{cm} & -4840\,\tfrac{kN}{cm} & -940.9\,\tfrac{kN}{cm} & -806.1\,\tfrac{kN}{cm} \\ 806.1\,\tfrac{kN}{cm} & 690.9\,\tfrac{kN}{cm} & 0 & -806.1\,\tfrac{kN}{cm} & -690.9\,\tfrac{kN}{cm} \\ -4840\,\tfrac{kN}{cm} & 0 & 10487\,\tfrac{kN}{cm} & 0 & 0 \\ -940.9\,\tfrac{kN}{cm} & -806.1\,\tfrac{kN}{cm} & 0 & 1826.5\,\tfrac{kN}{cm} & -80.0\,\tfrac{kN}{cm} \\ -806.1\,\tfrac{kN}{cm} & -690.9\,\tfrac{kN}{cm} & 0 & -80.0\,\tfrac{kN}{cm} & 7224\,\tfrac{kN}{cm} \end{bmatrix} \begin{bmatrix} \Delta_{Ax} \\ \Delta_{Ay} \\ \Delta_{Bx} \\ \Delta_{Dx} \\ \Delta_{Dy} \end{bmatrix} + \begin{matrix} 1 \\ 2 \\ 3 \\ 7 \\ 8 \end{matrix} \begin{bmatrix} 4 & 5 & 6 \\ 0 & 0 & 0 \\ 0 & 0 & 0 \\ 0 & -5647\,\tfrac{kN}{cm} & 0 \\ 0 & -886.1\,\tfrac{kN}{cm} & 886.1\,\tfrac{kN}{cm} \\ -5647\,\tfrac{kN}{cm} & 886.1\,\tfrac{kN}{cm} & -886.1\,\tfrac{kN}{cm} \end{bmatrix} \begin{bmatrix} 0 \\ 0 \\ 0.2\text{ cm} \end{bmatrix} \begin{matrix} 4 \\ 5 \\ 6 \end{matrix}$$

$$[\mathbf{D}_f] = [\mathbf{K}_{ff}]^{-1}([\mathbf{Q}_f] - [\mathbf{K}_{fs}][\mathbf{D}_s])$$

$$\begin{bmatrix} \Delta_{Ax} \\ \Delta_{Ay} \\ \Delta_{Bx} \\ \Delta_{Dx} \\ \Delta_{Dy} \end{bmatrix} = \begin{matrix} 1 \\ 2 \\ 3 \\ 7 \\ 8 \end{matrix} \begin{bmatrix} 5780\,\tfrac{kN}{cm} & 806.1\,\tfrac{kN}{cm} & -4840\,\tfrac{kN}{cm} & -940.9\,\tfrac{kN}{cm} & -806.1\,\tfrac{kN}{cm} \\ 806.1\,\tfrac{kN}{cm} & 690.9\,\tfrac{kN}{cm} & 0 & -806.1\,\tfrac{kN}{cm} & -690.9\,\tfrac{kN}{cm} \\ -4840\,\tfrac{kN}{cm} & 0 & 10487\,\tfrac{kN}{cm} & 0 & 0 \\ -940.9\,\tfrac{kN}{cm} & -806.1\,\tfrac{kN}{cm} & 0 & 1826.5\,\tfrac{kN}{cm} & -80.0\,\tfrac{kN}{cm} \\ -806.1\,\tfrac{kN}{cm} & -690.9\,\tfrac{kN}{cm} & 0 & -80.0\,\tfrac{kN}{cm} & 7224\,\tfrac{kN}{cm} \end{bmatrix}^{-1} \left(\begin{bmatrix} 0 \\ 0 \\ 0 \\ 0 \\ 0 \end{bmatrix} - \begin{bmatrix} 0 \\ 0 \\ 0 \\ 177.22\text{ kN} \\ -177.22\text{ kN} \end{bmatrix} \right)$$

$$\begin{bmatrix} \Delta_{Ax} \\ \Delta_{Ay} \\ \Delta_{Bx} \\ \Delta_{Dx} \\ \Delta_{Dy} \end{bmatrix} = \begin{bmatrix} 0\text{ cm} \\ -0.2333\text{ cm} \\ 0\text{ cm} \\ -0.2000\text{ cm} \\ 0\text{ cm} \end{bmatrix} = \begin{bmatrix} 0\text{ mm} \\ -2.3\text{ mm} \\ 0\text{ mm} \\ -2.0\text{ mm} \\ 0\text{ mm} \end{bmatrix} \begin{matrix} 1 \\ 2 \\ 3 \\ 7 \\ 8 \end{matrix}$$

Find the reactions:

Evaluation of Results:

Satisfied Fundamental Principles?

We expect all the reactions to be zero, so this serves as a check.

$$\begin{bmatrix} B_y \\ C_x \\ C_y \end{bmatrix} = \begin{matrix} 4 \\ 5 \\ 6 \end{matrix} \begin{bmatrix} 1 & 2 & 3 & 7 & 8 \\ 0 & 0 & 0 & 0 & -5647\,\tfrac{kN}{cm} \\ 0 & 0 & -5647\,\tfrac{kN}{cm} & -886.1\,\tfrac{kN}{cm} & 886.1\,\tfrac{kN}{cm} \\ 0 & 0 & 0 & 886.1\,\tfrac{kN}{cm} & -886.1\,\tfrac{kN}{cm} \end{bmatrix} \begin{bmatrix} 0\text{ cm} \\ -0.2333\text{ cm} \\ 0\text{ cm} \\ -0.2000\text{ cm} \\ 0\text{ cm} \end{bmatrix}$$

$$+ \begin{matrix} 4 \\ 5 \\ 6 \end{matrix} \begin{bmatrix} 4 & 5 & 6 \\ 5647\,\tfrac{kN}{cm} & 0 & 0 \\ 0 & 6533\,\tfrac{kN}{cm} & -886.1\,\tfrac{kN}{cm} \\ 0 & -886.1\,\tfrac{kN}{cm} & 886.1\,\tfrac{kN}{cm} \end{bmatrix} \begin{bmatrix} 0 \\ 0 \\ 0.2\text{ cm} \end{bmatrix}$$

$$\begin{bmatrix} B_y \\ C_x \\ C_y \end{bmatrix} = \begin{bmatrix} 0\text{ kN} \\ 177.22\text{ kN} \\ -177.22\text{ kN} \end{bmatrix} + \begin{bmatrix} 0\text{ kN} \\ -177.22\text{ kN} \\ 177.22\text{ kN} \end{bmatrix} = \begin{bmatrix} 0 \\ 0 \\ 0 \end{bmatrix} \checkmark$$

Find the internal force:

Evaluation of Results:

Satisfied Fundamental Principles?

We expect all the internal forces to be zero, so this too serves as a check. Let's check just two members.

Member AD:

$$q'^{AD}_F = 1631.2\,\frac{kN}{cm}[-0.7593 \quad -0.6508 \quad 0.7593 \quad 0.6508] \begin{bmatrix} 0 \\ -0.2333\text{ cm} \\ -0.2000\text{ cm} \\ 0 \end{bmatrix}$$

$$= 1631.2\,\frac{kN}{cm}(0 + 0.15183\text{ cm} - 0.15186\text{ cm} + 0)$$

$$= -0.05\text{ kN} \approx 0 \quad \checkmark$$

Since there are no applied loads, it is not obvious what a sufficiently small number is. However, we generated force magnitudes when we calculated the reactions: 177 kN. When we compare with those magnitudes, 0.05 kN is essentially zero.

Member *CD*:
Note that we include the support displacement.

$$q'^{CD}_F = 1772.3 \frac{kN}{cm} [0.7071 \quad -0.7071 \quad -0.7071 \quad 0.7071] \begin{bmatrix} 0 \\ 0.2000 \text{ cm} \\ -0.2000 \text{ cm} \\ 0 \end{bmatrix}$$

$$= 1772.3 \text{ kN/cm}(0 - 0.14142 \text{ cm} + 0.14142 \text{ cm} + 0)$$

$$= 0 \checkmark$$

Evaluation of Results:

Satisfied Fundamental Principles?
Compare the results from similar triangles with the results from the Direct Stiffness Method:

    Similar triangles: $\Delta_{Ay} = -2.33$ mm $(+\uparrow)$

    Direct Stiffness Method: $\Delta_{Ay} = -2.33$ mm $(+\uparrow)$

The predicted vertical displacements at *A* are the same to at least three significant figures. ✓

Conclusion:
    These results plus the checks on the reactions and internal forces strongly suggest that we have reasonable results.

Combine results:

Since we assume the structure remains linear elastic, we can superimpose the results for live load and support movement:

$$\begin{bmatrix} \Delta_{Ax} \\ \Delta_{Ay} \\ \Delta_{Bx} \\ \Delta_{Dx} \\ \Delta_{Dy} \end{bmatrix} = \begin{bmatrix} 0.1343 \text{ cm} \\ -1.3011 \text{ cm} \\ 0.0620 \text{ cm} \\ -0.5101 \text{ cm} \\ -0.1151 \text{ cm} \end{bmatrix} + \begin{bmatrix} 0 \text{ cm} \\ -0.2333 \text{ cm} \\ 0 \text{ cm} \\ -0.2000 \text{ cm} \\ 0 \text{ cm} \end{bmatrix} = \begin{bmatrix} 0.1343 \text{ cm} \\ -1.5344 \text{ cm} \\ 0.0620 \text{ cm} \\ -0.7101 \text{ cm} \\ -0.1151 \text{ cm} \end{bmatrix} \begin{matrix} 1 \\ 2 \\ 3 \\ 7 \\ 8 \end{matrix}$$

$$\begin{bmatrix} B_y \\ C_x \\ C_y \end{bmatrix} = \begin{bmatrix} 650 \text{ kN} \\ -0.1 \text{ kN} \\ -350 \text{ kN} \end{bmatrix} + \begin{bmatrix} 0 \\ 0 \\ 0 \end{bmatrix} = \begin{bmatrix} 650 \text{ kN} \\ -0.1 \text{ kN} \\ -350 \text{ kN} \end{bmatrix} \begin{matrix} 4 \\ 5 \\ 6 \end{matrix}$$

Notice that when we consider the full live load and the lift at the support, the net vertical displacement at *A* is larger than the allowable displacement:

$$\Delta^{Total}_{Ay} = 15.3 \text{ mm} > \Delta^{Allowable}_{Ay} = 15 \text{ mm}$$

Therefore, our team will need to redesign the truss to reduce the anticipated displacement. The Virtual Work Method we discussed in Section 5.3 is a great tool to help in that process.

Direct Stiffness Method with Combined Loading:

Our supervisor has challenged us to verify that superposition does produce the same results as considering both the live load and support lift simultaneously.

Section 14.5 Additional Loadings

Find the deformations:

$$\begin{bmatrix} 1 \\ 2 \\ 3 \\ 7 \\ 8 \end{bmatrix} \begin{bmatrix} 0 \\ -300 \text{ kN} \\ 0 \\ 0 \\ 0 \end{bmatrix} = \begin{bmatrix} 1 \\ 2 \\ 3 \\ 7 \\ 8 \end{bmatrix} \begin{bmatrix} 5780 \frac{\text{kN}}{\text{cm}} & 806.1 \frac{\text{kN}}{\text{cm}} & -4840 \frac{\text{kN}}{\text{cm}} & -940.9 \frac{\text{kN}}{\text{cm}} & -806.1 \frac{\text{kN}}{\text{cm}} \\ 806.1 \frac{\text{kN}}{\text{cm}} & 690.9 \frac{\text{kN}}{\text{cm}} & 0 & -806.1 \frac{\text{kN}}{\text{cm}} & -690.9 \frac{\text{kN}}{\text{cm}} \\ -4840 \frac{\text{kN}}{\text{cm}} & 0 & 10487 \frac{\text{kN}}{\text{cm}} & 0 & 0 \\ -940.9 \frac{\text{kN}}{\text{cm}} & -806.1 \frac{\text{kN}}{\text{cm}} & 0 & 1826.5 \frac{\text{kN}}{\text{cm}} & -80.0 \frac{\text{kN}}{\text{cm}} \\ -806.1 \frac{\text{kN}}{\text{cm}} & -690.9 \frac{\text{kN}}{\text{cm}} & 0 & -80.0 \frac{\text{kN}}{\text{cm}} & 7224 \frac{\text{kN}}{\text{cm}} \end{bmatrix} \begin{bmatrix} \Delta_{Ax} \\ \Delta_{Ay} \\ \Delta_{Bx} \\ \Delta_{Dx} \\ \Delta_{Dy} \end{bmatrix} + \begin{bmatrix} 1 \\ 2 \\ 3 \\ 7 \\ 8 \end{bmatrix} \begin{bmatrix} 0 & 0 & 0 \\ 0 & 0 & 0 \\ 0 & -5647 \frac{\text{kN}}{\text{cm}} & 0 \\ 0 & -886.1 \frac{\text{kN}}{\text{cm}} & 886.1 \frac{\text{kN}}{\text{cm}} \\ -5647 \frac{\text{kN}}{\text{cm}} & 886.1 \frac{\text{kN}}{\text{cm}} & -886.1 \frac{\text{kN}}{\text{cm}} \end{bmatrix} \begin{bmatrix} 0 \\ 0 \\ 0.2 \text{ cm} \end{bmatrix}$$

$$\begin{bmatrix} \Delta_{Ax} \\ \Delta_{Ay} \\ \Delta_{Bx} \\ \Delta_{Dx} \\ \Delta_{Dy} \end{bmatrix} = \begin{bmatrix} 1 \\ 2 \\ 3 \\ 7 \\ 8 \end{bmatrix} \begin{bmatrix} 5780 \frac{\text{kN}}{\text{cm}} & 806.1 \frac{\text{kN}}{\text{cm}} & -4840 \frac{\text{kN}}{\text{cm}} & -940.9 \frac{\text{kN}}{\text{cm}} & -806.1 \frac{\text{kN}}{\text{cm}} \\ 806.1 \frac{\text{kN}}{\text{cm}} & 690.9 \frac{\text{kN}}{\text{cm}} & 0 & -806.1 \frac{\text{kN}}{\text{cm}} & -690.9 \frac{\text{kN}}{\text{cm}} \\ -4840 \frac{\text{kN}}{\text{cm}} & 0 & 10487 \frac{\text{kN}}{\text{cm}} & 0 & 0 \\ -940.9 \frac{\text{kN}}{\text{cm}} & -806.1 \frac{\text{kN}}{\text{cm}} & 0 & 1826.5 \frac{\text{kN}}{\text{cm}} & -80.0 \frac{\text{kN}}{\text{cm}} \\ -806.1 \frac{\text{kN}}{\text{cm}} & -690.9 \frac{\text{kN}}{\text{cm}} & 0 & -80.0 \frac{\text{kN}}{\text{cm}} & 7224 \frac{\text{kN}}{\text{cm}} \end{bmatrix}^{-1} \left(\begin{bmatrix} 0 \\ -300 \text{ kN} \\ 0 \\ 0 \\ 0 \end{bmatrix} - \begin{bmatrix} 0 \\ 0 \\ 0 \\ 177.22 \text{ kN} \\ -177.22 \text{ kN} \end{bmatrix} \right)$$

$$\begin{bmatrix} \Delta_{Ax} \\ \Delta_{Ay} \\ \Delta_{Bx} \\ \Delta_{Dx} \\ \Delta_{Dy} \end{bmatrix} = \begin{bmatrix} 0.1342 \text{ cm} \\ -1.5345 \text{ cm} \\ 0.0619 \text{ cm} \\ -0.7102 \text{ cm} \\ -0.1151 \text{ cm} \end{bmatrix} = \begin{bmatrix} 1.3 \text{ mm} \\ -15.3 \text{ mm} \\ 0.6 \text{ mm} \\ -7.1 \text{ mm} \\ -1.2 \text{ mm} \end{bmatrix} \begin{matrix} 1 \\ 2 \\ 3 \\ 7 \\ 8 \end{matrix}$$

Find the reactions:

$$\begin{bmatrix} B_y \\ C_x \\ C_y \end{bmatrix} = \begin{matrix} 4 \\ 5 \\ 6 \end{matrix} \begin{bmatrix} 0 & 0 & 0 & 0 & -5647 \frac{\text{kN}}{\text{cm}} \\ 0 & 0 & -5647 \frac{\text{kN}}{\text{cm}} & -886.1 \frac{\text{kN}}{\text{cm}} & 886.1 \frac{\text{kN}}{\text{cm}} \\ 0 & 0 & 0 & 886.1 \frac{\text{kN}}{\text{cm}} & -886.1 \frac{\text{kN}}{\text{cm}} \end{bmatrix} \begin{bmatrix} 0.1342 \text{ cm} \\ -1.5345 \text{ cm} \\ 0.0619 \text{ cm} \\ -0.7102 \text{ cm} \\ -0.1151 \text{ cm} \end{bmatrix} + \begin{matrix} 4 \\ 5 \\ 6 \end{matrix} \begin{bmatrix} 5647 \frac{\text{kN}}{\text{cm}} & 0 & 0 \\ 0 & 6533 \frac{\text{kN}}{\text{cm}} & -886.1 \frac{\text{kN}}{\text{cm}} \\ 0 & -886.1 \frac{\text{kN}}{\text{cm}} & 886.1 \frac{\text{kN}}{\text{cm}} \end{bmatrix} \begin{bmatrix} 0 \\ 0 \\ 0.2 \text{ cm} \end{bmatrix}$$

$$\begin{bmatrix} B_y \\ C_x \\ C_y \end{bmatrix} = \begin{bmatrix} 650.0 \text{ kN} \\ 177.77 \text{ kN} \\ -527.3 \text{ kN} \end{bmatrix} + \begin{bmatrix} 0 \text{ kN} \\ -177.22 \text{ kN} \\ 177.22 \text{ kN} \end{bmatrix} = \begin{bmatrix} 650.0 \text{ kN} \\ 0.55 \text{ kN} \\ -350.1 \text{ kN} \end{bmatrix} \begin{matrix} 4 \\ 5 \\ 6 \end{matrix}$$

Evaluation of Results:
Satisfied Fundamental Principles?
Indeed, the results from superposition match the results from considering both loading cases simultaneously. ✓

14.5 Additional Loadings

Introduction
The Direct Stiffness Method works only when loads are acting at the nodes, but sometimes loading is along the member. In these cases, we need to convert the loading to equivalent loads at the nodes.

How-To
Most loading on a truss occurs at the nodes, but we still want to represent several loadings that do not (i.e., length change, temperature change). In order to use the Direct Stiffness Method, we will need to convert these load effects into equivalent loads that act only at the nodes.

The governing system of equations without the additional loading is represented by the expression

$$[Q] = [K][D]$$

We can interpret the right-hand side of the equation as the net loads at the nodes due to the deformations at the nodes. This interpretation is the key to figuring out how to incorporate the effects of loads between the nodes.

If we imagine that all degrees of freedom at all nodes start as supported, then any load between the nodes will cause reactions at these imaginary supports. These reactions are the nodal loads that are required to hold the elements in their original positions. So these would be net loads at the nodes that do not have any deformations.

We superimpose the loads due to deformations and the nodal equivalent loads in order to get the net effect. Before we can show that in an equation, we need additional symbols to distinguish between the net loads at the nodes and the nodal loads needed to hold the elements in place. Let's use a subscript to tell them apart:

q'_o = Nodal equivalent load; the load required at the end of one element to hold it in place (local coordinates)

q_o = Nodal equivalent load; the load required at the end of one element to hold it in place (global coordinates)

Q_o = Net nodal equivalent load; the sum of all q_o at a DOF for the structure (global coordinates)

The governing system of equations is then represented by the expression

$$[Q] = [K][D] + [Q_o]$$

We still need to break down the system into two subsets of equations, but all of the Q_o's are knowns. We start by finding the deformations:

$$[Q_f] = [K_{ff}][D_f] + [K_{fs}][D_s] + [Q_{fo}]$$

$$[D_f] = [K_{ff}]^{-1}([Q_f] - [K_{fs}][D_s] - [Q_{fo}])$$

We can then find the net reactions:

$$[Q_s] = [K_{sf}][D_f] + [K_{ss}][D_s] + [Q_{so}]$$

We can also find the net internal forces:

$$[q'] = [k'][T][d] + [q'_o]$$

$$q'_F = \frac{AE}{l}[-\cos\theta \quad -\sin\theta \quad \cos\theta \quad \sin\theta]\begin{bmatrix} d_{Nx} \\ d_{Ny} \\ d_{Fx} \\ d_{Fy} \end{bmatrix} + q'_{Fo}$$

Length Change

We design each member of a structure to be a certain length, but the actual length can be different for a variety of reasons. Prefabricated members (e.g., steel, aluminum, timber, precast concrete) have length tolerances. The fabricators guarantee that the length will be what was ordered plus or minus some amount. For example, the American Institute of Steel Construction

Code of Standard Practice for Steel Buildings and Bridges (AISC 2016) states that steel members no longer than 9 m can have a length variation of ±2 mm, and the Precast/Prestressed Concrete Institute recommends a tolerance for precast concrete I-beams of ±1 inch (PCI 2017). Sometimes, however, members arrive from the fabricator outside of tolerance. Rather than reject the member right away, the contractor might request that we evaluate whether the deviation can be tolerated. To answer that question, we will need to know the forces generated in the members and possibly the deformations.

Also, some materials change length over time. For example, concrete shrinks over time whether or not it is carrying load. Under sustained axial compression, concrete and timber get shorter over time. We call this phenomenon *creep*. Steel cables under sustained tension elongate; we call this phenomenon *relaxation*.

To find the nodal forces, we use the Force Method (Chapter 12). Let's consider elongation a positive change in length. To make the element determinate, we can change the far node support from a pin to a roller (Figure 14.25). The redundant is then a force applied at the far end in the x'-direction.

For compatibility, the applied length change and the elongation due to the redundant must add to zero. From mechanics of materials, we know how elongation and axial force are related:

$$\Delta l_{q'} = \frac{q'_{Fo} l}{AE}$$

When we combine them, we get the following compatibility equation:

$$\Delta l + \Delta l_{q'} = 0$$

$$\Delta l + \frac{q'_{Fo} l}{AE} = 0$$

We can solve this for the equivalent nodal load at the far end:

$$q'_{Fo} = -\frac{\Delta l}{l} AE$$

Equilibrium gives us the value at the near end:

$$q'_{No} = +\frac{\Delta l}{l} AE$$

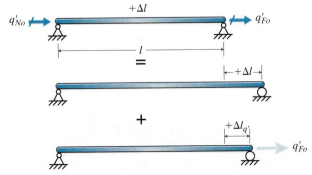

Figure 14.25 Pin-pin truss member with a length change can be decomposed into the length change and an applied axial force.

Temperature Change

When the temperature of an element changes, the length also changes. As an element gets hotter, it gets longer. Similarly, as it gets cooler, the element gets shorter. In a determinate structure, these length changes do not cause reactions or internal forces, but they do cause deformations. In indeterminate structures, temperature changes can cause all three: deformations, reactions, and internal forces. To predict the effects of temperature change, we need to calculate the nodal equivalent loads.

We can find the nodal equivalent loads by using the Force Method. Let's consider an increase in temperature as positive. As with the length change, let's change the far node support to a roller (Figure 14.26). Again, the redundant is the force at the far end in the x'-direction.

From mechanics of materials, we know how much an element changes length when the entire element undergoes a uniform temperature change:

$$\Delta l = \alpha \Delta T l$$

For compatibility, the two length changes must add to zero:

$$\Delta l + \Delta l_{q'} = 0$$

$$\alpha \Delta T l + \frac{q'_{Fo} l}{AE} = 0$$

Solving this equation for the equivalent nodal load at the far end gives us the following expression:

$$q'_{Fo} = -\alpha \Delta T A E$$

For equilibrium of the element, the equivalent nodal load at the near end must be equal and opposite:

$$q'_{No} = +\alpha \Delta T A E$$

Assembling the Global Vector

Before we can find the deformations, we need to convert the equivalent nodal loads from local to global coordinates and then combine them to create the vector of nodal loads for the structure.

The reactions make up $[\mathbf{q}'_o]$. To convert them to global coordinates, we use the transpose of the transformation matrix:

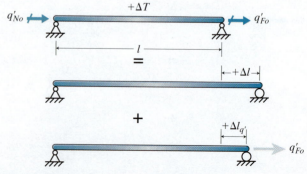

Figure 14.26 Pin-pin truss member with a temperature change can be decomposed into the length changes due to temperature and due to an applied axial force.

$$[\mathbf{q}_o] = [\mathbf{T}]^T[\mathbf{q}'_o]$$

where

$$[\mathbf{q}_o] = \begin{bmatrix} q_{Nxo} \\ q_{Nyo} \\ q_{Fxo} \\ q_{Fyo} \end{bmatrix} = \text{Internal forces at the ends of an element due to the nodal equivalent load (global coordinates)}$$

$$[\mathbf{T}]^T = \begin{bmatrix} \cos\theta & 0 \\ \sin\theta & 0 \\ 0 & \cos\theta \\ 0 & \sin\theta \end{bmatrix} = \text{Transpose of transformation matrix}$$

$$[\mathbf{q}'_o] = \begin{bmatrix} q'_{Nxo} \\ q'_{Fxo} \end{bmatrix} = \text{Nodal equivalent loads at the ends of an element (local coordinates)}$$

We can still use node coordinates to calculate $\cos\theta$ and $\sin\theta$, as we learned in Section 14.2.

Now we have the loads in global coordinates, but we still need to combine them for the structure:

$$[\mathbf{Q}_o] = \Sigma[\mathbf{q}_o]$$

To do that, we need to pay attention to degree-of-freedom numbers. The vector of Q_o contains the nodal equivalent loads for each degree of freedom for the structure. If multiple elements have nodal equivalent loads and meet at a DOF, then Q_o is the net load. Signs matter, so we must make sure that the process starts with $[\mathbf{q}'_o]$ based on the same x'-direction we used to create $[\mathbf{k}]$. If no element contributes a nodal equivalent load for a DOF, then $Q_o = 0$ for that DOF.

Loads at Supports

Different from nodal equivalent loads, sometimes we have loads applied directly to supported degrees of freedom (Figure 14.27). They are applied loads, so they are known, but they are at supported DOFs. In our system of equations, these are neither Q_f nor Q_s. To account for these loads, we put them in $[\mathbf{Q}_{so}]$:

$$[\mathbf{Q}_s] = [\mathbf{K}_{sf}][\mathbf{D}_f] + [\mathbf{K}_{ss}][\mathbf{D}_s] + [\mathbf{Q}_{so}]$$

If we consider a determinate truss with load applied to a supported DOF and then apply equilibrium (Figure 14.28), we find that the reaction, Q_s, is equal and opposite to the applied load. Therefore, we need to reverse the signs of the applied loads at supporting DOFs before adding them to $[\mathbf{Q}_{so}]$.

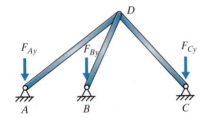

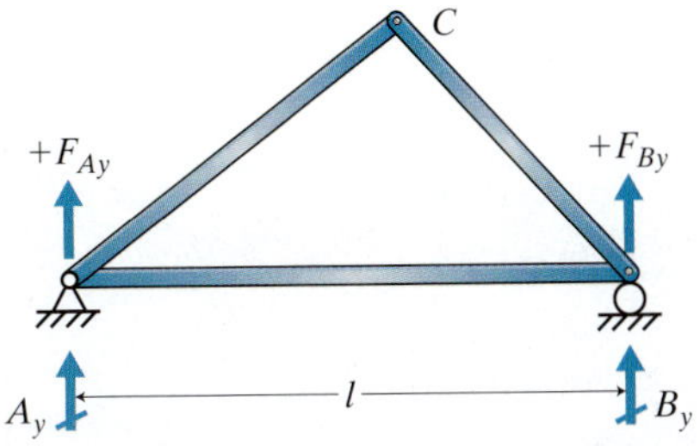

Figure 14.27 Loads applied at supported degrees of freedom.

$$\circlearrowright \Sigma M_B = 0 = A_y l + F_{Ay} l$$
$$A_y = -F_{Ay} \quad (+\uparrow)$$

$$+\uparrow \Sigma F_y = 0 = F_{Ay} - F_{Ay} + F_{By} + B_y$$
$$B_y = -F_{By} \quad (+\uparrow)$$

Figure 14.28 Reactions for a truss with loads applied at supported DOFs.

Section 14.5 Highlights

Additional Loadings

Additional Terms:

q'_o = Nodal equivalent load; the load required at the end of one element to hold it in place (local coordinates)
q_o = Nodal equivalent load; the load required at the end of one element to hold it in place (global coordinates)
Q_o = Net nodal equivalent load; the sum of all q_o at a DOF for the structure (global coordinates)

System of Equations:

$$[Q] = [K][D] + [Q_o]$$

where

$$[Q_o] = \Sigma [q_o]$$

Note: This is the full form of the Direct Stiffness Method. If there are no additional loadings, $[Q_o] = [0]$, so we get the expressions in Section 14.4.

Subset to find the deformations:

$$[Q_f] = [K_{ff}][D_f] + [K_{fs}][D_s] + [Q_{fo}]$$
$$[D_f] = [K_{ff}]^{-1}([Q_f] - [K_{fs}][D_s] - [Q_{fo}])$$

Subset to find the reactions:

$$[Q_s] = [K_{sf}][D_f] + [K_{ss}][D_s] + [Q_{so}]$$

Axial forces:

$$[q'] = [k'][T][d] + [q'_o]$$

$$q'_F = \frac{AE}{l}[-\cos\theta \quad -\sin\theta \quad \cos\theta \quad \sin\theta]\begin{bmatrix} d_{Nx} \\ d_{Ny} \\ d_{Fx} \\ d_{Fy} \end{bmatrix} + q'_{Fo}$$

Transforming Nodal Equivalent Loads:

$$\begin{bmatrix} q_{Nxo} \\ q_{Nyo} \\ q_{Fxo} \\ q_{Fyo} \end{bmatrix} = \begin{bmatrix} \cos\theta & 0 \\ \sin\theta & 0 \\ 0 & \cos\theta \\ 0 & \sin\theta \end{bmatrix} \begin{bmatrix} q'_{Nxo} \\ q'_{Fxo} \end{bmatrix}$$

Calculating Equivalent Nodal Loads:

Length change:

   Longer than planned $= +\Delta l$

$$q'_{No} = +\frac{\Delta l}{l}AE \quad q'_{Fo} = -\frac{\Delta l}{l}AE$$

Temperature change:

   Increase in temperature $= +\Delta T$
   Coefficient of thermal expansion $= \alpha$

$$q'_{No} = +\alpha \Delta T AE \quad q'_{Fo} = -\alpha \Delta T AE$$

Loads at supported DOFs:

   Added to initial support reactions $[Q_{so}]$
   Reaction added to $[Q_{so}]$ is the same magnitude but the opposite direction of the load applied to the supported DOF.

Example 14.5

Our firm has designed a large pipeline along the edge of a canal in Singapore. Our team has been tasked with designing a structure to support the pipe. Other engineers on the team have designed the truss members for strength.

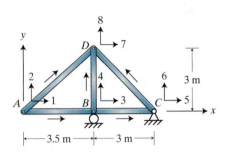

Figure 14.29

Because we are new to the firm, our supervisor wants to verify our knowledge of the Direct Stiffness Method. A determinate truss is a great way to do that.

In addition to the live load in the pipe, the truss might experience a temperature change in only the diagonal members as the sun heats it up in the late morning. A reasonable temperature increase to plan for is 10°C. To minimize the possibility of leaks in the pipeline, we need to make sure that the truss does not deflect more than 15 mm when the pipeline changes from empty to full and the temperature increases.

We found the vertical displacement due to the live load in Example 14.3. Our supervisor wants us to find the effect of just the temperature change. We can superimpose these results with the live load results to predict the total effect and then evaluate the adequacy of the design.

The coefficient of thermal expansion for steel is $11.7 \times 10^{-6}/°C$.

Estimated Solution:

Because the truss is determinate, the temperature change will not cause reactions or internal forces.

Increasing the temperature of members AD and CD will cause them to get longer. Considering the geometry of the truss and the support conditions, both lengthening members will cause node A to displace down.

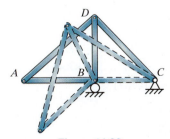

Figure 14.30

834   **Chapter 14**   Direct Stiffness Method for Trusses

Direct Stiffness Method:

Nodal equivalent loads:

Member AD – temperature change:

From Example 14.1,

$$\cos\theta = 0.7593$$
$$\sin\theta = 0.6508$$
$$A = 37.6 \text{ cm}^2$$

Nodal loads in local coordinates:

$$q'_{No} = +\alpha \Delta T A E$$
$$= (11.7 \times 10^{-6}/°C)(+10°C)(37.6 \text{ cm}^2)(20{,}000 \text{ kN/cm}^2)$$
$$= 87.98 \text{ kN}$$
$$q'_{Fo} = -87.98 \text{ kN}$$

Nodal loads in global coordinates:

$$\begin{bmatrix} q_{Nxo} \\ q_{Nyo} \\ q_{Fxo} \\ q_{Fyo} \end{bmatrix} = \begin{bmatrix} \cos\theta & 0 \\ \sin\theta & 0 \\ 0 & \cos\theta \\ 0 & \sin\theta \end{bmatrix} \begin{bmatrix} q'_{Nxo} \\ q'_{Fxo} \end{bmatrix}$$

$$= \begin{bmatrix} 0.7593 & 0 \\ 0.6508 & 0 \\ 0 & 0.7593 \\ 0 & 0.6508 \end{bmatrix} \begin{bmatrix} 87.98 \text{ kN} \\ -87.98 \text{ kN} \end{bmatrix}$$

$$= \begin{bmatrix} 66.80 \text{ kN} \\ 57.26 \text{ kN} \\ -66.80 \text{ kN} \\ -57.26 \text{ kN} \end{bmatrix} \begin{matrix} 1 \\ 2 \\ 7 \\ 8 \end{matrix}$$

Member CD – temperature change:

From Example 14.1,

$$\cos\theta = -0.7071$$
$$\sin\theta = 0.7071$$
$$A = 37.6 \text{ cm}^2$$

Nodal loads in local coordinates:

$$q'_{No} = (11.7 \times 10^{-6}/°C)(10°C)(37.6 \text{ cm}^2)(20{,}000 \text{ kN/cm}^2)$$
$$= 87.98 \text{ kN}$$
$$q'_{Fo} = -87.98 \text{ kN}$$

Nodal loads in global coordinates:

$$\begin{bmatrix} q_{Nxo} \\ q_{Nyo} \\ q_{Fxo} \\ q_{Fyo} \end{bmatrix} = \begin{bmatrix} -0.7071 & 0 \\ 0.7071 & 0 \\ 0 & -0.7071 \\ 0 & 0.7071 \end{bmatrix} \begin{bmatrix} 87.98 \text{ kN} \\ -87.98 \text{ kN} \end{bmatrix}$$

$$= \begin{bmatrix} -62.21 \text{ kN} \\ 62.21 \text{ kN} \\ 62.21 \text{ kN} \\ -62.21 \text{ kN} \end{bmatrix} \begin{matrix} 5 \\ 6 \\ 7 \\ 8 \end{matrix}$$

Net nodal loads:
$$Q_{io} = \Sigma q_{io}$$

$Q_{1o} = 66.80$ kN
$Q_{2o} = 57.26$ kN
$Q_{3o} = 0$
$Q_{4o} = 0$
$Q_{5o} = -62.21$ kN
$Q_{6o} = 62.21$ kN
$Q_{7o} = -66.80$ kN $+ 62.21$ kN $= -4.59$ kN
$Q_{8o} = -57.26$ kN $- 62.21$ kN $= -119.47$ kN

$$[\mathbf{Q}_o] = \begin{matrix} 1 \\ 2 \\ 3 \\ 4 \\ 5 \\ 6 \\ 7 \\ 8 \end{matrix} \begin{bmatrix} 66.80 \text{ kN} \\ 57.26 \text{ kN} \\ 0 \\ 0 \\ -62.21 \text{ kN} \\ 62.21 \text{ kN} \\ -4.59 \text{ kN} \\ -119.47 \text{ kN} \end{bmatrix}$$

$$[\mathbf{Q}_{fo}] = \begin{matrix} 1 \\ 2 \\ 3 \\ 7 \\ 8 \end{matrix} \begin{bmatrix} 66.80 \text{ kN} \\ 57.26 \text{ kN} \\ 0 \\ -4.59 \text{ kN} \\ -119.47 \text{ kN} \end{bmatrix}$$

$$[\mathbf{Q}_{so}] = \begin{matrix} 4 \\ 5 \\ 6 \end{matrix} \begin{bmatrix} 0 \\ -62.21 \text{ kN} \\ 62.21 \text{ kN} \end{bmatrix}$$

Find the deformations:
$$[\mathbf{Q}_f] = [\mathbf{K}_{ff}][\mathbf{D}_f] + [\mathbf{K}_{fs}]\underbrace{[\mathbf{D}_s]}_{[\mathbf{0}]} + [\mathbf{Q}_{fo}]$$

$$\begin{matrix} 1 \\ 2 \\ 3 \\ 7 \\ 8 \end{matrix} \begin{bmatrix} 0 \\ 0 \\ 0 \\ 0 \\ 0 \end{bmatrix} = \begin{matrix} \\ 1 \\ 2 \\ 3 \\ 7 \\ 8 \end{matrix} \begin{bmatrix} 1 & 2 & 3 & 7 & 8 \\ 5780 \frac{\text{kN}}{\text{cm}} & 806.1 \frac{\text{kN}}{\text{cm}} & -4840 \frac{\text{kN}}{\text{cm}} & -940.9 \frac{\text{kN}}{\text{cm}} & -806.1 \frac{\text{kN}}{\text{cm}} \\ 806.1 \frac{\text{kN}}{\text{cm}} & 690.9 \frac{\text{kN}}{\text{cm}} & 0 & -806.1 \frac{\text{kN}}{\text{cm}} & -690.9 \frac{\text{kN}}{\text{cm}} \\ -4840 \frac{\text{kN}}{\text{cm}} & 0 & 1087 \frac{\text{kN}}{\text{cm}} & 0 & 0 \\ -940.9 \frac{\text{kN}}{\text{cm}} & -806.1 \frac{\text{kN}}{\text{cm}} & 0 & 1826.5 \frac{\text{kN}}{\text{cm}} & -80.0 \frac{\text{kN}}{\text{cm}} \\ -806.1 \frac{\text{kN}}{\text{cm}} & -690.9 \frac{\text{kN}}{\text{cm}} & 0 & -80.0 \frac{\text{kN}}{\text{cm}} & 7224 \frac{\text{kN}}{\text{cm}} \end{bmatrix} \begin{bmatrix} \Delta_{Ax} \\ \Delta_{Ay} \\ \Delta_{Bx} \\ \Delta_{Dx} \\ \Delta_{Dy} \end{bmatrix} + \begin{matrix} 1 \\ 2 \\ 3 \\ 7 \\ 8 \end{matrix} \begin{bmatrix} 66.80 \text{ kN} \\ 57.26 \text{ kN} \\ 0 \\ -4.59 \text{ kN} \\ -119.47 \text{ kN} \end{bmatrix}$$

$$[\mathbf{D}_f] = [\mathbf{K}_{ff}]^{-1}([\mathbf{Q}_f] - [\mathbf{Q}_{fo}])$$

$$\begin{bmatrix} \Delta_{Ax} \\ \Delta_{Ay} \\ \Delta_{Bx} \\ \Delta_{Dx} \\ \Delta_{Dy} \end{bmatrix} = \begin{matrix} 1 \\ 2 \\ 3 \\ 7 \\ 8 \end{matrix} \begin{bmatrix} 1 & 2 & 3 & 7 & 8 \\ 5780 \frac{\text{kN}}{\text{cm}} & 806.1 \frac{\text{kN}}{\text{cm}} & -4840 \frac{\text{kN}}{\text{cm}} & -940.9 \frac{\text{kN}}{\text{cm}} & -806.1 \frac{\text{kN}}{\text{cm}} \\ 806.1 \frac{\text{kN}}{\text{cm}} & 690.9 \frac{\text{kN}}{\text{cm}} & 0 & -806.1 \frac{\text{kN}}{\text{cm}} & -690.9 \frac{\text{kN}}{\text{cm}} \\ -4840 \frac{\text{kN}}{\text{cm}} & 0 & 10487 \frac{\text{kN}}{\text{cm}} & 0 & 0 \\ -940.9 \frac{\text{kN}}{\text{cm}} & -806.1 \frac{\text{kN}}{\text{cm}} & 0 & 1826.5 \frac{\text{kN}}{\text{cm}} & -80.0 \frac{\text{kN}}{\text{cm}} \\ -806.1 \frac{\text{kN}}{\text{cm}} & -690.9 \frac{\text{kN}}{\text{cm}} & 0 & -80.0 \frac{\text{kN}}{\text{cm}} & 7224 \frac{\text{kN}}{\text{cm}} \end{bmatrix}^{-1} \left(\begin{bmatrix} 0 \\ 0 \\ 0 \\ 0 \\ 0 \end{bmatrix} - \begin{bmatrix} 66.80 \text{ kN} \\ 57.26 \text{ kN} \\ 0 \\ -4.59 \text{ kN} \\ -119.47 \text{ kN} \end{bmatrix} \right)$$

$$\begin{bmatrix} \Delta_{Ax} \\ \Delta_{Ay} \\ \Delta_{Bx} \\ \Delta_{Dx} \\ \Delta_{Dy} \end{bmatrix} = \begin{bmatrix} -0.0000 \text{ cm} \\ -0.16481 \text{ cm} \\ -0.0000 \text{ cm} \\ -0.07023 \text{ cm} \\ -0.0000 \text{ cm} \end{bmatrix} = \begin{bmatrix} 0.00 \text{ mm} \\ -1.6 \text{ mm} \\ 0.00 \text{ mm} \\ -0.70 \text{ mm} \\ 0.00 \text{ mm} \end{bmatrix} \begin{matrix} 1 \\ 2 \\ 3 \\ 7 \\ 8 \end{matrix}$$

Find the reactions:

$$[Q_s] = [K_{sf}][D_f] + [K_{ss}][D_s] + [Q_{so}]$$
$$\phantom{[Q_s] = [K_{sf}][D_f] + [K_{ss}]}[0]$$

$$\begin{bmatrix} B_y \\ C_x \\ C_y \end{bmatrix} = \begin{matrix} 4 \\ 5 \\ 6 \end{matrix} \begin{bmatrix} \overset{1}{0} & \overset{2}{0} & \overset{3}{0} & \overset{7}{0} & \overset{8}{-5647\,\frac{kN}{cm}} \\ 0 & 0 & -5647\,\frac{kN}{cm} & -886.1\,\frac{kN}{cm} & 886.1\,\frac{kN}{cm} \\ 0 & 0 & 0 & 886.1\,\frac{kN}{cm} & -886.1\,\frac{kN}{cm} \end{bmatrix} \begin{bmatrix} 0\,\text{cm} \\ -0.16481\,\text{cm} \\ 0\,\text{cm} \\ -0.07023\,\text{cm} \\ 0\,\text{cm} \end{bmatrix} + \begin{bmatrix} 0 \\ -62.21\,\text{kN} \\ 62.21\,\text{kN} \end{bmatrix}$$

$$= \begin{bmatrix} 0 + 0 \\ 62.23\,\text{kN} - 62.21\,\text{kN} \\ -62.23\,\text{kN} + 62.21\,\text{kN} \end{bmatrix} = \begin{bmatrix} 0 \\ 0.02\,\text{kN} \\ -0.02\,\text{kN} \end{bmatrix} \begin{matrix} 4 \\ 5 \\ 6 \end{matrix}$$

When compared with the components, these values are essentially zero.

$$\begin{bmatrix} B_y \\ C_x \\ C_y \end{bmatrix} = \begin{bmatrix} 0 \\ 0 \\ 0 \end{bmatrix} \begin{matrix} 4 \\ 5 \\ 6 \end{matrix}$$

Find the internal forces:

Member AB:

$$q'^{AB}_F = 4840\,\frac{kN}{cm}[-1\ \ 0\ \ 1\ \ 0]\begin{bmatrix} 0 \\ -0.16481\,\text{cm} \\ 0 \\ 0 \end{bmatrix} = 0$$

Member AD:

$$q'^{AD}_F = 1631.2\,\frac{kN}{cm}[-0.7593\ \ -0.6508\ \ 0.7593\ \ 0.6508]\begin{bmatrix} 0 \\ -0.16481\,\text{cm} \\ -0.07023\,\text{cm} \\ 0 \end{bmatrix}$$
$$+ (-87.98\,\text{kN}) = 0$$

Member BC:

$$q'^{BC}_F = 5647\,\frac{kN}{cm}[-1\ \ 0\ \ 1\ \ 0]\begin{bmatrix} 0 \\ 0 \\ 0 \\ 0 \end{bmatrix} = 0$$

Member BD:

$$q'^{BD}_F = 5647\,\frac{kN}{cm}[0\ \ -1\ \ 0\ \ 1]\begin{bmatrix} 0 \\ 0 \\ -0.07023\,\text{cm} \\ 0 \end{bmatrix} = 0$$

Member CD:

$$q'^{CD}_F = 1772.3\,\frac{kN}{cm}[0.7071\ \ -0.7071\ \ -0.7071\ \ 0.7071]\begin{bmatrix} 0 \\ 0 \\ -0.07023\,\text{cm} \\ 0 \end{bmatrix}$$
$$+ (-87.98\,\text{kN}) = 0.03\,\text{kN}$$

Section 14.5 Additional Loadings

Evaluation of Results:

Satisfied Fundamental Principles?
The reactions and internal forces are zero, as expected because the truss is determinate. ✓

The vertical displacement at *A* is downward, as expected: ✓

$$\Delta_{Ay} = -1.6 \text{ mm} \quad (+\uparrow)$$

Conculsion:
These results plus the checks on the reactions and internal forces strongly suggest that we have reasonable results.

Evaluation of Design:
The vertical displacement will not push the live load displacement beyond the limit.

$$\Delta_{Ay}^{Temp} = -1.6 \text{ mm} \quad (+\uparrow)$$
$$\Delta_{Ay}^{Total} = -1.6 \text{ mm} - 13.0 \text{ mm} = -14.6 \text{ mm} \quad (+\uparrow)$$

EXAMPLE 14.6

Our firm has designed a large pipeline along the edge of a canal in Singapore. Our team has been tasked with designing a structure to support the pipe. Other engineers on the team have designed the truss members for strength.

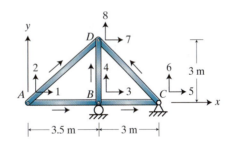

Figure 14.31

Because we are new to the firm, our supervisor wants to verify our knowledge of the Direct Stiffness Method. A determinate truss is a great way to do that.

Our final test from the supervisor is to analyze the truss to find the reactions due to just the self-weight of the truss. We are not concerned with the vertical displacement at *A* because the pipe will be attached after the truss displaces under its self-weight.

Member properties:

Diagonals:	Double angle	120 × 120 × 8	$w_o = 0.288$ kN/m
All others:	Universal columns	250 × 250 × 66.5	$w_o = 0.652$ kN/m

Loading:
We can convert the distributed load to loads at the nodes by taking the total load on each member and applying half to each end.

Member AB:
$l_{AB} = 3.50$ m
$W_{AB} = 3.500$ m $(0.652$ kN/m$) = 2.282$ kN
$P_{Ay}^{AB} = 2.282$ kN$/2 = 1.141$ kN $(+\downarrow)$
$P_{By}^{AB} = 1.141$ kN $(+\downarrow)$

Member AD:
$l_{AB} = 4.610$ m
$W_{AD} = 4.610$ m $(0.288$ kN/m$) = 1.328$ kN
$P_{Ay}^{AD} = 1.328$ kN$/2 = 0.664$ kN $(+\downarrow)$
$P_{Dy}^{AD} = 0.664$ kN $(+\downarrow)$

Member BC:
$l_{BC} = 3.000$ m
$W_{BC} = 3.000$ m $(0.652$ kN/m$) = 1.956$ kN
$P_{By}^{BC} = 1.956$ kN$/2 = 0.978$ kN $(+\downarrow)$
$P_{Cy}^{BC} = 0.978$ kN $(+\downarrow)$

Member BD:
$l_{BD} = 3.000$ m
$W_{BD} = 3.000$ m $(0.652$ kN/m$) = 1.956$ kN
$P_{By}^{BD} = 1.956$ kN$/2 = 0.978$ kN $(+\downarrow)$
$P_{Dy}^{BD} = 0.978$ kN $(+\downarrow)$

Member CD:
$l_{CD} = 4.243$ m
$W_{CD} = 4.243$ m $(0.288$ kN/m$) = 1.222$ kN
$P_{Cy}^{CD} = 1.222$ kN$/2 = 0.611$ kN $(+\downarrow)$
$P_{Dy}^{CD} = 0.611$ kN $(+\downarrow)$

Net forces on nodes:
$P_{Ay} = 1.141$ kN $+ 0.664$ kN $= 1.805$ kN $(+\downarrow)$
$P_{By} = 1.141$ kN $+ 0.978$ kN $+ 0.978$ kN $= 3.097$ kN $(+\downarrow)$
$P_{Cy} = 0.978$ kN $+ 0.611$ kN $= 1.589$ kN $(+\downarrow)$
$P_{Dy} = 0.664$ kN $+ 0.978$ kN $+ 0.611$ kN $= 2.253$ kN $(+\downarrow)$

Predicted Solution:

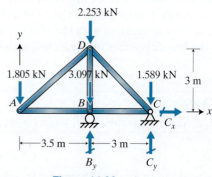

Figure 14.32

Apply equilibrium:

$$\circlearrowright \Sigma M_C = 0 = 1.805 \text{ kN}(6.5 \text{ m}) + 2.253 \text{ kN}(3 \text{ m}) + 3.097 \text{ kN}(3 \text{ m}) - B_y(3 \text{ m})$$
$$B_y = 9.261 \text{ kN} \quad (+\uparrow)$$

$$+\uparrow \Sigma F_y = 0 = -1.805 \text{ kN} - 2.253 \text{ kN} - 3.097 \text{ kN} + 9.261 \text{ kN} - 1.589 \text{ kN} + C_y$$
$$C_y = -0.517 \text{ kN} \quad (+\uparrow)$$

$$\xrightarrow{+} \Sigma F_x = 0 = C_x$$
$$C_x = 0$$

Direct Stiffness Method:

Assemble Q_f and Q_{so}:

$$[\mathbf{Q}_f] = \begin{matrix} 1 \\ 2 \\ 3 \\ 7 \\ 8 \end{matrix} \begin{bmatrix} 0 \\ -1.805 \text{ kN} \\ 0 \\ 0 \\ -2.253 \text{ kN} \end{bmatrix}$$

Since there are loads applied directly to the supported DOFs, we reverse the signs before adding them to $[\mathbf{Q}_{so}]$:

$$[\mathbf{Q}_{so}] = \begin{matrix} 4 \\ 5 \\ 6 \end{matrix} \begin{bmatrix} +3.097 \text{ kN} \\ 0 \\ +1.589 \text{ kN} \end{bmatrix}$$

Find the deformations:

$$[\mathbf{Q}_f] = [\mathbf{K}_{ff}][\mathbf{D}_f] + [\mathbf{K}_{fs}][\mathbf{D}_s]$$
$$\phantom{[\mathbf{Q}_f] = [\mathbf{K}_{ff}][\mathbf{D}_f] + [\mathbf{K}_{fs}]}[\mathbf{0}]$$

$$\begin{matrix} 1 \\ 2 \\ 3 \\ 7 \\ 8 \end{matrix} \begin{bmatrix} 0 \\ -1.805 \text{ kN} \\ 0 \\ 0 \\ -2.253 \text{ kN} \end{bmatrix} = \begin{matrix} 1 \\ 2 \\ 3 \\ 7 \\ 8 \end{matrix} \begin{bmatrix} 5780 \tfrac{\text{kN}}{\text{cm}} & 806.1 \tfrac{\text{kN}}{\text{cm}} & -4840 \tfrac{\text{kN}}{\text{cm}} & -940.9 \tfrac{\text{kN}}{\text{cm}} & -806.1 \tfrac{\text{kN}}{\text{cm}} \\ 806.1 \tfrac{\text{kN}}{\text{cm}} & 690.9 \tfrac{\text{kN}}{\text{cm}} & 0 & -806.1 \tfrac{\text{kN}}{\text{cm}} & -690.9 \tfrac{\text{kN}}{\text{cm}} \\ -4840 \tfrac{\text{kN}}{\text{cm}} & 0 & 10847 \tfrac{\text{kN}}{\text{cm}} & 0 & 0 \\ -940.9 \tfrac{\text{kN}}{\text{cm}} & -806.1 \tfrac{\text{kN}}{\text{cm}} & 0 & 1826.5 \tfrac{\text{kN}}{\text{cm}} & -80.0 \tfrac{\text{kN}}{\text{cm}} \\ -806.1 \tfrac{\text{kN}}{\text{cm}} & -690.9 \tfrac{\text{kN}}{\text{cm}} & 0 & -80.0 \tfrac{\text{kN}}{\text{cm}} & 7224 \tfrac{\text{kN}}{\text{cm}} \end{bmatrix} \begin{bmatrix} \Delta_{Ax} \\ \Delta_{Ay} \\ \Delta_{Bx} \\ \Delta_{Dx} \\ \Delta_{Dy} \end{bmatrix}$$

$$[\mathbf{D}_f] = [\mathbf{K}_{ff}]^{-1}([\mathbf{Q}_f])$$

$$\begin{bmatrix} \Delta_{Ax} \\ \Delta_{Ay} \\ \Delta_{Bx} \\ \Delta_{Dx} \\ \Delta_{Dy} \end{bmatrix} = \begin{matrix} 1 \\ 2 \\ 3 \\ 7 \\ 8 \end{matrix} \begin{bmatrix} 5780 \tfrac{\text{kN}}{\text{cm}} & 806.1 \tfrac{\text{kN}}{\text{cm}} & -4840 \tfrac{\text{kN}}{\text{cm}} & -940.9 \tfrac{\text{kN}}{\text{cm}} & -806.1 \tfrac{\text{kN}}{\text{cm}} \\ 806.1 \tfrac{\text{kN}}{\text{cm}} & 690.9 \tfrac{\text{kN}}{\text{cm}} & 0 & -806.1 \tfrac{\text{kN}}{\text{cm}} & -690.9 \tfrac{\text{kN}}{\text{cm}} \\ -4840 \tfrac{\text{kN}}{\text{cm}} & 0 & 10487 \tfrac{\text{kN}}{\text{cm}} & 0 & 0 \\ -940.9 \tfrac{\text{kN}}{\text{cm}} & -806.1 \tfrac{\text{kN}}{\text{cm}} & 0 & 1826.5 \tfrac{\text{kN}}{\text{cm}} & -80.0 \tfrac{\text{kN}}{\text{cm}} \\ -806.1 \tfrac{\text{kN}}{\text{cm}} & -690.9 \tfrac{\text{kN}}{\text{cm}} & 0 & -80.0 \tfrac{\text{kN}}{\text{cm}} & 7224 \tfrac{\text{kN}}{\text{cm}} \end{bmatrix}^{-1} \left(\begin{bmatrix} 0 \\ -1.805 \text{ kN} \\ 0 \\ 0 \\ -2.253 \text{ kN} \end{bmatrix} \right)$$

$$\begin{bmatrix} \Delta_{Ax} \\ \Delta_{Ay} \\ \Delta_{Bx} \\ \Delta_{Dx} \\ \Delta_{Dy} \end{bmatrix} = \begin{bmatrix} 8.077 \times 10^{-4} \text{ cm} \\ -8.693 \times 10^{-3} \text{ cm} \\ 3.728 \times 10^{-4} \text{ cm} \\ -3.468 \times 10^{-3} \text{ cm} \\ -1.0916 \times 10^{-3} \text{ cm} \end{bmatrix} \begin{matrix} 1 \\ 2 \\ 3 \\ 7 \\ 8 \end{matrix}$$

840 Chapter 14 Direct Stiffness Method for Trusses

Find the reactions:
$$[Q_s] = [K_{sf}][D_f] + [K_{ss}][D_s] + [Q_{so}]$$
$$[0]$$

$$\begin{bmatrix} B_y \\ C_x \\ C_y \end{bmatrix} = \begin{matrix} 4 \\ 5 \\ 6 \end{matrix} \begin{bmatrix} 1 & 2 & 3 & 7 & 8 \\ 0 & 0 & 0 & 0 & -5647\,\frac{kN}{cm} \\ 0 & 0 & -5647\,\frac{kN}{cm} & -886.1\,\frac{kN}{cm} & 886.1\,\frac{kN}{cm} \\ 0 & 0 & 0 & 886.1\,\frac{kN}{cm} & -886.1\,\frac{kN}{cm} \end{bmatrix} \begin{bmatrix} 8.077 \times 10^{-4}\,cm \\ -8.693 \times 10^{-3}\,cm \\ 3.728 \times 10^{-4}\,cm \\ -3.468 \times 10^{-3}\,cm \\ -1.0916 \times 10^{-3}\,cm \end{bmatrix} + \begin{bmatrix} +3.097\,kN \\ 0 \\ +1.589\,kN \end{bmatrix}$$

$$= \begin{bmatrix} 6.164\,kN + 3.097\,kN \\ 0.001\,kN - 0 \\ -2.106\,kN + 1.589\,kN \end{bmatrix} = \begin{bmatrix} 9.261\,kN \\ 0.001\,kN \\ -0.517\,kN \end{bmatrix} \begin{matrix} 4 \\ 5 \\ 6 \end{matrix}$$

Evaluation of Results:
Satisfied Fundamental Principles?
Compare the results from equilibrium with the results from the Direct Stiffness Method.

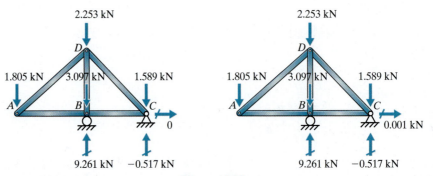

Figure 14.33

The results are identical to at least three significant figures. ✓
The horizontal reaction at C is within three significant figures compared with the magnitudes of the other reactions.

Conculsion:
This suggests that we have reasonable results.

References Cited

AISC (American Institute of Steel Construction). 2016. *Code of Standard Practice for Steel Buildings and Bridges (303–16)*. Chicago, IL: AISC.
PCI (Precast/Prestressed Concrete Institute). 2017. *PCI Design Handbook*, 8th ed. Chicago, IL: PCI.

Homework Problems

14.1 through 14.3 A local business owner has contacted our firm about an old hoist attached to its building. The hoist was used decades ago to unload deliveries. The owner would like to start using the hoist again but is concerned about the capacity of the aging hoist.

Our firm sent someone to inspect the hoist and take measurements. The lower member appears to be a pair of MC6 × 7 channels ($A = 4.18\text{ in}^2$), and the top member is a $\frac{1}{2}$-in.-diameter rod. Both members are steel ($E = 29{,}000$ ksi). The business owner has an automotive repair shop and wants to use the hoist structure to remove engines. An engineer on our team has done some research and determined that we should assume a 1000-lb point live load as a reasonable upper bound. We can apply the load factor to our results after the analysis.

The peak effect will be if the load is at the end of the hoist. In that case, the structure is effectively a determinate truss. Since we are new to the firm, our supervisor wants to use this project to evaluate our knowledge of the analysis method used in the firm's computer software: the Direct Stiffness Method.

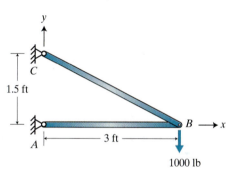

Probs. 14.1–14.3

14.1 Our first test is to create the structural stiffness matrix.

a. Calculate the element stiffness matrix, in global coordinates, for each of the truss members. Designate the near and far nodes for each member, and label all of the degrees of freedom.
b. Assemble the global stiffness matrix for the structure.

14.2 Our second test is to analyze the truss.

a. Guess which member will experience the largest axial force magnitude.
b. Find the support reactions and internal forces by hand using equilibrium.
c. Calculate the anticipated displacements of the free degrees of freedom using the Direct Stiffness Method. Note: to solve the system of equations, you may use computer software, but provide a printout of the inputs and outputs.
d. Use the displacements from part (c) to calculate the anticipated reactions. Verify equilibrium of the truss using these reactions.
e. Use the displacements from part (c) to calculate the anticipated internal force in each member.
f. Are your results from the Direct Stiffness Method reasonable? Make a comprehensive argument to justify your answer.
g. Using structural analysis software, determine the displacements of the free degrees of freedom, the reactions, and the internal forces. Comment on how these results compare with your results from parts (c) through (e).
h. Comment on why your guess in part (a) does or does not match what you discovered in part (b).

842   Chapter 14   Direct Stiffness Method for Trusses

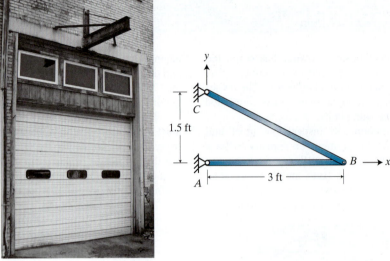

Probs. 14.1–14.3

14.3 Our third test is to determine the impact of temperature change on the truss. If the owner uses the hoist during the winter, it could be 50°F colder than the temperature when the hoist was installed. A potential concern is that point B will drop so much that the attached pulley might begin to slide off the end. Therefore, our supervisor wants to know how much vertical displacement to expect due to the 50°F drop in temperature. The coefficient of thermal expansion for steel is $6.5 \times 10^{-6}/°F$.

a. Guess whether point B will rise or drop with the temperature change.
b. Calculate the nodal equivalent loads in global coordinates.
c. Calculate the anticipated displacements of the free degrees of freedom using the Direct Stiffness Method. Note: to solve the system of equations, you may use computer software, but provide a printout of the inputs and outputs.
d. Use the displacements from part (c) to calculate the anticipated reactions. Verify equilibrium of the truss using these reactions.
e. Use the displacements from part (c) to calculate the anticipated internal force in each member.
f. Are your results from the Direct Stiffness Method reasonable? Make a comprehensive argument to justify your answer.
g. Using structural analysis software, determine the displacements of the free degrees of freedom, the reactions, and the internal forces due to the uniform temperature change.
h. Comment on why your guess in part (a) does or does not match what you discovered in part (c).

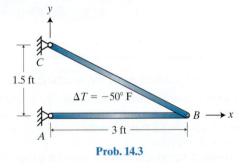

Prob. 14.3

14.4 through 14.7 We work for a steel fabricator as a connection designer. Our company just got a job from a local gym, which wants us to design and fabricate a pull-up bar that can be mounted on the wall. The owners have tried a rigid frame version but want something stiffer, like a truss. For the safety of the users, no members can be below the bar.

Our firm has purchased new structural analysis software, and our supervisor wants the software tested before we use it on projects. We have been tasked with analyzing the truss version of the pull-up bar by hand using the Direct Stiffness Method. Our results will be used to help verify the results of the new software.

Another team member has calculated the design load. For the purposes of this test, we assign different section properties to each of the members, and we assume the truss will be made of aluminum:

Member CD:	25 × 25 × 1.5	A = 139 mm²
Member BD:	25 × 25 × 2	A = 181 mm²
Member AD:	25 × 25 × 3	A = 259 mm²
Aluminum:	E = 69 GPa = 69 kN/mm²	

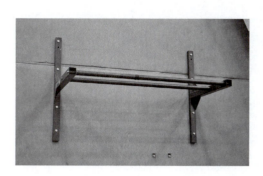
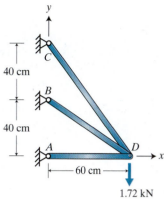

Probs. 14.4–14.7

14.4 Our first task is to create the structural stiffness matrix.
a. Calculate the element stiffness matrix, in global coordinates, for each of the truss members. Designate the near and far nodes for each member, and label all of the degrees of freedom.
b. Assemble the global stiffness matrix for the structure.

14.5 Our second task is to analyze the truss.
a. Guess which way node D will move in both the horizontal and vertical directions.
b. Using approximate analysis, find the support reactions and internal forces.
c. Calculate the anticipated displacements of the free degrees of freedom using the Direct Stiffness Method. Note: to solve the system of equations, you may use computer software, but provide a printout of the inputs and outputs.
d. Use the displacements from part (c) to calculate the anticipated reactions. Verify equilibrium of the truss using these reactions.
e. Use the displacements from part (c) to calculate the anticipated internal force in each member.
f. Are your results from the Direct Stiffness Method reasonable? Make a comprehensive argument to justify your answer.
g. Using structural analysis software, determine the displacements of the free degrees of freedom, the reactions, and the internal forces. Comment on how these results compare with your results from parts (c) through (e).
h. Comment on why your guess in part (a) does or does not match what you discovered in part (c).

14.6 When designing the truss, we must account for the possibility that the members are not exactly the length we planned. Therefore, our third task is to determine the impact if one of the members is fabricated short. For the purposes of testing our software, we will check only one possibility: member CD is fabricated 2 mm shorter than its planned length. We will consider this effect simultaneous with the live loading.

a. Guess which way node D will move in both the horizontal and vertical directions.
b. Using qualitative truss analysis, predict the sense of the axial force in each member. Consider the live load and the length change separately; then superimpose the qualitative results.
c. Calculate the nodal equivalent loads in global coordinates.
d. Calculate the anticipated displacements of the free degrees of freedom using the Direct Stiffness Method. Note: to solve the system of equations, you may use computer software, but provide a printout of the inputs and outputs.
e. Use the displacements from part (d) to calculate the anticipated reactions. Verify equilibrium of the truss using these reactions.
f. Use the displacements from part (d) to calculate the anticipated internal force in each member.
g. Are your results from the Direct Stiffness Method reasonable? Make a comprehensive argument to justify your answer.
h. Using structural analysis software, determine the displacements of the free degrees of freedom, the reactions, and the internal forces due to the live load and length change together. Comment on how these results compare with your results from parts (d) through (f).
i. Comment on why your guess in part (a) does or does not match what you discovered in part (d).

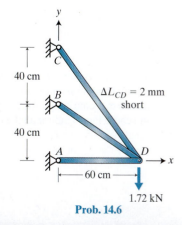

Prob. 14.6

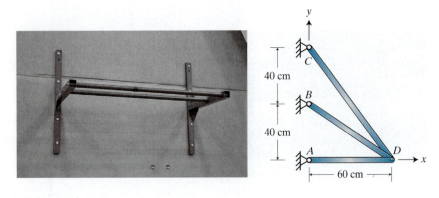

Probs. 14.4–14.7

14.7 Since the reactions are indeterminate for this truss, any movement in the connection to the wall will affect the deformations, internal forces, and reactions. To test our software, let's consider the possibility of the top support moving down 3 mm. We will consider this effect simultaneous with the live loading.

a. Guess which way node D will move in both the horizontal and vertical directions.
b. Using qualitative truss analysis, predict the sense of the axial force in each member. Consider the live load and the support movement separately; then superimpose the qualitative results.
c. Calculate the anticipated displacements of the free degrees of freedom using the Direct Stiffness Method. Note: to solve the system of equations, you may use computer software, but provide a printout of the inputs and outputs.
d. Use the displacements from part (c) to calculate the anticipated reactions. Verify equilibrium of the truss using these reactions.
e. Use the displacements from part (c) to calculate the anticipated internal force in each member.
f. Are your results from the Direct Stiffness Method reasonable? Make a comprehensive argument to justify your answer.
g. Using structural analysis software, determine the displacements of the free degrees of freedom, the reactions, and the internal forces due to the live load and support movement together. Comment on how these results compare with your results from parts (c) through (e).
h. Comment on why your guess in part (a) does or does not match what you discovered in part (c).

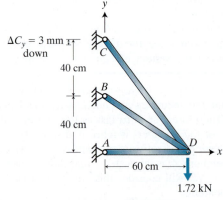

Prob. 14.7

14.8 through 14.12 Our firm is designing a rack for road maintenance equipment for a community in a zone that experiences moderate seismic activity. Our team leader chose an indeterminate truss to provide lateral stability for the structure. One of the load cases we must consider is the static equivalent loading from a seismic event.

Another engineer on our team has performed preliminary design of the truss and chose Pipe 3 Std for all the members:

Pipe 3 Std: $A = 2.08 \text{ in}^2$
Steel: $E = 29{,}000 \text{ ksi}$

Recently our firm upgraded the version of its structural analysis software. The firm's policy is to verify the software with several comprehensive tests cases, and our supervisor would like to use this truss as one of them.

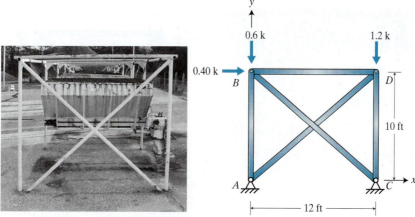

Probs. 14.8–14.12

14.8 Our first task is to create the structural stiffness matrix.
a. Calculate the element stiffness matrix, in global coordinates, for each of the truss members. Designate the near and far nodes for each member, and label all of the degrees of freedom.
b. Assemble the global stiffness matrix for the structure.

14.9 Our second task is to analyze the truss.
a. Guess the directions of each of the support reactions.
b. Using approximate analysis, find the support reactions and internal forces.
c. Calculate the anticipated displacements of the free degrees of freedom using the Direct Stiffness Method. Note: to solve the system of equations, you may use computer software, but provide a printout of the inputs and outputs.
d. Use the displacements from part (c) to calculate the anticipated reactions. Verify equilibrium of the truss using these reactions.
e. Use the displacements from part (c) to calculate the anticipated internal force in each member.
f. Are your results from the Direct Stiffness Method reasonable? Make a comprehensive argument to justify your answer.
g. Using structural analysis software, determine the displacements of the free degrees of freedom, the reactions, and the internal forces. Comment on how these results compare with your results from parts (c) through (e).
h. Comment on why your guess in part (a) does or does not match what you discovered in part (d).

14.10 An important test of the software is dealing with temperature change. It is reasonable that as the sun rises, one side of the truss gets warmer than the others. Therefore, let's determine the impact of a 15°F temperature increase for only member CD. The coefficient of thermal expansion for steel is $6.5 \times 10^{-6}/°F$.
a. Guess the directions of each of the support reactions.
b. Using qualitative truss analysis, predict the sense of the axial force in each member.
c. Calculate the nodal equivalent loads in global coordinates.
d. Calculate the anticipated displacements of the free degrees of freedom using the Direct Stiffness Method. Note: to solve the system of equations, you may use computer software, but provide a printout of the inputs and outputs.
e. Use the displacements from part (d) to calculate the anticipated reactions. Verify equilibrium of the truss using these reactions.
f. Use the displacements from part (d) to calculate the anticipated internal force in members AD and CD.
g. Are your results from the Direct Stiffness Method reasonable? Make a comprehensive argument to justify your answer.
h. Using structural analysis software, determine the displacements of the free degrees of freedom, the reactions, and the internal forces due to the temperature change. Comment on how these results compare with your results from parts (d) through (f).
i. Comment on why your guess in part (a) does or does not match what you discovered in part (e).

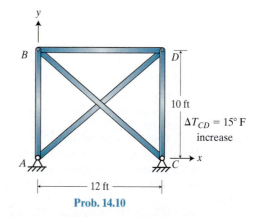

Prob. 14.10

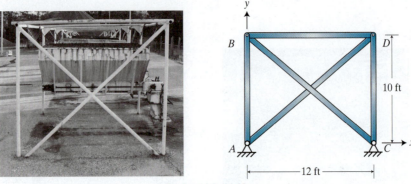

Probs. 14.8–14.12

14.11 Another test of the software is dealing with length change. Sometimes pieces arrive from the fabricator with a different length than planned. Most of the time, the effect on the structure is negligible. However, we need to account for the possibility. To test the software, let's determine the impact of member AD being ⅛ inch short.

a. Guess the directions of each of the support reactions.
b. Using qualitative truss analysis, predict the sense of the axial force in each member.
c. Calculate the nodal equivalent loads in global coordinates.
d. Calculate the anticipated displacements of the free degrees of freedom using the Direct Stiffness Method. Note: to solve the system of equations, you may use computer software, but provide a printout of the inputs and outputs.
e. Use the displacements from part (d) to calculate the anticipated reactions. Verify equilibrium of the truss using these reactions.
f. Use the displacements from part (d) to calculate the anticipated internal force in members AD and CD.
g. Are your results from the Direct Stiffness Method reasonable? Make a comprehensive argument to justify your answer.

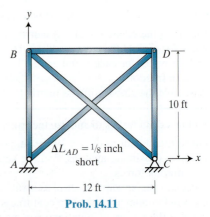

Prob. 14.11

h. Using structural analysis software, determine the displacements of the free degrees of freedom, the reactions, and the internal forces due to the length change. Comment on how these results compare with your results from parts (d) through (f).
i. Comment on why your guess in part (a) does or does not match what you discovered in part (e).

14.12 Differential settlement means that one foundation moves down more than the others over time. For a structure with indeterminate reactions, that causes internal forces and reactions. We want to know that the software can accurately predict this effect. Therefore, let's consider that node A moves ½ inch down relative to node C.

a. Guess the directions of each of the support reactions.
b. Using qualitative truss analysis, predict the sense of the axial force in each member.
c. Calculate the anticipated displacements of the free degrees of freedom using the Direct Stiffness Method. Note: to solve the system of equations, you may use computer software, but provide a printout of the inputs and outputs.

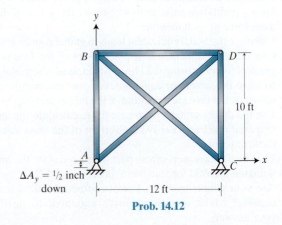

Prob. 14.12

d. Use the displacements from part (c) to calculate the anticipated reactions. Verify equilibrium of the truss using these reactions.
e. Use the displacements from part (c) to calculate the anticipated internal force in members AD and CD.
f. Are your results from the Direct Stiffness Method reasonable? Make a comprehensive argument to justify your answer.
g. Using structural analysis software, determine the displacements of the free degrees of freedom, the reactions, and the internal forces. Comment on how these results compare with your results from parts (c) through (e).
h. Comment on why your guess in part (a) does or does not match what you discovered in part (d).

14.13 through 14.16 Our firm has been selected to design a high-rise building that presents a unique challenge. Half of the building will hang over active railroad tracks. Therefore, there can be no foundations on one side. The principal in charge and the architect have decided to use a roof truss to transfer the load from one side of the building to the other.

The team is in the preliminary design phase, so the members have not been designed yet. In preparation for that design, the team leader wants to get a sense of how the truss will behave. Performing the Direct Stiffness Method by hand will provide quantitative information about the impact of each member; the team leader wants that information before refining the design.

The truss will be made of steel ($E = 200$ GPa $= 200$ kN/mm^2). An experienced engineer estimates that the members will have the following cross-sectional areas:

① $A_1 = 600$ cm^2
② $A_2 = 800$ cm^2
③ $A_3 = 500$ cm^2
④ $A_4 = 200$ cm^2
⑤ $A_5 = 100$ cm^2

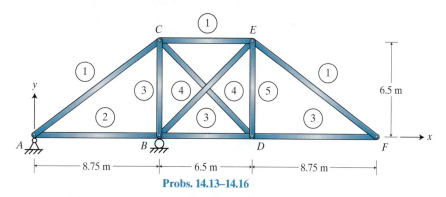

Probs. 14.13–14.16

14.13 We start by creating the structural stiffness matrix.
a. Calculate the element stiffness matrix, in global coordinates, for each of the truss members. Designate the near and far nodes for each member, and label all of the degrees of freedom.
b. Assemble the global stiffness matrix for the structure.

848 **Chapter 14** Direct Stiffness Method for Trusses

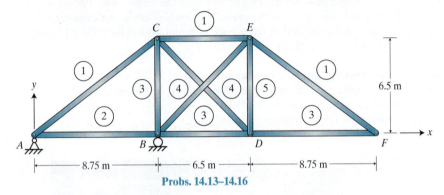

Probs. 14.13–14.16

14.14 Another team member has estimated the load each truss must carry.

a. Guess which member will carry the largest axial force (tension or compression).
b. Using approximate analysis, find the support reactions and internal forces. Which of these are upper bounds, lower bounds, or just approximations?
c. Calculate the anticipated displacements of the free degrees of freedom using the Direct Stiffness Method. Note: to solve the system of equations, you may use computer software, but provide a printout of the inputs and outputs.
d. Use the displacements from part (c) to calculate the anticipated reactions. Verify equilibrium of the truss using these reactions.
e. Use the displacements from part (c) to calculate the anticipated internal force in each member.
f. Are your results from the Direct Stiffness Method reasonable? Make a comprehensive argument to justify your answer.
g. Using structural analysis software, determine the displacements of the free degrees of freedom, the reactions, and the internal forces. Comment on how these results compare with your results from parts (c) through (e).
h. Comment on why your guess in part (a) does or does not match what you discovered in part (e).

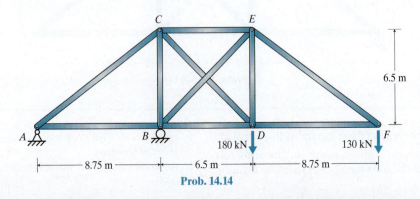

Prob. 14.14

14.15 As the sun rises, members *DE* and *EF* will warm up faster than the other members. Therefore, we want to predict the deformations at node *E* due to a temperature increase of 10°C for members *DE* and *EF* only. Let's consider the effect of the temperature change without any other loads; we can superimpose the results later if needed. The coefficient of thermal expansion for steel is $11.7 \times 10^{-6}/°C$.

a. Guess which way node *E* will move in both the horizontal and vertical directions.
b. Using qualitative truss analysis, predict the sense of the axial force in each member.
c. Calculate the nodal equivalent loads in global coordinates.
d. Calculate the anticipated displacements of the free degrees of freedom using the Direct Stiffness Method. Note: to solve the system of equations, you may use computer software, but provide a printout of the inputs and outputs.
e. Use the displacements from part (d) to calculate the vertical reaction at *B*. Comment on why calculating just this one reaction is enough to verify all of the reactions.
f. Use the displacements from part (d) to calculate the anticipated internal force in members *DE* and *EF*.
g. Are your results from the Direct Stiffness Method reasonable? Make a comprehensive argument to justify your answer.
h. Using structural analysis software, determine the displacements of the free degrees of freedom, the reactions, and the internal forces due to the temperature change. Comment on how these results compare with your results from parts (d) through (f).
i. Comment on why your guess in part (a) does or does not match what you discovered in part (d).

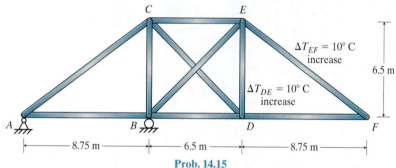

Prob. 14.15

14.16 Our project leader has pointed out that fabricated members have tolerances; therefore, we cannot expect all of the members to arrive from the shop with exactly the size designed. To anticipate the potential impact of a member that is within tolerance but different from the designed length, we have been asked to find out what will happen if member *CD* is fabricated 5 mm too long.

a. Guess which way node *F* will move in both the horizontal and vertical directions.
b. Using qualitative truss analysis, predict the sense of the axial force in each member.
c. Calculate the nodal equivalent loads in global coordinates.
d. Calculate the anticipated displacements of the free degrees of freedom using the Direct Stiffness Method. Note: to solve the system of equations, you may use computer software, but provide a printout of the inputs and outputs.
e. Use the displacements from part (d) to calculate the reactions at *A*. Comment on why calculating just these two reactions is enough to verify all of the reactions.
f. Use the displacements from part (d) to calculate the anticipated internal force in members *CD* and *DF*.
g. Are your results from the Direct Stiffness Method reasonable? Make a comprehensive argument to justify your answer.
h. Using structural analysis software, determine the displacements of the free degrees of freedom, the reactions, and the internal forces due to the length change. Comment on how these results compare with your results from parts (d) through (f).
i. Comment on why your guess in part (a) does or does not match what you discovered in part (d).

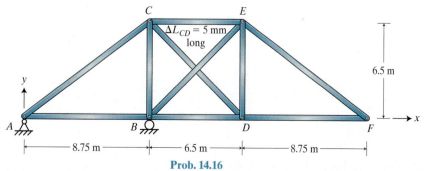

Prob. 14.16

14.17 through 14.20 Our team is designing the truss to support a large billboard. The senior engineer picked the configuration of the new truss based on the soil conditions at the site. Our firm just purchased new structural analysis software. In order to verify that the software is working properly for two-dimensional trusses, we have been tasked with analyzing the truss by hand using the Direct Stiffness Method.

For preliminary design purposes, an engineer on our team has chosen L4 × 4 × 3/8 galvanized steel angles:

L4 × 4 × 3/8 $A = 2.86 \text{ in}^2$

Steel: $E = 29{,}000 \text{ ksi}$

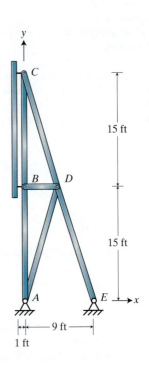

Probs. 14.17–14.20

14.17 Step 1 is to create the structural stiffness matrix.

a. Calculate the element stiffness matrix, in global coordinates, for each of the truss members. Designate the near and far nodes for each member, and label all of the degrees of freedom.

b. Assemble the global stiffness matrix for the structure.

14.18 Step 2 is to analyze the truss with the loads from the weight of the sign. Because the sign hangs out away from the truss, both vertical and horizontal forces are transferred to the truss.

a. Guess which member will experience the largest axial force magnitude.
b. Find the support reactions and internal forces by hand using equilibrium.
c. Calculate the anticipated displacements of the free degrees of freedom using the Direct Stiffness Method. Note: to solve the system of equations, you may use computer software, but provide a printout of the inputs and outputs.
d. Use the displacements from part (c) to calculate the anticipated reactions. Verify equilibrium of the truss using these reactions.
e. Use the displacements from part (c) to calculate the anticipated internal force in each member.
f. Are your results from the Direct Stiffness Method reasonable? Make a comprehensive argument to justify your answer.
g. Using structural analysis software, determine the displacements of the free degrees of freedom, the reactions, and the internal forces. Comment on how these results compare with your results from parts (c) through (e).
h. Comment on why your guess in part (a) does or does not match what you discovered in part (e).

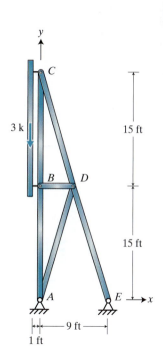
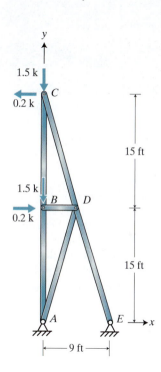

Prob. 14.18

852 Chapter 14 Direct Stiffness Method for Trusses

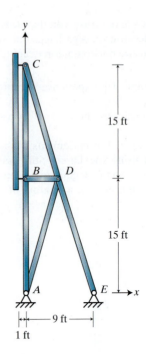

Probs. 14.17–14.20

14.19 An important test of the software is dealing with temperature change. As the sun rises, the billboard will shield member BC, but member AB will warm up. Let's determine the impact of a 20°F temperature increase for just member AB. We can ignore the load from the billboard for this check. The coefficient of thermal expansion for steel is $6.5 \times 10^{-6}/°F$.

a. Guess which way node C will move in both the horizontal and vertical directions.
b. Using qualitative truss analysis, predict the sense of the axial force in each member.
c. Calculate the nodal equivalent loads in global coordinates.
d. Calculate the anticipated displacements of the free degrees of freedom using the Direct Stiffness Method. Note: to solve the system of equations, you may use computer software, but provide a printout of the inputs and outputs.
e. Use the displacements from part (d) to calculate the reactions at A. Comment on why calculating just these two reactions is enough to verify all of the reactions.
f. Use the displacements from part (d) to calculate the anticipated internal force in members AB and CD.
g. Are your results from the Direct Stiffness Method reasonable? Make a comprehensive argument to justify your answer.
h. Using structural analysis software, determine the displacements of the free degrees of freedom, the reactions, and the internal forces due to the temperature change. Comment on how these results compare with your results from parts (d) through (f).
i. Comment on why your guess in part (a) does or does not match what you discovered in part (d).

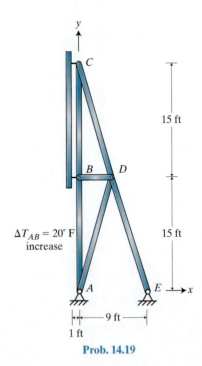

Prob. 14.19

14.20 While the truss is being constructed, we receive a call from the field telling us that member AD was fabricated ½ inch short. The contractor wants to know if the member can still be used even though it is outside the specified tolerance. To answer this question, we want to analyze the truss considering only the fabrication error. We can superimpose other loading results later. Since we analyzed the truss by hand before construction started, it may be faster for us to analyze the truss by hand again.

a. Guess the directions of each of the support reactions at A.
b. Using qualitative truss analysis, predict the sense of the axial force in each member.
c. Calculate the nodal equivalent loads in global coordinates.
d. Calculate the anticipated displacements of the free degrees of freedom using the Direct Stiffness Method. Note: to solve the system of equations, you may use computer software, but provide a printout of the inputs and outputs.
e. Use the displacements from part (d) to calculate the reactions at A. Comment on why calculating just these two reactions is enough to verify all of the reactions.
f. Use the displacements from part (d) to calculate the anticipated internal force in members AD and CD.
g. Are your results from the Direct Stiffness Method reasonable? Make a comprehensive argument to justify your answer.
h. Using structural analysis software, determine the displacements of the free degrees of freedom, the reactions, and the internal forces due to the length change. Comment on how these results compare with your results from parts (d) through (f).
i. Comment on why your guess in part (a) does or does not match what you discovered in part (e).

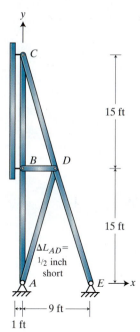

Prob. 14.20

CHAPTER 15

DIRECT STIFFNESS METHOD FOR FRAMES

Frames can satisfy geometry requirements that other structural systems cannot.

MOTIVATION

The Direct Stiffness Method is an extremely powerful tool for analyzing frames. We can use the method to find deformations at nodes, displaced shape, internal forces, and reactions. The concept is exactly the same for frames as it is for trusses. The only differences are the element stiffness matrix and the additional types of loads between nodes.

Even though the Direct Stiffness Method is a great tool for analyzing frames, we almost never use the method by hand. Instead, we typically implement the method in computer software. So, why do we study it? We identified three reasons at the beginning of Chapter 14: Direct Stiffness Method for Trusses that are also valid for frames. In addition, understanding how the method works empowers us to make wise modeling decisions. For example, a node in a three-dimensional frame has six degrees of freedom. Adding just one node to a frame that already has 200 nodes means that we expand the structural stiffness matrix by more than 14,000 coefficients. Understanding that cost and the potential benefit of the additional node is just one reason to practice the method by hand.

15.1 Element Stiffness Matrix in Local Coordinates

Introduction

At the core of the Direct Stiffness Method is the element stiffness matrix. Deriving the element stiffnesses in global coordinates would be a rather cumbersome process. Instead, we start with the element stiffness matrix in local coordinates.

How-To

Degrees of Freedom

In Chapter 14, we defined a degree of freedom (DOF) as a deformation associated with an internal force at one end of an element. For an element of a three-dimensional frame, there are six internal forces at one end: the axial force, two shears, torsion, and two moments (Figure 15.1). Those six internal forces are associated with six deformations: three displacements and three rotations. Therefore, each node in a three-dimensional frame has six degrees of freedom.

An element of a 2D frame has three internal forces at one end: the axial force, shear, and moment (Figure 15.2). These three internal forces are associated with three deformations: two displacements and one rotation (Figure 15.3). So, each node in a 2D frame has three degrees of freedom. For the purposes of this text, we limit ourselves to 2D frames. Entire books are dedicated to the Direct Stiffness Method; they cover the derivation of the 3D truss and frame elements.

We can number the DOFs in any sequence we want, but using a consistent sequence will allow us to create one formula for the element stiffness that we can use for every frame element. The typical numbering sequence is x-direction, y-direction, rotation. If we default to the xy-plane, which is most common, the rotation is about the z-axis. The standard practice is to use the right-hand rule to define a positive rotation. If we set the global x-direction as to the right and the global y-direction as up, a positive rotation will be counterclockwise for two-dimensional frames and the z-axis will point out of the page. Figure 15.4 shows the standard DOF numbering on an example frame.

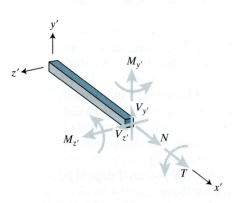

Figure 15.1 Internal forces at one end of a 3D frame element.

Figure 15.2 Internal forces at one end of a 2D frame element.

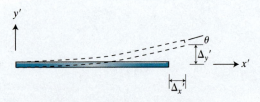

Figure 15.3 Deformations at one end of a 2D frame element.

Axial Deformation

The easiest way to derive the element stiffness matrix is to start in local coordinates and to consider one deformation at a time. Figure 15.5 shows a frame element with axial displacement at the near end and no other deformations.

This is the same situation we saw for truss elements in Section 14.2: Transformation and Element Stiffness Matrices. Axial deformation causes only axial force:

$$q'_{Nx} = \frac{AE}{l} d'_{Nx}$$

where
- A = Cross-sectional area
- E = Modulus of elasticity
- l = Length of the element

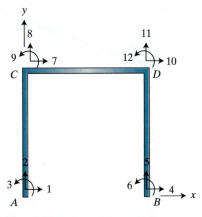

Figure 15.4 Standard numbering sequence for degrees of freedom for 2D frame nodes.

For the element to be in equilibrium, the axial force at the far end must be equal and opposite to the force at the near end, so we have

$$q'_{Fx} = -\frac{AE}{l} d'_{Nx}$$

All of the other internal forces at the ends of the element are zero. Because axial displacement does not cause shear, we say that the axial force and shear are *uncoupled*. Similarly, the axial force and moment are uncoupled.

If the axial displacement is at the far end (Figure 15.6), we still get only axial forces.

The relationships between axial displacement at the far end and axial forces at the two ends are

$$q'_{Nx} = -\frac{AE}{l} d'_{Fx}$$

$$q'_{Fx} = \frac{AE}{l} d'_{Fx}$$

Figure 15.5 Axial deformation at the near node of a frame element.

Figure 15.6 Axial deformation at the far node of a frame element.

Figure 15.7 Possible local y-axis directions for a 2D frame element.

Shear Deformation

Next, let's consider shear displacement at only the near end. Remember that we define the local x-axis as running from the near to the far node. That doesn't tell us which way the local y-axis points. We have two possibilities for the local y-axis (Figure 15.7).

The local z-axis is defined by the right-hand rule. So, if we choose down for the local y-axis in Figure 15.7, the local z-axis will point into the page. If we choose up for the local y-axis in Figure 15.7, the local z-axis will point out of the page, just like the global z-axis. Standard practice is actually to choose the local z-axis to match the global z-axis; then the local y-axis is set by default. For example, consider the frame element in Figure 15.8. The local x-axis is set by the definition of the near and far nodes. The local z-axis is set to be out of the page. In order to satisfy the right-hand rule, the local y-axis must be as shown.

Since we are considering only one deformation at a time, neither end can rotate when we apply shear displacement at the near end. In order for the frame element to take this shape, there must be both shear and moment at each end (Figure 15.9).

The frame element is 3° indeterminate, so we need to use the Force Method (Chapter 12) to find the reactions. Fortunately, someone has already done that for us. We can find those reactions inside the back cover of this text (Figure 15.10). Notice that no axial force is generated. We expect that because the axial force and shear are uncoupled. Both shear and moment end forces are generated, however, which means that shear and moment are coupled.

Figure 15.8 Example of how the local y-axis is set by the direction of the local x-axis and choosing the z-axis to be out of the page.

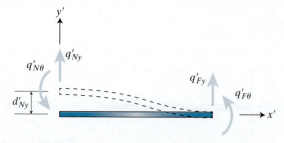

Figure 15.9 Shear deformation at the near node of a frame element.

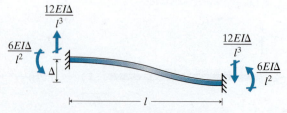

Figure 15.10 Reactions created by shear deformation at the near node of a frame element from inside the back cover of this text.

Once we adjust the signs to match our local axes, we get the following end forces due to shear displacement at the near end:

$$q'_{Ny} = \frac{12EI}{l^3} d'_{Ny}$$

$$q'_{N\theta} = \frac{6EI}{l^2} d'_{Ny}$$

$$q'_{Fy} = -\frac{12EI}{l^3} d'_{Ny}$$

$$q'_{F\theta} = \frac{6EI}{l^2} d'_{Ny}$$

where

I = Moment of inertia (about the z'-axis)

Shear displacement at the far end (Figure 15.11) creates the mirror image of the situation in Figure 15.10. The end forces in local coordinates are

$$q'_{Ny} = -\frac{12EI}{l^3} d'_{Fy}$$

$$q'_{N\theta} = -\frac{6EI}{l^2} d'_{Fy}$$

$$q'_{Fy} = \frac{12EI}{l^3} d'_{Fy}$$

$$q'_{F\theta} = -\frac{6EI}{l^2} d'_{Fy}$$

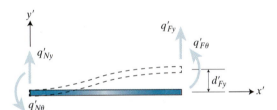

Figure 15.11 Shear deformation at the far node of a frame element.

Rotation

The final deformations are rotations at the ends. First, let's consider rotation at the near end (Figure 15.12). Remember that positive rotation is clockwise. In order for the frame element to take this shape, again there must be both shear and moment at each end. This observation has an

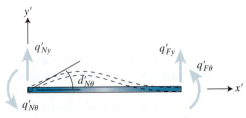

Figure 15.12 Rotation at the near node of a frame element.

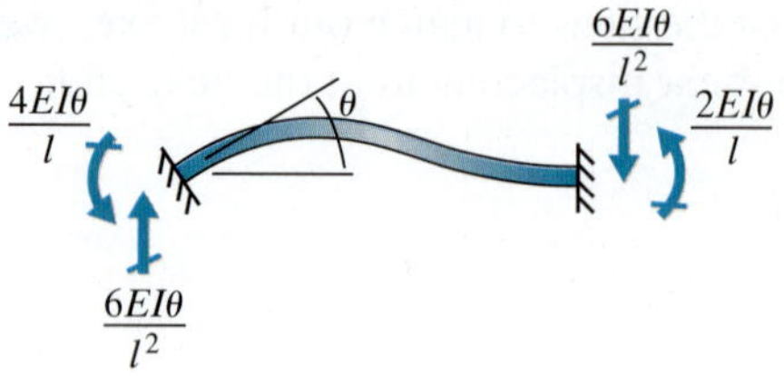

860 Chapter 15 Direct Stiffness Method for Frames

Figure 15.13 Reactions created by rotation at the near node of a frame element.

important implication: when two degrees of freedom are coupled, it does not matter which deforms—both experience internal forces.

Once again this is a 3° indeterminate member, and someone has already performed the analysis for us (Figure 15.13).

Once we adjust the signs to match our local axes, we get the following end forces due to rotation at the near node:

$$q'_{Ny} = \frac{6EI}{l^2} d'_{N\theta}$$

$$q'_{N\theta} = \frac{4EI}{l} d'_{N\theta}$$

$$q'_{Fy} = -\frac{6EI}{l^2} d'_{N\theta}$$

$$q'_{F\theta} = \frac{2EI}{l} d'_{N\theta}$$

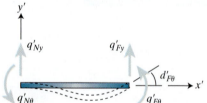

Figure 15.14 Rotation at the far node of a frame element.

Positive rotation at the far end produces the displaced shape in Figure 15.14. Reorienting the member in Figure 15.13 leads us to the following end forces due to rotation at the far node:

$$q'_{Ny} = \frac{6EI}{l^2} d'_{F\theta}$$

$$q'_{N\theta} = \frac{2EI}{l} d'_{F\theta}$$

$$q'_{Fy} = -\frac{6EI}{l^2} d'_{F\theta}$$

$$q'_{F\theta} = \frac{4EI}{l} d'_{F\theta}$$

Element Stiffness Matrix in Local Coordinates

The element stiffness matrix is the tool we use to calculate the internal forces generated by the nodal deformations. For a two-dimensional frame element, there are six internal forces and six nodal deformations, so the element stiffness matrix is 6 × 6. Each row is associated with a different internal force, so we assemble the matrix by superimposing the expressions for internal force that we derived in this section.

Section 15.1 Element Stiffness Matrix in Local Coordinates

$$[\mathbf{k}'] = \begin{array}{c} q'_{Nx} \\ q'_{Ny} \\ q'_{N\theta} \\ q'_{Fx} \\ q'_{Fy} \\ q'_{F\theta} \end{array} \begin{bmatrix} \dfrac{AE}{l} & 0 & 0 & -\dfrac{AE}{l} & 0 & 0 \\ 0 & \dfrac{12EI}{l^3} & \dfrac{6EI}{l^2} & 0 & -\dfrac{12EI}{l^3} & \dfrac{6EI}{l^2} \\ 0 & \dfrac{6EI}{l^2} & \dfrac{4EI}{l} & 0 & -\dfrac{6EI}{l^2} & \dfrac{2EI}{l} \\ -\dfrac{AE}{l} & 0 & 0 & \dfrac{AE}{l} & 0 & 0 \\ 0 & -\dfrac{12EI}{l^3} & -\dfrac{6EI}{l^2} & 0 & \dfrac{12EI}{l^3} & -\dfrac{6EI}{l^2} \\ 0 & \dfrac{6EI}{l^2} & \dfrac{2EI}{l} & 0 & -\dfrac{6EI}{l^2} & \dfrac{4EI}{l} \end{bmatrix} \begin{array}{c} d'_{Nx} \\ d'_{Ny} \\ d'_{N\theta} \\ d'_{Fx} \\ d'_{Fy} \\ d'_{F\theta} \end{array}$$

SECTION 15.1 HIGHLIGHTS

Element Stiffness Matrix in Local Coordinates

New and Modified Terms:

- d'_x = Axial displacement at the end of an element (local coordinates)
- d'_y = Shear displacement at the end of an element (local coordinates)
- d'_θ = Rotation at the end of an element (local coordinates)
- I = Moment of inertia (bending about the z'-axis)
- q'_x = Axial force at the end of an element (local coordinates)
- q'_y = Shear force at the end of an element (local coordinates)
- q'_θ = Moment at the end of an element (local coordinates)

Element Stiffnesses in Local Coordinates:

$$[\mathbf{k}'] = \begin{array}{c} q'_{Nx} \\ q'_{Ny} \\ q'_{N\theta} \\ q'_{Fx} \\ q'_{Fy} \\ q'_{F\theta} \end{array} \begin{bmatrix} \dfrac{AE}{l} & 0 & 0 & -\dfrac{AE}{l} & 0 & 0 \\ 0 & \dfrac{12EI}{l^3} & \dfrac{6EI}{l^2} & 0 & -\dfrac{12EI}{l^3} & \dfrac{6EI}{l^2} \\ 0 & \dfrac{6EI}{l^2} & \dfrac{4EI}{l} & 0 & -\dfrac{6EI}{l^2} & \dfrac{2EI}{l} \\ -\dfrac{AE}{l} & 0 & 0 & \dfrac{AE}{l} & 0 & 0 \\ 0 & -\dfrac{12EI}{l^3} & -\dfrac{6EI}{l^2} & 0 & \dfrac{12EI}{l^3} & -\dfrac{6EI}{l^2} \\ 0 & \dfrac{6EI}{l^2} & \dfrac{2EI}{l} & 0 & -\dfrac{6EI}{l^2} & \dfrac{4EI}{l} \end{bmatrix} \begin{array}{c} d'_{Nx} \\ d'_{Ny} \\ d'_{N\theta} \\ d'_{Fx} \\ d'_{Fy} \\ d'_{F\theta} \end{array}$$

15.2 Element Stiffness Matrix in Global Coordinates

Introduction

To convert the element stiffness matrix from local coordinates to global coordinates, we need to use the transformation matrix. The derivation process is basically the same as we used in Chapter 14: Direct Stiffness Method for Trusses, but the matrices have different sizes and entries than those for a truss.

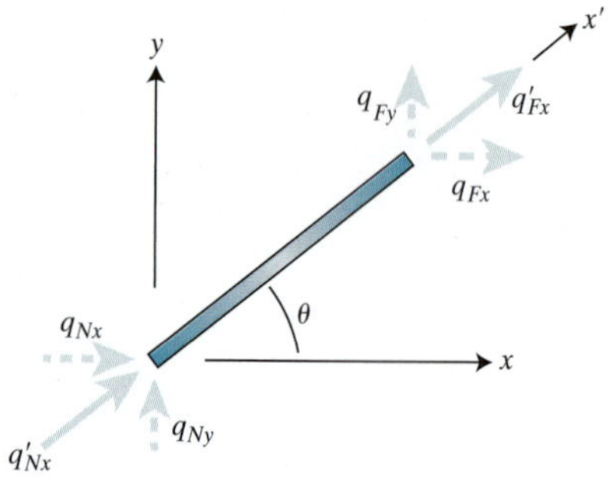

How-To

Transformation Matrix

The transformation matrix converts a vector from global coordinates to local coordinates. The transpose of the transformation matrix reverses the process; it converts from local to global coordinates. We need both in order to create the element stiffness matrix in global coordinates. Let's derive the transpose of the transformation matrix by considering internal forces. To do that, we can consider one internal force at a time and superimpose the results.

Figure 15.15 shows a frame element with only axial forces at the ends. We can break each axial force into its global x- and y-components using the angle between the global x-axis and the local x-axis.

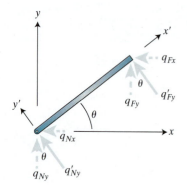

Figure 15.15 Frame element with axial forces broken into global coordinates.

These are the resulting components:

$$q_{Nx} = q'_{Nx} \cos\theta$$

$$q_{Ny} = q'_{Nx} \sin\theta$$

$$q_{Fx} = q'_{Fx} \cos\theta$$

$$q_{Fy} = q'_{Fx} \sin\theta$$

Similarly, we can consider only the shear forces at the ends (Figure 15.16). We can break each shear force into its global components, but notice that the global x-components are negative.

This process results in the following components of the shear forces:

$$q_{Nx} = q'_{Ny}(-\sin\theta)$$

$$q_{Ny} = q'_{Ny} \cos\theta$$

$$q_{Fx} = q'_{Fy}(-\sin\theta)$$

$$q_{Fy} = q'_{Fy} \cos\theta$$

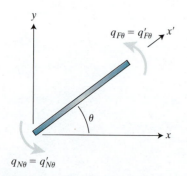

Figure 15.16 Frame element with shear forces broken into global coordinates.

In Figure 15.17, we see a frame element with a moment at each of the ends.

Figure 15.17 Frame element with a moment at each end.

Section 15.2 Element Stiffness Matrix in Global Coordinates

Because we chose to have the local z-axis point in the same direction as the global z-axis, the global moment is exactly the same as the local moment. That means we have the following transformation:

$$q_{N\theta} = q'_{N\theta}(1)$$
$$q_{F\theta} = q'_{F\theta}(1)$$

When we superimpose all of the transformation expressions, we get the following system of equations:

$$q_{Nx} = q'_{Nx}\cos\theta + q'_{Ny}(-\sin\theta)$$
$$q_{Ny} = q'_{Nx}\sin\theta + q'_{Ny}\cos\theta$$
$$q_{Fx} = q'_{Fx}\cos\theta + q'_{Fy}(-\sin\theta)$$
$$q_{Fy} = q'_{Fx}\sin\theta + q'_{Fy}\cos\theta$$
$$q_{N\theta} = q'_{N\theta}(1)$$
$$q_{F\theta} = q'_{F\theta}(1)$$

We can express this system of equations in matrix form as

$$[\mathbf{q}] = [\mathbf{T}]^T[\mathbf{q}']$$

where

$$[\mathbf{q}] = \begin{bmatrix} q_{Nx} \\ q_{Ny} \\ q_{N\theta} \\ q_{Fx} \\ q_{Fy} \\ q_{F\theta} \end{bmatrix} = \text{Internal forces for a frame element in global coordinates}$$

$$[\mathbf{T}]^T = \begin{bmatrix} \cos\theta & -\sin\theta & 0 & 0 & 0 & 0 \\ \sin\theta & \cos\theta & 0 & 0 & 0 & 0 \\ 0 & 0 & 1 & 0 & 0 & 0 \\ 0 & 0 & 0 & \cos\theta & -\sin\theta & 0 \\ 0 & 0 & 0 & \sin\theta & \cos\theta & 0 \\ 0 & 0 & 0 & 0 & 0 & 1 \end{bmatrix} = \text{Transpose of the transformation matrix for a frame element}$$

$$[\mathbf{q}'] = \begin{bmatrix} q'_{Nx} \\ q'_{Ny} \\ q'_{N\theta} \\ q'_{Fx} \\ q'_{Fy} \\ q'_{F\theta} \end{bmatrix} = \text{Internal forces for a frame element in local coordinates}$$

Here is the transformation matrix, which converts from global to local coordinates:

$$[\mathbf{T}] = \begin{bmatrix} \cos\theta & \sin\theta & 0 & 0 & 0 & 0 \\ -\sin\theta & \cos\theta & 0 & 0 & 0 & 0 \\ 0 & 0 & 1 & 0 & 0 & 0 \\ 0 & 0 & 0 & \cos\theta & \sin\theta & 0 \\ 0 & 0 & 0 & -\sin\theta & \cos\theta & 0 \\ 0 & 0 & 0 & 0 & 0 & 1 \end{bmatrix}$$

Element Stiffness Matrix in Global Coordinates

To convert the element stiffness matrix from local to global coordinates, we use both the transformation matrix and its transpose:

$$[\mathbf{k}] = [\mathbf{T}]^T [\mathbf{k}'] [\mathbf{T}]$$

After we multiply the matrices together, we get an expression for the element stiffness matrix for a two-dimensional frame element in global coordinates. Because the expression gets rather large, we typically use symbols for the directional cosines:

$$\lambda_x = \cos\theta$$

$$\lambda_y = \sin\theta$$

This results in the following stiffness matrix expression:

$$[\mathbf{k}] = \begin{array}{c} q_{Nx} \\ q_{Ny} \\ q_{N\theta} \\ q_{Fx} \\ q_{Fy} \\ q_{F\theta} \end{array} \begin{bmatrix} \left(\frac{AE}{l}\lambda_x^2 + \frac{12EI}{l^3}\lambda_y^2\right) & \left(\frac{AE}{l} - \frac{12EI}{l^3}\right)\lambda_x\lambda_y & -\frac{6EI}{l^2}\lambda_y & -\left(\frac{AE}{l}\lambda_x^2 + \frac{12EI}{l^3}\lambda_y^2\right) & -\left(\frac{AE}{l} - \frac{12EI}{l^3}\right)\lambda_x\lambda_y & -\frac{6EI}{l^2}\lambda_y \\ \left(\frac{AE}{l} - \frac{12EI}{l^3}\right)\lambda_x\lambda_y & \left(\frac{AE}{l}\lambda_y^2 + \frac{12EI}{l^3}\lambda_x^2\right) & \frac{6EI}{l^2}\lambda_x & -\left(\frac{AE}{l} - \frac{12EI}{l^3}\right)\lambda_x\lambda_y & -\left(\frac{AE}{l}\lambda_y^2 + \frac{12EI}{l^3}\lambda_x^2\right) & \frac{6EI}{l^2}\lambda_x \\ -\frac{6EI}{l^2}\lambda_y & \frac{6EI}{l^2}\lambda_x & \frac{4EI}{l} & \frac{6EI}{l^2}\lambda_y & -\frac{6EI}{l^2}\lambda_x & \frac{2EI}{l} \\ -\left(\frac{AE}{l}\lambda_x^2 + \frac{12EI}{l^3}\lambda_y^2\right) & -\left(\frac{AE}{l} - \frac{12EI}{l^3}\right)\lambda_x\lambda_y & \frac{6EI}{l^2}\lambda_y & \left(\frac{AE}{l}\lambda_x^2 + \frac{12EI}{l^3}\lambda_y^2\right) & \left(\frac{AE}{l} - \frac{12EI}{l^3}\right)\lambda_x\lambda_y & \frac{6EI}{l^2}\lambda_y \\ -\left(\frac{AE}{l} - \frac{12EI}{l^3}\right)\lambda_x\lambda_y & -\left(\frac{AE}{l}\lambda_y^2 + \frac{12EI}{l^3}\lambda_x^2\right) & -\frac{6EI}{l^2}\lambda_x & \left(\frac{AE}{l} - \frac{12EI}{l^3}\right)\lambda_x\lambda_y & \left(\frac{AE}{l}\lambda_y^2 + \frac{12EI}{l^3}\lambda_x^2\right) & -\frac{6EI}{l^2}\lambda_x \\ -\frac{6EI}{l^2}\lambda_y & \frac{6EI}{l^2}\lambda_x & \frac{2EI}{l} & \frac{6EI}{l^2}\lambda_y & -\frac{6EI}{l^2}\lambda_x & \frac{4EI}{l} \end{bmatrix}$$

(columns: d_{Nx}, d_{Ny}, $d_{N\theta}$, d_{Fx}, d_{Fy}, $d_{F\theta}$)

Just as with trusses, we need to be consistent about our choice of units when compiling the element stiffness matrix. The easiest way to do that is to choose one force unit (e.g., kN, lb, or k) and one length unit (e.g., cm, mm, or in.), and then use only the chosen units for each of the contributing properties: A, E, I, and l. The resulting deformations and internal forces will be in those units.

Section 15.2 Highlights

Element Stiffness Matrix in Global Coordinates

Transformation Matrix:

$$[d'] = [T][d] \quad [d] = [T]^T[d']$$
$$[q'] = [T][q] \quad [q] = [T]^T[q']$$

$$[T] = \begin{bmatrix} \cos\theta & \sin\theta & 0 & 0 & 0 & 0 \\ -\sin\theta & \cos\theta & 0 & 0 & 0 & 0 \\ 0 & 0 & 1 & 0 & 0 & 0 \\ 0 & 0 & 0 & \cos\theta & \sin\theta & 0 \\ 0 & 0 & 0 & -\sin\theta & \cos\theta & 0 \\ 0 & 0 & 0 & 0 & 0 & 1 \end{bmatrix}$$

$$[T]^T = \begin{bmatrix} \cos\theta & -\sin\theta & 0 & 0 & 0 & 0 \\ \sin\theta & \cos\theta & 0 & 0 & 0 & 0 \\ 0 & 0 & 1 & 0 & 0 & 0 \\ 0 & 0 & 0 & \cos\theta & -\sin\theta & 0 \\ 0 & 0 & 0 & \sin\theta & \cos\theta & 0 \\ 0 & 0 & 0 & 0 & 0 & 1 \end{bmatrix}$$

Element Stiffnesses in Global Coordinates:

$$[k] = \begin{matrix} q_{Nx} \\ q_{Ny} \\ q_{N\theta} \\ q_{Fx} \\ q_{Fy} \\ q_{F\theta} \end{matrix} \begin{bmatrix} \left(\frac{AE}{l}\lambda_x^2 + \frac{12EI}{l^3}\lambda_y^2\right) & \left(\frac{AE}{l} - \frac{12EI}{l^3}\right)\lambda_x\lambda_y & -\frac{6EI}{l^2}\lambda_y & -\left(\frac{AE}{l}\lambda_x^2 + \frac{12EI}{l^3}\lambda_y^2\right) & -\left(\frac{AE}{l} - \frac{12EI}{l^3}\right)\lambda_x\lambda_y & -\frac{6EI}{l^2}\lambda_y \\ \left(\frac{AE}{l} - \frac{12EI}{l^3}\right)\lambda_x\lambda_y & \left(\frac{AE}{l}\lambda_y^2 + \frac{12EI}{l^3}\lambda_x^2\right) & \frac{6EI}{l^2}\lambda_x & -\left(\frac{AE}{l} - \frac{12EI}{l^3}\right)\lambda_x\lambda_y & -\left(\frac{AE}{l}\lambda_y^2 + \frac{12EI}{l^3}\lambda_x^2\right) & \frac{6EI}{l^2}\lambda_x \\ -\frac{6EI}{l^2}\lambda_y & \frac{6EI}{l^2}\lambda_x & \frac{4EI}{l} & \frac{6EI}{l^2}\lambda_y & -\frac{6EI}{l^2}\lambda_x & \frac{2EI}{l} \\ -\left(\frac{AE}{l}\lambda_x^2 + \frac{12EI}{l^3}\lambda_y^2\right) & -\left(\frac{AE}{l} - \frac{12EI}{l^3}\right)\lambda_x\lambda_y & \frac{6EI}{l^2}\lambda_y & \left(\frac{AE}{l}\lambda_x^2 + \frac{12EI}{l^3}\lambda_y^2\right) & \left(\frac{AE}{l} - \frac{12EI}{l^3}\right)\lambda_x\lambda_y & \frac{6EI}{l^2}\lambda_y \\ -\left(\frac{AE}{l} - \frac{12EI}{l^3}\right)\lambda_x\lambda_y & -\left(\frac{AE}{l}\lambda_y^2 + \frac{12EI}{l^3}\lambda_x^2\right) & -\frac{6EI}{l^2}\lambda_x & \left(\frac{AE}{l} - \frac{12EI}{l^3}\right)\lambda_x\lambda_y & \left(\frac{AE}{l}\lambda_y^2 + \frac{12EI}{l^3}\lambda_x^2\right) & -\frac{6EI}{l^2}\lambda_x \\ -\frac{6EI}{l^2}\lambda_y & \frac{6EI}{l^2}\lambda_x & \frac{2EI}{l} & \frac{6EI}{l^2}\lambda_y & -\frac{6EI}{l^2}\lambda_x & \frac{4EI}{l} \end{bmatrix}$$

$$\lambda_x = \cos\theta = \frac{x_F - x_N}{\sqrt{(x_F - x_N)^2 + (y_F - y_N)^2}}$$

$$\lambda_y = \sin\theta = \frac{y_F - y_N}{\sqrt{(x_F - x_N)^2 + (y_F - y_N)^2}}$$

Special Considerations:

- Make sure all the stiffness coefficients have units that are consistent with the units for the deformations and internal forces we are calculating.
- Carry four significant figures in order to have adequate accuracy.

866 Chapter 15 Direct Stiffness Method for Frames

EXAMPLE 15.1

Our firm just purchased the newest version of structural analysis software from our preferred vendor. Standard procedure in the office is to test the software with several problems before putting it into general use.

We have been asked to use the Direct Stiffness Method by hand to find the deformations, reactions, and internal forces at the nodes for an asymmetric frame subjected to gravity load. Another engineer on our team is doing the same thing using the software. For the purposes of this test, we consider the supports to be fixed.

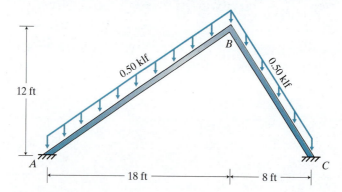

Figure 15.18

Our first task is to find the element stiffness matrices.

Member properties:

HSS8 × 3 × 3/8:	$A = 6.88$ in²	$I = 48.5$ in⁴
Steel:	$E = 29{,}000$ ksi	

Element Stiffnesses in Global Coordinates:

Choose the units:
   Let's use inches and kips for all units in order ensure consistency.
Define the near and far nodes:
   Let's work alphabetically across each element.
Define the DOF numbers:
   Let's use the standard numbering sequence at each node and work alphabetically across the structure.

Section 15.2 Element Stiffness Matrix in Global Coordinates

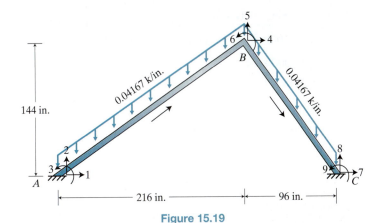

Figure 15.19

Element AB:

$$l_{AB} = \sqrt{(216 \text{ in.})^2 + (144 \text{ in.})^2} = 259.6 \text{ in.}$$

$$\lambda_x = \frac{216 \text{ in.} - 0 \text{ in.}}{\sqrt{(216 \text{ in.} - 0 \text{ in.})^2 + (144 \text{ in.} - 0 \text{ in.})^2}} = 0.8321$$

$$\lambda_y = \frac{144 \text{ in.} - 0 \text{ in.}}{\sqrt{(216 \text{ in.} - 0 \text{ in.})^2 + (144 \text{ in.} - 0 \text{ in.})^2}} = 0.5547$$

$$\frac{AE}{l} = \frac{(6.88 \text{ in}^2)(29{,}000 \text{ ksi})}{259.6 \text{ in.}} = 768.6 \text{ k/in.}$$

$$\frac{12EI}{l^3} = \frac{12(29{,}000 \text{ ksi})(48.5 \text{ in}^4)}{(259.6 \text{ in.})^3} = 0.9647 \text{ k/in.}$$

$$\frac{6EI}{l^2} = \frac{6(29{,}000 \text{ ksi})(48.5 \text{ in}^4)}{(259.6 \text{ in.})^2} = 125.22 \text{ k}$$

$$\frac{4EI}{l} = \frac{4(29{,}000 \text{ ksi})(48.5 \text{ in}^4)}{259.6 \text{ in.}} = 21{,}670 \text{ k} \cdot \text{in.}$$

$$\frac{2EI}{l} = \frac{2(29{,}000 \text{ ksi})(48.5 \text{ in}^4)}{259.6 \text{ in.}} = 10{,}836 \text{ k} \cdot \text{in.}$$

$$[\mathbf{k}_{AB}] = \begin{array}{c} \\ 1 \\ 2 \\ 3 \\ 4 \\ 5 \\ 6 \end{array} \begin{bmatrix} 1 & 2 & 3 & 4 & 5 & 6 \\ 532.5 \text{ k/in.} & 354.3 \text{ k/in.} & -69.46 \text{ k} & -532.5 \text{ k/in.} & -354.3 \text{ k/in.} & -69.46 \text{ k} \\ 354.3 \text{ k/in.} & 237.2 \text{ k/in.} & 104.20 \text{ k} & -354.3 \text{ k/in.} & -237.2 \text{ k/in.} & 104.20 \text{ k} \\ -69.46 \text{ k} & 104.20 \text{ k} & 21670 \text{ k} \cdot \text{in.} & 69.46 \text{ k} & -104.20 \text{ k} & 10836 \text{ k} \cdot \text{in.} \\ -532.5 \text{ k/in.} & -354.3 \text{ k/in.} & 69.46 \text{ k} & 532.5 \text{ k/in.} & 354.3 \text{ k/in.} & 69.46 \text{ k} \\ -354.3 \text{ k/in.} & -237.2 \text{ k/in.} & -104.20 \text{ k} & 354.3 \text{ k/in.} & 237.2 \text{ k/in.} & -104.20 \text{ k} \\ -69.46 \text{ k} & 104.20 \text{ k} & 10836 \text{ k} \cdot \text{in.} & 69.46 \text{ k} & -104.20 \text{ k} & 21670 \text{ k} \cdot \text{in.} \end{bmatrix}$$

Element BC:

$$l_{BC} = \sqrt{(96 \text{ in.})^2 + (144 \text{ in.})^2} = 173.07 \text{ in.}$$

$$\lambda_x = \frac{312 \text{ in.} - 216 \text{ in.}}{\sqrt{(312 \text{ in.} - 216 \text{ in.})^2 + (0 \text{ in.} - 144 \text{ in.})^2}} = 0.5547$$

$$\lambda_y = \frac{0 \text{ in.} - 144 \text{ in.}}{\sqrt{(312 \text{ in.} - 216 \text{ in.})^2 + (0 \text{ in.} - 144 \text{ in.})^2}} = -0.8321$$

$$\frac{AE}{l} = \frac{(6.88 \text{ in}^2)(29{,}000 \text{ ksi})}{173.07 \text{ in.}} = 1152.8 \text{ k/in.}$$

$$\frac{12EI}{l^3} = \frac{12(29{,}000 \text{ ksi})(48.5 \text{ in}^4)}{(173.07 \text{ in.})^3} = 3.256 \text{ k/in.}$$

$$\frac{6EI}{l^2} = \frac{6(29{,}000 \text{ ksi})(48.5 \text{ in}^4)}{(173.07 \text{ in.})^2} = 281.7 \text{ k}$$

$$\frac{4EI}{l} = \frac{4(29{,}000 \text{ ksi})(48.5 \text{ in}^4)}{173.07 \text{ in.}} = 32{,}510 \text{ k·in.}$$

$$\frac{2EI}{l} = \frac{2(29{,}000 \text{ ksi})(48.5 \text{ in}^4)}{173.07 \text{ in.}} = 16{,}254 \text{ k·in.}$$

$$[\mathbf{k}_{BC}] = \begin{array}{c} 4 \\ 5 \\ 6 \\ 7 \\ 8 \\ 9 \end{array} \begin{bmatrix} \overset{4}{357.0 \text{ k/in.}} & \overset{5}{-530.6 \text{ k/in.}} & \overset{6}{234.4 \text{ k}} & \overset{7}{-357.0 \text{ k/in.}} & \overset{8}{530.6 \text{ k/in.}} & \overset{9}{234.4 \text{ k}} \\ -530.6 \text{ k/in.} & 799.2 \text{ k/in.} & 156.26 \text{ k} & 530.6 \text{ k/in.} & -799.2 \text{ k/in.} & 156.26 \text{ k} \\ 234.4 \text{ k} & 156.26 \text{ k} & 32510 \text{ k·in.} & -234.4 \text{ k} & -156.26 \text{ k} & 16254 \text{ k·in.} \\ -357.0 \text{ k/in.} & 530.6 \text{ k/in.} & -234.4 \text{ k} & 357.0 \text{ k/in.} & -530.6 \text{ k/in.} & -234.4 \text{ k} \\ 530.6 \text{ k/in.} & -799.2 \text{ k/in.} & -156.26 \text{ k} & -530.6 \text{ k/in.} & 799.2 \text{ k/in.} & -156.26 \text{ k} \\ 234.4 \text{ k} & 156.26 \text{ k} & 16254 \text{ k·in.} & -234.4 \text{ k} & -156.26 \text{ k} & 32510 \text{ k·in.} \end{bmatrix}$$

15.3 Loads Between Nodes

Introduction

For trusses, most loads are applied at the nodes, but that is not true for frames. Beams often carry distributed loads from diaphragms. Columns sometimes experience lateral loads along their length. We deal with these loads between nodes the same way we dealt with temperature changes and length changes in Chapter 14: convert the loading to nodal equivalent loads.

How-To

In order to use the Direct Stiffness Method for frames that have loads between nodes, we superimpose end forces due to deformations with fixed-end forces:

$$[Q] = [K][D] + [Q_o]$$

This is the same process we followed for trusses in Section 14.5: Additional Loadings.

Fixed-end forces are the reactions at the ends of members that are fixed supported at both ends (e.g., Figure 15.20). The internal forces at the ends of the member are identical to the fixed-end forces, but we must pay attention to our local coordinate sign convention.

Section 15.3 Loads Between Nodes

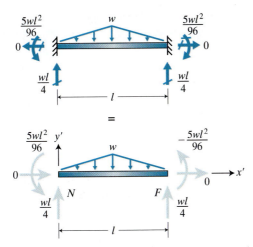

Figure 15.20 Fixed-end forces are reactions for a fixed-fixed supported member. They are identical to the internal forces at the ends of the member.

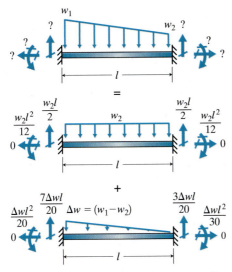

Figure 15.21 Trapezoidal distributed load is also the superposition of uniform and triangular distributed loads.

We start by calculating the fixed-end forces in local coordinates for each member that experiences loads between its nodes. To find the fixed-end forces, we can use the tables inside the back cover of this text for many common loadings, or we can use the Force Method (Chapter 12) for loadings that are not covered there. In some cases, we can break the actual loading into a superposition of multiple loadings that are inside the back cover of this text. For example, a trapezoidal distributed load is not included in the loadings inside the back cover. However, we can break that loading into a uniform distributed load and a triangular distributed load (Figure 15.21). Both of those are included inside the back cover.

If a frame member experiences a temperature change or a length change, the fixed-end forces are the same as we calculated in Section 14.5: Additional Loadings (Figure 15.22). This is because those two loadings cause an axial force that is uncoupled from shear and moment.

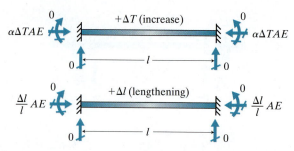

Figure 15.22 Fixed-end forces due to temperature change and length change.

Before we can consolidate all the fixed-end forces for the structure, we need to convert them to global coordinates. We use the transpose of the transformation matrix to do that:

$$[\mathbf{q}_o] = [\mathbf{T}]^T[\mathbf{q}'_o]$$

Once we have all the fixed-end forces in global coordinates, we add them, one DOF at a time:

$$[Q_o] = \Sigma q_o$$

Then we compile the Q_o for the entire structure, $[\mathbf{Q}_o]$.

SECTION 15.3 HIGHLIGHTS

Loads Between Nodes

Convert to nodal equivalent loads:

Use fixed-end forces
- Inside back cover
- Superposition
- Force Method

Temperature or length change

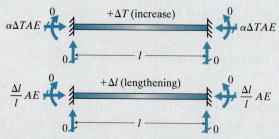

Convert to global coordinates and compile for structure:

$$[\mathbf{q}_o] = [\mathbf{T}]^T[\mathbf{q}'_o]$$

$$\begin{bmatrix} q_{Nxo} \\ q_{Nyo} \\ q_{N\theta o} \\ q_{Fxo} \\ q_{Fyo} \\ q_{F\theta o} \end{bmatrix} = \begin{bmatrix} \cos\theta & -\sin\theta & 0 & 0 & 0 & 0 \\ \sin\theta & \cos\theta & 0 & 0 & 0 & 0 \\ 0 & 0 & 1 & 0 & 0 & 0 \\ 0 & 0 & 0 & \cos\theta & -\sin\theta & 0 \\ 0 & 0 & 0 & \sin\theta & \cos\theta & 0 \\ 0 & 0 & 0 & 0 & 0 & 1 \end{bmatrix} \begin{bmatrix} q'_{Nxo} \\ q'_{Nyo} \\ q'_{N\theta o} \\ q'_{Fxo} \\ q'_{Fyo} \\ q'_{F\theta o} \end{bmatrix}$$

Compile by DOF number:

$$[\mathbf{Q}_o] = \Sigma[\mathbf{q}_o]$$

Example 15.2

Our firm just purchased the newest version of structural analysis software from our preferred vendor. Standard procedure in the office is to test the software with several problems before putting it into general use.

We have been asked to use the Direct Stiffness Method by hand to find the deformations, reactions, and internal forces at the nodes for an asymmetric frame subjected to gravity load. Another engineer on our team is doing the same thing using the software. For the purposes of this test, we consider the supports to be fixed.

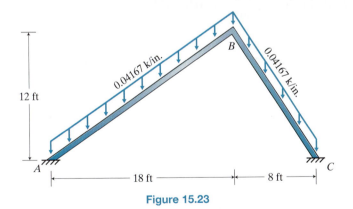

Figure 15.23

Before analyzing the frame, we need to find the nodal equivalent loads.

Nodal Equivalent Loads:

Member AB:
Use superposition to break down the loading:

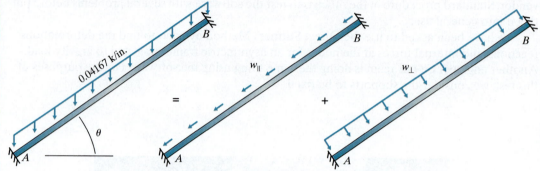

Figure 15.24

Consider the loading as a vector:

Figure 15.25

$$w_\| = w \sin\theta$$
$$= (0.04167 \text{ k/in.})\frac{144 \text{ in.}}{259.6 \text{ in.}} = 0.02311 \text{ k/in.}$$

$$w_\perp = w \cos\theta$$
$$= (0.04167 \text{ k/in.})\frac{216 \text{ in.}}{259.6 \text{ in.}} = 0.03467 \text{ k/in.}$$

Loading along the member:
Fixed-end forces are from the table inside the back cover.

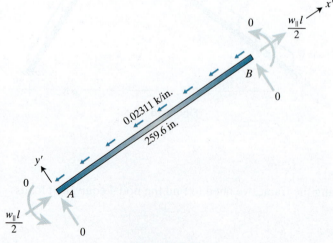

Figure 15.26

$$q'_{Nxo} = \frac{w_\| l}{2} = \frac{(0.02311 \text{ k/in.})(259.6 \text{ in.})}{2} = 3.000 \text{ k}$$

$$q'_{Nyo} = 0$$

$$q'_{N\theta o} = 0$$

$$q'_{Fxo} = \frac{w_\| l}{2} = \frac{(0.02311 \text{ k/in.})(259.6 \text{ in.})}{2} = 3.000 \text{ k}$$

$$q'_{Fyo} = 0$$

$$q'_{F\theta o} = 0$$

Loading perpendicular to the member:
Fixed-end forces are from the table inside the back cover.

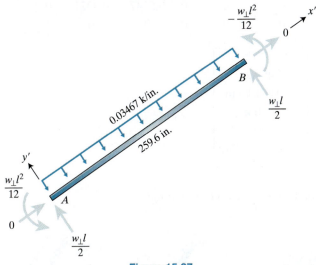

Figure 15.27

$$q'_{Nxo} = 0$$

$$q'_{Nyo} = \frac{w_\perp l}{2} = \frac{(0.03467 \text{ k/in.})(259.6 \text{ in.})}{2} = 4.500 \text{ k}$$

$$q'_{N\theta o} = \frac{w_\perp l^2}{12} = \frac{(0.03467 \text{ k/in.})(259.6 \text{ in.})^2}{12} = 194.71 \text{ k} \cdot \text{in.}$$

$$q'_{Fxo} = 0$$

$$q'_{Fyo} = \frac{w_\perp l}{2} = \frac{(0.03467 \text{ k/in.})(259.6 \text{ in.})}{2} = 4.500 \text{ k}$$

$$q'_{F\theta o} = -\frac{w_\perp l^2}{12} = -\frac{(0.03467 \text{ k/in.})(259.6 \text{ in.})^2}{12} = -194.71 \text{ k} \cdot \text{in.}$$

Superimpose the results:

$$\begin{bmatrix} q'_{Nxo} \\ q'_{Nyo} \\ q'_{N\theta o} \\ q'_{Fxo} \\ q'_{Fyo} \\ q'_{F\theta o} \end{bmatrix} = \begin{bmatrix} 3.000 \text{ k} \\ 0 \\ 0 \\ 3.000 \text{ k} \\ 0 \\ 0 \end{bmatrix} + \begin{bmatrix} 0 \\ 4.500 \text{ k} \\ 194.71 \text{ k} \cdot \text{in.} \\ 0 \\ 4.500 \text{ k} \\ -194.71 \text{ k} \cdot \text{in.} \end{bmatrix} = \begin{bmatrix} 3.000 \text{ k} \\ 4.500 \text{ k} \\ 194.71 \text{ k} \cdot \text{in.} \\ 3.000 \text{ k} \\ 4.500 \text{ k} \\ -194.71 \text{ k} \cdot \text{in.} \end{bmatrix}$$

Convert to global coordinates:

$$\cos\theta = \frac{216 \text{ in.} - 0 \text{ in.}}{\sqrt{(216 \text{ in.} - 0 \text{ in.})^2 + (144 \text{ in.} - 0 \text{ in.})^2}} = 0.8321$$

$$\sin\theta = \frac{144 \text{ in.} - 0 \text{ in.}}{\sqrt{(216 \text{ in.} - 0 \text{ in.})^2 + (144 \text{ in.} - 0 \text{ in.})^2}} = 0.5547$$

$$\begin{bmatrix} q_{Nxo} \\ q_{Nyo} \\ q_{N\theta o} \\ q_{Fxo} \\ q_{Fyo} \\ q_{F\theta o} \end{bmatrix} = \begin{bmatrix} 0.8321 & -0.5547 & 0 & 0 & 0 & 0 \\ 0.5547 & 0.8321 & 0 & 0 & 0 & 0 \\ 0 & 0 & 1 & 0 & 0 & 0 \\ 0 & 0 & 0 & 0.8321 & -0.5547 & 0 \\ 0 & 0 & 0 & 0.5547 & 0.8321 & 0 \\ 0 & 0 & 0 & 0 & 0 & 1 \end{bmatrix} \begin{bmatrix} 3.000 \text{ k} \\ 4.500 \text{ k} \\ 194.71 \text{ k} \cdot \text{in.} \\ 3.000 \text{ k} \\ 4.500 \text{ k} \\ -194.71 \text{ k} \cdot \text{in.} \end{bmatrix}$$

$$= \begin{bmatrix} 0 \text{ k} \\ 5.409 \text{ k} \\ 194.71 \text{ k} \cdot \text{in.} \\ 0 \text{ k} \\ 5.409 \text{ k} \\ -194.71 \text{ k} \cdot \text{in.} \end{bmatrix}$$

Member BC:

Use superposition to break down the loading:

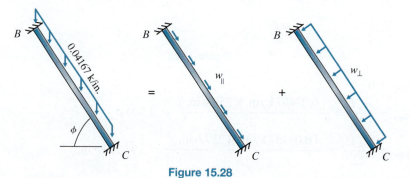

Figure 15.28

Consider the loading as a vector:

Figure 15.29

$w_{\parallel} = w \sin\phi$

$\quad = (0.04167 \text{ k/in.}) \dfrac{144 \text{ in.}}{173.07 \text{ in.}} = 0.03467 \text{ k/in.}$

$w_{\perp} = w \cos\phi$

$\quad = (0.04167 \text{ k/in.}) \dfrac{96 \text{ in.}}{173.07 \text{ in.}} = 0.02311 \text{ k/in.}$

Loading along the member:
Fixed-end forces are from the table inside the back cover.

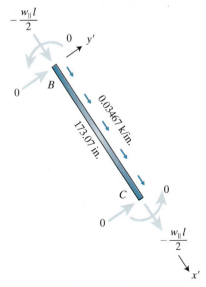

Figure 15.30

$q'_{Nxo} = -\dfrac{w_{\parallel}l}{2} = -\dfrac{(0.03467 \text{ k/in.})(173.07 \text{ in.})}{2} = -3.000 \text{ k}$

$q'_{Nyo} = 0$

$q'_{N\theta o} = 0$

$q'_{Fxo} = -\dfrac{w_{\parallel}l}{2} = -\dfrac{(0.03467 \text{ k/in.})(173.07 \text{ in.})}{2} = -3.000 \text{ k}$

$q'_{Fyo} = 0$

$q'_{F\theta o} = 0$

Loading perpendicular to the member:
Fixed-end forces are from the table inside the back cover.

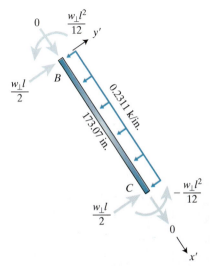

Figure 15.31

$$q'_{Nxo} = 0$$

$$q'_{Nyo} = \frac{w_\| l}{2} = \frac{(0.02311 \text{ k/in.})(173.07 \text{ in.})}{2} = 2.000 \text{ k}$$

$$q'_{N\theta o} = \frac{w_\perp l^2}{12} = \frac{(0.02311 \text{ k/in.})(173.07 \text{ in.})^2}{12} = 57.68 \text{ k} \cdot \text{in.}$$

$$q'_{Fxo} = 0$$

$$q'_{Fyo} = \frac{w_\| l}{2} = \frac{(0.02311 \text{ k/in.})(173.07 \text{ in.})}{2} = 2.000 \text{ k}$$

$$q'_{F\theta o} = -\frac{w_\perp l^2}{12} = -\frac{(0.02311 \text{ k/in.})(173.07 \text{ in.})^2}{12} = -57.68 \text{ k} \cdot \text{in.}$$

Superimpose the results:

$$\begin{bmatrix} q'_{Nxo} \\ q'_{Nyo} \\ q'_{N\theta o} \\ q'_{Fxo} \\ q'_{Fyo} \\ q'_{F\theta o} \end{bmatrix} = \begin{bmatrix} -3.000 \text{ k} \\ 0 \\ 0 \\ -3.000 \text{ k} \\ 0 \\ 0 \end{bmatrix} + \begin{bmatrix} 0 \\ 2.000 \text{ k} \\ 57.68 \text{ k} \cdot \text{in.} \\ 0 \\ 2.000 \text{ k} \\ -57.68 \text{ k} \cdot \text{in.} \end{bmatrix} = \begin{bmatrix} -3.000 \text{ k} \\ 2.000 \text{ k} \\ 57.68 \text{ k} \cdot \text{in.} \\ -3.000 \text{ k} \\ 2.000 \text{ k} \\ -57.68 \text{ k} \cdot \text{in.} \end{bmatrix}$$

Convert to global coordinates:

Note that θ in the formula for $[\mathbf{T}]^T$ is relative to the global axis, so it is not identical to the ϕ we labeled.

$$\cos \theta = \frac{312 \text{ in.} - 216 \text{ in.}}{\sqrt{(312 \text{ in.} - 216 \text{ in.})^2 + (0 \text{ in.} - 144 \text{ in.})^2}} = 0.5547$$

$$\sin \theta = \frac{0 \text{ in.} - 144 \text{ in.}}{\sqrt{(312 \text{ in.} - 216 \text{ in.})^2 + (0 \text{ in.} - 144 \text{ in.})^2}} = -0.8321$$

$$\begin{bmatrix} q_{Nxo} \\ q_{Nyo} \\ q_{N\theta o} \\ q_{Fxo} \\ q_{Fyo} \\ q_{F\theta o} \end{bmatrix} = \begin{bmatrix} 0.5547 & 0.8321 & 0 & 0 & 0 & 0 \\ -0.8321 & 0.5547 & 0 & 0 & 0 & 0 \\ 0 & 0 & 1 & 0 & 0 & 0 \\ 0 & 0 & 0 & 0.5547 & 0.8321 & 0 \\ 0 & 0 & 0 & -0.8321 & 0.5547 & 0 \\ 0 & 0 & 0 & 0 & 0 & 1 \end{bmatrix} \begin{bmatrix} -3.000 \text{ k} \\ 2.000 \text{ k} \\ 57.68 \text{ k} \cdot \text{in.} \\ -3.000 \text{ k} \\ 2.000 \text{ k} \\ -57.68 \text{ k} \cdot \text{in.} \end{bmatrix}$$

$$= \begin{bmatrix} 0 \text{ k} \\ 3.606 \text{ k} \\ 57.68 \text{ k} \cdot \text{in.} \\ 0 \text{ k} \\ 3.606 \text{ k} \\ -57.68 \text{ k} \cdot \text{in.} \end{bmatrix}$$

Assemble Nodal Equivalent Loads for Structure:

$[Q_o] = \Sigma [q_o]$

$$[Q_o] = \begin{array}{c} 1 \\ 2 \\ 3 \\ 4 \\ 5 \\ 6 \end{array} \begin{bmatrix} 0\,\text{k} \\ 5.409\,\text{k} \\ 194.71\,\text{k} \cdot \text{in.} \\ 0\,\text{k} \\ 5.409\,\text{k} \\ -194.71\,\text{k} \cdot \text{in.} \end{bmatrix} + \begin{array}{c} 4 \\ 5 \\ 6 \\ 7 \\ 8 \\ 9 \end{array} \begin{bmatrix} 0\,\text{k} \\ 3.606\,\text{k} \\ 57.68\,\text{k} \cdot \text{in.} \\ 0\,\text{k} \\ 3.606\,\text{k} \\ -57.68\,\text{k} \cdot \text{in.} \end{bmatrix} = \begin{array}{c} 1 \\ 2 \\ 3 \\ 4 \\ 5 \\ 6 \\ 7 \\ 8 \\ 9 \end{array} \begin{bmatrix} 0\,\text{k} \\ 5.409\,\text{k} \\ 194.71\,\text{k} \cdot \text{in.} \\ 0\,\text{k} \\ 9.015\,\text{k} \\ -137.03\,\text{k} \cdot \text{in.} \\ 0\,\text{k} \\ 3.606\,\text{k} \\ -57.68\,\text{k} \cdot \text{in.} \end{bmatrix}$$

15.4 Finding Deformations, Reactions, and Internal Forces

Introduction
One of the appeals of the Direct Stiffness Method is that the process for analyzing a determinate beam is exactly the same as the process for analyzing a highly indeterminate frame. The only change is the size of the system of equations that we must solve.

How-To
Once we have the element stiffness matrices in global coordinates, we assemble the structural stiffness matrix by summing all the coefficients. Remember that we add the coefficients based on degree-of-freedom numbers.

$$K_{ij} = \sum_{\substack{i,j=\text{all} \\ \text{DOF \#'s}}} k_{ij}$$

The assembled structural stiffness matrix, [K], should be symmetric because the Maxwell-Betti Reciprocal Theorem we discussed in Section 12.2: Multi-Degree Indeterminate Beams still applies. Verifying the symmetry about the diagonal is a great check against errors:

$$K_{ij} = K_{ji}, i \neq j$$

We can express the system of equilibrium equations with the following matrix notation:

$[Q] = [K][D] + [Q_o]$

Just as we discovered for trusses, we cannot solve this system of equations directly. Instead, we reorganize the equations to gather all of the free DOF Q together. We do that by shuffling the sequence of the equations. We also reorganize the terms in the equations to gather all of the free DOF D together. This gives us two subsets of equations that we can solve.

We solve the first subset for deformations:

$$[\mathbf{Q}_f] = [\mathbf{K}_{ff}][\mathbf{D}_f] + [\mathbf{K}_{fs}][\mathbf{D}_s] + [\mathbf{Q}_{fo}]$$
$$[\mathbf{D}_f] = [\mathbf{K}_{ff}]^{-1}([\mathbf{Q}_f] - [\mathbf{K}_{fs}][\mathbf{D}_s] - [\mathbf{Q}_{fo}])$$

Notice that the deformations are not limited to displacements. Frames have rotation degrees of freedom as well. Once we know the deformations, we can solve the second subset for reactions:

$$[\mathbf{Q}_s] = [\mathbf{K}_{sf}][\mathbf{D}_f] + [\mathbf{K}_{ss}][\mathbf{D}_s] + [\mathbf{Q}_{so}]$$

With the deformations, we can also find the internal forces at the ends of any element we want:

$$[\mathbf{q}'] = [\mathbf{k}'][\mathbf{T}][\mathbf{d}] + [\mathbf{q}'_o]$$

If we premultiply the $[\mathbf{k}']$ and $[\mathbf{T}]$ parts, we get the following expression for the internal forces at the ends of a two-dimensional frame element in local coordinates:

$$\begin{bmatrix} q'_{Nx} \\ q'_{Ny} \\ q'_{N\theta} \\ q'_{Fx} \\ q'_{Fy} \\ q'_{F\theta} \end{bmatrix} = \begin{bmatrix} \frac{AE}{l}\lambda_x & \frac{AE}{l}\lambda_y & 0 & -\frac{AE}{l}\lambda_x & -\frac{AE}{l}\lambda_y & 0 \\ -\frac{12EI}{l^3}\lambda_y & \frac{12EI}{l^3}\lambda_x & \frac{6EI}{l^2} & \frac{12EI}{l^3}\lambda_y & -\frac{12EI}{l^3}\lambda_x & \frac{6EI}{l^2} \\ -\frac{6EI}{l^2}\lambda_y & \frac{6EI}{l^2}\lambda_x & \frac{4EI}{l} & \frac{6EI}{l^2}\lambda_y & -\frac{6EI}{l^2}\lambda_x & \frac{2EI}{l} \\ -\frac{AE}{l}\lambda_x & -\frac{AE}{l}\lambda_y & 0 & \frac{AE}{l}\lambda_x & \frac{AE}{l}\lambda_y & 0 \\ \frac{12EI}{l^3}\lambda_y & -\frac{12EI}{l^3}\lambda_x & -\frac{6EI}{l^2} & -\frac{12EI}{l^3}\lambda_y & \frac{12EI}{l^3}\lambda_x & -\frac{6EI}{l^2} \\ -\frac{6EI}{l^2}\lambda_y & \frac{6EI}{l^2}\lambda_x & \frac{2EI}{l} & \frac{6EI}{l^2}\lambda_y & -\frac{6EI}{l^2}\lambda_x & \frac{4EI}{l} \end{bmatrix} \begin{bmatrix} d_{Nx} \\ d_{Ny} \\ d_{N\theta} \\ d_{Fx} \\ d_{Fy} \\ d_{F\theta} \end{bmatrix} + \begin{bmatrix} q'_{Nxo} \\ q'_{Nyo} \\ q'_{N\theta o} \\ q'_{Fxo} \\ q'_{Fyo} \\ q'_{F\theta o} \end{bmatrix}$$

Although the Direct Stiffness Method does not give us displaced shape or internal force diagrams directly, we can create them from the results we do obtain. Rarely do we generate those diagrams by hand from results of the Direct Stiffness Method; however, in order to program the Direct Stiffness Method, we need to know the process for developing the diagrams.

Since we know the loading on each element, we can use what we learned in Section 4.1: Internal Forces by Integration to develop expressions for the internal forces along each element. To find the constants of integration, we need internal force values at only one point in each element. The values at one end from $[\mathbf{q}']$ satisfy this requirement.

With equations for the moment along each element, we can use what we learned in Section 5.1: Double Integration Method to develop expressions for deformation along each element. We can find the constants of integration from the rotation and local y-displacement values at the near node:

$$[\mathbf{d}'] = [\mathbf{T}][\mathbf{d}]$$

The resulting expressions show how each element deforms between the nodes.

Section 15.4 Highlights

Finding Deformations, Reactions, and Internal Forces

Assembling the Structural Stiffness Matrix:

Add the stiffness coefficients from all elements, based on DOF numbers:

$$K_{ij} = \sum_{\substack{i,j=\text{all}\\ \text{DOF \#'s}}} k_{ij}$$

Finding Deformations:

Reorder the equations and terms to gather free and supported DOFs:

$$\begin{bmatrix} \mathbf{Q}_f \\ \mathbf{Q}_s \end{bmatrix} = \begin{bmatrix} \mathbf{K}_{ff} & \mathbf{K}_{fs} \\ \mathbf{K}_{sf} & \mathbf{K}_{ss} \end{bmatrix} \begin{bmatrix} \mathbf{D}_f \\ \mathbf{D}_s \end{bmatrix} + \begin{bmatrix} \mathbf{Q}_{fo} \\ \mathbf{Q}_{so} \end{bmatrix}$$

Solve the first subset of equations for deformations:

$$[\mathbf{Q}_f] = [\mathbf{K}_{ff}][\mathbf{D}_f] + [\mathbf{K}_{fs}][\mathbf{D}_s] + [\mathbf{Q}_{fo}]$$

$$[\mathbf{D}_f] = [\mathbf{K}_{ff}]^{-1}([\mathbf{Q}_f] - [\mathbf{K}_{fs}][\mathbf{D}_s] - [\mathbf{Q}_{fo}])$$

Finding Reactions:

Solve the second subset of equations for reactions:

$$[\mathbf{Q}_s] = [\mathbf{K}_{sf}][\mathbf{D}_f] + [\mathbf{K}_{ss}][\mathbf{D}_s] + [\mathbf{Q}_{so}]$$

Finding Internal Forces:

Use the deformations at the ends of the element and fixed-end forces to find the internal forces at the ends of an element.

Internal forces in local coordinates are typically calculated using deformations in global coordinates:

$$[\mathbf{q}'] = [\mathbf{k}'][\mathbf{T}][\mathbf{d}] + [\mathbf{q}'_o]$$

$$\begin{bmatrix} q'_{Nx} \\ q'_{Ny} \\ q'_{N\theta} \\ q'_{Fx} \\ q'_{Fy} \\ q'_{F\theta} \end{bmatrix} = \begin{bmatrix} \frac{AE}{l}\lambda_x & \frac{AE}{l}\lambda_y & 0 & -\frac{AE}{l}\lambda_x & -\frac{AE}{l}\lambda_y & 0 \\ -\frac{12EI}{l^3}\lambda_y & \frac{12EI}{l^3}\lambda_x & \frac{6EI}{l^2} & \frac{12EI}{l^3}\lambda_y & -\frac{12EI}{l^3}\lambda_x & \frac{6EI}{l^2} \\ -\frac{6EI}{l^2}\lambda_y & \frac{6EI}{l^2}\lambda_x & \frac{4EI}{l} & \frac{6EI}{l^2}\lambda_y & -\frac{6EI}{l^2}\lambda_x & \frac{2EI}{l} \\ -\frac{AE}{l}\lambda_x & -\frac{AE}{l}\lambda_y & 0 & \frac{AE}{l}\lambda_x & \frac{AE}{l}\lambda_y & 0 \\ \frac{12EI}{l^3}\lambda_y & -\frac{12EI}{l^3}\lambda_x & -\frac{6EI}{l^2} & -\frac{12EI}{l^3}\lambda_y & \frac{12EI}{l^3}\lambda_x & -\frac{6EI}{l^2} \\ -\frac{6EI}{l^2}\lambda_y & \frac{6EI}{l^2}\lambda_x & \frac{2EI}{l} & \frac{6EI}{l^2}\lambda_y & -\frac{6EI}{l^2}\lambda_x & \frac{4EI}{l} \end{bmatrix} \begin{bmatrix} d_{Nx} \\ d_{Ny} \\ d_{N\theta} \\ d_{Fx} \\ d_{Fy} \\ d_{F\theta} \end{bmatrix} + \begin{bmatrix} q'_{Nxo} \\ q'_{Nyo} \\ q'_{N\theta o} \\ q'_{Fxo} \\ q'_{Fyo} \\ q'_{F\theta o} \end{bmatrix}$$

EXAMPLE 15.3

Our firm just purchased the newest version of structural analysis software from our preferred vendor. Standard procedure in the office is to test the software with several problems before putting it into general use.

We have been asked to use the Direct Stiffness Method by hand to find the deformations, reactions, and internal forces at the nodes for an asymmetric frame subjected to gravity load. Another engineer on our team is doing the same thing using the software. For the purposes of this test, we consider the supports to be fixed.

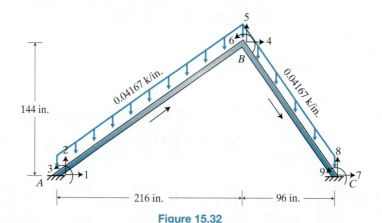

Figure 15.32

Given that we have already found the element stiffness matrices (Example 15.1) and the nodal equivalent loads (Example 15.2), we are ready to analyze the frame.

Estimated Solution:

Anticipated behavior:

Node B should move down under the downward loading.

Node B should rotate counterclockwise because of the longer span on the left.

The moment at A should be counterclockwise to hold the longer span in place.

The moment direction at C is difficult to predict because the rotation at B tends to require counterclockwise moment at C, but the uniform load on BC tends to require clockwise moment at C. These opposing effects suggest that M_C will be smaller magnitude than M_A.

Section 15.4 Finding Deformations, Reactions, and Internal Forces

Horizontal force reactions should point inward because of the arch action of the frame. Vertical force reactions should point upward to balance the gravity load.

Approximate analysis:
Approximate the frame as a flat beam. The situation is covered inside the back cover of this text. Convert the total load to an equivalent uniform distributed load.

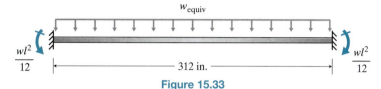

Figure 15.33

$$P_{total} = (0.04167 \text{ k/in.})(259.6 \text{ in.} + 173.07 \text{ in.}) = 18.029 \text{ k}$$

$$w_{equiv} = \frac{18.029 \text{ k}}{(216 \text{ in.} + 96 \text{ in.})} = 0.05779 \text{ k/in.}$$

$$M_{end} = \frac{(0.05779 \text{ k/in.})(312 \text{ in.})^2}{12} = 469 \text{ k} \cdot \text{in.}$$

These moment reactions are probably upper bounds because we have neglected the benefit of the arch action. We also shifted a little bit of load toward midspan, which has the effect of increasing the moment reactions.

Direct Stiffness Method:

Structural stiffness matrix:
Compile terms to form the structural stiffness matrix:

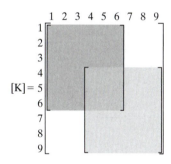

Two regions of [**K**] are made of zeros, since no element shares the DOF combinations: $K(1{:}3,7{:}9)$ and $K(7{:}9,1{:}3)$.

Only nine coefficients receive stiffness from both elements:

$K_{44} = 532.5 \text{ k/in.} + 357.0 \text{ k/in.} = 889.5 \text{ k/in.}$
$K_{45} = 354.3 \text{ k/in.} - 530.6 \text{ k/in.} = -176.5 \text{ k/in.}$
$K_{46} = 69.46 \text{ k} + 234.4 \text{ k} = 303.9 \text{ k}$

$K_{54} = 354.3 \text{ k/in.} - 530.6 \text{ k/in.} = -176.5 \text{ k/in.}$
$K_{55} = 237.2 \text{ k/in.} + 799.2 \text{ k/in.} = 1036.4 \text{ k/in.}$
$K_{56} = -104.20 \text{ k} + 156.26 \text{ k} = 52.06 \text{ k}$

$K_{64} = 69.46 \text{ k} + 234.4 \text{ k} = 303.9 \text{ k}$
$K_{65} = -104.20 \text{ k} + 156.26 \text{ k} = 52.06 \text{ k}$
$K_{66} = 21{,}670 \text{ k} \cdot \text{in.} + 32{,}510 \text{ k} \cdot \text{in.} = 54{,}180 \text{ k} \cdot \text{in.}$

882 Chapter 15 Direct Stiffness Method for Frames

Assembled matrix:

$$[\mathbf{K}] = \begin{array}{c} 1 \\ 2 \\ 3 \\ 4 \\ 5 \\ 6 \\ 7 \\ 8 \\ 9 \end{array} \begin{bmatrix} 532.5 \text{ k/in.} & 354.3 \text{ k/in.} & -69.46 \text{ k} & -532.5 \text{ k/in.} & -354.3 \text{ k/in.} & -69.46 \text{ k} & 0 & 0 & 0 \\ 354.3 \text{ k/in.} & 237.2 \text{ k/in.} & 104.20 \text{ k} & -354.3 \text{ k/in.} & -237.2 \text{ k/in.} & 104.20 \text{ k} & 0 & 0 & 0 \\ -69.46 \text{ k} & 104.20 \text{ k} & 21670 \text{ k} \cdot \text{in.} & 69.46 \text{ k} & -104.20 \text{ k} & 10836 \text{ k} \cdot \text{in.} & 0 & 0 & 0 \\ -532.5 \text{ k/in.} & -354.3 \text{ k/in.} & 69.46 \text{ k} & 889.5 \text{ k/in.} & -176.5 \text{ k/in.} & 303.9 \text{ k} & -357.0 \text{ k/in.} & 530.6 \text{ k/in.} & 234.4 \text{ k} \\ -354.3 \text{ k/in.} & -237.2 \text{ k/in.} & -104.20 \text{ k} & -176.5 \text{ k/in.} & 1036.4 \text{ k/in.} & 52.06 \text{ k} & 530.6 \text{ k/in.} & -799.2 \text{ k/in.} & 156.26 \text{ k} \\ -69.46 \text{ k} & 104.20 \text{ k} & 10836 \text{ k} \cdot \text{in.} & 303.9 \text{ k} & 52.06 \text{ k} & 54180 \text{ k} \cdot \text{in.} & -234.4 \text{ k} & -156.26 \text{ k} & 16254 \text{ k} \cdot \text{in.} \\ 0 & 0 & 0 & -357.0 \text{ k/in.} & 530.6 \text{ k/in.} & -234.4 \text{ k} & 357.0 \text{ k/in.} & -530.6 \text{ k/in.} & -234.4 \text{ k} \\ 0 & 0 & 0 & 530.6 \text{ k/in.} & -799.2 \text{ k/in.} & -156.26 \text{ k} & -530.6 \text{ k/in.} & 799.2 \text{ k/in.} & -156.26 \text{ k} \\ 0 & 0 & 0 & 234.4 \text{ k} & 156.26 \text{ k} & 16254 \text{ k} \cdot \text{in.} & -234.4 \text{ k} & -156.26 \text{ k} & 32510 \text{ k} \cdot \text{in.} \end{bmatrix}$$

Partition the system of equations:

Free DOFs: 4, 5, 6

Supported DOFs: 1, 2, 3, 7, 8, 9

$$[\mathbf{Q}_f] = [\mathbf{K}_{ff}][\mathbf{D}_f] + [\mathbf{K}_{fs}]\underbrace{[\mathbf{D}_s]}_{[\mathbf{0}]} + [\mathbf{Q}_{fo}]$$

$$\begin{array}{c} 4 \\ 5 \\ 6 \end{array} \begin{bmatrix} 0 \\ 0 \\ 0 \end{bmatrix} = \begin{array}{c} 4 \\ 5 \\ 6 \end{array} \begin{bmatrix} 889.5 \text{ k/in.} & -176.5 \text{ k/in.} & 303.9 \text{ k} \\ -176.5 \text{ k/in.} & 1036.4 \text{ k/in.} & 52.06 \text{ k} \\ 303.9 \text{ k} & 52.06 \text{ k} & 54180 \text{ k} \cdot \text{in.} \end{bmatrix} \begin{bmatrix} \Delta_{Bx} \\ \Delta_{By} \\ \theta_B \end{bmatrix} + \begin{array}{c} 4 \\ 5 \\ 6 \end{array} \begin{bmatrix} 0 \text{ k} \\ 9.015 \text{ k} \\ -137.03 \text{ k} \cdot \text{in.} \end{bmatrix}$$

$$[\mathbf{Q}_s] = [\mathbf{K}_{sf}][\mathbf{D}_f] + [\mathbf{K}_{ss}]\underbrace{[\mathbf{D}_s]}_{[\mathbf{0}]} + [\mathbf{Q}_{so}]$$

$$\begin{array}{c} 1 \\ 2 \\ 3 \\ 7 \\ 8 \\ 9 \end{array} \begin{bmatrix} A_x \\ A_y \\ M_A \\ C_x \\ C_y \\ M_C \end{bmatrix} = \begin{array}{c} 1 \\ 2 \\ 3 \\ 7 \\ 8 \\ 9 \end{array} \begin{bmatrix} -532.5 \text{ k/in.} & -354.3 \text{ k/in.} & -69.46 \text{ k} \\ -354.3 \text{ k/in.} & -237.2 \text{ k/in.} & 104.20 \text{ k} \\ 69.46 \text{ k} & -104.20 \text{ k} & 10836 \text{ k} \cdot \text{in.} \\ -357.0 \text{ k/in.} & 530.6 \text{ k/in.} & -234.4 \text{ k} \\ 530.6 \text{ k/in.} & -799.2 \text{ k/in.} & -156.26 \text{ k} \\ 234.4 \text{ k} & 156.26 \text{ k} & 16254 \text{ k} \cdot \text{in.} \end{bmatrix} \begin{bmatrix} \Delta_{Bx} \\ \Delta_{By} \\ \theta_B \end{bmatrix} + \begin{array}{c} 1 \\ 2 \\ 3 \\ 7 \\ 8 \\ 9 \end{array} \begin{bmatrix} 0 \text{ k} \\ 5.409 \text{ k} \\ 194.71 \text{ k} \cdot \text{in.} \\ 0 \text{ k} \\ 3.606 \text{ k} \\ -57.68 \text{ k} \cdot \text{in.} \end{bmatrix}$$

Find deformations:

$$[\mathbf{D}_f] = [\mathbf{K}_{ff}]^{-1}([\mathbf{Q}_f] - [\mathbf{Q}_{fo}])$$

$$\begin{bmatrix} \Delta_{Bx} \\ \Delta_{By} \\ \theta_B \end{bmatrix} = \begin{array}{c} 4 \\ 5 \\ 6 \end{array} \begin{bmatrix} 889.5 \text{ k/in.} & -176.5 \text{ k/in.} & 303.9 \text{ k} \\ -176.5 \text{ k/in.} & 1036.4 \text{ k/in.} & 52.06 \text{ k} \\ 303.9 \text{ k} & 52.06 \text{ k} & 54180 \text{ k} \cdot \text{in.} \end{bmatrix}^{-1} \left(\begin{bmatrix} 0 \\ 0 \\ 0 \end{bmatrix} - \begin{bmatrix} 0 \text{ k} \\ 9.015 \text{ k} \\ -137.03 \text{ k} \cdot \text{in.} \end{bmatrix} \right)$$

Solve the system of equations:

$$\begin{bmatrix} \Delta_{Bx} \\ \Delta_{By} \\ \theta_B \end{bmatrix} = \begin{bmatrix} -0.002715 \text{ in.} \\ -0.009288 \text{ in.} \\ 0.002553 \end{bmatrix} \begin{array}{c} 4 \\ 5 \\ 6 \end{array}$$

Find reactions:

$$[Q_s] = [K_{sf}][D_f] + [Q_{so}]$$

$$\begin{bmatrix} A_x \\ A_y \\ M_A \\ C_x \\ C_y \\ M_C \end{bmatrix} = \begin{matrix} 1 \\ 2 \\ 3 \\ 7 \\ 8 \\ 9 \end{matrix} \begin{matrix} 4 & 5 & 6 \\ \end{matrix} \begin{bmatrix} -532.5 \text{ k/in.} & -354.3 \text{ k/in.} & -69.46 \text{ k} \\ -354.3 \text{ k/in.} & -237.2 \text{ k/in.} & 104.20 \text{ k} \\ 69.46 \text{ k} & -104.20 \text{ k} & 10836 \text{ k} \cdot \text{in.} \\ -357.0 \text{ k/in.} & 530.6 \text{ k/in.} & -234.4 \text{ k} \\ 530.6 \text{ k/in.} & -799.2 \text{ k/in.} & -156.26 \text{ k} \\ 234.4 \text{ k} & 156.26 \text{ k} & 16254 \text{ k} \cdot \text{in.} \end{bmatrix} \begin{bmatrix} -0.002715 \text{ in.} \\ -0.009288 \text{ in.} \\ 0.002553 \end{bmatrix} + \begin{matrix} 1 \\ 2 \\ 3 \\ 7 \\ 8 \\ 9 \end{matrix} \begin{bmatrix} 0 \text{ k} \\ 5.409 \text{ k} \\ 194.71 \text{ k} \cdot \text{in.} \\ 0 \text{ k} \\ 3.606 \text{ k} \\ -57.68 \text{ k} \cdot \text{in.} \end{bmatrix}$$

Multiply the matrices:

$$\begin{bmatrix} A_x \\ A_y \\ M_A \\ C_x \\ C_y \\ M_C \end{bmatrix} = \begin{bmatrix} 4.559 \text{ k} \\ 8.840 \text{ k} \\ 223.2 \text{ k} \cdot \text{in.} \\ -4.557 \text{ k} \\ 9.189 \text{ k} \\ -18.27 \text{ k} \cdot \text{in.} \end{bmatrix} \begin{matrix} 1 \\ 2 \\ 3 \\ 7 \\ 8 \\ 9 \end{matrix}$$

Find internal forces:

Member AB:

From Example 15.1,

$$\lambda_x = 0.8321 \qquad \lambda_y = 0.5547$$

$$\frac{AE}{l} = 768.6 \text{ k/in.} \qquad \frac{12EI}{l^3} = 0.9647 \text{ k/in.}$$

$$\frac{6EI}{l^2} = 125.22 \text{ k} \qquad \frac{4EI}{l} = 21{,}670 \text{ k} \cdot \text{in.} \qquad \frac{2EI}{l} = 10{,}836 \text{ k} \cdot \text{in.}$$

From Example 15.2,

$$\begin{bmatrix} q'_{ABxo} \\ q'_{AByo} \\ q'_{AB\theta o} \\ q'_{BAxo} \\ q'_{BAyo} \\ q'_{BA\theta o} \end{bmatrix} = \begin{bmatrix} 3.000 \text{ k} \\ 4.500 \text{ k} \\ 194.71 \text{ k} \cdot \text{in.} \\ 3.000 \text{ k} \\ 4.500 \text{ k} \\ -194.71 \text{ k} \cdot \text{in.} \end{bmatrix}$$

Internal forces:

$$\begin{bmatrix} q'_{ABx} \\ q'_{ABy} \\ q'_{AB\theta} \\ q'_{BAx} \\ q'_{BAy} \\ q'_{BA\theta} \end{bmatrix} = \begin{bmatrix} 639.6 \text{ k/in.} & 426.3 \text{ k/in.} & 0 & -639.6 \text{ k/in.} & -426.3 \text{ k/in.} & 0 \\ -0.5351 \text{ k/in.} & 0.8027 \text{ k/in.} & 125.22 \text{ k} & 0.5351 \text{ k/in.} & -0.8027 \text{ k/in.} & 125.22 \text{ k} \\ -69.46 \text{ k} & 104.20 \text{ k} & 21670 \text{ k} \cdot \text{in.} & 69.46 \text{ k} & -104.20 \text{ k} & 10836 \text{ k} \cdot \text{in.} \\ -639.6 \text{ k/in.} & -426.3 \text{ k/in.} & 0 & 639.6 \text{ k/in.} & 426.3 \text{ k/in.} & 0 \\ 0.5351 \text{ k/in.} & -0.8027 \text{ k/in.} & -125.22 \text{ k} & -0.5351 \text{ k/in.} & 0.8027 \text{ k/in.} & -125.22 \text{ k} \\ -69.46 \text{ k} & 104.20 \text{ k} & 10836 \text{ k} \cdot \text{in.} & 69.46 \text{ k} & -104.20 \text{ k} & 21670 \text{ k} \cdot \text{in.} \end{bmatrix} \begin{bmatrix} 0 \\ 0 \\ 0 \\ -0.002715 \text{ in.} \\ -0.009288 \text{ in.} \\ 0.002553 \end{bmatrix} + \begin{bmatrix} 3.000 \text{ k} \\ 4.500 \text{ k} \\ 194.71 \text{ k} \cdot \text{in.} \\ 3.000 \text{ k} \\ 4.500 \text{ k} \\ -194.71 \text{ k} \cdot \text{in.} \end{bmatrix}$$

$$= \begin{bmatrix} 8.696 \text{ k} \\ 4.826 \text{ k} \\ 223.2 \text{ k} \cdot \text{in.} \\ -2.696 \text{ k} \\ 4.174 \text{ k} \\ -138.61 \text{ k} \cdot \text{in.} \end{bmatrix}$$

Member BC:
From Example 15.1,

$$\lambda_x = 0.5547 \qquad \lambda_y = -0.8321$$

$$\frac{AE}{l} = 1152.8 \text{ k/in.} \qquad \frac{12EI}{l^3} = 3.256 \text{ k/in.}$$

$$\frac{6EI}{l^2} = 281.7 \text{ k} \qquad \frac{4EI}{l} = 32{,}510 \text{ k·in.} \qquad \frac{2EI}{l} = 16{,}254 \text{ k·in.}$$

From Example 15.2,

$$\begin{bmatrix} q'_{BCxo} \\ q'_{BCyo} \\ q'_{BC\theta o} \\ q'_{CBxo} \\ q'_{CByo} \\ q'_{CB\theta o} \end{bmatrix} = \begin{bmatrix} -3.000 \text{ k} \\ 2.000 \text{ k} \\ 57.68 \text{ k·in.} \\ -3.000 \text{ k} \\ 2.000 \text{ k} \\ -57.68 \text{ k·in.} \end{bmatrix}$$

Internal forces:

$$\begin{bmatrix} q'_{BCx} \\ q'_{BCy} \\ q'_{BC\theta} \\ q'_{CBx} \\ q'_{CBy} \\ q'_{CB\theta} \end{bmatrix} = \begin{bmatrix} 639.5 \text{ k/in.} & -959.2 \text{ k/in.} & 0 & -639.5 \text{ k/in.} & 959.2 \text{ k/in.} & 0 \\ 2.709 \text{ k/in.} & 1.8061 \text{ k/in.} & 281.7 \text{ k} & -2.709 \text{ k/in.} & -1.8061 \text{ k/in.} & 281.7 \text{ k} \\ 234.4 \text{ k} & 156.26 \text{ k} & 32510 \text{ k·in.} & -234.4 \text{ k} & -156.26 \text{ k} & 16254 \text{ k·in.} \\ -639.5 \text{ k/in.} & 959.2 \text{ k/in.} & 0 & 639.5 \text{ k/in.} & -959.2 \text{ k/in.} & 0 \\ -2.709 \text{ k/in.} & -1.8061 \text{ k/in.} & -281.7 \text{ k} & 2.709 \text{ k/in.} & 1.8061 \text{ k/in.} & -281.7 \text{ k} \\ 234.4 \text{ k} & 156.26 \text{ k} & 16254 \text{ k·in.} & -234.4 \text{ k} & -156.26 \text{ k} & 32510 \text{ k·in.} \end{bmatrix} \begin{bmatrix} -0.002715 \text{ in.} \\ -0.009288 \text{ in.} \\ 0.002553 \\ 0 \\ 0 \\ 0 \end{bmatrix} + \begin{bmatrix} -3.000 \text{ k} \\ 2.000 \text{ k} \\ 57.68 \text{ k·in.} \\ -3.000 \text{ k} \\ 2.000 \text{ k} \\ -57.68 \text{ k·in.} \end{bmatrix}$$

$$= \begin{bmatrix} 4.173 \text{ k} \\ 2.695 \text{ k} \\ 138.59 \text{ k·in.} \\ -10.173 \text{ k} \\ 1.305 \text{ k} \\ -18.27 \text{ k·in.} \end{bmatrix}$$

Summary of Results:

$$\begin{bmatrix} \Delta_{Bx} \\ \Delta_{By} \\ \theta_B \end{bmatrix} = \begin{bmatrix} -0.002715 \text{ in.} \\ -0.009288 \text{ in.} \\ 0.002553 \end{bmatrix} \begin{matrix} (+\uparrow) \\ (\xrightarrow{+}) \\ (\curvearrowright) \end{matrix} \qquad \begin{bmatrix} A_x \\ A_y \\ M_A \\ C_x \\ C_y \\ M_C \end{bmatrix} = \begin{bmatrix} 4.559 \text{ k} \\ 8.840 \text{ k} \\ 223.2 \text{ k·in.} \\ -4.557 \text{ k} \\ 9.189 \text{ k} \\ -18.27 \text{ k·in.} \end{bmatrix} \begin{matrix} (+\uparrow) \\ (\xrightarrow{+}) \\ (\curvearrowright) \\ (+\uparrow) \\ (\xrightarrow{+}) \\ (\curvearrowright) \end{matrix}$$

$$\begin{bmatrix} q'_{ABx} \\ q'_{ABy} \\ q'_{AB\theta} \\ q'_{BAx} \\ q'_{BAy} \\ q'_{BA\theta} \end{bmatrix} = \begin{bmatrix} 8.696 \text{ k} \\ 4.826 \text{ k} \\ 223.2 \text{ k·in.} \\ -2.696 \text{ k} \\ 4.174 \text{ k} \\ -138.61 \text{ k·in.} \end{bmatrix} \qquad \begin{bmatrix} q'_{BCx} \\ q'_{BCy} \\ q'_{BC\theta} \\ q'_{CBx} \\ q'_{CBy} \\ q'_{CB\theta} \end{bmatrix} = \begin{bmatrix} 4.173 \text{ k} \\ 2.695 \text{ k} \\ 138.59 \text{ k·in.} \\ -10.173 \text{ k} \\ 1.305 \text{ k} \\ -18.27 \text{ k·in.} \end{bmatrix}$$

Section 15.4 Finding Deformations, Reactions, and Internal Forces

Evaluation of Results:
Observed Expected Features?
Comparison with predictions:
  Node B moves down under the downward loading. ✓
  Node B rotates counterclockwise because of the longer span on the left. ✓
  The moment reaction at A is counterclockwise to hold the longer span in place. ✓
  The moment reaction at C is smaller magnitude than M_A. ✓
  Horizontal force reactions point inward because of the arch action of the frame. ✓
  Vertical force reactions point upward to balance the gravity load. ✓

Satisfied Fundamental Principles?
Verify equilibrium:

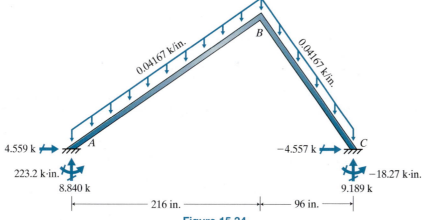

Figure 15.34

$\pm \rightarrow \Sigma F_x = 4.559 \text{ k} - 4.557 \text{ k} = 0.002 \text{ k} \approx 0$ ✓
$+\uparrow \Sigma F_y = 8.840 \text{ k} - 0.04167 \text{k/in.}(259.6 \text{ in.} + 173.07 \text{ in.}) + 9.189 \text{ k}$
$\qquad = -4 \times 10^{-4} \text{ k} \approx 0$ ✓
$\circlearrowright \Sigma M_C = 223.2 \text{ k} \cdot \text{in.} - 8.840 \text{ k}(312 \text{ in.}) - 18.27 \text{ k} \cdot \text{in.}$
$\qquad + (0.04167 \text{ k/in.})(259.6 \text{ in.})(216 \text{ in.}/2 + 96 \text{ in.})$
$\qquad + (0.04167 \text{ k/in.})(173.07 \text{ in.})(96 \text{ in.}/2) = -0.2 \text{ k} \cdot \text{in.} \approx 0$ ✓

Approximation Predicted Outcomes?
Comparison with approximation:
  Both moment reactions, 223.2 k·in. and 18.27 k·in., are less than the upper bound approximation of 469 k·in. ✓

Conclusion:
  Together these checks strongly suggest that we have reasonable results from the Direct Stiffness Method. We can use these results to verify the new software.

Homework Problems

15.1 through 15.4 Our firm has won the contract to design a new two-span overpass. Our team has finished preliminary design of the girders. The firm has purchased bridge analysis and design software to help in the design. Before we use that software to complete the design, our policy is to verify the software with by hand calculations.

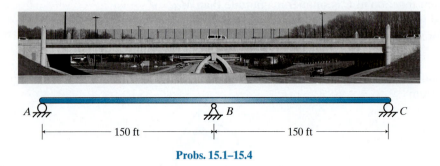

Probs. 15.1–15.4

15.1 From preliminary design, we expect one lane of the bridge to have these composite section properties:

$$A = 3750 \text{ in}^2 \quad I = 2{,}220{,}000 \text{ in}^4$$

Concrete: $E = 5000$ ksi

Our first task is to create the structural stiffness matrix.

a. Calculate the element stiffness matrix, in global coordinates, for each of the beam members. Designate the near and far nodes for each member, and label all of the degrees of freedom.
b. Assemble the global stiffness matrix for the structure.

15.2 One of the load cases we must consider is uniform dead load.

a. Guess the displaced shape under this loading.
b. Using approximate analysis, find the moment at midspan of each span. You can do so by assuming the two spans are each simply supported; this is the same as assuming the inflection point is at B. Is this an upper bound, a lower bound, or just an approximation?
c. Calculate the nodal equivalent loads for the structure in global coordinates.
d. Calculate the anticipated deformations of the free degrees of freedom using the Direct Stiffness Method. Note: to solve the system of equations, you may use computer software, but provide a printout of the inputs and outputs.
e. Use the deformations from part (d) to calculate the anticipated reactions. Verify equilibrium of the bridge using these reactions.
f. Use the deformations from part (d) to calculate the anticipated internal forces at the ends of member BC. Also find the moment at midspan of member BC.
g. Are your results from the Direct Stiffness Method reasonable? Make a comprehensive argument to justify your answer.
h. Using structural analysis software, determine the deformations of the free degrees of freedom, the reactions, and the internal forces at the ends of the members. Comment on how these results compare with your results from parts (d) through (f).
i. Print out the displaced shape from your computer aided analysis results. Identify at least three features that suggest you have a reasonable result.
j. Comment on why your guess in part (a) does or does not match what you discovered in part (i).

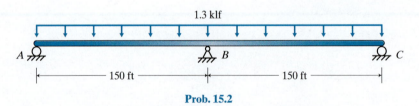

Prob. 15.2

15.3 To finalize the design, we need to find the largest internal forces generated by the design truck at different locations along the bridge. The design truck we are most concerned with is a fully loaded, five-axle semi with a relatively short span. Although there would actually be five point loads from the truck, for the purpose of verifying the software we can combine the tandem loads.

The bridge deck will distribute the point forces to the various bridge girders across the width. Another engineer on our team will use the Distribution Factor Method to calculate that distribution after we finish our analyses. One of the load cases is to put the middle axle at midspan.

a. Guess the directions of each of the reactions under this loading.
b. Using approximate analysis, find the moment at midspan of span AB. You can do so by assuming the two spans are each simply supported; this is the same as assuming the inflection point is at B. Is this an upper bound, a lower bound, or just an approximation for the midspan moment?
c. Calculate the nodal equivalent loads for the structure in global coordinates.
d. Calculate the anticipated deformations of the free degrees of freedom using the Direct Stiffness Method. Note: to solve the system of equations, you may use computer software, but provide a printout of the inputs and outputs.
e. Use the deformations from part (d) to calculate the anticipated reactions. Verify equilibrium of the bridge using these reactions.
f. Use the deformations from part (d) to calculate the anticipated internal forces at the right end of member AB. Also find the moment at midspan of member AB.
g. Are your results from the Direct Stiffness Method reasonable? Make a comprehensive argument to justify your answer.
h. Using structural analysis software, determine the deformations of the free degrees of freedom, the reactions, and the internal forces. Comment on how these results compare with your results from parts (d) through (f).
i. Comment on why your guess in part (a) does or does not match what you discovered in part (e).

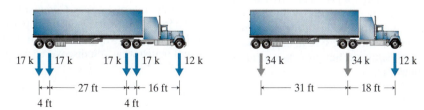

Prob. 15.3a

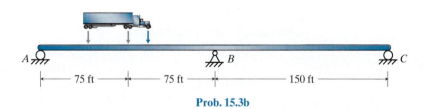

Prob. 15.3b

888 Chapter 15 Direct Stiffness Method for Frames

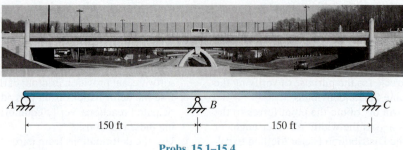

Probs. 15.1–15.4

15.4 We must also consider the possibility of differential settlement. For example, support A may settle 1 inch more than the others.

a. Guess the directions of each of the reactions under this loading.
b. Using approximate analysis, find the moment at B. You can do that by treating member AB as a cantilever that displaces 1 inch due to a point load. Displacement formulas are available inside the front cover of this text. Is this an upper bound, a lower bound, or just an approximation?
c. Calculate the anticipated deformations of the free degrees of freedom for the entire bridge using the Direct Stiffness Method. Note: to solve the system of equations, you may use computer software, but provide a printout of the inputs and outputs.
d. Use the deformations from part (c) to calculate the anticipated reactions. Verify equilibrium of the bridge using these reactions.
e. Use the deformations from part (c) to calculate the anticipated moment at B.
f. Are your results from the Direct Stiffness Method reasonable? Make a comprehensive argument to justify your answer.
g. Using structural analysis software, determine the deformations of the free degrees of freedom, the reactions, and the internal forces. Comment on how these results compare with your results from parts (c) through (e).
h. Comment on why your guess in part (a) does or does not match what you discovered in part (d).

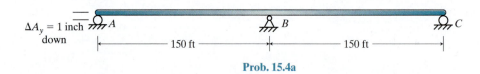

Prob. 15.4a

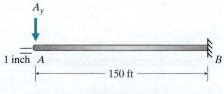

Prob. 15.4b

15.5 through 15.7 A pedestrian walkway must be suspended over a maintenance corridor. To do that, our firm has chosen to support the walkway on a frame fixed to the adjacent building and on a very stiff column at the other end. Because we are new to the firm, our supervisor wants to test our understanding of the Direct Stiffness Method.

Another engineer on the team has selected the members based on preliminary analysis:

W27 × 94:	$A = 27.6 \text{ in}^2$	$I_x = 3270 \text{ in}^4$
W14 × 61:	$A = 17.9 \text{ in}^2$	$I_y = 107 \text{ in}^4$
Steel:	$E = 29{,}000 \text{ ksi}$	

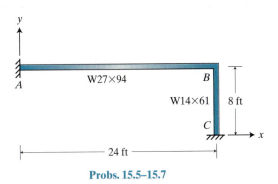

Probs. 15.5–15.7

15.5 Our first task is to create the structural stiffness matrix.

a. Calculate the element stiffness matrix, in global coordinates, for each of the frame members. Designate the near and far nodes for each member, and label all of the degrees of freedom.

b. Assemble the global stiffness matrix for the structure.

890 Chapter 15 Direct Stiffness Method for Frames

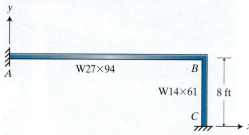

Probs. 15.5–15.7

15.6 Our second test is to analyze the frame experiencing the full live load on the pedestrian bridge.

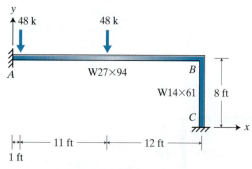

Prob. 15.6a

a. Guess the directions of the moment reactions under this loading.
b. Using approximate analysis, estimate the reactions. You can do so by assuming the girder to be fixed at both ends. Using the fixed-end forces tabulated inside the back cover of this text can help. Which of these are upper bounds, lower bounds, or just approximations?

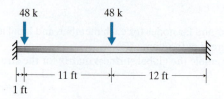

Prob. 15.6b

c. Calculate the nodal equivalent loads for the structure in global coordinates.

d. Calculate the anticipated deformations of the free degrees of freedom for the full frame using the Direct Stiffness Method. Note: to solve the system of equations, you may use computer software, but provide a printout of the inputs and outputs.

e. Use the deformations from part (d) to calculate the anticipated moment reactions.

f. Use the deformations from part (d) to calculate the anticipated internal forces at the ends of member AB.

g. Are your results from the Direct Stiffness Method reasonable? Make a comprehensive argument to justify your answer.

h. Using structural analysis software, determine the deformations of the free degrees of freedom, the moment reactions, and the internal forces at the ends of member AB. Comment on how these results compare with your results from parts (d) through (f).

i. Comment on why your guess in part (a) does or does not match what you discovered in part (e).

15.7 If the connection at A is not perfectly rigid, the deformations, reactions, and internal forces will be affected. Our third test is to predict the effect of rotation of the support at A.

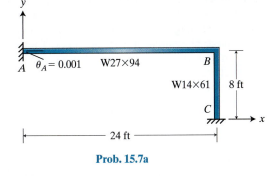

Prob. 15.7a

a. Guess the directions of the moment reactions under this loading.
b. Using approximate analysis, find the moment at A. You can do that by treating member AB as fixed at one end and pin supported at the other. We found the relationship between M and θ in Section 13.2: Fixed-End Moments and Distribution Factors. Is this moment an upper bound, a lower bound, or just an approximation?

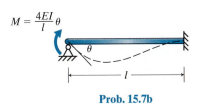

Prob. 15.7b

c. Using approximate analysis, find the moment at A. You can do that by treating member AB as simply supported with an applied couple moment. You can use the relationship between the applied moment and rotation from inside the front cover. Is this an upper bound, a lower bound, or just an approximation?

Prob. 15.7c

d. Calculate the anticipated deformations of the free degrees of freedom for the full frame using the Direct Stiffness Method. Note: to solve the system of equations, you may use computer software, but provide a printout of the inputs and outputs.
e. Use the deformations from part (d) to calculate the anticipated moment reactions.
f. Use the deformations from part (d) to calculate the anticipated internal forces at the ends of member AB.
g. Are your results from the Direct Stiffness Method reasonable? Make a comprehensive argument to justify your answer.
h. Using structural analysis software, determine the deformations of the free degrees of freedom, the moment reactions, and the internal forces at the ends of member AB. Comment on how these results compare with your results from parts (d) through (f).
i. Comment on why your guess in part (a) does or does not match what you discovered in part (e).

15.8 through 15.11 A school installed a lightweight frame to mark the start and finish lines at the track. For a special event coming up, the organizers want to hang a single heavy piece of equipment from the frame. They contacted our firm to evaluate the frame to determine whether they can safely hang the equipment from it.

One of our concerns is the possibility of a seismic event occurring while the equipment is hanging on the frame. Another engineer on our team has calculated the static equivalent seismic load.

To model the behavior, we can calculate effective section properties to represent the frame. Based on field measurements, we know that each section of the frame is made with four aluminum pipes. Each pipe has a 50-mm outside diameter and is 3 mm thick. When we consider the spacing of the pipes, we get the properties.

Another engineer on our team has analyzed the frame using commercial software. However, the results don't make sense. To check the software, we will perform the Direct Stiffness Method by hand.

One pipe:	$A = 229$ mm^2	Spacing $= 350$ mm
Frame:	$A = 916$ mm^2	$I = 2.80 \times 10^7$ mm^4
Aluminum:	$E = 69$ GPa $= 69$ kN/mm^2	

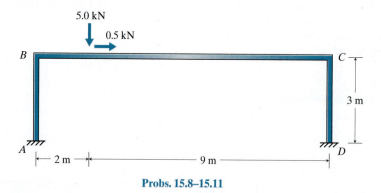

Probs. 15.8–15.11

15.8 We start by creating the structural stiffness matrix.

a. Calculate the element stiffness matrix, in global coordinates, for each of the frame members. Designate the near and far nodes for each member, and label all of the degrees of freedom.

b. Assemble the global stiffness matrix for the structure.

15.9 Our second task is to analyze the frame experiencing the static equivalent seismic load.

a. Guess the directions of the moment reactions under this loading.
b. Using approximate analysis, estimate the reactions. You can use the approaches from Chapter 9: Approximate Analysis of Rigid Frames. Which reactions are upper bounds, lower bounds, or just approximations?
c. Calculate the nodal equivalent loads for the structure in global coordinates.
d. Calculate the anticipated deformations of the free degrees of freedom using the Direct Stiffness Method. Note: to solve the system of equations, you may use computer software, but provide a printout of the inputs and outputs.
e. Use the deformations from part (d) to calculate the anticipated moment reactions.
f. Use the deformations from part (d) to calculate the anticipated internal forces at the ends of member BC.
g. Are your results from the Direct Stiffness Method reasonable? Make a comprehensive argument to justify your answer.
h. Using structural analysis software, determine the deformations of the free degrees of freedom, the moment reactions, and the internal forces at the ends of member BC. Comment on how these results compare with your results from parts (d) through (f).
i. Comment on why your guess in part (a) does or does not match what you discovered in part (e).

15.10 Since we are concerned about the performance of the software, let's check other loadings as well. The peak temperature drop the frame will experience is likely 30°C compared with the temperature when it was installed. The coefficient of thermal expansion for aluminum is $23.1 \times 10^{-6}/°C$.

a. Guess the directions of the moment reactions with just this uniform temperature drop for all members.
b. Using approximate analysis, estimate the moment reactions. You can do so by considering the columns to be fixed at both ends, with the tops moving based on how much the beam shortens. Using the fixed-end forces tabulated inside the back cover of this text can help. Are these moment reactions upper bounds, lower bounds, or just approximations?
c. Calculate the nodal equivalent loads for the structure in global coordinates.
d. Calculate the anticipated deformations of the free degrees of freedom using the Direct Stiffness Method. Note: to solve the system of equations, you may use computer software, but provide a printout of the inputs and outputs.
e. Use the deformations from part (d) to calculate the anticipated moment reactions.
f. Use the deformations from part (d) to calculate the anticipated internal forces at the ends of member CD.
g. Are your results from the Direct Stiffness Method reasonable? Make a comprehensive argument to justify your answer.
h. Using structural analysis software, determine the deformations of the free degrees of freedom, the moment reactions, and the internal forces at the ends of member CD. Comment on how these results compare with your results from parts (d) through (f).
i. Comment on why your guess in part (a) does or does not match what you discovered in part (e).

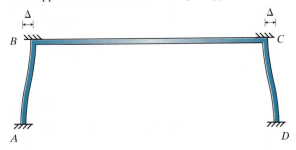

Prob. 15.10

894 Chapter 15 Direct Stiffness Method for Frames

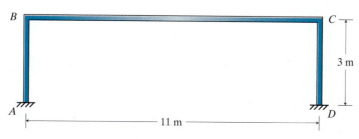

Probs. 15.8–15.11

15.11 Another check we want to make of the software is support movement. The column members are held to the foundations with bolts through baseplates. The holes are slightly larger than the bolt diameters; therefore, the baseplates could move. Let's consider the effect of the baseplate at A moving 5 mm to the right.

a. Guess the directions of the moment reactions with just the baseplate movement.
b. Using approximate analysis, estimate the moment reactions. You can do so by considering the columns to be fixed at both ends, with the bottoms each moving 2.5 mm inward. Using the fixed-end forces tabulated inside the back cover of this text can help. Are these moment reactions upper bounds, lower bounds, or just approximations?
c. Calculate the anticipated deformations of the free degrees of freedom using the Direct Stiffness Method. Consider only A to move the full 5 mm to the right. Note: to solve the system of equations, you may use computer software, but provide a printout of the inputs and outputs.
d. Use the deformations from part (c) to calculate the anticipated moment reactions.
e. Use the deformations from part (c) to calculate the anticipated internal forces at the ends of member CD.
f. Are your results from the Direct Stiffness Method reasonable? Make a comprehensive argument to justify your answer.
g. Using structural analysis software, determine the deformations of the free degrees of freedom, the moment reactions, and the internal forces at the ends of member CD. Comment on how these results compare with your results from parts (c) through (e).
h. Comment on why your guess in part (a) does or does not match what you discovered in part (d).

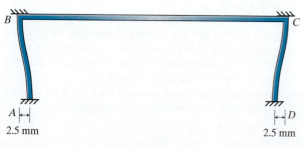

Prob. 15.11

15.12 through 15.15 The owners of a building in Europe that has a below-grade window well would like to put on a roof in order to turn the space into a greenhouse. Originally, the design team considered making the roof removable, but the potential for wind uplift made that option not practical. Instead, our team has chosen to use fixed supports on the two ends. The team leader anticipates that we will use 45 × 145 wood members as a rigid frame ($E = 10$ GPa, $A = 65.2$ cm^2, $I = 1143$ cm^4).

An engineer on our team has estimated the dead load on the frame.

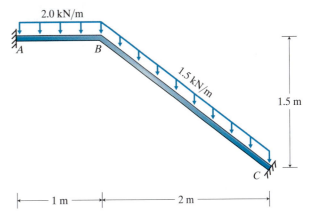

Probs. 15.12–15.15

As we start to work on the project, we get a new updated version of our structural analysis software. Company policy is to verify each new version of software, so our team leader has chosen this project for the verification.

15.12 Our first task is to create the structural stiffness matrix.

a. Calculate the element stiffness matrix, in global coordinates, for each of the frame members. Designate the near and far nodes for each member, and label all of the degrees of freedom.

b. Assemble the global stiffness matrix for the structure.

15.13 Our second task is to analyze the frame experiencing unfactored dead load.

a. Guess the three directions that node B will move under this loading: Δ_{Bx}, Δ_{By}, and θ_B.

b. Using approximate analysis, estimate the vertical reactions. You can approximate the frame as a fixed-end beam with an equivalent uniform load. Which vertical reactions are upper bounds, lower bounds, or just approximations?

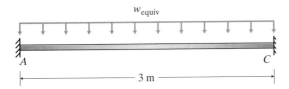

Prob. 15.13

c. Calculate the nodal equivalent loads for the structure in global coordinates for the full frame.

d. Calculate the anticipated deformations of the free degrees of freedom for the full frame using the Direct Stiffness Method. Note: to solve the system of equations, you may use computer software, but provide a printout of the inputs and outputs.

e. Use the deformations from part (d) to calculate the anticipated vertical reactions.

f. Use the deformations from part (d) to calculate the anticipated internal forces at the ends of member BC.

g. Are your results from the Direct Stiffness Method reasonable? Make a comprehensive argument to justify your answer.

h. Using structural analysis software, determine the deformations of the free degrees of freedom, the vertical reactions, and the internal forces at the ends of member BC. Comment on how these results compare with your results from parts (d) through (f).

i. Comment on why your guess in part (a) does or does not match what you discovered in part (d).

15.14 Temperature change is one of the load cases we analyze using the software. Since this is a greenhouse roof, we should consider the possibility that the frame experiences a 20°C increase compared with the temperature when it was installed. The coefficient of thermal expansion for fir, a common European wood, is $3.7\times10^{-6}/°C$.

a. Guess the directions of the moment reactions with just this uniform temperature increase for both members.
b. Using approximate analysis, estimate the horizontal reactions. You can do that by approximating the frame as a fixed-end beam. Is this an upper bound, a lower bound, or just an approximation?

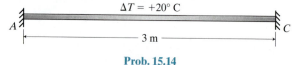

Prob. 15.14

c. Calculate the nodal equivalent loads for the structure in global coordinates for the full frame.

d. Calculate the anticipated deformations of the free degrees of freedom for the full frame using the Direct Stiffness Method. Note: to solve the system of equations, you may use computer software, but provide a printout of the inputs and outputs.
e. Use the deformations from part (d) to calculate the anticipated horizontal reactions.
f. Use the deformations from part (d) to calculate the anticipated internal forces at the ends of member BC.
g. Are your results from the Direct Stiffness Method reasonable? Make a comprehensive argument to justify your answer.
h. Using structural analysis software, determine the deformations of the free degrees of freedom, the moment reactions, and the internal forces at the ends of member BC. Comment on how these results compare with your results from parts (d) through (f).
i. Comment on why your guess in part (a) does or does not match what you discovered in part (e).

Probs. 15.12–15.15

15.15 Another load case we use the software to analyze is support movement. The geotechnical engineer working on the project says the house could still settle 10 mm more than the outer wall.

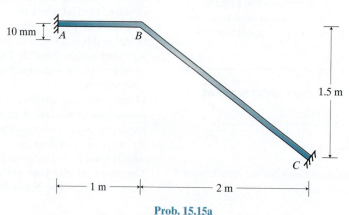

Prob. 15.15a

a. Guess the three directions that node B will move under this loading: Δ_{Bx}, Δ_{By}, and θ_B.
b. Using approximate analysis, estimate the moment reactions. You can do that by approximating the frame as a fixed-end beam. Using the fixed-end forces tabulated inside the back cover of this text can help. Are these moment reactions upper bounds, lower bounds, or just approximations?
c. Calculate the anticipated deformations of the free degrees of freedom for the full frame using the Direct Stiffness Method. Note: to solve the system of equations, you may use computer software, but provide a printout of the inputs and outputs.
d. Use the deformations from part (c) to calculate the anticipated moment reactions.
e. Use the deformations from part (c) to calculate the anticipated internal forces at the ends of member AB.

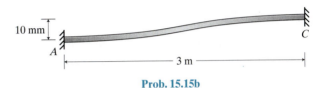

Prob. 15.15b

f. Are your results from the Direct Stiffness Method reasonable? Make a comprehensive argument to justify your answer.
g. Using structural analysis software, determine the deformations of the free degrees of freedom, the moment reactions, and the internal forces at the ends of member BC. Comment on how these results compare with your results from parts (c) through (e).
h. Comment on why your guess in part (a) does or does not match what you discovered in part (c).

15.16 through 15.18 We are being considered for a position at a structural engineering firm and our interview is tomorrow morning. When we arrive at the hotel, we find a welcome basket and a note:

Welcome to our city. We are excited to meet with you tomorrow. In preparation for our meeting, we would like you to demonstrate some of your skills. The sketch is of a project we recently completed. Please perform the analysis tasks listed and bring your answers with you in the morning.

The structure is a frame that provides shade for picnic tables. It appears to carry a uniform dead load along both members. Although the frame is determinate, the firm wants us to analyze it by hand using the Direct Stiffness Method.

The problem description includes a caution: this combination of geometry and section properties is very sensitive to roundoff error when the Direct Stiffness Method is used. We are to carry four significant figures in our calculations but recognize that our results might not match computer results to within 1%.

The paper says both members are round HSS sections with the following properties:

HSS6 × 0.280:	A = 4.69 in²	I = 19.3 in⁴
Steel:	E = 29,000 ksi	

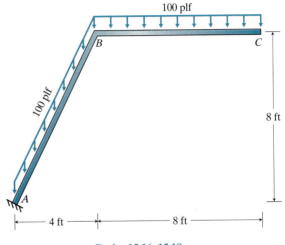

Probs. 15.16–15.18

898   Chapter 15   Direct Stiffness Method for Frames

15.16 Our first test is to create the structural stiffness matrix.

a. Calculate the element stiffness matrix, in global coordinates, for each of the frame members. Designate the near and far nodes for each member, and label all of the degrees of freedom.

b. Assemble the global stiffness matrix for the structure.

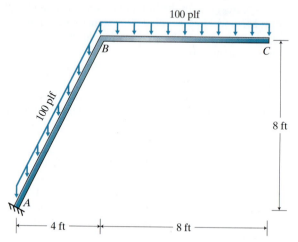

Probs. 15.16–15.18

15.17 Our second test is to analyze the frame experiencing unfactored dead load.

a. Guess the shape of the moment diagram.
b. Using approximate analysis, estimate the vertical displacement at C. You can approximate the frame as a cantilevered beam with an equivalent uniform load. Will this be an upper bound, a lower bound, or just an approximation?

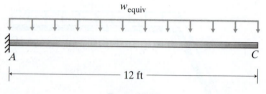

Prob. 15.17

c. Calculate the nodal equivalent loads for the structure in global coordinates for the full frame.
d. Calculate the anticipated deformations of the free degrees of freedom for the full frame using the Direct Stiffness Method. Note: to solve the system of equations, you may use computer software, but provide a printout of the inputs and outputs.
e. Use the deformations from part (d) to calculate the anticipated reactions. Verify equilibrium.
f. Use the deformations from part (d) to calculate the anticipated internal forces at the ends of member AB.
g. Are your results from the Direct Stiffness Method reasonable? Make a comprehensive argument to justify your answer.
h. Using structural analysis software, determine the deformations of the free degrees of freedom, the reactions, and the internal forces at the ends of member AB. Comment on how these results compare with your results from parts (d) through (f).
i. Using structural analysis software, print out the moment diagram for the frame. Identify at least three features that suggest you have a reasonable result.
j. Comment on why your guess in part (a) does or does not match what you discovered in part (i).

15.18 The notes for the problem say that expecting the unexpected is an important part of the job. The last test is to consider someone doing exercises on the end of the structure while the foundation rotates.

Prob. 15.18b

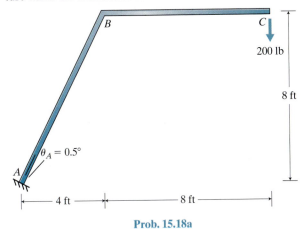

Prob. 15.18a

a. Guess the shape of the moment diagram.
b. Using approximate analysis, estimate the vertical displacement at C. You can do that by superimposing the two load cases. You can use trigonometry to predict the vertical displacement due to the foundation rotation, since the frame moves as a rigid body. For the effect of the live load, you can approximate the frame as a cantilever beam. Is this an upper bound, a lower bound, or just an approximation?
c. Calculate the anticipated deformations of the free degrees of freedom for the full frame using the Direct Stiffness Method. Note: to solve the system of equations, you may use computer software, but provide a printout of the inputs and outputs.
d. Use the deformations from part (c) to calculate the anticipated reactions. Verify equilibrium.
e. Use the deformations from part (c) to calculate the anticipated internal forces at the ends of member AB.
f. Are your results from the Direct Stiffness Method reasonable? Make a comprehensive argument to justify your answer.
g. Using structural analysis software, determine the deformations of the free degrees of freedom, the reactions, and the internal forces at the ends of member AB. Comment on how these results compare with your results from parts (c) through (e).
h. Using structural analysis software, print out the moment diagram for the frame. Identify at least three features that suggest you have a reasonable result.
i. Comment on why your guess in part (a) does or does not match what you discovered in part (h).

INDEX

A

Abutments, 71, 155, 787
Additional loadings for Direct Stiffness Method
    global vector assembly, 830–831
    length change, 828–829
    loads at supports, 831
    loads between nodes, 868–877
    temperature change, 830
Analysis errors with computer aided analysis, 339–340
    assumptions in method, 339–340
    roundoff errors, 340
Applied moments, effect on internal forces, 174–175
Approximate analysis definition, 437
Approximate analysis of beams
    change the boundary conditions, 98, 243
    change the loading, 98, 101, 178, 186–187, 190
    change the structure shape, 98, 100, 181–182, 253
Approximate analysis of braced frames, 384–434
    braced frames with gravity loads, 401–416
        assumptions, 406
        distributed loading, 403–404
        process of approximate analysis, 405
        truss approximation, 401–403
    braced frames with lateral loads, 384–400
        assumptions, 384–386
        axial force members, 387
        nonidentical braces, 386
        transfer of lateral force to braced bays, 388
        process of approximate analysis, 388
        purpose of each member, 387–388
Approximate analysis of determinate trusses, 264–265
Approximate analysis of indeterminate trusses, 374–383
    more than two diagonals, 376–379
    process of approximate analysis, 376–377
Approximate analysis of rigid frames, 436–513
    Cantilever Method for lateral loads, 473–489
        applicability, 478
        assumptions, 473
        neutral axis of frame, 474
        process of approximate analysis, 478
    change the structure shape, 208, 273
    combined gravity and lateral loads, 490–496
        process of approximate analysis, 492

    Gravity Load Method, 438–457
        assumptions, 439–440
        inflection points, 439
        nonuniform load, 443–444
        pin supports, 440–443
        process of approximate analysis, 444
    Portal Method for lateral loads, 458–472
        applicability, 462
        assumptions, 459–460
        process of approximate analysis, 462
Approximate lateral displacements for braced frames, 516–542
    Story Drift Method, 516–526
        assumptions, 519
        bays with different brace stiffnesses, 517–519, 524–526
        implications, 519
        limitations, 519
    Virtual Work Method, 526–542
        assumptions, 527
        implications, 527
        limitations, 527
Approximate lateral displacements for rigid frames, 542–564
    Stiff Beam Method, 542–549
        assumptions, 545
        implications, 545
        limitations, 545
        pinned base, 544
        shear in each column, 544–545
        total drift, 544
    Virtual Work Method, 550–564
        approximate analysis derivation, 550–552
        assumptions, 552
        implications, 552
        limitations, 552
        pinned base, 552
Approximate lateral displacements for walls, 565–582
    multistory, 575–582
        assumptions, 578
        flexural contribution, 577
        implications, 578
        limitations, 578
        shear contribution, 575–576
        story drift, 577–578, 582
        total drift, 577–582

Index **901**

single story, 565–574
    assumptions, 570
    coupled walls, 566–567, 570–574
    flexural contribution, 566
    implications, 570
    openings, 568
    rotation at base, 568–570
    shear contribution, 565
Arches, 152–162
    order-of-magnitude estimates, 159
    static equivalent seismic load, 159
    three-hinge arch, 153–162
    thrust arch with fixed supports, 152, 159
    thrust arch with pinned supports, 152
    tied arch with simple supports, 152, 153
    tie force, 155
Area load, 36
Axial deformation for Direct Stiffness Method for frames, 857
Axial force diagrams, 194–195

B

Bays with different brace stiffness and Story Drift Method, 517–519
Beam, 28
Beams analyzed with Moment Distribution Method, 734–753
    assumptions, 742
    carryover moments, 738, 739
    convergence, 738–739
    finding reactions, internal forces, and deformations, 741–742
    fixed-end moments, 736–737
    moments distributing into members, 736–739
    net moment, 739–741
    process of analysis, 742
    unbalanced moment, 737, 740
Bounding the solution, 98–101
    approximate answers, 98
    lower bound, 98–99, 100, 102, 107, 253, 264, 273, 568–570, 573
    order-of-magnitude errors, 98
    overprediction, 98–99, 101
    underprediction, 98–99, 101
    upper bound, 63, 98–99, 102–103, 105–106, 181–182, 568, 573–574, 618, 687, 694, 748, 759–760, 881
Brace, 29
Braced frames with gravity loads, 401–416
    assumptions, 406
    distributed loading, 403–404
    process of approximate analysis, 405
    truss approximation, 401–403

Braced frames with lateral loads, 384–400
    assumptions, 384–386
    axial force members, 387
    nonidentical braces, 386
    transfer of lateral force to braced bays, 388
    process of approximate analysis, 388
    purpose of each member, 387–388

C

Cables with point loads, 120–139
    analysis methods, 126–130
        General Cable Theorem, 129–130
        Method of Joints, 124, 127–128
        system of nonlinear equations, 126–129
    connection types, 120–121
        pin, 121
        pulley, 121
    first-order analysis, 123
    geometric nonlinearity, 121–124
    second-order analysis, 123
    sag, 125–126
Cables with uniform loads, 140–151
    catenary shape, 146
    self-weight of cables, 144–146
    uniform load, 140–142
    uniform load plus point load, 143–144
Camber, 110, 235
Cantilever Method for approximate analysis of rigid frames with lateral loads, 473–489
    applicability, 478
    assumptions, 473
    neutral axis of frame, 474
    process of approximate analysis, 478
Carryover moments, 738, 739
Catenary shape, 146
Center of rigidity for distribution of lateral loads, 620–622
Chevron braced frame, 384, 387, 402–403, 405, 527
Chord forces, 646–647
Collector beams, 617, 638
Collector forces for in plane shear, 638–640
Column, 29
Columns pinned at base with Virtual Work Method, 552
Combined gravity and lateral loads for approximate analysis of rigid frames, 490–496
Compatibility boundary conditions, 239–240
Compatibility for analysis of braced frames, 517–519
Compatibility for analysis of coupled walls, 566–567
Compatibility for Force Method
    indeterminate beams, one degree, 677–679
    indeterminate beams, multi-degree, 691–692
    indeterminate trusses, 710

902 Index

Compiling the system of equations for Direct Stiffness Method for trusses, 807–815
  deformation vector, 811–812
  degrees of freedom, 807
  load vector, 810–811
  structural stiffness matrix, 807–810
Computer aided analysis, 336–370
  analysis errors, 339–340
    assumptions in method, 339–340
    roundoff errors, 340
  features of solution check, 350–352
    displaced shape review, 350–351
    internal force diagrams review, 351–352
  fundamental principles check, 342–344
    horizontal equilibrium, 343
    moment equilibrium, 344
    vertical equilibrium, 344
  human mistakes, 340
  identifying mistakes, 340–342
    evaluation of results, 341
    prediction of results, 341
  input errors, 338–339
    loading, 339
    material behavior, 338
    structural idealization, 339
Conjugate Beam Method, 247–257
Connection, 29, 31–33, 35
Continuous walls, 50–52
Corrected Equivalent Beam Model, 648
Coupled internal forces, 858
Coupled walls, 566–567, 568
Cross braced frames, 384, 386–387, 401, 516, 640

D

Dead load, 4, 12, 235
Dead load densities, common construction materials, 5
Dead load pressures, common construction materials, 5–6
Deduction Method for internal force diagrams, 192–205
  axial force diagrams, 194–195
  shear and moment diagrams, 192–194
Deformations, 234–294
  Conjugate Beam Method, 247–257
    assumptions, 250
    conjugate support conditions, 249
    procedure, 250
  Double Integration Method, 236–247
    assumptions, 240
    deriving governing equation, 236–239
    peak values, 240
    procedure, 240
    using governing equation, 239–240

  Virtual Work Method, 257–276
    assumptions, 263
    deriving governing equation, 258–262
    procedure, 263
    using governing equation, 262–263
Deformation vector for Direct Stiffness Method, 811–812
Degree of indeterminacy, 376, 438, 440, 442, 677, 691, 699–700
Degrees of freedom for Direct Stiffness Method
  frames, 856
  trusses, 807
Densities, common construction materials, 5
Deriving governing equation
  Double Integration Method, 236–239
  Virtual Work Method, 258–262
Design value, 12, 20–21, 23–24, 27
Detailed analysis methods—definition, 675
Diagonally braced frames, 384–386, 401
Diaphragm chords, 617, 646–647
Diaphragms, 59–66, 616–673
  collector beams, 617, 638
  diaphragm chords, 617, 646–647
  distribution of lateral loads by flexible diaphragm, 59–66
    assumption, 61
    approximation, 61
  distribution of lateral loads by rigid diaphragm, 618–633
    assumptions, 625
    center of rigidity, 620–622
    rotation, 622–624
    translation, 619–620
  drag struts, 617
  flexible diaphragm—definition, 617
  in plane moment, 645–660
    chord forces, 646–647
    Corrected Equivalent Beam Model, 648
    Equivalent Beam Model, 647
    flexible diaphragm M_{max} calculation, 647
    rigid diaphragm M_{max} calculation, 648–649
    semirigid diaphragm M_{max} calculation, 650
  in plane shear, 633–645
    analysis of collector forces, 638–640
    calculating in plane shear, 635–636
    collector beams, 638
    evaluating shear capacity, 636–638
    transfer length, 639
  rigid diaphragm—definition, 617
Direct Stiffness Method for frames, 854–899
  element stiffness matrix in global coordinates, 862–868
    transformation matrix, 862–864

element stiffness matrix in local coordinates, 856–861
    axial deformation, 857
    degrees of freedom, 856
    rotation, 859–860
    shear deformation, 858–859
finding deformations, reactions, and internal forces, 877–885
loads between nodes, 868–877
Direct Stiffness Method for trusses, 792–853
  additional loadings, 827–840
    global vector assembly, 830–831
    length change, 828–829
    loads at supports, 831
    temperature change, 830
  compiling the system of equations, 807–815
    deformation vector, 811–812
    degrees of freedom, 807
    load vector, 810–811
    structural stiffness matrix, 807–810
  element stiffness matrix, 795–806
    element stiffness in global coordinates, 801–803
    element stiffness in local coordinates, 796–797
    local coordinate system, 795
    transformation matrix, 797–801
  finding deformations, reactions, and internal forces, 815–827
    partitioning the system of equations, 815–817
    solving for deformations, 816–817
    solving for internal forces, 818
    solving for reactions, 817
  overview, 794–795
    elements, 794
    element stiffness, 794
    nodes, 794
Displaced shape review in features of solution check, 350–351
Distributed loading for braced frames with gravity loads, 403–404
Distributed loads normal to a member, internal forces, 172–174
Distributed loads parallel to a member, internal forces, 175–176
Distribution factor with Moment Distribution Method, 731–735
  moment-rotation relationship, 731–732
  special cases, 734–735
  stiffness—definition, 731
Distribution of lateral loads by flexible diaphragm, 59–66
  flexible diaphragm, 59–61
  lateral stiffness, 59
  Equivalent Beam Model, 61, 66

Distribution of lateral loads by rigid diaphragm, 618–633
  assumptions, 625
  center of rigidity, 620–622
  rotation, 622–624
  translation, 619–620
Double Integration Method, 236–247
  assumptions, 240
  deriving governing equation, 236–239
  peak values, 240
  procedure, 240
  using governing equation, 239–240
Drag struts, 617
Drift—definition, 515

E

Earthquake as load, 8, 159
Elastic buckling, 90
Elements, 794
Element stiffness—definition, 794
Element stiffness in global coordinates, 801–803, 862–868
Element stiffness in local coordinates, 796–797, 860–861
Element stiffness for frames, 857–860
  axial deformation, 857
  rotation, 859–860
  shear deformation, 858–859
Element stiffness matrix for Direct Stiffness Method
  frames, 856–868
    element stiffness matrix in global coordinates, 862–868
    transformation matrix, 862–864
    element stiffness matrix in local coordinates, 856–861
      axial deformation, 857
      degrees of freedom, 856
      rotation, 859–860
      shear deformation, 858–859
  trusses, 795–806
    element stiffness in global coordinates, 801–803
    element stiffness in local coordinates, 796–797
    local coordinate system, 795
    transformation matrix, 797–801
Equation intervals, 176, 239
Equation of compatibility for Force Method, 676
Equivalent Beam Model, 61, 66
Errors, 338–340
  analysis error, 339–340
  human mistakes, 340
  input error, 338–339
Evaluation of results, 341–342
  features of solution check, 350–352
  fundamental principles check, 342–344

F

Features of solution check, 350–352
  displaced shape review, 350–351
  internal force diagrams review, 351–352
Finding deformations, reactions, and internal forces with Direct Stiffness Method for frames, 877–885
Finding deformations, reactions, and internal forces with Direct Stiffness Method for trusses, 815–827
  partitioning the system of equations, 815–817
  solving for deformations, 816–817
  solving for internal forces, 818
  solving for reactions, 817
First-order analysis, 94, 119, 123, 131, 153
Fixed-end moments for Moment Distribution Method, 728, 730–734, 736–737
  cantilevered end, 730–731
  definition, 728, 730
  step in Moment Distribution Method process, 736–737
Fixed support, 30
Flexible diaphragm—definition, 59, 60, 617
Flexibility coefficients for Force Method, 679–680
Flexible diaphragm M_{max} calculation, 647
Flexural deformation for lateral displacement of walls, 566, 577
Floor diaphragm, 29
Fluid as load, 8
Force Method, 674–725
  indeterminate trusses, 699–710
    assumptions, 705
    degree of indeterminacy, 699–700
    predicting deformations for multi-degree indeterminate, 703–705
    predicting deformations for one degree indeterminate, 701–703
    redundants, 705
  multi-degree indeterminate beams, 691–699
    assumptions, 693
    compatibility, 691–692
    degree of indeterminacy, 691
    Maxwell-Betti's reciprocal theorem, 692
    process of analysis, 693
    redundants, 691–692
    solving for redundants, 692
  one degree indeterminate beams, 676–690
    assumptions, 680
    compatibility, 677–679
    degree of indeterminacy, 677
    equation of compatibility, 676
    flexibility coefficients, 679–680
    internal moment redundant, 678–679
    overview, 676–677
    process of analysis, 680
    redundants, 677
Fundamental principles check, 342–344
  horizontal equilibrium, 343
  moment equilibrium, 344
  vertical equilibrium, 344

G

General Cable Theorem for analysis of cables with point loads, 129–130
Geometric nonlinearity for analysis of cables with point loads, 121–124
Girder, 28
Girt, 54
Global vector assembly for loads between nodes with Direct Stiffness Method, 830–831
Grade beam, 504, 599–600, 785–786
Gravity load, 4
Gravity load application, 35–48
  area load, 36
  conversion of gravity pressure, 36–41
  distributed loads, 37–48
  idealizing the loading, 36–41
  line loads, 37
  load path, 36
  one-way action, 37
  one-way diaphragm, 37
  tributary width, 37
  two-way action, 37
  two-way diaphragm, 37–41, 44–45
Gravity Load Method for approximate analysis of rigid frames with gravity loads, 438–457
  assumptions, 439–440
  inflection points, 439
  nonuniform load, 443–444
  pin supports, 440–443
  process of approximate analysis, 444

H

Hardy Cross, 727
Horizontal projection, 6
Human mistakes, 340
  identifying mistakes, 340–342
  evaluation of results, 341
  performance of analysis, 341
  prediction of results, 341

I

Idealized connection, 29, 31–33, 35
Idealized load
  gravity load, 36–48
  lateral load distribution by flexible diaphragm, 59–66

flexible diaphragm, 59–61
lateral stiffness, 59
Equivalent Beam Model, 61, 66
lateral load distribution by rigid diaphragm, 618–633
assumptions, 625
center of rigidity, 620–622
rotation, 622–624
translation, 619–620
lateral load to diaphragms, 49–58
Idealized support, 29–30, 33, 34
Idealized wall diaphragms, 49–52
Idealizing the structure, 28–33
Identifying mistakes, 340–342
Inconsistency discounting, 83, 341
Indeterminate trusses
approximate analysis, 374–383
process, 376
more than two diagonals, 376
predicting deformations for multi-degree indeterminate, 703–705
predicting deformations for one degree indeterminate, 701–703
Inelastic material behavior, 338
Inelastic response drift, 571, 589, 594, 613
Inflection points, 437
Influence lines, 296–335
Müller-Breslau Method, 312–321
Table-of-Points Method, 298–312
using, 322–328
Information overload, 83, 341
In plane moment: diaphragm chords, 645–660
chord forces, 646–647
Corrected Equivalent Beam Model, 648
Equivalent Beam Model, 647
flexible diaphragm M_{max} calculation, 647
rigid diaphragm M_{max} calculation, 648–649
semirigid diaphragm M_{max} calculation, 650
In plane shear: diaphragm collectors, 633–645
analysis of collector forces, 638–640
calculating in plane shear, 635–636
collector beams, 638
evaluating shear capacity, 636–638
transfer length, 639
Input errors with computer aided analysis, 338–339
loading, 339
material behavior, 338
structural idealization, 339
Integration Method for internal force diagrams, 172–191
applied moments, 174–175
distributed loads normal to a member, 172–174
distributed loads parallel to a member, 175–176
equation intervals, 176
finding peak values, 176–177

point loads normal to a member, 174–175
point loads parallel to a member, 176
Internal force diagrams, 170–233
deduction, 192–205
axial force diagrams, 194–195
shear and moment diagrams, 192–194
frames, 205–215
integration, 172–191
applied moments, 174–175
distributed loads normal to a member, 172–174, 175–176
equation intervals, 176
finding peak values, 176–177
point loads normal to a member, 174–175
point loads parallel to a member, 176
Internal force diagrams review in features of solution check, 351–352

L
Lateral load application, 49–58
continuous walls, 50–52
idealized distributed load, 52
idealized lateral load, 53, 55, 57, 58
idealized wall diaphragms, 51
load path, 49
pressure on wall diaphragm, 49
Lateral load-resisting systems, 59, 384, 458, 565
Length change with Direct Stiffness Method, 828–829, 870
Linear behavior, 94–95
Linear elastic behavior, 94, 238, 258, 260, 322, 473, 491, 575, 678, 755
Line loads, 37
Live load, 4–6, 7, 12, 35, 104, 235, 297, 298–300, 322, 344, 490
Live load pressures, 7
Load combinations, 11–27
design value, 12, 20–21, 23–24, 27
load factors, 12
maximum likely load, 11
probability distribution of peak load, 11
ultimate value, 12
Load factors, 12
Loading input errors, 339
Loading to maximize displacement, 97
Load path, 36, 49
Load types, 4–10
dead, 4
dead load densities, common construction materials, 5
dead load pressures, common construction materials, 5–6
earthquake, 8
fluid, 8
gravity, 4

Load types (*Continued*)
  horizontal projection, 6
  live, 4
  live load pressures, 7
  snow, 6–7
  snow drift, 37, 117, 223
  soil, 8
  superimposed dead load, 4
  units, 8
  wind, 7
Load idealization, 35–66
  gravity load application, 35–48
    area load, 36
    conversion of gravity pressure, 36–41
    distributed loads, 37–41, 42–43, 44–45, 46–48
    idealizing the loading, 36–41
    line loads, 37
    load path, 36
    one-way action, 37
    one-way diaphragm, 37
    tributary width, 37
    two-way action, 37
    two-way diaphragm, 37–41, 44–45
  lateral load application
    continuous walls, 50–52
    idealized distributed load, 52
    idealized wall diaphragms, 51
    load path, 49
    pressure on wall diaphragm, 49
    static equivalent seismic load, 159
  lateral load distribution by flexible diaphragm, 59–66
    flexible diaphragm, 59–61
    lateral stiffness, 59
    Equivalent Beam Model, 61, 66
  lateral load distribution by rigid diaphragm, 618–633
    assumptions, 625
    center of rigidity, 620–622
    rotation, 622–624
    translation, 619–620
Loads at supports with Direct Stiffness Method, 831
Loads between nodes with Direct Stiffness Method, 827–840, 868–877
Load vector for Direct Stiffness Method, 810–811
Local coordinate system for Direct Stiffness Method, 795
Lower bound, 98–99, 100, 102, 107, 253, 264, 273, 568–570, 573

M

Material behavior input errors, 338
Maximum inelastic response drift, 571, 589, 594, 613
Maximum likely load, 11
Maxwell-Betti's reciprocal theorem, 692, 809–810, 877
Method of Joints, 84–87, 89, 124, 127–128, 376, 388, 390–391, 403, 405

Method of Sections, 84–85, 88–89, 129, 376–379, 381–383, 388, 403, 405
Moment Distribution Method, 726–791
  beams, 734–753
    assumptions, 742
    carryover moments, 738, 739
    convergence, 738–739
    finding reactions, internal forces, and deformations, 741–742
    fixed-end moments, 736–737
    moments distributing into members, 736–739
    net moment, 739–741
    process of analysis, 742
    unbalanced moment, 737, 740
  convergence, 738–739
  distribution factors, 731–735
    moment-rotation relationship, 731–732
    special cases, 734–735
    stiffness—definition, 731
  fixed-end moment, 728, 730–731, 736–737
    cantilevered end, 730–731
    definition, 728, 730
    step in Moment Distribution Method process, 736–737
  node, 728
  overview, 728–729
    fixed-end moment, 728
    node, 728
    unbalanced moment, 728
  sidesway inhibited frames, 734–753
    assumptions, 742
    carryover moments, 738, 739
    convergence, 738–739
    finding reactions, internal forces, and deformations, 741–742
    fixed-end moments, 736–737
    moments distributing into members, 736–739
    net moment, 739–741
    process of analysis, 742
    unbalanced moment, 737, 740
  sidesway frames, 735, 754–776
    assumptions, 756
    multistory frames, 756–758
    scaling factor, 756
    single-story frames, 755–756
  unbalanced moment, 728, 737, 740
Moment-rotation relationship, 31–32, 731–732
Moments distributing into members, 736–739
Müller-Breslau Method, 312–321
Multi-degree indeterminate beams
  Force Method, 691–699
    assumptions, 693
    compatibility, 691–692
    degree of indeterminacy, 691
    Maxwell-Betti's reciprocal theorem, 692
    process of analysis, 693

Index **907**

redundants, 691–692
solving for redundants, 692
Moment Distribution Method, 734–753
assumptions, 742
carryover moments, 738, 739
convergence, 738–739
finding reactions, internal forces, and deformations, 741–742
fixed-end moments, 736–737
moments distributing into members, 736–739
net moment, 739–741
process of analysis, 742
unbalanced moment, 737, 740

N

Net moments with Moment Distribution Method, 739–741
Node, 728, 794
Nonidentical braces for braced frames with lateral loads, 386
Nonlinear behavior, 94, 338
Nonlinear equations for analysis of cables with point loads, 126–129
Nonuniform load with Gravity Load Method for approximate analysis of rigid frames, 443–444

O

One degree indeterminate beams with Force Method, 676–690
assumptions, 680
compatibility, 677–679
degree of indeterminacy, 677
equation of compatibility, 676
flexibility coefficients, 679–680
internal moment redundant, 678–679
overview, 676–677
process of analysis, 680
redundants, 677
One-way action, 37
One-way diaphragm, 37
Open-web joist, 110, 223, 368
Order-of-magnitude errors, 98
Order-of-magnitude estimates, 159
Overprediction, 98–99, 101, 182, 208, 564, 568, 573, 647

P

Partitioning the system of equations with Direct Stiffness Method, 815–817
Pin connection, 31–33
for cables, 121
Pin support, 30, 31
with Stiff Beam Method, 544
Point loads
normal to a member, internal forces, 174–175
parallel to a member, internal forces, 176
with uniform loads on cables, 143–144

Point loads
normal to a member, internal forces, 174–175
parallel to a member, internal forces, 176
with uniform loads on cables, 143–144
Point of fixity, 230, 292, 363, 497, 499, 502, 503, 511, 594, 786, 788
Portal Method for approximate analysis of rigid frames with lateral loads, 458–472
applicability, 462
assumptions, 459–460
process of approximate analysis, 462
Predicting results, 82–108, 341
approximating loading conditions, 102–108
average load, 106–107
equivalent loads, 102–104, 108
bounding the solution, 98–101
approximate answers, 98
order-of-magnitude errors, 98
overprediction, 98–99, 101, 182, 208, 564, 568, 573, 647
underprediction, 98–99, 101, 338, 519, 527, 545, 552, 568, 570, 577–578
identifying mistakes, 341
inconsistency discounting, 83, 341
information overload, 83, 341
qualitative truss analysis, 84–93
Method of Joints, 84–87, 89
Method of Sections, 84–85, 88–89
zero-force members, 84–85
superposition principle, 94–97
linear behavior, 94–95
linear elastic behavior, 94
linear load versus displacement plot, 94
loading to maximize displacement, 97
nonlinear behavior, 94
Pressure on wall diaphragm, 49
Probability distribution of peak load, 11
Process of approximate analysis
for braced frames with gravity loads, 405
for braced frames with lateral loads, 388
for indeterminate trusses, 376–377
for rigid frames with gravity and lateral loads, 492
for rigid frames with gravity loads, 444
for rigid frames with lateral loads, Cantilever Method, 478
for rigid frames with lateral loads, Portal Method, 462
Pulley connections for cables, 121
Purlin, 54, 75

Q

Qualitative truss analysis, 84–93
Method of Joints, 84–87, 89
Method of Sections, 84–85, 88–89
zero-force members, 84–85

R

Redundants with Force Method
  indeterminate trusses, 705
  multi-degree indeterminate beams, 691–692
  one degree indeterminate beams, 677
Rigid connection, 31–33
Rigid diaphragm—definition, 617
Rigid diaphragm M_{max} calculation, 648–649
Rigid frames
  approximate analysis with gravity loads, 438–457
  approximate analysis with lateral loads
    Cantilever Method, 473–489
    Portal Method, 458–472
  detailed analysis
    Direct Stiffness Method, 854–899
    Force Method, 674–725
    Moment Distribution Method, 726–791
Rigidity, 618
Roller connection, 31
Roller support, 30
Rotation for distribution of lateral loads, 622–624
Rotation with Direct Stiffness Method for frames, 859–860
Roundoff errors, 340, 692, 739, 802

S

Sag with cables, 125–126
Scaling factor with Moment Distribution Method for sidesway frames, 756
Second-order analysis, 119, 123–124, 129
Second-order effect, 94
Seismic loads, 8, 159, 469, 487, 571, 634, 759
Self-weight of cables, 144–146
Semirigid diaphragm M_{max} calculation, 650
Shear and moment diagrams, 192–194
Shear capacity evaluation for in plane shear, 636–638
Shear deformation for lateral displacement of walls, 565, 575–576
Shear deformation with Direct Stiffness Method for frames, 858–859
Shear distribution to columns with Stiff Beam Method, 544–545
Sidesway frame—definition, 734–735
Sidesway frames analyzed with Moment Distribution Method, 735, 754–776
  assumptions, 756
  multistory frames, 756–758
  scaling factor, 756
  single-story frames, 755–756
Sidesway inhibited frames analyzed with Moment Distribution Method, 734–753
  assumptions, 742
  carryover moments, 738–739
  convergence, 738–739
  finding reactions, internal forces, and deformations, 741–742
  fixed-end moments, 736–737
  moments distributing into members, 736–739
  net moment, 739–741
  process of analysis, 742
  unbalanced moment, 737, 740
Slab-on-grade, 49–52, 53, 56, 58, 63, 75
Snow as load, 6–7
Soil as load, 8
Solid walls
  multistory, approximate lateral displacement, 575–582
  single story, approximate lateral displacement, 565–574
Solving for deformations with Direct Stiffness Method, 816–817
Solving for internal forces with Direct Stiffness Method, 818
Solving for reactions with Direct Stiffness Method, 817
Solving for redundants for multi-degree indeterminate beams with Force Method, 692
Static equivalent seismic load, 159, 469, 487, 571, 634, 759
Stiff Beam Method for rigid frames and approximate lateral displacement, 542–549
  assumptions, 545
  implications, 545
  limitations, 545
  pinned base, 544
  shear in each column, 544–545
  total drift, 544
Stiffness
  beams
    flexural, 569
    rotational, 731–732
  braces, 386, 516–517
  cantilevers, 460, 734
  columns
    exterior, 459
    interior, 459
  element
    frame in global coordinates, 864
    frame in local coordinates, 857–861
    truss in global coordinates, 801–802
    truss in local coordinates, 796–797
  floors
    in plane, 59
    out of plane, 29
  frames, 59
  lateral, 59, 619
  relative, 385, 402, 439, 458, 727, 736
  torsional of floor, 569
  truss member, 796–797
  walls
    in plane, 29, 568
    out of plane, 565

Story Drift Method for braced frames
    and approximate lateral displacement,
    516–526
  assumptions, 519
  bays with different brace stiffnesses, 517–519, 524–526
  implications, 519
  limitations, 519
Structural idealization input errors, 339
Structural system parts, 28–29
  beam, 28
  brace, 29
  column, 29
  floor diaphragm, 29
  girder, 28
  structural wall, 29
Structural stiffness matrix for Direct Stiffness Method,
    807–810, 877
Structural wall, 29
Structure idealization, 28–35
  connection, 29–33, 35
  member, 28–29
  support, 29–30, 33–34
  typical idealizations of structural supports and
    connections, 30–31
Superimposed dead load, 4
Superposition principle, 94–97
  linear behavior, 94–95
  linear elastic behavior, 94
  linear load versus displacement plot, 94
  loading to maximize displacement, 97
  nonlinear behavior, 94
Superposition process with Moment Distribution
    Method for multistory frames, 756–758
Superposition process with Moment Distribution
    Method for single-story frames, 755–756
Support, 29–30, 33, 34
System of nonlinear equations for analysis of cables
    with point loads, 126–129

T

Table-of-Points Method, 298–312
Temperature change, 154, 827, 830, 832,
    833–837, 869–870
  Direct Stiffness Method, 830, 870
Three-hinge arch, 153–162
Thrust arch
  fixed supports, 152, 159
  pinned supports, 152
Tied arch with simple supports, 152, 153
Tie force, 155
Total drift, 517, 526, 545, 552, 570, 577
Transfer girder, 110, 186, 203
Transfer length for in plane shear, 639

Transfer of lateral force to braced bays, 388
Transformation matrix for Direct Stiffness Method
  frame, 862–864
  truss, 797–801
Translation for distribution of lateral loads, 619–620
Tributary height, 55, 58
Tributary width, 37–40, 42–45, 47, 55, 61, 63
Truss approximation for braced frames
    with gravity loads, 401–403
Two-way action, 37
Two-way diaphragm, 37–38
Typical idealizations of structural supports and
    connections, 30–31
  fixed, 30
  pin, 30, 31
  rigid, 31
  roller, 30, 31

U

Ultimate value, 12
Unbalanced moment, 728, 737, 740
Uncoupled internal forces, 857
Underprediction, 98–99, 101, 338, 519, 527, 545,
    552, 568, 570, 577–578
Unfactored internal forces, 16–20
Unfactored reactions, 22–23, 25–27
Units, based on type of load, 8
Upper bound, 63, 98–99, 102–103, 105–106, 181–182,
    568, 573–574, 618, 687, 694, 748, 759–760, 881

V

Virtual Work Method, 257–276
  assumptions, 263
  deriving governing equation, 258–262
  procedure for analysis, 263
  using governing equation, 262–263
Virtual Work Method for braced frames and
    approximate lateral displacement, 526–542
  assumptions, 527
  implications, 527
  limitations, 527
Virtual Work Method for rigid frames and approximate
    lateral displacement, 550–564
  approximate analysis derivation, 550–552
  assumptions, 552
  implications, 552
  limitations, 552
  pinned base, 552

W

Wind as load, 7

Z

Zero-force members, 84–85

Fixed-Fixed Beams

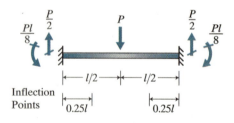

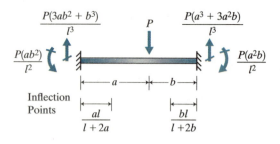

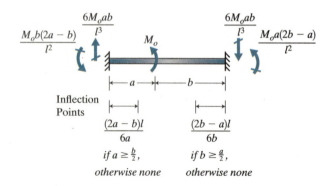

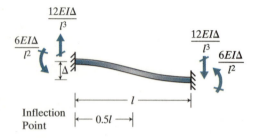